Technology, Law, and the Working Environment

Technology, Law, and the Working Environment

Nicholas A. Ashford

Charles C. Caldart

Massachusetts Institute of Technology

VNR VAN NOSTRAND REINHOLD
New York

Library of Congress Catalog Number 90-38160
ISBN 0-442-23926-2

Printed in the United States of America

Van Nostrand Reinhold
115 Fifth Avenue
New York, New York 10003

Chapman and Hall
2-6 Boundary Row
London, SE1 8HN, England

Thomas Nelson Australia
102 Dodds Street
South Melbourn 3205
Victoria, Australia

Nelson Canada
1120 Birchmount Road
Scarborough, Ontario M1K 5G4, Canada

16 15 14 13 12 11 10 9 8 7 6 5 4 3 2 1

Library of Congress Cataloging-in-Publication Data

Ashford, Nicholas Askounes.
    Technology, law, and the working environment / by Nicholas A.
Ashford and Charles C. Caldart.
        p.   cm.
    includes index.
    ISBN 0-442-23926-2
    1. Industrial hygiene—Law and legislation—United States.
    2. Industrial safety—Law and legislation—United States.
    3. Technology and law.   I. Caldart, Charles C.   II. Title.
    KF3570.A78   1991
    344.73'0465—dc20
    [347.304465]   90-38160
                                                            CIP

Στήν Ἀνδρονίκη με ἀγάπη και ὑπερηφάνεια.

N.A.A.

To my parents, Clyde and Cora, for their values,
support, and example, and to Anna, for her unflagging
patience as this book was being written.

C.C.C.

Contents in Brief

Preface   xxi

Acknowledgments   xxiii

CHAPTER 1   Technology, Work, and Health   1

CHAPTER 2   Administrative Law   39

CHAPTER 3   The Occupational Safety and Health Act of 1970   87

CHAPTER 4   The Toxic Substances Control Act of 1976   189

CHAPTER 5   Economic Issues in Occupational Health and Safety   221

CHAPTER 6   Regulation of Labor-Management Relations Under the National Labor Relations Act   249

CHAPTER 7   Toxics Information Transfer in the Workplace   299

CHAPTER 8   Retaliatory Employment Practices   363

CHAPTER 9   Compensation for Occupational Injury and Disease   453

APPENDIX A   The Occupational Safety and Health Act   509

APPENDIX B   The Toxic Substances Control Act   541

Contents

Preface   xxi

Acknowledgments   xxiii

CHAPTER 1

Technology, Work, and Health   1

A. Introduction   1
B. The Early Period: 1900–1945   3
    1. Background   3
    2. Technology and the Organization of Work   4
    3. The Structure of Industry and Occupations   5
    4. Labor-Management Relations   6
    5. Health and Safety   8
C. Postwar Changes: 1945–Present   10
    1. Technological Change   10
    2. The Organization of Work   12
    3. Changes in the Structure of Industries and Occupations   16
    4. Labor-Management Relations   21
        a. Unionization   21
        b. Worker Satisfaction   23
        c. Innovations in Labor-Management Relations   24
            (i) *Expansion of Individual Rights*   24
            (ii) *Worker Participation*   25
    5. Health and Safety   27
        a. The Health and Safety Crisis of the 1960s   27
        b. After OSHA   29
            (i) *The Role of Unions*   29
            (ii) *Continuing Problems*   30

D. Conclusion   32
    References   34

CHAPTER 2

Administrative Law   39

A. Questions to Consider When Analyzing a Regulatory Framework   40
B. The Constitutional Basis for Labor, Health, and Safety Regulation   41
    1. Direction from the Legislative Branch   41
        a. The Statutory Mandate   41
    Notes   42
        b. The Commerce Clause   42
        c. The Delegation Doctrine   42
        d. The Procedural Mandate   43
        e. Interpreting the Statutory Mandate   44
        f. Statutory Amendment and Informal Controls   45
    2. Direction from the Executive Branch   46
    Schultz, W., and Vladeck, D., An Obstacle to Public Safety, Washington Post, May
    10, 1988, p. 20 46
        Notes   48
    3. Direction from the Judicial Branch   49
    Ferguson, J. H., and McHenry, D. E., Constitutional Courts from The American
    System of Government, pp. 441-453 (1981) 50
        Notes   55
    Llewellyn, K. N., The Bramble Bush, pp. 41-45 (1951) 55
C. Administrative Rulemaking   58
    1. The Distinction between Rulemaking, Adjudication, and Enforcement   58
    2. A General Look at Rulemaking under the Administrative Procedure Act   59
    3. Informal Rulemaking by OSHA and EPA   60
    Ashford, N., Advisory Committees in OSHA and EPA: Their Use in Regulatory
    Decisionmaking, 9 Science, Technology & Human Values 72 (1984) 60
D. Citizen or Corporate Access to the Administrative Process   61
    1. Initiation of Rulemaking   61
    2. Access to Agency Proceedings and Records   62
    3. Access to Advisory Committees   62
    4. Access to the Courts   63
    Calumet Industries v. Brock, 807 F.2d 225 (D.C. Cir. 1986) 65
    5. Monetary Limitations on the Availability of Review   67
    6. Bypassing the Agency: Citizen Enforcement through the Private Right of
        Action   68

E.   The Role of the Courts in Reviewing Agency Action   69
     1. Five Judicial Limitations on Agency Authority   69
     Marshall v. Barlow's, Inc., 436 U.S. 307 (1978) 69
     2. The Scope of Factual Review   73
     Motor Vehicle Manufacturers' Association v. State Farm Mutual Automobile
     Insurance Company, 463 U.S. 29 (1983) 75
          Notes   78
     3. Judicial Review of Agency Decisions Not to Act   79
F.   Federal Preemption and the Role of State and Local Laws   79
     Appendix: Administrative Procedure Act, 5 U.S.C. 551, *et seq.*   80

CHAPTER 3

The Occupational Safety and Health Act of 1970   87

A.   Introduction   87
     1. The History of Federal Regulation   87
     2. The Occupational Safety and Health Act of 1970   88
B.   Description and Summary of the OSHAct   89
          Notes   96
C.   A Brief History of Standard Setting under the OSHAct   99
     1. 6(a) and 6(b) Standards   99
     2. The Generic Cancer Standard   100
D.   Judicial Interpretation of Standard Setting under the OSHAct   102
     1. The Early Cases: Weight of the Evidence and the Meaning of Feasibility   102
     Industrial Union Department, AFL-CIO v. Hodgson, 499 F.2d 467 (D.C. Cir.
     1974) 102
          Notes   113
     The Society of the Plastics Industry, Inc., v. Occupational Safety and Health
     Administration, 509 F.2d 1301 (2d Cir. 1975) 114
          Notes   120
     2. The Limits of Technology Forcing   121
     American Iron and Steel Institute v. OSHA, 577 F.2d 825 (3d Cir. 1978) 121
          Notes   121
     3. The Significant Risk Requirement   122
     Industrial Union Department v. American Petroleum Institute, 448 U.S. 607
     (1980) 123
          Notes   136
     4. Standard Setting After the Benzene Decision: Feasibility and Cost-Benefit
     Analysis   137

American Textile Manufacturers Institute, Inc. v. Donovan, 542 U.S. 490 (1981) 137

    Notes 145

United Steel Workers of America, AFL-CIO-CLC v. Marshall and Bingham, 647 F.2d 1189 (D.C. Cir. 1980) 145

    Notes 157

5. Emergency Temporary Standards 157

Asbestos Information Association/North America v. OSHA, 727 F. Supp. 415 (5th Cir. 1984) 157

    Notes 163

6. Short-Term Exposure Limits 164

Public Citizen Health Research Group v. Brock, 823 F.2d 626 (D.C. Cir. 1987) 164

    Notes 166

7. Protection of Workers from Exposure to Noise: A Special Case or a Troubling Precedent? 166

Forging Industry Association v. Secretary of Labor, 748 F.2d 210 (4th Cir. 1984) 167

    Notes 171

8. Medical Removal Protection 171

United Steelworkers of America, AFL-CIO-CLC v. Marshall and Bingham, 647 F.2d 1189 (D.C. Cir. 1980) 171

    Notes 179

E.   Evolution of the General Duty Obligation 179

Ashford, N., Crisis in the Workplace: Occupational Disease and Injury 180

International Union, United Automobile, Aerospace and Agricultural Implement Workers of America v. General Dynamics Land Systems Div., 815 F.2d 626 (D.C. Cir. 1987) 182

    Notes 186

Discussion Questions 186

CHAPTER 4

The Toxic Substances Control Act of 1976   189

A.   Introduction 189

   1. Purposes 190

   2. Definitions 190

   3. Authority to Regulate 190

      a. Rulemaking under Section 6 191

      b. Imminent Hazards 191

      c. Regulating Carcinogens, Mutagens, and Teratogens 192

        d. Premanufacturing Notification   192

        e. Significant New Use Rule   193

        f. PMN Exemption   193

    4. Authority to Require Testing   193

    5. Recordkeeping and Reporting Requirements   193

    6. Relationship to Other Acts   194

    7. Citizen Suits and Civil Petitions   194

    8. Judicial Review   195

B.   The Implementation of TSCA   195

    1. Regulation   196

        a. Section 6 Regulations   196

        b. Imminent Hazard Regulation (Section 7)   197

        c. Section 4(f) Actions   197

        d. PMN Activity   197

    2. Testing   197

    3. Reasons for Slow Implementation   199

    Notes   200

C.   Judicial Interpretations of the Act   200

    1. Unreasonable Risk   200

    EDF v. EPA, 636 F.2d 1267 (D.C. Cir. 1980)   200

    2. Imminent Hazard   207

    United States v. Commonwealth Edison Co., 620 F. Supp. 1404 (D.C. IL 1985)   207

    3. Testing Rules   209

    NRDC v. Costle, 14 Environment Reporter Cases 1858 (S.D.N.Y. 1980)   209

    NRDC v. EPA, 595 F. Supp. 1255 (S.D.N.Y. 1984)   211

    Notes   215

D.   The Reluctance to Regulate: The Regulatory History of Formaldehyde   215

    Ashford, N., et al., A Hard Look at Federal Regulation of Formaldehyde: A Departure from Reasoned Decisionmaking, Harvard Environmental Law Review 7(2):297 (1983)   215

    Discussion Questions   219

CHAPTER 5

Economic Issues in Occupational Health and Safety   221

A.   Introduction   221

B.   The Market for Health and Safety   222

    1. The Idealized Job Market   223

    2. Performance of the Job Market: The U.S. Experience   224

C.   Market Imperfections as the Basis for Government Intervention   226

   1. Imperfections in the Job Market   226

      a. Imperfect Information   226

      b. Externalities   228

      c. Imperfect Competition   230

      d. Market-transmitted Injustices or Inequities   231

   2. Types of Government Intervention   233

D.   Cost-Benefit Analysis   236

   1. Problems in Estimating Public Policy Benefits   238

   2. Problems in Estimating Public Policy Costs   241

   3. Problems of Equity and Ethics   243

   4. Other Misuses and Abuses   244

   5. Conclusions   246

   References   247

CHAPTER 6

Regulation of Labor-Management Relations Under the National Labor Relations Act   249

A.   The Role of Unions in the United States   250

Kochan, T., et al., The Transformation of American Industrial Relations, Basic Books, New York, 1986, pp. 47–50   251

B.   The Essential Rights and Obligations   253

   1. Unfair Labor Practices   253

   2. The Organizing Process   254

Weiler, P., Promises to Keep: Securing Workers' Rights to Self-organization under the NLRA, 96 Harv. L. Rev. 1769, 1793–1795 (1983)   254

   Note   255

C.   The Collective Bargaining Process   256

   1. General Overview   256

Dunlop, J. T., Past and Future Tendencies in American Labor Organizations, Daedalus, "A New America?" Winter 1978, Vol. 107, No. 1   256

   2. The Duty to Engage in Collective Bargaining   258

   Notes   259

   3. Bargaining over Safety and Health   260

National Labor Relations Board v. Gulf Power Company, 384 F.2d 822 (5th Cir. 1967)   260

Robinson, J., Labor Union Involvement in Occupational Safety and Health,

1957–1987, Journal of Health Politics, Policy and Law 13(3):453–468
(1988) 262

    Notes 265

4. The Timing of the Duty to Bargain: "Decision" vs. "Effects" Bargaining 266

First National Maintenance Corporation v. National Labor Relations Board, 452
U.S. 66 (1981) 266

    Notes 273

Kohler, T., Distinctions without Differences: Effects Bargaining in Light of First
National Maintenance, 5 Ind. Rel. Law J. 402, 421–423 (1983) 274

5. Collective Bargaining over Issues of Technological Change 275

Ashford, N., and Ayers, C., Changes and Opportunities in the Environment
for Technology Bargaining, 62 Notre Dame L. Rev. 810, 818–819, 842–855
(1987) 275

    Notes 282

6. Arbitration Clauses and the Right to Strike 282

    a. Types of Strikes 282
    b. The No-Strike Clause 283

Boys Markets, Inc., v. Retail Clerks Union, Local 770, 398 U.S. 235 (1970) 283

    c. Arbitration 285

    Notes 286

D. Labor-Management Health and Safety Committees 287

    Notes 288

E. The Dissenting Employee 289

1. Impermissible Discrimination and the Duty of Fair Representation 289

2. Must Employees Join the Union? 290

    Notes 290

F. The Role of the NLRB 291

NLRB, A Guide to Basic Law and Procedures, Washington, D.C., U.S. Government
Printing Office (1987), pp. 39–44 291

Appendix to Chapter 6 293

CHAPTER 7

Toxics Information Transfer in the Workplace   299

A. Introduction 299

B. The Right to Know 300

Caldart, C., Promises and Pitfalls of Workplace Right-to-Know, 1 Seminars in
Occupational Medicine 81–86, 89 (1986) 300

C.   Employee and Workplace Monitoring   305

Ashford, N., et al., Monitoring the Worker for Exposure and Disease: Scientific, Legal and Ethical Considerations in the Use of Biomarkers (1990) 306

D.   Federal Information Transfer Laws   310

1. The OSHAct   310

a. The Authority to Require the Generation of Information   310

Ashford, N., et al., Monitoring the Worker for Exposure and Disease: Scientific, Legal and Ethical Considerations in the Use of Biomarkers (1990) 310

b. The OSHA Hazard Communication Standard   313

United Steelworkers of America, AFL-CIO-CLC v. Auchter, 763 F.2d 728 (3d Cir. 1985) 313

(i) *Purpose of the Hazard Communication Standard*   319

(ii) *Definitions*   320

(iii) *Requirements*   320

(iv) *Employee Information and Training*   321

(v) *Trade Secrets*   321

(vi) *"Tailored" Provisions*   321

c. The OSHA Medical Access Rule   322

Ashford, N., et al., Monitoring the Worker for Exposure and Disease: Scientific, Legal and Ethical Considerations in the Use of Biomarkers (1990) 322

2. The Toxic Substances Control Act   325

Ashford, N., et al., Monitoring the Worker for Exposure and Disease: Scientific, Legal and Ethical Considerations in the Use of Biomarkers (1990) 325

3. The National Labor Relations Act   327

Winona Industries, Inc., and International Chemical Workers Union, Local 893, AFL-CIO, case-18-Ca 6509 Before the National Labor Relations Board Division of Judges 327

Mentzer, M., Union's Right to Information about Occupational Health Hazards under the National Labor Relations Act, Industrial Relations Law Journal Vol. 5: 247, 255–256 330

Oil, Chemical & Atomic Workers Local Union No. 6-418, AFL-CIO v. National Labor Relations Board, 711 F.2d 348 (D.C. Cir 1983) 331

Notes   334

4. The Emergency Planning and Community Right to Know Act   335

5. High Risk Notification   336

E.   State and Local Information Transfer Laws   338

New Jersey State Chamber of Commerce v. Hughey, 774 F.2d 587 (3d Cir. 1985) 339

Notes   344

F.   Potential Barriers to Disclosure   345

1. Employee Confidentiality   345

Ashford, N., et al., Monitoring the Worker for Exposure and Disease: Scientific, Legal and Ethical Considerations in the Use of Biomarkers (1990) 346

2. Trade Secrets   351

Ashford, N., and Caldart, C., The Right to Know: Toxics Information Transfer in the Workplace, Ann. Rev. Public Health 6: 383–401, 1985   351

Ruckleshaus v. Monsanto Co., 467 U.S. 986 (1984)   352

    Notes   356

United Steelworkers of America, AFL-CIO v. Auchter, 763 F.2d 728 (3d Cir. 1985)   356

New Jersey State Chamber of Commerce v. Hughey, 600 F. Supp. 606 (D.N.J. 1985)   357

    Notes   358

3. Restrictions on Disclosure Once Access Has Been Granted   358

Lawlor v. Shannon, U.S. District Court, Massachusetts Civ. Action No. 86-2516-Mc, August 29, 1988   359

CHAPTER 8

Retaliatory Employment Practices   363

A. A Starting Point: Four Potential Workplace Freedoms   363

B. The Decline of the Employment at Will Doctrine   364

    Feinman, J. M., The Development of the Employment at Will Rule, 20 Am. J. Legal Hist. 118 (1976)   365

    Hoerr, J., et al., Beyond Unions: A Revolution in Employee Rights Is in the Making, Business Week, July 8, 1985, 72   367

        Notes   367

C. The Right to Refuse Hazardous Work under Federal Labor Statutes   372

    1. The National Labor Relations Act   372

        a. Generally   372

    Ashford, N.A., and Katz, J. I., Unsafe Working Conditions: Employee Rights under the Labor Management Relations Act and the Occupational Safety and Health Act, 52 Notre Dame Lawyer 802 (1977)   372

    National Labor Relations Board v. Tamara Foods, Inc., 692 F.2d 1171 (8th Cir. 1982)   376

        Notes   382

        b. "Concerted Activity" under Section 7   382

        Notes   385

    2. The Occupational Safety and Health Act   385

    Whirlpool Corporation v. Marshall, 445 U.S. 1 (1980)   386

        Notes   391

D.   Protections from Other Retaliatory Employment Practices   392

　1. Section 11(c) of the OSHAct   392

　　a. The Scope of the Protection   392

　Note   393

　　b. Proof of Discrimination   393

Marshall v. Commonwealth Aquarium, 611 F.2d 1 (1st Cir. 1979) 393

　Notes   395

　　c. Procedural Requirements   396

Donovan v. Square D Company, 709 F.2d 335 (5th Cir. 1983)   396

　Notes   396

Chadsey v. United States of America, 11 OSHC 1198 (D. Ore. 1983) 399

　2. Section 8 of the NLRA   400

Weiler, P., Promises to Keep: Securing Workers' Rights to Self-Organization under the NLRA, 96 Harv. L.R. 1769 (1983) 401

　Notes   405

　3. Federal Environmental and Safety Statutes   406

Section 210 of the Energy Reorganization Act, 42 U.S.C. § 5851   406

　Notes   407

　4. General Whistleblower Statutes   407

　5. Common Law Protections   408

　　a. Violation of Personnel Policies   408

　　b. Breach of Implied Promise of Job Security   409

　　c. Breach of Implied Covenant of Good Faith and Fair Dealing   409

　　d. Violation of Public Policy   409

　　　(i) *Discharge for Carrying Out Some Articulated Public Policy*   410

Lally v. Copygraphics, 85 N.J. 668 (1981) 410

　Notes   412

Palmateer v. International Harvester Co., 85 Ill.2d 124, 421 N.E.2d 876 (1981)   412

　　　(ii) *Discharge for Refusing to Carry Out an Action That Would Have Contravened an Articulated Public Policy*   415

Pierce v. Ortho Pharmaceutical Corp., 84 N.J. 61 (1980) 415

　Notes   420

E.   Exclusion from the Workplace Because of Presumed Susceptibility to Workplace Hazards   420

　1. Exclusion of Pregnant or Fertile Female Employees in Order to Protect the Fetus   421

　　a. The OSHAct   421

Oil, Chemical and Atomic Workers International Union v. American Cyanamid Company, 741 F.2d 444 (D.C. Cir. 1984) 421

　Notes   425

　　b. The Civil Rights Act   425

United Auto Workers v. Johnson Controls, Inc., 886 F.2d 871 (7th Cir. 1989) 426

　Notes   446

2. Exclusion of Employees on the Basis of Human Monitoring Data   447

Ashford, N.A., et al., Monitoring the Worker for Exposure and Disease: Scientific, Legal and Ethical Considerations in the Use of Biomarkers (1990) 447

Notes   451

CHAPTER 9

Compensation for Occupational Injury and Disease   453

A.   State Workers' Compensation   453

Office of Technology Assessment, U.S. Congress, Preventing Illness and Injury in the Workplace (207–209) 1985 454

Boden, L., Workers' Compensation, from Levy and Wegman, eds., Occupational Health: Recognizing and Preventing Work-related Disease 149–162 (1988) 455

B.   The Level of Compensation for Occupational Injury and Disease   461

U.S. Department of Labor, An Interim Report to Congress on Occupational Diseases, Washington, D.C.: U.S. Government Printing Office, 1980 461

Caldart, C., Are Workers Adequately Compensated for Injury Resulting from Exposure to Toxic Substances? An Overview of Worker Compensation and Suits in Tort, Regulation and Compliance, 92 (1985), Basel: S. Karger AG, 1985, pp. 92–98 463

Notes   465

C.   The Use of Tort Law to Augment Workers' Compensation   466

1. Suits against Employers   467

Johns-Manville Products Corporation v. Contra Costa Superior Court (Reba Rudkin, Real Party in Interest), 165 Cal. Rptr. 858, 612 P.2d 948 (1980) 467

Blankenship v. Cincinnati Milacron Chemicals, 69 Ohio St.2d 608 (1982) 470

Notes   474

2. Suits against Manufacturers   476

Barker v. Lull Engineering Co., 20 Cal.3d 413, 573 P.2d 443 (1978) 476

Beshada v. Johns-Manville Products Corporation, 90 N.J. 191, 447 A.2d 539 (1982) 477

Notes   482

3. Suits against Other Third Parties   484

a. The Government   484

Cunningham v. United States, 786 F.2d 1445 (9th Cir. 1986) 485

Notes   486

b. Unions   486

United Steel Workers of America v. Rawson,—U.S.—S.Ct 1904 (1990) 487

Notes   495

c. Health and Safety Contractors   495

Canipe v. National Loss Control Service Corp., 736 F.2d 1055 (5th Cir. 1984) 495

Notes   498

d. Supervisory Personnel   499

Streeter v. Sullivan, 509 So.2d 268 (Fla. 1987) 499

e. Medical Professionals   501

Ashford, N., Caldart, C., The Control of Reproductive Hazards in the Workplace: A Prescription for Prevention, 5 Ind. Rel. Law J. 523–563, 558 (1983) 501

Notes   502

D. Proof of Causation in Occupational Disease Cases Growing Out of Exposure to Toxic Substances   503

Ashford, N., et al., Monitoring Workers for Exposure and Disease: Scientific, Legal and Ethical Considerations in the Use of Biomarkers (1990) 503

Notes   506

APPENDIX A

The Occupational Safety and Health Act   509

APPENDIX B

The Toxic Substances Control Act   541

Index   593

Index of Cases   612

Preface

Technology is the mainstay of the modern industrial state. New developments in materials, manufacturing processes, final products, and work organization are critical to a dynamic economy. However, the development and utilization of technology often has negative consequences for workers, such as injury and disease, and deskilling and displacement. The legal system has responded in many ways to these problems, and that response affects employers, workers, and a variety of workplace professionals.

Managers, engineers, and scientists are central to the success of the modern industrial firm. They are the professionals who manage existing technological systems and they also are ultimately responsible for changing technology to meet environmental, health, and safety needs. They play crucial roles in stimulating and planning the development of new technology and production systems. Thus, understanding health and safety regulation and the laws that govern industrial relations has become increasingly important to successful management and technological planning.

This book is an outgrowth of a graduate-level course developed and taught at the Massachusetts Institute of Technology, as well as at several law schools and schools of public health, over the last 13 years. The book addresses a variety of scientific, legal, and policy issues growing out of the technological transformation of the American workplace.

The text begins with a discussion of the evolution of technology, work, and health since the turn of the century (Chapter 1). This chapter traces the economic and political forces that spurred the development of modern workplace law. A brief primer in administrative law—the law guiding governmental agencies—is provided in Chapter 2 in order to place in proper perspective the obligations, powers, and limitations of the Occupational Safety and Health Administration (OSHA), the Environmental Protection Agency (EPA), the National Labor Relations Board (NLRB), the Equal Employment Opportunity Commission (EEOC), the Office of Management and Budget (OMB), and the other governmental bodies whose actions affect the workplace.

The direct regulation of technologies affecting the health and safety of workers is explored through an analysis of the Occupational Safety and Health Act of 1970 (Chapter 3) and the Toxic Substances Control Act of 1976 (Chapter 4). Economic issues are fundamental to an analysis of these two regulatory systems. Chapter 5 addresses: (1) the costs of occupational disease and injury, (2) market alternatives to regulating health and safety, (3) the role of economic considerations in standard setting, and (4) the usefulness of economic analysis in regulatory decision making.

Much has been written about problems of labor commitment to productivity improvement. Employer attitudes toward health and safety in the workplace signal the extent employers care about their workers. Ideally, such evidence of concern should stimulate cooperation between management and labor which, in turn, should lead to improved workplace technology that is both safe and productive. Chapter 6 looks at the most important *formal* structure within which such cooperation takes place: the collective bargaining process of the National Labor Relations Act.

Regulatory regimes, market-based approaches, and collective bargaining all require adequate and accessible information to function optimally. In the context of employee health and safety, this information includes the identification of chemicals used in production, measurement results of ambient air monitoring, and data gathered from the biomonitoring of workers. Chapter 7 addresses the transfer of such information through right to know laws and explores its importance as an instrument for achieving statutory and societal goals.

Because employees' attempts to improve their working conditions have at times resulted in retaliation by employers, Chapter 8 explores the legal avenues available for worker redress. State workers' compensation systems and worker tort suits represent legal avenues for remediation where workplace health and safety efforts fail. They may also promote injury prevention and technology redesign. These topics are addressed in Chapter 9.

This text is intended for students at all levels. It is written for students in business, law, and engineering because they are likely to be the future managers of technological companies. The book is also meant to assist the professionals who have the knowledge and opportunity to redesign workplace technology and to ameliorate its effect on working conditions—scientists, particularly chemists and material scientists, public health professionals, occupational physicians, and process engineers. Finally, this book is written for working men and women. In the last analysis, they may be the most important agents for change.

This text may be used as part of a larger course on health and safety, industrial relations, or business law, or it may stand alone. The basic legal skills and familiarity with legal materials provided by this book should also benefit students regardless of their ultimate career choice.

A comment is in order regarding the way in which citations and references appear in the text: There are three different conventions followed, reflecting the different styles used in legal, social science, and economic writing. In the legal chapters, the legal references appear in the text. In Chapter 1, citations appear in the text as to author, year, and page, with fuller references at the end of the chapter. In Chapter 5, we utilize numbered footnotes, which appear at the bottom of the page, for material that is essential or useful for understanding the text. The mixed conventions reflect the multidisciplinary nature of the book. In excerpted material from both cases and articles, footnotes are often omitted.

Acknowledgments

Many persons and institutions were important in making this book possible. Richard Kazis and Robert F. Stone contributed most of the writing in Chapters 1 and 5, respectively. Kathleen Rest contributed original writing, the excerpting of case material, organizational assistance, and a sharp editing hand to the manuscript. Christine White assisted in the preparation of copy.

The National Institute for Occupational Safety and Health and the Kellogg Foundation provided financial support. The Massachusetts Institute of Technology afforded us the opportunity to develop and teach a course on the material that stimulated the writing of this book.

We are grateful to all of the above for making this book possible.

About the Authors

Nicholas A. Ashford, Ph.D., J.D., is Associate Professor of Technology and Policy at the Massachusetts Institute of Technology. He was Chairman of the National Advisory Committee on Occupational Safety and Health, served on the EPA Science Advisory Board, is a Fellow of the American Association for the Advancement of Science, and is currently Chairman of the Committee on Technology Innovation and Economics of the EPA National Advisory Council for Environmental Policy and Technology. He is the author of *Crisis in the Workplace: Occupational Disease and Injury* (1976), and co-author of *Evaluating Chemical Regulations: Trade-off Analysis and Impact Assessment for Environmental Decision-making* (1980), *Analyzing the Benefits of Health, Safety, and Environmental Regulations* (1982), *Monitoring the Worker for Exposure and Disease: Scientific and Ethical Considerations in the Use of Biomarkers* (1990), and *Chemical Exposures: Low Levels and High Stakes* (1991), Van Nostrand Reinhold. Dr. Ashford holds both a Doctorate in Chemistry and a Law Degree from the University of Chicago.

Charles C. Caldart, J.D., M.P.H., is a Lecturer in law and technology in the Department of Civil Engineering at the Massachusetts Institute of Technology, and a Research Associate at MIT's Center for Technology, Policy, and Industrial Development. Mr. Caldart is also the director of litigation at the Boston office of the National Environmental Law Center, a non-profit public interest law and policy organization. He is a co-author of *Monitoring the Worker for Exposure and Disease: Scientific, Legal and Ethical Considerations in the Use of Biomarkers* (1990). Mr. Caldart received his law degree from the University of Washington, and his M.P.H. from the Harvard University School of Public Health.

CHAPTER 1

Technology, Work,

and Health

A. INTRODUCTION

The laws, legal processes, and institutions discussed in this book have a common focus: broadly defined, they shape the conditions of work in America. Labor law, occupational health and safety protections, regulation of chemical production, right-to-know laws, rules against discrimination on the job, and compensation for harm—each provides workers with some form of protection or avenue for redress. One of the aims of this book is to illustrate how the legal system can be used to control existing technology and to redirect technological development into less hazardous and health-threatening directions, thereby making our nation's workplaces healthier, safer, and more desirable environments.

Before beginning a detailed exploration of specific topics, it is important to examine how the evolution of America's industries and workplaces has led to pressure for the enactment of the various laws and legal remedies that are the core subject of this book. In this chapter, readers are provided a brief historical overview of how work has changed in the United States in the past century and how economic growth and development have triggered a variety of efforts to limit the costs to and increase the protections for American workers. Particular emphasis is placed on the period since 1945, the postwar period during which most of the statutes and evolving legal protections discussed below were enacted or established.

The authors are indebted to Richard Kazis for assistance with this chapter.

What are the characteristics of an industrial system at any particular moment in time? What are the factors that define our day-to-day experience of work? In this chapter, we focus on five factors, each of which helps shape the conditions and the experience of work. These are the *technology* used to produce and distribute goods and services; the *mix of industries and occupations* in the economy; the *way work tasks are organized and allocated; the relations between management and employees,* in union and nonunion settings; and *the kinds of dangers to health and safety* that workers face on the job.

Each of these five factors colors our experience of and attitude toward work. Each is potentially a source of conflict, conflict that at times has led to regulation and legislation designed to alter the balance of power and responsibility in the workplace.

The *technology* used in a given job or workplace—and the *scale, scope, and speed of technological change*—are major determinants of the level of complexity, productivity, employee satisfaction, security, and risks to health and safety associated with that work (Colton and Bruchey 1987; Cyert and Mowery 1987; Guile and Brooks 1987; Piore and Sabel 1984; Office of Technology Assessment 1988).

The *structure of industry and of occupations* determines what kinds of jobs and in what kinds of environments we are likely to spend our working lives. In the early 1800s, for example, Americans were most likely to be employed in agriculture or skilled trades; today, new entrants to the work force are far more likely to work in an office than on a production line (Hartmann 1987; Johnston 1987; Leontief and Duchin 1985; Singlemann 1978; Department of Labor 1988).

The *organization of work* also shapes the way we experience and relate to our jobs. If the workplace hierarchy is steep and rigid and employee initiative is discouraged, job satisfaction is likely to be lower than if creativity and initiative are encouraged. If one works for a small firm, one's day-to-day experience is very different from that of someone working for General Motors or du Pont (Bowles et al. 1983; Braverman 1974; Chandler 1977; Gordon et al. 1982; Hirschhorn 1984; Zuboff 1988).

Labor-management relations, including the role of unions, is concerned with how people interact in an industrial system. This system is shaped by internal dynamics but also by technological change, historical patterns of work organization, and the kinds of industries and occupations that are growing and contracting. One's experience of work is deeply affected by patterns of labor-management relations at the workplace—how employers treat employees and how workers react (Bowles et al., 1983; Brody 1980; Freeman and Medoff 1984; Green 1980; Heckscher 1988; Kochan et al. 1986; Osterman 1988).

Finally, *workplace health and safety* is a critical determinant of job satisfaction and commitment. Moreover, the level of concern for health and safety in the workplace is also an important indicator of a society's overall commitment to its members' well-being (Ashford 1976; Bureau of National Affairs 1987; Nelkin and Brown 1984; Noble 1986; OTA 1985a; OTA 1985b; Weil 1987, 1991a and 1991b).

Technology and technological change, the industrial and occupational structure, the organization of work, and the nature of labor-management relations all play a role in shaping working conditions and all have an impact on employee health, safety, and welfare. The literature that explores the many ways in which these factors interrelate is vast. For the purposes of this book, the precise chain of causality is less important than the interconnectedness. Changes in one element of the system can ripple through the various parts of the whole. Regulation of one element of the industrial system inevitably affects the others.

This chapter outlines historical trends and changes in technology, the structure of industry and occupations, the organization of work, labor-management relations, and workplace health and safety that have provided the background and impetus for the major legal and regulatory initiatives addressing the conditions of work. The intent is to provide sufficient background to enable readers to appreciate the timing and nature of management, labor, government, and societal responses to industrial change as they are played out in legislation, collective bargaining, and other institutional frameworks.

B. THE EARLY PERIOD: 1900–1945

1. Background

The first half of this century, through World War II, was the period of formation for the U.S. industrial system as we know it. Many of the giant corporations that dominate the U.S. economy were created in a wave of mergers and consolidations in the years between 1890 and 1920, including U.S. Steel, Standard Oil, Nabisco, International Harvester, AT&T, ALCOA, and Eastman Kodak (Chandler 1977). Huge multiunit firms, coordinated by new and sophisticated systems of administrative management, established themselves as the dominant force in many manufacturing and processing industries, such as tobacco, soap, meatpacking, sewing machinery, farm equipment, chemicals, oil, copper, steel, and shoes. These firms adopted mass production methods that enabled them to lower per-unit costs, and they pioneered mass distribution networks that could guarantee sufficiently large markets. In this period, many older industries died and new ones, such as chemicals and automobiles, became increasingly central to the economy.

As the technology of production and the organization of work changed, so did labor relations in much of American industry. The new power of large industrial firms and the new layers of authority within the firm, especially the increased shop floor power of foremen, increased the range and intensity of pressures facing workers at the same time as it rendered the dominant craft model of unionization inadequate. Strikes reached a new peak in 1919, after the end of World War I. During the Depression, with unemployment reaching as high as 70 percent in some cities and desperate firms cutting wages to stay afloat, labor unrest exploded. By the end of the 1930s, a new legal framework for industrial relations had been created in the form of the Wagner Act, the federal labor legislation that remains the basis for U.S. industrial relations to this day.

The revolution in American manufacturing also forced issues of workplace safety and health into public consciousness during the first half of this century. With the scale and pace of production increasing and new mechanical technologies injuring workers at unprecedented rates in the early decades of the 1900s, firms initiated voluntary safety programs, workers found courts more receptive to liability suits, and state governments enacted a series of workers' compensation laws to stabilize the situation. In general, the era in which the giant corporation evolved and achieved dominance can be characterized as one in which the system of production grew far more rapidly and unpredictably than the mechanisms of regulation and control, either in the private or the public sector. Evidence of this can be seen in the state and federal regulatory initiatives of the Progressive era, as well as the extensive federal labor and social welfare legislation of the New Deal years. The following pages examine in greater detail the economic and social forces that shaped the way work was experienced in this period.

We focus on the five factors outlined above: technology, the organization of work, the structure of industry and occupations, industrial relations, and workplace health and safety.

2.   Technology and the Organization of Work

Beginning in the late 1800s, the American economy underwent a period of phenomenal growth and change. Modern transportation and communications networks, such as transcontinental railroads and the telegraph, made it possible to coordinate the production and sale of great quantities of goods. In all areas of U.S. manufacturing, mechanized production processes were invented and introduced that enabled producers to take advantage of the expanding national market and the potential for volume production. In some industries, machines replaced manual labor; in others, they also integrated processes, greatly increasing output. In the first decade of the new century, manufacturers, financiers, and investors eager to profit from the control of large-scale industrial enterprises initiated a wave of consolidations and mergers that created the modern business enterprises that have been central to the production and distribution of goods in the United States since that time. As these firms developed and grew, they became capital-intensive, energy-consuming giants producing in large volume for mass markets. Production and administrative technologies were introduced with this single goal in mind (Chandler 1977; Gordon et al. 1982).

In mass production firms, specialized machines are operated by workers who have narrow job definitions and the responsibility of performing specific operations at high speeds and with frequent repetitions. The high volumes of standardized goods enable firms to invest in expensive capital equipment designed to do a very particular operation. While it would be wrong to think that by the 1930s most manufacturing facilities in the United States were organized along the lines of Henry Ford's assembly plants, it would not be incorrect to say that mass production had triumphed by that time. That is, the principles of mass production served as the basis for the manufacturing strategy of the nation's most advanced firms and the model to which smaller and less advanced firms aspired.

The triumph of mass production had dramatic implications for the organization of work. Between 1890 and 1929, as mass production spread across diverse industries, craft skills were replaced by machines and semiskilled operative work became the typical production job. Economist David Gordon and his coauthors (1982, p. 128) have labeled these years the period of the "homogenization of labor," during which production jobs became more and more alike. Homogenization of the nature of industrial work was accompanied by increasingly impersonal conditions of work as the size of many plants and firms grew significantly. As plant size grew, production workers came increasingly under the supervision of foremen, who were given wide discretion by managers who found they could no longer run the firm and still maintain personal relations with all employees.

For industrial workers, these changes meant a reduction in responsibility, security, and control of their work. As craft skills were replaced and supervision tightened, workers were treated more and more like an appendage to the machine, interchangeable with others, needing little in the way of education and training. The shift can be seen dramatically in the changing mix of skilled, semiskilled, and unskilled jobs at Ford between 1910 and 1917, before and after the introduction of the assembly line.

During that period, skilled workers dropped from 31 to 21 percent of total employment, unskilled laborers dropped from 38 to 16 percent of the work force, and semiskilled operatives jumped from 29 to 62 percent of total employment (Gordon et al. 1982, p. 133). Changes in technology, work organization, and worker discretion and initiative were inextricably intertwined. The "scientific management" of industrial production, as championed by Frederick W. Taylor and other industrial engineers, entailed a number of related organizational innovations: centralized planning and routing of all parts of the production process; detailed and extensive analysis and simplification of job tasks; allotment to workers of specific, predetermined tasks; and close supervision of performance. Consider this description of shoe production in a New England factory in the 1930s:

> In the modern, largely mechanized shoe factory, jobs have been extensively simplified, routinized and standardized. In the process of manufacture, a pair of shoes, instead of remaining largely in the hands of one or a very few experts, now passes through the hands of a large number of operatives, each of whom performs only a small part of the manufacturing process. In addition, many of the jobs consist primarily in feeding materials to machines which perform the actual processes. . . . The exceedingly limited scope of the technological activity permitted the modern shoe operative makes it relatively easy for a foreman to maintain discipline because he can immediately detect any deviation from the permitted patterns of behavior (Warner and Low 1947, p. 174).

3. The Structure of Industry and Occupations

During the early part of this century, manufacturing became the engine driving the U.S. economy. Agriculture had already begun its steady decline as an employer of American workers, shrinking from about 37 percent of the work force in 1900 to 27 percent in 1920 and 17 percent in 1940. From 1900 to 1920, the national labor force grew by almost 50 percent, and much of the growth was concentrated in manufacturing. Manufacturing employment rose from 2.5 million in 1870 to more than 11 million in 1920, with textiles employing the largest group of manufacturing workers. By 1920, almost four out of ten American workers were employed in manufacturing. From an agrarian nation dominated by self-employed workers, the United States had been transformed into an urban, industrial society in which 80 percent of the work force toiled for others. As one historian put it, "The typical male worker in 1920 left home early in the morning and spent the next eight to ten hours laboring in a mill or factory among scores, or in some cases thousands, of fellow workers" (Dubofsky 1975, p. 4).

While manufacturing and manufacturing-related employment moved to center stage during this period, the nation's industrial structure remained quite diverse. Mining and construction employed twice as many workers in 1920 as did iron and steel production. And through the first few decades of the 1900s, domestic servants out-numbered all other categories of employees (Dubofsky 1975). In the 1920s, clerical work became the fastest-growing occupation as large corporations required more secretaries, managers, and accountants to manage mass production and distribution systems. By 1930, more than 2.2 million women worked in offices (Green 1980, p. 105).

As the number of jobs in manufacturing and services expanded, new sources of workers were tapped. One source was workers who could no longer make a living off the land: almost 11 million workers left the nation's farms between 1870 and 1920, and millions more followed in the 1930s and 1940s. A second, critically important source of workers, particularly for manufacturing, was the mass of European peasants who came to America in successive waves of immigration beginning in the late 1800s and continuing through the 1924 limitation of immigration by Congress. During the first two decades of the century, new immigrants numbered close to one million a year. The nation's women provided a third source of new workers. Women increased their labor force participation over 35 percent between 1890 and 1930, to 22 percent of the total workforce. Finally, after the abrupt end to open immigration in the 1920s, American blacks began to enter the industrial work force in significant numbers.

Different segments of the labor force tended to be concentrated in particular industries and occupations. Relatively unskilled and uneducated immigrants joined the ranks of the growing industrial work force. Outside of the apparel industry, women did not benefit greatly from the growth of manufacturing employment. In 1930, 79 percent of women workers were employed as teachers, clerical workers, retail sales workers, apparel operatives, or domestics (Gordon et al. 1982, p. 151). Although black industrial employment began to rise in the 1920s, most blacks remained outside the industrial economy until World War II. In 1910, blacks accounted for only 2.2 percent of nonagricultural employment in the North and West. Black inroads into nonagricultural employment included work in tobacco and construction and, as a result of their introduction as strikebreakers, coal mining, meatpacking, and iron-ore mining. When blacks were employed in manufacturing, they were isolated in the most low-skilled, and generally the most dangerous, jobs. As labor historian David Brody writes, "The sharecroppers and farm laborers took their places at the bottom of the ladder vacated by the Slavic and Italian workers" (Brody 1980, p. 18). In 1928, 80 percent of black and Mexican employees of United States Steel's Gary Works held unskilled jobs; 85 to 95 percent of skilled and semi-skilled jobs were held by whites (Green 1980, p. 125).

4. Labor-Management Relations

Accompanying the rapid, dramatic changes in the technology and organization of work in the early 1900s was an intensification of industrial conflict. This was followed by a period of accommodation in the 1920s as employers devised responses to worker unrest, which was followed in turn by a resurgence of conflict in the 1930s during the hard times of the Great Depression. Union membership rose 300 percent between 1897 and 1904, but stopped abruptly in the face of intense employer counterattacks in the period before World War I.

Across many manufacturing industries, new technologies and patterns of work organization challenged existing labor-management relations. Some crafts, such as glassblowing, disappeared almost entirely. Others were deskilled, leading to battles between threatened workers and employers over work pacing, wages, work rules, and union recognition. Employers organized in trade associations and pressed their advantage in efforts to keep unions from expanding their membership as manufacturing industries and firms grew. By and large, their efforts were successful. Unions retreated from steel mills, packing houses, and auto plants in the years before World War I. While skilled craft workers were shrinking in numbers and the ranks of semi-skilled operatives were expanding, the existing craft structure of U.S. labor unions (coupled with intense opposition from employers) kept unions from making great strides in membership or representation. Efforts to respond to the changing size and organization of

American industry by building industrial unions led to three large-scale organizing efforts in the years preceding the Depression, in apparel and mining, which were largely successful, and in steel, which was crushed.

During World War I, labor shortages and a concern with maintaining high levels of production made conditions for organizing more favorable. The American Federation of Labor, which had long resisted government assistance in its fight for membership and recognition, became a partner in the war effort, for which it was rewarded with government protection for organizing efforts in major war industries. Union membership rose from 2.8 million in 1916 to 5 million in 1920, though most of the gains were in craft unions in transportation, machine shops, and construction.

After the war, though, employers again took the offensive. The campaign had elements of both the carrot and the stick. On the one hand, employers united in a campaign to break the nation's most powerful unions by abolishing union shops. On the other hand, the nation's largest and most profitable employers instituted safety programs, community funds, health plans, and other benefits for their workers. The anti-union campaign was quite successful. By the end of 1922, unions in meatpacking, steel, lumber, and the maritime industry were virtually nonexistent. The United Mine Workers lost five out of six members in the 1920s. By the eve of the Depression, American unions represented only 12 percent of the work force. At the same time, labor-management relations in many large firms were being reshaped by programs of "welfare capitalism," a recognition by employers of some responsibility for their workers beyond wages that led many employers to initiate group insurance, mutual aid societies, and even pensions for their workers. Paternalism without unions spread as the new model of labor-management relations, made possible in part by the changes in technology, work organization, and employer power that characterized the transformation of the U.S. economy in the early 1900s. However, as David Brody points out, it was the prosperity of the 1920s that made the welfare capitalism model work. After 1929, with production at a virtual standstill, the largesse—and the ethos of cooperation that went with it—could not be sustained.

The Great Crash of 1929 and the ensuing Depression years reshaped labor-management relations in the United States. As a result of both increased labor militance and New Deal reforms, by the end of World War II union membership had risen above 30 percent of the work force and a system of industrial relations had been institutionalized and formalized at the national level.

In the first few years of the Depression, as unemployment reached 60 to 80 percent in some industrial cities and as business credibility plummeted, worker militance grew in cities and industries across the nation. Spurred by passage of the National Industrial Recovery Act in 1933, which recognized the right of workers to "organize unions of their own choosing" and to bargain collectively with employers, the United Mine Workers and other unions demanded recognition and contracts. In 1935, after the NIRA was declared unconstitutional, Congress passed the National Labor Relations Act, also known as the Wagner Act, which recognized the right to organize and established the National Labor Relations Board as the enforcement mechanism to oversee union elections and to regularize the unionization and collective bargaining processes. Workers and unions across the United States again pushed for recognition and protection.

This surge of activism was propelled by the diffusion of a successful tactic: the sit-down strike. Pioneered in the rubber factories of Akron, sitdowns spread to include auto workers and others in heavy industry, as well as hospital workers, garbage collectors, and sales workers. By April 1937, union membership—split between unions of the American Federation of Labor and the Congress of Industrial Organizations—had jumped to seven million.

In the late 1930s, America was an industrial battleground and the permanence of labor's gains and of the Wagner Act framework were still in question. However, World War II and the first few postwar years settled many of the outstanding questions. Labor shortages and the need for national unity during the war years led to the consolidation and strengthening of American unions, greatly increased membership (up to 14.7 million), and recognition by many of the more intransigent large employers, such as Ford and Bethlehem Steel. By the end of the war, the Wagner Act framework had become institutionalized and labor unions had assumed an important role in the political arena. In the first few postwar years, *a modus vivendi* was arrived at between the leaders of corporate America and its unions. Labor unions would pursue wage and benefit gains but would not challenge management prerogative in areas of corporate decisionmaking. Unions would also limit the role of labor radicals in their organizations. In exchange, the Wagner Act framework of industrial relations—with amendments outlawing sitdown strikes and some other tactics that were perceived to give labor serious disruptive power—would be preserved and productivity gains would be shared. Union membership continued to increase. Labor had become a recognized institutional and political force.

5.   Health and Safety

The rapid changes in the nature of production in the United States in the early 1900s had a direct and adverse impact on health and safety on the job. Mechanization led to the intensification of the pace of work for many workers. Although hard data are difficult to find, the increased speed of work and the introduction of large new machinery clearly contributed to a jump in industrial accidents. One study of industrial accidents in New York State in the years 1911 through 1914 concluded that power machinery was by far the most significant cause of nonlethal work accidents. A journalist writing in 1908 estimated that 35,000 workers were killed and 536,000 injured each year in a work force of only 30 million. In August 1910, the Cleveland Citizen editorialized that the United States had become an "industrial slaughterhouse." The average number of industrial accidents during the period 1926 through 1929, when production was high and union organization and other kinds of worker protections weak, was almost twice that of the postwar years between 1950 and 1970.

While industrial accidents were the most visible threat to health and safety on the job, work-related diseases were also taking their toll: stone cutters weakened and died of lung disease; cigar and tobacco workers suffered heart and respiratory ailments; hat makers contracted nerve disorders from inhaling the mercury used to treat furs and felt; women who painted clock faces with radium had high incidence of fatal radiation poisoning. Yet, throughout the first half of the century, little attention was paid to the problem of occupational disease and little protection was offered workers who might be most at risk.

Reformers in and outside industry paid some attention to industrial safety during the go-go years of the early 1900s. Prompted by middle-class Progressive reformers, state legislatures in 26 states enacted legislation in the first decade of the 1900s that made it easier for workers injured or killed in industrial accidents to win compensation in court. These employer liability laws removed the employers' defense, known as the "fellow servant" rule, that had allowed employers to escape liability if the employer could show that a fellow worker was responsible for the injury. As these laws proliferated, employers began to fear they would have to bear the costs of accidents that for years they had avoided. Led by the National Civic Federation, employers pushed for a workers' compensation system that would recognize employer liability but stabilize the cost of compensation. By 1920, all but six southern states had passed workers' compensation laws.

While these statutes redressed a longstanding wrong, they also had an unfortunate effect on employer attitudes toward workplace safety. Before passage of workers' compensation laws, railroads, U.S. Steel and other large employers had begun to develop serious safety programs designed to reduce accidents. With the costs of accidents made more predictable, employers lost some of the economic incentive to avoid accidents. As a result, safety programs languished (Berman 1978, p. 77).

During the Depression, health and safety concerns took a back seat to simple survival. Employers were unable to afford safe work places, and they were aware that workers had few options but to take whatever jobs they could find. Thus, although the total number of industrial accidents declined during the Depression, the decline was caused by the precipitous drop in production, not by increased protection of worker health and safety. In fact, for those who were employed, dangers abounded. At the 1936 United Auto Workers convention, one doctor reported that there had been 13,000 cases of lead poisoning in auto factories since 1929. A worker in a Dodge plant described conditions in 1932: "[There] was no attempt to ventilate the work areas or to take the pollutants out of the air. . . . It was an accepted fact that thousands of metal finishers in the auto industry suffered from lead poisoning" (Kazis and Grossman 1982, p. 174). Across manufacturing, mining, construction, and other dangerous trades conditions were similar.

The federal government tried to respond during the 1930s through the health and safety activities of the newly created Division of Labor Standards. This agency helped local and state governments and labor officials deal with specific hazards and was responsible for developing safety and health standards. It also provided training for state health and safety inspectors. During the Depression, however, little progress was made.

During World War II, the demands of wartime production and the enlistment of millions of new workers into the manufacturing economy precipitated a new surge in industrial accidents. During the first three years of the war, more Americans were killed and injured in work-related accidents than on the battlefield (Markowitz and Rosner 1986, p. 343). In a climate of national unity and sacrifice—and an economy short on available labor—this was unacceptable. Through the Division of Labor Standards, the federal government took a more active role in plant inspections, training for labor and management, and education on industrial safety. Immediately after the war, though, the combination of a reaction against the New Deal labor laws and the return of male workers into the labor force resulted in the effective dismantling of these programs, leaving American workers with few protections for occupational health and safety. This situation continued until the passage of the Mine Safety and Health Act and the Occupational Safety and Health Act in 1969 and 1970, respectively.

C. POSTWAR CHANGES: 1945–PRESENT

There is a temptation to look back on the first three decades after World War II as the glory years of the American economy. Challenges from overseas competitors were nonexistent for much of the period. American industry seemed an invincible giant. Manufacturing productivity was high. Labor-relations stabilized, with the relative acceptance of a permanent, if limited, role for responsible trade unions. The United States seemed to be the initiator of technological innovations that were copied by other industrial nations forced to play catch-up. In the 1950s and into the 1960s, America seemed destined to be the unchallenged economic and political leader of the non-Communist world.

This dominance, however, was shorter-lived and far less stable than most people imagined in the late 1950s. For both internal and external reasons, the relative stability and remarkable prosperity of the first three postwar decades has been replaced by a far more difficult and challenging set of circumstances and conditions that have had a very significant impact on work in the United States. The globalization of the economy together with the technological revolution represented by the diffusion of microelectronics-based technologies, have had a dramatic effect. As a result, the technology and organization of work, the structure of occupations and industries in the United States, the nature of labor-management relations, and the challenges to health and safety on the job have all changed so greatly that the reality and experience of work in the late 1980s differ markedly from those of even 15 years ago.

It is useful to think of the postwar era as two very different periods: (1) 1945 to the mid-1970s, which was a period in which relative prosperity and stability hid the build-up of serious grievances until they became visible in the 1960s; and (2) late-1970s to the present, a period of intense dislocation and change that has shattered many of our assumptions about work and working. During the first period, the tensions generated at work—by technology, work organization, human relations, and health and safety hazards on the job—were primarily a result of tendencies within the U.S. industrial system that had been building since before World War II and that accelerated in the 1960s. During the second period, the forces shaping the working life of Americans have been both domestic and international. In each period, as in the first half of the 1900s, rapid changes in the production system generated a host of work-related problems that prompted demand for legislation, regulation, and other avenues of redress and protection.

1. Technological Change

In the first few decades of the postwar period, the dynamic technologies were not the computer-related technologies that surround us today. Rather, they were the chemical, electronic, and mechanical technologies that had been at the heart of industrial production for decades. Added to these, though, were technologies developed or exploited more fully during the war, such as transistors, synthetic chemicals, and nuclear technologies.

Much of the technological change in this period was designed to make mass production more efficient through mechanization of jobs that had previously been done manually. Thus, the early 1950s saw the beginning of major investment in expensive factory automation in auto, steel, rubber, textiles, and other industries.

Automation could lead to greatly increased output per worker. Ford found that one worker could run a transfer machine performing 500 operations that were once done by as many as 70 workers. Similarly, two workers using automatic machinery could assemble 1000 radios a day, replacing 200 workers who had previously done the job by hand (Green 1980, p. 213). New technology designed to increase output was not restricted to assembly lines. In mining, for example, the introduction of continuous mining machinery revolutionized underground mining. These changes in the technology of production prompted broad concern about the displacement of workers by machines. Congressional hearings were called. A federal commission was established. However, in the context of a growing economy, displaced production workers had relatively modest income loss or unemployment spells before finding new work of comparable status.

Productivity-enhancing innovations were an important element in the success of American industry during the 1950s and 1960s. There was, however, a dark side to these changes. In addition to the problem of worker dislocation, the continued and accelerated degradation of work for many industrial workers and the increased speed of production using new machines and new procedures—both of which are discussed in greater detail below—contributed to the seriousness of the occupational health and safety crisis that led to demands for protection in the late 1960s and early 1970s.

In addition to technological change designed to mechanize production, the first few decades of the postwar period also witnessed a phenomenal increase in the use of new petrochemical-based materials in both industrial and consumer products. Before World War II, U.S. production of synthetic organic chemicals totaled fewer than one billion pounds a year. By 1976, production had increased to 162.9 billion pounds a year. The number of commercially available synthetic compounds rose from about 17,000 in 1958 to 58,000 in 1971 to over 70,000 in the mid-1980s (Odell 1980, p. 132). The proliferation of chemical compounds and chemically based products—such as polyesters and nylons, detergents, plastics, lubricants, insecticides, pesticides, and herbicides—made chemicals one of the major growth industries of the postwar period and introduced chemical products into every nook and cranny of U.S. industry.

The rapid surge in the use of petrochemicals resulted from the substitution of chemicals for natural products: "plastics for paper, wood, and metals; detergents for soap; nitrogen fertilizer for soil, organic matter and nitrogen-fixing crops (the natural sources of nitrogen); pesticides for the insect's natural predators," according to biologist Barry Commoner (1987, p. 59). Between 1940 and 1960, fertilizer production increased fourfold. Plastic production more than doubled in the ten years between 1955 and 1965.

Unregulated and untested for negative environmental and health impacts, the proliferation of synthetic chemical products during the postwar years, like the introduction of productivity-enhancing mechanization and automation, proved to be a major problem in the 1960s. According to Commoner (1987, p. 60), the changes in the technology of production that have occurred since World War II, "turned the nation's factories, farms, vehicles and shops into seedbeds of pollution: nitrates from fertilizers; phosphates from detergents; toxic residues from pesticides; smog and carcinogenic exhaust from vehicles; the growing list of toxic chemicals and the mounds of undegradable plastic containers, wrappings, and gewgaws from the petrochemical industry." By the 1960s, in communities and workplaces, the negative effects of this technological change were difficult to ignore.

In the past two decades, and particularly since the mid-1970s, a new era of technological development has begun, based on the development and diffusion of microelectronic-based products and processes. The silicon chip has improved information and communication technologies in ways that are affecting all aspects of U.S. society and economy. As labor scholar Steven Deutsch (1988, p. 5) notes, "In the 1950s, it took millions of dollars to automate a car factory. Today, automating a travel agency, bank, insurance, office, or nearly any workplace may cost only a few thousand dollars." Today, one in six U.S. workers uses a computer on the job. One half the U.S. labor force is employed in workplaces where microelectronic technology has been applied or will be applied in the next few years. (Deutsch 1988.) According to the 1988 study, Technology and the American Economic Transition, prepared by the Congressional Office of Technology Assessment, "An overwhelming body of evidence suggests . . . that new technologies for collecting, storing, manipulating, and communicating information have the potential to revolutionize the structure and performance of the national economy" (Office of Technology Assessment 1988, p. 16).

New information technologies have facilitated the globalization of production by reducing the cost and increasing the speed of international coordination of economic activity. They have led to dramatic changes in the organization of production, making it possible to reorganize manufacturing away from dominant standardized long-run mass production systems toward more flexible, shorter-run, niche strategies. They have had widespread impact on the structure of industry and occupations and on the nature of work in the American economy. They have also created their own set of occupational health hazards (LaDou 1985).

The pace of change has been phenomenal. Over 40 percent of all new investment in plant and equipment is now in "information technology," which includes computers, copying machines, facsimile machines, and the like. This is twice the 1978 share of total investment accounted for by such purchases (OTA 1988, p. 16). As the OTA report notes, "Computing power once confined to specially equipped, air-conditioned rooms has already insinuated itself into everything from automobile carburetors to greeting cards and teddy bears." While it is too early to tell whether the dislocation and uncertainty that this wave of technological change has triggered is a one-time disequilibrium or the beginning of a period of endemic instability, there is no question that microelectronics-based technologies are changing the way most Americans work and experience work.

2. The Organization of Work

The postwar period has seen both the maturation of mass production and the beginning of its disintegration as the dominant form of work organization in the United States. During the first few postwar decades, into the 1970s, most American manufacturing industries continued to work on developing ways to improve their ability to produce large quantities of standardized goods with great efficiency. The mass production model and the narrowing of job content and worker initiative that accompanied it reached their most highly developed form in industries such as automobiles and steel. In Detroit's auto plants, for example, one worker would weld a fender, a second inspect the weld, a third order more fenders from storage, a fourth maintain the welding gun, a fifth supervise, and a sixth (the industrial engineer) determine the content of the other five jobs. Prior to a postwar effort to simplify the steel industry's job classification system, there were between 45,000 and 50,000 job titles at United States Steel (Cole 1979, p. 105).

This method of organizing work was not confined to the shopfloor. Skilled and professional work was also organized so that employees developed narrow, though deep, skills. In the auto industry, for example, there are engineers whose sole function is to design door locks. The commercial aircraft industry has a bewildering range of engineering specialists, each with extensive knowledge of one aspect of building airplanes but few with a full understanding of the production process.

What has this organizational structure meant for workers in American manufacturing? According to a pioneering 1952 sociological study of assembly line work, is has meant mechanical pacing of work, repetitiveness, minimum skill requirements, minute subdivision of product worked on, and surface mental attention (Walker and Guest 1952). For the semiskilled operatives who became the backbone of American manufacturing, mass production meant limited involvement in production and, often, limited satisfaction with their underutilization. Here are some typical comments of production workers during the heyday of American mass production (Walker and Guest 1952, pp. 54, 138).

> There is nothing more discouraging than having a barrel beside you with 10,000 bolts in it and using them all up. Then you get a barrel with another 10,000 bolts, and you know every one of those 10,000 bolts has to be picked up and put in exactly the same place as the last 10,000 bolts.

> One of the main things wrong with this job is that there is no figuring for yourself; no chance to use my brain.

> You're just a number to them. They number the stock, and they number you.

In the past decade, there are signs that the mass production model is giving way to new forms of work organization, as firms adjust to new microelectronics-based technologies and search for ways to compete in a global marketplace in which product cycles have become much shorter and responsiveness and quality have, in many cases, replaced standardization and price as the criteria for success. Learning from their Japanese and European competitors, for whom Taylorism was never so dominant and work was never organized in such narrow, formalized jobs, American firms have begun to experiment with a variety of ways to broaden job responsibilities, achieve more flexible deployment of human and physical capital, and restructure the way work is organized (Piore and Sabel 1984; Osterman 1988; Zuboff 1988). Richard Walton of the Harvard Business School calls this trend moving from "control to commitment in the workplace" and has described the kinds of changes it entails in a range of areas, including job design, performance expectations, management organization, compensation policies, employment assurances, employee voice policies, and labor-management relations (see Table 1-1) (Walton 1985).

Some firms are experimenting with decentralizing their production process, giving more responsibility to subcontractors and playing a greater role in helping them meet new quality demands. As a recent National Academy of Sciences report notes (Cyert and Mowery 1987, p. 125):

> The advantages of the large firm, which were rooted for many years in the low costs of intrafirm communications, are being eroded in some industries by the rapid decline in the costs of interfirm communication, in addition to other factors. Together with intensified international competition, this development has led U.S. firms in some industries to rely on external sources for administrative and support services, which often results in "spinning off" portions of these activities to other organizations.

Table 1-1. Work-force strategies

	Control	Transition	Commitment
Job Design Principles	Individual attention limited to performing individual job.	Scope of individual responsibility extended to upgrading system performance, via participative problem-solving groups in QWL,* EI,** and quality circle programs.	Individual responsibility extended to upgrading system performance.
	Job design deskills and fragments work and separates doing and thinking.	No change in traditional job design or accountability.	Job design enhances content of work, emphasizes whole task, and combines doing and thinking.
	Accountability focused on individual.		Frequent use of teams as basic accountable unit.
	Fixed job definition.		Flexible definition of duties, contingent on changing conditions.
Performance Expectations	Measured standards define minimum performance. Stability seen as desirable.		Emphasis placed on higher, "stretch objectives," which tend to be dynamic and oriented to the marketplace.
Management Organization: Structure, Systems, and Style	Structure tends to be layered, with top-down controls.	No basic changes in approaches to structure, control, or authority.	Flat organization structure with mutual influence systems.
	Coordination and control rely on rules and procedures.		Coordination and control based more on shared goals, values, and traditions.
	More emphasis on prerogatives and positional authority.		Management emphasis on problem solving and relevant information and expertise.
	Status symbols distributed to reinforce hierarchy.	A few visible symbols change.	Minimum status differentials to de-emphasize inherent hierarchy.
Compensation Policies	Variable pay where feasible to provide individual incentive.	Typically no basic changes in compensation concepts.	Variable rewards to create equity and to reinforce group achievements: gain sharing, profit sharing.
	Individual pay geared to job evaluation.		Individual pay linked to skills and mastery.
	In downturn, cuts concentrated on hourly payroll.	Equality of sacrifice among employee groups.	Equality of sacrifice.
Employment Assurances	Employees regarded as variable costs.	Assurances that participation will not result in loss of job.	Assurances that participation will not result in loss of job.

Table 1-1. continued

	Control	Transition	Commitment
		Extra effort to avoid layoffs.	High commitment to avoid or assist in reemployment.
			Priority for training and retaining existing work force.
Employee Voice Policies	Employee input allowed on relatively narrow agenda. Attendant risks emphasized. Methods include open-door policy, attitude surveys, grievance procedures, and collective bargaining in some organizations.	Addition of limited, ad hoc consultation mechanisms. No change in corporate governance.	Employee participation encouraged on wide range of issues. Attendant benefits emphasized. New concepts of corporate governance.
	Business information distributed on strictly defined "need to know" basis.	Additional sharing of information.	Business data shared widely.
Labor-Management Relations	Adversarial labor relations; emphasis on interest conflict.	Thawing of adversarial attitudes; joint sponsorship of QWL or EI; emphasis on common fate.	Mutuality in labor relations; joint planning and problem solving on expanded agenda.
			Unions, management, and workers redefine their respective roles.

*Quality of working life.
**Employee involvement.
Source: Walton, p. 81. Reprinted by permission of *Harvard Business Review.* Copyright © 1985 by the President and Fellows of Harvard College; all rights reserved.

In many firms, management is pushing decisionmaking authority closer to the point where production or customer service decisions are made in an effort to improve productivity, cut waste, and increase competitiveness. This strategy, which is often accompanied by increases in expenditure on employee education and training and by recomposition of the skill levels of many jobs, is in many ways antithetical to the principles of mass production that had dominated U.S. industrial thinking for so long.

New technologies are also having an important, though sometimes contradictory, impact on the organization of work. Computer technologies can reduce the amount of skill required to do some jobs just as they can increase the amount and change the kind of skills required for others. In large part, the outcome depends on managerial choice rather than anything inherent in the technology itself (Spenner 1985; Adler 1984). There are certainly instances of deskilling. Just as certainly, though, upskilling results and "multiskilled" workers are needed to make use of the full capabilities of new technologies. However, the introduction of some computer technologies, such as computer-integrated manufacturing or flexible manufacturing systems, appears to give an edge to firms that change their organizational structure in ways that reduce barriers between functional areas and require greater communications, coordination, and planning.

Taken together, trends in technology, product markets, and intensified competition are placing new demands on American firms. While there is great variation among firms and industries, they heyday of mass production has passed, and American firms are experimenting continually and often in contradictory ways with new forms of work organization, some of which are a decided break from the past.

3.   Changes in the Structure of Industries and Occupations

In the postwar period, the structure of the U.S. economy has shifted dramatically. Many talk of the coming of a "postindustrial" economy. Manufacturing employment has contracted. Almost all new jobs today are being created in service industries. White collar and professional work has been growing while blue collar occupations have been in decline (see Table 1-2).

From the 1940s through the 1960s, mechanization of agriculture continued to push workers, particularly Southern blacks, off the farm and into manufacturing and service jobs. Farming employment as a percentage of total employment dropped to 8 percent in 1960 and 3 percent in 1980. For much of the postwar period, manufacturing employment expanded, though at a slower rate than in the first half of the century and at a slower rate than in the service sector. As a result, manufacturing employment as a percentage of total private nonfarm employment slipped from 36 percent in 1966 to 28 percent in 1979 (Cyert and Mowery 1987, p. 79). Since 1979, however, manufacturing employment has declined both relatively and absolutely. Estimates vary, but one figure places the net loss of manufacturing jobs in the United States between 1979 and 1987 at 1.9 million jobs (Mishel and Simon 1988, p. 25). Some industries were hit much harder than others: steel, automobiles, tires, textiles and apparel, and footwear all suffered

Table 1-2. Employment growth by sector, 1979–1987

Industry Sector	Employment		Job Growth	Industry Share of Job Growth	Median Weekly Earnings (1987)
	1979	1987			
	(1)	(2)	(3)	(4)	(5)
	(1,000s)	(1,000s)	(1,000s)		
Goods producing	26,461	24,885	−1,576	−12.8%	$389
Mining	958	742	−216	−1.8	508
Construction	4,463	5,032	569	4.6	402
Manufacturing	21,040	19,112	−1,928	−15.7	398
Durable Goods	12,760	11,235	−1,525	−12.4	408
Nondurable Goods	8,280	7,876	−404	−3.3	349
Service producing	63,363	77,219	13,856	112.8	344
Transp., Comm., Util.	5,136	5,377	241	2.0	485
Wholesale	5,204	5,797	593	4.8	401
Retail	14,989	18,262	3,273	26.6	258
Fin., Ins., Real Est.	4,975	6,588	1,613	13.1	370
Services	17,112	24,136	7,024	57.2	327
Government	15,947	17,063	1,116	9.1	425
Total	89,823	102,105	12,282		372

Source: Mishel and Simon, p. 25.

heavy employment losses, while aircraft and missiles, instrumentation, computers, and radio and TV equipment registered employment gains (U.S. Department of Labor 1988, p. 39).

Services have grown rapidly and steadily and have become by far the most important provider of new jobs in the economy (see Figure 1-1). Services employed about 59 percent of the labor force in 1950, but now provide three out of every four jobs. Today, hospitals employ more people than auto and steel combined; business services employ more workers than construction; jobs in retail food stores employ more than agriculture (Sweeney and Nussbaum 1989). Nine out of ten jobs created in the past 25 years have been in service industries. The lion's share of these new jobs—80 percent—were concentrated in three industries: retail trade, health, and business services. Finance, insurance, and real estate, which provide relatively high-paying service jobs, also grew rapidly in the past two decades.

While the proportion of service jobs has increased at the expense of manufacturing, many of these jobs depend upon the health of the manufacturing sector. According to political economists John Zysman and Stephen Cohen, "The employment of . . . 40, 50, or even 60 million Americans, half to three-quarters of whom are counted as service workers, depends directly upon manufacturing production" (Zysman and Cohen 1987, p. 58). Design and engineering services, payroll and accounting services, finance and insurance, repair and maintenance of plant and machinery, training and recruitment, testing and lab work, industrial waste disposal, advertising, and many other services are directly linked to manufacturing. According to one estimate, 60 percent of service

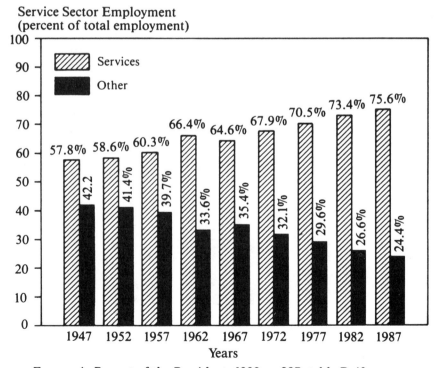

Source: *Economic Report of the President, 1988,* p. 297, table B-43.

Figure 1-1. Growth of the service sector, 1947–1987. p. f-33. (Source: *Sweeney, John J. and Karen Nussbaum,* Solutions for the New Work Force, *Washington, DC: Seven Locks Press, 1989, p. 194.*)

industry workers are involved in packaging, financing, insuring, distributing and selling manufactured products (Sweeney and Nussbaum 1989, p. 196).

Paralleling the change in the broad industrial mix during the postwar period as a whole and the past decade in particular has been a shift in the occupational mix of available jobs. In general, this shift has been away from jobs requiring less skill and education to those requiring more. White collar workers constituted about 36 percent of the labor force in 1950; by the early 1980s, more than half of all workers held white collar jobs. Between 1972 and 1986, employment of executive, administrative, and managerial workers grew twice as fast as did total employment. The most rapid growth was in employment of technicians and related support workers, an occupational category that includes health technologists and technicians. Employment of professional workers grew faster than total employment. In contrast, the number of private household workers dropped 11 percent, the number of fabricators, operators, and laborers fell by more than 9 percent and the number of farming, forestry, and fishing workers also fell. While the number of precision production, craft, and repair workers rose between 1972 and 1986, it did so at a much slower rate than total employment (U.S. Department of Labor 1988, p. 44). Even within industries, the mix of occupations is changing. In manufacturing, 13 percent of workers are employed in service or clerical occupations (Sweeney and Nussbaum 1989).

Since World War II, the composition of the work force and the distribution of job opportunities available to different population groups has changed significantly (see Table 1-3). Women, blacks, and other racial and ethnic groups have entered the work force in growing numbers and have been able to move into occupations and industries in which they were poorly represented in earlier decades. As jobs in occupations that tended to employ women—such as education, health, and clerical jobs—expanded in number, the number of women working in those jobs increased accordingly. By 1970, six million women worked as nurses, teachers, or health technicians, compared

Table 1-3. Females and blacks continue to be employed in low-paid occupations

Occupation	Total Who Are Female		Total Who Are Black	
	1960	1984	1960	1984
Total labor force	33%	44%	11%	10%
White-collar	42	55	4	7
Professional and technical	36	48	4	7
Managers, officials, and proprietors	16	34	3	5
Sales	40	48	2	5
Clerical	68	80	5	10
Blue-collar	15	18	12	11
Craftworkers and foremen	3	9	5	7
Operatives	28	26	12	14
Nonfarm laborers	2	18	27	15
Service workers	65	61	27	18
Private household	98	96	50	30
Farm workers	18	16	16	8

Source: Bezold et al., p. 26.

to 800,000 in 1920. In the past decade, women have accounted for 60 percent of labor force growth. Women have moved into white collar positions from which they had long been restricted—in law, medicine, finance, and other industries. In lesser numbers, women have also been able to move into traditionally male construction, repair, and manufacturing jobs. However, job segregation by sex in the labor market remains substantial, with negative effects on earnings and career mobility (Hartmann 1987). Cashiers, waitresses, bookkeepers, secretaries, and typists continue to dominate the ranks of working women.

Occupational segregation by race and ethnic group also persists. As blacks were forced off the farm by the mechanization of agriculture, many moved to northern and midwestern cities. But the decentralization and suburbanization of manufacturing in the postwar period left many blacks without work or with few options but low paying jobs in peripheral manufacturing industries and in services. In the 1960s, black workers made some earnings and employment gains in hospitals, government, services, and in core industries such as auto and steel. A black middle class has developed. However, much of the relative improvement was checked in the recessions of the 1970s and 1980s.

In the 1980s, the problem for blacks was not simply occupational segregation and the related limits to income and mobility, but a severe lack of employment experience and prospects for a large percentage of urban blacks, particularly for black males. Thus, while the percentage of black males in white collar jobs rose from 5 percent in 1940 to 27 percent in 1983, the proportion of employed black men dropped from 80 percent in 1930 to 56 percent in 1983. In the 1970s, black male labor force participation rates dropped below that of white males for all age groups. In 1979, with overall unemployment below 6 percent, 34 percent of black male teenagers were without work (Wilson 1987). For Hispanic workers in the United States, the pattern is similar, though less extreme.

One more trend in the structure of job opportunities in the United States must be noted: the dramatic increase in the number of people who comprise the "contingent" work force of part-time, temporary, and contract employees. A decade of corporate restructuring has led many firms to seek flexibility (and lower costs) through hiring on a part-time or temporary basis rather than creating new permanent jobs (see Table 1-4). Half of all new jobs in the 1980s were filled by part-time or temporary employees. (See Table 1-5 for variation in part-time work by industry in 1987.) Groups at the bottom of the employment hierarchy—women, minorities, youth, the elderly—are all over-

Table 1-4. Growth of the contingent work force

| | (millions of workers) | | Percent |
	1980	1988	Change
Part-time	14.3	20.0	40.0
Temporary	0.6	1.3	117.0
Business services	2.5	4.0	60.0
Self-employed	8.4	9.8	16.7
Total contingent work force	25.8	35.1	36.0
Total civilian work force	105.5	120.9	14.6

Source: Sweeney and Nussbaum, p. 56.

represented in the contingent work force. And while most contigent workers hold low-level retail, clerical, and service jobs, many white-collar employees such as accountants, engineers, and even doctors and managers work on a contract basis. While some of this flexible employment is by choice, involuntary part-time work is increasing far more rapidly than voluntary part-time employment (see Figure 1-2) (Sweeney and Nussbaum 1989; Mishel and Simon 1988).

Table 1-5. Rate of part-time work by industry, 1987

Industry Sector	Involuntary	Voluntary	Total Part-Time
Goods Producing	3.5%	3.4%	6.9%
Mining	3.9	2.2	6.1
Construction	6.9	4.6	11.5
Manufacturing	2.5	3.1	5.6
Durable goods	1.6	2.3	3.9
Nondurable goods	3.9	4.3	8.2
Service Producing	5.0	16.3	21.3
Transp., Comm., Util.	3.2	5.3	8.5
Wholesale trade	2.2	6.2	8.4
Retail trade	8.9	27.0	35.8
Fin., Ins., Real Est.	1.9	9.5	11.4
Public administration	1.3	5.1	6.4
Education	3.7	21.2	24.9
Service industries	5.3	17.0	22.3

Source: Mishel and Simon, p. 25.

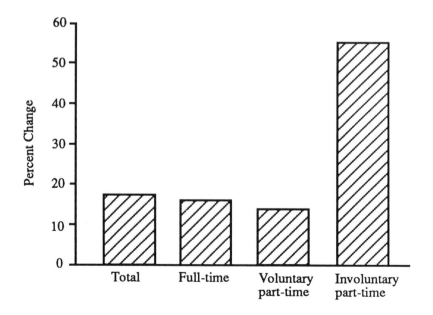

Figure 1-2. Growth in jobs by type, 1979–1987. p. f-37. (Source: *Mishel, Lawrence and Jaqueline Simon,* The State of Working America, *Washington, DC: Economic Policy Institute, 1988, p. 22.*)

4.   Labor-Management Relations

The postwar period has seen the rise of union membership and acceptance followed first by gradual and then by dramatic declines in union membership as well as in other measures of union strength, popularity, and acceptance. Through the 1970s, the change was gradual. By the mid-1980s, however, labor-management relations in one industry after another changed dramatically. On the one hand, the ability of unions to represent their members and the mass of unorganized workers weakened. On the other hand, a range of innovations in labor-management relations—from Employee Stock Ownership Plans to worker participation on corporate boards to quality circles and other forms of shopfloor participation to joint decisionmaking on a host of issues—were being introduced. In this section, we survey postwar trends in unionization, worker satisfaction, and innovations in labor-management relations.

(a) Unionization

After a few years of instability and conflict immediately after the war, a relative calm was established in labor-management relations. Although employers never abandoned the desire to operate without unions, collective bargaining and the system of industrial relations established by New Deal legislation became increasingly institutionalized and accepted. While employers pushed for limitations to the Wagner Act in the 1947 Taft-Hartley and 1959 Landrum-Griffin Acts and tried both to limit union influence and to defeat unionization drives, labor-management relations assumed greater predictability and day-to-day accommodation. In the 1950s, employer offensives against unions could no longer destroy the nation's trade unions as they had in the 1870s, 1890s, and 1920s. By the late 1950s, industrial relations professionals such as John Dunlop of Harvard were writing books describing the contours of the nascent U.S. "industrial relations system."

The high-water mark of union membership was 1954, when unions represented 39 percent of the private nonagricultural work force (34.7 percent of all U.S. workers). After that time, while union membership continued to rise absolutely, its growth did not keep pace with overall employment growth. Union representation of blue collar workers declined slightly through 1975 as a percentage of the total blue collar work force and unions failed to make significant gains in representation of white collar employees. Public sector and hospital worker unionization was one of the few bright spots for labor union membership in the 1960s and 1970s, sparked by the extension of the Wagner Act to hospital workers in 1964 and changes in state labor laws that legalized public sector unions. Public sector unionization rose from 11.6 percent in 1953 to 39.5 percent in 1975.

During the 1960s and 1970s, even as collective bargaining seemed to be increasingly stable and accepted in unionized settings, a significant nonunion sector emerged in almost every industry in which unions had been dominant. Moreover, whole new industries grew so rapidly that unionization did not penetrate. Unionization slipped from 90 to 60 percent of the blue collar work force in petroleum refining between 1960 and the 1980s. The nonunion segment of the underground bituminous coal mining industry rose from zero to 20 percent. In construction, nonunion shops first captured a significant portion of residential and light commercial construction and have since won major industrial and office jobs. In deregulated industries such as trucking and airlines, unionized carriers have gone bankrupt and have been replaced by smaller, unorganized companies. Even within the same company, the percentage of unionized facilities has been sliding. The number of nonunion facilities at companies such as Monsanto, 3M, and General Electric has been increasing steadily since the 1960s. In addition, in manufacturing industries such as computers and high tech electronics and service industries such as insurance and finance, unions have been unable to gain any significant presence (Kochan et al. 1986).

In the late 1970s, as international competition accelerated and American industry entered a period of intense restructuring, the position of American trade unions eroded rapidly. Between 1979 and 1984, union representation among blue collar workers fell from 44 to 32 percent in manufacturing, from 38 to 27 percent in construction, from 47 to 20 percent in mining and agriculture and from 62 to 46 percent in transportation, communications, and public utilities (Robinson 1987, p. 15). Union representation of the private nonagricultural work force has fallen below 16 percent, less than half its 1954 peak (Heckscher 1988, p. 3).

There are many explanations for the rapidity of this decline. They include the failure of American unions to organize the unorganized, the intensity of employers' legal and illegal union avoidance strategies, the emergence of a nonunion human resources management strategy that provided an alternative to collective bargaining, the inability of unions to respond to the pressures on management to cut costs in the face of international competition or to stand their ground in industries in which nonunion segments weakened union bargaining and strike power. Whatever the reason or reasons, the gradual and then precipitous decline in union membership in the United States has been an important factor in the evolution of workplace rights and protections and in the reshaping of labor-management relations. Management has taken the initiative in bargaining and in pursuing union-avoidance strategies. Many innovations in labor-management relations in recent years have been initiated in the

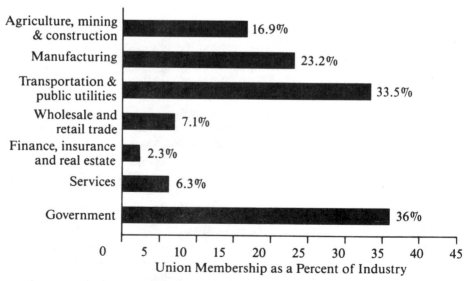

Source: Bureau of Labor Statistics. "Union Membership in 1987," News Release, USDL 88-27, January 22, 1988. Table 2.

Figure 1-3. Union membership by major industry, 1987. p. f-42. (Source: *Sweeney, John J. and Karen Nussbaum,* Solutions for the New Work Force, *Washington, DC: Seven Locks Press, 1989, p. 191.*)

nonunion sector, as part of concession bargaining in the union sector, or in the arena of legislated individual rights.

By 1986, according to one estimate, the proportion of union membership had dropped further in many industries: manufacturing, 24 percent unionized; transportation and utilities, 35 percent; agriculture, mining, and construction, 13.8 percent; services 6.3 percent; finance, insurance, and real estate, 2.6 percent. Government employment is one of the few areas where unionization has increased in the past 20 years. In 1986, 36 percent of government workers were represented by a labor union (Sweeney and Nussbaum 1989). (See Figure 1-3 for 1987.)

(b) Worker satisfaction

After an initial wave of strikes and difficult negotiations in the late 1940s, the 1950s brought a welcome lull in labor-management conflict. Workers seemed to be relatively happy with their jobs and lives. This "myth of the happy worker," as author Harvey Swados called it in an influential 1959 article, was punctured in the 1960s and 1970s as worker dissatisfaction became a basic fact of life in postwar industrial America.

In the mid-1960s, labor-management relations became more conflictual than they had been in more than 15 years. Industrial workers, like students and blacks during that tumultuous decade, were not content with the status quo. As young workers with more education entered the industrial work force only to find their jobs routinized and their creative potential checked by the dictates of mass production work, they chafed. In a period of social turbulence, that discontent evolved into a challenge to authority—a challenge targeted both at employers, who were responsible for degrading working conditions and authoritarian management, and at unions, who resisted demands to put anything other than wages on the bargaining table. In 1964, auto workers in Detroit launched their own campaign to "humanize working conditions." The incidence of strikes rose to a ten-year high in 1966 and the frequency of wildcat strikes, usually seen as an indication of discontent with working conditions, jumped between 1967 and 1969 (Noble 1986, p. 70). Sabotage, slowdowns, absenteeism and other forms of individual and collective protest also increased. In 1972, the "labor revolt" received national attention when workers at General Motors's newest, most productive plant in Lordstown, Ohio, walked off the job in response to a production speedup and over-regimentation of the work process. A year later, a special task force of the Department of Health, Education and Welfare published *Work in America*, a report that brought concern about alienation at work into the mainstream. According to the report, because work had not "changed fast enough to keep up with the rapid and widespread changes in worker attitudes, aspirations and values," many American workers were "dissatisfied with their working life" (Green 1980, p. 221).

Apparently, worker satisfaction did not improve in the 1970s and 1980s. For blue collar workers in declining mass production industries, the threat of unemployment and disruptive technological change has obviously fueled discontent. However, what pollster Daniel Yankelovich has labeled the "commitment gap" affects many other occupational groups as well (Yankelovich and Immerwahr 1983). The growing ranks of semiprofessional workers—the middle managers, technicians, sales representatives, and others who are playing an increasingly important role in the U.S. economy—are a good example. For these workers, the widespread and continual restructuring of white collar and management staffs in U.S. corporations has planted the seeds for significant disenchantment. The Opinion Research Corporation (ORC), which has surveyed close to 200 companies periodically during the past 30 years, found that managers' attitudes toward their firms has deteriorated. Perceptions of fairness, job security, and responsiveness to employees have all changed for the worse. Clerical workers are another group deeply troubled by today's work environment. According to ORC surveys, from the 1950s through 1979, clerical workers' approval of company fairness dropped from 70 to 20 percent (Heckscher 1988, p. 241). Other groups—such as part-time, temporary, and other contingent workers—are also under great stress in the current period of economic insecurity.

Many workers today do not see their efforts leading to gains for them. Their educational attainment is higher, yet their jobs do not seem to be changing rapidly enough in terms of discretion and authority. There remains a serious disjuncture between what people are being asked to do and what they feel they are capable of doing (Walton 1985). According to Charles Heckscher (1988, p. 241), "It is . . . fair to estimate that more than three-quarters of the work force is undergoing significant disruption of expectations." According to a quantitative model developed by economist Samuel Bowles and his colleagues, deteriorating worker motivation (as measured by trends in real spendable hourly earnings, work safety, and job satisfaction) was an important contributing factor in the slowdown of U.S. productivity growth in the 1970s (Bowles et al. 1984).

(c) Innovations in labor-management relations

As union strength and density has weakened, efforts to redress worker concerns and advance worker protections have taken two very different forms: (1) expansion through legislation of the rights of individual workers; and (2) new forms of employee participation and labor-management cooperation both outside and within the context of collective bargaining and union representation.

(i) Expansion of Individual Rights. Beginning in the early 1960s and accelerating in the 1970s, there has been a significant increase at the federal and state level in legislative and judicial actions to increase the protection of individual workers vis à vis their employers. The first innovation of this type was Title VII of the 1964 Civil Rights Act, which protects minority workers against discipline or discharge for reasons other than work performance. Following a costly suit against AT&T in the early 1970s, many corporations voluntarily changed their personnel policies and discipline systems so as to eliminate arbitrary managerial practices and avoid consequential legal actions.

In the 1970s, a number of federal statutes followed the Title VII precedent and restricted employer discretion in dealing with individual workers. Charles Hecksher (1988, pp. 159-160) provides a good summary:

> Due process protection was extended to people over the age of forty by the Age Discrimination in Employment Act. Among federal contractors, the Vocational Rehabilitation Act prohibited discharge or discrimination based on physical handicaps or mental limitations. The Occupational Safety and Health Act provided workers important rights to enforce standards of safety in the workplace. And a long series of regulatory statutes have included specific protections for workers. Before 1974, a company could fire employees to prevent them from vesting in pension funds, but the Employment Retirement Income Security Act of 1974 put a stop to that. Polluters, after the passage of the Clean Air Act Amendments of 1977, could no longer punish whistleblowers in their ranks. The list goes on: at least twenty-six major new statutes now contain similar protections.

At the state level, legislative initiatives to protect individual rights have been even more extensive and varied. Restrictions on the use of polygraph tests have been passed in many states, as have whistleblower protections and statutes outlawing employment discrimination on the basis of sexual preference, marital status, political activity, and other worker characteristics. At present, in a number of states the long-standing doctrine of "employment-at-will," which enables employers to fire workers without cause if and when they please, has been weakened by a number of court rulings (Stieber 1985). Together, these state and federal extensions of worker rights at the expense of management prerogative have dramatically changed the nature of labor-management relations in a number of arenas. And it is likely that this trend toward elaboration of particular individual rights will continue. Unfortunately, the lack of coherence and the unevenness of protections from state to state make it a system of protections that is inefficient, often unfair, and based on litigation and governmental intervention in labor-management relations rather than on an expansion of negotiation and bargaining between employers and their employees.

(ii) Worker Participation. Also starting in the 1960s, American firms have begun to experiment with various ways to increase worker participation in workplace decision making. One form of worker participation is through ownership. Employee Stock Ownership Plans have become more common as a way to motivate and give a stake to workers. In general, these plans tend to separate decisionmaking from ownership, leaving workers as owners but with little input regarding production, investment, or other decisions. In some cases, though, employee ownership has been a vehicle for the integration of management and ownership and has involved workers and their representatives in all aspects of running their firm.

Another form of worker participation that has had some, albeit limited, impact in the United States is the inclusion of union representatives on corporate boards of directors. While widespread in West Germany and some other countries, this cooperation at the top, often called codetermination, is quite rare in the United States. Moreover, it is a limited form of participation. Union representation on corporate boards appears to have little impact on policy and decisions. Benefits to workers and management to date have not been far-reaching.

The most common form of labor-management participation and cooperation is on the shopfloor level. These innovations are initiated with differing levels of worker representation. In nonunion companies, they are usually instituted from on-high. In union settings, as in the auto and communications industries, they can involve union leaders and members in program development, though usually outside the collective bargaining framework and the traditional union hierarchy. Called quality circles, employee involvement, labor-management teams, socio-technical work systems, and other names, these informal participation processes at the workplace have often been initiated to address the motivational costs of mass production and repetitive work. In addition, health and safety concerns have led to the creation of thousands of joint labor-management health and safety committees, particularly in unionized companies. In 1986, 49 percent of collective bargaining contracts had provisions for labor-management safety committees; among manufacturing firms, the proportion was 62 percent (Ruttenberg 1988).

During the past two decades, these initiatives have become increasingly more inclusive and ambitious, involving more workers and providing more responsibility than earlier efforts. From job enlargement through job enrichment to problem-solving groups and autonomous work groups, firms have tried to respond to workers' growing dissatisfaction with narrow job tasks and their greater interest in taking control over their work lives with strategies to increase worker participation (Heckscher 1988). These efforts can be beneficial to both labor and management. Workers given greater responsibility tend to take more pride and derive more satisfaction from their work. This bolsters creativity and enthusiasm in the workplace and enhances performance, which in turn can translate into increased productivity, increased product quality and reliability, and increased contribution to new products and production systems (Bowles et al. 1984).

Of course, after decades of work organization designed to limit worker initiative and autonomy, this shift does not come easily. Unions have mistrusted management motives, often for good reason. Middle and lower management have often resisted efforts ordered from above. Both labor and management must assume new risks and responsibilities, some of which are difficult to take on unless there is a prior relationship of trust and cooperation. Progress is slow. As a result, the pattern has been one of recurrent waves of innovation that seem promising for worker satisfaction and productivity, but that do not seem to diffuse broadly (Heckscher 1988, p. 117).

A 1982 survey found that at least one-third of Fortune 500 firms, some union and some nonunion, had instituted some form of participative management or quality of work life program. However, this figure reflects a wide range of participative programs, from suggestion boxes to state-of-the-art experiments in worker and union involvement in areas of decisionmaking that had traditionally been restricted to management. Some of the most ambitious efforts involve unionized firms—AT&T and the Communications Workers and International Brotherhood of Electrical Workers, Xerox and the Amalgamated Clothing and Textile Workers Union, several steel companies and the United Steel Workers of America, the United Auto Workers with both Ford and with the GM-Toyota NUMMI venture (U.S. Department of Labor 1986).

While the level of participation and the results are often impressive—for instance, of 1150 proposals offered by hourly workers for changes in design and production methods for three Ford small trucks, more than 700 were adopted—these efforts do not fit easily into the existing framework of industrial relations established by the Wagner Act and successive labor laws. Thus, while voluntary participation initiatives are proliferating, changing the climate and practice of labor-management relations in some firms and industries, there are ways in which these efforts conflict with and are hampered by the legalistic and adversarial framework of American industrial relations. The future of quality of work life experiments is unclear. It does appear, though, that a majority of workers in nonunion and union settings alike believe they have a right to greater say in decisions that affect them on the job (Heckscher 1988, p. 242). Labor-management cooperation at the shopfloor level will always be an important element of any effort to address that need.

An additional avenue for worker participation can be seen in the advances made by a number of unions, including the Auto Workers, the Machinists, and the Communications Workers, in bargaining over the introduction of new technology (Ashford and Ayres 1987). It is not clear whether even major changes in workplace technology will be recognized as mandatory subjects of collective bargaining under current labor law (see Chapter Six). However, with rapid and complex technological change being so much a part of the experience of work today, labor may be expected to assume a more active role in examining current technology and adopting new technology in any event.

5. Health and Safety

The decade of the 1970s witnessed the enactment of a number of laws that significantly restructured the rights and responsibilities of employees and their employers in the area of occupational health and safety. Before a National Labor Relations Board ruling in *Gulf Light v. IBEW* in 1966, companies could—and usually did—claim that the entire issue of health and safety on the job was part of management prerogative and outside the scope of collective bargaining (Robinson 1988, p. 454). A year later, in *NLRB v. Gulf Power Co.*, the Fifth Circuit Court reinforced the NLRB ruling and concluded that safety rules and work practices were mandatory subjects for collective bargaining (Ashford and Ayres 1987). Further, with the passage of the Mine Safety and Health Act of 1969 and the Occupational Safety and Health Act of 1970 Congress began to establish a comprehensive set of institutions, procedures, and regulations designed to increase protection of health and safety on the job. This effort has continued through the 1970s and 1980s with enactment of the Toxic Substances Control Act and other statutes and regulations.

In retrospect, this concentrated burst of legislative activity after years of inaction is not surprising. By the late 1960s, it was increasingly clear that the existing collective bargaining and legislative mechanisms for protecting workers had failed to provide significant protection of health and safety on the job. New materials, technologies, and production processes were generating a variety of quantitatively and qualitatively different health and safety problems. The inclusion of health and safety as an issue for collective bargaining was limited to only a few unions representing workers in industries where hazardous work was common. Even in these industries, such as auto, coal, and electrical work, health and safety bargaining was quite restricted (Robinson 1987; Ashford and Ayres 1987). State and federal laws were nonexistent or unenforceable. By the 1960s, the bill for decades of rapid growth without adequate attention to the workplace and community environments had come due. Congress and the Nixon administration responded to very real workplace and community environmental problems. And they responded with more than incremental tinkering because of the political organization and volatility of the environmental and labor constituencies. In a period of social unrest, legislative action was taken to address a growing health and safety crisis and, at the same time, to defuse for what might otherwise have become a more acute political crisis.

(a) The health and safety crisis of the 1960s

Evidence of both a safety and a health crisis could not be ignored by the late 1960s. From 1961 to 1970, the incidence of industrial accidents rose 29 percent (Ashford 1976, p. 85). For most of the postwar period, the accident rate had risen and fallen predictably with the business cycle. During periods of sluggish growth, accidents declined in frequency and usually in severity. During periods of economic expansion, as firms hired inexperienced workers, opened new plants, introduced new equipment, and ran existing facilities at full capacity, accident rates rose. Beginning in 1966, however, the accident rate rose faster than the business cycle alone would explain. According to one analysis, employers facing declining profits began to intensify the pace of work (Bowles et al. 1983). At the same time, the Vietnam War buildup put additional strain on many segments of American manufacturing. Accident rates rose most sharply in war-related industries, such as primary and fabricated metals and machinery. In ordnance industries, for example, the rate of increase in accidents doubled (Noble 1986, p. 64).

Occupational illness, though far more difficult to verify statistically, also was emerging as an increasingly serious and visible problem. Changes in the raw materials used in American industry and in the technology and organization of production in the decades of prosperity that followed World War II generated a broad range of new health and safety challenges and threats that were also much in evidence by the late 1960s. In coal mines, for example, the introduction of continuous mining machines in the 1940s not only sharply reduced the number of miners needed to produce coal but also increased the amount of coal dust miners were inhaling, increasing the likelihood that miners would develop black lung disease. The rapid increase in uranium mining for the nuclear weapons program put another group of miners at risk, exposing them to radiation without any provisions for protection. In many industries, new petrochemical-based substances and processes were exposing workers to a broad range of health hazards. In 1961, for example, more than one in ten workers at Uniroyal's Eau Claire, Wisconsin, plant came down with a respiratory illness that was eventually traced to a new vulcanizing process. During the 1960s, scientific evidence began to mount that occupational disease was indeed a serious problem. An extended study of workers in a Paterson, New Jersey, asbestos plant found unusually high rates of lung cancer, asbestosis, and mesothelioma. Studies of cancer deaths among residents of Allegheny County, Pennsylvania, in the 1960s found high rates of cancer among steelworkers exposed to coke oven emissions. In the coal fields, studies of mine workers found what they and their relatives already knew: black lung killed. Throughout the 1960s, invisible occupational killers were becoming more visible. And the results were reaching an environment- and health-conscious public.

Toward the end of the decade, the connection between the growing environmental movement and issues of occupational health and safety was increasingly clear. Two events in particular added urgency to the public's concern and outrage. On November 20, 1968, a mine explosion in Farmington, West Virginia killed 78 miners. Two months later, an oil rig off the Santa Barbara coast exploded, covering 13 miles of ocean beach in viscous, black oil. Both disasters received extensive TV coverage. Both events touched the public nerve and added to the momentum for change.

These events raised both environmental protection and occupational health and safety to the top of the domestic policy agenda. In a climate of rank-and-file discontent over the conditions of work, increased public concern about the environment, consumer product safety, and the quality of life, the beginnings of organized union and public health activism on occupational safety and health issues, and the political interest first of the Johnson administration and then of President Nixon to address both labor and middle-class quality of life concerns, legislative action became all but inevitable.

In 1969, 43,000 miners shut down West Virginia's mines for three weeks and marched on the state capital to demand workers' compensation for black lung disease.

The threat of an expansion of miner activism pushed Congress to pass the Coal Mine Health and Safety Act of 1969 (expanded to the Mine Safety and Health Act in 1977) and propelled action on the Occupational Safety and Health Act of 1970, workplace parallels to the flurry of environmental laws—including the National Environmental Policy Act, the Clean Air Act, and the Federal Water Pollution Control Act—that were enacted or strengthened during Richard Nixon's first term.

(b) After OSHA

The passage of the Occupational Safety and Health Act (OSHAct) changed the basic institutional framework for addressing occupational health and safety problems in this country. One result has been a significant increase in union activity in the area of protecting worker health and safety. Yet, despite the creation of a health and safety infrastructure and the passage of other workplace health and safety-related laws since 1970, serious problems persist. Moreover, new problems have been identified and other new concerns have been created in emerging era of microelectronics based technologies.

(i) The Role of Unions. Unions have played an important role in the protection of health and safety on the job, despite their shrinking share of the work force. According to economist James Robinson, workers appear to turn to union representation as one strategy for dealing with unsafe and unhealthy working conditions. Union membership and pro-union sentiment are much higher in hazardous jobs than in safe jobs. Although there has been substantial erosion of union strength even in the most dangerous employments since the late 1970s, unions remain the one organization that can play an intervening role in occupational health and safety conflicts (Robinson 1988; Weil 1987 1991a and 1991b).

The percentage of union contracts that contain clauses regarding safety equipment, the creation of joint safety committees, hazard pay, the right to refuse hazardous work, and other health and safety protections has risen steadily since the 1950s, accelerating in the 1970s and 1980s (Robinson 1988). Moreover, union representation makes a difference in the ability of any group of workers to advance health and safety protections. Organization and resources enable workers to force employers to take safety seriously. The Philadephia Project on Occupational Safety and Health (PHILAPOSH) assembled a list of union-initiated interventions in the Philadelphia area that gives a sense of what is possible. The list includes campaigns by a variety of different union locals that resulted in improved spray booths at a pump plant, new handling procedures for PCBs, removal of asbestos from ceilings in a city clinic, reduced workloads to relieve stress of welfare department workers, an end to the use of a fiberglass-coated wire that was causing rashes and itching at an assembly plant, a new grievance procedure and health and safety training program at a large plant to address OSHA findings of excess levels of lead and other heavy metals, installation of a fan and hood after workers tested the air and found dangerously high vapor levels, and many other similar changes (Noble 1986, pp. 135-136).

Not only can union representation afford workers the resources and mechanisms necessary for a stronger negotiating position, union representation also increases the effectiveness of federal health and safety rules and regulations. Union representation is correlated with a greater probability of OSHA inspection, a higher number of citations per inspection, and, in manufacturing, greater monetary penalties per violation. According to a recent empirical study, "Unions in the manufacturing, construction, and service sectors increase both the quantity (number of inspections) and quality of inspections (measured in inspection duration and scope of inspection activity) over their nonunion counterparts. Additionally, unions reduce the time required to abate violations of OSHA standards and limit employer alterations of those abatement periods" (Weil 1987, p. 334). Unions create incentives for employers to improve health and safety conditions voluntarily by increasing the threat of regulatory enforcement. Given the inadequacy of OSHA's budget and staffing throughout its history and particularly since the deep cuts suffered during the Reagan years, union representation provides perhaps the best way for workers to improve their chances of working in a safe and healthy environment.

(ii) Continuing Problems. The enactment of a law and the creation of a bureaucracy does not make a problem disappear. And OSHA itself has provided only limited relief to America's workers, improving conditions significantly in some industries such as textiles and industries where explosions were common, but having little impact in many other dangerous and less dangerous trades. Not surprisingly, therefore, health and safety problems at work persist. According to the National Institute for Occupational Safety and Health the ten leading work-related diseases and injuries are:

1. Occupational lung diseases: asbestosis, byssinosis, silicosis, coal workers' pneumoconiosis, lung cancer, occupational asthma

2. Musculoskeletal injuries: disorders of the back, trunk, upper extremity, neck, lower extremity; traumatically induced Raynaud's phenomenon

3. Occupational cancers (other than lung): leukemia; mesothelioma; cancers of the bladder, nose, and liver

4. Amputations, fractures, eye loss, lacerations, and traumatic deaths

5. Cardiovascular diseases: hypertension, coronary artery disease, acute myocardial infarction

6. Disorders of reproduction: infertility, spontaneous abortion, teratogenesis

7. Neurotoxic disorders: peripheral neuropathy, toxic encephalitis, psychoses, extreme personality changes (exposure-related)

8. Noise-induced loss of hearing

9. Dermatologic conditions: dermatoses, burns (scaldings), chemical burns, contusions (abrasions)

10. Psychologic disorders: neuroses, personality disorders, alcoholism, drug dependency

The conditions listed under each category are *selected examples,* not comprehensive definitions of the category (NIOSH 1983). Further, most suspect carcinogens still remain unregulated, as do reproductive hazards and neurotoxins. (The reader is referred to Levy and Wegman 1988 and Rom 1983 for extensive reviews of occupational disease.)

As in the earlier half of the century, the risk of illness and death on the job is not

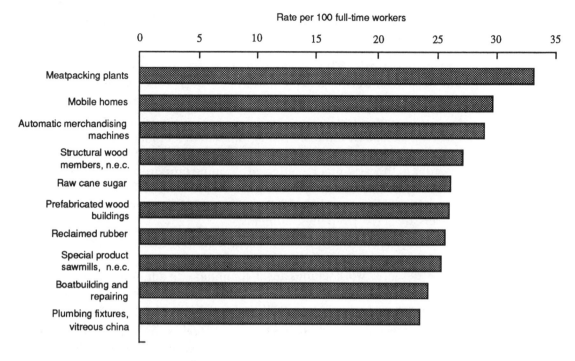

Rate per 100 full-time workers

Figure 1-4. Injury and illness incidence rates, total cases, high-risk manufacturing industries, 1986 BLS Annual Survey, p. f-59. n.e.c. = not elsewhere classified.

evenly distributed. Certain industries are far more dangerous than others (see Figure 1-4). Farm work is among the most dangerous. The National Safety Council has estimated that the death rate in agriculture is 66 per 100,000, far higher than the industrial average of 18 per 100,000. Although few statistics are available, one government estimate is that 80,000 to 90,000 field workers get sick and 800 to 1000 die each year from pesticide exposure (Pollack and Grozuczak 1984). Minorities are concentrated in the most dangerous industries, including dry cleaning, foundries, hospitals, farm work, textiles, and tobacco. Studies of occupational segregation in textile, steel, and other industries have shown that within those industries minorities hold the most dangerous jobs. The increased exposure to dangerous and unhealthy work has quite serious costs. While the life expectancy of the average American is over 70 years, the life expectancy of a migrant farmworker is only 49 years. Black workers face a 37 percent greater risk of illness and a 20 percent greater risk of death due to their jobs than white workers, according to the Urban Environmental Conference (Pollack and Grozuczak 1984). Black workers are one and a half times more likely to be severely disabled from job injuries and illnesses.

In addition to the problems generated by the technologies of yesterday, emerging technologies in manufacturing and office work are generating a set of new problems that are only beginning to be addressed. Of course, new technology can eliminate dangerous work, as has often been the case in U.S. manufacturing. The use of robots to replace workers in auto paint shops is one good example. At the same time, though, new unintended problems are often created. Steven Deutsch (1988) notes five areas of concern emerging from the proliferation of microelectronics-based technologies. These are as follows:

1. **Chemical hazards.** The solvents and other chemicals used in the manufacture of semiconductors and computers have been associated with a host of skin diseases and other adverse effects including central nervous system problems.

2. **Musculoskeletal problems.** New equipment that is not ergonomically well designed can lead to problems such as "carpal tunnel syndrome," a nerve disorder in the wrist that has affected alarming proportions of grocery clerks, clerical workers, postal workers, and others. The National Institute for Occupational Safety and Health estimates that 15 to 20 percent of American workers are at risk from these ailments.

3. **Video display terminal hazards.** There is growing evidence of arm, back, wrist, and leg problems associated with VDT use. Vision problems due to poorly designed work stations are also common. In addition, there remain serious questions as to the possibility of radiation hazards of VDT work, particularly for pregnant women (Bureau of National Affairs 1987).

4. **Stress.** In recent years, there has been a noticeable rise in stress-related worker compensation claims and a general increased awareness of stress-related illness. Workers on microelectronics-based equipment suffer increased stress resulting from the pacing of work, computer monitoring by employers, invasion of privacy, and pressure for production (OTA 1987).

5. **Job loss and fear of job loss.** It is well documented in the literature that job loss and fear of displacement manifest themselves in increased levels of depression, alcohol and drug abuse, cardiovascular disease, gastrointestinal disorders and other physical problems. New technologies and the pace of technological change in the past decade have reduced the job security of tens of millions of Americans and have resulted in the elimination of jobs for as many as a million employed workers per year since 1979 (Cyert and Mowery 1987). This increased instability is an occupational hazard that is becoming more widespread.

In addition to this list of new problems, widespread classical problems such as lead and other heavy metal poisoning, systemic toxicity, and dust-induced respiratory disease persist. Chemical sensitivities causing allergy-like problems are also receiving increased attention (Ashford and Miller 1991; Cullen 1987).

D. CONCLUSION

This chapter has surveyed the changing nature of work and the manner in which it is experienced by American workers. Focusing on five interrelated aspects of working life—technology, the organization of work, the structure of industry and occupations, labor-management relations, and workplace health and safety—we have explored the kinds of changes and challenges that workers have had to face during the course of the twentieth century. If anything is clear from this survey, it is that the conditions of work are constantly changing and that adjustment to change is often costly in terms of employment stability, health and safety, and other indicators of job quality and satisfaction. Moreover, it appears that the past decade of industrial restructuring has posed particularly difficult choices and challenges to firms, unions, and workers.

This chapter has focused on the past, trying to explain the antecedents and the context for much of the protective legislation, regulation, and rules that govern the workplace today. One of the conclusions we draw from this historical survey is that unless workers, and the institutions that provide representation for workers, are involved at an early stage in discussions about and planning for technological and other organizational changes in the workplace, the costs of adjustment will remain higher than they need be or should be.

In 1987, the Panel on Technology and Employment of the Committee on Science, Engineering and Public Policy jointly established by the National Academies of Sciences and Engineering and the Institute of Medicine published *Technology and Employment,* a careful look at the impact of technological change on employment. The panel's principal finding is worth quoting in its entirety:

> Technological change is an essential component of a dynamic, expanding economy. Recent and prospective levels of technological change will not produce significant increases in total unemployment, although individuals will face painful and costly adjustments. The modern U.S. economy, in whch international trade plays an increasingly important role, must generate and adopt advanced technologies rapidly in both the manufacturing and nonmanufacturing sectors if growth in U.S. employment and wages is to be maintained. Rather than producing mass unemployment, technological change will make its maximum contribution to higher living standards, wages, and employment levels if appropriate public and private policies are adopted to support the adjustment to new technologies (Cyert and Mowery 1987, p. 3).

Recognizing that the health and safety implications of the current wave of technological and organizational changes are extremely complex, the panel proposed that a major interdisciplinary study be conducted of the consequences of technological change and the appropriate regulatory structure for ensuring protection of health and safety on the job. Three broad areas of study were proposed: (1) workplace hazards created by technological change; (2) the potential for technology to reduce workplace hazards; and (3) the impact of changes in the work environment—the structure of industries and occupations, the organization of work, and other factors—on health and safety.

The panel's recommendations focus on easing adjustment for displaced workers, existing employees, and new entrants to the workforce. An important conclusion mirrors our own assessment of how best to deal with issues of technological change. That is, not only should workers be prepared to deal with the effects of technological change, but they should also be involved in the decisionmaking process about the adoption and use of new technologies. The panel recommended "that management give advance notice of and consult with workers about job redesign and technological change" (Cyert and Mowery 1987, p. 14). We agree. Only through policies that engender greater worker commitment will the costs of technical change be minimized and its benefits maximized.

Over a decade ago, one of the authors of the present text wrote a treatise analyzing the state of occupational health and safety in the United States a few years after the enactment of the Occupational Safety and Health Act. The book, *Crisis in the Workplace*, concluded with this observation:

> From a practical point of view, increased *responsibility* of the worker for his own work environment can only come with a corresponding increase in both his *knowledge* and his *control* of that work environment. . . . To the extent that collective bargaining is being used to settle occupational health and safety issues by extracting better *employer* performance, there is little shift in power. However, to the extent that workers are gaining participatory rights in the process, important changes may be occuring. . . . While the policy instruments cited above must be further developed and aggressively wielded, they will nonetheless be insufficient even when working in concert. Workers themselves must assume a greater role in self-monitoring the quality of their own work environments. . . . While progress can be expected in the courts and hearings—and in the laboratory and classrooms—much more is likely if workers and their representatives are able to function responsibly and effectively in fruitful interaction with employers, and as a necessary check on even the best-intentioned employers (Ashford 1976, pp. 537-539).

In effect, that book ended where this one begins—with a call for increased worker participation and responsibility in decisions about technology, work organization, labor-management relations, and, in particular, health and safety. In the following chapters, we analyze how existing labor law, occupational health and safety protections, chemical production regulations, right-to-know laws, rules against discrimination, and workers' compensation and tort laws—the basic framework for protection from and redress of inequities in the workplace—can be used by workers and their representatives, not only to compensate victims, but in a proactive way to avoid some of the most significant costs of technological and organizational changes on the job.

REFERENCES

Adler, P. 1984. Rethinking the Skill Requirements of New Technologies. Boston: Graduate School of Business Administration, Harvard University.

Ashford, N. A. 1976. *Crisis in the Workplace: Occupational Disease and Injury*. Cambridge: MIT Press.

Ashford, N. A. and C. Ayres. 1987. Changes and Opportunities in the Environment for Technology Bargaining. *Notre Dame Law Review* 62(5):810–858.

Ashford, N. A. and C. S. Miller. 1991. *Chemical Exposures: Low Levels and High Stakes*. New York: Van Nostrand Reinhold.

Berman, D. 1978. *Death on the Job*. New York: Monthly Review Press.

Bezold, C., R. J. Carlson, and J. C. Peck. 1986. *The Future of Work and Health*. Dover: Auburn House Publishing.

Bluestone, B. and B. Harrison. 1982. *The Deindustrialization of America*. New York: Basic Books.

Bowles, S., D. Gordon, and T. Weisskopf. 1983. *Beyond the Wasteland*. Garden City, NY: Anchor Press.

Bowles, S. D. Gordon, and T. Weisskopf. 1984. A Social Model for U.S. Productivity Growth. *Challenge.*

Braverman, H. 1974. *Labor and Monopoly Capital.* New York: Monthly Review Press.

Brody, D. 1980. *Workers in Industrial America.* New York: Oxford University Press.

Bureau of National Affairs. 1987. *Stress in the Workplace.* Washington, DC: BNA.

Chandler, A. D., Jr. 1977. *The Visible Hand.* Cambridge: Harvard University Press.

Cole, R. 1979. *Work, Mobility and Participation.* Berkeley: University of California Press.

Colton, J. and S. Bruchey. 1987. *Technology, the Economy, and Society: The American Experience.* New York: Columbia University Press.

Commoner, B. 1987. A Reporter at Large: The Environment. *The New Yorker,* June 15: 46–71.

Cullen, M. 1987. The Worker with Multiple Chemical Sensitivities: An Overview, in *Workers with Multiple Chemical Sensitivities, Occupational Medicine: State of the Art Reviews,* (Cullen, ed.). Philadelphia: Hanley & Belfus 2(4):655–662.

Cyert, R. M. and D. C. Mowery, eds. 1987. *Technology and Employment.* Washington, DC: National Academy of Sciences.

Deutsch, S. 1988. New Technology and the Work Environment: Problems and Solutions. *LOHP Monitor,* July-September:5–8.

Dubofsky, M. 1975. *Industrialism and the American Worker, 1865–1920.* Arlington Heights: AHM Publishing.

Freeman, R. B. and J. L. Medoff. 1984. *What Do Unions Do?* New York: Basic Books.

Ginzberg, E. 1982. The Mechanization of Work *Scientific American* 247(3):48–55.

Gordon, D. M., R. Edwards, and M. Reich. 1982. *Segmented Work, Divided Workers.* New York: Cambridge University Press.

Green, J. R. 1980. *The World of the Worker.* New York: Hill and Wang.

Guile, B. and H. Brooks, eds. 1987. *Technology and Global Industry.* Washington, DC: National Academy Press.

Hartmann, H., ed. 1987. *Computer Chips and Paper Clips.* Washington, DC: National Academy of Sciences.

Heckscher, C. C. 1988. *The New Unionism.* New York: Basic Books.

Hirschhorn, L. 1984. *Beyond Mechanization.* Cambridge: MIT Press.

Howard, R. 1985. *Brave New Workplace.* New York: Viking Press.

Johnston, W. B. 1987. *Workforce 2000.* Indianapolis: Hudson Institute.

Kazis, R. and R. L. Grossman. 1982. *Fear at Work.* New York: Pilgrim Press.

Kochan, T. A., H. C. Katz, R. B. McKersie. 1986. *The Transformation of American Industrial Relations.* New York: Basic Books.

LaDou, J., (ed.) 1985. The Microelectronics Industry, in *Occupational Medicine: State of the Art Reviews.* Philadelphia: Hanley and Belfus, 1(1):1–197.

Leontief, W. and F. Duchin. 1985. *The Future Impact of Automation on Employment.* New York: Oxford University Press.

Levy, B.S. and D.H. Wegman, (eds.). 1988. *Occupational Health: Recognizing and Preventing Work-Related Disease.* Boston: Little, Brown and Co.

Levitan, S. A. and C. M. Johnson. 1982. *Second Thoughts on Work.* Kalamazoo: The W.E. Upjohn Institute for Employment Research.

Markowitz, G. and D. Rosner. 1986. More than Economism: The Politics of Workers' Safety and Health, 1932–1947. *The Milbank Quarterly* 64(3):331–354.

Mishel, L. and J. Simon. 1988. *The State of Working America.* Washington, DC: Economic Policy Institute.

National Institute for Occupational Safety and Health. 1983. Leading Work-Related Disease and Injuries—United States. *Morbidity and Mortality Weekly Report,* January 21, 1983.

Nelkin, D. and M. S. Brown. 1984. *Workers at Risk.* Chicago: University of Chicago Press.

Noble, C. 1986. *Liberalism at Work.* Philadelphia: Temple University Press.

Odell, R. 1980. *Environmental Awakening.* Cambridge: Ballinger.

Osterman, P. 1988. *Employment Futures.* New York: Oxford University Press.

Personick, M. E. and K. Taylor-Shirley. 1989. Profiles in Safety and Health: Occupational Hazards of Meatpacking. *Monthly Labor Review.*

Piore, M. J. and C. F. Sabel. 1984. *The Second Industrial Divide.* New York: Basic Books.

Pollack, S. and J. Grozuczak. 1984. *Reagan, Toxics, and Minorities.* Washington, DC: Urban Environment Conference.

Robinson, J. C. 1987. *Labor Union Involvement in Occupational Health and Safety.* Manuscript. Berkeley: *School of Public Health.*

Robinson, J.C., 1988. Labor Union Involvement in Occupational Safety and Health, 1957–1987. *Journal of Health Politics, Policy and Law* 13(3):453–468.

Rom, W., (ed.) 1983. *Environmental and Occupational Medicine* Boston: Little, Brown and Co.

Ruttenberg, R. 1988. *The Role of Labor-Management Committees in Safeguarding Worker Safety and Health.* Washington, DC: U.S. Department of Labor, Bureau of Labor-Management Relations and Cooperative Programs.

Singlemann, J. 1978. *From Agriculture to Services.* Beverly Hills: Sage.

Spenner, K. 1985. The Upgrading and Downgrading of Occupations. *Review of Educational Research* 55:125–154.

Stellman, J. and M. S. Henifin. 1983. *Office Work May Be Hazardous to Your Health.* New York: Pantheon.

Stieber, J. 1985. Recent Developments in Employment-at-Will. *Labor Law Journal,* August:557–563.

Sweeney, J. J. and K. Nussbaum. 1989. *Solutions for the New Work Force.* Washington, DC: Seven Locks Press.

U.S. Congress. Office of Technology Assessment. 1985a. *Preventing Illness and Injury in the Workplace.* Washington, DC: U.S. Government Printing Office.

U.S. Congress. Office of Technology Assessment. 1985b. *Reproductive Health Hazards in the Workplace.* Washington, DC: U.S. Government Printing Office.

U.S. Congress. Office of Technology Assessment. 1986. *Technology and Structural Unemployment.* Washington, DC: U.S. Government Printing Office.

U.S. Congress. Office of Technology Assessment. 1987. *The Electronic Supervisor.* Washington, DC: U.S. Government Printing Office.

U.S. Congress. Office of Technology Assessment. 1988. *Technology and the American Economic Transition.* Washington, DC: U.S. Government Printing Office.

U.S. Department of Labor. 1980. *Protecting People at Work.* Washington, DC: U.S. Government Printing Office.

U.S. Department of Labor. 1986. *U.S. Labor Law and the Future of Labor-Management Cooperation.* Washington, DC: U.S. Government Printing Office.

U.S. Department of Labor. 1988. *Projections 2000.* Washington, DC: U.S. Government Printing Office.

Walker, C. R. and R. H. Guest. 1952. *The Man on the Assembly Line.* Cambridge: Harvard University Press.

Walton, R. 1985. From Control to Commitment in the Workplace. *Harvard Business Review,* March-April.

Warner, W. L. and J.O. Low. 1947. *The Social Structure of the Modern Factory.* New Haven: Yale University Press.

Weil, D. 1987. Government and Labor at the Workplace. *Proceedings of the 1987 Industrial Relations Research Association,* 332–335.

Weil D. 1991a. Labor Unions and the Implementation of the Occupational Safety and Health Act. *Industrial Relations,* forthcoming Winter 1991.

Weil, D. 1991b. Building Safety: The Role of Construction Unions in the Enforcement of the Occupational Safety and Health Act. *Journal of Labor Research,* forthcoming 1991.

Wilson, W. J. 1987. *The Truly Disadvantaged.* Chicago: University of Chicago Press.

Yankelovich, D. and J. Immerwahr. 1983. *Putting the Work Ethic to Work.* New York: Public Agenda Foundation.

Zuboff, S. 1988. *In the Age of the Smart Machine.* New York: Basic Books.

Zysman, J. and S. Cohen. 1987. The Myth of a Post-Industrial Economy. *Technology Review,* February-March.

CHAPTER 2

Administrative Law

Much of the "law" referred to in the title of this text stems from federal regulatory statutes—such as the Occupational Safety and Health Act and the Toxic Substances Control Act—that govern various aspects of the workplace and its technology. The key to an understanding of how these regulatory systems function is administrative law.

In essence, *administrative law* is the body of law that governs the way in which administrative agencies make and implement decisions. Federal administrative law is grounded in the U.S. Constitution and in various federal statutes. Although administrative law can appear to be little more than a series of seemingly arcane structural and procedural rules, a basic understanding of administrative law is essential to an understanding of how the regulatory system works. It is administrative law that allows us to "push" the regulatory system in one direction or another—to propose a regulation that we feel is needed, to challenge a regulation that we feel is too stringent (or not stringent enough), to obtain information from the records of an administrative agency, to provide input to the standard-setting process, and to do a host of other things that give us some measure of control over the workplace environment. In this light, administrative law can be seen as the useful tool it often is.

The purpose of this chapter is to provide an overview of the administrative system—highlighting the relationships between Congress, the agencies, and the courts—and then to explore some of the issues that are of particular importance to the regulation of the workplace. In addition, this chapter provides a brief introduction to the legal system for those unfamiliar with its workings.

A. QUESTIONS TO CONSIDER WHEN ANALYZING A REGULATORY FRAMEWORK

Before delving into the theories and practice of administrative law, it is useful to reflect for a moment on the broader picture. What is it that can be gained from a study of administrative law and the administrative system? Certainly, it should provide some familiarity with the details of administrative procedure. Beyond the technical details, however, administrative law gives us a conceptual basis from which we can analyze, and actually come to *understand*, a particular regulatory system. The following questions provide a logical focus for conducting such an analysis in the area of health, safety, and environmental regulation. In each case, an understanding of administrative law will help ferret out the appropriate answer.

1. How is the regulatory "problem" defined? What is being regulated, and how is it delineated?

2. What "risk formulation" is embodied in the statutory standard? For example, what level of disease prevention or pollution abatement must the implementing agency attain and/or maintain?
   - Is the standard health-based? Technology-based?
   - To what extent—if at all—is economics to be taken into consideration?
     - Is the standard tied to *economic feasibility*?
     - Is some form of social *cost-benefit* analysis required (or permitted)?
     - Does the standard look to the economics of the *firm*, or to the economics of the entire *industry*?

3. How has the agency interpreted and carried out its statutory mandate?

4. Through what procedures are the agency's regulations promulgated? To what extent are those outside the agency able to place issues onto the agency's regulatory agenda and participate in the agency's rulemaking procedures?

5. Who has the statutory burden of proof on the various issues of importance? In a broad sense, is the burden on industry to prove that its product or process is "acceptable," or is the burden on the regulatory agency to prove that it is not?

6. What health or safety testing requirements—if any—does the regulatory framework impose?

7. What avenues are available for appealing the agency's decision to take (or not to take) a particular regulatory action? Is there an express "citizen suit" provision that allows affected citizens to sue the agency or individual violators of the statute?

8. What enforcement options are available to the agency?

9. What role is left for *state* statutes or common law in this area? To what extent—if at all—does the federal regulatory system *preempt* state action?

With these general inquiries in mind, we now turn to the administrative system itself. The conceptual basis for administrative law is familiar to any student who has had a high school civics course. Indeed, it is *so* familiar that it is often taken for granted or even overlooked completely. Almost all questions of administrative law in the United States are grounded in the tripartite model of government embodied in the Constitution: the separation of powers among the legislative, executive, and judicial branches of government. In general, it is this model that judges have in their mind's eye—either explicitly or implicitly—when they approach issues of administrative law.

This chapter is designed to guide the reader through this tripartite system. Although the chapter emphasizes the federal system, the broad concepts are largely applicable to the various state administrative and legal systems as well. There are, however, important differences between the federal system and many state systems. Any attempt to understand or use a particular state's administrative process should be preceded by careful attention to the specific features of that state's system.

B.   THE CONSTITUTIONAL BASIS FOR LABOR, HEALTH, AND SAFETY REGULATION

Typically, administrative agencies are created by the legislative branch, are run on a day-to-day basis by the executive branch, and are subjected to periodic review by the judicial branch. Thus, a functional analysis of agency behavior—one that asks both what it is that agencies do and how their activities can be influenced—must examine the manner in which each of the branches directs and controls agency behavior.

1.   Direction from the Legislative Branch

The legislative branch can direct and influence agency behavior through a series of formal and informal controls.

(a) The statutory mandate

The administrative process begins with an act of Congress that either *creates a new agency* to deal with a particular area of concern, or *grants new powers and responsibilities* to an existing agency to deal with that area of concern. We will refer in this text to a statute that gives such authority to a new or existing agency as the agency's *originating statute* (otherwise known as *enabling legislation*) in that area. It is the originating statute that gives the agency its *statutory mandate*—its formal directive from Congress with regard to the subject matter in question. Obviously, an agency often administers more than one originating statute, and the demarcation between subject areas is not always clear. For example, the Environmental Protection Agency has been given the directive to deal with toxic air pollutants under the Clean Air Act, toxic water pollutants under the Clean Water Act, toxic wastes deposited into the ground under the Resource Conservation and Recovery Act, and toxic chemicals generally under the Toxic Substances Control Act. A close examination of each of these originating statutes reveals that EPA not only has a variety of statutory directives from Congress, but that it could use any one of these statutes, especially the latter, in such a way as to limit the discharge of toxic substances into *all* environmental media. Further, there are situations in which two or more agencies have overlapping statutory mandates. For example, both OSHA and EPA have the authority to regulate worker exposure to harmful chemicals.

◇ NOTES

1. Inconveniently enough, the sections of many federal statutes are known by two different numbers. When an act is passed by Congress, its sections are numbered sequentially, beginning with 1. Thus, the Occupational Safety and Health Act is numbered from Section 1 through Section 34. (Longer statutes, such as the Clean Air Act, commonly are divided into "subchapters" or "titles." Within each subchapter or title, the various sections are numbered in sequential fashion: 101, 102, 201, 202, etc,.) It is by these original section numbers that the various sections of a statute usually become known to those who work with them on a regular basis. However, all federal statutes are grouped by subject matter area and placed into the United States Code. Both the National Labor Relations Act and the Occupational Safety and Health Act are found in Title 28 of the Code, for example, because this is the area of the Code dealing generally with workplace issues. Federal statutes are cited by their title and section numbers. Thus, Section 6 of the OSHAct becomes 28 U.S.C. §655. This is the "official" citation.

2. EPA is an exception to the general way administrative agencies are created. It was not created by an act of Congress but by a presidential order in 1970.

(b) The commerce clause

Congressional power to grant such authority to administrative agencies flows from the U.S. Constitution. By now, it is well settled that Congress has broad powers under the commerce clause of the Constitution to regulate in the general areas of health, safety, and the environment.

The Constitution grants Congress the power "to regulate commerce . . . among the States." (Article I, Section 8) This commerce power allows Congress to regulate activities that affect interstate commerce, a power that has broadened considerably over the years through judicial interpretation. For example, in upholding the Civil Rights Act of 1964, the Supreme Court accepted as valid an asserted relationship between interstate commerce and racial discrimination in restaurants. It reasoned that restaurants who refused to serve blacks sold fewer interstate goods, obstructed interstate travel by blacks, and generally affected the free flow of commerce across state lines. If the commerce clause could be used to sustain such social regulation, it is not difficult to see how it was extended to the environmental and occupational arena as well.

To discourage any challenge to a statute enacted under its commerce powers on the grounds that it is not sufficiently connected to interstate commerce, Congress generally makes explicit reference to the effect on commerce in the language of the statute itself.

(c) The delegation doctrine

Many statutory mandates to agencies—especially in the areas of health, safety, and the environment—are strikingly broad and nonspecific. For example, Section 6 of the Toxic Substances Control Act directs EPA to regulate chemicals that pose "an unreasonable risk of injury to health or the environment." Beyond directing EPA to "consider the environmental, social, and economic impact of any action," TSCA leaves the agency considerable discretion to determine which risks will be deemed "unreasonable" and which will not.

Earlier in this century, such a broad grant of authority to an administrative agency might well have been considered unconstitutional. The relevant constitutional principle is known as the "delegation doctrine." This doctrine stems from the classic understanding that Congress—as the duly elected representative of the American public—is the repository of all federal legislative power. According to the delegation doctrine Congress cannot delegate this legislative power to another party, such as an administrative agency, because the agency has *not* been elected by the people. Under a strict application of the doctrine, Congress is required to provide reasonably clear and specific statutory standards to guide agency decisionmaking.

The delegation doctrine reached its apex in 1936, when the U.S. Supreme Court struck down two separate statutes on the grounds that they granted improperly broad decisionmaking authority to administrative agencies. Since that time, however, the number of administrative agencies has increased dramatically, and agency decisionmaking has become the principal means of federal regulation. Administrative agencies have even been said to comprise the "fourth branch" of government. In an apparent acquiescence to political reality, the courts have also relaxed the delegation doctrine considerably. Broad delegations of substantive authority to administrative agencies have become very much the rule rather than the exception. When an especially expansive statutory mandate *is* challenged, the courts have responded, not by invoking the delegation doctrine to strike down the statute, but by either: (1) giving a narrower interpretation to the statutory language, or (2) ordering the agency to develop its own standards for interpreting the statutory language. Indeed, the Supreme Court has not invoked the delegation doctrine to invalidate a statute since 1936.

This has led some commentators to conclude that the delegation doctrine is no more than a historical artifact. Such a conclusion may be premature, however. On two occasions, Justice William Rehnquist has authored a separate opinion in which he has taken the position that Section 6(b) of the OSHAct—the provision that directs OSHA to set standards for workplace exposures to toxic substances—violates the delegation doctrine and should be invalidated. That section requires OSHA to "set the standard which most adequately assures, to the extent feasible . . . that no employee will suffer material impairment of health or functional capacity." In failing to be more specific, concludes Justice Rehnquist, Congress has unconstitutionally delegated to an agency the responsibility for making the "fundamental policy decisions" that must properly be made by Congress itself. *Industrial Union Dept., AFL-CIO v. American Petroleum Institute*, 448 U.S. 607, 671 (1980) (Rehnquist, J., concurring). See also *American Textile Manufacturers Institute v. Donovan*, 452 U.S. 490, 543 (1981) (Rehnquist, J., dissenting). Although the position has not been adopted by other current members of the Court, it could signal a resurgence of the delegation doctrine. If so, many of the broad statutory mandates in the area of health, safety, and the environment could be called into question.

(d) The procedural mandate

In addition to following the substantive mandates of originating statutes, federal agencies are also required to adhere to the more general procedural directives of several other statutes. Where the substantive mandate provides guidance to the agency on *which* decisions it should make, the procedural mandate provides guidance on *how* those decisions should be made. The most important procedural statute for federal agencies is the Administrative Procedure Act (APA), 5 U.S.C. §551. Passed in 1946, this statute remains the chief means through which Congress controls the procedures of the various federal agencies. The APA sets forth regularized procedures for agency rulemaking and adjudication, for judicial review of administrative decisionmaking, and for citizen access to these administrative and judicial processes.

A variety of other statutes also provide important procedural directives. For our purposes, the most important of these are the Freedom of Information Act, which requires agencies to make most of their internal documents available to the public; the Government in the Sunshine Act, which requires agencies to make many of their proceedings open to the public; the Federal Advisory Committee Act, which specifies procedures for an agency's use of outside advisors; the Privacy Act, which gives private citizens access to agency documents and information concerning them; and the National Environmental Policy Act, which requires federal agencies to prepare an environmental impact statement before approving a major action significantly affecting the environment.

The originating statute may specify procedural requirements as well, and conflicts can arise between the procedural directives of the originating statute and the directives of one or more of these general procedural statutes mentioned above. The OSHAct, for example, contains specifications for rulemaking that differ from those found in the Administrative Procedure Act. In general, the more specific requirements set forth in the originating statute will control. However, when the originating statute is silent on a point, the agency will be required to follow the "generic" requirements set forth in the APA.

(e) Interpreting the statutory mandate

A central task for the agency—and for the courts in reviewing the agency's decisions—is interpreting the statutory mandate. Often, this is far from easy. The broader and less specific the substantive mandate, the more difficult it is to divine the intent of Congress.

To determine legislative intent, the agency and the courts start logically with the statutory language itself. This basic principle has been well stated by the Supreme Court. "First, always, is the question whether Congress has directly spoken to the precise question at issue. If the intent of Congress is clear, that is the end of the matter; for the court, as well as the agency, must give effect to the unambiguously expressed intent of Congress." *Chevron, U.S.A., v. NRDC,* 467 U.S. 837, 842-843 (1984). Thus, where the language of the statute is sufficiently clear and unambiguous, the agency need not, and *may not,* look any further. It must carry out the intent of Congress as expressed in the language of the statute.

In many cases, however, the intent of Congress is not clear from the bare language of the statute. Here, one must look behind the language to the statute's "legislative history." There are three basic sources of this history. Prior drafts of the statute—the early House and Senate bills—can help reveal what Congress chose *not* to include in the final statute. This allows one to draw logical inferences about the language that did become law. Reports of the congressional committees that helped draft the language can also be helpful. These reports—the House Report, the Senate Report, and the

Conference Report (which is written by representatives of both committees when they meet to work out the compromise language for the final statute)—contain explanatory comments on the final statutory language. As the Conference Report represents something of a consensus document, the courts generally consider it to be the most instructive and influential. The third and generally least influential source of legislative history is the record of the Congressional floor debates on the various versions of the statute. Such commentary is most useful when it elucidates positions on both sides of a particular issue and in so doing sheds light on why a particular provision was rejected in favor of another. Quite often, however, senators or representatives have a particular ax to grind and offer comments on a bill with the obvious intent of influencing court interpretations at a later time. Unless such commentary provides evidence of widespread Congressional support for the proffered interpretation, it is given little weight by the courts.

Finally, courts look to how the agency has interpreted the statute. Unless the agency's interpretation conflicts with the language or legislative history of the statute, the courts generally defer to the agency.

(f) Statutory amendment and informal controls

If Congress believes that an agency is pursuing regulatory policies that run counter to legislative intent or directives (or if it believes that reviewing courts have taken the implementation of a statute in a direction not in concert with Congress's current desires), it has several ways to remedy the problem. The most direct course of action would be for Congress to formally amend the statute to clarify its mandate to the agency. Although most direct, this may not be accomplished easily. Congress does not often speak with a unified voice. The passage of a major piece of legislation usually requires considerable time and political compromise. Indeed, the very breadth of the statutory mandate may have arisen from the need to strike such a compromise. Any attempt to inject further specificity into a piece of legislation by amending it may well face a long and difficult battle. This is not to say that statutes are never changed in response to Congressional dissatisfaction with agency behavior. The Solid Waste Disposal Act (also known as the Resource Conservation and Recovery Act), for example, was substantially modified in 1984 in response to a widespread perception in Congress that EPA was not moving swiftly enough to regulate the disposal of hazardous waste. In common practice, however, Congress often uses other means to influence agency decisionmaking.

Congress has a number of informal, more broadly "political" controls at its disposal. For example, members of Congress are free to make statements—on the floor of Congress and in other public fora—that criticize how an agency handles a particular matter. Especially if they receive media attention, such commentaries can effectively spur the agency to consider a change in direction. Congress also can use committee hearings to question and verbally admonish recalcitrant agency officials. These include both "oversight hearings," which permit close and often harsh questioning of agency officials, and budget hearings.

During the annual budget period, top agency officials come before Congress to explain and defend the administration's funding proposal for their agency. This gives members of Congress an opportunity to influence behavior by suggesting an increase or threatening a decrease in the agency's overall funding. In the early years of the

Reagan administration, for example, both EPA and OSHA were called to task for requesting a budget that many members of Congress believed was too small to fulfill their statutory mandates. In addition, Congress can designate specific line items in an agency's authority for special funding.

2. Direction from the Executive Branch

Although usually created by the legislative branch (EPA is an exception), administrative agencies sit within the executive branch. Accordingly, the executive also exercises considerable control over agency decisionmaking. Much of the executive's influence over the direction of an agency stems from the President's control of the appointment process. Most statutes that create an administrative agency also permit the President to appoint the agency's top decisionmakers (the so-called political appointments), subject to the approval of the Senate. The power to appoint includes the power to remove from office, along with all the more subtle means of persuasion that lie between the two. The underlying theory, presumably, is that each new administration should be free, within the bounds of the applicable statutory mandates, to chart the direction of the agencies that operate within its purview. However, this approach often entails an inherent conflict because the direction favored by the administration frequently differs from that favored by Congress. This appears to be an accepted part of the political process.

The executive branch also wields considerable influence over the agencies through the budget process. Although final approval of the national budget rests with Congress, the budget is shaped, in large part, by the proposed budget submitted to Congress by the executive. Even more directly than Congress, then, the executive branch can use its grip on the national purse strings to expand the size of those regulatory programs it favors and to contract the size of those it does not. Further, since 1980 the President has used the Office of Management and Budget (OMB) to oversee an economic analysis of all proposed major regulations. This has had a significant effect on the regulatory initiatives proposed by OSHA and EPA, and, as the following commentary indicates, has generated substantial controversy.

An Obstacle to Public Safety

William B. Schultz and David C. Vladeck

Before President Ronald Reagan was elected, the principal function of the Office of Management and Budget was to manage the federal budget. But in the past seven years, OMB has taken on a new role that has had a chilling effect on regulations designed to protect consumers and workers.

OMB's authority comes from an executive order issued less than a month after Reagan took office. It requires that OMB review all major federal regula-

Source: From the *Washington Post,* May 10, 1988, p. 20.

tory decisions and do an economic analysis of the costs of implementing each proposal. This little-noticed order from the White House has accomplished perhaps the most significant change in administrative law in the past 50 years.

Take the case of asbestos. In the early 1960s, asbestos was identified as a hazard that killed thousands of people annually, and in 1984 the Environmental Protection Agency proposed to phase out this substance over five to 15 years. But the proposal, like every important regulation issued by federal health and safety agencies during the Reagan administration, had a major hurdle to overcome: the Office of Management and Budget.

OMB performed a cost-benefit analysis, balancing the lives that would be lost if asbestos were permitted to be used in products such as insulation against the cost to industry of a ban. Its officials decided that a life is worth $1 million but then used an economist's tool called "discounting" to adjust for their expectation that most people would not die from asbestos-induced cancer until many years after their initial exposure. Using discounting, OMB's economists calculated the adjusted value of a human life at $208,000. OMB found that the regulation was not justified because its costs exceeded the value of human lives saved, and sent it back to EPA for revision. Now, EPA's final decision is not expected until this summer.

The pace of standard-setting at the Occupational Safety and Health Administration (OSHA), which was always slow, is now glacial. Created in 1970, OSHA's mandate is to foster a safe environment for American workers. One of its principal tools is the strict limitations that it places on toxic substances in the workplace.

Although OSHA had been averaging two to three health regulations per year, the agency did not issue a single standard during the first 2½ years of the Reagan administration, and it has issued only six standards during the past seven years. Four of these —ethylene oxide, benzene, formaldehyde and field sanitation—were issued only after a court order setting a specific deadline for agency action.

In the process, OMB succeeded in delaying standards for several years or more. A good example is ethylene oxide (EtO), a highly toxic and carcinogenic gas widely used in hospitals to sterilize medical equipment. In 1981, OSHA estimated that at exposure levels then permitted in the workplace, from 6 to 10 percent of 75,000 exposed hospital workers would get cancer over the course of their lives.

In 1983, a federal appeals court found that OSHA had illegally delayed the stronger EtO standard and ordered the agency to act. OSHA drafted a standard that was generally acceptable to labor and consumer groups, but the hospital industry objected to the part of the regulation that limited the amount of EtO a worker could receive in a single burst.

Having lost at OSHA, the industry took its case to OMB, which adopted the industry's view and overruled OSHA. That decision was also reversed by the court of appeals, and, nearly two years later, the new regulation was issued in March.

OSHA's benzene regulation suffered a similar fate; it was delayed three years before being issued in 1987.

One of the most troubling instances of current OMB interference involves cadmium, a metal used in electroplating and extensively in industrial processing. OSHA estimates that more than 213,000 workers are exposed to very high levels of cadmium. As a result, health officials estimate, there may be 1,106 excess cancer deaths per 10,000 workers, affecting 11 percent of the work force. Even greater numbers could suffer kidney damage, according to OSHA.

In the past, OSHA acted very quickly to curb exposures. Today, after factoring in OMB review, OSHA estimates that it will take three years to issue a cadmium standard. The agency projects that for each year it delays, nearly 500 workers could contract cadmium-induced lung cancer.

The Food and Drug Administration, one of the oldest federal regulatory agencies, is charged with regulating foods, drugs, and cosmetics. It has always been seen as relatively non-political. Yet today, every important FDA decision must survive a political review at OMB.

Unlike OSHA standards, most of the regulations that the FDA ultimately issues have not been significantly changed by OMB. But when important public health issues are at stake, OMB has delayed and indirectly blocked FDA regulations.

Take the case of aspirin and Reye's syndrome. Reye's syndrome is a rare but sometimes fatal disease that in the late 1970s was killing several hundred children a year. In the fall of 1981, the federal Centers for Disease Control, supported by four separate studies, identified a link between Reye's syndrome and the use of aspirin by children with flu and chicken pox.

No one suggested that aspirin be taken off the market, but the FDA drafted a proposed regulation that would have required a warning label.

The aspirin industry immediately began lobbying OMB. The president of its trade association, the Aspirin Foundation, met with a high-level OMB official, who, as he later recounted in a sworn deposition, reviewed the FDA's scientific data concerning the link between aspirin and Reye's Syndrome. Within a few days, he rejected the work that it had taken scientists at the FDA more than six months to complete. Shortly thereafter, the FDA decided to kill the proposed regulation until an additional study was completed.

In February 1986, almost four years later, the FDA at last issued a final regulation requiring the warning label on aspirin products, and the incidence of Reye's syndrome, which also had received considerable publicity, has since declined significantly.

OMB's impact on health and safety regulation is not limited to highly publicized cases such as aspirin. Often, FDA officials choose not to issue important regulations because they know that OMB will not give its approval. Usually, the public never learns about these efforts that are not pursued, but one extraordinary example, urethane in alcohol, has recently come to light.

Urethane is a carcinogen found in many alcoholic beverages. Canada, at the end of 1985, set limits in wine and liquor. While there are many uncertainties in applying data from animal research to humans, one study concluded that daily consumption of the amount of urethane in two shots of many brands of bourbon sold in the U.S. might cause cancer in one in 200 people. Gary Flamm, director of FDA's Office of Toxicological Sciences, ranks urethane among the top three carcinogens that should be feared. Concerned about these risks, the FDA and the industry have tested about 1,000 products to determine their urethane levels, and about 100 have been found to have levels that exceed the Canadian limits. The

Center for Science in the Public Interest has petitioned the FDA to follow Canada's lead and regulate this carcinogen.

The agency seriously considered issuing a regulation, but, instead, entered into voluntary agreements with the wine and liquor industry under which the manufacturers would not meet the Canadian limits until 1995. The agreements, moreover, are not enforceable by the FDA.

The problem with OMB review is that it allows economists who have little contact with the regulatory agency and virtually no technical expertise to evaluate essentially scientific decisions. OMB compounds the damage by leaving no paper trail, so the public often blames other agencies for delays or inadequate regulations that were the fault of OMB.

Until now, most of the criticism of OMB has come from the public interest community. Industry lobbyists have kept quiet as they successfully use OMB to overrule agency decisions that displease their clients. But these same lobbyists privately have expressed concerns about OMB review. They realize that their success will come back to haunt them if there is a shift in the political winds.

◇ NOTES

1. For a more detailed, scholarly review that comes to the same general conclusions about OMB's role, see Olson, The Quiet Shift of Power: Office of Management and Budget Supervision of Environmental Protection Agency Rulemaking Under Executive Order 12, 291, 4 Virginia Journal of Natural Resources Law, 1 (1984). For two such articles coming to a contrary conclusion, see Bernstein, The Presidential Role in Administrative Rulemaking: Improving Policy Directives: One Vote for Not Tying the President's Hands, 56 Tul.L.Rev. 818 (1982); and Comment, Capitalizing on a Congressional Void: Executive Order 12, 291, 31 Am.U.L.Rev. 613 (1981).

2. President Reagan's 1981 Executive Order 12291 (the substance of which remains in effect) specifies that regulatory action may not be taken unless the potential benefits to society outweigh the potential costs, where such benefits and costs are to be quantified in monetary terms to the extent possible. A precursor of this Executive Order was President Ford's 1974 Executive Order 11821, which requires that all regulations by executive branch agencies be accompanied by an inflationary impact statement, where "inflationary" was defined by the Council on Wage and Price Stability as a situation in which the costs of the regulation exceeded the benefits. However, it did not require that the regulation not be inflationary but only that the inflationary impacts be evaluated. Similarly, President Carter's Executive Order 12044 required federal agencies to analyze the economic consequences of significant regulations (those with an annual effect on the economy of $100 million or more) and their alternatives, but it imposes no cost-benefit requirement. In essence, President Reagan's Executive Order imposed the cost-benefit criterion (when compared to the baseline) as a prerequisite to promulgation of federal regulations. Under the Order, OMB is empowered to review proposed and final regulations of federal agencies, and the accompanying cost-benefit analysis, in advance of publication. As noted in the Schultz and Vladeck article, it has used this power to delay the promulgation of several regulations. (For a more detailed exploration of cost-benefit analysis as a guide for decisionmaking, see Chapter 6.)

3. OMB has sought to impose the cost-benefit criterion on agency decisionmaking even when the underlying statute has required that the regulation be promulgated according to criteria *other* than cost-benefit balancing. What issue of constitutional law does this raise? What arguments might OMB make to support its role in the face of a conflicting Congressional mandate?

4. OMB has also sought to influence the content of agency regulations under authority granted to it in the federal Paperwork Reduction Act (PRA). However, the Supreme Court has held that the PRA does not authorize OMB to interfere with substantive policy decisions made by other agencies in carrying out their own statutory mandates. See *Dole v. United Steelworkers of America* _ _ _ U.S. _ _ _, 110 S. Ct. 929 (1990).

3.   Direction from the Judicial Branch

Absent a statutory or constitutional amendment, the ultimate arbiters of the meaning of a particular statute or constitutional provision are the courts. This principle flows from the venerable case of *Marbury v. Madison,* 1 Cranch 137, 2 L. Ed.60 (1803), in which the Supreme Court held that a court could invalidate an act of Congress if it found the statutory language to be in violation of the Constitution. In a very real sense then, what an agency can or must do is what the courts *say* it can or must do. Congress can amend a statute to circumvent a judicial interpretation that it does not like, but even the new statutory language will face potential scrutiny by a reviewing court.

As powerful as the judicial branch is, however, it has at least one major Achilles heel. Unless someone brings a lawsuit, even the nine justices of the nation's most powerful judicial body—the United States Supreme Court—can do absolutely nothing to correct an agency action or decision that they believe to be unconstitutional or in violation of the agency's statutory mandate. Even if a lawsuit *is* filed, the Supreme Court cannot act on the matter until the case winds its way up from the lower courts, a process that can take years. (There are special circumstances in which a case can *originate* in the Supreme Court, but these are limited and are not relevant to most cases involving labor, health, or safety regulation.)

Familiarity and comfort with the court system takes time and cannot be gained through a few readings in a textbook. The following excerpt describes briefly what courts do and how the federal judicial system is organized.

Constitutional Courts

J. H. Ferguson and D. E. McHenry

The judicial article of the Constitution, Article III, is amazingly brief. It consists of but six paragraphs, the reading of which provides little understanding of our judicial system. The key to understanding is the opening sentence: "The judicial power of the United States, shall be vested in one supreme Court, and in such inferior Courts as the Congress may from time to time ordain and establish." ...

Source: From *The American System of Government*, J.H. Ferguson and D.E. McHenry, pp. 441–453. New York: McGraw Hill Publishing Company, © 1981. Reprinted by permission.

FEDERAL JUDGES

The Omnibus Judgeship Act of 1978 provided additional judgeships, creating positions in 117 district courts and 35 courts of appeals—the largest single increase in Federal judgeships in American History, and the first since 1970. The filling of those 152 positions will bring the total number of sitting judges to about 650, more than three-fourths of whom preside at the district-court level. The new legislation also authorized the President to promulgate "standards and guidelines" for the selection of new judges on the basis of merit. Moreover, it stipulated that in making nominations the President should

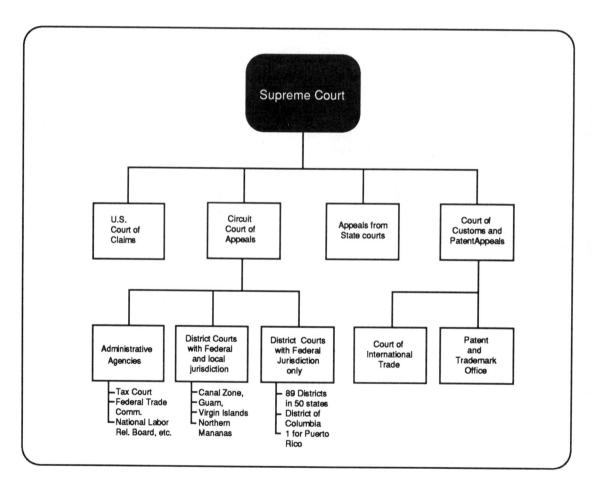

Figure 2-1. The Federal judicial system, like that in each of the states, consists of trial and appellate courts.

"give due consideration to qualified women, blacks, Hispanics and other minority individuals."

Selection and Appointment

All Federal judges are appointed by the President with the advice and consent of the Senate for terms of "good behavior." No qualifications are stated in the Constitution; hence the President is free to appoint anyone whom the Senate agrees to confirm. Although the President appoints judges, the rule of senatorial courtesy has traditionally required that a name submitted be acceptable to one or both senators, when of the President's party, in whose state the vacancy exists. Critics have for years claimed that the system unduly politicizes judicial selection; slows down the appointing process; fosters discrimination for reasons of race, national origin, creed, partisanship, and sex; and reduces the likelihood of having a judiciary worthy of commanding the confidence and respect required during the uncertain times ahead. . . .

Judicial salaries are fixed by Congress; they can be raised at any time, but they cannot be lowered during the incumbency of any particular judge. At the age of seventy, or at sixty-five with fifteen years on the bench, judges may retire or resign at full pay. If they retire, they are eligible for special assignments of a judicial character. Unlike most pension plans, the Federal one for judges requires no financial contributions from their own earnings during the course of employment.

After assuming office, judges are required by recently adopted canons of judicial ethics to refrain from such political activity as holding office in a political organization; making speeches for, or publicly endorsing, a political organization or candidate; soliciting funds, paying an assessment, or making a contribution to a political organization or candidate; attending political gatherings or purchasing tickets for political party dinners or other functions; and running for office without first resigning.

Removal from Office

The fact that no definite time limit is placed on judicial tenure has been understood to mean that terms run for life or until a judge chooses to resign or retire. It has also been understood that forcible removal was possible only by the impeachment process, although the Constitution does not explicitly state this to be the case. . . .

JURISDICTION

The term "jurisdiction" refers to the types of disputes and issues which may be taken to Federal courts for decision.

Constitutional courts may deal only with "cases" and "controversies." This explains why they will not give "advisory opinions" when asked by the President or Congress to clarify legal issues. Cases and controversies arise from contests between litigants who have rights and interests at stake and standing to sue in Federal courts. The rights and interests must be real and substantial, not hypothetical or trivial. . . .

The classes of cases and controversies which may come before Federal courts are shown graphically below. Some of these raise *Federal questions,* i.e., they involve the Constitution, acts of Congress, treaties, or vessels on navigable waters. Others reach Federal courts because of the *character* or *citizenship of the parties* involved [Figure 2-2].

Item 4, diversity of citizenship, presents difficulties. After much confusion, two rules now govern: (1) United States courts decide cases involving citizens of different states according to the same rules of law as would govern the case in a state court. (2) In such cases, Federal procedures are followed.

Item 6 has been qualified by the Eleventh Amendment, Congress, and the courts, in keeping with the precept that a sovereign cannot be sued unless consent is given. States may now be sued in Federal courts without their consent only by another state or by the Federal government. If an alien, citizen of another state, or citizen of the same state wishes to sue a state, this can be done only with the consent of the state involved, and then only in state courts. States may, however, initiate suits in Federal courts against aliens, citizens of other states, and foreign governments; although disputes with foreign governments are often settled by diplomatic negotiation.

Federal jurisdiction over the types of cases and controversies mentioned is not exclusive, however. Rather, Congress is free to distribute jurisdiction over most of them as it sees fit. Indeed, Congress may completely divest Federal courts of jurisdiction in certain instances. As matters stand, Federal courts have *exclusive* jurisdiction over some of them, have *concurrent* jurisdiction over others, and are totally *denied* consideration of still others. The division of responsibility is given in [Figure 2-3] below. . . .

Federal Question	1. Cases arising under the Constitution
	2. Cases involving Federal laws and treaties
	3. Admiralty and maritime cases
Character of Citizenship of Parties	1. Cases affecting ambassadors, other public ministers, and foreign consuls
	2. Controversies in which the United States is a party
	3. Controversies between two or more states
	4. Controversies between citizens of different states (diverse citizenship)
	5. Controversies between citizens of the same state claiming lands under grants of different states
	6. Controversies between a state or its citizens, and foreign states or their citizens or subjects

Figure 2-2. Types of cases and controversies.

Exclusively Federal	Civil actions in which states are parties (subject to exceptions noted)
	All suits and proceedings brought against (but not necessarily those initiated by) ambassadors, others possessing diplomatic immunity, and foreign consuls
	All cases involving Federal criminal laws
	All admiralty, maritime, patent-right, copyright, and bankruptcy cases
	All civil cases against the Federal government where consent to sue has been granted
Denied Federal	Civil suits involving citizens of different states where the amount at issue is less than $10,000
Concurrent with States	Civil suits involving citizens of different states where the amount at issue is $10,000 or more

Figure 2-3. Jurisdiction of Federal courts.

SUPREME COURT

Standing at the pinnacle of the Federal court system is the United States Supreme Court. Launched by the Judiciary Act of 1789, the Court held its first two terms on Wall Street in New York City, but in neither term were there any cases. Its next two terms were held in Philadelphia; thereafter it met in Washington. As first constituted it consisted of a Chief Justice and five associates. Congress by statute reduced the membership to five in 1801; increased it to seven in 1807; increased it to nine in 1827 and ten in 1863; reduced it to seven in 1866; and fixed it at nine in 1869. . . .

Cases that do not originate in the Supreme Court come to it by what is technically known as an appeal or by writ of certiorari. Appeals are allowed as a matter of right in cases involving Federal and state powers which obviously require a ruling by the highest Court. On petitions for certiorari the Court has the option of granting or denying review. If granted, as comparatively few are, lower Federal or state courts are directed to send up the entire record and proceeding for review or retrial. Typically, over three-fourths of the Supreme Court's business arises from petitions for certiorari. . . .

Decisions and Opinions

Several hundred cases and controversies reach the Supreme Court each year. A large number of appeals and petitions for writs of certiorari are disposed of without serious consideration for want of jurisdiction or merit. Many petitions for certiorari are merely granted or denied upon comparatively short briefs without oral argument. Others involving rather well-settled points of law about which the Court can come to a decision without oral hearings are disposed of as brief *per curiam* decisions. The remainder are decided after oral argument and in comparatively long written opinions setting forth reasons and justifications.

A decision may be unanimous or divided. If divided, both *majority* and *dissenting* opinions are usually written. Often one justice or more than one agrees with the conclusion reached in the majority or the dissenting opinion but for different reasons, in which case *concurring* opinions may be written. Thus, in a case involving complicated and controversial issues, there may be a majority opinion, a dissenting opinion, and concurring opinions. Six justices constitute a quorum, and at least a majority must concur before a decision is reached. *Obiter dicta* appear from time to time. Those are passing remarks, or observations, which are not binding on the case at hand but may have a bearing upon how opinions are interpreted, enforced, or decided at some time in the future. Published opinions once bore such titles as Dallas, Cranch, Wheaton, Wallace, and Otto, reflecting the names of Court reporters; since 1882 they have appeared as *United States Reports*.

COURTS OF APPEALS

Immediately below the Supreme Court stand the courts of appeals, created in 1891. There is one for the District of Columbia and eleven others. The eleventh, including Alabama, Georgia, and Florida, was added in 1980, by splitting the fifth. . . .

Each court has four to twenty-six permanent judges and usually hears cases in divisions consisting of three judges, but all judges may sit. The justices of the Supreme Court may also sit within circuits to

---

Warrant—commands appearance, arrest, search, or seizure.

Summons—directs plaintiff in civil suit to appear and answer complaint.

Writ of execution—directs defendant to satisfy judgment awarded in civil suit.

Writ of ejectment—ejects defendant from property held which court has found belongs to plaintiff.

Injunction—directs person to do, or more commonly, not to do, damaging act.

Mandamus—orders public official to perform act required by law.

Certiorari—orders public official, especially inferior judicial tribunal, to send up record for review.

Figure 2-4. Common judicial writs.

which each has been assigned, but time prevents them from "riding circuit," as they did in the early days of the Republic. District judges may also be assigned to serve in the appeals courts, although they may not judge cases which were before them on the lower bench. In some circuits, court is always held in the same city; in others, it may be held in two or more designated cities. The courts sit at irregular intervals in buildings owned or leased by the Federal government. Appeals judges are appointed by the President with the advice and consent of the Senate for terms of good behavior.

The appeals courts have slight original jurisdiction; they are primarily appellate courts. With few exceptions, cases decided in the district courts, special constitutional courts, legislative courts, and quasi-judicial boards and commissions go next to the appeals courts. Only the Supreme Court reviews the decisions of the appeals courts. . . .

DISTRICT COURTS

Eighty-nine district courts are located in the states. An additional one serves the District of Columbia. Districts have from one to twenty-seven judges; in a few instances, one judge serves two or more districts. The judges are appointed by the President with Senate approval for terms of good behavior.

A small state may itself constitute a district; otherwise, districts are arranged with regard for population, distance, and volume of business. Congress establishes district configuration.

Except for those assigned to the District of Columbia, district judges must reside in the district, or one of the districts, for which they are appointed. A permanent office must be maintained at a principal city, but court is usually held at regular intervals in various cities within each district.

Most cases and controversies start in district courts. Theirs is chiefly original jurisdiction; no cases come to them on appeal; cases begun in state courts are occasionally transferred to them. In the district courts nearly all accused of committing Federal crimes are tried. Ordinarily, cases are tried with only one judge presiding, but three judges must sit in certain types of cases.

Traditionally, the many bankruptcy cases arising under Federal law originated in the district courts, where they were handled by court-appointed "referees." In 1978, Congress changed that system when it approved the first major revision of the nation's bankruptcy laws in nearly forty years. After debating

whether primary responsibility should be retained by the district courts or shifted to the circuit courts, Congress decided in favor of the former and elevated the rank and status of referees to bankruptcy judges.

Under the new system, bankruptcy judgeships were placed on a permanent basis rather than at the discretion of the United States Judicial Conference. The number of judgeships was increased substantially; they were made appointive by the President with Senate approval for fourteen-year terms; salaries and benefits were raised; and jurisdiction was broadened. Provision for merit selection was made by authorizing circuit-court counsels to recommend candidates, although, as in other instances when merit lists are supplied, the President is not required to nominate any of the candidates named.

Although bankruptcy judges remain adjuncts of the district courts, they have both constitutional and legislative status. They now have all the jurisdictional authority over bankruptcy proceedings that a district judge has in civil cases, but their terms are limited to a stated number of years, as are those of legislative-court judges, as noted below. Most appeals are taken to district courts, although appeals may be heard by circuit courts with the consent of all parties, or to panels of three bankruptcy judges designated by the circuit courts. . . .

FEDERAL MAGISTRATES

Until recent times, minor judicial functions were performed by commissioners appointed by district courts for terms of four years and paid from fees. Legal training was not a prerequisite, and partisan considerations usually determined who would be appointed. Legislation enacted in 1968 provided the first reform in more than a hundred years. The name "magistrate" was substituted for "commissioner"; legal training was required; and appointments were made by district courts for terms of eight years; salaries were substituted for fees; and trial jurisdiction was broadened.

Further upgrading resulted from legislation passed in 1979. That legislation allowed magistrates to preside, provided [litigants] consent in writing, at jury and nonjury civil trials; criminal misdemeanor trials; and juvenile trials which do not permit incarceration. Magistrates continue to make eight-year appointments, although merit selection procedures promulgated by the United States Judicial Conference must now be used. Direct appeals can be taken to United States

Circuit Courts of Appeals unless litigants consent before trial to appeal to United States District Courts. As presently constituted, magistrate courts play about the same role in the Federal system of justice as do the lowest state trial courts presided over by officers bearing such titles as magistrates, justices of the peace, or community court judges.

COURT OFFICERS

Attached to each Federal court are the usual clerks, reporters, stenographers, bailiffs, and other aides. Appointments are usually made by the courts themselves using merit standards set by the Administrative Office of the United States Courts.

◇ **NOTES**

1. In addition to the sources of exclusive federal jurisdiction mentioned here, certain federal statutes require that cases involving those statutes be brought in federal court. The federal Clean Water Act, for example, specifies that a "citizen suit" against an alleged violator of the Act must be brought in the federal district court in which the alleged violation occurred. See 33 U.S.C. §1365(c).

2. In addition to the 12 Courts of Appeal mentioned here, there is a Court of Appeals for the Federal Circuit, which hears appeals from such specialized tribunals as the Court of Claims, the Patent and Trademark Office, and the Court of International Trade.

3. There is also a United States Tax Court, a trial level court that presides over certain disputes arising under the Internal Revenue Code.

◇

The following excerpt is from a lecture given by Professor Karl Llewellyn to law students at Columbia University. First published in 1930, Professor Llewellyn's observations remain perhaps the classic statement on the difficulty, and importance, of that arcane branch of literature known as the "judicial opinion."

This Case System: What to Do with the Cases

K.N. Llewellyn

I have now sketched for you in charcoal outline—an unkind critic might remark, in bastard caricature—the part that law and law's minions play in our society, and the general history of a case at law. All this, for your ordained affairs, is background; you will be getting restive. Indeed, in delivering these lectures orally, I have felt the need, before the sec-

Source: From K.N. Llewellyn, *The Bramble Bush* (Oceana Pub. Co., 1981), pp. 41-45.

ond hour, of setting to work to dynamite the foreground stumps: cases in casebooks have been assigned to you; what, then, are you to do with them?

Now the first thing you are to do with an opinion is to read it. Does this sound commonplace? Does this amuse you? There is no reason why it should amuse you. You have already read past seventeen expressions of whose meaning you have no conception. So hopeless is your ignorance of their meaning that you have no hard-edged memory of having seen unmeaning symbols on the page. You have applied to the court's opinion the reading technique that you use upon the Satevepost. Is a word unfamiliar? Read on that much more quickly! Onward and upward—we must not hold up the story.

That will not do. It is a pity, but you must learn to *read*. To read each word. To understand *each* word. You are outlanders in this country of the law. You do not know the speech. It must be learned. Like any other foreign tongue, it must be learned: by seeing words, by using them until they are familiar; meantime, by constant reference to the dictionary. What, dictionary? Tort, trespass, trover, plea, assumpsit, nisi prius, venire de novo, demurrer, joinder, traverse, abatement, general issue, tender, mandamus, certiorari, adverse possession, dependent relative revocation, and the rest. Law Latin, law French, aye, or law English—what do these strange terms mean to you? Can you rely upon the crumbs of language that remain from school? Does *Cattle levant and couchant* mean *cows getting up and lying down*? Does *nisi prius* mean *unless before*? Or *traverse* mean an upper gallery in a church? I fear a dictionary is your only hope—a law dictionary—the one-volume kind you can keep ready on your desk. Can you trust the dictionary, is it accurate, does it give you what you want? Of course not. No dictionary does. The life of words is in the using of them, in the wide network of their long associations, in the intangible something we denominate their feel. But the bare bones to work with, the dictionary offers; and without those bare bones you may be sure the feel will never come.

The first thing to do with an opinion, then, is read it. The next thing is to get clear the actual decision, the judgment rendered. Who won, the plaintiff or defendant? And watch your step here. You are after in first instance the plaintiff and defendant *below*, in the trial court. In order to follow through what happened you must therefore first know the outcome *below;* else you do not see what was appealed from, nor by whom. You now follow through in order to see exactly what *further* judgment has been rendered on appeal. The stage is then cleared of form—although of course you do not yet know all that these forms mean, that they imply. You can turn now to what you want peculiarly to know. Given the actual judgments below and above as your indispensable framework—what has the case decided, and what can you derive from it as to what will be decided later?

You will be looking, in the opinion, or in the preliminary matter plus the opinion, for the following: a statement of the facts the court assumes; a statement of the precise way the question has come before the court—which includes what the plaintiff

wanted below, and what the defendant did about it, the judgment below, and what the trial court did that is complained of; then the outcome on appeal, the judgment; and finally the reasons this court gives for doing what it did. This does not look so bad. But it is much worse than it looks.

For all our cases are decided, all our opinions are written, all our predictions, all our arguments are made, on certain four assumptions. They are the first presuppositions of our study. They must be rutted into you till you can juggle with them standing on your head and in your sleep.

1) *The court must decide the dispute that is before it.* It cannot refuse because the job is hard, or dubious, or dangerous.

2) *The court can decide* only *the particular dispute which is before it.* When it speaks to that question it speaks ex cathedra, with authority, with finality, with an almost magic power. When it speaks to the question before it, it announces *law*, and if what it announces is new, it legislates, it *makes* the law. But when it speaks to any other question at all, it says mere words, which no man needs to follow. Are such words worthless? They are not. We know them as judicial *dicta;* when they are wholly off the point at issue we call them *obiter dicta*—words dropped along the road, wayside remarks. Yet even wayside remarks shed light on the remarker. They may be very useful in the future to him, or to us. But he will not feel bound to them, as to his ex cathedra utterance. They came not hallowed by a Delphic frenzy. He may be slow to change them; but not so slow as in the other case.

3) *The court can decide the particular dispute only according to a* general *rule which covers a whole class of like disputes.* Our legal theory does not admit of single decisions standing on their own. If judges are free, are indeed forced, to decide new cases for which there is no rule, they must at least make a new rule as they decide. So far, good. But how wide, or how narrow, is the general rule in this particular case? That is a troublesome matter. The practice of our case-law, however, is I think fairly stated thus: it pays to be suspicious of general rules which look too wide; it pays to go slow in feeling *certain* that a wide rule has been laid down at all, or that, if seemingly laid down, it will be followed. For there is a fourth accepted canon:

4) *Everything, everything, big or small, a judge may say in an opinion, is to be read with primary reference to the particular dispute, the particular question before him.* You are not to think that the words mean what they might

if they stood alone. You are to have your eye on the case in hand, and to learn how to interpret all that has been said *merely* as a reason for deciding *that* case *that* way. At need.

Now why these canons? The first, I take it, goes back to the primary purpose of law. If the job is in first instance to settle disputes which do not otherwise get settled, then the only way to do it is to do it. And it will not matter so much *how* it is done, in a baffling instance, so long as it is done at all.

The third, that cases must be decided according to a general rule, goes back in origin less to purpose than to superstition. As long as law was felt as something ordained of God, or even as something inherently right in the order of nature, the judge was to be regarded as a mouthpiece, not as a creator; and a mouthpiece of the general, who but made clear an application to the particular. Else he broke faith, else he was arbitrary, and either biased or corrupt. Moreover, justice demands, wherever that concept is found, that like men be treated alike in like conditions. Why, I do not know; the fact is given. That calls for general rules, and for their even application. So, too, the "separation of powers" comes in powerfully to urge that general rules are made by the Legislature or the system, not the judges, and that the judge has but to act *according* to the general rules there are. Finally, a philosophy even of expediency will urge the same. Whatever may be the need of shaping decision to individual cases in the juvenile court, or in the court of domestic relations, or in a business man's tribunal for commercial cases—still, when the supreme court of a state speaks, it speaks first to clear up a point of general interest. And the responsibility for formulating general policy forces a wider survey, a more thorough study of the policies involved. So, too, we gain an added guarantee against either sentimentalism or influence in individual cases. And, what is not to be disregarded, we fit with the common notion of what justice calls for. —All of which last, I may say in passing, seemed to me once too pat to be convincing. In point of fact, the clearing up by courts of points of general interest does not require any constant strain to make a *rule;* and the felt need to make a rule that would be safe has often led our courts to overcaution. Responsibility for formulating general policy is not, as European practice shows, essential to fine, firm and conscientious work of judges, nor does it guarantee with us the absence of disturbing factors. And, finally, the man in the street can understand so little of the distinctions lawyers take that his reliance for justice is now just what it would be under a different system: the traditional respect for and training of the bench, the character of the personnel, the substantial soundness of its output. —I take time to say this because I deem it important that early, very early in this game you meet some counterweight against what I may call the unconscious snobbery of social institutions: against the bland assumption that because things social are, therefore they must be; against the touching faith that the current rationalizations of an institution, first, fit the facts, second, exhaust the subject, third, negate other, negate better possibilities. Nowhere more than in law do you need armor against that type of ethnocentric and chronocentric snobbery—the smugness of your own tribe and your own time. *We* are the Greeks; *all* others are barbarians. For the partial cure which is provided in the arts, in business organization, and most magnificently in science, by the international character of the data, of their application, their communication —these in law are no more. Law, as against the other disciplines, is like a tree. In its own soil it roots, and shades one spot alone. Atoms and x-rays seem to behave indifferent to whether the laboratory be in Paris or Chicago. A German appendix acts much like an American; the knife can speak a tongue more international than Volapluk. But the day when the Roman sources as modernized by Bartolus "and the other old mastiffs of the law" were serviceable equally in Italy and Germany and France —that day is past. The day is over when a dispute in Bristol could be referred to two Italian merchants, because the Italian cities were the root and flower of commercial customs of the world. If you would keep perspective, then—if you would really *see* the common law, you must beware of letting it submerge you. Perhaps when you see it, when you weigh it against other possibilities, you will look upon its work again and find it good. That, as common lawyer, I may hope, and do. But I should think it a cheap valuation of our wayward, wilful, charming Mistress to feel that she must be kept from comparison, or even scrutiny, lest her charm should fail.

Back, if I may now, to the why of the two canons I have left: that the court *can* decide only the particular dispute before it; that all that is said is to be read with eyes on that dispute. Why these? I do believe, gentlemen, that here we have as fine a deposit of slow-growing wisdom as ever has been laid down through the centuries by the unthinking social sea. Here, hardened into institutions, carved out and given line by rationale. What is this wisdom? Look to your own discussion, look to any argument. You know where you would go. You reach, at random if hurried, more carefully if not, for a foundation, for a major premise. But never for itself. Its interest lies in leading to the conclusion you are headed for. You shape its words, its content, to an end decreed. More, with your mind upon your object you use words, you bring in illustrations, you deploy and advance and concentrate again. When you have done, you have said much you did not mean. You did not mean, that is, *except* in reference to your point. You have brought generalization after generalization up, and discharged it at your goal; all, in the heat of argument, were over-stated. None would you stand to, if your opponent should urge them to *another* issue.

So with the judge. Nay, more so with the judge. He is not merely human, as are you. He is, as well, a lawyer; which you, yet, are not. A lawyer, and as such skilled in manipulating the resources of persuasion at his hand. A lawyer, and as such prone without thought to twist analogies, and rules, and instances, to his conclusion. A lawyer, and as such peculiarly prone to disregard the implications which do not bear directly on his case.

More, as a practiced campaigner in the art of exposition, he has learned that one must prepare the way for argument. You set the mood, the tone, you lay the intellectual foundation—all with the case in mind, with the conclusion—all, because those who hear you also have the case in mind, without the niggling criticism which may later follow. You wind up, as a pitcher will wind up—and as in the pitcher's case, the wind-up often is superfluous. As in the pitcher's case, it has been known to be intentionally misleading.

With this it should be clear, then, why our canons thunder. Why we create a class of dicta, of unnecessary words, which later readers, their minds now on quite other cases, can mark off as not quite essential to the argument. Why we create a class of *obiter dicta*, the wilder flailings of the pitcher's arms, the wilder motions of his gum-ruminant jaws. Why we set about, as our job, to crack the kernel from the nut, to find the true rule the case in fact decides: the *rule of the case.*

C. ADMINISTRATIVE RULEMAKING

In the broad sense, an agency carries out its statutory mandate by promulgating administrative regulations. These regulations are first published in the *Federal Register,* and are later codified in the *Code of Federal Regulations*. Unless they are challenged successfully in court or in Congress, they have the force of law. This section explores the process by which agency rulemaking is effectuated.

1. The Distinction between *Rulemaking, Adjudication* and *Enforcement*

Agencies perform legislative, judicial, and enforcement functions. When an agency engages in "rulemaking," it is doing many of the things we normally associate with the legislative process, and it is making policy. Within the confines of its statutory mandate, the agency is laying out a policy that all who fall within its ambit will have to follow. When OSHA sets a standard for a specific workplace exposure, it is engaging in rulemaking and is establishing a requirement to which all employers covered by the OSHAct will have to adhere.

But OSHA does more than develop and promulgate administrative rules; it also *enforces* them. In so doing, the agency performs a function that looks more like police work than legislation. This is the agency's "enforcement" function.

Finally, many regulatory statutes create some kind of administrative tribunal to adjudicate disputes that arise when the agency uses its enforcement power. The

OSHAct, for example, created the Occupational Safety and Health Review Commission. Employers cited for violating the Act or some regulation promulgated under the Act can appeal the citation to the Review Commission. The Commission hears the facts and determines whether a violation of the Act has occurred. It does not issue broad policy declarations, but rather issues an interpretation of the law *as applied to the facts at hand.* In so doing, the Review Commission is engaging in "adjudication," not rulemaking.

2. A General Look at Rulemaking under the Administrative Procedure Act

When an agency engages in rulemaking, Section 4 of the Administrative Procedure Act (APA) requires the agency to publish a "general notice of proposed rulemaking" in the Federal Register, setting forth: 1) a statement of the time, place, and nature of the public rulemaking proceedings; 2) reference to the legal authority under which the rule is proposed; and 3) either the terms or substance of the proposed rule or a description of the subjects and issues involved. Beyond this, the APA specifies two general procedures that can be used by agencies who are promulgating administrative regulations: formal rulemaking and informal (or notice and comment) rulemaking.

Formal rulemaking, defined in Sections 7 and 8 of the APA, is to be performed only if it is specifically required by Congress in the originating statute. As it can be a lengthy and cumbersome process, full formal rulemaking has been required in only a few modern regulatory statutes. The requisites of the formal rulemaking process include a trial-type hearing before an impartial presiding officer, an opportunity to present evidence and to cross examine witnesses, a decision made *on the record*, an opportunity to submit proposed findings, exceptions, and supporting reasons, and a statement of findings and conclusions and the reasons therefore. As with a court trial, a *verbatim* record of the entire proceeding must be maintained. If an agency engages in formal rulemaking, the APA directs the reviewing court to set aside the agency's action if it is "found to be . . . unsupported by substantial evidence [on the record as a whole]."

Unless the originating statute specifies otherwise, informal rulemaking is to be used. This form of rulemaking, defined in Section 4 of the APA, is often called notice and comment rulemaking. It requires that, after giving the required notice in the Federal Register, the agency give "interested parties an opportunity to participate in the rulemaking through submission of written data, views, or arguments with or without opportunity for oral presentation." It does *not* require a hearing and allows the agency to consider materials other than those brought forward at any hearing the agency choses to conduct. Unlike formal rulemaking, informal rulemaking allows the agency to look beyond the hearing record in making rules. When courts review agency action under informal rulemaking, agencies are not held to the "substantial evidence" test noted above, but rather to an arguably less stringent requirement that their determinations not be "arbitrary" or "capricious."

Sections 4, 7, 8, and 10 of the APA are reprinted in the appendix to this chapter. Many originating statutes contain procedural requirements in addition to—or in lieu of—those specified in the APA. Close attention to the particular requirements of each statutory scheme is important.

3. Informal Rulemaking by OSHA and EPA

Consistent with general Congressional preference, rulemaking by OSHA and EPA is conducted according to the informal notice and comment process. No formal, trial-type proceeding is required. In both the Occupational Safety and Health Act (OSHAct) and the Toxic Substances Control Act (TSCA), however, Congress has gone beyond the bare requirements of the Administrative Procedure Act and has provided additional opportunities for public input. Moreover, both EPA and OSHA are held to the substantial evidence test when promulgating rules under TSCA and the OSHAct. This mixture of formal and informal rulemaking procedures is often characterized as *hybrid rulemaking*.

Both the OSHAct and TSCA provide for an informal public hearing as a part of the rulemaking process. In promulgating an "occupational safety or health standard" under Section 6(b) of the OSHAct, OSHA must hold a "public hearing" if requested to do so by "any interested person." Although OSHA need not provide an opportunity for the cross-examination of witnesses at such a hearing, it usually does so as a matter of policy. In promulgating a rule regulating the production, use, or disposal of a hazardous chemical under Section 6 of TSCA, EPA similarly must provide interested persons an opportunity for an "informal hearing" in each case. In addition, however, Section 6 of TSCA explicitly requires EPA to allow the cross-examination of witnesses to the extent necessary "for a full and true disclosure with respect to [the relevant] issues."

Both agencies are also authorized to submit certain regulatory issues to advisory committees made up of persons outside the agency. These advisory committees do not supplant the agencies' regulatory power or responsibility, but they do provide input on selected technical *and* policy issues that arise in the course of agency deliberations. The following excerpt describes the relevant OSHA and EPA committees.

Advisory Committees in OSHA and EPA: Their Use in Regulatory Decisionmaking

N.A. Ashford

Section 7(a) of the Occupational Safety and Health Act establishes a mandatory, permanent, "balanced" advisory committee known as the National Advisory Committee on Occupational Safety and Health (NACOSH) to advise the Secretary of Labor and the Secretary of Health and Human Services on general issues pertaining to the administration of the OSH Act. The committee's twelve members are drawn from management, labor, occupational safety and health professionals, and the public, and are selected "upon the basis of their experience and competence in the field of occupational safety and health."

Source: From *9 Science, Technology & Human Values* 72 (1984). Copyright © 1984 by Sage Publications, West Newbury, CA. Reprinted by permission of Sage Publications, Inc.

NACOSH is required to hold no fewer than two meetings during each calendar year.

In addition, the OSH Act makes provision for the creation of limited-lifetime, "balanced," *ad hoc* advisory committees to consider specific subjects related to the setting of occupational standards. The authority for such committees is found in Section 7(b) of the Act, which empowers, but does not require, the Secretary of Labor to appoint special advisory committees "to assist him in his standard-setting function under Section 6. Each of these committees is to have no more than fifteen members, and is to be balanced equally between "persons qualified by experience and affiliation to present the viewpoint of the employers involved" and "persons similarly qualified to present the viewpoint of the workers involved." In addition, an *ad hoc* committee must include at least one representative of a state health and safety agency, and may include other persons "who are qualified by knowledge and experience to make a useful contribution," as long as the number of such representatives does not exceed the number from federal or state agencies.

Although the Environmental Protection Agency established a Science Advisory Board (SAB) administratively in 1974, that Board was subsequently established by statute in the Environmental Research, Development, and Demonstration Authorization Act. The original Board consisted of five standing committees and an executive committee. Under the Reagan administration, however, the EPA reduced the number of standing committees to three. By statute, these committees are to meet periodically with the EPA administrator "to provide advice . . . on the scientific and technical aspects of environmental problems and issues." Their membership is to be "a body of independent scientists and engineers of sufficient size and diversity to provide a range of expertise required to assess the scientific and technical aspects of environmental issues," and additional *ad hoc* committees may be drawn from the members.

D. CITIZEN OR CORPORATE ACCESS TO THE ADMINISTRATIVE PROCESS

The affected public has considerable opportunity to influence agency rulemaking. At the very least, notice and an opportunity to comment must be provided before the promulgation of any substantive regulation. Although most people neither read the Federal Register nor take time to comment, those who do participate tend to reflect the broader populace. Typically, the Federal Register notice of proposed rulemaking in the area of workplace health or safety attracts the attention of labor unions, public interest groups, and representatives of the affected industries. They likely *will* participate in the process, and in so doing they will present the views of their respective constituencies. Thus, although the outcome is still determined by the agency, a rough form of "administrative democracy" is at work.

Further, other avenues of access provide additional opportunity to influence the outcome of administrative proceedings.

1.   Initiation of Rulemaking

Section 4(e) of the APA requires every agency to "give an interested person the right to petition for the issuance, amendment, or repeal of a rule." Thus, although this section does not require the agency to actually *take* the requested action, it does give the interested public a formal opportunity to prod administrative consideration of a particular issue. Further, unless the request is patently frivolous, the agency may be required to provide a statement of its rationale if it declines to act on the petition.

Some originating statutes provide even greater leverage to citizens seeking to initiate agency rulemaking. Section 21 of TSCA, for example, contains specific provisions authorizing citizen petitions. In addition, it requires EPA to "either grant or deny" the petition within 90 days, authorizes the agency to hold a public hearing on the petition if "appropriate," and requires it to publish its rationale in the Federal Register if it denies the petition. Finally, a citizen whose petition for rulemaking is denied has a right under Section 21 to appeal the matter to a federal district court for a *de novo* proceeding. In such an appeal, the court is required by the statute to evaluate the petition *on its merits* and may not simply defer to the discretion of the agency.

2.   Access to Agency Proceedings and Records

Much of what an agency does is a matter of public record, and is generally accessible to the public. Three statutes, added to the APA in the 1970s, are particularly useful. The Government in the Sunshine Act, 5 U.S.C. §552b, requires that, except for certain enumerated exceptions, "every portion of every meeting of an agency shall be open to public observation." The exceptions are designed primarily to provide protection for personal privacy, trade secrets, agency enforcement efforts, and internal agency personnel rules and practices. A companion provision pertaining to agency records is the Freedom of Information Act (FOIA), 5 U.S.C. §552, which requires each agency to make all of its records "promptly available to any person," subject to a similar set of exceptions. Under the FOIA, an agency is required to respond to any request for records within ten working days. Anyone who is denied access to records to which he or she is entitled under the FOIA may take the agency to federal district court to secure access and may recover attorneys fees and the costs of the suit. Finally, there is the Privacy Act, 5 U.S.C. §552a, which provides broad access—with certain narrow exceptions—to agency records pertaining to oneself. The Act provides a right to correct inaccurate references to oneself in agency records and limits an agency's disclosure of personalized information.

3.   Access to Advisory Committees

The proceedings and records of agency advisory committees also are generally open to the public. The Federal Advisory Committee Act (FACA), 5 U.S.C. App. I, requires that, again with certain exceptions, the meetings of advisory committees be open to the general public and the records of these committees be made available to the general public. The exceptions generally follow those found in the Government in the Sunshine Act and the Freedom of Information Act. FACA also provides the public with some control over the composition of advisory committees. It requires the membership of an advisory committee to be "fairly balanced in terms of the points of view presented and the functions to be performed." This provision was designed to prevent advisory committees from being unduly biased toward any particular viewpoint on important policy or technical issues. The extent to which an agency determination based on an "unbalanced" advisory committee recommendation may be challenged in court, however, has yet to be decided.

An important feature of FACA is that it pertains to almost all formal and informal advisory bodies convened by the agency, regardless of whether the agency refers to them as "advisory committees." Thus, if an agency meets with a group of industry scientists with regard to a proposed chemical regulation, but fails to include experts with other viewpoints (or, conversely, if it meets with a group of scientists from the public interest community without including representatives of industry), it may well be in violation of the fair balance requirement of FACA. Further, if it holds such meetings and refuses admission to other interested parties, it may well be in violation of the public access provisions of the Act. Careful attention to FACA, then, can be a useful tool for those seeking to have input to agency decisionmaking, especially on issues of science and technology.

In addition to these general rights of access, the originating statute may provide for other routes of access to agency proceedings or records. Conversely, in some cases the originating statute will contain provisions that *limit* the applicability of one or more of these laws.

4.   Access to the Courts

Finally, the public has a limited right of access to the courts to seek judicial review of agency performance. The standards by which courts evaluate agency decisions are discussed in the next section. Before actually *getting* to court, however, there are a few practical and procedural hoops through which one must be able to jump.

In general, a court does not have the power to review an agency decision unless there is statutory authority for judicial review. To determine if a court can review an agency decision, one usually looks to the APA and the particular regulatory statute under which the agency decision was made.

Section 10(c) of the APA provides for judicial review in situations in which the originating statute is silent on the point. But not every agency decision is reviewable. To come under the ambit of this provision, the decision must be one that can be characterized as a "final agency action." The term *agency action* is defined elsewhere in the Act as "the whole or a part of an agency rule, order, license, sanction, relief, or the equivalent or denial thereof, or failure to act." Importantly, then, an agency decision not to take regulatory action in a particular area is reviewable under this provision. The word *final* is not defined in the APA, but the clear intent is to preclude review under this section before the regulatory issue in question has had an opportunity to make its way through the regulatory decisionmaking process.

Section 10 also provides that an agency decision that would be otherwise reviewable under the APA will not be reviewable if either another statute precludes judicial review or the decision is one that is "committed to agency discretion by law." Thus, one must always look to the originating statute to determine whether it contains language that would place the decision into one of these categories. Originating statutes may also *expand* the availability of judicial review. Indeed, many health, safety, and environmental statutes have sections that specify direct access to the courts for review of certain agency decisions. TSCA and the OSHAct both contain such provisions.

Beyond this determination, there are certain *threshold limitations on the availability of judicial review.* Three doctrines of administrative law—commonly dubbed "standing to sue," "exhaustion of administrative remedies," and "ripeness"—sometimes pose threshold problems. The *standing doctrine,* discussed in detail below, deals with the question of whether the person seeking judicial review is among those who are entitled to seek such review. The latter two address the question of whether the administrative decision in question has "matured" sufficiently—whether it has reached that point in the administrative process at which it is deemed reviewable. *Exhaustion of administrative remedies* is the rule that all intra-agency appeals must be utilized before one can seek judicial review. In large part, this doctrine is consistent with that portion of the APA that provides review only for final agency actions. The exhaustion doctrine can be modified, however, if the court finds that additional proceedings in the agency would be futile or would work irreparable harm. *Ripeness,* a related doctrine, is a rule of constitutional derivation. Under the constitution, federal courts can only entertain cases or controversies that are real and present or imminent. Thus, if the agency is still considering the regulatory matter in question, or has not yet taken contemplated regulatory action, a court may find that the matter is not yet ripe for judicial review.

Among the three, standing to sue probably is the most likely to be an issue. Under the constitution, one cannot invoke the jurisdiction of the federal courts unless one has a "personal stake in the outcome of the controversy" sought to be resolved. *Baker v. Carr,* 369 U.S. 186, 204 (1962). The standing doctrine flows logically from this limitation. Section 10 of the Administrative Procedure Act provides as follows:

> A person suffering legal wrong because of agency action, or adversely affected by agency action within the meaning of a relevant statute, is entitled to judicial review thereof.

Thus, before one is entitled to judicial review of an agency decision—before one has standing to sue—one must be able to establish that one will be injured by that decision. What kind of "injury" is required, and how close must the nexus be between that injury and the person seeking review?

The leading modern case on this issue is *Sierra Club v. Morton,* 405 U.S. 727 (1972). There, the Sierra Club (a nonprofit environmental group) sought to block the construction of a large resort in the Mineral King Valley in California, directly adjacent to Sequoia National Park. The group alleged that the Secretary of Interior's decisions to allow such construction "would destroy or otherwise affect the scenery, natural and historic objects and wildlife of the park and would impair the enjoyment of the park for future generations." It did not allege that any of its members actually used the park, but rather alleged a general interest in protecting the environment on behalf of all members of the public. The Court agreed that the kind of injury alleged here—which it characterized as aesthetic rather than economic—"may amount to an 'injury in fact' sufficient to lay the groundwork for standing under Section 10 of the APA." However, the Court held that the Sierra Club did not have standing in this case because it had not alleged any *specific* aesthetic injury to any specific Sierra Club member. Standing, noted the Court, "requires that the party seeking review be himself among the injured."

In practice, this criterion generally is not difficult to meet. If the administrative decision involves an environmental issue, one must be sure to locate someone who has a direct economic or aesthetic interest in the environmental amenity in question. If it involves a workplace issue, one must locate an affected employer or employee. And once standing to sue is established by one having a direct interest in the outcome, that party may raise the broader interests of the public-at-large.

Sometimes more difficult to establish, however, is that the interest in question is one that is "within the meaning of a relevant statute" as required by Section 10 of the APA. As noted in *Sierra Club v. Morton,* standing is a actually a two-pronged test:

> persons [have] standing to obtain judicial review of agency action under Section 10 of the APA where they [have] alleged that the challenged action [has] caused them an 'injury in fact,' and where the alleged injury [is] to an interest arguably within the zone of interests to be protected or regulated by the statutes that the agencies [are] claimed to have violated.

The following case illustrates the importance of this second requirement.

Calumet Industries v. Brock

807 F.2d 225 (D.C. Cir 1986)

SILBERMAN, Circuit Judge.

Petitioners Calumet Industries, Macmillan Ring-Free Oil Co., and Seaview Lubricants seek review of a Notice of Interpretation ("Notice") issued by the Occupational Safety and Health Administration ("OSHA"). The Notice seeks to draw a bright line between "carcinogenic" lubricating oils and "noncarcinogenic" lubricating oils. In these consolidated cases, petitioners contend that the Notice is invalid because OSHA failed to follow the notice and comment procedures of the Administrative Procedure Act, 5 U.S.C. §551 *et seq.* (1982) ("APA") and the Occupational Safety and Health Act, (required procedures codified at 29 U.S.C. §665(b) (1982) ("OSH Act"); and that OSHA's action in promulgating the Notice was "arbitrary and capricious" and therefore unlawful, 5 U.S.C. §706(a)(A). We do not reach the merits of these claims because we are convinced that petitioners lack standing to challenge OSHA's issuance of the Notice.

The genesis of this lawsuit occurred on November 25, 1983, when, pursuant to his authority under the OSH Act, the Secretary of Labor issued the Hazard Communication Standard, 29 C.F.R. 1910.1200 (1985) ("Standard"), scheduled to take effect two years later (November 25, 1985). The Standard requires chemical manufacturers and importers to assess the hazards of the chemicals they produce or import and affix labels with appropriate warnings to the containers of all chemicals deemed to be hazardous. It further specifies that carcinogenic chemicals be labeled as health hazards. A listing in one of several authoritative sources, including the International Agency for Research on Cancer (IARC) *Monographs*, establishes that a chemical is carcinogenic.

In April, 1984, IARC issued a Monograph that examined the carcinogenity of lubricating oils. In its study, IARC divided lubricating oils into eight classes, including the four classes pertinent here: vacuum distillates, acid-treated oils, solvent-refined oils, and hydrotreated oils. The Monograph concluded: (1) all vacuum distillates and acid-treated oils are carcinogenic; (2) "mildly" solvent-refined oils are carcinogenic, but "severely" solvent-refined oils are *not* carcinogenic; and (3) "mildly" hydrotreated oils are carcinogenic, but the evidence concerning "severely" hydrotreated oils does not permit a conclusion that such oils either are or are not carcinogenic. IARC did not,

however, provide any definition of either "mild" and "severe" solvent-refining or "mild" and "severe" hydrotreating.

Affected oil refiners (those who made hydrotreated and solvent-refined lubricating oils) were quite perturbed by this series of events. Taken together, the Standard and the IARC Monograph required labeling "mildly" solvent-refined and hydrotreated oils. But where, they asked, was the line distinguishing "mild" from "severe?" As the effective date of the Standard drew near, numerous refiners and other interested parties urged OSHA to clarify the labeling requirements to identify the distinction between "mildly" and "severely" hydrotreated oils.[1]

In an effort to do so, OSHA examined several research papers that had formed the basis of IARC's study, seeking to identify those factors that distinguished the two forms of hydrotreating. Although OSHA was unable to find a definitive distinction between the two, it concluded from its research that temperature and pressure were the two most significant factors distinguishing "mild" from "severe" hydrotreating. Accordingly, OSHA decided to issue a "Notice of Interpretation," which took the form of a definition of "mild" hydrotreated for labeling purposes under the Standard. On December 20, 1985, OSHA published its Notice of Interpretation, which said: "OSHA will . . . regard an oil as mildly hydrotreated if it has been processed at a pressure of 800 pounds per square inch (psi) or less, at temperatures up to 800°F, independent of other process parameters." 50 Fed. Reg. 51854 (Dec. 20, 1985). The Notice did not address the distinction between "mildly" and "severely" solvent-refined oils.

Petitioners do not make either hydrotreated or solvent-refined lubricating oils; Calumet and Seaview manufacture vacuum distillate oils, and Macmillan makes vacuum distillate and acid-treated oils. All three petitioners acknowledge that the Standard clearly requires them to label all their products and that they have complied with that requirement. Despite the fact that IARC's 1984 Monograph did not create any confusion about *their* labelling obligation, petitioners nevertheless participated in the efforts described above to convince OSHA to more pre-

---

[1] Apparently, the imprecise distinction between "mildly" and "severely" solvent-refined oils was not the major focus of refiners' complaints.

cisely interpret the Standard's labelling requirements as they pertained to hydrotreated oils. But while directly affected refiners merely requested clarification of the labelling requirements, petitioners urged OSHA to adopt a more stringent labelling policy whereby any product for which health evidence was inconclusive must be labelled as hazardous. Petitioners argued this approach would, among other things, reduce the availability of unlabelled lubricating oils, thus protecting the market position of those refiners (like petitioners) who were already required to label their products. OSHA rejected that argument, choosing (as we have noted) to permit its unclarified Standard to go into effect as scheduled on November 25, 1985.

*   *   *

Conceding that the Notice does not pertain to the products they make, petitioners nevertheless contend they have standing to bring this complaint because they are losing sales and profits to producers of hydrotreated and solvent-refined oils who are *not* required to label their products under the Standard as interpreted by the Notice. Petitioners argue that the line drawn between "mildly" and "severely" hydrotreated oils is underinclusive and should be redrawn to require labelling additional hydrotreated oils. They also imply that manufacturers of carcinogenic solvent-refined oils are exploiting the continuing impreciseness in the Standard regarding solvent-refined oils to avoid labelling their products. Petitioners seek reversal of OSHA's Notice and a stay of the Standard, as it pertains to lubricating oils, until the labelling requirements of the Standard are "adequately" clarified.

II

Petitioners contend they have met both the constitutional and prudential requirements of standing. Although we very much doubt petitioners could establish constitutional standing,[2] we need not reach that issue because it is, in our view, abundantly clear that petitioners do not meet non-constitutional pru-

dential standing criteria. When a litigant seeks judicial review of an agency decision, he must show that "the interest sought to be protected by the complaint is arguably within the zone of interests to be protected or regulated by the statute . . . in question." *Association of Data Processing Service Organizations, Inc. v. Camp*, 397 U.S. 150, 153 (1970). There must be "some indicia—however slight—that the litigant before the court was intended to be protected, benefited or regulated by the statute under which suit is brought." *Copper & Brass Fabricators Council, Inc. v. Department of the Treasury*, 679 F.2d 951, 952 (D.C. Cir. 1982). Petitioners gain standing under the APA not because their interests are merely "implicated" by the agency action but because "Congress intended . . . that class to be relied upon to challenge agency disregard of the law." *Block v. Community Nutrition Institute*, 467 U.S. 340, 347 (1984).

Petitioners would have standing under the "protected interests" leg of the zone of interest test if their complaint asserted the infringement of an interest that is protected by the OSH Act. However, if is now well settled in other Circuits that the interest to be protected by the OSH Act is worker safety, *see, e.g., R. T. Vanderbilt Co. v. OSHRC*, 728 F.2d 815, 818 [11 OSHC 1829, 1831] (6th Cir. 1984); *R. T. Vanderbilt Co. v. OSHRC*, 708 F.2d 570, 577 [11 OSHC 1545, 1549-50] (11th Cir. 1983); *Fire Equipment Manufacturers' Association, Inc. v. Marshall*, 679 F.2d 679, 682 [10 OSHC 1649, 1650-51] (7th Cir. 1982), *cert. denied*, 459 U.S. 1105 [11 OSHC 1096] (1983), and *not* business profits, *see, e.g., R. T. Vanderbilt*, 708 F.2d at 578 [11 OSHC at 1550] (manufacturer's economic injury not sufficient to obtain judicial review); *Fire Equipment Manufacturers'*, 679 F.2d at 682 [10 OSHC at 1549-50] (manufacturers' lost profits not within zone of interest protected by OSH Act). We agree with these courts. As petitioners here do not come before us as protectors of worker safety, but instead as entrepreneurs seeking to protect their competitive interests, we think it plain they lack standing under the "protected interests" leg of the zone of interests test.

Nevertheless, petitioners contend that although not arguably protected by the OSH Act, they are

[2]To have standing under the Constitution, "[a] plaintiff must allege personal injury fairly traceable to the defendant's allegedly unlawful conduct and likely to be redressed by the requested relief." *Valley Forge Christian College v. Americans for Separation of Church and State, Inc.*, 454 U.S. 464, 472 (1982). If the Notice that petitioners challenge were invalidated, then the Standard for manufacturers of hydrotreated oils would again be ambiguous; arguably, petitioners' competitors would exploit this

ambiguity to avoid labeling their products. It is difficult to see how that result could enhance petitioners' competitive position. Indeed, petitioners appear to complain about that very situation with respect to manufacturers of solvent-refined oils. Petitioners would therefore seem to run afoul of the causation requirement reflected in both the "fairly traceable" and "redressible" aspects of the standing test.

certainly included within the Act's "regulatory scheme" and therefore have standing under the "regulated interests" leg of the zone of interests test. To be sure, petitioners are regulated by the OSH Act, which gives the Labor Secretary authority to promulgate a wide range of occupational safety and health standards. And petitioners are also regulated by the Hazard Communication Standard, which requires them, as manufacturers of hazardous chemicals, to label their lubricating oils. But petitioners are concededly *not* regulated by the agency action they challenge—the Notice. We have held that to have standing under the "regulated interests" leg, it is not enough that petitioners be regulated merely by the statute upon which the agency action is based. *Tax Analysts and Advocates v. Blumenthal,* 566 F2d 130, 143–45 (D.C. Cir. 1977), *cert. denied,* 434 U.S. 1086 (1978) (taxpayer regulated by Internal Revenue Code cannot challenge tax credit provision not pertaining to him). Granting "regulated interest" standing to petitioners to challenge OSHA's Notice simply because they happen to be regulated by the OSH Act in general and by other aspects of the Standard would "stretch[ ] the concept of regulation [as used in the zone of interests test] to implausible limits." *Tax Analysts,* 566 F2d at 143.

Arguing that a competitor oil manufacturer not subject to an OSHA regulation has standing to allege that another oil manufacturer was treated too leniently by that regulation is simply another way of contending that a competitor's market position is protected by the OSH Act—a proposition we, and other courts, soundly reject. Instead, to gain standing under the "regulated interests" leg of the zone of interests test, petitioners must show they are directly, even if only minimally, regulated by the agency action they challenge.[3] Since they do not manufacture hydrotreated oils (or, for that matter solvent-refined oils), petitioners are merely challenging the Standard's application (as interpreted by the Notice) to third-party competitors. To paraphrase our statement in *Tax Analysts,* where we denied standing in a similar situation, "[petitioners are] not directly regulated by the rulings being challenged in this case. Rather, a more appropriate description is that [they] operate [ ] in an industry which is regulated by the rulings but do[ ] not operate in that sphere of the industry which is the object of the regulation." *Tax Analysists,* 566 F2d 143 n.82. Accordingly, we hold that petitioners lack standing; their petition is therefore, *denied.*

[3]The only litigants who would have "regulated interests" standing to challenge OSHA's Notice are manufacturers of hydrotreated oils processed at a pressure of 800 psi or less, at temperatures up to 800°F, who are required by virtue of the Notice to label their products. Moreover, such litigants would have standing only to contest the burden the Notice imposes on *them,* and not to contest the Notice's failure to similarly burden others.

5. Monetary Limitations on the Availability of Review

Legal considerations are not the only important factors in determining whether judicial review will be available. There are monetary limitations on the availability of review. Of primary *practical* concern is the fact that lawsuits cost money. Not only does one have to hire a lawyer to bring the case, but it is often also necessary to employ technical consultants to analyze the subject matter in question and to critique the agency's approach to technical issues. Fortunately, Congress has provided some assistance in this area. The Equal Access to Justice Act, 28 U.S.C. §2412, passed in 1984, provides for an award of "fees and other expenses" to any "prevailing party" in a lawsuit (other than a tort suit) with an agency of the United States, unless such an award is specifically prohibited by another statute. The Act defines "fees and expenses" rather broadly; they include:

> the reasonable expenses of expert witnesses, the reasonable cost of any study, analysis, engineering report, test, or project which is found by the court to be necessary for the preparation of the party's case, and reasonable attorney fees.

Thus, one who is successful in securing judicial reversal or remand of an agency decision can obtain reimbursement—generally at something less than market rates—for the principal costs of such litigation. In practice, this may make it possible for individuals to retain an attorney who will agree to defer the receipt of compensation on the expectation that it will be received from the government when the case is won. (This is often referred to as taking a case on a contingent fee basis.)

There are limitations on the availability of fees and expenses under the Equal Access to Justice Act. For-profit corporations, and individuals with a net worth in excess of one million dollars, are not eligible under the Act. Further, one must be a "prevailing" party to qualify. In general, this means that one must be successful on at least one major part of the case. Finally the court may deny or reduce an award to the extent that it finds that the position of the agency was "substantially justified," that the party seeking the award "engaged in conduct which unduly and unreasonably protracted the final resolution of the matter in controversy," or that "special circumstances make an award unjust."

In addition to the Equal Access to Justice Act, specific judicial review provisions found in some originating statutes also provide for an award of fees and expenses. Section 19(d) of TSCA, for example, specifies that a court may award "costs of suit and reasonable fees for attorneys and expert witnesses" in cases challenging a TSCA standard. There is no similar provision in Section 6(f) of the OSHAct, which governs judicial review of OSHA standards.

6. Bypassing the Agency: Citizen Enforcement through the Private Right of Action

As discussed, the major functions of administrative agencies are to set regulatory standards and then to enforce them. What role does the citizen have when an agency does not follow through on its enforcement function, either because of a lack of resources or because of a lack of will? In the absence of a statute specifically authorizing citizen enforcement, that role probably is limited.

Under the Administrative Procedure Act, a citizen can seek a court order—a *writ of mandamus*—to compel the agency to carry out its responsibilities under the law. In general, however, the courts tend to defer to an agency's "prosecutorial discretion," and to give the agency wide latitude in determining when to enforce the law and whom to enforce it against.

A citizen can also seek to bypass the agency, and to enforce the law directly against the violator. In doing so, the citizen must persuade the court that Congress intended that there be what is called a "private right of action" to enforce the statute in question. Where the statute is silent on the point, one must argue that there is an *implied* private right of action. This is an uphill battle, as the criteria for establishing such an implied right of action are difficult to meet. See *California v. Sierra Club*, 451 U.S. 287 (1981). However, the "citizen suit" provisions found in most federal environmental statutes— including TSCA—contain specific authorization for private rights of action. In general, these provisions allow affected citizens to bring suit against violators of the act if certain preconditions are met, and to recover attorneys fees and expenses if the suit is successful. These are extremely powerful tools for the citizen activist. Although the citizen's right of enforcement is clearly subordinate to that of the agency, these citizen suit provisions authorize aggressive private enforcement in situations in which the agency declines to act. By coupling the private right of action with a specific authorization for the recovery of attorneys fees and expenses from the violator, Congress here created real incentives for citizen participation in the environmental enforcement process. By and large, however, Congress has not chosen to create such incentives with regard to workplace health and safety, as the OSHAct does not provide for private rights of action.

E. THE ROLE OF THE COURTS IN REVIEWING AGENCY ACTION

Once one has secured access to the courts for review of an agency decision, the issue of the scope of judicial review remains. That is, to what extent will the courts delve into the actual details of the agency decisionmaking process in their review of that process? When one explores this issue, one is also exploring the broader issue of the nature of the agency's responsibilities under the law. For, absent further action by Congress, the question of what an agency is authorized to do, or what an agency must or must not do, is up to the courts to decide.

1. Five Judicial Limitations on Agency Authority

In general, there are five sources of law to which courts can turn in evaluating agency action. A court can reverse the agency's decision, or remand it for further consideration, if it finds that the agency has violated any of these sources of law. The first is the *originating statute* itself. Agencies cannot violate their principal statutory mandates. Thus, although it is never appropriate for the reviewing court to substitute *its* policy judgments for those of the agency, it is the court's duty to remand the agency's decision if the agency has taken policy positions that offend the policy choices made by Congress in the originating statute. Secondly, the courts look to *other applicable legislation,* including various procedural statutes such as the APA. To the extent they apply to the agency's action, all other statutes must be followed as well. A third source of law is *the agency's own rules and procedures.* Although not always fatal, an agency's failure to adhere to the procedures it has developed for promulgating regulations may result in a court remand. A fourth limitation on agency behavior could be termed *administrative common law*—substantive or procedural requirements developed by the federal courts in their review of agency decisions. In theory, the courts have no authority to require more of the agencies than is required by statute. In practice, however, the courts take enough leeway in interpreting those statutes that they effectively impose requirements of their own. Finally, agency behavior is limited by the *Constitution.* As the following case illustrates, agency action that offends the Constitution is unlawful even when it is specifically authorized by statute.

Marshall v. Barlow's, Inc.

436 U.S. 307 (1978)

Mr. Justice WHITE delivered the opinion of the Court.

Section 8(a) of the Occupational Safety and Health Act of 1970 (OSHA) empowers agents of the Secretary of Labor (the Secretary) to search the work area of any employment facility within the Act's jurisdiction. The purpose of the search is to inspect for safety hazards and violations of OSHA regulations. No search warrant or other process is expressly required under the Act.

On the morning of September 11, 1975, an OSHA inspector entered the customer service area of Barlow's, Inc., an electrical and plumbing installation business located in Pocatello, Idaho. The president and general manager, Ferrol G. "Bill" Barlow, was on hand; and the OSHA inspector, after showing his credentials, informed Mr. Barlow that he wished to conduct a search of the working areas of the business. Mr. Barlow inquired whether any complaint had been received about his company. The inspector answered no, but that Barlow's, Inc. had simply turned up in the agency's selection process. The inspector again asked to enter the nonpublic area of the business; Mr. Barlow's response was to inquire whether the inspector had a search warrant. The inspector had none. Thereupon, Mr. Barlow refused the inspector admission to the employee area of his business. He said he was relying on his rights as guaranteed by the Fourth Amendment of the United States Constitution.

Three months later, the Secretary petitioned the United States District Court for the District of Idaho to issue an order compelling Mr. Barlow to admit the inspector. The requested order was issued on December 30, 1975, and was presented to Mr. Barlow on January 5, 1976. Mr. Barlow again refused admission, and he sought his own injunctive relief against the warrantless searches assertedly permitted by OSHA. A three-judge court was convened. On December 30, 1976, it ruled in Mr. Barlow's favor. 424 F. Supp. 437 [4 OSHC 1887]. Concluding that *Camara v. Municipal Court*, 387 U.S. 523, 528-529 (1967), and *See v. City of Seattle*, 387 U.S. 541, 543 (1967), controlled this case, the court held that the Fourth Amendment required a warrant for the type of search involved here and that the statutory authorization for warrantless inspections was unconstitutional. An injunction against searches or inspections pursuant to §8(a) was entered. The Secretary appealed, challenging the judgment, and we noted probable jurisdiction. 430 U.S. 964.

I

The Secretary urges that warrantless inspections to enforce OSHA are reasonable within the meaning of the Fourth Amendment. Among other things, he relies on §8(a) of the Act, 29 U.S.C. §657(a), which authorizes inspection of business premises without a warrant and which the Secretary urges represents a congressional construction of the Fourth Amendment that the courts should not reject. Regretfully, we are unable to agree.

The Warrant Clause of the Fourth Amendment protects commercial buildings as well as private homes. To hold otherwise would belie the origin of that Amendment, and the American colonial experience. An important forerunner of the first 10 Amendments to the United States Constitution, the Virginia Bill of Rights, specifically opposed "general warrants, whereby an officer or messenger may be commanded to search suspected places without evidence of a fact committed." The general warrant was a recurring point of contention in the colonies immediately preceding the Revolution. The particular offensive-

ness it engendered was acutely felt by the merchants and businessmen whose premises and products were inspected for compliance with the several Parliamentary revenue measures that most irritated the colonists'. "[T]he Fourth Amendment's commands grew in large measure out of the colonists' experience with the writs of assistance ... [that] granted sweeping power to customs officials and other agents of the King to search at large for smuggled goods." *United States v. Chadwick*, 433 U.S. 1, 7-8 (1977). See also *G.M. Leasing Corporation v. United States*, 429 U.S. 338, 355 (1977). Against this background, it is untenable that the ban on warrantless searches was not intended to shield places of business as well as of residence.

This Court has already held that warrantless searches are generally unreasonable, and that this rule applies to commercial premises as well as homes. In *Camara v. Municipal Court*, 387 U.S. 523, 528-529 (1967), we held:

[E]xcept in certain carefully defined classes of cases, a search of private property without proper consent is 'unreasonable' unless it has been authorized by a valid search warrant.

On the same day, we also ruled:

As we explained in *Camara*, a search of private houses is presumptively unreasonable if conducted without a warrant. The businessman, like the occupant of a residence, has a constitutional right to go about his business free from unreasonable official entries upon his private commercial property. The businessman, too, has that right placed in jeopardy if the decision to enter and inspect for violation of regulatory laws can be made and enforced by the inspector in the field without official authority evidenced by a warrant. *See v. City of Seattle*, 387 U.S. 541, 543 (1967).

These same cases also held that the Fourth Amendment prohibition against unreasonable searches protects against warrantless intrusions during civil as well as criminal investigations. *See v. City of Seattle, supra*, at 543. The reason is found in the "basic purpose of this Amendment ... [which] is to safeguard the privacy and security of individuals against arbitrary invasions by governmental officials." *Camara, supra*, at 528. If the government intrudes on a person's property, the privacy interest suffers whether the government's motivation is to investigate violations of criminal laws or breaches of other statutory or regulatory standards. It therefore appears that unless some recognized exception to the warrant requirement applies, *See v. City of Seattle, supra,* would require a warrant to conduct the inspection sought in this case.

The Secretary urges that an exception from the search warrant requirement has been recognized for "pervasively regulated business[es]," *United States v. Biswell,* 406 U.S. 311, 316 (1972), and for "closely regulated" industries "long subject to close supervision and inspection." *Colonnade Catering Corp. v. United States,* 297 U.S. 72, 74, 77, (1970). These cases are indeed exceptions, but they represent responses to relatively unique circumstances. Certain industries have such a history of government oversight that no reasonable expectation of privacy, see *Katz v. United States,* 380 U.S. 347, 351-352 (1967), could exist for a proprietor over the stock of such an enterprise. Liquor (*Colonnade*) and firearms (*Biswell*) are industries of this type; when an entrepreneur embarks upon such a business, he has voluntarily chosen to subject himself to a full arsenal of governmental regulation.

Industries such as these fall within the "certain carefully defined classes of cases." referenced in *Camara, supra,* at 528. The element that distinguishes these enterprises from ordinary businesses is a long tradition of close government supervision, of which any person who chooses to enter such a business must already be aware. "A central difference between those cases [*Colonnade* and *Biswell*] and this one is that businessmen engaged in such federally licensed and regulated enterprises accept the burdens as well as the benefits of their trade, whereas the petitioner here was not engaged in any regulated or licensed business. The businessman in a regulated industry in effect consents to the restrictions placed upon him." *Almeida-Sanchez v. United States,* 413 U.S. 266, 271 (1973).

The clear import of our cases is that the closely regulated industry of the type involved in *Colonnade* and *Biswell* is the exception. The Secretary would make it the rule. Invoking the Walsh-Healy Act of 1936, 41 U.S.C. §35 *et seq.,* the Secretary attempts to support a conclusion that all businesses involved in interstate commerce have long been subjected to close supervision of employee safety and health conditions. But the degree of federal involvement in employee working circumstances has never been of the order of specificity and pervasiveness that OSHA mandates. It is quite unconvincing to argue that the imposition of minimum wages and maximum hours on employers who contracted with the government under the Walsh-Healy Act prepared the entirety of American interstate commerce for regulation of working conditions to the minutest detail. Nor can any but the most fictional sense of voluntary consent to later searches be found in the single fact that one conducts a business affecting interstate commerce; under current practice and law, few businesses can be conducted without having some effect on interstate commerce.

The Secretary also attempts to derive support for a *Colonnade-Biswell*-type exception by drawing analogies from the field of labor law. In *Republic Aviation Corp. v. NLRB,* 324 U.S. 793 (1945), this Court upheld the rights of employees to solicit for a union during nonworking time where efficiency was not compromised. By opening up his property to employees, the employer had yielded so much of his private property rights as to allow those employees to exercise §7 rights under the National Labor Relations Act. But this Court also held that the private property rights of an owner prevailed over the intrusion of non-employee organizers, even in nonworking areas of the plant and during nonworking hours. *NLRB v. Babcock & Wilcox Co.,* 351 U.S. 105 (1956).

The critical fact in this case is that entry over Mr. Barlow's objection is being sought by a Government agent. Employees are not being prohibited from reporting OSHA violations. What they observe in their daily functions is undoubtedly beyond the employer's reasonable expectation of privacy. The Government inspector, however, is not an employee. Without a warrant he stands in no better position than a member of the public. What is observable by the public is observable, without a warrant, by the Government inspector as well. The owner of a business has not, by the necessary utilization of employees in his operation, thrown open the areas where employees alone are permitted to the warrantless scrutiny of Government agents. That an employee is free to report, and the Government is free to use, any evidence of noncompliance with OSHA that the employee observes furnishes no justification for federal agents to enter a place of business from which the public is restricted and to conduct their own warrantless search.

II

The Secretary nevertheless stoutly argues that the enforcement scheme of the Act requires warrantless searches, and that the restrictions on search discretion contained in the Act and its regulations already protect as much privacy as a warrant would. The secretary thereby asserts the actual reasonableness of OSHA searches, whatever the general rule against warrantless searches might be. Because "reasonableness is still the ultimate standard," *Camara v. Municipal Court, supra,* at 539, the Secretary suggests that the Court decide whether a warrant is needed by arriving at a sensible balance between the administrative necessities of OSHA inspections and the incremental protection of privacy of business owners a warrant would afford. He suggests that only a decision exempting OSHA inspections from the Warrant Clause would give "full recognition to the competing public and private interests here at stake." *Camara v. Municipal Court, supra,* at 539.

The Secretary submits that warrantless inspections are essential to the proper enforcement of OSHA because they afford the opportunity to inspect without prior notice and hence to preserve the advantages of surprise. While the dangerous conditions outlawed by the Act include structural defects that cannot be quickly hidden or remedied, the Act also regulates a myriad of safety details that may be amendable to speedy alteration or disguise. The risk is that during the interval between an inspector's initial request to search a plant and his procuring a warrant following the owner's refusal or permission, violations of this latter type could be corrected and thus escape the inspector's notice. To the suggestion that warrants may be issued *ex parte* and executed without delay and without prior notice, thereby preserving the element of surprise, the Secretary expresses concern for the administrative strain that would be experienced by the inspection system, and by the courts, should *ex parte* warrants issued in advance become standard practice.

We are unconvinced, however, that requiring warrants to inspect will impose serious burdens on the inspection system or the courts, will prevent inspections necessary to enforce the statute, or will make them less effective. In the first place, the great majority of businessmen can be expected in normal course to consent to inspection without warrant; the Secretary has not brought to this Court's attention any widespread pattern of refusal. In those cases where an owner does insist on a warrant, the Secretary argues that inspection efficiency will be impeded by the advance notice and delay. The Act's penalty provisions for giving advance notice of a search, 29 U.S.C. §666(f), and the Secretary's own regulations, 29 CFR §1903.6, indicate that surprise searches are indeed contemplated. However, the Secretary has

also promulgated a regulation providing that upon refusal to permit an inspector to enter the property or to complete his inspection, the inspector shall attempt to ascertain the reasons for the refusal and report to his superior, who shall "promptly take appropriate action, including compulsory process, if necessary." 29 CFR §1903.4. The regulation represents a choice to proceed by process where entry is refused and on the basis of evidence available from present practice, the Act's effectiveness has not been crippled by providing those owners who wish to refuse an initial requested entry with a time lapse while the inspector obtains the necessary process. Indeed, the kind of process sought in this case and apparently anticipated by the regulation provides notice to the business operator. If this safeguard endangers the efficient administration of OSHA, the Secretary should never have adopted it, particularly when the Act does not require it. Nor is it immediately apparent why the advantages of surprise would be lost if, after being refused entry, procedures were available for the Secretary to seek an *ex parte* warrant and to reappear at the premise without further notice to the establishment being inspected.

Whether the Secretary proceeds to secure a warrant or other process, with or without prior notice, his entitlement to inspect will not depend on his demonstrating probable cause to believe that conditions in violation of OSHA exist on the premises. Probable cause in the criminal law sense is not required. For purposes of an administrative search such as this, probable cause justifying the issuance of a warrant may be based not only on specific evidence of an existing violation but also on a showing that "reasonable legislative or administrative standards for conducting an . . . inspection are satisfied with respect to a particular [establishment]," *Camara v. Municipal Court, supra,* at 538. A warrant showing that a specific business has been chosen for an OSHA search on the basis of a general administrative plan for the enforcement of the Act derived from neutral sources such as, for example, dispersion of employees in various types of industries across a given area, and the desired frequency of searches in any of the lesser divisions of the area, would protect an employer's Fourth Amendment rights. We doubt that the consumption of enforcement energies in the obtaining of such warrants will exceed manageable proportions.

Finally, the Secretary urges that requiring a warrant for OSHA inspectors will mean that, as a practical matter, warrantless search provisions in other regulatory statutes are also constitutionally infirm. The reasonableness of a warrantless search, however, will depend upon the specific enforcement needs and privacy guarantees of each statute. Some of the statutes cited apply only to a single industry, where regulations might already be so pervasive that a *Colonnade-Biswell* exception to the warrant requirement could apply. Some statutes already envision resort to federal court enforcement when entry is refused, employing specific language in some cases and general language in others. In short, we base today's opinion on the facts and law concerned with OSHA and do not retreat from a holding appropriate to that statute because of its real or imagined effect on other, different administrative schemes.

Nor do we agree that the incremental protections afforded the employer's privacy by a warrant are so marginal that they fail to justify the administrative burdens that may be entailed. The authority to make warrantless searches devolves almost unbridled discretion upon executive and administrative officers particularly those in the field, as to when to search and whom to search. A warrant, by contrast, would provide assurance from a neutral officer that the inspection is reasonable under the Constitution, is authorized by statute, and is pursuant to an admin-istrative plan containing specific neutral criteria. Also, a warrant would then and there advise the owner of the scope and objects of the search, beyond which limits the inspector is not expected to proceed. These are important functions for a warrant to perform, functions which underlie the Court's prior decisions that the Warrant Clause applies to inspections for compliance with regulatory statutes. *Camara v. Municipal Court, supra; See v. City of Seattle, supra.* We conclude that the concerns expressed by the Secretary do not suffice to justify warrantless inspections under OSHA or vitiate the general constitutional requirement that for a search to be reasonable a warrant must be obtained.

III

We hold that Barlow was entitled to a declaratory judgment that the Act is unconstitutional insofar as it purports to authorize inspections without warrant or its equivalent and to an injunction enjoining the Act's enforcement to that extent. The judgment of the District Court is therefore affirmed.

[Dissenting opinion of STEPHENS, J. (in which BLACKMAN, J. and RHENQUIST, J., joined) omitted. BRENNAN, J., took no part in the case.]

2. The Scope of Factual Review

Clearly, a court is empowered to strike down an agency decision that violates a relevant statutory or constitutional provision. But even when there is no strictly "legal" issue of this nature, courts play an important role in reviewing—and ultimately shaping—the decisions that agencies make. The Administrative Procedure Act, and in many cases the originating statute itself, direct the reviewing court to examine the *factual* basis for the agency's decision as well.

As discussed previously, the APA employs a two-fold standard to define the scope of factual review. If the agency decision was made according to informal notice and comment rulemaking, the court is empowered to "hold unlawful and set aside agency action, findings, and conclusions found to be . . . arbitrary, capricious, [or] an abuse of discretion." But if the agency decision was made according to formal rulemaking, the court is empowered to set it aside if the decision is found to be "unsupported by substantial evidence." The original intention of Congress appears clear. If the agency employed a formal rulemaking process—with a full hearing, cross examination, and the other trappings of a trial-type proceeding—there would be an extensive record for the court to review. It thus made sense to Congress to require that the court examine that record carefully to ascertain that each important aspect of the agency's decision was supported by "substantial evidence" in the record. With the much sparser record expected to be generated by informal rulemaking, however, it made sense to expect a less thorough review. Hence, the less demanding arbitrary and capricious standard.

In practice, however, the distinction between the two standards of review has blurred considerably. There are two principal reasons for this, and both may be traced to the actions of Congress. The first is that Congress began sending mixed signals to the courts with the language of originating statutes passed well after the APA. In the 1970s, federal statutes were passed that required agencies to conduct their decisionmaking according to informal rulemaking, but required the courts to review those decisions according to the substantial evidence standard. (As noted, this is true of both the OSHAct and TSCA.) In conducting their review under these so-called hybrid rulemaking statutes, the courts began to require that the agencies develop more extensive administrative records under informal rulemaking than had been the accepted practice. In addition, with the extensive delegation of policymaking authority to agencies that occurred during the 1970s, the federal courts found that they were spending considerably more of their time reviewing agency decisionmaking. Perhaps because they recognized the importance of the social policy issues inherent in the decisions they were asked to review, the courts began to require more of the agencies than they had in the past, even when conducting review under the arbitrary and capricious standard.

The Supreme Court case that opened the proverbial floodgates was *Citizens to Preserve Overton Park v. Volpe*, 401 U.S. 402 (1971). Here, national and local environmental groups challenged a decision by the Department of Transportation to authorize the expenditure of federal funds to construct a six-lane highway through a public park in Memphis, Tennessee. Initial judicial review was secured in U.S. District Court under the Administrative Procedure Act, and the statutory scope of review was the APA's arbitrary and capricious standard. Concluding that the administrative record had not been reviewed in sufficient depth by the lower court, the Supreme Court sent the case back to the district court for further deliberation. In doing so, the Court noted that reviewing courts applying the arbitrary and capricious standard must engage in "a substantial inquiry . . . a thorough, probing, in-depth review." It directed reviewing courts to determine: 1) whether the agency acted within the scope of its statutory authority; 2) whether the agency's decision "was based on a consideration of the relevant factors and whether there has been a clear error of judgment;" and 3) whether the agency "followed the necessary procedural requirements."

Overton Park enunciated a much broader standard for judicial review of informal agency decisionmaking than had been assumed to exist at that time. Many courts, especially the influential District of Columbia Court of Appeals, took this as authorization to look much more closely at the factual basis for agency rulemaking and to remand agency decisions that appeared to the court to be at odds with the underlying factual record. What the courts must demand from agencies, the D.C. Circuit noted, was "reasoned decisionmaking." An agency practices reasoned decisionmaking, said the court, when it: 1) takes a "hard look . . . at the relevant issues"; 2) deliberates "in a manner calculated to negate the danger of arbitrariness and irrationality"; 3) violates "no law"; and 4) provides an "articulated justification" that makes a "rational connection between the facts found and the choice made." See, for example, *Action for Children's Television v. FCC*, 564 F.2d 458, 472 n. 24, 479 (D.C. Cir. 1977).

But the viability of *Overton Park*, and thus of the "reasoned decisionmaking" standard generally, was called into question by *Vermont Yankee Power Corp. v. Natural Resources Defense Council*, 435 U.S. 519 (1978), where the Supreme Court overturned the D.C. Circuit's reversal of a Nuclear Regulatory Commission (NRC) decision to grant licenses for two nuclear power plants. The court of appeals had, in effect, required the

NRC to provide an opportunity for cross examination in its informal rulemaking process. As the relevant statutes contain no such requirement, the Supreme Court reversed, cautioning reviewing courts "against engrafting their own notions of proper procedures upon agencies entrusted with substantive functions by Congress." Some commentators read this case as a retreat from the "substantial inquiry" standard enunciated in *Overton Park*, and many interpreted it as a "slap on the wrist" to the D.C. Circuit for having taken an overly aggressive approach to judicial review.

This proved to be an overreaction. Five years later, in *Motor Vehicle Manufacturers' Association v. State Farm Mutual Automobile Insurance Company*, 463 U.S. 29 (1983), the Supreme court clarified its *Vermont Yankee* opinion, and left little doubt that it had not intended in that case to signal a retreat from thoroughgoing judicial review. *State Farm* dealt with a decision by the Department of Transportation (through the National Highway Traffic Safety Administration, or NHTSA) to rescind its earlier regulation ("Standard 208") requiring "passive restraints" (airbags or detachable automatic seatbelts) in new model motor vehicles. The agency based this decision on its finding that the installation of detachable automatic seatbelts would not bring about "even a five percentage point increase" in seatbelt usage. All nine members of the Court agreed that the recision itself was arbitrary and capricious and a five-member majority held that the agency's factual finding was unsupported by the record.

Motor Vehicle Manufacturers' Association v. State Farm Mutual Automobile Insurance Company

463 U.S. 29 (1983)

Mr. Justice WHITE delivered the opinion of the Court.

* * *

The Department of Transportation accepts the applicability of the "arbitrary and capricious" standard. It argues that under this standard, a reviewing court may not set aside an agency rule that is rational, based on consideration of the relevant factors and within the scope of the authority delegated to the agency by the statute. We do not disagree with this formulation. The scope of review under the "arbitrary and capricious" standard is narrow and a court is not to substitute its judgment for that of the agency. Nevertheless, the agency must examine the relevant data and articulate a satisfactory explanation for its action including a "rational connection between the facts found and the choice made." *Burlington Truck Lines v. United States*, 371 U.S. 156, 168 (1962). In reviewing that explanation, we must "consider whether the decision was based on a consideration of the relevant factors and whether there has been a clear error of judgment." *Bowman Transp. Inc. v. Arkansas-Best Freight System, Citizens to Preserve Overton Park v.*

Volpe, supra, at 416. Normally, an agency rule would be arbitrary and capricious if the agency has relied on factors which Congress has not intended it to consider, entirely failed to consider an important aspect of the problem, offered an explanation for its decision that runs counter to the evidence before the agency, or is so implausible that it could not be ascribed to a difference in view or the product of agency expertise. The reviewing court should not attempt itself to make up for such deficiencies: "We may not supply a reasoned basis for the agency's action that the agency itself has not given." *SEC v. Chenery Corp.*, 332 U.S. 194, 196 (1947). "We will, however, uphold a decision of less than ideal clarity if the agency's path may reasonably be discerned." *Bowman Transp. Inc. v. Arkansas-Best Freight Systems, supra,* at 286. See also *Camp v. Pitts*, 411 U.S. 138, 142-143 (1973) (per curiam).

V

The ultimate question before us is whether NHTSA's rescission of the passive restraint requirement of Standard 208 was arbitrary and capricious. We conclude, as did the Court of Appeals, that it was. We also conclude, but for somewhat different reasons, that further consideration of the issue by the agency is therefore required. We deal separately with the rescission as it applies to airbags and as it applies to seatbelts.

A

The first and most obvious reason for finding the rescission arbitrary and capricious is that NHTSA apparently gave no consideration whatever to modifying the Standard to require that airbag technology be utilized. Standard 208 sought to achieve automatic crash protection by requiring automobile manufacturers to install either of two passive restraint devices: airbags or automatic seatbelts. There was no suggestion in the long rulemaking process that led to Standard 208 that if only one of these options were feasible, no passive restraint standard should be promulgated. Indeed, the agency's original proposed standard contemplated the installation of inflatable restraints in all cars. Automatic belts were added as a means of complying with the standard because they were believed to be as effective as airbags in achieving the goal of occupant crash protection. 36 Fed. Reg. 12,858, 12,859 (July 8, 1971). At that time, the passive belt approved by the agency could not be detached. Only later, at a manufacturer's behest, did the agency approve of the detachability feature—and only after assurances that the feature would not compromise the safety benefits of the restraint. Although it was then foreseen that 60% of the new cars would contain airbags and 40% would have automatic seatbelts, the ratio between the two was not significant as long as the passive belt would also assure greater passenger safety.

The agency has now determined that the detachable automatic belts will not attain anticipated safety benefits because so many individuals will detach the mechanism. Even if this conclusion were acceptable in its entirety . . . standing alone it would not justify any more than an amendment of Standard 208 to disallow compliance by means of the one technology which will not provide effective passenger protection. It does not cast doubt on the need for a passive restraint standard or upon the efficacy of airbag technology. In its most recent rule-making, the agency again acknowledged the life-saving potential of the airbag:

> The agency has no basis at this time for changing its earlier conclusions in 1976 and 1977 that basic airbag technology is sound and has been sufficiently demonstrated to be effective in those

vehicles in current use. . . . NHTSA Final Regulatory Impact Analysis (RIA) at XI-4 (App. 264).

Given the effectiveness ascribed to airbag technology by the agency, the mandate of the Safety Act to achieve traffic safety would suggest that the logical response to the faults of detachable seatbelts would be to require the installation of airbags. At the very least this alternative way of achieving the objectives of the Act should have been addressed and adequate reasons given for its abandonment. . . .

Petitioners also invoke our decision in *Vermont Yankee Nuclear Power Corp.* v. *NRDC,* 435 U.S. 519 (1977), as though it were a talisman under which any agency decision is by definition unimpeachable. Specifically, it is submitted that to require an agency to consider an airbags-only alternative is, in essence, to dictate to the agency the procedures it is to follow. Petitioners both misread *Vermont Yankee* and misconstrue the nature of the remand that is in order. In *Vermont Yankee,* we held that a court may not impose additional procedural requirements upon an agency. We do not require today any specific procedures which NHTSA must follow. Nor do we broadly require an agency to consider all policy alternatives in reaching decision. It is true that a rulemaking "cannot be found wanting simply because the agency failed to include every alternative device and thought conceivable by the mind of man . . . regardless of how uncommon or unknown that alternative may have been. . . ." 435 U.S., at 551. But the airbag is more than a policy alternative to the passive restraint standard; it is a technological alternative within the ambit of the existing standard. We hold only that given the judgment made in 1977 that airbags are an effective and cost-beneficial life-saving technology, the mandatory passive-restraint rule may not be abandoned without any consideration whatsoever of an airbags-only requirement.

B

Although the issue is closer, we also find that the agency was too quick to dismiss the safety benefits of automatic seatbelts. NHTSA's critical finding was that, in light of the industry's plans to install readily detachable passive belts, it could not reliably predict "even a 5 percentage point increase as the minimum level of expected usage increase." 46 Fed. Reg., at 53,423. The Court of Appeals rejected this finding because there is "not one iota" of evidence that Modified Standard 208 will fail to increase nationwide seatbelt use by at least 13 percentage points, the level of increased usage necessary for the standard to justify its cost. Given the lack of probative evidence, the court held that "only a well-justified refusal to seek more evidence could render rescission non-arbitrary." 680 F2d, at 232.

Petitioners object to this conclusion. In their view, "substantial uncertainty" that a regulation will accomplish its intended purpose is sufficient reason, without more, to rescind a regulation. We agree with petitioners that just as an agency reasonably may decline to issue a safety standard if it is uncertain about its efficacy, an agency may also revoke a standard on the basis of serious uncertainties if supported by the record and reasonably explained. Rescission of the passive restraint requirement would not be arbitrary and capricious simply because there was no evidence in direct support of the agency's conclusion. It is not infrequent that the available data [do] not settle a regulatory issue and the agency must then exercise its judgment in moving from the facts and probabilities on the record to a policy conclusion. Recognizing that policymaking in a complex society must account for uncertainty, however, does not imply that it is sufficient for an agency to merely recite the terms "substantial uncertainty" as a justification for its actions. The agency must explain the evidence which is available, and must offer a "rational connection between the facts found and the choice made." *Burlington Truck Lines, Inc. v. United States, supra,* at 168. Generally one aspect of that explanation would be a justification for rescinding the regulation before engaging in a search for further evidence.

In this case, the agency's explanation for rescission of the passive restraint requirement is *not* sufficient to enable us to conclude that the rescission was the product of reasoned decisionmaking. To reach this conclusion, we do not upset the agency's view of the facts, but we do appreciate the limitations of this record in supporting the agency's decision. We start with the accepted ground that if used, seatbelts unquestionably would save many thousands of lives and would prevent tens of thousands of crippling injuries. Unlike recent regulatory decisions we have reviewed, *Industrial Union Department v. American Petroleum Institute,* 448 U.S. 607 (1980); *American Textile Manufactures Inst., Inc. v. Donovan,* 452 U.S. 490 (1981), the safety benefits of wearing seatbelts are not in doubt and it is not challenged that were those benefits to accrue, the monetary costs of implementing the standard would be easily justified. We move next to the fact that there is no direct evidence in support

of the agency's finding that detachable automatic belts cannot be predicted to yield a substantial increase in usage. The empirical evidence on the record, consisting of surveys of drivers of automobiles equipped with passive belts, reveals more than a doubling of the usage rate experienced with manual belts. Much of the agency's rulemaking statement—and much of the controversy in this case—centers on the conclusions that should be drawn from these studies. The agency maintained that the doubling of seatbelt usage in these studies could not be extrapolated to an across-the-board mandatory standard because the passive seatbelts were guarded by ignition interlocks and purchasers of the tested cars are somewhat atypical. Respondents insist these studies demonstrate that Modified Standard 208 will substantially increase seat belt usage. We believe that it is within the agency's discretion to pass upon the generalizability of these field studies. This is precisely the type of issue which rests within the expertise of NHTSA, and upon which a reviewing court must be most hesitant to intrude.

But accepting the agency's view of the field tests on passive restraints indicates only that there is no reliable real-world experience that usage rates will substantially increase. To be sure, NHTSA opines that "it cannot reliably predict even a 5 percentage point increase as the minimum level of increased usage." Notice 25, 46 Fed. Reg., at 53,423. But this and other statements that passive belts will not yield substantial increases in seatbelt usage apparently take no account of the critical difference between detachable automatic belts and current manual belts. A detached passive belt does require an affirmative act to reconnect it, but—unlike a manual seat belt— the passive belt, once reattached, will continue to function automatically unless again disconnected. Thus, inertia—a factor which the agency's own studies have found significant in explaining the current low usage rates for seatbelts—works in *favor* of, not *against,* use of the protective device. Since 20 to 50% of motorists currently wear seatbelts on some occasions, there would seem to be grounds to believe that seatbelt use by occasional users will be substantially increased by the detachable passive belts. Whether this is in fact the case is a matter for the agency to decide, but it must bring its expertise to bear on the question.

[Concurring and dissenting opinion of RHENQUIST, J. (in which BURGER, C. J., POWELL, J. and O'CONNOR, J. joined) omitted.]

This decision certainly can be read as a reaffirmation of the principles laid down in *Overton Park*. (Clearly, the Court was not shy about examining the factual basis for the agency's decision.) To borrow the Supreme Court's language from that earlier case, however, the question of just how "substantial" an inquiry, or how "in-depth" a review a court must conduct remains somewhat open today. This is especially true where, as is often the case with issues of occupational health and safety, the agency's decision is based on the evaluation of highly technical data.

Some commentators argue that lay judges are simply not qualified to understand and interpret such data. Thus, their argument runs, courts should defer to an agency's specialized expertise in these matters except in cases of obvious errors in reasoning. Others, however, argue that reviewing courts shirk their responsibility under the APA if they do not examine carefully the factual basis for the regulatory choices made by the agency. If the underlying data and methodologies are complex, this argument proceeds, it is incumbent on the agency and the other parties to the litigation to elucidate them for the court. Support for both of these philosophies can be found in judicial decisions, and it remains rather difficult to predict which one of them the courts will rely upon in any particular case. Suffice it to say that whenever an agency fails to provide a clear, well-reasoned factual analysis in support of a regulatory decision, it runs the risk of judicial reversal.

◇ NOTES

1. It is important to reiterate that, no matter how carefully a reviewing court scrutinizes the details of the agency's reasoning, the court may not overturn an administrative decision merely because it disagrees with the agency's policy determinations. In the administrative arena, policy decisions are the province of Congress, not the courts. Thus, policies are to be set in accordance with the statutory mandate.

2. The authors of this text have argued that, in the area of health and safety regulation, reviewing courts will fail to recognize the important *policy* decisions inherent in an agency's assessment of particular risks—and thus will be unable to determine whether the agency is carrying out the policy directives set forth in its statutory mandate—unless they look carefully at the risk assessment methodologies that lie behind the regulatory decision. See Ashford, Ryan, and Caldart, A Hard Look at Federal Regulation of Formaldehyde: A Departure from Reasoned Decisionmaking, 7 Harvard Environmental Law Review 297 (1983). For a somewhat different viewpoint, see: McGarity, Beyond the Hard Look: A New Standard for Judicial Review? 2 Natural Resources and the Environment 32 (1986).

3. The OSHAct cases reprinted in Chapter 3 provide an excellent look at how far various courts have been willing to go in analyzing the technical data underlying agency decisionmaking and at how well they have performed the analysis.

4. For a look at how various members of the D.C. Circuit Court of Appeals have addressed the issue of judicial review of "science-based" agency decisionmaking, see Wald, Making "Informed" Decisions on the District of Columbia Circuit, 50 George Washington Law Review 135 (1982); Bazelon, Science and Uncertainty: A

Jurist's View, 5 Harvard Environmental Law Review 211 (1981); Bazelon, Coping with Technology Through the Legal Process, 62 Cornell Law Review 822 (1977); Leventhal, Environmental Decisionmaking and the Role of the Courts, 122 University of Pennsylvania Law Review 511 (1974); Wright, The Courts and the Rule-making Process: The Limits of Judicial Review, 59 Cornell Law Review 375 (1974).

3. Judicial Review of Agency Decisions Not to Act

A final issue of importance is the scope of judicial review when the agency declines to take rulemaking action. Traditionally, courts have given greater deference to agency discretion in such situations. The D.C. Circuit Court of Appeals has set forth six principal reasons for affording such deference: (1) that the issues involved may turn on "factors not inherently susceptible to judicial resolution," such as the management of budget and personnel and the balancing of competing policies within a broad statutory framework; (2) that there may be "such rapid technological development that regulations would be outdated by the time they could become effective"; (3) that the currently available data may be an inadequate basis for regulation; (4) that "the circumstances in the regulated industry may be evolving in such a way that could vitiate the need for regulation"; (5) that the agency may not yet possess "the expertise necessary for effective regulation" in the area in question; and (6) that the record on review may not be "narrowly focused on the particular rule advocated by [the party seeking review]." *Natural Resources Defense Council v. SEC,* 606 F.2d 1031, 1046 (D.C. Cir. 1979). Nonetheless, the courts clearly are empowered under Section 10 of the APA to review decisions not to act, and to "compel agency action unlawfully withheld or unreasonably delayed." They have done so on several occasions, some of which are described in the chapters on the OSHAct and TSCA that follow. (See, particularly, the discussion of *Public Citizen Health Research Group v. Brock* in Chapter 3 and *NRDC v. Costle* in Chapter 4.)

F. FEDERAL PREEMPTION AND THE ROLE OF STATE AND LOCAL LAWS

Once the federal government has chosen to take regulatory action in a particular area, what role is left for state and local governments? By and large, the answer is that they may play whatever role the federal government allows them to play. Due to the so-called supremacy clause of the U.S. Constitution, which characterizes the Acts of Congress as the "supreme law of the land," Congress can *preempt* state and local laws. In many cases, then, a federal statute or regulation will explicitly provide that state laws governing the same subject matter are prohibited. The OSHAct, for example, has a version of such a provision, as does the National Labor Relations Act. Other statutes, such as the Toxic Substances Control Act, explicitly disavow any intent to preempt, thus leaving the way clear for state regulation. Where a federal statute is silent on the subject of preemption, it is up to the courts—as usual—to divine the intent of Congress. In general, courts will not find an intent to preempt unless: 1) the federal law and the state law are so much in conflict that it is impossible to comply with both; 2) the federal government has so pervasively regulated the subject matter area that it can be said to have "preempted the field"; or 3) the state statute or provision can rightly be said to be "an obstacle to the accomplishment of the purposes of Congress." See *Silkwood v. Kerr-McGee Corporation,* 464 U.S. 238, 248 (1984).

From time to time, OSHA has used its power under the OSHAct to preempt conflicting state and local laws governing workplace safety and health. As discussed in Chapter 7, for example, OSHA's Hazard Communication Standard has preempted many state and local toxic substance "right-to-know" laws. Preemption has also been raised as a defense by employers who face *criminal* charges as a result of their treatment of their employees. In a few states, county prosecutors have brought charges under state criminal law against employers alleged to have intentionally or recklessly exposed employees to workplace conditions that were manifestly unsafe. Invariably, the defendants have argued that such prosecutions are preempted by the OSHAct, as that statute provides its own punishment (a jail sentence of up to six months, a fine of up to $10,000, or both) for employers when a worker is killed as a result of the violation of an OSHA standard. This provision of the OSHAct, the argument runs, signals a Congressional intent to limit the criminal liability of employers for occupational safety and health infractions. The use of state criminal law to impose higher penalties, it is argued, conflicts with this Congressional purpose and must be deemed preempted.

Thus far, this argument has met with little success in state appellate courts. The Illinois Supreme Court has held that the OSHAct does not preempt the State of Illinois from prosecuting a corporate employer *and its officers* for aggravated battery and reckless conduct for allegedly exposing workers to toxic substances. *People v. Chicago Magnet Wire Corporation,* 126 Ill. 2d 356 (1989). Similarly, the Supreme Court of Michigan has cleared the way for the involuntary manslaughter prosecution of a supervisor for the death of a worker who died of carbon monoxide poisoning while working in a company-owned van. *People v. Hegedus,* 432 Mich. 598 (1989). If this trend continues, it could signal the emergence of state criminal law as an important factor in regulating occupational safety and health.

Appendix to Chapter 2

ADMINISTRATIVE PROCEDURE ACT, 5 U.S.C. §551, *et seq.*

§553. Rule Making [Section 4]

(a) This section applies, according to the provisions thereof, except to the extent that there is involved—
  (1) a military or foreign affairs function of the United States; or
  (2) a matter relating to agency management or personnel or to public property, loans, grants, benefits, or contracts.

(b) General notice of proposed rule making shall be published in the Federal Register, unless persons subject thereto are named and either personally served or otherwise have actual notice thereof in accordance with law. The notice shall include—

(1) a statement of the time, place, and nature of public rule making proceedings;

(2) reference to the legal authority under which the rule is proposed; and

(3) either the terms or substance of the proposed rule or a description of the subjects and issues involved.

Except when notice or hearing is required by statute, this subsection does not apply—

(A) to interpretative rules, general statements of policy, or rules of agency organization, procedure, or practice; or

(B) when the agency for good cause finds (and incorporates the finding and a brief statement of reasons therefor in the rules issued) that notice and public procedure thereon are impracticable, unnecessary, or contrary to the public interest.

(c) After notice required by this section, the agency shall give interested persons an opportunity to participate in the rule making through submission of written data, views, or arguments with or without opportunity for oral presentation. After consideration of the relevant matter presented, the agency shall incorporate in the rules adopted a concise general statement of their basis and purpose. When rules are required by statute to be made on the record after opportunity for an agency hearing, sections 556 and 557 of this title apply instead of this subsection.

(d) The required publication or service of a substantive rule shall be made not less than 30 days before its effective date, except—

(1) a substantive rule which grants or recognizes an exemption or relieves a restriction;

(2) interpretative rules and statements of policy; or

(3) as otherwise provided by the agency for good cause found and published with the rule.

(e) Each agency shall give an interested person the right to petition for the issuance, amendment, or repeal of a rule.

§556. Hearings; Presiding Employees; Powers and Duties; Burden of Proof; Evidence; Record as Basis of Decision [Section 7]

(a) This section applies, according to the provisions thereof, to hearings required by section 553 or 554 of this title to be conducted in accordance with this section.

(b) There shall preside at the taking of evidence—

(1) the agency;

(2) one or more members of the body which comprises the agency; or

(3) one or more administrative law judges appointed under section 3105 of this title. This subchapter does not supersede the conduct of specified classes of proceedings, in whole or in part, by or before boards of other employees specially provided for by or designated under statute. The functions of presiding employees and of employees participating in decisions in accordance with section 557 of this title shall be conducted in an impartial manner. A presiding or participating employee may at any time disqualify himself. On the filing in good faith of a timely and sufficient affidavit of personal bias or other disqualification of a presiding or participating employee, the agency shall determine the matters as a part of the record and decision in the case.

(c) Subject to published rules of the agency and within its powers, employees presiding at hearings may—

(1) administer oaths and affirmations;

(2) issue subpoenas authorized by law;

(3) rule on offers of proof and receive relevant evidence;

(4) take depositions or have depositions taken when the ends of justice would be served;

(5) regulate the course of the hearing;

(6) hold conferences for the settlement or simplification of the issues by consent of the parties;

(7) dispose of procedural requests or similar matters;

(8) make or recommend decisions in accordance with section 557 of this title; and

(9) take other action authorized by agency rule consistent with this subchapter.

(d) Except as otherwise provided by statute, the proponent of a rule or order has the burden of proof. Any oral or documentary evidence may be received, but the agency as a matter of policy shall provide for the exclusion of irrelevant, immaterial, or unduly repetitious evidence. A sanction may not be imposed or rule or order issued except on consideration of the whole record or those parts thereof cited by a party and supported by and in accordance with the reliable, probative, and substantial evidence. The agency may, to the extent consistent with the interests of justice and the policy of the underlying statutes administered by the agency, consider a violation of section 557(d) of this title sufficient grounds for a decision adverse to a party who has knowingly committed such violation or knowingly caused such violation to occur. A party is entitled to present his case or defense by oral or documentary evidence, to submit rebuttal evidence, and to conduct such cross-examination as may be required for a full and true disclosure of the facts. In rule making or determining claims for money or benefits or applications for initial licenses an agency may, when a party will not be prejudiced thereby, adopt procedures for the submission of all or part of the evidence in written form.

(e) The transcript of testimony and exhibits, together with all papers and requests filed in the proceeding, constitutes the exclusive record for decision in accordance with section 557 of this title and, on payment of lawfully prescribed costs, shall be made available to the parties. When an agency decision rests on official notice of a material fact not appearing in the evidence in the record, a party is entitled, on timely request, to an opportunity to show the contrary.

§557. Initial Decisions; Conclusiveness; Review by Agency; Submissions by Parties; Contents of Decisions; Record [Section 8]

(a) This section applies, according to the provisions thereof, when a hearing is required to be conducted in accordance with section 556 of this title.

(b) When the agency did not preside at the reception of the evidence, the presiding employee or, in cases not subject to section 554(d) of this title, an employee qualified to preside at hearings pursuant to section 556 of this title, shall initially decide the case unless the agency requires, either in specific cases or by general rule, the entire record to be certified to it for decision. When the presiding employee makes an initial decision, that decision then becomes the decision of the agency without further proceedings unless there is an appeal to, or review on motion of, the agency within time provided by rule. On appeal from or review of the initial decision, the agency has all the powers which it would have in making the initial decision except as it may limit the issues on notice or by rule. When the agency makes the decision without having presided at the reception of the evidence, the presiding employee or an employee qualified to preside at hearings pursuant to section 556 of this title shall first recommend a decision, except that in rule making or determining application for initial licenses—

(1) instead thereof the agency may issue a tentative decision or one of its responsible employees may recommend a decision; or

(2) this procedure may be omitted in a case in which the agency finds on the record that due and timely execution of its functions imperatively and unavoidably so requires.

(c) Before a recommended, initial, or tentative decision, or a decision on agency review of the decision of subordinate employees, the parties are entitled to a reasonable opportunity to submit for the consideration of the employees participating in the decisions—

(1) proposed finding and conclusions; or

(2) exceptions to the decisions or recommended decisions of subordinate employees or to tentative agency decisions; and

(3) supporting reasons for the exceptions or proposed findings or conclusions.

The record shall show the ruling on each finding, conclusion, or exception presented. All decisions, including initial, recommended, and tentative decisions, are a part of the record and shall include a statement of—

(A) findings and conclusions, and the reasons or basis therefor, on all the material issues of fact, law, or discretion presented on the record; and

(B) the appropriate rule, order, sanction, relief, or denial thereof.

(d)(1) In any agency proceeding which is subject to subsection (1) of this section, except to the extent required for the disposition of ex parte matters as authorized by law—

(A) no interested person outside the agency shall make or knowingly cause to be made to any member of the body comprising the agency, administrative law judge, or other employee who is or may reasonably be expected to be involved in the decisional process of the proceeding, an ex parte communication relevant to the merits of the proceeding;

(B) no member of the body comprising the agency, administrative law judge, or other employee who is or may reasonably be expected to be involved in the decisional process of the proceeding, shall make or knowingly cause to be made to any interested person outside the agency an ex parte communication relevant to the merits of the proceeding;

(C) a member of the body comprising the agency, administrative law judge, or other employee who is or may reasonably be expected to be involved in the decisional process of such proceeding who receives, or who makes or knowingly causes to be made, a communication prohibited by this subsection shall place on the public record of the proceeding:

(i) all such written communications;

(ii) memoranda stating the substance of all such oral communications; and

(iii) all written response, and memoranda stating the substance of all oral responses, to the materials described in clauses (i) and (ii) of this subparagraph;

(D) upon receipt of a communication knowingly made or knowingly caused to be made by a party in violation of this subsection, the agency, administrative law judge, or other employee presiding at the hearing may, to the extent consistent with the interests of justice and the policy of the underlying statutes, require the party to show cause why his claim or interest in the proceeding should not be dismissed, denied, disregarded, or otherwise adversely affected on account of such violation; and

(E) the prohibitions of this subsection shall apply beginning at such time as the agency may designate, but in no case shall they begin to apply later than the time at which a proceeding is noticed for hearing unless the person responsible for the communication has knowledge that it will be noticed, in which case the prohibitions shall apply beginning at the time of his acquisition of such knowledge.

(2) This subsection does not constitute authority to withhold information from Congress.

JUDICIAL REVIEW [Section 10]

§701. Application; Definitions

(a) This chapter applies, according to the provisions thereof, except to the extent that—

(1) statutes preclude judicial review; or

(2) agency action is committed to agency discretion by law.

(b) For the purpose of this chapter—

(1) "agency" means each authority of the Government of the United States, whether or not it is within or subject to review by another agency, but does not include—

(A) the Congress;

(B) the courts of the United States;

(C) the governments of the territories or possessions of the United States;

(D) the government of the District of Columbia;

(E) agencies composed of representatives of the parties or of representatives of organizations of the parties to the disputes determined by them;

(F) courts martial and military commissions;

(G) military authority exercised in the field in time of war or in occupied territory; or

(H) functions conferred by sections 1738, 1739, 1743, and 1744 of title 12; chapter 2 of title 41; or sections 1622, 1884, 1891-1902, and former section 1641(b)(2), of title 50, appendix; and

(2) "person," "rule," "order," "license," "sanction," "relief," and "agency action" have the meanings given them by section 551 of this title.

§702. Right of Review

A person suffering legal wrong because of agency action, or adversely affected or aggrieved by agency action within the meaning of a relevant statute, is entitled to judicial review thereof. An action in a court of the United States seeking relief other than money damages and stating a claim that an agency or an officer or employee thereof acted or failed to act in an official capacity or under color of legal authority shall not be dismissed nor relief therein be denied on the ground that it is against the United States or that the United States is an indispensable party. The United States may be named as a defendant in any such action, and a judgment or decree may be entered against the United States: *Provided,* That any mandatory or injunctive decree shall specify the Federal officer or officers (by name or by title), and their successors in office, personally responsible for compliance. Nothing herein (1) affects other limitations on judicial review or the power or duty of the court to dismiss any action or deny relief on any other appropriate legal or equitable ground; and (2) confers authority to grant relief if any other statute that grants consent to suit expressly or impliedly forbids the relief which is sought.

§703. Form and Venue of Proceeding

The form of proceeding for judicial review is the special statutory review proceeding relevant to the subject matter in a court specified by statute or, in the absence or inadequacy thereof, any applicable form of legal action, including actions for declaratory judgments or writs of prohibitory or mandatory injunction or habeas corpus, in a court of competent jurisdiction. If no special statutory review proceeding is applicable, the action for judicial review may be brought against the United States, the agency by its official title, or the appropriate officer. Except to the extent that prior, adequate, and exclusive opportunity for judicial review is provided by law, agency action is subject to judicial review in civil or criminal proceedings for judicial enforcement.

§704. Actions Reviewable

Agency action made reviewable by statute and final agency action for which there is no adequate remedy in a court are subject to judicial review. A preliminary, procedural, or intermediate agency action or ruling not directly reviewable is subject to review on the review of the final agency action. Except as otherwise expressly required by statute, agency action otherwise final is final for the purposes of this section whether or not there has been presented or determined an application for a declaratory order, for any form of reconsideration, or, unless the agency otherwise requires by rule and provides that the action meanwhile is inoperative, for an appeal to superior agency authority.

§705. Relief Pending Review

When an agency finds that justice so requires, it may postpone the effective date of action taken by it, pending judicial review. On such conditions as may be required and to the extent necessary to prevent irreparable injury, the reviewing court, including the court to which a case may be taken on appeal from or on application for certiorari or other writ to a reviewing court, may issue all necessary and appropriate process to postpone the effective date of an agency action or to preserve status or rights pending conclusion of the review proceedings.

§706. Scope of Review

To the extent necessary to decision and when presented, the reviewing court shall decide all relevant questions of law, interpret constitutional and statutory provisions, and determine the meaning or applicability of the terms of an agency action. The reviewing court shall—
   (1) compel agency action unlawfully withheld or unreasonably delayed; and

(2) hold unlawful and set aside agency action, findings, and conclusions found to be—

(A) arbitrary, capricious, an abuse of discretion, or otherwise not in accordance with law;

(B) contrary to constitutional right, power, privilege, or immunity;

(C) in excess of statutory jurisdiction, authority, or limitations, or short of statutory right;

(D) without observance of procedure required by law;

(E) unsupported by substantial evidence in a case subject to section 556 and 557 of this title or otherwise reviewed on the record of an agency hearing provided by statute; or

(F) unwarranted by the facts to the extent that the facts are subject to trial de novo by the reviewing court.

In making the foregoing determinations, the court shall review the whole record or those parts of it cited by a party, and due account shall be taken of the rule of prejudicial error.

CHAPTER 3

The Occupational Safety

and Health Act of 1970

A. INTRODUCTION

Although the existence of occupational disease and injury has been documented since ancient times, general awareness of their significance for U.S. workers has waxed and waned since the nation's first period of industrial expansion in the late 1800s. Most early occupational health and safety efforts were undertaken at the state level. Massachusetts enacted the first worker safety law in 1877 and, by 1900, most heavily industrialized states had some redimentary legislation requiring employers to control certain workplace hazards. However, most state programs were weak, poorly funded, understaffed, and highly fragmented.

1. The History of Federal Regulation

Early federal regulation focused on a small number of dangerous occupations, those of merchant seamen, railroad workers, and miners. Other federal efforts early in the twentieth century focused more broadly on research. The federal Office of Industrial Hygiene and Sanitation in the U.S. Public Health Service and the U.S. Bureau of Labor Standards looked at lead poisoning, radium poisoning in the watch industry, hazards in brass foundries, phosphorous-induced disease, and the dangers of the dusty trades. Broad-based regulation, however, was slow in coming. Only after the public witnessed numerous spectacular workplace disasters and experienced many more private and personal tragedies brought on by workplace injury and disease was the federal government prompted to take the problem of workplace health and safety more seriously.

In 1934, under Franklin D. Roosevelt, Secretary of Labor Frances Perkins created the Bureau of Labor Standards, the first permanent federal agency with a mandate to promote health and safety for the U.S. workforce. The Walsh-Healey Public Contracts Act of 1936 created a federal regulatory role by directing the Department of Labor to ensure that federal contractors met minimum health and safety standards. In the late 1950s and early 1960s, federal involvement in standard setting intensified, culminating in the passage of the Federal Coal Mine Health and Safety Act of 1969. This act helped set the stage for more generally focused legislation to be enacted one year later.

By the late 1960s, the National Safety Council was estimating that more that 14,000 workers were being killed on the job every year and that two million more were being injured and disabled annually. At the same time, the environmental movement was gaining public prominence in its efforts to protect human health and the environment from toxic substances contained in air and water pollution. A fortunate and reasonable spillover of these activities focused on the environment of the workplace, where individuals faced significantly higher exposures to toxic substances on a daily basis.

In 1968, President Lyndon Johnson proposed and submitted to Congress a bill to establish the nation's first comprehensive occupational health and safety program. Opposed by industry, the bill never reached the House or Senate floor. This effort was continued during the Nixon administration and, after a period of considerable conflict and compromise, the 91st Congress passed the Occupational Safety and Health Act of 1970 (OSHAct) which was signed into law on December 29, 1970.

2.   The Occupational Safety and Health Act of 1970

The OSHAct applies to most private sector employers and extends both health and safety protections to their workers. Workplaces already regulated under the Atomic Energy Act of 1954 or under other specific health and safety statutes, such as the Mine Safety and Health Act, are exempt from coverage to the extent that these other regulatory agencies exercise their statutory authority to prescribe or enforce health and safety standards.

The OSHAct does not reach public employees directly. An executive order of the President provides protection for federal employees. State and municipal government employees are not covered unless the state in which they work assumes responsibility for occupational health and safety activities under a state plan approved by the Occupational Safety and Health Administration (OSHA).

The Act provides a clear statement of its purpose and statutory mandate: "... to assure so far as possible every working man and woman in the Nation safe and healthful working conditions...." Congress placed the primary responsibility for achieving this end squarely on the shoulders of employers, who would henceforth be obligated to comply with specific health and safety standards and fulfill a general duty "to each of his employees" to provide a safe and healthful workplace. In imposing such a statutory duty, Congress was clear in its intentions. Because employers control the conditions of employment, the onus of protecting workers from occupational disease and injury was to rest with them as well.

Any serious effort to protect and promote worker health and safety must begin with a solid understanding of the OSHAct. Fortunately, it is an exceptionally well-crafted piece of legislation that is clearly organized, easily understood, and often eloquent. The courts have provided further elaboration and interpretation of the Act as industry and labor have waged ongoing battles during the standard setting process. This chapter is designed to: (1) familiarize readers with the Act itself; (2) help them understand the evolution of the Act through court interpretation; and (3) alert them to the continuing policy issues and debates involved in protecting worker health and safety.

B.   DESCRIPTION AND SUMMARY OF THE OSHACT

The best way to understand the Act is to read it with pencil in hand. The Act is included in its entirety in Appendix A. The following sections, adapted from *Crisis in the Workplace*,[a] describe and discuss the most salient features of the Act, but cannot substitute for a careful reading of the Act itself.

1.   Duties of Employers and Employees

Under Section 5(a)(2), each employer has a specific duty to comply with occupational safety and health standards promulgated under the OSHAct. In all cases not covered by specific standards, the employer has a general duty "to furnish to each of his employees employment and a place of employment which are free from recognized hazards that are causing or are likely to cause death or serious physical harm to his employees." (Section 5(a)(1)). The latter requirement, commonly referred to as the "general duty clause," has been the basis of many citations issued against employers, and has been subject to evolving legal interpretation.

The duties of employees are specified in Section 5(b) of the act: "Each employee shall comply with occupational safety and health standards and all rules, regulations, and orders issued pursuant to this Act which are applicable to his own actions and conduct."

Since the purpose of the Act is to provide for safe workplaces, the duties imposed under the Act do not directly affect any workers' compensation law. Furthermore, the Act does not "diminish or affect in any other manner" the common-law or statutory rights, *duties,* or liabilities of employers and employees under any law with respect to injuries, diseases, or death of employees arising from the course of employment. (Section 4(b)(4)).

The Act specifically prohibits discrimination against employees for the filing of OSHA complaints or any similar exercise of rights under the Act (Section 11(c)). The federal district courts have jurisdiction over suits alleging such discrimination.

2.   Administration

Three agencies have been set up within the federal government to administer and enforce the OSHAct. The agency responsible for administering most of the provisions of the Act is the Occupational Safety and Health Administration (OSHA), located within the Department of Labor. OSHA is required to set occupational safety and health standards and to conduct inspections of workplaces covered under the Act for

[a]N. A. Ashford. 1976. *Crisis in the Workplace: Occupational Disease and Injury,* Cambridge: MIT press, pp. 143–152.

the purpose of assuring compliance with standards and the general duty obligation. The agency has the power to issue citations against employers and to assess penalties for violations.

The Occupational Safety and Health Review Commission (OSHRC) is an independent quasi-judicial review board; it consists of three members appointed to six-year terms by the President. The Commission rules upon all challenged enforcement actions of the Occupational Safety and Health Administration.

To assist the Secretary of Labor in enforcing the OSHAct, Congress raised to national institute status the former Bureau of Occupational Safety and Health, and renamed it the National Institute for Occupational Safety and Health (NIOSH). A research agency within the Department of Health and Human Services, NIOSH is located in Atlanta and is part of the Centers for Disease Control. NIOSH is responsible for developing and recommending occupational safety and health standards. The agency is specifically required to publish a list of all known toxic substances and the concentrations at which these substances exhibit toxic effects. In order to accomplish these tasks, NIOSH is authorized to conduct research and experimental programs in occupational health and safety, promote the training of an adequate supply of research personnel, and conduct hazard evaluations.

In addition to these three agencies, Section 7 of the OSHAct establishes the National Advisory Committee on Occupational Safety and Health (NACOSH). NACOSH consists of "representatives of management, labor, occupational safety, and occupational health professions, and of the public." NACOSH is required to advise, consult with, and make recommendations to the Secretary of Labor and the Secretary of Health and Human Services on matters relating to the administration of the Act. The Secretary of Labor may also appoint other advisory committees to aid in the setting of standards. (There was also continued in existence a previously established federal advisory committee dealing primarily with safety in federal employment.)

3. Occupational Safety and Health Standards

The primary function of OSHA is to promulgate and enforce standards governing the work environment in businesses under its jurisdiction. Three types of standards are mandated by the OSHAct: (1) *interim* (initial) standards consisting of federal standards from other Acts and national consensus standards existing at the time of implementation of the OSHAct; (2) *permanent* standards to replace or augment the interim standards; and (3) *temporary emergency* standards which may be issued immediately upon the finding of grave danger to employees.

Under Section 6(a) of the Act, OSHA was given two years, until April 28, 1973, within which to promulgate interim standards. Congress was of the opinion that many basic safety and health standards had already been subject to public debate and could be promulgated immediately to protect employees. Accordingly, the Act required the Secretary to adopt as interim occupational safety and health standards any appropriate standards which had been promulgated by other federal agencies. Such standards were explicitly exempted from the rulemaking procedures ordinarily required by the Administrative Procedure Act and were allowed to be made effective immediately upon publication in the *Federal Register.* In addition, the Secretary could adopt and promulgate national consensus standards produced by such recognized organizations as the National Fire Protection Association or the American National Standards

Institute. The Act required the Secretary to promulgate as interim standards only those national consensus standards for which there was substantial agreement among affected parties, thus eliminating standards which unduly favored either employers or employees. The Secretary announced an initial package of standards covering most industries soon after the Act became effective, and a 90-day grace period was afforded employers to allow them time to comply.

Under Section 6(b) of the OSHAct, the Secretary of Labor may—upon the basis of information submitted by the Secretary of Health and Human Services, NIOSH, advisory committees, or other groups—revise, modify, or revoke existing standards, and promulgate new permanent ones. Whenever the Secretary concludes that a permanent standard is appropriate, he or she may either publish the proposed standard directly in the *Federal Register* and ask for comments, or may first request the recommendations of an advisory committee which may then be appointed for that purpose. Once a proposed standard has been published in the *Federal Register,* interested parties are afforded at least 30 days in which to comment and/or to request a public hearing on objections to the standard. Within thirty days of the close of the comment period or the hearing, the Secretary must issue a final standard. Once a standard has been adopted, any person adversely affected by the standard may challenge its validity by petitioning the appropriate U.S. Circuit Court of Appeals within 60 days. [See Section 6(f).]

If the Secretary finds that employees are exposed to grave danger from an insufficiently controlled hazard, a temporary emergency standard may be issued under Section 6(c) of the Act. Such a standard is effective immediately upon publication in the *Federal Register,* but may be challenged thereafter in the Court of Appeals. A temporary emergency standard must be "necessary" to protect employees from such danger. It may remain in effect for six months; thereafter, the Secretary must adopt a permanent standard governing the hazard.

Individual variances from occupational safety and health standards may be issued to employers by the Secretary for two reasons. Employers may apply for a temporary variance for the purpose of obtaining additional time if they cannot come into compliance for such reasons as unavailability of equipment or personnel (Section 6(b)6(A)). Permanent variances may be granted if the Secretary finds that the safety measures used by an employer are as safe as those specifically required by a standard (Section 6(d)).

The Act also provides for experimental variances (Section 6(b)(6)(C)) and national defense variances (Section 16).

4. Enforcement

When enacted in 1970, the OSHAct gave inspectors the right to enter any establishment covered by the Act at reasonable times, without delay, and without prior notification for the purpose of inspecting the premises and all pertinent conditions, structures, machines, and apparatus within. As discussed in Chapter 2, however, the Supreme Court held in 1978 that this provision of the Act violates the Fourth Amendment. Thus, inspectors are now required to have a search warrant. (See Note 2 on page 96.) The Act also allows inspectors to question privately any employer, owner, operator, agent, or employee.

The inspector must present proper credentials to the employer or employer representative in charge of the workplace. Procedures are provided in Section 15 of the Act to assure the confidentiality of trade secrets obtained by OSHA compliance officers. Employees and employers have the right to accompany the inspector during the inspection (the so-called "walkaround" right). Where no "authorized employee representative" exists, the inspector must consult with a reasonable number of employees.

After completing the inspection, the inspecting officer (called the "compliance officer" in OSHA regulations) ordinarily reports any violation to the Area Director, who decides whether to issue a citation. In some cases, the inspector may issue it directly. A citation must describe the specific nature of the violation and establish a reasonable time for abatement of the condition in question. A copy of the citation must be prominently posted at or near the place of violation. Notices, in lieu of citations, may be issued for *de minimis* violations, i.e., ones which have no direct or immediate relationship to safety or health. A citation must be issued with reasonable promptness, but in any case within six months following the inspection.

If OSHA finds any conditions or practices in a place of employment which could reasonably be expected to cause death or serious physical harm before normal enforcement procedures could eliminate the hazard, it may seek a restraining order in the U.S. District Court under Section 11 of the Act. (In state occupational health and safety programs, there may be quicker relief. In California, for example, a state safety inspector can stop operations on the spot.) If the agency fails to seek such relief from an imminent danger, the only recourse for an affected party is to seek a writ of mandamus to compel the Secretary of Labor to do so.

Within a reasonable time after issuance of a citation, the Secretary must notify the employer by certified mail of any proposed penalties. The employer has fifteen working days from the receipt of the notice in which to contest the proposed penalty or citation. If the employer fails to notify the Secretary within this allotted time, the citation and the assessed penalty are deemed the final order of the Review Commission, barring employee objections to the abatement period, and are not subject to review by any court or agency.

If the employer contests the citation or the penalty, the case goes before an administrative law judge (hearing officer) of the Review Commission. The judge's decision is considered final unless, within 30 days, one of the three members of the Review Commission exercises his or her discretionary right to "direct" review. If a review is directed, the case must be decided by the full three-member Review Commission. Any aggrieved party (not just the employer) may seek judicial review in the U.S. Court of Appeals within 60 days of the administrative law judge's decision or, if a review is directed, of the Review Commission's final disposition.

5. The Abatement Period

A reasonable period for abatement of a hazard must be prescribed in the citation issued by OSHA. The abatement period is to begin immediately or up to fifteen days after issuance of a citation (although some have begun later) if the employer does not contest the citation. If the employer files a notice of contest in good faith, and not solely for delay or avoidance of penalties, the abatement period begins after the final administrative disposition by the Review Commission. In general, commencement of appeal proceedings in the U.S. Court of Appeals does not operate as a stay of the order of the Review Commission, unless the court so orders.

Any employer who fails to correct a violation for which a citation has been issued within the period permitted for abatement may be assessed a civil penalty of not more than $1000 a day for each day the violation continues. The Secretary of Labor must notify the employer in writing of this proposed penalty. This notice and assessment is final unless the employer contests the same by notice to the Secretary within fifteen days. The Secretary must afford an opportunity for a hearing if the abatement has not been completed because of factors beyond the employer's control and the employer shows he has made a good-faith effort to comply with the abatement requirements of a citation.

6. Penalties for Violations

The OSHAct provides for a range of monetary penalties for violation of the Act or rules promulgated under it. Any employer who willfully or repeatedly violates the general duty clause or any standard, rule, or order may be assessed a civil penalty of up to $10,000 for each violation. If an employer receives a citation for a serious violation, he must be assessed a civil penalty not to exceed $1000 per violation. A serious violation exists where there is a substantial probability that death or serious physical harm could result. Even if a violation is specifically determined not to be of a serious nature, an employer may receive a civil penalty of up to $1000 per violation.

An employer who commits a willful violation which results in the death of an employee is liable upon conviction for a penalty of up to $10,000 and/or up to six months imprisonment. Subsequent violations can increase the criminal penalty to $20,000 and/or one year imprisonment.

Penalties can also be assessed for interference with enforcement activities. A person giving unauthorized advance notification of an inspection may be liable for a fine of up to $1000 and/or imprisonment for up to six months. Anyone convicted of knowingly making false statements can be assessed a fine of up to $10,000 and/or six months imprisonment. Failure to post official notification required under the Act may result in a civil penalty of up to $1000 per violation.

7. Record-Keeping Requirements

Section 2(b)(12) of the OSHAct requires each employer to maintain a record-keeping system for the purpose of obtaining accurate and uniform information about workplace injuries and illnesses. The system provides inspectors with on-site records of occupational accidents and illnesses. The maintenance of records is also expected to heighten employers' awareness of occupational safety and health problems.

An injury is deemed recordable if it involves medical treatment, loss of consciousness, restriction of work or motion, or transfer to another job. Employers must also maintain accurate records of employee exposure to toxic or physically harmful substances for which specific standards have been promulgated. The Secretary of Labor, working with the Secretary of HHS, is required to promulgate regulations to govern the monitoring of these toxic substances and to provide employees or their representatives with access to monitoring records and records of their own exposure to these substances. (See Chapter 7 for a fuller discussion of access to medical and exposure information under the Act.)

8.   Statistics

The record-keeping system is the basic source of data for the statistical program required by the OSHAct. The Secretary of Labor, in conjunction with the Secretary of HHS, must compile accurate statistics on work injuries and illnesses, including all disabling, serious, or significant injuries and illnesses. To carry out these requirements, grants to or contracts with private organizations, states, or local political subdivisions may be used. Employers can be required to file reports of work injuries and illnesses for this statistical program.

9.   Research

The National Institute for Occupational Safety and Health is required to carry out the research and educational functions assigned to the Secretary of HHS under the OSHAct. The Institute is authorized to recommend criteria for occupational safety and health standards and to conduct research and experimental programs determined by the Institute Director to be necessary for developing or improving job safety and health standards.

A prime function of the Institute is to generate and circulate information in the fields of occupational health and safety. For this purpose, the Institute may conduct medical monitoring of employees and monitoring of the workplace environment to measure potentially toxic substances or harmful physical agents. The Institute must update annually its list of known toxic substances and the concentrations at which toxicity is known to occur. In addition, industry-wide studies on chronic or low-level exposure to a broad variety of industrial materials and processes are mandated.

10.   Education and Training Programs

The Secretary of HHS, in conjunction with the Secretary of Labor, is required to conduct educational programs to provide an adequate supply of qualified persons to enforce the Act. Programs are also authorized for instruction in the proper use of safety and health equipment. Under Section 21, the Secretary of Labor has prime responsibility in the following areas: short-term training of personnel enforcing the Act; general programs for employers and employees on the recognition, avoidance, and prevention of unsafe or unhealthful working conditions; and programs on effective means of preventing occupational injuries or illnesses.

11.   Assistance from the Small Business Administration

Section 28 of the OSHAct amends sections of the Small Business Act to provide for financial assistance to small businesses for the purpose of altering equipment, methods of operation, or facilities to comply with standards and duties required by the Act. Loans are available to a small business if OSHA determines that compliance with the Act is likely to cause substantial economic injury.

12.   Employee Rights

Perhaps the most important employee right under the Act is the right to request an OSHA inspection. Any employee or employee representative who believes that a violation of a safety or health standard, or of the general duty obligation, has occurred which threatens physical harm or which is imminently dangerous, may request in writing an inspection by the Occupational Safety and Health Administration. A signed letter must be sent to OSHA detailing specific information about the alleged violation. The employee need not notify the employer of the inspection request to OSHA. Moreover, although OSHA gives a copy of the letter to the employer prior to inspection, the employee's name must be withheld by the Secretary if the employee so requests. A special investigation must be made if the Secretary determines that there are reasonable grounds to believe that a violation or danger exists. However, if the Secretary finds that there are not reasonable grounds to believe that a violation exists, notification of this finding must be given to the complainant.

The following comprehensive list of employee rights under the Act has been compiled by the United Steelworkers of America:

1. The right to petition the Secretary to commence the procedure for promulgating standards (Section 6(b)(1)).

2. The right to standards which most adequately and *feasibly* assure that no employee will suffer any impairment of health or functional capacity, or diminished life expectancy, even if such employee has regular exposure to toxic or harmful materials (Section 6(b)(5)).

3. The right to standards which prescribe, where necessary, the labeling of hazardous substances, protective equipment, and monitoring (Section 6(b)(7)).

4. The right to medical examinations so as to determine whether exposure is adversely affecting health (Section 6(b)(7)).

5. The right to have the results of one's medical examination transmitted to an employee's physician (Section 6(b)(7)).

6. The right to notification of an employer's request for variance and to participate in the hearing called to evaluate the request (Section 6(d)).

7. The right to employee representation on the standard-setting advisory committees and on NACOSH.

8. The right to have employers keep them informed of their rights under the Act (Section 8(c)(1)).

9. The right to observe the monitoring equipment and have access to the records thereof (Section 8(c)(3)).

10. The right to be notified when concentration levels are being exceeded, and to be informed of measures being taken to correct such a situation (Section 8(c)(3)).

11. The right to accompany the OSHA inspector (Section 8(e)).

12. The right to request a special inspection; to be able to inform the inspector of any alleged violation; and to receive written explanation of failure by the inspector to issue a citation of violation and to receive an informal review thereof (Section 8(f)(1-2)).

13. The right to demand a hearing if an unreasonable (long) time is set for the abatement of a hazardous situation by the Secretary (Section 10(c)).

14. The right to participate in hearings to determine whether an employer should receive a modification of the original abatement order (Section 10(c)).

15. The right to protection against disciplinary action for exercising rights under the Act (Section 11(c)(1-3)).

16. The right to request HHS determination of the toxicity of any substance normally found in the workplace (Section 20(a)(6)).

◇ NOTES

1. Section 5(b) imposes a duty on *employees* to comply with occupational health and safety standards, as well as with all applicable rules and regulations issued under the Act. However, the Act contains no mechanism to enforce these employee-centered duties. How can one reconcile this absence of enforcement with the duty imposed in 5(b)? Very likely, Congress relied on the power relationship of employment to provide a private enforcement mechanism. Employers are free to take disciplinary action against employees who disregard health and safety rules. But this does not absolve the employer from ultimate responsibility. Indeed, employers are often held responsible for a violation even when it was caused by an employee's contravention of the employer's health and safety rules.

2. OSHA's inspection powers have been steadily eroded since the Act's passage. In *Marshall v. Barlow's, Inc.*, excerpted in Chapter 2, the Supreme Court ruled that the Fourth Amendment requires OSHA to obtain a search warrant, which need not be based on probable cause, before it can enter and inspect a worksite without the employer's consent. OSHA's inspection powers have been limited even further in some jurisdictions. At least two appeals courts have ruled that OSHA must obtain a warrant or subpoena before it can examine an employer's occupational health and safety records. See *Brock v. Emerson Electric Co.*, 834 F.2d 994 (11th Cir. 1987); *McLaughlin v. Kings Island, Division of Taft Broadcasting Co.*, 849 F.2d 990 (6th Cir. 1988). These courts have ruled that the OSHA regulation that requires employers to provide OSHA with their injury and illness logs is unconstitutional in that it violates the Fourth Amendment's prohibition of unreasonable search and seizures. The Fourth Circuit has ruled otherwise. *McLaughlin v. A.B. Chance Co*, 842 F.2d 724 (4th Cir. 1988).

3. While Section 8(f)(1) protects the confidentiality of employees who request OSHA inspections, one can reasonably ask if it is always in the employee's best interest to request anonymity. Understandably, workers may fear retaliation if OSHA informs their employer of their identity when following up on a complaint. Protection against retaliation is available under Section 11(c), which provides that employers cannot discharge or otherwise discriminate against employees for exercising their rights under the Act. (See Chapter 8 for more detail on 11(c)). If retaliation does occur, any discrimination case might be stronger if the worker could show that the employer was aware of his or her identity from the beginning. Hence anonymity may not always be in the worker's interest.

4. While Section 8(e) provides that an employee representative must be allowed to accompany a compliance officer during a workplace inspection, it is silent on the issue of walkaround pay—i.e., the issue of whether employees are entitled to be paid for the time spent accompanying the inspector. OSHA's policy has changed several times.

The issue was first raised in 1971, when a labor union contended that an employer's refusal to pay for walkaround time constituted discrimination under Section 11(c) of the Act. The Solicitor General disagreed, however, and OSHA declined to take the matter to court. Subsequently, the D.C. Court of Appeals rejected the argument that failure to provide walkaround pay was a violation of the Fair Labor Standards Act.

In 1977, Assistant Secretary Eula Bingham reversed OSHA policy and, without prior notice and comment, issued an "interpretative rule" stating that an employer's refusal to provide walkaround pay constituted discrimination under 11(c). Although upheld by the district court, the rule was vacated by the D.C. Court of Appeals on procedural grounds. Concluding that the agency was not really "interpreting" the Act, but rather was formulating new policy to implement the Act, the Court held that the agency must comply with the notice and comment procedures of the Administrative Procedure Act. OSHA promptly instituted the required rulemaking procedure. In January 1981, the agency issued a rule requiring employers to provide walkaround pay. It based that rule not on discrimination, but on OSHA's authority to provide for effective walkaround inspections under 8(e).

Four months later, under the new Reagan Administration, Assistant Secretary Thorne Auchter revoked the rule. Using essentially the same record used by Bingham, he concluded that (1) the vast majority of employee representatives are compensated for walkaround time either by their employer or their union; (2) lack of pay has not served as a disincentive to employee participation even when compensation was not provided; and (3) inspectors can always rely on consultation with individual employees as authorized under 8(c). The revocation has not been challenged. [b]

5. Various definitions of "repeated violations" have been offered by the courts and by the Occupational Safety and Health Review Commission (OSHRC). In 1976, the Third Circuit Court of Appeals concluded that a violation is "repeated" if it occurs more than twice. *Bethlehem Steel v. OSHRC,* 540 F.2d 157 (3d Cir. 1976). The Fourth and Ninth Circuits have held that a "repeated violation" can be cited where there is one prior violation. *George Hyman Constr. Co. v. OSHRC,* 582 F.2d 834 (4th Cir. 1978); *Todd Shipyards Corp. v. Secretary,* 566 F.2d 1327 (9th Cir. 1977). In 1979, the OSHRC offered a broad interpretation by stating that "a violation is repeated under Section 17(a) of the Act if, at the time of the alleged repeated violation, there was a Commission final order against the same employer for a substantially similar violation." *Potlatch Corp,* 7 OSHC 1061 (1979). The OSHRC acknowledged the potential difficulty of showing similarity under certain circumstances.

OSHA imposes both geographical and time limitations on the determinations of whether violations are "repeated." OSHA's 1986 Field Operations Manual states that for employers with "fixed establishments" (e.g., factories, terminals, stores), OSHA will look only to violations occurring at the cited establishment. For employers with nonfixed establishments (e.g., construction companies), it will look to prior violations occurring anywhere within the jurisdiction of the OSHA area office. The manual also states that repeated violations cannot be cited if they occur more than three years after the previous violation became final or within three years of the final abatement date, whichever is later.

For companies within the jurisdiction of the Third Circuit Court of Appeals, the first repeat citation cannot be issued until there are two final orders issued to that company for the same standard. *Bethlehem Steel v. OSHRC,* cited above.

---

[b]B. W. Mintz. 1984. *OSHA: History, Law, and Policy.* Washington: The Bureau of National Affairs pp. 546–556.

6. Through the use of appropriations "riders," Congress can use the appropriations process to make changes in the way agencies operate, most often to limit their activities. OSHA has not been immune from this tactic. The fiscal 1989 appropriations included eight provisos that restrict OSHA's use of funds. Most have been included in OSHA appropriations since the mid-1970s. For example, one rider exempts from scheduled safety inspections businesses with ten or fewer employees in industries with low injury and illness rates. Another exempts farms with ten or fewer employees from OSHA coverage entirely, unless the farms have a temporary labor camp. This rider effectively excludes from OSHA protection more than one million workers in an industry with the second highest lost work-day injury rate in the nation. Sometimes riders prompt regulation. For example, in 1974, Congress enacted an appropriations rider that exempted employers with ten or fewer employees from OSHA record keeping requirements. OSHA eventually issued a regulation to that effect (29 CFR 1904.15) and the rider was dropped from subsequent appropriations bills.

7. Section 21(c) of the OSHAct authorizes OSHA to establish education and training programs for employers and employees. OSHA's most notable education and training effort was the New Directions program, instituted in 1978 to provide grants to employee, employer, educational, and nonprofit organizations. New Directions adopted a "peer" approach to education and training in that it sought to provide individuals with the necessary knowledge and skills through trained peers.

New Directions was originally conceived as a grant program with a five-year funding cycle. In an effort to build institutional competence, each grantee would receive funds for a five-year period. New grantees would be added each year for six years, when Round I grantees would be phased out and Round VI grantees would commence. The program did not work out as planned. Round I began in 1979 with a budget of about three million dollars and 86 grantees (27 labor unions, 14 employer associations, 32 educational institutions, and 13 nonprofit organizations). As expected, Round II followed in 1980 with 69 new grantees. No new awards were made in 1981, the first year of the Reagan Administration, nor were new awards made in 1982 and 1983. A third round of grants was made in 1984, but only 26 new grantees were funded, none of which was an educational institution or nonprofit organization. No new awards were made in 1985 and 1986, but a fourth round was funded in 1987 with 18 new grantees.

The New Directions budget hit 14.1 million in 1981, when the Reagan Administration agreed to maintain funding for Round I and II grantees, but then declined drastically to a low of 1.1 million in 1986. From 1979-1987, OSHA expended a total of $60.8 million on New Directions. This was supplemented by $14 million from the National Cancer Institute from 1979-1983. OSHA estimates that 570,000 individuals have been reached through their New Directions program to date.

8. If you have studied the OSHAct, you should be able to discuss the following concepts and terms as they relate to the Act.

+ General duty clause, §5(a)(1)

+ Recognized hazard, §5(a)(1)

+ Consensus standard, §6(a)

- Permanent standard, §6(b)
- Emergency temporary standard, §6(c)
- Feasibility, §6(b)(5)
- Standard of judicial review, §6(f), 11(a)
- Variances, §6(b)6A,6(b)6(c),6(d), 16
- Employee access to information, §6(b)(7)
- Employee access to records, §8(3)
- Imminent danger, §13
- Discrimination protection, §11(c)(1)
- Trade secrets, §15
- Record keeping, §8(c)(1)
- Willful violation, §17(a)
- State plan, §18

C.   A BRIEF HISTORY OF STANDARD SETTING UNDER THE OSHACT

1.   6(a) and 6(b) Standards

As noted, Section 6(a) required OSHA to adopt existing national consensus standards within two years of the Act's passage without further rulemaking. OSHA moved quickly and, in 1971, adopted thousands of consensus standards. Most pertained to safety (rather than health) and many were already outdated, overly specific, or inconsequential (such as the standard regulating the shape of toilet seats). Approximately 400 of the standards pertained to health. These included threshold limit values (TLVs) recommended as guidelines by the American Conference of Governmental Industrial Hygienists (ACGIH), and exposure levels for toxic substances as recommended by the American National Standards Institute (ANSI). These standards put permissible exposure limits (PELs) into place but included no requirements for exposure monitoring, medical surveillance, record keeping, hazard warning, or employee education. Moreover, they were set at a level thought to protect the "average" worker, not the sensitive worker or the worker with prior exposure or existing disease.

Although, many of the guidelines were subsequently revised by the organizations that developed them, the corresponding Section 6(a) standards remained as originally adopted by OSHA. Recognizing the need to update these start-up health standards, OSHA proposed a plan in 1988 to revise them largely *en masse* by replacing them with the current TLVs recommended by the ACGIH and, for 15 substances, by adopting more stringent NIOSH recommendations. The final standard, incorporating a total of 376 PELs, was promulgated on January 19, 1989. The revised standard has drawn both support and criticism, and is, as this text goes to press, under challenge in the courts. From a worker protection perspective, the revision would be something of a mixed bag. It would certainly reduce exposure to many toxic substances without the lengthy and cumbersome process of individual rulemaking required under Section 6(b). However,

Table 3-1. Chemical substances for which the Occupational Safety and Health Administration (OSHA) has promulgated health standards 1972–1988

2-Acetylaminofluorine	4-Dimethylaminoazobenzene
Acrylonitrile	Ethylene oxide
4-Aminodiphenyl	Etyleneimine
Arsenic (inorganic)	Formaldehyde
Asbestos	Lead (inorganic)
Benzene	Methyl chloromethyl ether
Benzidine	α-Naphthylamine
bis-Chloromethyl ether	β-Naphthylamine
Coke oven emissions	4-Nitrobiphenyl
Cotton dust	N-Nitrosodimethylamine
1,2-Dibromo-3-chloropropane	β-Propiolactone
3,3′-Dichlorobenzidine (and its salts)	Vinyl chloride

Source: Adapted from 29 C.F.R. 1910, Subpart Z (1988). OSHA has initiated rule making for nine other substances, including 1,3-butadiene, cadmium dust and fume, 2-ethoxyethanol (Cellosolve), 2-ethoxyethyl acetate, ethylene dibromide, methyl Cellosolve, methyl Cellosolve acetate, methylene chloride and 4,4′-methylenedianiline [53 *Fed. Reg.* 21,246 (June 7, 1988)].

it would also repeat many of the mistakes of the past, by (1) adopting many inadequately protective exposure limits (OSHA disregarded NIOSH recommendations for stricter limits for 68 specific substances)[c], (2) authorizing the use of respirators for up to five years to meet the new PELs, and (3) omitting requirements for exposure monitoring and medical surveillance.

OSHA also has promulgated 24 health standards under Section 6(b), which are listed in Table 3-1. Each standard has a unique story behind it. Taken together, they illustrate the sad plight of OSHA rulemaking. It is a process characterized by frustration, delay, and missed opportunity. The court cases discussed in this chapter provide a fascinating look at the process of setting standards for individual toxic substances.

2. The Generic Cancer Standard

In addition to the substance-specific standards, OSHA has promulgated two "generic" standards for toxic substances. One is the hazard communication standard which is discussed in detail in Chapter 7. The other is OSHA's generic cancer standard.

In order to increase its speed and efficiency in dealing with occupational carcinogens, OSHA promulgated a generic standard in 1980. The standard established (1) scientifically based criteria for identifying and classifying potential occupational carcinogens; (2) procedures and timetables for this purpose; and (3) model standards for the setting of standards for these substances. By resolving certain scientific issues *as a matter of OSHA policy* (thus foreclosing their being raised in future regulatory proceedings as a general issue) and standardizing the form of carcinogen regulation, the agency sought

[c]G. E. Ziem and B. I. Castleman. 1989. Threshold Limit Values: Historical Perspectives and Current Practice, *Journal of Occupational Medicine*, 31(1): 910–918.

to create a more efficient and effective rulemaking process. Among the important science policy issues "resolved" by the OSHA generic standards were the following:

1. Positive results in human epidemiological studies alone, or positive results in at least one animal bioassay meeting certain minimum standards of quality and in the presence of supporting evidence, will suffice to establish the inference that a substance is a carcinogen.

2. Negative results in animal bioassays or human epidemiological studies will not be considered in deciding whether a chemical is a carcinogen if positive results have been obtained in either an animal bioassay or epidemiological study, unless the negative studies meet stringent quality standards.

3. In animal bioassays, any tumor produced (including benign tumors), or an increase in the incidence of tumors which occur spontaneously, or the occurrence of tumors at the site of application of a substance by the oral, inhalation, or dermal routes, will establish the inference that the substance is a carcinogen.

4. Tumors induced in animals at high dosage levels (including the maximum tolerated dose) will be considered evidence of carcinogenicity of that substance, unless strong evidence can be supplied that the metabolism of the substance in question is such that at a low dose, no carcinogenic metabolites are produced.

In general, these positions appear to represent a reasonable consensus on the state of the art in carcinogenicity testing.

The generic standard classified as "Class I" suspect carcinogens those substances with strong supporting evidence of carcinogenicity (i.e., data from human studies and positive results from one animal study plus other supporting data such as positive results from short-term in vitro tests.). Any resulting permanent standard for a Class I suspect carcinogen would require reduction of exposure to the maximum extent feasible; no exposure would be allowed at all when a less hazardous substitute was available. The standard classified as Class II suspect carcinogens those substances with less conclusive evidence, such as unconfirmed results from a single animal study. Any resulting permanent standard would limit exposure "as appropriate and consistent with statutory requirements."

The standard required OSHA to publish, every six months, two priority lists of Class I and Class II carcinogens, each containing about ten substances, selected from prior lists of candidates for regulation. These priority lists constituted a nonbinding indication of those substances that OSHA intended to regulate in the near future.

The model standards contained in the generic standard recommended that permanent standards for both Class I and Class II carcinogens include descriptions of work and hygiene practices, maximum exposure limits, and provisions for personal protective equipment, exposure monitoring, and medical recordkeeping.

After the promulgation of the generic cancer standard, however, the Supreme Court held that OSHA can promulgate permanent standards for chemical hazards only where the chemical poses a "significant risk" of harm. (See *Marshall v. American Petroleum Institute* discussed later in this chapter.) This decision called into question many of the assumptions underlying the generic cancer policy, and necessitated revisions in the policy. OSHA published amendments to the generic standard in

1981, but the agency withdrew them shortly thereafter. In 1982, OSHA issued an advance notice of proposed rulemaking for a revised proposal and, in 1986, published its intent to revise the policy by January 1987. Thus far, however, this has not occurred.

Although it is still on the books, the original generic cancer policy is not used by the agency. Nonetheless, talk of reopening and revising the standard continues to be heard, and the generic standard may yet resurface as a viable regulatory tool. The resolution of recurring science policy issues through a generic standard could greatly reduce the time and resources required of OSHA during regulatory proceedings. Besides the remaining scientific issues related to the determination of carcinogenicity, however, OSHA would still be required to determine what level of control is economically and technologically feasible.

D. JUDICIAL INTERPRETATION OF STANDARD SETTING UNDER THE OSHACT

As noted, OSHA rulemaking is a time-consuming and cumbersome process. Most of the health standards promulgated by OSHA have been challenged in the courts, and the resulting judicial interpretations have shaped both the scientific and procedural dimensions of OSHA's standard setting process.

This section presents excerpts from the key decisions of the federal courts with regard to permanent and temporary emergency standards.

1. The Early Cases: Weight of the Evidence and the Meaning of Feasibility

In this first case, which deals with OSHA's promulgation of a permanent asbestos standard, the D.C. Circuit Court of Appeals addressed the reach of the substantial evidence test in the course of judicial review, the weight to be given NIOSH recommendations, and the meaning of feasibility under the OSHAct.

Industrial Union Department, AFL-CIO v. Hodgson

499 F.2d 467 (D.C. Cir. 1974)

McGOWAN, Circuit Judge.

This direct review proceeding presents a classic case of what Judge Friendly has aptly termed "a new form of uneasy partnership" between agency and court that results whenever Congress delegates decision making of a legislative character to the one, subject to review by the other. *Associated Industries v. United States Dept. of Labor*, 487 F.2d 342, 354 (2nd Cir. 1973). The angularity of this relationship is only sharpened when, as here, Congress—with no apparent awareness of anomaly—has explicitly combined an informal agency procedure with a standard of review traditionally conceived of as suited to formal adjudication or rule-making. The federal courts, hard pressed

as they are by the flood of new tasks imposed upon them by Congress, surely have some claim to be spared additional burdens deriving from the illogic of legislative compromise. At the least, it would have been helpful if there had been some recognition by Congress that the quick answer it gave to a legislative stalemate posed serious problems for a reviewing court, and that there would inevitably have to be some latitude accorded it to surmount those problems consistently with the legislative purposes. The duty remains, in any event, to decide the case before us in accordance with our statutory mandate, however dimly the rationale, if any, underlying it can be perceived.

The petition before us seeks review of standards promulgated by the Secretary of Labor under the

Occupational Safety and Health Act of 1970, 29 U.S.C. Secs. 651 *et seq.*, (hereinafter OSHA). The standards in question regulate the atmospheric concentrations of asbestos dust in industrial workplaces. Petitioners are unions whose members are affected by the health hazards of asbestos dust. They challenge the timetable established by the standards for the achievement of permissible levels of concentration, and object to portions of the standards concerning methods of compliance, monitoring intervals and techniques, cautionary labels and notices, and medical examinations and records. We remand two of such issues to the Secretary for further consideration. In all other respects, the petition is denied.

I

A. The Occupational Safety and Health Act

Technological progress in industry appears not to have been accompanied uniformly by corresponding reductions in the health hazards of industrial working conditions. More than 2.2 million persons are disabled on the job each year, and in 1967 the Surgeon General estimated that approximately 400,000 new cases of occupational disease would occur in each succeeding year.[1] The chairman of the Committee on Labor and Public Welfare summarized the problem as follows:

> Not only are occupational diseases which first came to light at the beginning of the Industrial Revolution still undermining the health of workers, but new substances, new processes, and new sources of energy are presenting health problems of ever-increasing complexity.

Foreword, Legislative History of the Occupational Safety and Health Act of 1970 (hereinafter Legis. Hist.)

OSHA, the first comprehensive attempt by Congress to deal with these problems,[2] covers every employer whose business affects interstate commerce.[3] Eschewing any attempt to establish substantive provisions to control all these various employers, the

Act erects a general framework to govern the development of regulations, and delegates the task for formulating particular health and safety standards to the Secretary of Labor. Civil and criminal sanctions are provided to enforce compliance.

OSHA specifies the procedure to be followed in the promulgation of standards, and provides for the establishment of a research institute and the appointment of advisory committees to assist the Secretary.[4] The substantive provisions of the Act impose a general obligation upon employers to provide safe working conditions. 29 U.S.C. Sec. 654(a)(1) (1970). The Secretary is required to promulgate standards to control particular health hazards that come to his attention. Certain types of controls, including monitoring, medical examinations, warnings, record keeping, and specific protective measures are specified by the statute itself, but the decision as to when and how they should be required with regard to particular health hazards is left to the Secretary.

*   *   *

B. Asbestos

Asbestos is a generic term applicable to a number of fibrous, inorganic, silicate minerals that are incombustible in air. Its commercial value is high, and its uses are many and varied. Asbestos can be woven into cloth, used in powder form, or incorporated into materials of various shapes and consistencies. Almost one million tons of asbestos are used in this country annually; and, for many purposes, it cannot easily be replaced with other substances.

Unfortunately, asbestos is as hazardous to health as it is useful to industry. During its production and use, tiny asbestos fibers are released as a dust in the air, and, over the course of this century, thousands of workers have been killed or disabled by the effects of inhaling these fibers. There are no precise figures concerning the number of workers involved, but it is estimated that three to five million workers are exposed to some extent to asbestos fibers in the

---

[1] House Comm. on Education and Labor, Occupational Safety and Health Act, H.R. Rep. No. 91-1291, 91st Cong., 2d Sess. 14 (1970); Cohen, The Occupational Safety and Health Act: A Labor Lawyer's Overview, 33 Ohio St. L. J. 788, 789–90 (1972).

[2] For summaries of prior regulation, *see* Cohen, note 1 *supra*, 788-89; Comment, OSHA; Employer Beware, 10 Houston L. Rev. 426, 426-28 (1973).

[3] The United States, and state and local governments, are exempted from OSHA by 29 U.S.C. Sec. 652(5) (1970).

[4] The Act establishes within the Department of Health, Education, and Welfare a National Institute for Occupational Safety and Health (NIOSH) which is authorized to "develop and establish recommended occupational safety and health standards" for transmission to the Secretary of Labor, 29 U.S.C. Sec. 671. The Secretary of Labor may also appoint an advisory committee to assist him in his standard-setting functions. In this instance a 5-member Advisory Committee on Asbestos Standards was constituted, consisting of two employer and two labor members, and the representative of the public.

building construction and shipyard industries alone. While OSHA was under consideration in Congress, the health hazards of the asbestos industry were among the examples used to stress the need for legislation.[7]

C. Proceedings before the Secretary

Within a few months of the effective date of OSHA, petitioners requested the Secretary to establish an emergency standard to control concentrations of asbestos dust.[8] The Secretary promptly issued a temporary standard and set in motion the procedure for establishment of a permanent standard. Notice of the proposed rulemaking was published, and interested persons were invited to submit their views. NIOSH submitted its recommendations, as did the Advisory Committee. These were made public, and the Secretary conducted a hearing at which various representatives and experts appeared on behalf of interested parties. On the basis of these recommendations and a formidable record of documents and oral testimony, including highly technical statements by expert witnesses, the Secretary established the standards in question.[9] His statement of reasons

---

[7]Legislative History at 319, 412–13, 1002. The Secretary expressed the problem as follows (37 ER. 11318):

> No one has disputed that exposure to asbestos of high enough intensity and long enough duration is causally related to asbestosis and cancers. The dispute is as to the determination of a specific level below which exposure is safe.

[8]Under 29 U.S.C. Sec. 655(c), the Secretary can issue an immediately effective emergency standard without observing the procedural requirements for a permanent standard if (1) employees are exposed to grave danger, and (2) the emergency standard is necessary to protect them. The Secretary must then publish a permanent standard within six months.

[9]A qualified hearing examiner presided over the four days consumed by the public hearing. At the close his only function was to certify the record to the Secretary, which consisted of the written statements and comments on the proposed standards received prior to the hearing in response to the notice of rulemaking, the transcript of the hearing itself, and many exhibits received during the hearing and in a further period allowed after the hearing for this purpose. The Joint Appendix filed in this court contains over 1100 pages, of which over 400 are from the hearing transcript. The testimonial pattern generally was for the witnesses to read long statements at the close of which they were subject to cross-examination. The questions actually asked tended to be few, sporadic, and perfunctory, and the record resembles nothing so much as that of a typical legislative committee hearing.

covers some four and one-half pages of the Federal Register.[10]

Petitioners allege no procedural errors in the promulgation of these standards, but they characterize them as inadequate to protect the health of employees as required by the Act. They attack the Secretary's interpretation of OSHA in certain particulars, as well as the enforcement measures he has selected.

II

OSHA is a self-contained statute in the sense that it does not depend upon reference to the Administrative Procedure Act for specification of the procedures to be followed. It prescribes that the process of promulgating a standard is to be initiated by the publication of a proposed rule. Interested persons are given a period of 20 days thereafter within which to submit written data or comments. Within this period any interested person may submit written objections, and may request a public hearing thereon. In such event, the Secretary shall publish a notice specifying the particular standard involved and stating the time and place of the hearing. Within 60 days after the completion of such hearing, the Secretary shall make his decision. Judicial review by the courts of appeals is provided.[11]

This procedure is characteristic of the informal rulemaking contemplated by Section 4 of the APA, 5 U.S.C. Sec. 553, and it was so understood by the Congress. By regulation, however, the Secretary, although describing it as "legislative in type," has provided that the oral hearing called for in the

---

[10]The statutory direction is that the Secretary, whenever he promulgates a standard, "shall include a statement of reasons for such action. . . ." 29 U.S.C. Sec. 655(e). The Secretary has by regulation, 29 C.F.R. Sec. 1911.19(b), as amended, 37 ER. 8655 (1972), defined this task in these terms:

> Any rule or standard adopted . . . shall incorporate a concise general statement of its basis and purpose. The statement is not required to include specific and detailed findings and conclusions of the kind customarily associated with formal proceedings. However, the statement will show the significant issues which have been faced, and will articulate the rationale for their solution.

Petitioners have not challenged the propriety of this formulation.

[11]29 U.S.C. Sec. 655(f) reads in relevant part: "The determinations of the Secretary shall be conclusive if supported by substantial evidence in the record as a whole."

statute shall contain some elements normally associated with the adjunctory or formal rulemaking model. As indicated in the text of the regulations, set forth in the margin,[12] the Secretary apparently concluded that this was necessary because of the necessity of having a record to which the statutorily mandated substantial evidence test could be meaningfully applied by a reviewing court. The only controversy we have in this case as to the procedural requirements of the statute is not with respect to the manner in which the rulemaking was done by the Secretary, but as to the reach of the substantial evidence test in the course of judicial review.

The substantial evidence test has customarily been directed to adjudicatory proceedings or formal rulemaking.[13] The hybrid nature of OSHA in this respect can be explained historically, if not logically,

[12]In 29 C.F.R. Sec. 1911.15 ("Nature of Hearing"), the Secretary stated in relevant part:

(a)(2) Section 6(b)(3) provides an opportunity for a hearing on objections to proposed rule making, and section 6(f) provides in connection with the judicial review of standards, that determinations of the Secretary shall be conclusive if supported by substantial evidence in the record as a whole. Although these sections are not read as requiring a rule making proceeding within the meaning of the last sentence of 5 U.S.C. 553(c) requiring the application of the formal requirements of 5 U.S.C. 556 and 557, they do suggest a Congressional expectation that the rule making would be on the basis of a record to which a substantial evidence test, where pertinent, may be applied in the event an informal hearing is held.

(3) The oral hearing shall be legislative in type. However, fairness may require an opportunity for cross-examination on crucial issues. The presiding officer is empowered to permit cross-examination under such circumstances. . . .

(b) Although any hearing shall be informal and legislative in type, this part is intended to provide more than the bare essentials of informal rule making under 5 U.S.C. 553. The additional requirements are the following:

(1) The presiding officer shall be a hearing examiner appointed under 5 U.S.C. 3105.

(2) The presiding officer shall provide an opportunity for cross-examination on crucial issues.

(3) The hearing shall be reported verbatim, and a transcript shall be available to any interested person on such terms as the presiding officer may provide.

[13]See Camp v. Pitts, 411 U.S. 138 (1973); City of Chicago v. FPC, 458 F.2d 731, 744 (D.C. Cir. 1971), cert. denied, 405 U.S. 1074 (1972); Automotive Parts and Accessories Ass'n v. Boyd, 407 F.2d 330, 334-38 (D.C. Cir. 1968); Wirtz v. Baldor Electric Co., 337 F.2d 518, 525-28 (D.C. Cir. 1963).

as a legislative compromise. The Conference Report reflects that the Senate bill called for informal rulemaking, but the House version specified formal rulemaking and substantial evidence review. The House receded on the procedure for promulgating standards, but the substantial evidence standard of review was adopted.[14]

*   *   *

Faced with the fact that his determinations were commanded by Congress to be reviewed under a substantial evidence standard, the Secretary did voluntarily move his procedures significantly towards the formal model. He directed that (1) a qualified hearing examiner should preside over the oral hearing, (2) cross-examination should be permitted, and (3) a verbatim transcript made. The total record in this case was in part created under the conditions that obtain in a formal proceeding. In substantial remaining part, however, it consists of a melange of written statements, letters, reports, and similar materials received outside the bounds of the oral hearing and untested by anything approaching the adversary process.

Thus, in some degree the record approaches the form of one customarily conceived of as appropriate for substantial evidence review. In other respects, it does not. On a record of this mixed nature, when the facts underlying the Secretary's determinations are susceptible of being found in the usual sense, that must be done, and the reviewing court will weigh them by the substantial evidence standard. But, in a statute like OSHA where the decision making vested in the Secretary is legislative in character, there are areas where explicit factual findings are not possible, and the act of decision is essentially a prediction based upon pure legislative judgment, as when a Congressman decides to vote for or against a particular bill.

OSHA sets forth general policy objectives and

[14]This combination is made even more confusing by a statement in the report that seems to indicate that the Conference Committee thought the substantial evidence standard was less exacting than the standard of rationality ordinarily applicable to the results of informal rulemaking. For a more detailed discussion of these legislative events, see Associated Industries of NY State v. Department of Labor [487 F.2d 342 (2dCir. 1973)], where the Second Circuit said that the Congressional intention was clear "to adopt the substantial evidence test for review as a trade-off for the House's abandoning its insistence on rulemaking on the record. . . ." 487 F.2d at 329 [OSHC at 1344].

establishes the basic procedural framework for the promulgation of standards, but the formulation of specific substantive provisions is left largely to the Secretary.[17] The Secretary's task thus contains "elements of both a legislative policy determination and an adjudicative resolution of disputed facts." *Mobil Oil Corp. v. FPC*, 483 F. 2d 1238, 1257 (D.C. Cir. 1973). Although in practice these elements may so intertwine as to be virtually inseparable, they are conceptually distinct and can only be regarded as such by a reviewing court.

From extensive and often conflicting evidence, the Secretary in this case made numerous factual determinations. With respect to some of those questions, the evidence was such that the task consisted primarily of evaluating the data and drawing conclusions from it. The court can review that data in the record and determine whether it reflects substantial support for the Secretary's findings. But some of the questions involved in the promulgation of these standards are on the frontiers of scientific knowledge, and consequently as to them insufficient data is presently available to make a fully informed factual determination. Decision making must in that circumstance depend to a greater extent upon policy judgments and less upon purely factual analysis.[18] Thus, in addition to currently unresolved factual issues, the formulation of standards involves choices that by their nature require basic policy determinations rather than resolution of factual controversies. Judicial review of inherently legislative decisions of this sort is obviously an undertaking of different dimensions.[19]

For example, in this case the evidence indicated that reliable data is not currently available with respect to the precisely predictable health effects of various levels of exposure to asbestos dust; nevertheless, the Secretary was obligated to establish some specific level as the maximum permissible exposure. After considering all the conflicting evidence, the Secretary explained his decision to adopt, over strong employer objection, a relatively low limit in terms of

the severe health consequences which could result from over-exposure. Inasmuch as the protection of the health of employees is the overriding concern of OSHA, this choice is doubtless sound, but it rests in the final analysis on an essentially legislative policy judgment, rather than a factual determination, concerning the relative risks of underprotection as compared to overprotection.

Regardless of the manner in which the tasks of judicial review is articulated, policy choices of this sort are not susceptible to the same type of verification or refutation by reference to the record as are some factual questions. Consequently, the court's approach must necessarily be different no matter how the standards of review are labeled. That does not mean that such decisions escape exacting scrutiny, for as this court has stated in a similar context:

> This exercise need be no less searching and strict in its weighing of whether the agency has performed in accordance with the Congressional purposes, but, because it is addressed to different materials, it inevitably varies from the adjudicatory model. The paramount objective is to see whether the agency, given an essentially legislative task to perform, has carried it out in a manner calculated to negate the dangers of arbitrariness and irrationality in the formulation of rules for general application in the future.

Automotive Parts & Accessories Association v. Boyd, 407 F.2d 330, 338 (1968).

We do not understand Congress to have in this instance nullified this approach for all purposes by directing substantial evidence review. As noted above, that provision is important as an indication of how we should approach certain kinds of questions and what kind of record we should demand of the Secretary. But it is surely not to be taken as a direction by Congress that we treat the Secretary's decision making under OSHA as something different from what it is, namely, the exercise of delegated power to make within certain limits decisions that Congress normally makes itself, and by processes, as the courts have long recognized and accepted, peculiar to itself. A due respect for the boundaries between the legislative and the judicial function dictates that we approach our reviewing tasks with a flexibility informed and shaped by sensitivity to the diverse origins of the determinations that enter into a legislative judgment.

What we are entitled to at all events is a careful identification by the Secretary, when his proposed standards are challenged, of the reasons why he chooses to follow one course rather than another.

---

[17]For a comparison of this aspect of OSHA with the National Labor Relations Act, which defines specifically prohibited practices, *see* Cohen, *supra* note 1, at 798-800.

[18]Where existing methodology or research in a new area of regulation is deficient, the agency necessarily enjoys broad discretion to attempt to formulate a solution to the best of its ability on the basis of available information. Permian Basin Area Rate Case, 390 U.S. 747, 811 (1968).

[19]*See Automotive Parts and Accessories Ass'n v. Boyd*, 407 F.2d 330, 336 (D.C. Cir. 1968); *Body & Tank Corp. v. NLRB*, 339 F.2d 76, 78-79 (2d Cir. 1964).

Where that choice purports to be based on the existence of certain determinable facts, the Secretary must, in form as well as substance, find those facts from evidence in the record. By the same token, when the Secretary is obligated to make policy judgments where no factual certainties exist or where facts alone do not provide the answer, he should so state and go on to identify the considerations he found persuasive.

* * *

In the case of OSHA, the Secretary has wisely acted by regulation to go beyond the minimum requirements of the statute and to expand his capacity to find facts by providing an evidentiary hearing in which cross-examination is available. We think it equally the part of wisdom and restraint on our part to show a comparable flexibility, and to be always mindful that at least some legislative judgments cannot be anchored securely and solely in demonstrable fact. Such a principle, far from being destructive of the Congressional contemplation, is essential to its preservation.

* * *

III

Before addressing the specific challenges made to the Secretary's action, we examine two further problems raised by petitioners involving statutory construction. One has to do with the weight to be accorded by the Secretary to the NIOSH recommendations. The other relates to whether the Secretary may take economic considerations into account.

With respect to the former, the statute directs NIOSH to develop criteria documents that describe safe levels of exposure, and the Secretary is to promulgate standards that insure that employees are protected. The language employed by Congress in these two mandates is essentially identical except that the Secretary must consider elements of feasibility.[21] From this similarity petitioners argue that the determinations of NIOSH are meant to be conclu-

sive on the question of what exposure levels adequately protect health, and that the Secretary may deviate from the NIOSH document only to the extent dictated by feasibility.

The Act merely says that the Director of NIOSH shall immediately forward recommended standards to the Secretary without specifying how the Secretary is to use them, but the procedure prescribed for the formulation of standards militates against petitioners' position. It is the Secretary rather than NIOSH who conducts hearings and receives the comments of interested persons. The Secretary may also appoint a special advisory committee to assist him in his standard-setting functions, and receive recommendations from it, as he did here.

The Act, or so it seems to us, must be taken as contemplating that the Secretary may consider all of this information as well as that received from NIOSH. Petitioners' argument would restrict the advisory committees and interested parties to comments relating solely to feasibility, a role petitioners themselves clearly—and, we think, legitimately—exceeded at the hearing and in their arguments before this court. The NIOSH recommendation was undoubtedly important in the eyes of Congress as an aid to the Secretary, but we cannot see that it was intended as more than that.

In connection with the second issue, we note that the statutory authority for the promulgation of standards reads in relevant part:

> The Secretary . . . shall set the standard which most adequately assures, *to the extent feasible,* on the basis of the best available evidence, that no employee will suffer material impairment of health or functional capacity. . . . 29 U.S.C. Sec. 655(b)(5) (emphasis supplied).

The standards as promulgated retain the concentration level specified by the temporary emergency standard until 1976 when a lower permanent standard becomes effective. The Secretary explained his decision to delay, two years longer than the period suggested by NIOSH, implementation of the tougher standard as "necessary to allow employers to make

---

[21]*Compare* 29 U.S.C. Sec. 669(a)(3):
[NIOSH] shall develop criteria dealing with toxic materials and harmful physical agents and substances which will describe exposure levels that are safe for various periods of employment, including but not limited to exposure levels at which no employee will suffer impaired health or functional capacities or diminished life expectancy as a result of his work experience.

with 29 U.S.C. Sec. 655(b)(5):
The Secretary, in promulgating standards dealing with toxic materials or harmful physical agents under this subsection, shall set the standard which most adequately assures, *to the extent feasible,* on the basis of the best available evidence, that no employee will suffer material impairment of health or functional capacity even if such employee has regular exposure to the hazard dealt with by the standard for the period of his working life. . . . (Emphasis supplied.)

the needed changes for coming into compliance," and petitioners argue that the Secretary improperly considered economic factors entering into the Secretary's conclusion could properly include problems of economic feasibility.

There can be no question that OSHA represents a decision to require safeguards for the health of employees even if such measures substantially increase production costs. This is not, however, the same thing as saying that Congress intended to require immediate implementation of all protective measures technologically achievable without regard for their economic impact. To the contrary, it would comport with common usage to say that a standard that is prohibitively expensive is not "feasible."[22] Senator Javits, author of the amendment that added the phrase in question to the Act, explained it in these terms:

> As a result of this amendment the Secretary, in setting standards, is expressly required to consider feasibility of proposed standards. This is an improvement over the Daniels bill, which might be interpreted to require absolute health and safety in all cases, regardless of feasibility, and the Administration bill, which contains no criteria for standards at all.

S. Rep. No. 91-1282, 91st Cong., 2d Sess., p. 58; Legis. Hist. at 197. The thrust of these remarks would seem to be that practical considerations can temper protective requirements. Congress does not appear to have intended to protect employees by putting their employers out of business—either by requiring protective devices unavailable under existing technology or by making financial viability generally impossible.

This qualification is not intended to provide a route by which recalcitrant employers or industries may avoid the reforms contemplated by the Act. Standards may be economically feasible even though, from the standpoint of employer, they are financially burdensome and affect profit margins adversely. Nor does the concept of economic feasibility necessarily guarantee the continued existence of individual employers. It would appear to be consistent with the purposes of the Act to envisage the economic demise of an employer who has lagged behind the rest of the industry in protecting the health and safety of employees and is consequently financially unable to comply with new standards as quickly as other employers.[23] As the effect becomes more widespread within an industry, the problem of economic feasibility becomes more pressing. For example, if the standard requires changes that only a few leading firms could quickly achieve, delay might be necessary to avoid increasing the concentration of that industry. Similarly, if the competitive structure or posture of the industry would be otherwise adversely affected—perhaps rendered unable to compete with imports or with substitute products[24]—the Secretary could properly consider that factor. These tentative examples are offered not to illustrate concrete instances of economic unfeasibility but rather to suggest the complex elements that may be relevant to such a determination.[25]

With the aid of the foregoing analytic background of the procedural and substantive provisions of the Act, we turn to the specific objections raised by petitioners to the standards.

---

[22] A discussion of some of the costs of dust control is found in Hills, Economics of Dust Control, 132 *Annals of the New York Academy of Sciences*, 322 (1965). Several industry representatives testified in detail concerning the cost of attempting to meet the standards. *Cf. H & H Tire Co. v. United States Dept. of Transportation*, 471 F.2d 350 (7th Cir. 1972); *Chrysler Corporation v. Department of Transportation*, 472 F.2d 659 (6th Cir. 1972). These cases support the proposition that "practical" as employed in the Automobile Safety Act of 1966, 15 U.S.C. Sec. 1392(a), includes economic considerations, but the legislative history of that statute, unlike the history of OSHA, is more explicit on that point.

[23] Temporary variances may be obtained when timely compliance is technologically impossible.

[24] Testimony of industry representatives predicted both of these results.

[25] Since technological progress is here linked to objectives other than the traditional competitive, profit-oriented concerns of industry, accommodation of both sets of values will sometimes involve novel economic problems. *International Harvester Co. v. Ruckelshaus*, 478 F.2d 615 (D.C. Cir. 1973), illustrates some of these problems in the context of the automobile emissions standards of the Clean Air Act. 42 U.S.C. Secs. 1857 *et seq.* In the highly concentrated automobile industry the court deemed it likely that, by virtue of their size and importance to the economy, any one of the three major companies could obtain a relaxation of the automobile emissions standards if it could not meet them. If this occurred after other manufacturers had prepared to comply with the standard, the technological laggard would enjoy a competitive advantage because installation of the control devices renders the vehicles less efficient to operate. This circumstance justified insuring that the standards could be met by all major producers before they became effective.

1. Effective Date for the Two Fiber Standard. The most important aspect of setting the standards was the determination of an acceptable dust concentration level. Under the emergency standards, the eight hour time-weighted average airborne concentration of asbestos dust had been limited to five fibers greater than five microns in length per milliliter of air (hereinafter "the five fiber standard").[26] A principal issue at the hearings on the permanent standards was whether the standard should remain at five fibers or be lowered to two. Proponents of standards ranging from zero to 12 fibers appeared, and it is fair to say that the evidence did not establish any one position as clearly correct. The Secretary decided to resolve this doubt in favor of greater protection of the health of employees, and established the two fiber standard recommended by NIOSH and his Advisory Committee as the level ultimately to be achieved.

Industry representatives testified that they simply could not reduce concentrations to the two fiber level in the foreseeable future. In the course of formulating its proposal, NIOSH had undertaken a limited analysis of industry's capacity to comply and had recommended delaying the effective date of the two fiber standard for two years, *i.e.*, July 1, 1974. The Secretary decided to retain the five fiber standard for approximately four years (July 1, 1976) before requiring the reduction to two fibers, in order to give employers time to prepare for the lower limit. Petitioners assert that the four year delay permitted by the Secretary is too long because (1) the health of employees is endangered thereby, and (2) employers do not need that much time.

a. Health Hazards Occasioned by the Delay. The Secretary solicited the views of several experts on the question of the predictable health effects of maintaining a five fiber standard until 1976. The experts differed sharply in some of their opinions, but their responses are generally cautious and reflect deficiencies in available data concerning the relationship between exposure to asbestos dust and the likelihood of disease. The record indicates that no precise prediction of increased harm can be made at this time.[27]

The Secretary must establish those standards that most adequately insure that no employee will suffer material impairment of health. We cannot say, on the basis of the conflicting testimony in the record, that the Secretary erred in his prediction of the health effect of the four year delay, but neither can we say that employees are not exposed to some additional risk of disease because of greater exposure. In view of the Act's express allowance for problems of feasibility, the Secretary's decision to allow a four year delay is not irrational with regard to those industries that require that long to meet the standard. It is appropriate to allow sufficient time to permit an orderly industry-wide transition since, in those cases, the indeterminate degree of risk involved is counterbalanced by considerations of feasibility; it is not, however, a risk to which employees should be needlessly exposed.

*   *   *

2. Monitoring Requirements. Within six months of publication of the standards, all employers are required to monitor workplaces to determine whether the concentrations of asbestos dust are within the allowable limits; thereafter, monitoring must in all cases be of such pattern and frequency as to identify accurately the levels of exposure. Monitoring must occur no less frequently than once every six months where the concentrations may reasonably be foreseen to exceed the standards. Petitioners object that these provisions are inadequate to protect the health of employees.[33]

The monitoring provisions are especially impor-

---

[26]Although size and shape of fibers are relevant to this propensity to cause harm, the court is limited to fibers longer than five microns for practical rather than medical reasons. The most accurate sampling technique that can feasibly be employed, the membrane filter method, does not measure smaller particles.

[27]The Secretary directed his inquiries to Marcus M. Key, M.D., Assistant Surgeon General, Director of NIOSH; George W. Wright, M.D., Head of the Medical Research Division of St. Luke's Hospital, Cleveland, Ohio; W. Clark Cooper, M.D., Professor in Residence, University of California at Berkeley; and Duncan A. Holaday, Research Professor, Mount Sinai School of Medicine, City University of New York. As examples of the responses, Dr. Key anticipated some increase in significant proportions. With regard to potential carcinogenic effects, he said data was insufficient to make a prediction but that safety required proceeding on the assumption that exposure should be minimized.

[33][Section 6(B)(F)] requires that the standards where appropriate shall provide for monitoring or measuring employee exposure . . . as may be necessary for the protection of employees.

tant because the results of that process often determine when and what protective measures are required. Petitioners argue that the reasonable foreseeability qualification vests control of this key provision in the discretion of the employer.

*   *   *

They suggest that an employer may evade the controls imposed by the standards simply by deciding that no violations can reasonably be foreseen at his plant, thereby exempting his business from the requirement that monitoring be conducted once every six months. Although it is true that the standards require the employer to exercise some judgment concerning the likelihood of a violation, we do not give the foreseeability exception the broad construction feared by petitioners.

*   *   *

Thus we consider the qualification based on reasonable foreseeability to be a narrow one applicable only when considerable information regarding the workplace is available. Even then occasional monitoring would be required in order to insure that no changes had occurred, but the six months schedule would not apply.

Some jobs pose particularly difficult monitoring problems, and petitioners argue that monitoring once every six months is inadequate to control such situations effectively. That may be true, but the problem should be adequately handled by the general requirement that monitoring be tailored to reflect exposure levels accurately. Where excessive exposure may reasonably be foreseen, monitoring is required *at least* once every six months and more frequently where appropriate.

Petitioners seem to fear that, because of the imprecise language of this requirement, the six months maximum interval between monitoring samples will in practice become the minimum as well, and they consider that inappropriate. The imprecision is perhaps unavoidable, however, in view of the multitude of diverse industrial situations involved. The most effective manner in which to deal with problems of this sort would appear to be to invoke the procedures for detecting and correcting violations in particular workplaces.

*   *   *

3. Labels and Warnings. In order to insure that employees handle materials likely to produce asbestos dust carefully and to prevent persons from entering areas where they will be exposed to such dust needlessly, cautionary labels and warning signs are required by the Secretary's standards. Petitioners assert that the language selected by the Secretary is not strong enough for these purposes, and they have suggested alternatives that they consider better suited to the task.[43]

Although the language favored by petitioners was recommended by NIOSH and was included in the standards as originally proposed, we cannot say it is required by OSHA. After examination of the standards promulgated in light of the terms of the statute, we conclude that they are within the range of the Secretary's discretion.

4. Medical Examinations. The standards require that all employers provide a medical examination (1) when an employee is first assigned to a job involving exposure to asbestos, (2) when he leaves that job, and (3) annually during his employment in the position. Petitioners do not attack this timetable or the substantive requirements governing the nature of the examination, but they do assert that the examinations should be given by a physician of the employee's choice, and that the results should be given to the employer only if the employee decides to do so.[44]

Petitioners argue that the standards violate the principle of physician/patient confidentiality because they would allow the physical examinations to be conducted by company doctors and would make the

[43]The statutory mandate regarding labeling is as follows:
  Any standard promulgated under this subsection shall prescribe the use of labels or other appropriate forms of warning as are necessary to insure that employees are apprised of all hazards to which they are exposed....
  The language favored by intervenors would use the terms "Danger" and "warning" and would make specific mention of particular health hazards such as cancer and asbestosis.

[44]Petitioners also argue that the examination should be available to retired as well as to active employees. The statute requires that where appropriate the standards "prescribe the type and frequency of medical examinations or other tests which shall be made available, by the employer or at his cost, to employees exposed to hazards in order to most effectively determine whether the health of such employees is adversely affected by such exposure." 29 U.S.C. Sec. 655(b)(7). By its terms the protection of this provision extends only to employees and would terminate with that status. Although that language might conceivably be read to include former employees if such a construction were necessary to effectuate the purposes of the Act, we are not persuaded that the measures adopted by the Secretary have been shown to be inadequate by reference to the language of the statute.

results available to the employer. Confidentiality is necessary, they argue, to avert the possibility that, in hiring and discharging employees, employers will discriminate against those with symptoms of asbestos-related disease or prior histories of exposure to asbestos dust. The Secretary recognized this potential problem, and stated that uses of the records would be scrutinized carefully. However, he did not consider the possibility of such abuse sufficient to outweigh the opposing consideration.

The standards require the employer to take into consideration the result of an employee's most recent physical examination in making assignments to jobs requiring the use of respirators. An employee who cannot safely perform such a job is to be reassigned without loss of seniority or wages. The Secretary reasoned that the salutary purposes of this provision could not be fulfilled if employers were denied access to the medical records.[45]

Since the results are to be made available to the employer, allowing the employer to select the physician has the advantages of both convenience and efficiency. Some employers already maintain an industrial medical staff skilled in dealing with asbestos-related medical problems, and the requirements of the standards as promulgated appear likely to encourage the development of additional staffs of this sort. The employee would then benefit from the expertise associated with specialization, and the state of medical knowledge concerning these diseases should likewise be improved. All of these factors, identified in the record, operate to make the Secretary's decision on this point a non-arbitrary one.

5. The Recordkeeping Requirements. Many of the problems the Secretary faced in establishing standards regarding asbestos dust were directly attributable to the lack of reliable information concerning asbestos-related diseases. The Act attempts to correct this deficiency by requiring that:

> Each employer shall make, keep and preserve, and make available to the Secretary or the Secretary of Health, Education, and Welfare, such records regarding his activities relating to this chapter as the Secretary . . . may prescribe by regulation as necessary or appropriate for the enforcement of this chapter or for developing information regarding the cases and prevention

[45]The ultimate choice remains in the hands of the employees in any event since the examinations provided by the employer are entirely voluntary insofar as employees are concerned.

of occupational accidents and illnesses. 29 U.S.C. Sec. 657 (c) (1).

Under the standards promulgated by the Secretary to implement this provision of the statute, employers must retain for twenty years records of each employee's required medical examinations. The standards also require that employers keep records of personal and environmental monitoring, but these records need only be kept for the most recent three years.

Petitioners challenge both requirements as inadequate to protect the health of employees and to advance the present state of knowledge in this area. They argue that, in view of the long latency period associated with many asbestos-related diseases, all records, beginning with initial exposure and continuing throughout the life of each employee, should be retained. Since the recordkeeping requirements for medical examinations and monitoring differ not only in retention period but also in purpose, we treat each separately and reach different conclusions with regard to the two standards.

a. Medical Records. Records of an employee's prior physical examinations may assist a physician in the detection, diagnosis and treatment of that employee's illnesses. Although completeness would be desirable, it may be that the most critical medical records are the most recent ones, perhaps reflecting the examinations of the past two to five years. Since Congress expressed the desire to avoid unnecessary recordkeeping and to impose the "minimum burden upon employers," 29 U.S.C. Sec. 657(d), the Secretary's decision to require retention of only records from the most recent twenty years was permitted him under the Act.

Records of past medical examinations may also be useful in research; for example, they may facilitate the plotting of the longitudinal course of disease or the evaluation of the long-term effects of various levels of exposure to asbestos fibers. These purposes might require a longer retention period than would often be critical for treatment of particular patients; however, even for research, records of some effects of exposure are detectable in the examinations. Thus, even though that often will not occur until several years after initial exposure, the Secretary could reasonably conclude that the records of twenty years would ordinarily encompass the period during which disease could be detected and studied.

Further, we note that twenty years simply marks the minimum time which records must be maintained by employers. If retention is required for a longer period of time for purposes of research, the research organization can obtain copies of the records from the employer and keep them. The standards require an employer to supply these records to the Government upon request, and the twenty year period appears reasonably adequate for these purposes. This arrangement seems most practical since the records of every employee may not be needed for research. Ordinarily, sampling techniques would be employed to select limited numbers of representative records. After research groups have had a reasonable opportunity to obtain the records, and once the other records have been digested in statistical summaries, the remaining records arguably are of little utility.

b. Records of Exposure Levels Detected by Monitoring. The three year retention period for the monitoring records seems surprisingly short in comparison to the twenty years requirement for medical records—especially in view of their respective functions. Whereas the medical records of primary importance may be those beginning with the first manifestations of a disorder, a complete record of an employee's history of exposure to asbestos prior to the development of disease would be important to research. At this point in time, relatively little is known about the causal relationship between exposure to asbestos dust and various diseases. Persons manifesting disorders are being studied, but that research is hindered by the lack of information concerning the exposure levels at industrial workplaces in the past.

As the Secretary observed:

> [W]e have now evidence of the consequences of exposure, but we do not have, in general, accurate measures of the levels of exposure occurring 20 or 30 years ago, which have given rise to these consequences.

For this reason, in the proceedings before the Advisory Committee two members of the committee suggested that a twenty year period for exposure records and a five year requirement for medical records would be appropriate.

The Secretary did not explain his decision to require only a three year retention period for exposure records, but the Government has suggested several justifications in its brief. Passing over the problem of considering a rationale advanced by counsel rather than by the Secretary, we have examined those arguments and conclude that they do not adequately clarify the Secretary's action.

The Government notes that the Act specifies that citations for violations must be issued within six months of the occurrence of the violation. Thus it would appear that a period of three years is more than adequate for enforcement purposes. This argument is accurate, but it does not speak to the Secretary's obligation under OSHA to require retention of those records necessary to develop information concerning the causes of disease.

The second justification offered is that it is the responsibility of NIOSH, the Secretary, and HEW to collect the exposure data necessary to re-evaluate the standards. 29 U.S.C. Secs. 671, 673. Employers must supply records on request,[52] and the Government asserts that three years is an adequate period of time to allow for the agencies to make such requests.

A single permanent storage center for all these records might well be the best solution. Since employees may change jobs and employers may cease operations, continuity may be best achieved if the agency collects and retains the records. If such a program is in fact implemented, the three year period may be acceptable, but there is no indication that current agency procedure includes compilation of such records. Further, the following footnote from the Government's brief raises some question concerning the interpretation of the purposes of recordkeeping:

> These records are not relevant concerning individual records of exposure. Monitoring of employee exposure is not continuously performed, therefore, worklife exposure is never truly obtainable. Unless specific employees are linked with exposure data these records are not relevant for purposes of determining cumulative employee exposure. Because these records are not relevant to total employee exposure there is no need to transfer monitoring records if a company goes out of business.

As noted above, data concerning prior exposure is considered very important in determining the causal relationship between exposure to asbestos dust and disease. To the extent that the note quoted above suggests that collection of such data is unnecessary, it reflects a misinterpretation of the Act. It may be that an adequate basis for research may be

---

[52]Records shall be maintained for a period of at least 3 years and shall be made available upon request to the Assistant Secretary of Labor for Occupational Safety and Health, the Director of the National Institute for Occupational Safety and Health, and to authorized representatives of either. 29 C.F.R. 1910. 93a(i)(1).

established on the basis of exposure levels generally prevalent in the industry without linking each employee to the particular samples taken from his workplace, but the purposes of the recordkeeping requirements cannot be fulfilled without providing some adequate means of relating health records to exposure levels.

The Secretary has provided no explanation of the relatively short retention period for monitoring results, and we find no adequate assurance in the record that the requirement as promulgated will provide the data needed for research into the causes and prevention of asbestos related disease. Consequently, we remand the recordkeeping requirements to the Secretary for such modification or clarification as may be necessary to insure that the statutory objectives will be fulfilled.

Except for the remand we order for reexamination of (1) the uniform application of the 1976 effective date for the two fiber standard and (2) the three year retention period for monitoring records, the petition for review is denied. All of the challenged features of the standards appear to partake of an essentially legislative type of decision making by the Secretary in the performance of the broad delegation made to him by Congress. Had any one of these decisions been made in the first instance by Congress itself and embodied in the statute, its vulnerability to judicial scrutiny would have been dubious indeed. In this context, therefore, judicial review inevitably runs the risk of becoming arbitrary supervision and revision of the Secretary's efforts to effectuate the legislative purposes in an area where variant responses might each be legitimate in the sight of Congress.

What, in our view, differentiates the two provisions we have remanded from those we have left untouched is that the record, examined closely in relation to the relevant concerns of the Act, leaves nagging questions —even for the inexpert observer—as to the reason and rationale for the Secretary's particular choices. However the statutory standard for our review may be characterized, we consider that our dispositions fall within it.

◇ NOTES

1. In this case, the court allows the Secretary of Labor to set "a relatively low limit" for worker exposure to asbestos, even though "reliable data [are] not currently available" for precisely predictable health effects at various levels of exposure. Could the court just as easily have reasoned that the Secretary did not meet his burden of proof? With what consequence?

2. The court describes the factual determinations in the case to be "on the frontiers of scientific knowledge." What guidance does this provide for the resolution of evidentiary conflicts in future cases involving the setting of health standards?

3. The court devotes much discussion to the boundaries and definition of feasibility. How much, if any, of the court's interpretation at this point in time could be considered *dicta* rather than a part of the court's holding in the case?

4. A subsequent case, *SOCMA v. Brennan,* 503 F.2d 1155 (3d Cir. 1974), concerned an OSHA standard setting exposure levels for 14 carcinogens. For some of these substances, the only evidence of carcinogenicity had come from tests performed on animals. The Third Circuit Court of Appeals upheld the standard. Quoting first from OSHA's written explanation of the standard, the court concluded that the Secretary's reliance on solid animal evidence was permissible under the Act as a "legal" determination:

> "We think it improper to afford less protection to workers when exposed to substances found to be carcinogenic only in experimental animals. Once the carcinogenicity of a substance has been demonstrated in animal experiments, the practical regulatory

alternatives are to consider them either noncarcinogenic or carcinogenic to humans, until evidence to the contrary is produced. The first alternative would logically require, not relaxed controls on exposure, but exclusion from regulation. The other alternative logically leads to the treatment of a substance as if it was known to be carcinogenic in man.

We agree with the director of NIOSH and the report of the Ad Hoc Committee on the Evaluation of Low Levels of Environmental Chemical Carcinogens to the Surgeon General, U.S. Public Health Service, April 22, 1970 that the second alternative is the responsible and correct one."

It seems to us that what the Secretary has done [here] in extrapolating from animal studies to humans is to make a legal rather than a factual determination. He has said in effect that if carcinogenicity in two animal species is established, as a matter of law Secs. 6(a) and 6(b)(5) require that they be treated as carcinogenic in man.

If the court had not upheld the standard, how might it have justified such a result?

5. NIOSH had consistently recommended that asbestos be regulated more stringently as a carcinogen, instead of as a classical lung toxicant. OSHA ignored these recommendations for many years. Finally, in 1986 OSHA did acknowledge asbestos as a carcinogen and revised the asbestos standard to reflect this. (See *Asbestos Information Association v. OSHA,* discussed later in this chapter.)

In hearing a challenge to OSHA's newly promulgated vinyl chloride standard, the Second Circuit Court of Appeals addressed the issue of sufficiency of scientific evidence. Looking to both human and animal data, the court upheld the Secretary's determination that no detectable exposure would be allowed because no safe level of exposure could be determined for vinyl chloride. The court also discussed the dimensions of feasibility from a technological, rather than economic perspective.

The Society of the Plastics Industry, Inc. v. Occupational Safety and Health Administration

509 F.2d 1301 (2d Cir. 1975)

Mr. Justice CLARK.

*   *   *

II. BACKGROUND

There are three basic components of the vinyl chloride industry. First, there are the manufacturers of vinyl chloride itself. A gas at ambient temperatures and pressure, vinyl chloride monomer (VCM) is primarily synthesized by the oxychlorination of ethylene in a handful of large, outdoor production plants which resemble oil refineries. Shell, Dow, and Goodrich are the leading producers, accounting for some 50% of the 5.2 billion pounds annually available in the United States. Because of the high degree of automation involved in this manufacturing process, only some 1,500 workers are employed in VCM production. VCM plants are open-air facilities, primarily in the South.

Second, there are the manufacturers of polyvinyl chloride (PVC). Virtually all vinyl chloride is polymerized into thermoplastic PVC resin which serves as the basis for a wide variety of useful plastic products. Goodrich is by far the largest single producer, producing some 20% of the country's 5.4 billion pounds annually, though in total there are only 21 companies operating the 37 PVC plants. Historically, PVC production has been a "batch" or noncontinuous operation carried out in relatively small (2,000–6,000 gallon) "reactors" which require frequent cleaning; the trend, however, is towards substantially larger reactors. PVC plants are not open-air facilities and are generally located in colder climates than VCM plants. They employ some 5,000 workers.

Third, and finally, there are the fabricators of products which utilize PVC resins. Innumerable firms throughout the country, employing thousands of workers, compound PVC with plasticizers, heat stabilizers, lubricants, light stabilizers, flame retardants, or impact modifiers to produce an astounding variety of wares, such as pipes and conduits for building and construction, flooring, wire and cable, furniture, phonograph records, and packaging. In fabrication, residual VCM that has been entrapped in the PVC resin escapes during the heating process, and in this way workers in the fabricating industry are also exposed to vinyl chloride.

It is now clear that the workers in all components of the vinyl chloride industry are subjected to a serious health risk from VCM. Although conclusive proof of the carcinogenic and, in turn, fatal character of VCM did not emerge until early in 1974 when the deaths of three workers in Goodrich's PVC plant at Louisville were reported, strong warning signals had appeared long before. As early as 1949, when the vinyl chloride industry had barely reached its tenth anniversary, a study conducted among vinyl chloride workers in the Soviet Union found liver damage in 15 of 48 workers studied, and in 1958 and 1959, Dow Chemical scientists elicited liver irregularities in rats and rabbits at a 100 ppm concentration of VCM. Although Dow recommended a 50 ppm allowable level in 1961, the industry adhered to its previous 500 ppm standard.

We need not outline in detail the morbid "Vinyl Chloride Chronology," published by an industry spokesman, the Manufacturing Chemists Association (MCA), in a 1974 press release in order to illustrate the mounting evidence of VCM's carcinogenicity. Indeed, the record shows what can only be described as a course of continued procrastination on the part of the industry to protect the lives of its employees. In 1967, when the industry had not reached its thirtieth anniversary, upon receiving recurring reports of the softening of the finger tips and bone of VCM/PVC workers, the Manufacturing Chemists Association had the University of Michigan study the causes of this abnormality. Three years later, in 1970, when it was advised that research could not

pinpoint the cause of the malady *but* recommend a VCM/PVC ceiling of 50 ppm VCM, nothing was done. And in March of the same year, Dr. P.L. Viola of the Regina Elena Institute for Cancer Research in Rome, Italy, published a report that 30,000 ppm VCM exposure for four hours per day, five days per week for a year caused cancer of the skin, lung, and bones of rats, and a few months later at the Tenth International Cancer Congress, described observations of malignant tumors in the ear canals of rats subjected to the same exposure. Apparently relying on Dr. Viola's comment at the close of his abstract that: "no implications to human pathology can be extrapolated from the experimental model reported in the paper," the industry did nothing.

In 1971, MCA began to inquire by letter regarding the conduct of toxicological studies on laboratory animals with VCM and in May of 1971 heard a detailed presentation by Dr. Viola regarding his earlier studies as well as other studies then unreported. The industry began talking about raising funds for epidemiological research on VCM carcinogenicity, but not until March 30, 1972 did 17 U.S. companies agree to finance such a study. By then, the first deaths of U.S. workers due to VCM exposure were being recorded.

Months were consumed in 1972 by negotiations among the participating company representatives, and it was not until February of 1973 that a protocol was agreed upon and a research contract for animal exposure studies signed. Meanwhile, startling results from European experiments were filtering back to the industry. By January of 1973 it was discovered that European experiments with rats had not only found tumors of the ear canal (which Dr. Viola had reported as early as 1970), but also of the kidneys and liver at concentrations as low as 250 ppm VCM, but not at 50 ppm. This discovery, however, was kept confidential, and it was not even revealed to the National Institute of Occupational Safety and Health (NIOSH) until July 17, 1973.

Earlier, on September 27, 1971, a patient of Dr. J.L. Creech, Jr., plant physician of the B.F. Goodrich Chemical Company in Louisville, Kentucky, died. The patient had been employed for 15 years as a helper and operator in Goodrich's Louisville PVC plant. When first hospitalized, a tentative diagnosis was made of a bleeding duodenal ulcer, but upon re-admittance and after an exploratory laparotomy and biopsy was carried out, he was found to have angiosarcoma of the liver, an exceptionally rare and irreversible cancer which strikes only 1 person in some 50,000, no more than 20-30 persons a year. Eighteen months later, on March 3, 1973, another former employee of the Louisville plant died, and a third died on December 19, 1973, again of angiosarcoma of the liver. Recognizing the rarity of the tumor and learning that all three had worked in the Goodrich PVC plant, Dr. Creech brought the matter to the attention of Goodrich, and then on January 22, 1974, to the attention of the National Institute of Occupational Safety and Health (NIOSH).

News of other deaths followed swiftly. On January 29, 1974, Goodrich reported the death of a fourth former employee from angiosarcoma; a report of the death of a fifth employee followed on February 15th. Six days later, Union Carbide advised NIOSH of the death of one of its PVC workers from liver angiosarcoma. Goodyear Tire and Rubber Company announced a vinyl chloride worker fatality from liver angiosarcoma on March 1, 1974, and reported two more such deaths from the same cause on March 22nd. Goodrich reported cases of liver angiosarcoma in two of its living employees. On April 16, 1974, Firestone Plastics announced the death of one of its employees from the same disease. Finally, on May 10, 1974, the National Cancer Institute diagnosed another Union Carbide VCM worker as a victim of the same disease. In all, the deaths of 13 workers in the PVC and fabricating industries were reported.

III. THE SECRETARY'S ACTIONS

Two days after Goodrich made its report to NIOSH of its first three VCM worker deaths, an inspection of its plant by NIOSH indicated considerable exposure of workers to VCM, and control procedures as well as precautionary monitoring were recommended. NIOSH alerted other federal agencies and after additional investigation soon concluded that VCM was the suspect agent for a new occupational cancer. On April 5, 1974, the Assistant Secretary of Labor, acting on behalf of the Secretary, held a hearing and promulgated an emergency temporary standard of 50 ppm TWA[1] in lieu of the prevailing 500 ppm one. At the hearing, industry itself showed concern, and evidence was presented that both VCM and PVC plants could reduce concentrations below 50 ppm TWA. As in the past, Dow Chemical spoke out for worker safety and urged that industry exposure be reduced to 50 ppm TWA by operational and engi-

neering changes and that appropriate respiratory protection be given where such level was not attainable. The Assistant Secretary concluded that the evidence demonstrated VCM to be carcinogenic for man. Monitoring and housekeeping requirements were also imposed, and a requirement was included that, if the 50 ppm level was breached, appropriate respirators equipment be furnished workers. The emergency order was limited to six months "during which time the whole question of possible exposure of humans to VC would be reconsidered more fully and in the light of more information, including experiments which are underway at the time."

On April 9, 1974, the Industrial Bio-Test Laboratory notified MCA that its preliminary findings in their animal exposure study showed that angiosarcoma of the liver was produced in mice at a level of 50 ppm of vinyl chloride, and MCA so informed the federal government. On May 10, 1974, the Assistant Secretary again acted. Issuing a notice of proposed permanent rulemaking, he drew attention to the MCA study results and noted that: "the question of a safe level of exposure for humans cannot be determined at this time, and may continue as a matter for scientific deliberation for many years." In the interim, he concluded, it was necessary to abandon the 50 ppm emergency temporary standard and to establish it at as low a level as can be detected using methodologies sensitive to 1 ppm plus-or-minus 50%, i.e., the so-called "no-detectable" level. This level was to be reached through the "institution of engineering controls and work practices as soon as feasible," but respiratory protection was to be afforded where the goal could not otherwise be achieved.

On May 24, 1974, notice of a hearing on the proposed standard to be held before an Administrative Law Judge was given. Eight days of hearings were held during June and July of 1974, and additional materials were received from interested parties until September 25th. In addition, the record of the previous hearing on February 25, 1974, was ordered to be included.

On October 1, 1974, the final standard was promulgated, effective January 1, 1975.[2] The main provisions are as follows:

(1) The standard applies to manufacturers of VCM and PVC and to fabricators of PVC, but excludes those merely handling or using already-fabricated products.

---

[1]Time Weighted Average (TWA) represents a worker's cumulative exposure to a toxic substance during a 9-hour shift.

[2]Petitioners' application for stay of the January 1st deadline was granted pending our disposition of the matter.

(2) In place of the potentially obscure "no detectable level" standard, a permissible exposure limit not greater than 1 ppm averaged over an eight-hour period is set, but allows for peaks of VCM exposure up to 5 ppm during periods not exceeding 15 minutes.

(3) All employers are required to conduct an initial program of monitoring and measurement of exposure levels, but need not continue monitoring if initial levels are below the so-called "action level" of 0.5 ppm averaged over the eight-hour work day.

(4) The standard requires that "feasible engineering and work practice controls" be employed to reduce exposure below the permissible exposure where ever possible or to the lowest practicable level if not possible, supplemented by respiratory protection.

(5) Where respiratory protection is required under the standard, suitable equipment, as indicated by a chart in the standard, shall be provided, and use shall be required of employees unless the VCM level is below 25 ppm measured over any 15-minute period, in which case the use of respiratory equipment is optional with the employee until January 1, 1976.

(6) Employees working in certain hazardous operations, especially those involved in physically cleaning the interiors of PVC manufacturing reactors, shall be provided protective garments and respiratory equipment.

(7) A medical surveillance program is required for all employees exposed to VCM in excess of the action level of 0.5 ppm.

(8) Finally, in addition to certain record-keeping requirements, all entrances, work-areas, and containers related to VCM or PVC manufacture and PVC fabrication are required to be labelled with the warning legend: "Cancer-Suspect Agent." 29 C.F.R. Sec. 1910.93q(1).

IV. PETITIONERS' CONTENTIONS

(a)

Petitioner's initial claim is that the available scientific and medical evidence does not establish that the 1 ppm exposure level adopted by the Secretary is required by health or safety considerations. They claim that no proof exists in the record to justify such a low standard, since all of the medical witnesses testified that no one can say whether exposure to VCM at low levels was safe or unsafe. Further, they point to studies of employee health by Dow Chemical over a number of years which concluded that exposure to VCM below 200 ppm did not lead to any adverse effect.

We find, however, that the evidence is quite sufficient to warrant the Secretary's choice. First, it must be remembered that we are dealing here with human lives, and the record reveals that 11 manufacturing plant workers have already died from the effects of this potent chemical. Moreover, the animal exposure study, sponsored by MCA, the industry's own trade association, identified fatal liver angiosarcoma and other kidney and liver diseases at the 50 ppm level. None of the physicians or scientists who testified could identify a safe level of exposure to VCM, nor the precise mechanism by which it produces cancer; yet expert after expert recommended that this "very virulent" carcinogen be restricted to the lowest detectable level. Indeed, as one witness, Dr. Kraybill of the National Cancer Institute, testified:

> Certainly, there is little margin for safety if a response can be expected at a level below 50 parts per million which in truth is now only a fiftyfold safety factor in terms of the proposed standard. According to toxicological principles, were this compound a noncarcinogen, then to establish a tolerance or safe level, there would have to be a 100 to 1 margin of safety in terms of a no-effect level, and from the experimental data on the animals we don't even know what the no-effect level is. Obviously, this would put the allowable level at a small fraction of a given standard—of the given standard.

As in *Industrial Union Department, AFL-CIO v. Hodgson*, [499 F.2d 467 (D.C. Civ. 1974)], the ultimate facts here in dispute are "on the frontiers of scientific knowledge" and, though the factual finger points, it does not conclude. Under the command of OSHA, it remains the duty of the Secretary to act to protect the working-man, and to act even in circumstances where existing methodology or research is deficient. The Secretary, in extrapolating the MCA study's finding from mouse to man, has chosen to reduce the permissible level to the lowest detectable one. We find no error in this respect.

(b)

Failing in this contention, petitioners strongly urge that the Secretary breached his statutory mandate to insure that the standard selected is a "feasible" one.[3]

[3]29 U.S.C. Sec. 655(b)(5) provides in part: The Secretary ... shall set the standard which most adequately assures, to the extent feasible, on the basis of the best available evidence, that no employee will suffer material impairment of health or functional capacity. ...

Relying on the so-called Snell Report,[4] petitioners claim that VCM and PVC manufacturers will never be able to reduce levels of exposure to 1 ppm through engineering means. They point to the conclusion reached by the Snell Report that:

> The costs of compliance increase rapidly with decreasing VCM target levels and represent significant engineering uncertainty or infeasibility beyond 10 ppm ceiling and 2-5 ppm TWA for the VCM industry and 15-25 ppm ceiling and 10-15 ppm TWA for the PVC industry.

According to the report, "[b]ased on the industry surveys and Snell's independent assessments of the state-of-the-art of the technology" the standard price of VCM would only rise from 7.41¢/lb. at present to 7.69¢/lb at a target level of 2-5 ppm TWA, but would supposedly soar to 12.71¢/lb. at the "no-detectable" level.

In his statement of reasons in support of the standard, the Assistant Secretary acknowledged the industry contention and the Snell conclusion about the infeasibility of the 1 ppm level, but noted that: "Labor union spokesmen and the Health Research Group, Inc., however, have suggested that such a level is attainable." The Assistant Secretary went on to say:

> Since there is no actual evidence that any of the VC or PVC manufacturers have already attained a 1 ppm level or in fact instituted all available engineering and work practice controls, any estimate as to the lowest feasible level attainable must necessarily involve subjective judgment. Likewise, the projections of industry, labor, and others concerning feasibility are essentially conjectural. Indeed, as Firestone has suggested, it is not possible to accurately predict the degree of improvement to be obtained from engineering changes until such changes are actually implemented.

> We agree that the PVC and VC establishments will not be able to attain a 1 ppm TWA level for all job classifications in the near future. We do believe, however, that they will, in time, be able to

[4]This report, "Economic Impact Studies of the Effects of Proposed OSHA Standards for Vinyl Chloride," was prepared in September of 1974 by Foster D. Snell, Inc., an independent consultant, at the request of the Secretary of Labor and estimates the cost to the industry of complying with various exposure levels between 50 ppm and 0-1 ppm. Plant and industry visits were the principal means of information gathering.

attain levels of 1 ppm TWA for most job classifications most of the time. It is apparent that reaching such levels may require some new technology and work practices. It may also be necessary to utilize technology presently used in other industries. In any event the VC and PVC industries have already made great strides in reducing exposure levels. (See testimony of Dow Chemical Co.) For example, B.F. Goodrich testified that it has reduced average exposure levels in several PVC plants from 35-40 ppm early this year to 12-13 ppm at the time of the hearing. We are confident that industry will continue to do so.

We cannot agree with petitioners that the standard is so clearly impossible of attainment. It appears that they simply need more faith in their own technological potentialities, since the record reveals that, despite similar predictions of impossibility regarding the emergency 50 ppm standard, vast improvements were made in a matter of weeks, and a variety of useful engineering and work practice controls have yet to be instituted. In the area of safety, we wish to emphasize, the Secretary is not restricted by the status quo. He may raise standards which require improvements in existing technologies or which require the development of new technology, and he is not limited to issuing standards based solely on devices already fully developed. *Cf. Chrysler Corp. v. Dept. of Transportation*, 472 F.2d 659, 673 (6th Cir. 1972); *Natural Resources Defense Council, Inc. v. E.P.A.*, 489 F.2d 390, 401 (5th Cir. 1974).

There is much testimony in the record, especially in the Snell Report, indicating that VCM concentration can be easily pinpointed and largely corrected. For example, many of the companies engaged in PVC manufacture still perform the cleaning of batch reactors by opening the vessel and having the worker physically enter it. In chipping off the accretion from the walls of the vessel, the worker is thus exposed to a high concentration of VCM. Yet other, less hazardous methods are currently available and in use, in which the vessels are cleaned by machinery, emulsions or simply water under high pressure. Other sources of exposure are encountered in filling tank cars, measuring, testing and repairing pipe joints or other connections. The Snell Report indicates that much of this may be alleviated.

But whether it can or not, the Secretary's compliance scheme does not rest only on engineering and work practice controls. He does mandate that the industry use such technology to the extent feasible, but, more importantly, he requires that, in addition, respiratory protection be used if engineering means cannot bring the VCM level down to the permissible limit.

To be sure, respirators have their drawbacks. These problems were detailed at the hearings and recognized by the Assistant Secretary. Self-contained and air-hose type breathing equipment is bulky, expensive, and infeasible for full-time use, as well as potentially hazardous in terms of tripping, restricted mobility, and over-exhaustion of workers. But the fact remains that they effectively eliminate exposure to VCM, and they are already being used by some PVC companies in the cleaning process and at other points in production with good success.

Like the industry's claims about the impossibility of achieving compliance through technological means, petitioners' claims of dire consequences from the requirement of respiratory protection are exaggerated. It does not appear that full-time use of respirators is necessary, and the Snell Report points this out. Furthermore, lightweight, inexpensive cartridge or canister-type respirators, which can effectively filter out VCM at low levels, are now available and acceptable. Contrary to petitioners' assertions, the Snell Report indicates that a variety of respirators are reasonably available.

(c)

We find that the Secretary's directions are clear, definite, and certain and that they are also entirely feasible, since the goal of the lowest detectable level can definitely be attained through the combination of technological means and respirators. Our conclusion in this regard is buttressed by the fact that only some 6,500 of the workers coming under the standard —those in the VCM and PVC manufacturing field— are potentially exposed to high or constant VCM concentrations, and only a small percentage of them are actually subject to excessive concentrations of the chemical during manufacture. The remaining hundreds of thousands of workers are in the fabrication field where exposure is already so low that some of the petitioners claim that the fabricators should not even be under the standard; these, of course, have been included because of their work

with PVC resin which in final form contains residual amounts of VCM, and the record shows that two deaths have in fact occurred among fabricators.

If, in the future, the monitoring under the standard indicates that the VCM level among fabricators is sufficiently low and that the sporadic, unpredictable exposure of these workers to residual VCM has been controlled, the fabricators might well be excluded or included in a separate, less rigorous standard. Certainly, if PVC producers were able to eliminate all residual VCM from the resin before it is delivered to the fabricator, there would seem to be no reason for their inclusion. But this would be for the Secretary to decide on appropriate application or on his own motion, and, as the record presently stands, their inclusion is amply justified.

In any event, compliance is time-phased and if the petitioners find that they cannot comply for reasons beyond their control, OSHA permits the amendment of standards. Upon application and sufficient proof of such a situation, we feel certain that the petitioners would obtain relief at the hands of the Secretary. This is especially true of the "cancer suspect agent" labelling requirements which petitioners urge are beyond statutory contemplation. We think that the Secretary is simply "fighting fire with fire" and using the labels to bring the danger of vinyl chloride forcibly to the attention of the workers. *Cf. Synthetic Organic Chemical Mfrs. Assn. v. Brennan*, 503 F.2d 1155 (3d Cir. 1974). They deserve no less treatment. We have also considered the other arguments of the petitioners and find them untenable.

V

It is our conclusion that the challenged aspects of the Secretary's vinyl chloride standard are supported by substantial evidence in the record and that the petitions for review must be denied. Taking into account the delay occasioned by these petitions, we think that a reasonable "lead time" is appropriate and therefore order that the Secretary's regulations, rather than being effective January 1, 1975, shall become effective sixty days after the date of this order and that the time requirement as to respiratory protection contained in 29 C.F.R. Sec. 1910.93q(g)(1) is rescheduled accordingly.

◇ NOTES

1. Evidence of a rare form of liver cancer in vinyl chloride-exposed workers prompted OSHA to set an emergency temporary standard (ETS) of 50 ppm while it moved to promulgate a new, more protective permanent standard. The ACGIH TLV of 500 ppm, adopted in 1971 along with 400 other TLVs as permissible exposure limits (PELs) under Section 6(a) of the OSHAct, was based on classical toxicity data, not carcinogenic effects.

2. The court made short shrift of the negative Dow study reporting no excess cancer in workers regularly exposed to 200 ppm of vinyl chloride. How should OSHA balance negative and positive evidence of carcinogenicity? Would the court have been at liberty to give more weight to the Dow study?

3. The existence of both human and animal data indicating carcinogenicity of vinyl chloride was used by OSHA, along with the assertion that no safe level to a carcinogen is known, to justify a standard requiring a "no detectable" level of 1 ppm. No risk assessment for exposures at 1 ppm (or higher) was performed. Did OSHA discharge its burden to determine a "safe level"? Compare the Supreme Court's benzene decision (*Industrial Union Department v. American Petroleum Institute* on page 123).

4. In requiring that "the factual finger" point, but not requiring that it necessarily conclude, the court appears to be adopting the reasoning of the D.C. Circuit in the asbestos case. Is the court saying that OSHA's decision is one of fact, but with the evidentiary requirements reduced, or that the agency's decision is one of law? Compare *SOCMA v. Brennan*, discussed in Note 4 on page 113.

5. The proposed standard included not only a PEL of 1 ppm over 8 hours, but also a short term exposure limit (STEL) of 5 ppm over 15 minutes and an "action level" of 0.5 ppm. What are the purposes of these additional requirements?

6. OSHA requires compliance with the vinyl chloride standard first through engineering controls and then by the use of personal protective equipment (i.e., respirators) if necessary. This is true of the asbestos standard as well. Is this approach statutorily justified? It would be cheaper for the employer to be able to comply *only* through the use of respirators. What is the rationale for requiring both?

7. The court agrees with OSHA's position that a standard is "feasible" even if it would require the development of new technology (technological innovation) or the use of technology borrowed from other industries (technological diffusion). This is the origin of OSHA's "technology forcing" authority. Is there statutory justification for this authority? Is the court endorsement of the authority a holding of the case or *dicta*? Why?

8. The industry argued that the increased production cost necessitated by the 1 ppm standard made compliance impossible to achieve. Is the industry claiming technological or economic infeasibility? What is the relationship between technological and economic feasibility?

2. The Limits of Technology Forcing

OSHA promulgated an exposure standard for coke oven emissions created in the production of steel, because workers at the coke ovens were found to have a significantly higher incidence of leukemia. In reviewing this standard, the Third Circuit Court of Appeals also endorsed the technology forcing authority given OSHA under the OSHAct, but limited the burden that OSHA could place on each employer.

American Iron and Steel Institute v. OSHA

577 F.2d 825 (3d Cir. 1978)

ROSENN, Circuit Judge.

* * *

B. REQUIRED RESEARCH AND DEVELOPMENT

The Secretary's standard requires that if, after implementation of all the required controls, the permissible exposure limit has not been met by January 20, 1980,

> employers shall research, develop and implement any other engineering and work practice controls necessary to reduce exposure to or below the permissible exposure limit.

41 Fed. Reg. 46785.

Petitioners attack this requirement contending that requiring employers to engage in unlimited research and development is not authorized by the Act. Although the Secretary may "raise standards which require improvement in existing technologies or which require the development of new technology," *Society of Plastics Industry, Inc. v. OSHA* 509 F.2d 1301, 1309 (2d Cir. 1975), petitioners note that the Secretary's technology-forcing power is limited to technology that "looms on today's horizon." *American Federation of Labor v. Brennan*, 530 F.2d 109, 131 (3d Cir. 1975). The petitioners also contend that the

requirement is fatally vague, providing no indication of the limit or magnitude of the employer's obligation. The Government maintains that the requirement is valid "technology-forcing" and that it is not fatally vague.

29 U.S.C. §665(b)(5) grants authority to the Secretary to develop and promulgate standards dealing with toxic materials or harmful agents "based upon research, demonstrations, experiments, and such other information as may be appropriate." Under the same statutory provision the Secretary is directed to consider the latest scientific data in the field. As we have construed the statute, the Secretary can impose a standard which requires an employer to implement technology "looming on today's horizon," and is not limited to issuing a standard solely based upon technology that is fully developed today. Nevertheless, the statute does not permit the Secretary to place an affirmative duty on each employer to research and develop new technology. Moreover, the speculative nature of the research and development provisions renders any assessment of feasibility practically impossible. In holding that the Secretary lacks statutory authorization to promulgate the research and development provision, we note in passing that we need not reach petitioners' challenge to the provision as fatally vague. Accordingly, we hold the research and development provision of the standard to be invalid and unenforceable.

◇ NOTES

1. Note that OSHA's technology forcing authority, which could have been interpreted as *dicta* in the vinyl chloride case, becomes a holding, and hence law, in this case. Industry did not question OSHA's authority to set standards that required new technology. Industry merely questioned OSHA's authority to require each and every employer to undertake an affirmative research and development (R and D) effort to develop that technology.

2. By using the words "technology that looms on today's horizon," is the court placing stricter limits on OSHA's technology forcing authority than did the Second Circuit Court of Appeals in the vinyl chloride case?

3. Does the Secretary's determination of feasibility have to be supported by "substantial evidence on the record as a whole"?

4. The OSHA coke oven emissions standard was promulgated after receiving the report of an appointed coke oven advisory committee authorized under Section 6(b)(1) of the OSHAct. What are the merits of using an advisory committee on standards?

5. In this case (in a portion of the opinion not reprinted above), the court commented on the standard of judicial review as follows:

> As we examine the regulation at issue in this case, it is imperative to distinguish between determinations bottomed on factual matters, and nonfactual, legislative-like policy decisions. It is only the former that we subject to the "substantial evidence" test. The evidence in support of a fact-finding is "substantial" when "from it [the evidence] an inference of the fact may be drawn reasonably." B. Schwartz, *Administrative Law* 595 (1977). In such a case, the reviewing court must uphold the finding "even though [it] would justifiably have made a different choice had the matter been before it de novo."

Note that the court subjects only factual matters to the substantial evidence test. How does this reasoning differ from the reasoning of the D.C. Circuit and the Second Circuit in the asbestos and vinyl chloride cases? Can the two approaches be reconciled? Compare the earlier discussion of *SOCMA v. Brennan* in Note 4 on page 113.

6. The court here concludes that OSHA's determination that there is no absolutely safe level of exposure to a carcinogen is supported by substantial evidence on the record. Is the determination a legal one or a factual one? Are the conclusion and reasoning consistent with the excerpt quoted in Note 5 above, which articulates the court's approach to judicial review?

3. The Significant Risk Requirement

Encouraged by the favorable reception that its standard setting had received in the courts of appeal, OSHA issued its proposed generic cancer policy on October 4, 1977. Although the generic cancer policy was not technically in effect at the time OSHA issued a final rule for benzene, it was clear that the approach set forth in the generic standard had been followed in setting the standard for benzene.

The American Petroleum Institute (API) challenged the validity of the benzene standard in the Fifth Circuit Court of Appeals on the grounds that OSHA had not met its alleged statutory burden to show that the standard was "reasonably necessary or appropriate to provide safe and healthful employment" as required by Section 3(8) of the Act. Even though economic and technology feasibility was not contested, the API argued that a cost-benefit analysis was required before OSHA could issue a standard. The court of appeals agreed, invalidating the standard.

On appeal, the Supreme Court, in its first review of an OSHA standard, declined to reach the question of cost-benefit analysis. Instead it articulated "significant risk" requirement for Section 6(6) standards, and caused the agency to revise its approach to the regulation of carcinogens.

Industrial Union Department v. American Petroleum Institute

448 U.S. 607 (1980)

Mr. Justice STEVENS announced the judgment of the Court and delivered an opinion in which the Chief Justice and Mr. Justice STEWART join and in Parts I, II, III-A-C and E of which Mr. Justice POWELL joins.

This case concerns a standard promulgated by the Secretary of Labor to regulate occupational exposure to benzene, a substance which has been shown to cause cancer at high exposure levels. The principal question is whether such a showing is a sufficient basis for a standard that places the most stringent limitation on exposure to benzene that is technologically and economically possible.

The Act delegates broad authority to the Secretary to promulgate different kinds of standards. The basic definition of an "occupational safety and health standard" is found in §3 (8), which provides:

> The term "occupational safety and health standard," means a standard which requires conditions, or the adoption or use of one or more practices, means, methods, operations, or processes, reasonably necessary or appropriate to provide safe or healthful employment and places of employment.

Where toxic materials or harmful physical agents are concerned, a standard must also comply with §6 (b) (5).[1]

*   *   *

Wherever the toxic material to be regulated is a carcinogen, the Secretary has taken the position that no safe exposure level can be determined and that §6(b) (5) requires him to set an exposure limit at the lowest technologically feasible level that will not impair the viability of the industries regulated. In this case, after having determined that there is a causal connection between benzene and leukemia (a cancer of the white blood cells), the Secretary set an exposure limit on airborne concentrations of benzene of one part benzene per million parts of air (1 ppm), regulated dermal and eye contact with solutions containing benzene, and imposed complex monitoring and medical testing requirements on employers whose workplaces contain 0.5 ppm or more of benzene.

*   *   *

On pre-enforcement review . . . the United States Court of Appeals for the Fifth Circuit held the regulation invalid. *American Petroleum Institute et al. v. OSHA* 581 F.2d 493 (5th Cir. 1978). The court concluded that OSHA had exceeded its standard-setting authority because it had not shown that the new benzene exposure limit was "reasonably necessary or appropriate to provide safe or healthful employment" as required by §3(8), and because §6(b) (5) does "not give OSHA the unbridled discretion to adopt standards designed to create absolutely risk-

---

[1]The second and third sentences of this section, which impose feasibility limits on the Secretary and allow him to take into account the best available evidence in developing standards, may apply to all health and safety standards. This conclusion follows if the term "subsection" used in the second sentence refers to the entire subsection 655(b) (which sets out procedures for the adoption of all types of health and safety standards), rather than simply to the toxic materials subsection, §655(b) (5). While Mr. Justice Marshall . . . and respondents agree with this position, see Brief for Respondents, at 39; see also Currie, OSHA, 1976 *Am. Bar Foundation Research J.* 1107, 1137, n. 151, the Government does not, see Brief for Federal Parties, at 58; see also Berger & Riskin, "Economic and Technological Feasibility in Regulating Toxic Substances under the Occupational Safety and Health Act," 7 *Ecol. L. Q.* 285, 294 (1978). There is no need for us to decide this issue in this case.

free workplaces regardless of costs." Reading the two provisions together, the Fifth Circuit held that the Secretary was under a duty to determine whether the benefits expected from the new standard bore a reasonable relationship to the costs that it imposed. *Id.,* at 503 [6 OSHC at 1965]. The court noted that OSHA had made an estimate of the costs of compliance, but that the record lacked substantial evidence of any discernible benefits.[5]

We agree with the Fifth Circuit's holding that §3(8) requires the Secretary to find, as a threshold matter, that the toxic substance in question poses a significant health risk in the workplace and that a new, lower standard is therefore "reasonably necessary or appropriate to provide safe or healthful employment and places of employment." Unless and until such a finding is made, it is not necessary to address the further question whether the Court of Appeals correctly held that there must be a reasonable correlation between costs and benefits, or whether, as the Government argues, the Secretary is then required by §6(b)(5) to promulgate a standard that goes as far as technologically and economically possible to eliminate the risk.

Because this is an unusually important case of first impression, we have reviewed the record with special care. In this opinion, we (1) describe the benzene standard, (2) analyze the Agency's rationale for imposing a 1 ppm exposure limit, (3) discuss the controlling legal issues, and (4) comment briefly on the dermal contact limitation.

I

Benzene is a familiar and important commodity. It is a colorless, aromatic liquid that evaporates rapidly under ordinary atmospheric conditions. Approximately 11 billion pounds of benzene were pro-

duced in the United States in 1976. Ninety-four percent of that total was produced by the petroleum and petrochemical industries, with the remainder produced by the steel industry as a byproduct of coking operations. Benzene is used in manufacturing a variety of products including motor fuels (which may contain as much as 2% benzene), solvents, detergents, pesticides, and other organic chemicals.

* * *

The entire population of the United States is exposed to small quantities of benzene, ranging from a few parts per billion to 0.5 ppm, in the ambient air. . . . Over one million workers are subject to additional low-level exposures as a consequence of their employment. The majority of these employees work in gasoline service stations, benzene production (petroleum refineries and coking operations), chemical processing, benzene transportation, rubber manufacturing and laboratory operations.

Benzene is a toxic substance. Although it could conceivably cause harm to a person who swallowed or touched it, the principal risk of harm comes from inhalation of benzene vapors. When these vapors are inhaled, the benzene diffuses through the lungs and is quickly absorbed into the blood. Exposure to high concentrations produces an almost immediate effect on the central nervous system. Inhalation of concentrations of 20,000 ppm can be fatal within minutes; exposures in the range of 250 to 500 ppm can cause vertigo, nausea, and other symptoms of mild poisoning. . . . Persistent exposures at levels above 25-40 ppm may lead to blood deficiencies and diseases of the blood-forming organs, including aplastic anemia, which is generally fatal.

* * *

In 1969 the American National Standards Institute adopted a national consensus standard of 10 ppm averaged over an eight-hour period with a ceiling concentration of 25 ppm for 10-minute periods or a maximum peak concentration of 50 ppm. In 1971, after the Occupational Health and Safety Act was passed, the Secretary adopted this consensus standard as the federal standard, pursuant to 29 U.S.C. §655(a).[7]

As early as 1928, some health experts theorized

---

[5]"The lack of substantial evidence of discernable benefits is highlighted when one considers that OSHA is unable to point to any empirical evidence documenting leukemia risk at 10 ppm even though that has been the permissible exposure limit since 1971. OSHA's assertion that benefits from reducing the permissible exposure limit from 10 ppm to 1 ppm are likely to be appreciable, an assumption based only on inferences drawn from studies involving much higher exposure levels rather than on studies involving these levels or sound statistical projections from the high-level studies, does not satisfy the reasonably necessary requirement limiting OSHA's action. *Aqua Slide* requires OSHA to estimate the extent of expected benefits in order to determine whether those benefits bear a reasonable relationship to the standard's demonstrably high costs." [6 OSHC at 1966].

[7]Section 6(a) of the Act, 29 U.S.C. §655(a), provides that Without regard to chapter 5 of Title 5 or to the other subsections of this section, the Secretary shall, as soon as practicable during the period beginning with the effective date of this chapter and ending two years after such date, by rule promulgate as an occupational safety

that there might also be a connection between benzene in the workplace and leukemia.

* * *

Between 1974 and 1976 additional studies were published which tended to confirm the view that benzene can cause leukemia, at least when exposure levels are high. In an August 1976 revision of its earlier commendation, NIOSH stated that these studies provided "conclusive" proof of a causal connection between benzene and leukemia. Although it acknowledged that none of the intervening studies had provided the dose-response data it had found lacking two years earlier. NIOSH nevertheless recommended that the exposure limit be set as low as possible.

* * *

In October 1976 NIOSH sent another memorandum to OSHA, seeking acceleration of the rulemaking process and "strongly" recommending the issuance of an emergency temporary standard for benzene and two other chemicals believed to be carcinogens. NIOSH recommended that a 1 ppm exposure limit be imposed for benzene.[14] Apparently because of the NIOSH recommendation, OSHA asked its consultant to determine the cost of complying with a 1 ppm standard instead of with the "minimum feasible" standard. It also issued voluntary guidelines for benzene, recommending that exposure levels be limited to 1 ppm on an 8-hour time-weighted average basis wherever possible.

or health standard any national consensus standard, and any established Federal standard, unless he determines that the promulgation of such a standard would not result in improved safety or health for specifically designated employees. In the event of conflict among any standards, the Secretary shall promulgate the standard which assures the greatest protection of the safety or health of the affected employees.

[14]At the hearing on the permanent standard NIOSH representatives testified that they had selected 1 ppm initially in connection with the issuance of a proposed standard for vinyl chloride. In that proceeding they had discovered that 1 ppm was approximately the lowest level detectable through the use of relatively unsophisticated monitoring instruments. With respect to benzene, they also thought that 1 ppm was an appropriate standard because any lower standard might require the elimination of the small amounts of benzene (in some places up to 0.5 ppm) that are normally present in the atmosphere. NIOSH's recommendation was *not* based on any evaluation of the feasibility, either technological or economic, of eliminating all exposures above 1 ppm.

* * *

OSHA did issue an emergency standard, effective May 21, 1977, reducing the benzene exposure limit from 10 ppm to 1 ppm, the ceiling for exposures of up to 10 minutes from 25 ppm to 5 ppm, and eliminating the authority for peak concentrations of 50 ppm. In its explanation accompanying the emergency standard, OSHA stated that benzene had been shown to cause leukemia at exposures below 25 ppm and that, in light of its consultant's report, it was feasible to reduce the exposure limit to 1 ppm.

* * *

On May 19, 1977, the Court of Appeals for the Fifth Circuit entered a temporary restraining order preventing the emergency standard from taking effect. Thereafter, OSHA abandoned its efforts to make the emergency standard effective and instead issued a proposal for a permanent standard patterned almost entirely after the aborted emergency standard.

* * *

II

The critical issue at this point in the litigation is whether the Court of Appeals was correct in refusing to enforce the 1 ppm exposure limit on the ground that it was not supported by appropriate findings.

* * *

In the end OSHA's rationale for lowering the permissible exposure limit to 1 ppm was based, not on any finding that leukemia has ever been caused by exposure to 10 ppm of benzene and that it will *not* be caused by exposure to 1 ppm, but rather on a series of assumptions indicating that some leukemias might result from exposure to 10 ppm and that the number of cases might be reduced by reducing the exposure level to 1 ppm. In reaching that result, the Agency first unequivocally concluded that benzene is a human carcinogen.[36] Second, it concluded

[36]"The evidence in the record conclusively establishes that benzene is a human carcinogen. The determination of benzene's leukemogenicity is derived from the evaluation of all the evidence in totality and is not based on any one particular study. OSHA recognizes, as indicated above, that individual reports vary considerably in quality, and that some investigations have significant methodological deficiencies. While recognizing the strengths and weaknesses in individual studies, OSHA nevertheless concludes that the benzene record as a whole clearly establishes a causal relationship between benzene and leukemia."

that industry had failed to prove that there is a safe threshold level of exposure to benzene below which no excess leukemia cases would occur. In reaching this conclusion OSHA rejected industry contentions that certain epidemiological studies indicating no excess risk of leukemia among workers exposed at levels below 10 ppm were sufficient to establish that the threshold level of safe exposure was at or above 10 ppm.[37] It also rejected an industry witness' testimony that a dose-response curve could be constructed on the basis of the reported epidemiological studies and that this curve indicated that reducing the permissible exposure limit from 10 to 1 ppm would prevent at most one leukemia and one other cancer death every six years.[38]

Third, the Agency applied its standard policy with respect to carcinogens,[39] concluding that, in the absence of definitive proof of a safe level, it must be assumed that *any* level above zero presents *some*

increased risk of cancer.[40] As the Government points out in its brief, there are a number of scientists and public health specialists who subscribe to this view, theorizing that a susceptible person may contract cancer from the absorption of even one molecule of a carcinogen like benzene.[41]

Fourth, the Agency reiterated its view of the Act, stating that it was required by §6(b) (5) to set the standard either at the level that has been demon-

[37]In rejecting these studies, OSHA stated that: "Although the epidemiological method can provide strong evidence of a causal relationship between exposure and disease in the case of positive findings, it is by its very nature relatively crude and an insensitive measure." After noting a number of specific ways in which such studies are often defective, the Agency stated that it is ". . . OSHA's policy when evaluating negative studies, to hold them to a higher standard of methodological accuracy." Viewing the studies in this light, OSHA concluded that each of them had sufficient methodological defects to make them unreliable indicators of the safety of low-level exposures to benzene.

[38]OSHA rejected this testimony in part because it believed the exposure data in the epidemiological studies to be inadequate to formulate a dose-response curve. It also indicated that even if the testimony was accepted—indeed as long as there was any increase in the risk of cancer—the agency was under an obligation to "select the level of exposure which is most protective of exposed employees."

[39]In his dissenting opinion, Mr. Justice Marshall states that the Agency did not rely "blindly on some draconian carcinogen 'policy' " in setting a permissible exposure limit for benzene. He points to the large number of witnesses the Agency heard and the voluminous record it compiled as evidence that it relied instead on the particular facts concerning benzene. With all due respect, we disagree with Mr. Justice Marshall's interpretation of the Agency's rationale for its decision. After hearing the evidence, the Agency relied on the same policy view it had stated at the outset, namely that, in the absence of clear evidence to the contrary, it must be assumed that no safe level exists for exposure to a carcinogen. The Agency also reached the entirely predictable conclusion that industry had not carried its concededly impossible burden of proving that a safe level of exposure exists for benzene. As the Agency made clear later in its proposed

generic cancer policy, it felt compelled to allow industry witnesses to go over the same ground in each regulation dealing with a carcinogen, despite its policy view. The generic policy, which has not yet gone into effect, was specifically designed to eliminate this duplication of effort in each case by foreclosing industry from arguing that there is a safe level for the particular carcinogen being regulated.

[40]"As stated above, the positive studies on benzene demonstrate the causal relationship of benzene to the induction of leukemia. Although these studies, for the most part involve high exposure levels, it is OSHA's view that once the carcinogenicity of a substance has been established qualitatively, any exposure must be considered to be attended by risk when considering any given population. OSHA therefore believes that occupational exposure to benzene at low levels poses a carcinogenic risk to workers."

[41]The so-called "one-hit" theory is based on laboratory studies indicating that one molecule of a carcinogen may react in the test tube with one molecule of DNA to produce a mutation. The theory is that, if this occurred in the human body, the mutated molecule could replicate over a period of years and eventually develop into a cancerous tumor. See OSHA's Proposed Rule on the Identification, Classification and Regulation of Toxic Substances Posing a Potential Carcinogenic Risk. Industry witnesses challenged this theory, arguing that the presence of several different defense mechanisms in the human body make it unlikely that a person would actually contract cancer as a result of absorbing one carcinogenic molecule. Thus, the molecule might be detoxified before reaching a critical site, damage to a DNA molecule might be repaired, or a mutated DNA molecule might be destroyed by the body's immunological defenses before it could develop into a cancer.

In light of the improbability of a person contracting cancer as a result of a single hit, a number of the scientists testifying on both sides of the issue agreed that every individual probably does have a threshold exposure limit below which he or she will not contract cancer. The problem, however, is that individual susceptibility appears to vary greatly and there is at present no way to calculate each and every person's threshold. Thus, even industry witnesses agreed that if the standard must ensure with absolute certainty that every single worker is protected from any risk of leukemia, only a zero exposure limit would suffice.

strated to be safe or at the lowest level feasible, whichever is higher. If no safe level is established, as in this case, the Secretary's interpretation of the statute automatically leads to the selection of an exposure limit that is the lowest feasible.[42] Because of benzene's importance to the economy, no one has ever suggested that it would be feasible to eliminate its use entirely, or to try to limit exposures to the small amounts that are omnipresent. Rather, the Agency selected 1 ppm as a workable exposure level, see n. 14, *supra,* and then determined that compliance with that level was technologically feasible and that "the economic impact . . . [compliance] will not be such as to threaten the financial welfare of the affected firms or the general economy. It therefore held that 1 ppm was the minimum feasible exposure level within the meaning of §6(b)(5) of the Act.

Finally, although the Agency did not refer in its discussion of the pertinent authority to any duty to identify the anticipated benefits of the new standard, it did conclude that some benefits were likely to result from reducing the exposure limit from 10 ppm to 1 ppm. This conclusion was based, again, not on evidence, but rather on the assumption that the risk of leukemia will decrease as exposure levels decrease. Although the Agency found it impossible to construct a dose-response curve that would predict with any accuracy the number of leukemias that could be expected to result from exposures at 10 ppm, at 1 ppm, or at any intermediate level, it nevertheless "determined that the benefits of the proposed standard are likely to be appreciable." In light of the Agency's disavowal of any ability to determine the numbers of employees likely to be adversely affected by exposure of 10 ppm, the Court of Appeals held this finding to be unsupported by the record. 581 F.2d, at 503.[44]

It is noteworthy, that at no point in its lengthy explanation did the Agency quote or even cite §3(8) of the Act. It made no finding that any of the provisions of the new standard were "reasonably necessary or appropriate to provide safe or healthful employ-ment and places of employment." Nor did it allude to the possibility that any such finding might have been appropriate.

III

Our resolution of the issues in this case turns, to a large extent, on the meaning of and the relationship between §3(8), which defines a health and safety standard as a standard that is "reasonably necessary and appropriate to provide safe or healthful employment," and §6(b)(5), which directs the Secretary in promulgating a health and safety standard for toxic materials to "set the standard which most adequately assures, to the extent feasible, on the basis of the best available evidence, that no employee will suffer material impairment of health or functional capacity. . . . "

In the Government's view, §3(8)'s definition of the term "standard" has no legal significance or at best merely requires that a standard not be totally irrational. It takes the position that §6(b)(5) is controlling and that it requires OSHA to promulgate a standard that either gives an absolute assurance of safety for each and every worker or that reduces exposures to the lowest level feasible. The Government interprets "feasible" as meaning technologically achievable at a cost that would not impair the viability of the industries subject to the regulation. The respondent industry representatives, on the other hand, argue that the Court of Appeals was correct in holding that the "reasonably necessary and appropriate" language of §3(8), along with the feasibility requirement of §6(b)(5), requires the Agency to quantify both the costs and the benefits of a proposed rule and to conclude that they are roughly commensurate.

In our view, it is not necessary to decide whether either the Government or industry is entirely correct. For we think it is clear that §3(8) does apply to all permanent standards promulgated under the Act and that it requires the Secretary, before issuing any standard, to determine that it is reasonably necessary and appropriate to remedy a significant risk of material health impairment. Only after the Secretary has made the threshold determination that such a risk exists with respect to a toxic substance, would it be necessary to decide whether §6(b)(5) requires him to select the most protective standard he can consistent with economic and technological feasibility, or whether, as respondents argue, the benefits of the regulation must be commensurate with the costs of its implementation. Because the Secretary did not make the required threshold finding in this case, we have no occasion to determine whether costs must be weighed against benefits in an appropriate case.

[42]"There is no doubt that benzene is a carcinogen and must, for the protection and safety of workers, be regulated as such. Given the inability to demonstrate a threshold or establish a safe level, it is appropriate that OSHA prescribe that the permissible exposure to benzene be reduced to the lowest level feasible."

[44]The court did, however, hold that the Agency's other conclusions—there there is *some* risk of leukemia at 10 ppm and that the risk would decrease by decreasing the exposure limit to 1 ppm—were supported by substantial evidence. 581 F.2d, at 503.

Under the Government's view, §3(8), if it has any substantive content at all,[45] merely requires OSHA to issue standards that are reasonable calculated to produce a safer or more healthy work environment. Apart from this minimal requirement of rationality, the Government argues that §3(8) imposes no limits on the Agency's power, and thus would not prevent it from requiring employers to do whatever would be "reasonably necessary" to eliminate all risks of any harm from their workplaces.[46] With respect to toxic substances and harmful physical agents, the Govern-

ment takes an even more extreme position. Relying on §6(b)(5)'s direction to set a standard "which most adequately assures . . . that no employee will suffer material impairment of health or functional capacity," the Government contends that the Secretary is required to impose standards that either guarantee workplaces that are free from any risk of material health impairment, however small, or that come as close as possible to doing so without ruining entire industries.

If the purpose of the statute were to eliminate completely and with absolute certainty any risk of serious harm, we would agree that it would be proper for the Secretary to interpret §§3(8) and 6(b)(5) in this fashion. But we think it is clear that the statute was not designed to require employers to provide absolutely risk-free workplaces whenever it is technologically feasible to do so, so long as the cost is not great enough to destroy an entire industry. Rather, both the language and structure of the Act, as well as its legislative history, indicate that it was intended to require the elimination, as far as feasible, of significant risks of harm.

B

By empowering the Secretary to promulgate standards that are "reasonably necessary or appropriate to provide safe or healthful employment and places of employment," the Act implies that, before promulgating any standard, the Secretary must make a finding that the workplaces in question are not safe. But "safe" is not the equivalent of "risk-free." There are many activities that we engage in every day—such as driving a car or even breathing city air—that entail some risk of accident or material health impairment; nevertheless, few people would consider these activities "unsafe" unless it threatens the workers with a significant risk of harm.

Therefore, before he can promulgate *any* permanent health or safety standard, the Secretary is required to make a threshold finding that a place of employment is unsafe—in the sense that significant risks are present and can be eliminated or lessened by a change in practices. This requirement applies to permanent standards promulgated pursuant to §6(b)(5), as well as to other types of permanent standards. For there is no reason why §3(8)'s definition of a standard should not be deemed incorporated by reference into §6(b)(5). The standards promulgated pursuant to §6(b)(5) are just one species of the genus of standards governed by the basic requirement. That section repeatedly uses the term "standard" without suggesting any exception from, or qualification of,

---

[45]We cannot accept the argument that §3(8) is totally meaningless. The Act authorizes the Secretary to promulgate three different kinds of standards—national consensus standards, permanent standards and temporary emergency standards. The only substantive criteria given for two of these—national consensus standards and permanent standards for safety hazards not covered by §6(b)(5)—are set forth in §3. While it is true that §3 is entitled "definitions," that fact does not drain each definition of substantive content. For otherwise, there would be no purpose in defining the critical terms of the statute. Moreover, if the definitions were ignored, there would be no statutory criteria at all to guide the Secretary in promulgating either national consensus standards or permanent standards other than those dealing with toxic materials and harmful physical agents. We may not expect Congress to display perfect craftmanship, but it is unrealistic to assume that it intended to give no direction whatsoever to the Secretary in promulgating most of his standards.

The structure of the separate subsection describing emergency temporary standards . . . supports this conclusion. It authorizes the Secretary to bypass the normal procedures for setting permanent standards if he makes two findings: (A) that employees are exposed to "grave danger" from exposure to toxic substances and (B) that an emergency standard is "necessary" to protect the employees from that danger. Those findings are to be compared with those that are implicitly required by the definition of the permanent standard—(A) that there be a significant—as opposed to a "grave"—risk, and (b) that additional regulation is "reasonably necessary or appropriate"—as opposed to "necessary." It would be anomalous for Congress to require specific findings for temporary standards but to give the Secretary a *carte blanche* for permanent standards.

[46]The Government does not concede that the feasibility requirement in the second sentence of §6(b)(5) applies to health and safety standards other than toxic substances standards. See n. 1, *supra*. However, even if it did, the Government's interpretation of the term "feasible," when coupled with its view of §3(8), would still allow the Agency to require the elimination of even insignificant risks at great cost, so long as an entire industry's viability would not be jeopardized.

the general definition; on the contrary, it directs the Secretary to select "*the* standard"—that is to say, one of various possible alternatives that satisfy the basic definition in §3(8)—that is most protective. Moreover, requiring the Secretary to make a threshold finding of significant risk is consistent with the scope of the regulatory power granted to him by §6(b)(5), which empowers the Secretary to promulgate standards, not for chemicals and physical agents generally, but for "*toxic* chemicals" and "*harmful* physical agents."

This interpretation of §§3(8) and 6(b)(5) is supported by the other provisions of the Act. Thus, for example, §6(g) provides in part that

> In determining the priority for establishing standards under this section, the Secretary shall give due regard to the urgency of the need for mandatory safety and health standards for particular industries, trades, crafts, occupations, businesses, workplaces or work environments.

The Government has expressly acknowledged that this section requires the Secretary to undertake some cost-benefit analysis before he promulgates any standard, requiring the elimination of the most serious hazards first.[49] If such an analysis must precede the promulgation of any standard, it seems manifest that Congress intended, at a bare minimum, that the Secretary find a significant risk of harm and therefore a probability of significant benefits before establishing a new standard.

*   *   *

In the absence of a clear mandate in the Act, it is unreasonable to assume that Congress intended to give the Secretary the unprecedented power over American industry that would result from the Government's view of §§3(8) and 6(b)(5), coupled with OSHA's cancer policy. Expert testimony that a substance is probably a human carcinogen—either be-

cause it has caused cancer in animals or because individuals have contracted cancer following extremely high exposures—would justify the conclusion that the substance poses some risks of serious harm no matter how minute the exposure and no matter how many experts testified that they regarded the risk as insignificant. That conclusion would in turn justify pervasive regulation limited only by the constraint of feasibility. In light of the fact that there are literally thousands of substances used in the workplace that have been identified as carcinogens or suspect carcinogens, the Government's theory would give OSHA power to impose enormous costs that might produce little, if any, discernible benefit.

If the Government were correct in arguing that neither §3(8) nor §6(b)(5) requires that the risk from a toxic substance be quantified sufficiently to enable the Secretary to characterize it as significant in an understandable way, the statute would make such a "sweeping delegation of legislative power" that it might be unconstitutional. . . . A construction of the statute that avoids this kind of open-ended grant should certainly be favored.

C

The legislative history also supports the conclusion that Congress was concerned, not with absolute safety, but with the elimination of significant harm. The examples of industrial hazards referred to in the committee hearings and debates all involved situations in which the risk was unquestionably significant.

*   *   *

In its reply brief the Government argues that . . . the Secretary is not required to eliminate threats of insignificant harm; it argues that §6(b)(5) still requires the Secretary to set standards that ensure that not even one employee will be subject to any risk of serious harm—no matter how small that risk may be. This interpretation is at odds with Congress' express recognition of the futility of trying to make all workplaces totally risk-free. Moreover, not even OSHA follows this interpretation of §6(b)(5) to its logical conclusion. Thus, if OSHA is correct that the only no-risk level for leukemia due to benzene exposure is zero and if its interpretation of §6(b)(5) is correct, OSHA should have set the exposure limit as close to zero as feasible. But OSHA did not go about its tasks in that way. Rather, it began with a 1 ppm level, selected at least in part to ensure that employers would not be required to eliminate ben-

[49]"First, 29 U.S.C. §655(g) requires the Secretary to establish priorities in setting occupational health and safety standards so that the more serious hazards are addressed first. In setting such priorities the Secretary must, of course, consider the relative costs, benefits and risks." The Government argues that the Secretary's setting of priorities under this section is not subject to judicial review. While we agree that a court cannot tell the Secretary which of two admittedly significant risks he should act to regulate first, this section, along with §§3(8) and 6(b)(5), indicates that the Act does limit the Secretary's power to requiring the elimination of significant risks.

zene concentrations that were little greater than the so-called "background" exposures experienced by the population at large. See n. 14, *supra*. Then, despite suggestions by some labor unions that it was feasible for at least some industries to reduce exposures to well below 1 ppm,[56] OSHA decided to apply the same limit to all, largely as a matter of administrative convenience.

* * *

OSHA also deviated from its own interpretation of §6(b) (5) in adopting an action level of 0.5 ppm below which monitoring and medical examinations are not required.

* * *

OSHA's concession to practicality in beginning with a 1 ppm exposure limit and using an action level concept implicitly adopt an interpretation of the statute as not requiring regulation of insignificant risks.[58] It is entirely consistent with this interpretation to hold that the Act also requires the Agency to limit its endeavors in the standard-setting area to eliminating significant risks of harm.

Finally, with respect to the legislative history, it is important to note that Congress repeatedly expressed its concern about allowing the Secretary to have too much power over American industry. Thus, Congress refused to give the Secretary the power to shut down plants unilaterally because of an imminent danger, see *Whirlpool Corp. v. Marshall*, 445 U.S. 1, and narrowly circumscribed the Secretary's power to issue temporary emergency standards.[59] This effort

by Congress to limit the Secretary's power is not consistent with a view that the mere possibility that some employee somewhere in the country may confront some risk of cancer is a sufficient basis for the exercise of the Secretary's power to require the expenditure of hundreds of millions of dollars to minimize that risk.

D

Given the conclusion that the Act empowers the Secretary to promulgate health and safety standards only where a significant risk of harm exists, the critical issue becomes how to define and allocate the burden of proving the significance of the risk in a case such as this, where scientific knowledge is imperfect and the precise quantification of risks is therefore impossible. The Agency's position is that there is substantial evidence in the record to support its conclusion that there is no absolutely safe level for a carcinogen and that, therefore, [the] burden is properly on industry to prove, apparently beyond a shadow of a doubt, that there *is* a safe level for benzene exposure. The Agency argues that, because of the uncertainties in this area, any other approach would render it helpless, forcing it to wait for the leukemia deaths that it believes are likely to occur[60] before taking any regulatory action.

We disagree. As we read the statute, the burden was on the Agency to show, on the basis of substantial evidence, that it is at least more likely than not that long-term exposure to 10 ppm of benzene presents a significant risk of material health impairment. Ordinarily, it is the proponent of a rule or order who has the burden of proof in administrative proceedings.

[56]One union suggested a 0.5 ppm permissible exposure limit for oil refineries and a 1 ppm ceiling (rather than a time-weighted average) exposure for all other industries, with no use of an action level. Another wanted a 1 ppm ceiling limit for all industries.
[58]The Government also states that it is OSHA's policy to attempt to quantify benefits wherever possible. While this is certainly a reasonable position, it is not consistent with OSHA's own view of its duty under §6(b)(5). In light of the inconsistencies in OSHA's position and the legislative history of the Act, we decline to defer to the Agency's interpretation.
[59]In *Florida Peach Growers Assn., Inc. v. Dept. of Labor*, 489 F.2d 120, 130, and n. 16 (CA5 1974), the court noted that Congress intended to restrict the use of emergency standards, which are promulgated without any notice or hearing. It held that, in promulgating an emergency standard, OSHA must find not only a danger *of* exposure or even some danger *from* exposure, but a grave danger

from exposure necessitating emergency action. Accord, *Dry Colors Mfrs. Assn., Inc. v. Dept. of Labor*, 486 F.2d 98, 100 (CA3 1973) (an emergency standard must be supported by something more than a possibility that a substance may cause cancer in man).

Congress also carefully circumscribed the Secretary's enforcement powers by creating a new, independent board to handle appeals from citations issued by the Secretary for noncompliance with health and safety standards. See 29 U.S.C. §§659-661.
[60]As noted above, OSHA acknowledged that there was no empirical evidence to support the conclusion that there was any risk whatsoever of deaths due to exposures at 10 ppm. What OSHA relied upon was a theory that, because leukemia deaths had occurred at much higher exposures, some (although fewer) were also likely to occur at relatively low exposures. The Court of Appeals specifically held that its conclusion that the number was "likely" to be appreciable was unsupported by the record.

In some cases involving toxic substances, Congress has shifted the burden of proving that a particular substance is safe onto the party opposing the proposed rule. The fact that Congress did not follow this course in enacting OSHA indicates that it intended the Agency to bear the normal burden of establishing the need for a proposed standard.

In this case OSHA did not even attempt to carry its burden of proof. The closest it came to making a finding that benzene presented a significant risk of harm in the workplace was its statement that the benefits to be derived from lowering the permissible exposure level from 10 to 1 ppm were "likely" to be "appreciable." The Court of Appeals held that this finding was not supported by substantial evidence. Of greater importance, even if it were supported by substantial evidence, such a finding would not be sufficient to satisfy the Agency's obligations under the Act.

*   *   *

Contrary to the Government's contentions, imposing a burden on the Agency of demonstrating a significant risk of harm will not strip it of its ability to regulate carcinogens, nor will it require the Agency to wait for deaths to occur before taking any action. First, the requirement that a "significant" risk be identified is not a mathematical straitjacket. It is the Agency's responsibility to determine, in the first instance, what it considers to be a "significant" risk. Some risks are plainly acceptable and others are plainly unacceptable. If, for example, the odds are one in a billion that a person will die from cancer by taking a drink of chlorinated water, the risk clearly could not be considered significant. On the other hand, if the odds are one in a thousand that regular inhalation of gasoline vapors that are two percent benzene will be fatal, a reasonable person might well consider the risk significant and take appropriate steps to decrease or eliminate it. Although the Agency has no duty to calculate the exact probability of harm, it does have an obligation to find that a significant risk is present before it can characterize a place of employment as "unsafe."[62]

Second, OSHA is not required to support its finding that a significant risk exists with anything approaching scientific certainty. Although the Agency's findings must be supported by substantial evidence, 29 U.S.C. §655(f), §6(b) (5) specifically allows the Secretary to regulate on the basis of the "best available evidence." As several courts of appeals have held, this provision requires a reviewing court to give OSHA some leeway where its findings must be made on the frontiers of scientific knowledge. See *Industrial Union Dept., AFL-CIO v. Hodgson,* 162 U.S. App. D.C. 331, 499 F.2d 467, 476 (1974); *Society of the Plastics Industry, Inc. v. OSHA,* 509 F.2d 1301, 1308 [2 OSHC 1496] (CA2 1975), cert. denied, 421 U.S. 962. Thus, so long as they are supported by a body of reputable scientific thought, the Agency is free to use conservative assumptions in interpreting the data with respect to carcinogens, risking error on the side of over-protection rather than under-protection.[63]

Finally, the record in this case and OSHA's own rulings on other carcinogens indicate that there are a number of ways in which the Agency can make a rational judgment about the relative significance of the risks associated with exposure to a particular carcinogen.

*   *   *

[Concurring opinions of Justices REHNQUIST, BURGER and POWELL omitted.]

*   *   *

[62]In his dissenting opinion, . . . Mr. Justice Marshall states that "when the question involves determination of the acceptable level of risk, the ultimate decision must necessarily be based on considerations of policy as well as empirically verifiable facts. Factual determinations can at most define the risk in some statistical way; the judgment whether that risk is tolerable cannot be based solely on a resolution of the facts." We agree. Thus, while the Agency must support its finding that a certain level of risk exists by substantial evidence, we recognize that its determination that a particular level of risk is "significant" will be based largely on policy consideration. At this point we have no need to reach the issue of what level of scrutiny a reviewing court should apply to the latter type of determination.

[63]Mr. Justice Marshall states that, under our approach, the agency must either wait for deaths to occur or must "deceive the public" by making a basically meaningless determination of significance based on totally inadequate evidence. Mr. Justice Marshall's view, however, rests on the erroneous premise that the only reason OSHA did not attempt to quantify benefits in this case was because it could not do so in any reasonable manner. As the discussion of the Agency's rejection of an industry attempt at formulating a dose-response curve demonstrates, however, the Agency's rejection of methods such as dose-response curves was based at least in part on its view that nothing less than absolute safety would suffice.

Dissenting Opinion

Mr. Justice MARSHALL with whom Mr. Justice BRENNAN, Mr. Justice WHITE, and Mr. Justice BLACKMUN join, dissenting.

In cases of statutory construction, this Court's authority is limited. If the statutory language and legislative intent are plain, the judicial inquiry is at an end. Under our jurisprudence, it is presumed that ill-considered or unwise legislation will be corrected through the democratic process; a court is not permitted to distort a statute's meaning in order to make it conform with the Justices' own views of sound social policy. . . .

Today's decision flagrantly disregards these restrictions on judicial authority. The plurality ignores the plain meaning of the Occupational Safety and Health Act of 1970 in order to bring the authority of the Secretary of Labor in line with the plurality's own views of proper regulatory policy. The unfortunate consequence is that the Federal Government's efforts to protect American workers from cancer and other crippling diseases may be substantially impaired.

*   *   *

In this case, the Secretary of Labor found, on the basis of substantial evidence, that (1) exposure to benzene creates a risk of cancer, chromosomal damage, and a variety of nonmalignant but potentially fatal blood disorders, even at the level of 1 ppm; (2) no safe level of exposure has been shown; (3) benefits in the form of saved lives would be derived from the permanent standard; (4) the number of lives that would be saved could turn out to be either substantial or relatively small; (5) under the present state of scientific knowledge, it is impossible to calculate even in a rough way the number of lives that would be saved, at least without making assumptions that would appear absurd to much of the medical community; and (6) the standard would not materially harm the financial condition of the covered industries. The court does not set aside any of these findings. Thus, it could not be plainer that the Secretary's decision was fully in accord with his statutory mandate "most aedequately [to] assure[ ] . . . that no employee will suffer material impairment of health or functional capacity. . . . "

The plurality's conclusion to the contrary is based on its interpretation of 29 U.S.C. §652(8), which defines an occupational safety and health standard as one "which requires conditions . . . reasonably necessary or appropriate to provide safe or healthful employment. . . . " According to the plurality, a stand-ard is not "reasonably necessary or appropriate" unless the Secretary is able to show that it is "at least more likely than not," that the risk he seeks to regulate is a "significant" one. Nothing in the statute's language or legislative history, however, indicates that the "reasonably necessary or appropriate" language should be given this meaning. Indeed, both demonstrate that the plurality's standard bears no connection with the acts or intentions of Congress and is based only on the plurality's solicitude for the welfare of regulated industries. And the plurality uses this standard to evaluate not the agency's decision in this case, but a strawman of its own creation.

Unlike the plurality, I do not purport to know whether the actions taken by Congress and its delegates to ensure occupational safety represent sound or unsound regulatory policy. The critical problem is cases like the one at bar is scientific uncertainty. While science has determined that exposure to benzene at levels above 1 ppm creates a definite risk of health impairment, the magnitude of the risk cannot be quantified at the present time. The risk at issue has hardly been shown to be insignificant; indeed, future research may reveal that the risk is in fact considerable. But the existing evidence may frequently be inadequate to enable the Secretary to make the threshold finding of "significance" that the Court requires today. If so, the consequence of the plurality's approach would be to subject American workers to a continuing risk of cancer and other fatal diseases, and to render the Federal Government powerless to take protective action on their behalf. Such an approach would place the burden of medical uncertainty squarely on the shoulders of the American worker, the intended beneficiary of the Occupational Safety and Health Act. It is fortunate indeed that at least a majority of the Justices reject the view that the Secretary is prevented from taking regulatory action when the magnitude of a health risk cannot be quantified on the basis of current techniques. . . .

Because today's holding has no basis in the Act, and because the Court has no authority to impose its own regulatory policies on the Nation, I dissent.

I

*   *   *

The legislative history of the Act reveals Congress' particular concern for health hazards of "unprecedented complexity" that had resulted from chemicals whose toxic effects "are only now being discovered." "Recent scientific knowledge points to hitherto unsuspected cause-and-effect relationships between occupational exposures and many of the so-called chronic diseases —cancer, respiratory ailments, allergies, heart disease, and others." *Ibid.* Members of Congress made repeated references to the dangers posed by carcinogens and to the defects in our knowledge of their operation and effect. One of the primary purposes of the Act was to ensure regulation of these "insidious 'silent' killers."

* * *

The authority conferred by §655(b)(5), however, is not absolute. The subsection itself contains two primary limitations. The requirement of "material" impairment was designed to prohibit the Secretary from regulating substances that create a trivial hazard to affected employees. Moreover, all standards promulgated under the subsection must be "feasible."

* * *

II

The plurality's discussion of the record in this case is both extraordinarily arrogant and extraordinarily unfair. It is arrogant because the plurality presumes to make its own factual findings with respect to a variety of disputed issues relating to carcinogen regulation. It should not be necessary to remind the Members of this Court that they were not appointed to undertake independent review of adequately supported scientific findings made by a technically expert agency.[9] And the plurality's discussion is unfair

[9]I do not, of course, suggest that it is appropriate for a federal court reviewing agency action blindly to defer to the agency's findings of fact and determinations of policy. Under *Citizens to Preserve Overton Park, Inc. v. Volpe,* 401 U.S. 402, 416 (1971), courts must undertake a "searching and careful" judicial inquiry into these factors. Such an inquiry is designed to require the agency to take a "hard look," *Kleppe v. Sierra Club,* 427 U.S. 390, 410 (1976) (citation omitted), by considering the proper factors and weighing them in a reasonable manner. There is also room for especially rigorous judicial scrutiny of agency decisions under a rationale akin to that offered in *United States v. Carolene Products, Inc.,* 304 U.S. 144, 152, n. 4 (1938). See *Environmental Defense Fund v. Ruckelshaus,* 142 U.S. App. D.C. 74, 439 F.2d 584 (1971).

I see no basis, however, for the approach taken by the plurality today, which amounts to nearly *de novo* review of questions of fact and of regulatory policy on behalf of institutions that are by no means unable to protect themselves in the political process. Such review is especially inappropriate when the factual questions at issue are ones about which the Court cannot reasonably be expected to have expertise.

because its characterization of the Secretary's report bears practically no resemblance to what the Secretary actually did in this case. Contrary to the plurality's suggestion, the Secretary did not rely blindly on some draconian carcinogen "policy." If he had, it would have been sufficient for him to have observed that benzene is a carcinogen, a proposition that respondents do not dispute. Instead, the Secretary gathered over 50 volumes of exhibits and testimony and offered a detailed and even-handed discussion of the relationship between exposure to benzene at all recorded exposure levels and chromosomal damage, aplastic anemia, and leukemia. In that discussion he evaluated, and took seriously, respondents' evidence of a safe exposure level.

The hearings on the proposed standard were extensive, encompassing 17 days from July 19 through August 10, 1977. The 95 witnesses included epidemiologists, toxicologists, physicians, political economists, industry representatives, and members of the affected work force. Witnesses were subjected to exhaustive questioning by representatives from a variety of interested groups and organizations.

* * *

Costs and benefits. The Secretary offered a detailed discussion of the role that economic considerations should play in his determination. He observed that standards must be "feasible," both economically and technologically. In his view the permanent standard for benzene was feasible under both tests. The economic impact would fall primarily on the more stable industries, such as petroleum refining and petro-chemical production. These industries would be able readily to absorb the costs or to pass them on to consumers. None of the twenty affected industries involving 157,000 facilities and 629,000 exposed employees, would be unable to bear the required expenditures. He concluded that the compliance costs were "well within the financial capability of the covered industries." An extensive survey of the national economic impact of the standard, undertaken by a private contractor, found first-year operating costs of between $187 and $205 million, recurring annual costs of $34 million, and investment in engineering controls of about $266 million.[22] Since respondents have not attacked the Secretary's basic

[22]The plurality's estimate of the amount of expenditure per employee is highly misleading. Most of the costs of the benzene standard would be incurred only once and would thus protect an unascertainable number of employees in the future; that number will be much higher than the number of employees currently employed.

conclusions as to costs, the Secretary's extensive discussion need not be summarized here.

Finally, the Secretary discussed the benefits to be derived from the permanent standard. During the hearings, it had been argued that the Secretary should estimate the health benefits of the proposed regulation. To do this he would be required to construct a dose-response curve showing, at least in a rough way, the number of lives that would be saved at each possible exposure level. Without some estimate of benefits, it was argued, the Secretary's decisionmaking would be defective. During the hearings an industry witness attempted to construct such a dose-response curve. Restricting himself to carcinogenic effects, he estimated that the proposed standard would save two lives every six years and suggested that this relatively minor benefit would not justify the regulation's costs.

The Secretary rejected the hypothesis that the standard would save only two lives in six years. This estimate, he concluded, was impossible to reconcile with the evidence in the record. He determined that, because of numerous uncertainties in the existing data, it was impossible to construct a dose-response curve by extrapolating from those data to lower exposure levels. More generally, the Secretary observed that it had not been established that there was a safe level of exposure for benzene. Since there was considerable testimony that the risk would decline with the exposure level, the new standard would save lives. The number of lives saved "may be appreciable," but there was no way to make a more precise determination.[25] The question was "on the frontiers of scientific knowledge."

The Secretary concluded that, in light of the scientific uncertainty, he was not required to calculate benefits more precisely. In any event he gave "careful consideration" to the question of whether the admittedly substantial costs were justified in light of the hazards of benzene exposure. He concluded that those costs were "necessary" in order to promote the purposes of the Act.

[25]At one point the Secretary did indicate that appreciable benefits were "likely" to result. The Court of Appeals held that this conclusion was unsupported by substantial evidence. The Secretary's suggestion, however, was made in the context of a lengthy discussion intended to show that appreciable benefits "may" be predicted but that their likelihood could not be quantified. The suggestion should not be taken as a definitive statement that appreciable benefits were more probable than not.

For reasons stated *infra,* there is nothing in the Act to prohibit the Secretary from acting when he is unable to conclude that appreciable benefits are more probable than not.

III

A

* * ، *

The plurality is insensitive to three factors which, in my view, make judicial review of occupational safety and health standards under the substantial evidence test particularly difficult. First, the issues often reach a high level of technical complexity. In such circumstances the courts are required to immerse themselves in matters to which they are unaccustomed by training or experience. Second, the factual issues with which the Secretary must deal are frequently not subject to any definitive resolution. Often "the factual finger points, it does not conclude." *Society of Plastics Indus., Inc. v. OSHA,* 509 F.2d 1301, 1308 (CA2) (Mr. Justice Clark), cert. denied, 421 U.S. 992 [3 OSHC 1195] (1975). Causal connections and theoretical extrapolations may be uncertain. Third, when the question involves determination of the acceptable level of risk, the ultimate decision must necessarily be based on considerations of policy as well as empirically verifiable facts. Factual determinations can at most define the risk in some statistical way; the judgment whether that risk is tolerable cannot be based solely on a resolution of the facts.

The decision to take action in conditions of uncertainty bears little resemblance to the sort of empirically verifiable factual conclusions to which the substantial evidence test is normally applied. Such decisions were not intended to be unreviewable; they too must be scrutinized to ensure that the Secretary has acted reasonably and within the boundaries set by Congress. But a reviewing court must be mindful of the limited nature of its role. See *Vermont Yankee Nuclear Power Corp. v. NRDC,* 435 U.S. 519 (1978). It must recognize that the ultimate decision cannot be based solely on determinations of fact, and that those factual conclusions that have been reached are ones which the courts are ill-equipped to resolve on their own.

Under this standard of review, the decision to reduce the permissible exposure level to 1 ppm was well within the Secretary's authority. The Court of Appeals upheld the Secretary's conclusions that benzene causes leukemia, blood disorders, and chromosomal damage even at low levels, that an exposure level of 10 ppm is more dangerous than one of 1 ppm, and that benefits will result from the proposed standard. It did not set aside his finding that the number of lives that would be saved was not subject to quantification. Nor did it question his conclusion that the reduction was "feasible."

In these circumstances, the Secretary's decision was reasonable and in full conformance with the statutory language requiring that he "set the standard which most adequately assures, to the extent feasible, on the basis of the best available evidence, that no employee will suffer material impairment of health or functional capacity even if such employee has regular exposure to the hazard dealt with by such standard for the period of his working life." On this record, the Secretary could conclude that regular exposure above the 1 ppm level would pose a definite risk resulting in material impairment to some indeterminate but possibly substantial number of employees. Studies revealed hundreds of deaths attributable to benzene exposure. Expert after expert testified that no safe level of exposure had been shown and that the extent of the risk declined with the exposure level. There was some direct evidence of incidence of leukemia, nonmalignant blood disorders, and chromosomal damage at exposure levels of 10 ppm and below. Moreover, numerous experts testified that existing evidence required an inference that an exposure level of about 1 ppm was hazardous. We have stated that "well-reasoned expert testimony—based on what is known and uncontradicted by empirical evidence—may in and of itself be 'substantial evidence' when first-hand evidence on the question . . . is unavailable." *FPC v. Florida Power & Light Co.,* 404 U.S. 453, 464-465 (1972). Nothing in the Act purports to prevent the Secretary from acting when definitive information as to the quantity of a standard's benefits is unavailable.[26] Where, as here, the deficiency in knowledge relates to the extent of the benefits rather than their existence, I see no reason to hold that the Secretary has exceeded his statutory authority.

[26]This is not to say that the Secretary is prohibited from examining relative costs and benefits in the process of setting priorities among hazardous substances, or that systematic consideration of costs and benefits is not to be attempted in the standard-setting process. Efforts to quantify costs and benefits, like statements of reasons generally, may help to promote informed consideration of decisional factors and facilitate judicial review. See *Dunlop v. Bachowski,* 421 U.S. 560, 571-574 (1975). The Secretary indicates that he has attempted to quantify costs and benefits in the past.

It is not necessary in the present case to say whether the Secretary must show a reasonable relation between costs and benefits. Discounting for the scientific uncertainty, the Secretary expressly—and reasonably—found such a relation here.

B

* * *

The plurality suggests that under the "reasonably necessary" clause, a workplace is not "unsafe" unless the Secretary is able to convince a reviewing court that a "significant" risk is at issue. . . . That approach is particularly embarrassing in this case, for it is contradicted by the plain language of the Act. The plurality's interpretation renders utterly superfluous the first sentence of §655(b) (5), which, as noted above, requires the Secretary to set the standard "which most adequately assures . . . that no employee will suffer material impairment of health." Indeed, the plurality's interpretation reads that sentence out of the Act. By so doing, the plurality makes the test for standards regulating toxic substances and harmful physical agents substantially identical to the test for standards generally—plainly, the opposite of what Congress intended. And it is an odd canon of construction that would insert in a vague and general definitional clause a threshold requirement that overcomes the specific language placed in a standard-setting provision. The most elementary principles of statutory construction demonstrate that precisely the opposite interpretation is appropriate. In short, Congress could have provided that the Secretary may not take regulatory action until the existing scientific evidence proves the risk at issue to be "significant,"[27] but it chose not to do so.

[27]It is useful to compare the Act with other regulatory statutes in which Congress has required a showing of a relationship between costs and benefits or of an "unreasonable risk." In some statutes Congress has expressly required cost-benefit analysis or a demonstration of some reasonable relation between costs and benefits. See 33 U.S.C. §701(a) (Flood Control Act of 1936); 42 U.S.C. §7545(c) (2) (B) (Clean Air Act): 33 U.S.C. §1314(b) (4) (B) (Clean Water Act). In others Congress has imposed two independent requirements: that administrative action be "feasible" and justified by a balancing of costs and benefits, e.g., 43 U.S.C. §1347(b) (Outer Continental Shelf Land Act); 42 U.S.C. §6295(a) (4) (D) (Energy Policy and Conservation Act). This approach demonstrates a legislative awareness of the difference between a feasibility constraint and a constraint based on weighing costs and benefits. In still others Congress has authorized regulation of "unreasonable risk," a term which has been read by some courts to require a balancing of costs and benefits. See, e.g., *Aqua Slide "N" Dive Corp. v. CPSC,* 569 F.2d 831 (CA5 1978) (construing 15 U.S.C. §2058(c) (2) (A) [Consumer Product Safety Act]); *Forester v. CPSC,* 182 U.S. App. D.C. 153, 559 F.2d 774 (1977) (construing 15 U.S.C. §1261 (s) [Child Protection and Toy Safety Act]).

The plurality's interpretation of the "reasonably necessary or appropriate" clause is also conclusively refuted by the legislative history. While the standard-setting provision that the plurality ignores received extensive legislative attention, the definitional clause received *none at all.* An earlier version of the Act did not embody a clear feasibility constraint and was not restricted to toxic substances or to "material" impairments. The "reasonably necessary or appropriate" clause was contained in this prior version of the bill, as it was at all relevant times. In debating this version, Members of Congress repeatedly expressed concern that it would require a risk-free universe. The definitional clause was not mentioned at all, an omission that would be incomprehensible if Congress intended by that clause to require the Secretary to quantify the risk he sought to regulate in order to demonstrate that it was "significant."

*   *   *

In short, today's decision represents a usurpation of decisionmaking authority that has been exercised by and properly belongs with Congress and its authorized representatives. The plurality's construction has no support in the statute's language, structure, or legislative history. The threshold finding that the plurality requires is the plurality's own invention. It bears no relationship to the acts or intentions of Congress, and it can be understood only as reflecting the personal views of the plurality as to the proper allocation of resources for safety in the American workplace.

C

*   *   *

Though it is difficult to see how a future Congress could be any more explicit on the matter than was the Congress that passed the Act in 1970, it is important to remember that today's decision is subject to legislative reversal. Congress may continue to believe that the Secretary should not be prevented from protecting American workers from cancer and other fatal diseases until scientific evidence has progressed to a point where he can convince a federal court that the risk is "significant." Today's decision is objectionable not because it is final, but because it places the burden of legislative inertia on the beneficiaries of the safety and health legislation in question in this case. By allocating the burden in this fashion, the Court requires the American worker to return to the political arena and to win a victory that he won once before in 1970. I am unable to discern any justification for that result.

*   *   *

◇ NOTES

1. While a majority of five Justices vacated the benzene standard, no majority agreed on the rationale. At times only three Justices (out of nine) joined the plurality opinion. Would the principles of the case be easy to reverse by a subsequent Supreme Court? Does Justice Marshall's dissent provide a way?

2. Is the Court correct in its interpretation of Section 3(8) of the OSHAct? Is Justice Marshall persuasive in his assertion that the plurality rewrote the OSHAct?

3. The Court gave guidelines on the nature of "significant risk," suggesting that a risk of death of one in a billion is not significant while a risk of one in a thousand is. Is this helpful? The Court was also careful to say that significant risk did not need to be proven with anything approaching scientific certainty. The Agency must simply establish that it was "more likely than not" that harm would result without regulation. The Court would not place OSHA in a "statistical straitjacket." However, OSHA is required to find significant risk before regulating. Would it be difficult for OSHA to prove significant risk for a carcinogenic substance?

4. After the benzene decision, OSHA's proposed revision to its generic cancer policy contained two major changes: (1) it required that only "significant risks" be regulated and (2) it removed the automatic triggering of temporary emergency and permanent standards.

5. Subsequent scientific studies confirmed that human cancer does indeed result from exposures to benzene at levels lower than 10 ppm. OSHA has recently promulgated a 1 ppm standard using these studies. Should the standard withstand challenge in the courts?

6. The effect of the benzene decision was seen in OSHA's approach to the regulation of formaldehyde. OSHA acknowledged formaldehyde to be a carcinogen, but it set the standard at 1 ppm even though 0.5 ppm was "feasible." OSHA argued that below 1 ppm, the number of workers affected did not constitute "a significant risk of material impairment."[d] The D.C. Circuit Court of Appeals was dissatisfied with OSHA's explanation for refusing to impose a lower limit and remanded the standard to OSHA for reconsideration on June 9, 1989.

4.   Standard Setting After the Benzene Decision: Feasibility and Cost-Benefit Analysis

On June 23, 1978, four months after issuing its benzene standard, OSHA promulgated a 6(b) standard regulating exposure to cotton dust. OSHA revised the previous 6(a) standard and established different lower limits for different parts of the cotton industry. The standard was challenged in the D.C. Circuit Court of Appeals by both industry and labor. While the D.C. Circuit was considering the standard, the Fifth Circuit Court of Appeals issued its opinion striking down the benzene standard. In its opinion, the D.C. Circuit criticized the reasoning of the Fifth Circuit and disagreed with that court's interpretation of the OSHAct regarding the use of cost-benefit analysis.

The Supreme Court granted review and sided with the D.C. Circuit on the cost-benefit issue.

[d]K. Rest and N. A. Ashford. 1988. Regulation and Technological Options: The Case of Occupational Exposure to Formaldehyde, *Harvard Journal of Law and Technology,* 1:63–96.

American Textile Manufacturers Institute v. Donovan

452 U.S. 490 (1981)

BRENNAN, J., delivered the opinion of the Court, in which WHITE, MARSHALL, BLACKMUN, and STEVENS, JJ., joined. STEWART, J., filed a dissenting opinion. REHNQUIST, J., filed a dissenting opinion, in which BURGER, C.J., joined. POWELL, J., took no part in the decision of the cases.

*   *   *

In 1978, the Secretary, acting through the Occupational Safety and Health Administration (OSHA),[1] promulgated a standard limiting occupational expo-

[1]This opinion will use the terms OSHA and the Secretary interchangeably when referring to the agency, the Secretary of Labor, or the Assistant Secretary for Occupational Safety and Health. The Secretary of Labor has delegated the authority to promulgate occupational safety and health standards to the Assistant Secretary.

sure to cotton dust, an airborne particle byproduct of the preparation and manufacture of cotton products, exposure to which induces a "constellation of respiratory effects" known as "byssinosis." This disease was one of the expressly recognized health hazards that led to passage of the Occupational Safety and Health Act of 1970.

* * *

Petitioners in these consolidated cases, representing the interests of the cotton industry,[2] challenged the validity of the "Cotton Dust Standard" in the Court of Appeals for the District of Columbia pursuant to §6(f) of the Act. They contend in this Court, as they did below, that the Act requires OSHA to demonstrate that its Standard reflects a reasonable relationship between the costs and benefits associated with the Standard. Respondents, the Secretary of Labor and two labor organizations,[3] counter that Congress balanced the costs and benefits in the Act itself, and that the Act should therefore be construed not to require OSHA to do so. They interpret the Act as mandating that OSHA enact the most protective standard possible to eliminate a significant risk of material health impairment, subject to the constraints of economic and technological feasibility. The Court of Appeals held that the Act did not require OSHA to compare costs and benefits. 617 F.2d 636 (1979). We granted certiorari,—U.S.— (1980), to resolve this important question, which was presented but not decided in last Term's *Industrial Union Department v. American Petroleum Institute,* 448 U.S. 607 (1980),[4] and to decide other issues related to the Cotton Dust Standard.[5]

I

Byssinosis, known in its more severe manifestations as "brown lung" disease, is a serious and potentially disabling respiratory disease primarily caused by the

inhalation of cotton dust.[6] . . . Byssinosis is a "continuum . . . disease," that has been categorized into four grades.[8] In its least serious form, byssinosis produces both subjective symptoms, such as chest tightness, shortness of breath, coughing, and wheezing, and objective indications of loss of pulmonary functions. In its most serious form, byssinosis is a chronic and irreversible obstructive pulmonary disease, clinically similar to chronic bronchitis or emphysema, and can be severely disabling. At worst, as is true of other respiratory diseases including bronchitis, emphysema, and asthma, byssinosis can create an

---

[2]Petitioners in No. 79–1429 include 12 individual cotton textile manufacturers, and the American Textile Manufacturers Institute, Inc. (ATMI), a trade association representing approximately 175 companies. In No. 79–1583, petitioner is the National Cotton Council of America, a nonprofit corporation chartered for the purpose of increasing the consumption of cotton and cotton products.

[3]The two labor organizations are the American Federation of Labor and Congress of Industrial Organizations, Industrial Union Department, AFL-CIO, and the Amalgamated Clothing & Textile Workers Union, AFL-CIO. In the Court of Appeals, the labor organizations challenged the Cotton Dust Standard as not sufficiently stringent.

[4]Justice Powell, concurring in part and in the judgment, was the only member of the Court to decide the cost-benefit issue expressly. Justice Powell concluded that the statute "requires the agency to determine that the economic effects of its standard bear a reasonable relationship to the expected benefits." *Industrial Union Department v. American Petroleum Institute,* 448 U.S. 607, 667 (Powell, J., concurring in part and in the judgment). Justice Marshall, dissenting, joined by Justice Brennan, Justice White, and Justice Blackmun, indicated that the statute did not contemplate cost-benefit analysis. See *id.,* 448 U.S. at 607 n. 30, 32.

[5]In addition to the cost-benefit issue, the other questions presented and addressed are (1) whether substantial evidence in the record as a whole supports OSHA's determination that the Cotton Dust Standard is economically feasible; and (2) whether OSHA has the authority under the Act to require that employers guarantee the wages and benefits of employees who are transferred to other positions because of their inability to wear respirators.

[6]Cotton dust is defined as

dust present in the air during the handling or processing of cotton, which may contain a mixture of many substances including ground up plant matter, fiber, bacteria, fungi, soil, pesticides, non-cotton plant matter and other contaminants which may have accumulated with the cotton during the growing, harvesting and subsequent processing or storage periods. Any dust present during the handling and processing of cotton through the weaving or knitting of fabrics, and dust present in other operations or manufacturing processes using new or waste cotton fibers or cotton fiber byproducts from textile mills are considered cotton dust.

29 CFR §1910.1043(b)(1980)(Cotton Dust Standard).

[8]Known generally as the Schilling classification grades, they include: "[Grade] 1/2: slight acute effect of dust on ventilatory capacity; no evidence of chronic ventilatory impairment.

"[Grade] 1: definite acute effect of dust on ventilatory capacity; no evidence of chronic ventilatory impairment.

"[Grade] 2: evidence of slight to moderate irreversible impairment of ventilatory capacity.

"[Grade] 3: evidence of moderate to severe irreversible impairment of ventilatory capacity."

additional strain on cardiovascular functions and can contribute to death from heart failure ("there is an association between mortality and the extent of dust exposure"). One authority has described the increasing seriousness of byssinosis as follows:

> In the first few years of exposure [to cotton dust], symptoms occur on Monday, or other days after absence from the work environment; later, symptoms occur on other days of the week; and eventually, symptoms are continuous, even in the absence of dust exposure. A. Bouhuys, Byssinosis in the United States.[9]

*   *   *

Estimates indicate that at least 35,000 employed and retired cotton mill workers, or 1 in 12 such workers, suffers from the most disabling form of byssinosis.[11]

[9]Descriptions of the disease by individual mill workers presented in hearings on the Cotton Dust Standard before an administrative law judge, are more vivid: "When they started speeding the looms up the dust got finer and more and more people started leaving the mill with breathing problems. My mother had to leave the mill in the early fifties. Before she left, her breathing got so short she just couldn't hold out to work. My stepfather left the mill on account of breaching [sic] problems. He had coughing spells til he couldn't breath, like a child's whooping cough. Both my sisters who work in the mill have breathing problems. My husband had to give up his job when he was only fifty-four years old because of the breathing problem." . . .

"I suppose I had a breathing problem since 1973. I just kept on getting sick and began losing time at the mill. Every time that I go into the mill I get deathly sick, choking and vomiting losing my breath. It would blow down all that lint and cotton and I have clothes right here where I have wore and they have been washed several times and I would like for you all to see them. That will not come out in washing.

"I am only fifty-seven years old and I am retired and I can't even get to go to church because of my breathing. I get short of breath just walking around the house or dressing [or] sometimes just watching T.V. I cough all the time."

". . . I had to quit because I couldn't lay down and rest without oxygen in the night and my doctor told me I would have to get out of there . . . I couln't [sic] even breathe, I had to get out of the door so I could breathe and he told me not to go back in [the mill] under any circumstances."

Byssinosis is not a newly discovered disease, having been described as early as in the 1820s in England and observed in Belgium in a study of 2,000 cotton workers in 1845.

*   *   *

Not until the early 1960's was byssinosis recognized in the United States as a distinct occupational hazard associated with cotton mills.

*   *   *

In 1966, the American Conference of Governmental Industrial Hygienists (ACGIH), a private organization, recommended that exposure to total cotton dust[14] be limited to a "threshold limit value" of 1,000 micrograms per cubic meter of air (1000 $\mu g/m^3$) averaged over an 8-hour workday. The United States Government first regulated exposure to cotton dust in 1968, when the Secretary of Labor, pursuant to the Walsh-Healey Act, 41 U.S.C. §35(e), promulgated airborne contaminant threshold limit values, applicable to public contractors, that included the 1000 $\mu g/m^3$ limit for total cotton dust. 34 Fed. Reg. 7953 (1969).[15] Following passage of the Act in 1970, the 1000 $\mu g/m^3$ standard was adopted as an "established Federal standard" under §6(a) of the Act, a provision designed to guarantee immediate protection of workers for the period between enactment of the statute and promulgation of permanent standards.[16]

In 1974, ACGIH, adopting a new measurement of unit respirable rather than total dust, lowered its previous exposure limit recommendation to 200 $\mu g/m^3$ measured by a vertical elutriator, a device that measures cotton dust particles 15 microns or less in diameter.[17]

[11]The criterion of disability used for the 35,000 worker estimate was a Forced Expiratory Volume (FEV1) measurement of pulmonary function of 1.2 liters or less. An FEV1 of 1.2 liters "is a small fraction of the pulmonary performance of a normal lung."
[14]"Total dust" includes both respirable and nonrespirable cotton dust.
[15]The Secretary of Labor adopted the threshold limit values contained in a list that had been prepared by the ACGIH.
[16]Section 6(a) of the Act, provides in pertinent part:
[T]he Secretary shall, as soon as practicable during the period beginning with the effective date of this chapter and ending two years after such date, by rule promulgate as an occupational safety or health standard . . . any established Federal standard, unless he determines that the promulgation of such a standard would not result in improved safety or health for specifically designated employees.
[17]In many cotton preparation and manufacturing operations, including opening, picking and carding, 1000 $\mu g/m^3$ of total dust is roughly equivalent to 500 $\mu g/m^3$ of respirable dust.

* * *

On December 28, 1976, OSHA published a proposal to replace the existing Federal standard on cotton dust with a new permanent standard, pursuant to §6(b) (5) of the Act. The proposed standard contained a PEL of 200 $\mu g/m^3$ of vertical elutriated lint-free respirable cotton dust for all segments of the cotton industry. It also suggested an implementation strategy for achieving the PEL that relied on respirators for the short-term and engineering controls for the long-term. OSHA invited interested parties to submit written comments within a 90-day period.[20]

* * *

OSHA issued its final Cotton Dust Standard—the one challenged in the instant case—on June 23, 1978. Along with an accompanying statement of findings and reasons, the Standard occupied 69 pages of the Federal Register.

The Cotton Dust Standard promulgated by OSHA establishes mandatory PEL's over an 8-hour period of 200 $\mu g/m^3$ for yarn manufacturing,[21] 750 $\mu g/m^3$ for slashing and weaving operations, and 500 $\mu g/m^3$ for all other processes in the cotton industry.[22] These levels represent a relaxation of the proposed PEL of 200 $\mu g/m^3$ for all segments of the cotton industry.

[20]The Act specifies an informal rulemaking procedure to accompany the promulgation of occupational safety and health standards. See 29 U.S.C. §655(b) (2), (3), (4).

[21]The Standard provides that exposure to lint-free respirable cotton dust may be measured by a vertical elutriator, with its 15-micron particle size cutoff, or a "method of equivalent accuracy and precision."

[22]The manufacturing of cotton textile products is divided into several different stages. (1) In the operations of *opening, picking, carding, drawing,* and *roving,* raw cotton is cleaned and prepared for spinning into yarn. (2) In the operations of *spinning, twisting, winding, spooling* and *warping,* the prepared cotton is made into yarn and readied for weaving and other processing. (3) In *slashing* and *weaving,* the yarn is manufactured into a woven fabric. The Cotton Dust Standard defines "yarn manufacturing" to mean "all textile mill operations from opening to, but not including, slashing and weaving."

The nontextile industries covered by the Standard's 500 $\mu g/m^3$ PEL include, but are not limited to, "warehousing, compressing of cotton lint, classing and marketing, using cotton yarn (i.e. knitting), reclaiming and marketing of textile manufacturing waste, delinting of cottonseed, marketing and converting of linters, reclaiming and marketing of gin motes and batting, yarn felt manufacturing using waste cotton fibers and by products.

OSHA chose an implementation strategy for the Standard that depended primarily on a mix of engineering controls, such as installation of ventilation systems,[23] and work practice controls, such as special floor sweeping procedures. Full compliance with the PELs is required within 4 years, except to the extent that employers can establish that the engineering and work practice controls are infeasible. During this compliance period, and at certain other times, the Standard requires employers to provide respirators to employees. Other requirements include monitoring of cotton dust exposure, medical surveillance of all employees, annual medical examinations, employee education and training programs, and the posting of warning signs. A specific provision also under challenge in the instant case requires employers to transfer employees unable to wear respirators to another position, if available, having a dust level at or below the Standard's PELs, with "no loss of earnings or other employment rights or benefits as a result of the transfer."

On the basis of the evidence in the record as a whole, the Secretary determined that exposure to cotton dust represents a "significant health hazard to employees," and that "the prevalence of byssinosis should be significantly reduced: by the adoption of the Standard's PELs." In assessing the health risks from cotton dust and the risk reduction obtained from lowered exposure, OSHA relied particularly on data showing a strong linear relationship between the prevalence of byssinosis and the concentration of lint-free respirable cotton dust. At 200 $\mu g/m^3$ PEL, OSHA found that the prevalence of at least Grade 1/2 byssinosis would be 13% of all employees in the yarn manufacturing sector.

In enacting the Cotton Dust Standard, OSHA interpreted the Act to require adoption of the most stringent standard to protect against material health impairment, bounded only by technological and economic feasibility.

* * *

The Court of Appeals upheld the Standard in all major respects.[24] The court rejected the industry's

[23]Ventilation systems include general controls, such as central air conditioning, and local exhaust controls, which capture emissions of cotton dust as close to the point of generation as possible.

[24]The court remanded to the agency that portion of the Standard dealing with the cottonseed oil industry, after concluding that the record failed to establish adequately the Standard's economic feasibility.—U.S. App. D.C.—, 617 F2d 636, 669, 677 (1979).

claim that OSHA failed to consider its proposed alternative or give sufficient reasons for failing to adopt it. 617 F2d, at 652-654. The court also held that the Standard was "reasonably necessary and appropriate" within the meaning of §3(8) of the Act, because of the risk of material health impairment caused by exposure to cotton dust. 617 F2d, at 654-655, 654, n. 83. Rejecting the industry position that OSHA must demonstrate that the benefits of the Standard are proportionate to its costs, the court instead agreed with OSHA's interpretation that the standard must protect employees against material health impairment subject only to the limits of technological and economic feasibility. *Id.*, at 662-666. The court held that "Congress itself struck the balance between costs and benefits in the mandate to the agency" under §6(b) (5) of the Act, and that OSHA is powerless to circumvent that judgment by adopting less than the most protective feasible standard. 617 F2d at 663. Finally, the court held that the agency's determination of technological and economic feasibility was supported by substantial evidence in the record as a whole. *Id.*, at 655-662.

We affirm in part, and vacate in part.[25]

II

The principal question presented in this case is whether the Occupational Safety and Health Act requires the Secretary, in promulgating a standard

pursuant to §6(b)(5) of the Act, to determine that the costs of the standard bear a reasonable relationship to its benefits. Relying on §§6(b) (5) and 3(8) of the Act petitioners urge not only that OSHA must show that a standard addresses a significant risk of material health impairment, see *Industrial Union Department v. American Petroleum Institute, supra,* 448 U.S. at 639 (plurality opinion), but also that OSHA must demonstrate that the reduction in risk of material health impairment is significant in light of the costs of attaining that reduction.[26] Respondents on the other hand contend that the Act requires OSHA to promulgate standards that eliminate or reduce such risks "to the extent such protection is technologically and economically feasible."

*   *   *

Secretary presumed that no safe exposure level existed for carcinogenic substances. *Industrial Union Department v. American Petroleum Institute, supra,* 448 U.S. at 620, 624, 635–636, n. 39 and 40 (plurality opinion). Following this Court's decision, OSHA deleted those provisions of the Cancer Policy which required the "automatic setting of the lowest feasible level" without regard to determinations of risk significance.

In distinct contrast with its Cancer Policy, OSHA expressly found that "exposure to cotton dust presents a significant health hazard to employees," and that "cotton dust produced significant health effects at low levels of exposure." In addition, the agency noted that "grade 1/2 byssinosis and associated pulmonary function decrements are significant health effects in themselves and should be prevented in so far as possible." In making its assessment of significant risk, OSHA relied on dose response curve data (the Merchant Study) showing that 25% of employees suffered at least Grade 1/2 byssinosis at a 500 g/m^3 PEL, and that 12.7% of all employees would suffer byssinosis at the 200 g/m^3 PEL standard. Examining the Merchant Study in light of other studies in the record, the agency found that "the Merchant study provides a reliable assessment of health risk to cotton textile workers from cotton dust." OSHA concluded that the "prevalence of byssinosis should be significantly reduced" by the 200 μg/m^3 PEL ("200 μg/m^3 represents a significant reduction in the number of affected workers"). It is difficult to imagine what else the agency could do to comply with this Court's decision in *Industrial Union Department v. American Petroleum Institute.*

[26]Petitioners ATMI et al. express their position in several ways. They maintain that OSHA "is required to show that a reasonable relationship exists between the risk reduction benefits and the cost of its standards." Petitioners also suggest that OSHA must show that "the standard is expected to achieve a *significant reduction in* [the significant risk of material health impairment]" based on "an assessment of the costs of achieving it." Allowing that

---

[25]The post-argument motions of the several parties for leave to file supplemental memoranda are granted. We decline to adopt the suggestion of the Secretary of Labor that we should "vacate the judgment of the court of appeals and remand the case so that the record may be returned to the Secretary for further consideration and development." We also decline to adopt the suggestion of petitioners that we should "hold these cases in abeyance and . . . remand the record to the court of appeals with an instruction that the record be remanded to the agency for further proceedings."

At oral argument, and in a letter addressed to the Court after oral argument, petitioners contended that the Secretary's recent amendment of OSHA's so-called "Cancer Policy" in light of this Court's decision in *Industrial Union Department v. American Petroleum Institute,* 448 U.S. 607, was relevant to the issues in the present case. We disagree.

OSHA amended its Cancer Policy to "carry out the Court's interpretation of the Occupational Safety and Health Act of 1970 that consideration must be given to the significance of the risk in the issuance of a carcinogen standard and that OSHA must consider all relevant evidence in making these determinations." Previously, although lacking such evidence as dose response data, the

To resolve this debate, we must turn to the language, structure, and legislative history of the Occupational Safety and Health Act.

A

* * *

Although their interpretations differ, all parties agree that the phrase "to the extent feasible" contains the critical language in §6(b) (5) for purposes of this case.

The plain meaning of the word "feasible" supports respondents' interpretation of the statute. According to Webster's Third New International Dictionary of the English Language, "feasible" means "capable of being done, executed, or effected," Accord, The Oxford English Dictionary ("Capable of being done, accomplished or carried out"); Funk & Wagnalls New "Standard" Dictionary of the English Language ("That may be done, performed or effected"). Thus, §6(b) (5) directs the Secretary to issue the standard that "most adequately assures . . . that no employee will suffer material impairment of health," limited only by the extent to which this is "capable of being done." In effect then, as the Court of Appeals held,

"[t]his does not mean that OSHA must engage in a rigidly formal cost-benefit calculation that places a dollar value on employee lives or health." Petitioners describe the required exercise as follows: "First, OSHA must make a responsible determination of the cost and risk reduction benefits of its standard. Pursuant to the requirement of Section 6(f) of the Act, this determination must be factually supported by substantial evidence in the record. The subsequent determination whether the reduction in health risk is 'significant' (based upon the factual assessment of costs and benefits) is a judgment to be made by the agency in first instance."

Respondent disputes petitioners' description of the exercise, claiming that any meaningful balancing must involve "placing a [dollar] value on human life and freedom from suffering," Brief for Respondent Secretary of Labor, at 59, and that there is no other way but through formal cost-benefit analysis to accomplish petitioners' desired balancing. Cost-benefit analysis contemplates "systematic enumeration of all benefits and all costs, tangible and intangible, whether readily quantifiable or difficult to measure, that will accrue to all members of society if a particular project is adopted." E. Stokey and R. Zeckhauser, A Primer for Policy Analysis 134 (1978); see National Academy of Sciences, Decision Making for Regulating Chemicals in the Environment 38 (1975). See generally E. Mishan, Cost-Benefit Analysis in 300 Economic Journal 683 (1965). Whether petitioners' or respondent's characterization is correct, we will sometimes refer to petitioners' proposed exercise as "cost-benefit analysis."

Congress itself defined the basic relationship between costs and benefits, by placing the "benefit" of worker health above all other considerations save those making attainment of this "benefit" unachievable. Any standard based on a balancing of costs and benefits by the Secretary that strikes a different balance than that struck by Congress would be inconsistent with the command set forth in §6(b)(5). Thus, cost-benefit analysis by OSHA is not required by the statute because feasibility analysis is.[29] See *Industrial Union Department v. American Petroleum Institute, supra,* 448 U.S. at 718 (Marshall, J., dissenting).

When Congress has intended that an agency engage in cost-benefit analysis, it has clearly indicated such intent on the face of the statute.

* * *

Certainly in light of its ordinary meaning, the word "feasible" cannot be construed to articulate such congressional intent. We therefore reject the argument that Congress required cost-benefit analysis in §6(b) (5).

B

Even though the plain language of §6(b)(5) supports this construction, we must still decide whether §3(8), the general definition of an occupational safety and health standard, either alone or in tandem with §6(b) (5), incorporates a cost-benefit requirement for standards dealing with toxic materials or harmful physical agents. Section 3(8) of the Act (emphasis added), provides:

The term "occupational safety and health standard" means a standard which requires conditions, or the adoption or use of one or more practices,

[29]In this case we are faced with the issue whether the Act requires OSHA to balance costs and benefits in promulgating a *single* toxic material and harmful physical agent standard under §6(b)(5). Petitioners argue that without cost-benefit balancing, the issuance of a single standard might result in a "serious misallocation of the finite resources that are available for the protection of worker safety and health," given the other health hazards on the workplace. . . . This argument is more properly addressed to other provisions of the Act which may authorize OSHA to explore costs and benefits for deciding between issuance of several standards regulating different varieties of health and safety hazards, e.g., §6(g) of the Act; see *Industrial Union Department v. American Petroleum Institute, supra,* 448 U.S. at 644; see also Case Comment, 60 B. U. L. R. 115, 122, n. 52, or [sic] for promulgating other types of standards not issued under §6(b)(5). We express no view on these questions.

means, methods, operations, or processes, *reasonably necessary or appropriate* to provide safe or healthful employment and places of employment.

Taken alone, the phrase "reasonably necessary or appropriate" might be construed to contemplate some balancing of the costs and benefits of a standard. Petitioners urge that, so construed, §3(8) engrafts a cost-benefit analysis requirement on the issuance of §6(b) (5) standards, even if §6(b) (5) itself does not authorize such analysis. We need not decide whether §3(8), standing alone, would contemplate some form of cost-benefit analysis. For even if it does, Congress specifically chose in §6(b)(5) to impose separate and additional requirements for issuance of a subcategory of occupational safety and health standards dealing with toxic materials and harmful physical agents: it required that those standards be issued to prevent material impairment of health *to the extent feasible*. Congress could reasonably have concluded that *health* standards should be subject to different criteria than *safety* standards because of the special problems presented in regulating them. See *Industrial Union Department v. American Petroleum Institute, supra*, 448 U.S. at 649, n. 54 (plurality opinion).

Agreement with petitioners' argument that §3(8) imposes an additional and overriding requirement of cost-benefit analysis on the issuance of §6(b) (5) standards would eviscerate the "to the extent feasible" requirement. Standards would inevitably be set at the level indicated by cost-benefit analysis, and not at the level specified by §6(b)(5). For example, if cost-benefit analysis indicated a protective standard of 1000 μg/m^3 PEL, while feasibility analysis indicated a 500 μg/m^3 PEL, the agency would be forced by the cost-benefit requirement to chose the less stringent point.[31] We cannot believe that Congress intended the general terms of §3(8) to countermand the specific feasibility requirement of §6(b) (5). Adoption of petitioners' interpretation would effectively write §6(b) (5) out of the Act. We decline to render Congress' decision to include a feasibility requirement nugatory, thereby offending the well-settled rule that all parts of a statute, if possible, are to be given effect.

*   *   *

Congress did not contemplate any further balancing by the agency for toxic material and harmful physical agents standards, and we should not "impute to Congress a purpose to paralyze with one hand what it sought to the promote with the other." *Weinberger v. Hynson, Westcott & Dunning, Inc., supra*, at 631, quoting *Clark v. Uebersec Finanz-Korporation*, 332 U.S. 480, 489 (1947).[32]

C

The legislative history of the Act, while concededly not crystal clear, provides general support for respondents' interpretation of the Act. The congressional reports and debates certainly confirm that Congress meant "feasible" and nothing else in using that term. Congress was concerned that the Act might be thought to require achievement of absolute safety, an impossible standard, and therefore insisted that health and safety goals be capable of economic and technological accomplishment. Perhaps most telling is the absence of any indication whatsoever that Congress intended OSHA to conduct its own cost-benefit analysis before promulgating a toxic material or harmful physical agent standard.

*   *   *

Not only does the legislative history confirm that Congress meant "feasible" rather than "cost-benefit" when it used the former term, but it also shows that

---

[31] In addition, as the legislative history makes plain, see *infra*, at 25-26, any standard that was not economically or technologically feasible would *a fortiori* not be "reasonably necessary or appropriate" under the Act. See *Industrial Union Department v. Hodgson,*—U.S. App. D.C.—, 499 F.2d 467, 478 (1974) ("Congress does not appear to have intended to protect employees by putting their employers out of business").

[32] This is not to say the §3(8) might not require the balancing of costs and benefits for standards promulgated under provisions other than §6(b)(5) of the Act. As a plurality of this Court noted in *Industrial Union Department,* if §3(8) had no substantive content, "there would be no statutory criteria at all to guide the Secretary in promulgating either national consensus standards or permanent standards other than those dealing with toxic material and harmful physical agents." Slip op., at 29–30, n. 45 [8 OSHC at 1597]. Furthermore, the mere fact that a 6(b)(5) standard is "feasible" does not mean that §3(8)'s "reasonably necessary or appropriate" language might not impose additional restraints on OSHA. For example, all §6()(5) standards must be addressed to "significant risks" of material health impairment. *Id.,* at 32 [at 1598]. In addition, if the use of one respirator would achieve the same reduction in health risk as the use of five, the use of five respirators was "technologically and economically feasible," and OSHA thus insisted on the use of five, then the "reasonably necessary or appropriate" limitation might come into play as an additional restriction on OSHA to choose the one-respirator standard. In this case we need not decide all the applications that §3(8) might have, either alone or together with §6(b)(5).

Congress understood that the Act would create substantial costs for employers, yet intended to impose such costs when necessary to create a safe and healthful working environment.[37] Congress viewed the costs of health and safety as a cost of doing business. Senator Yarborough, a cosponsor of the Williams bill, stated: "We know the costs would be put into consumer goods but that is the price we should pay for the 80 million workers in America." He asked:

> One may well ask too expensive for whom? Is it too expensive for the company who for lack of proper safety equipment loses the services of its skilled employees? Is it too expensive for the employee who loses his hand or leg or eyesight? Is it too expensive for the widow trying to raise her children on meager allowance under workmen's compensation and social security? And what about the man—a good hardworking man—tied to a wheel chair or hospital bed for the rest of his life? That is what we are dealing with when we talk about industrial safety. . . . We are talking about people's lives, not the indifference of some cost accountants.

Senator Eagleton commented that "[t]he costs that will be incurred by employers in meeting the standards of health and safety to be established under this bill are, in my view, *reasonable and necessary costs of doing business.*" Legis. Hist. 1150–1151 (emphasis added).[38]

Other Members of Congress voiced similar views.[39] Nowhere is there any indication that Congress contemplated a different balancing by OSHA of the benefits of worker health and safety against the costs of achieving them. Indeed Congress thought that the *financial costs* of health and safety problems in the workplace were as large or larger than the *financial costs* of eliminating these problems. In its statement of findings and declaration of purpose encompassed in the Act itself, Congress announced that "personal injuries and illnesses arising out of work situations impose a substantial burden upon, and are a hindrance to, interstate commerce in terms of lost production, wage loss, medical expenses, and disability compensation payment." The Senate was well aware of the magnitude of these costs:

> [T]he economic impact of industrial deaths and disability is staggering. Over $1.5 million is wasted in lost wages, and the annual loss to the Gross National Product is estimated to be over $8 billion. Vast resources that could be available for productive use are siphoned off to pay workmen's compensation benefits and medical expenses.

Senator Eagleton summarized, "Whether we, as individuals, are motivated by simple humanity or by simple economics, we can no longer permit profits to be dependent upon an unsafe or unhealthy worksite."

[Dissenting Opinions of Justices STEWART and REHNQUIST are omitted.]

---

[37]Because the costs of compliance would weigh particularly heavily on small businesses, Congress provided in §28 of the Act an amendment to the Small Business Act, 15 U.S.C. §636, making small businesses eligible for economic assistance through the Small Business Administration to comply with standards promulgated by the Secretary. Senator Dominick explained: "There is a provision in the bill which recognizes the impact that this particular legislation may have on small businesses. . . . It permits the Secretary to make loans to small businesses wherever the standards that are set by the National Government are so severe as to have caused a real and substantial economic injury. Under those circumstances the Secretary is entitled, through the Small Business Administration, to make loans to those businesses to get them over the hump, because of the need for new equipment, or because of new conditions within the shop, which would permit them to continue in operation. "I think that is a very significant and important provision for minimizing economic injury which could occur if the bill resulted in situations which would have very serious effects on businesses."

[38]Congress was concerned that some employers not obtain

a competitive advantage over others by declining to invest in worker health and safety:

> Although many employers in all industries have demonstrated an exemplary degree of concern for health and safety in the workplace, their efforts are too often undercut by those who are not so concerned. Moreover, the fact is that many employers—particularly smaller ones—simply cannot make the necessary investment in health and safety, and survive competitively, unless all are compelled to do so.

[39]See, e.g., Legis. Hist. 1030–1031 (remarks of Congressman Dent):

> Although I am very much disturbed over adding new costs to the operation of our production facilities because of the threats from abroad, I would say there is a greater concern and that must be for the production men who do the producing—the men who work in the service industries and the men and women in this country who daily go out and keep the economy moving and make it safe for all of us to live and to work and to be able to prosper in it.

NOTE

1. Given this interpretation of the Act, what is the propriety of the cost-benefit analysis performed as a part of OMB's review of proposed OSHA standards? (See Chapter 2 for a discussion of OMB's role in standard setting.) See Chapter 5 for a discussion of the use of cost-benefit analysis as a tool in standard setting.

◇

After the Supreme Court decided the benzene case, but before it affirmed the cotton dust decision of the D.C. Circuit Court of Appeals, the D.C. Circuit reviewed OSHA's occupational standard for lead.

OSHA had revised the prior 6(a) standard, and had reduced the permissible ambient air concentration from 200 $\mu g/m^3$ over an eight-hour period to 50 $\mu g/m^3$. The revised standard also included biological monitoring and medical surveillance provisions that required removing workers, with pay, whenever their blood level was above 50 $\mu g/dl$ blood (or for other medical reasons). The standard placed different time requirements for compliance on different parts of the lead industry.

Addressing the requirements laid down by the Supreme Court in the benzene decision, the Court found that subclinical effects of lead exposure did indeed constitute "material impairment," that OSHA properly determined that the risk of lead exposure was "significant," and that the new standard was "reasonably necessary or appropriate" for reducing this risk. The Court also made important clarifying contributions to the meaning of feasibility.

United Steelworkers of America, AFL-CIO-CLC v. Marshall and Bingham

647 F.2d 1189 (D.C. Cir. 1980)

WRIGHT, Chief Judge.

In November 1978 the Occupational Safety and Health Administration (OSHA), exercising its authority and responsibility under Section 6 of the Occupational Safety and Health Act, issued new rules designed to protect American workers from exposure to airborne lead in the workplace.[1] In these consolidated appeals petitioners representing both labor union and industry interests challenge virtually every aspect of the new lead standard and the massive rulemaking from which it emerged.[2] The unions[3] claim that OSHA has failed to carry out its statutory duty to ensure that "no employee will suf-

[1]The new standard appears at 43 Fed. Reg. 53007 (1978), with minor amendments at 44 Fed. Reg. 5446 (1979), and at 29 C.F.R. §1910.1025 (1979). The Preamble to the standard appears at 43 Fed. Reg. 52952–53007 (1978), with Attachments, which we shall refer to as part of the Preamble, at 43 Fed. Reg. 54354–54509 (1978). In the interests of simplicity, we shall cite the Preamble only by page and column number for Volume 43 of the Federal Register, and the final lead standard only by section number for Title 29 of the 1979 edition of the Code of Federal Regulations.

The Secretary of Labor has delegated his authority to set standards under 29 U.S.C. §655 (1976) to the Assistant Secretary of Labor for Occupational Safety and Health, who is the head of OSHA. For purposes of this opinion, the words "Secretary," "agency," and "OSHA" are interchangeable.

[2]By order of March 1, 1979 this court agreed to stay certain portions of the new lead standard pending outcome of this appeal. Our summary at the end of this opinion identifies parts of the stay which shall remain in effect for certain industries while OSHA reconsiders certain issues on remand.

[3]Briefs in opposition to portions of the standard were filed by the United Steelworkers of America (USWA) and intervenor Oil, Chemical and Atomic Workers International (OCAW). USWA, however, joined by the United Automobile Workers, also filed a reply brief defending the new lead standard against the challenges posed by the industry parties. Two *amici* briefs supporting employee interests were also filed, in [sic] behalf of California state

fer material impairment of health. . . . "[4] The industry parties[5] charge OSHA with almost every procedural sin of which an agency can be guilty in informal rulemaking, attack some of the most important substantive provisons of the standard as exceeding OSHA's statutory authority, and assert that the agency has failed to present substantial evidence to support the factual bases of the standard. Though the numerous challenges to the standard and the size and complexity of the rulemaking require of us a lengthy analysis of the issues, we affirm most of the new occupational lead standard, remanding to the agency for reconsideration only the question of the feasibility of the standard for a number of the affected industries.[6]

* * *

employment agencies and women's rights and civil rights groups, both briefs solely devoted to supporting the OSHA provision for medical removal protection.
[4]Because we find the union arguments that attack the standard without any notable merit, while the industry challenges pose extremely substantial questions of law and fact, we address the union arguments only very briefly near the end of this opinion.
[5]The central industry brief opposing the standard was filed by Lead Industries Association, Inc. (LIA), representing all affected industries except those which OSHA has placed in the category of "other industries." Industry briefs challenging OSHA were also filed by the American Iron & Steel Institute (AISI), representing some of other "other industries;" the National Association of Recycling Industries, Inc. (NARI), representing the secondary lead smelters as well as recyclers of non-lead metals; the three major domestic automobile manufacturers—Chrysler, Ford, and General Motors; and the Bell System. *Amicus* Capital Legal Foundation filed a brief opposing the medical removal provision. Intervenor National Construction Association filed a brief supporting OSHA's decision to exempt the construction industry from the new lead standard. This opinion will frequently refer to LIA as the source of industry arguments made by that party as well as other industry parties.
[6]The summary at the end of this opinion explains the terms of the remand and lists the industries for which the question of feasibility remains open on remand.

The dissenting opinion takes strong opposition to our views on such important issues as the role of the consultants in the rulemaking, the adequacy of notice in the rulemaking, the statutory validity of the medical removal protection program, and the technological and economic feasibility of the standard. As we acknowledge at several points in this opinion, OSHA's procedures and evidence-gathering were less than perfect, and a number of important questions on appeal are very close. We believe, however, that the Supreme Court's unanimous

II. SCOPE OF REVIEW

In our recent decision in the cotton dust case, *AFL-CIO v. Marshall*, 617 F.2d 636 (D.C. Cir. No. 78-1562, decided October 24, 1979), we dealt at length with the criteria for judicial review appropriate to so-called "hybrid rulemaking" in general, and to cases under the OSHAct in particular. In the present case we feel no need to reinvent the wheel by recounting the relevant legislative and judicial history of the OSHAct and the general background of hybrid rulemaking. Rather, we incorporate our analysis in the cotton dust case as the established and proper interpretation of our scope of review for OSHA cases. However, we summarize that analysis very briefly here.

Though the OSHAct adopts the "substantial evidence" test for judicial review, 29 U.S.C. §655(f)(1976), rulemaking under that Act remains essentially informal. *AFL-CIO v. Marshall, supra,* 617 F.2d at 650:

> The tasks of this reviewing court are thus to ensure that the agency has (1) acted within the scope of its authority; (2) followed the procedures required by statute and by its own regulations; (3) explicated the bases for its decision; [and] (4) adduced substantial evidence in the record to support its determinations.

(Footnotes omitted.) Of course, we must rigorously review the agency's interpretations of the substantive provisions of its statutory mandate. Moreover, we must ensure that the agency has lived up to statutory and constitutional standards in its rulemaking procedure—a subject we address in the next part of this opinion. These, however, are conventional problems of judicial review. The peculiar problem of reviewing the rules of agencies like OSHA lies in applying the substantial evidence test to regulations which are essentially legislative and rooted in inferences from complex scientific and factual data, and which often necessarily involve highly speculative projections of technological development in areas wholly lacking in scientific and economic certainty. We noted in the

opinion in *Vermont Yankee Nuclear Power Corp. v. Natural Resources Defense Council, Inc.,* 435 U.S. 519 (1978), requires us to be especially wary of imposing on the agency any procedural or evidentiary constraints beyond those explicitly established by Congress in the OSHAct. Therefore, though we recognize that the dissent has raised serious questions about a number of aspects of the rulemaking, we disagree with its conclusions. However, in as complex and unwieldy a case as this, we think it impractical and unnecessary to respond to the dissent in a point-by-point fashion. Rather, we believe our extensive discussions of the key issues adequately explain our reasons for rejecting the dissent's views. . . .

cotton dust case that we do not pretend to have the competence or the jurisdiction to resolve technical controversies in the record, 617 F.2d at 651-652, or, where the rule requires setting a numerical standard, to second-guess an agency decision that falls within a "zone of reasonableness," *id.*, at n.60, *quoting Hercules, Inc. v. EPA*, 598 F.2d 91, 107 (D.C. Cir. 1978); *see Industrial Union Dep't. AFL-CIO v. American Petroleum Institute,*—U.S.—,—, 48 U.S. L. Week 5022, 5035 (July 2, 1980) (herinafter cited to Law Week pages only) (plurality opinion); *id.*, at 5037 [at 1606-07] (Burger, C.J., concurring). Rather, our task is to "ensure public accountability," 617 F.2d at 651, by requiring the agency to identify relevant factual evidence, to explain the logic and the policies underlying any legislative choice, to state candidly any assumptions on which it relies, and to present its reasons for rejecting significant contrary evidence and argument. Generalization cannot usefully take us further. We will discuss other aspects of the proper scope of review as the need arises in our analysis of distinct issues in the case.

*   *   *

V. PERMISSIBLE EXPOSURE LIMIT

OSHA's creation of a permissible exposure limit for lead raises intertwined statutory and evidentiary issues. To resolve these difficult issues, we must address the relationship between and the respective language of the statutory sections under which OSHA must operate, and examine the sequence of reasoning which led OSHA to a PEL of 50 $\mu g/m^3$.

A. The Threshold Question: "Reasonable Necessity" and "Significant Risk"

The statutory criteria governing OSHA's creation of permissible exposure limits for toxic substances appear in two sections of the Act. Section 3(8) generally defines a "standard":

The term "occupational safety and health standard" means a standard which requires conditions, or the adoption or use of one or more practices, means, methods, operations, or processes, reasonably necessary or appropriate to provide safe and healthful employment and places of employment.

Section 6(b)(5) of the Act then sets out the Secretary's duties in creating a standard for dangerous substances:

The Secretary, in promulgating standards dealing with toxic material or harmful physical agents

under this subsection, shall set the standard which most adequately assures, to the extent feasible, on the basis of the best available evidence, that no employee will suffer material impairment of health or functional capacity even if such employee has regular exposure to the hazard dealt with by such standard for the period of his working life. Development of standards under this subsection shall be based upon research, demonstrations, experiments, and such other information as may be appropriate. In addition to the attainment of the highest degree of health and safety protection for the employee, other considerations shall be the latest available scientific data in the field, the feasibility of the standards, and experience gained under this and other health and safety laws.

Though the first of these provisions might appear to be definitional only, with the second specifically announcing the protective goal at which OSHA must aim and the resources on which it must draw, a plurality of the Supreme Court, in a decision issued after argument in the present case, has stated that Section 3(8) establishes a threshold responsibility OSHA must carry out before it exercises its authority under Section 6(b)(5). In *Industrial Union Dep't. AFL-CIO v. American Petroleum Institute, supra,* the plurality stated that before OSHA creates a new exposure limit for a toxic agent, it must first "find, as a threshold matter, that the toxic substance in question poses a significant health risk in the workplace and that a new, lower standard is therefore "reasonably necessary or appropriate to provide safe or healthful employment and places of employment.'" 48 U.S. L. Week at 5024.[84] In so construing this section, the plurality made clear its view that Congress had not mandated OSHA to seek an absolutely

[84]The *American Petroleum Institute* decision affirmed the judgment of the Fifth Circuit, which had invalidated the benzene standard for its failure to adhere to the implicit requirements of §3(8). *American Petroleum Institute v. OSHA*, 581 F.2d 493 [6 OSHC 1959] (5th Cir. 1978). But the lack of a majority opinion in the decision and the complex overlapping among the five separate opinions leave the Supreme Court's view of OSHA's statutory responsibilities in some doubt. Indeed, the Court reached no decision on the question that had proved decisive in the Fifth Circuit: whether §3(8) requires OSHA to weigh the costs of a new standard against its benefits.

Justice Stevens wrote the plurality opinion, and was joined completely by Justice Stewart and the Chief Justice, *Industrial Union Dep't, AFL-CIO v. American Petro-*

risk-free workplace, or to require industry to eliminate even insignificant risks of harm so long as the effort is not technologically impossible or financially ruinous. *Id.*, at 5031 [at 1598]. Rather, the Secretary must examine the current exposure level of the toxic substance to determine whether it leaves workers subject to a significant risk, and whether

leum Institute, supra . . . though the latter also wrote a separate concurring opinion emphasizing that the courts must continue to grant OSHA great leeway in its essentially legislative decisions on regulatory policy, 48 U.S. L. Week at 5036-5037.

Justice Powell concurred in Parts I, II, III-A, III-B, III-C, and III-E of Justice Stevens' opinion, and wrote a separate concurring opinion which took the view that §3(8) imposed more stringent requirements on OSHA than the plurality opinion had acknowledged. *Id.*, at 5037. Justice Powell found that in addition to invoking its presumption about carcinogens—a presumption for which he saw insufficient support, *id.*—OSHA had relied on the "fall-back position" that the benzene standard was based on specific evidence of harm. *Id.* In this view, OSHA had indeed *attempted* to carry its burden under §3(8) by finding that the benefits of lowering the benzene PEL to 1 ppm were likely to be "appreciable." *See id.* at §5034 (plurality opinion). But Justice Powell took that statutory burden to be greater than Justice Stevens had been willing to hold in this case, and concluded that OSHA had failed to meet it. Justice Powell apparently believed that OSHA had not presented substantial evidence of significant harm from benzene at the current PEL. *Id.*, at 5038 (Powell, J., concurring). But he concluded in any event that §3(8) also required OSHA to determine that the costs of the new standard bore a reasonable relationship to its expected benefits. *Id.* In asserting that an OSHA standard is invalid if its economic effect is "wholly disproportionate" to its likely health and safety benefits, *id.* Justice Powell, alone among his colleagues, took a partial step toward approving the Fifth Circuit's position that OSHA must attempt a quantitative comparison of the costs and benefits of a new standard to prove the standard "reasonably necessary" under §3(8). *American Petroleum Institute v. OSHA, supra* 581 F.2d at 502-503.

Justice Rehnquist concurred in the judgment only, stating in his separate opinion that the first sentence of §6(b)(5), mandating OSHA to "set the standard which most adequately assures, to the extent feasible, that no employee will suffer material impairment of health," was unconstitutional. *Id.*, at 5039. Finding the language in this crucial sentence merely precatory, *id.* at 5041, Justice Rehnquist asserted that it failed so utterly to inform OSHA as to what costs the agency could impose on industry in regulating a substance for which no safe level could be found that it violated the nondelegation doctrine established by *Schechter Poultry Corp. v. United States,* 295 U.S. 495 (1935), and *Panama Refining Co. v. Ryan,* 293 U.S. 388 (1935).

that risk can be eliminated or lessened by lowering the limit. *Id.*[85]

*   *   *

In creating the new lead standard, OSHA has clearly met the Section 3(8) threshold test of proving "significant harm" described by the *American Petroleum Institute* plurality. OSHA nowhere relied on categorical assumptions about the effects of lead poisoning; indeed, since the lead standard does not rest on the carcinogenic effects of lead, OSHA did not even have available to it any general policy dictating that there is no safe level of lead. Nor did OSHA rest on evidence of the dangers of lead at very high exposure levels and then simply infer that those dangers would decrease as the PEL was lowered. Rather, OSHA amassed voluminous evidence of the specific harmful effects of lead at particular blood-lead levels with air-lead levels. By this means OSHA was able to describe the actual harmful effects of lead on a worker population at both the current PEL and the new PEL. In its proof of significant harm from lead at the current PEL and its careful measurement of the likely reduction in that harm at the new PEL, the lead standard stands in marked contrast to the benzene standard struck down by the Supreme Court.

*   *   *

B. The Section 6(b)(5) Question: "Material Impairment"

Having carried its burden under Section 3(8), OSHA must still meet the demands of Section 6(b)(5). *Industrial Union Dep't, AFL-CIO v. American Petroleum Institute, supra,* 48 U.S. L. Week at 5031 n. 48 (plural-

Nevertheless, in stating that §3(8) of the Act did not, as the plurality believed, impose a general check on OSHA's exercise of its standardsetting duty under §6(b)(5), *id.* Justice Rehnquist actually created a five-person majority, along with the dissenters, Justices Marshall, Brennan, White, and Blackmun, *id.*, at 5043 (Marshall, J., dissenting), for the view that §3(8) does not place on OSHA any threshold burden of proving "significant harm."

[85]In mentioning this second aspect of the threshold requirement—whether the significant harm at the current level can be eliminated or lessened by a change in the PEL—the plurality implies that §3(8) requires OSHA to prove by specific evidence the level of risk at the *new* PEL, as well as the current PEL. The plurality leaves this point somewhat unclear, but in any event a requirement of such proof would seem to follow from the second statutory provision governing OSHA's toxic agent standards, §6(b)(5). *See* Part V-B *infra.*

ity opinion). To repeat, the key language of that section reads:

> The Secretary ... shall set the standard which most adequately assures, to the extent feasible, on the basis of the best available evidence, that no employee will suffer material impairment of health or functional capacity even if such employee has regular exposure to the hazard dealt with by such standard for the period of his working life.

This language is neither precise nor artful. In particular, we note the rather troublesome phrase "most adequately assures." "Adequately," both in normal use and as a contemporary legal cliché, means "suitably" or "passably" or "just barely." Literally read, "most adequately" would seem to require OSHA to set the *least demanding* standard which offers a tolerable degree of protection. That reading, of course, makes little sense in the context of the legislative history and other language in the statute, and we readily assume that Congress used "most adequately" to mean "best." We stress the point, however, to demonstrate the difficulty a court faces in reviewing the work of an agency acting under a statute whose vague language at times seems to reflect more invocational rhetoric than measured legislative mandate.

That difficulty becomes more acute when we consider the two limits this section places on OSHA's power to protect workers. We take up later the most vexing of these limits—the requirement that a standard be "feasible." *See* Part VI *infra*. But in reviewing the PEL we must address the other limit—OSHA's mandate to eliminate "*material* impairment" of health and safety. The word "material," like "adequately," may be mere rhetoric. On the other hand, it may qualify OSHA's power and discretion.[95] If so, "material" requires careful construction. Does it simply mean something beyond the "trivial" or the "*de minimis*"? Or does it mean the physically overt or manifest? If it means the latter, as the industry argues—and indeed *American Petroleum Institute* suggests that it does not mean the former, *see* text and notes at notes 84-85 *supra*—then from how remote a vantage must OSHA act to protect workers from overt impairment? Must OSHA wait until impairment is overt, or can it act earlier to prevent the impairment from ever becoming overt?

The essential question under Section 6(b)(5) for this case is whether OSHA acted within the limits of its mandate to establish "material" impairment of health when it set a standard designed to protect workers from the subclinical effects of lead. As a statutory matter, after examining precedent and legislative history, we hold that Section 6(b)(5) empowers OSHA to set a PEL that prevents the subclinical effects of lead that lie on a continuum shared with overt lead disease.

LIA's argument that the OSHAct bars the agency from acting to prevent the subclinical effects of lead exposure is *virtually* foreclosed by our recent decision in the cotton dust case. *AFL-CIO v. Marshall, supra.* The textile industry argued there that neither Section 6(b)(5) nor Section 3(8) (allowing OSHA to adopt means "reasonably necessary or appropriate" to protecting workers) permitted the agency to set a PEL designed to prevent the acute but reversible symptoms of byssinosis, since those symptoms were not a "material impairment of health." 617 F.2d at 654.

We acknowledged there that the statute did not define "material," but found it unnecessary to decide whether these early symptoms were indeed "material" impairments. 617 F.2d at 654, n.83. Rather, we assumed that chronic byssinosis was itself a material impairment, and that OSHA could control the early symptoms as a means of preventing the chronic disease from developing. 617 F.2d at 655. Though OSHA conceded there were serious gaps in medical knowledge about the causal relationship between the acute and the chronic phases of the disease, it had some evidence that the acute symptoms weakened the worker's pulmonary system and made him more susceptible to the chronic phase. 617 F.2d at 655.

Here, of course, OSHA is acting to prevent, not overt early symptoms of a disease, but subclinical effects. Nevertheless, the reasoning of our cotton dust holding applies here. We held there that OSHA need not wait until symptoms of a disease appeared, but could act to "reduce the risk" of serious material impairment. 617 F.2d at 654-655.

*   *   *

VI. FEASIBILITY—DEFINING THE STANDARD

The feasibility issue illustrates better than any other in this case the difficulty both court and agency face in working under very general statutory language.[102]

---

[95]The legislative history suggests that the phrase "material impairment" was substituted for "any impairment" to ensure that OSHA would not have the power to eliminate every possible hazard to worker health. . . .

[102]LIA argued in its brief that as part of its economic analysis OSHA must undertake formal cost-benefit analysis to ensure that the costs of the new standard bear a reasonable relation to the benefits the standard would

Section 6(b)(5) of the OSH Act simply says that the agency is to set a standard which best protects workers "to the extent feasible," and that "feasibility" must be a factor in any choice of a standard. On this crucial matter, which in this case was the subject of thousands of pages of record evidence and comments and testimony from 150 parties, Congress has said no more. The conventional arts of statutory construction are of little help in understanding such scant language, especially where the legislative history is virtually silent.[103] What guidance we have in our examination of the feasibility of the lead standard comes from two sources: judicial glosses on "feasibility," and OSHA's fairly detailed explanation of its approach to feasibility in the Preamble to the lead standard.

---

yield. However, in the cotton dust case, *AFL-CIO v. Marshall, supra*. . . , we squarely held that nothing in the statute requires such an analysis, and that in fact cost-benefit analysis would contravene the congressional goal of protecting worker health and safety within the limits of economic possibility. 617 F.2d at 659. In *American Petroleum Institute v. OSHA, supra* note 84, the Fifth Circuit took the opposite view of this question, but in affirming the Fifth Circuit's judgment, *Industrial Union Dep't, AFL-CIO v. American Petroleum Institute, supra,* the Supreme Court expressed no view on the cost-benefit question. The plurality took the view that because OSHA's benzene standard did not pass the threshold test of "significant health risk" pursuant to §3(8) of the statute, the Court had no need to consider in that case the question whether OSHA must also prove a reasonable correlation between the costs and benefits of a new standard. 48 U.S. L. Week at 5024. Only Justice Powell went on record as finding such a requirement in §3(8), while Justice Rehnquist and the four dissenting Justices concluded that §3(8) imposed no general check on OSHA's standardsetting discretion at all. *See* note 84 *supra.* The Court, however, will address the cost-benefit question anew. *Republic Steel Corp v. OSHA,* 577 F.2d 825 [6 OSHC 1451] (3d Cir. 1978), *cert. granted.,*—U.S.—, 48 U.S. L. Week 3851 (June 24, 1980); *American Iron & Steel Institute v. OSHA, supra.* . . .

[103]We have little beyond Senator Javits' remark in his separate statement appended to the Senate Committee report on the OSHA bill:

As a result of this amendment the Secretary, in setting standards, is expressly required to consider feasibility of proposed standards. This is an improvement over the Daniels bill, which might be interpreted to require absolute health and safety in all cases, regardless of feasibility, and the Administration bill, which contains no criteria for standards at all.

. . . The word "feasible," as it finally appears in §6(b)(5), was placed there at the urging of Senator Dominick, but the opinions in *Industrial Union Dep't, AFL-CIO v. American*

A. Judicial Interpretation

1. *The meaning of feasibility.* The judicial history at least establishes clearly that there are two types of feasibility—technological and economic. *American Iron & Steel Institute v. OSHA,* 577 F.2d 825, 832 (3rd Cir. 1978), *cert. granted,*—U.S.—, 48 U.S. L. Week 3851 (June 24, 1980).[104] And the cases in this circuit and other circuits generate some useful criteria for measuring both technological and economic feasibility. But the factual complexity of the record in this case demands that we pursue the meaning of feasibility further than have most earlier courts. Moreover, as we shall discuss in detail below, unprecedented language in the "Means of Compliance" section of the lead standard—establishing an express presumption that the standard is feasible without resort to respirators—creates new problems in determining the feasibility of this standard. We believe, however, that our analysis of this language may ultimately illuminate the meaning of feasibility for OSHA standards generally.

The oft-stated view of technological feasibility under the OSH Act is that Congress meant the statute to be "technology-forcing." *AFL-CIO v. Brennan,* 530 F.2d 109, 121 (3d Cir. 1975). This view means, at the very least, that OSHA can impose a standard which only the most technologically advanced plants in an industry have been able to achieve—even if only in some of their operations some of the time. *American Iron & Steel Institute v. OSHA, supra,* 577 F.2d at 832–835. But under this view OSHA can also force industry to develop and diffuse new technology. *Society of Plastics Industries, Inc. v. OSHA,* 509 F.2d 1301, 1309 (2d Cir.), *cert. denied.* 421 U.S. 992 (1975). At least where the agency gives industry a reasonable time to develop new technology, OSHA is not bound to the technological status quo. *Id.* So long as it presents substantial evidence that companies acting vigorously and in good faith can develop the technology, OSHA can require industry to meet PEL's never attained anywhere.[105]

Petroleum Institute, supra note 54, are greatly divided and ultimately inconclusive on the significance of the final language. *See* 48 U.S. L. Week at 5033 n.53 (plurality opinion); *id.* at 5041 & n.4 (Rehnquist, J., concurring); *id.* at 5045 & n.8 (Marshall, J., dissenting).

[104]OSHA must in any event prepare economic impact analyses of proposed regulations. Executive Order No. 11821, 3 C.F.R. 926 (1971–75 compilation).

[105]In *American Iron & Steel Institute v. OSHA, supra.* . . , 577 F.2d at 838, the court held that OSHA had acted beyond its statutory power in explicitly and affirmatively requir-

The most useful general judicial criteria for economic feasibility come from Judge McGowan's opinion in *Industrial Union Dep't, AFL-CIO v. Hodgson, supra*. A standard is not infeasible simply because it is financially burdensome, 499 F.2d at 478, or even because it threatens the survival of some companies within an industry:

> Nor does the concept of economic feasibility necessarily guarantee the continued existence of individual employers. It would appear to be consistent with the purposes of the Act to envisage the economic demise of an employer who has lagged behind the rest of the industry in protecting the health and safety of employees and is consequently financially unable to comply with new standards as quickly as other employers.

Id. (footnote omitted).[106] A standard is feasible if it does not threaten "massive dislocation" to, *AFL-CIO v. Brennan, supra*, 530 F.2d at 123, or imperil the existence of, *American Iron & Steel Institute v. OSHA, supra*, 577 F.2d at 836, the industry.[107] No matter how initially frightening the projected total or annual costs of compliance appear, a court must examine those costs in relation to the financial health and profitability of the industry and the likely effect of such costs on unit consumer prices. *Id.* More specifically, *Industrial Union Dep't, AFL-CIO v. Hodgson, supra*, teaches us that the practical question is whether the standard threatens the competitive stability of an industry, 499 F.2d at 478, or whether any intra-industry or inter-industry discrimination in the standard might wreck such stability or lead to undue concentration. *Id.*, at 478, 481. Granting companies reasonable time to comply with new PEL's might not only enhance economic feasibility generally, but, where the agency makes compliance deadlines uniform for competing segments of industry, can also prevent such injury to competition. *Id.*, at 479–481.[108]

2. Proving feasibility. Having this general guidance on the *meaning* of feasibility, we still need to know how OSHA is to *prove* feasibility. Within the rather generous constraints of the substantial evidence test, and given our necessary deference to the agency when it is making an essentially legislative decision, *see* Part II *supra*, we must decide what to require of OSHA in the matter of types and quantity of data, precision of cost estimates, and certainty or prediction. On this question the cases offer some modest help.

As for technological feasibility, we know that we cannot require of OSHA anything like certainty. Since "technology-forcing" assumes the agency will make highly speculative projections about future technology, a standard is obviously not infeasible solely because OSHA has no hard evidence to show that the standard has been met. More to the point here, we cannot require OSHA to prove with any certainty that industry will be able to develop the necessary technology, or even to identify the single technological means by which it expects industry to meet the PEL. OSHA can force employers to invest all reasonable faith in their own capacity for technological innovation, *Society of Plastics Industries, Inc. v. OSHA, supra*, 509 F.2d at 1309 and can thereby shift to industry some of the burden of choosing the best strategy for compliance. OSHA's duty is to show that modern technology has at least conceived some industrial strategies or devices which are likely to be capable of meeting the PEL and which the industries are generally capable of adopting.

Our view finds support in the statutory require-

---

ing industry to perform research and development of an undefined nature and quantity. But the court there made clear that OSHA could set a standard that will require technology still only on the scientific horizon, *id.*, and which would thus likely require research and development as a practical matter.

[106]Although many employers in all industries have demonstrated an exemplary degree of concern for health and safety in the workplace, their efforts are too often undercut by those who are not so concerned. Moreover, the fact is that many employers—particularly smaller ones—simply cannot make the necessary investment in health and safety, and survive competitively, unless all are compelled to do so. The competitive disadvantage of the more conscientious employer is especially evident where there is a long period between exposure to a hazard and manifestation of an illness. In such instances a particular employer has no economic incentive to invest in current precautions, not even in the reduction of workmen's compensation costs, because he will seldom have to pay for the consequences of his own neglect. . . .

[107]The Third Circuit has even hinted that OSHA could find a hazardous industrial activity of such insignificant redeeming social utility that it could absolutely ban the activity where there was no technologically feasible means of abating the hazard. *AFL-CIO v. Brennan*, 530 F.2d 109, 1212 (3d Cir. 1975).

[108]The Preamble to the lead standard shows that OSHA carefully considered the dangers of destroying competition. At one point OSHA expressed doubt that destruction of "subgroups" of an industry that threatens the industry's competitive structure might not make a standard infeasible, but since it never relied on this rationale in finding the standard feasible for any industry we need not consider its validity.

ment that OSHA act according to the "best *available* evidence." 29 U.S.C. §655(b)(5) (1976) (emphasis added). OSHA cannot let workers suffer while it awaits the Godot of scientific certainty. It can and must make reasonable predictions on the basis of "credible sources of information," whether data from existing plants or expert testimony. *AFL-CIO v. Marshall, supra,* 617 F.2d at 650; *see Industrial Union Dep't. AFL-CIO v. American Petroleum Institute, supra,* 48 U.S. L. Week at 5035 (plurality opinion). Our role is to ensure that the agency has developed substantial evidence in meeting that task.

The ironic truth is that "technology-forcing" makes the agency's standard of proof somewhat circular. Since the agency must hazard some prediction about experimental technology, it may not be able to determine the success of new means of compliance until industry implements them. *Society of Plastics Industries, Inc. v. OSHA, supra,* 509 F.2d at 1309 (citing OSHA's explanation of the vinyl chloride standard). Conversely, OSHA or the courts may discover that a standard is infeasible only after industry has exerted all good faith efforts to comply. *Atlantic & Gulf Stevedores, Inc. v. OSHRC, supra,* 534 F.2d at 550. Both the agency, in issuing, and the court, in upholding, a standard under this principle obviously run the risk that an apparently feasible standard will prove technologically impossible in the future. But, while we will address below the complexities of this issue, we note the general agreement in the courts that such flexible devices as variance proceedings can correct erroneous predictions about feasibility when the error manifests itself, and thus can reduce this risk. *E.g., Society of Plastics Industries, Inc. v. OSHA, supra,* 509 F.2d at 1310.

Proving economic feasibility presents different problems. We know from the cotton dust case that OSHA need not specifically weigh industry costs against worker benefits. . . . But when the agency has proved technological feasibility by making reasonable predictions about experimental means of compliance, the court probably cannot expect hard and precise estimates of costs.[109] Nevertheless, the agency must of course provide a reasonable assessment of the likely range of costs of its standard, and the likely effects of those costs on the industry. The practical question is what form that assessment must take.

The one case to address this question in some

detail is the cotton dust decision. Judge Bazelon clearly implied that the agency need not engage in massive data collection with its own staff, but can rely on data and estimates produced by consultants and even the industries themselves. OSHA, in effect, can then produce its "own" estimate by modifying the consultants' or industries' estimates where it finds specific faults in their calculations. More important, OSHA can correct such "double-counting" errors as failing to allow for investment tax credits or productivity benefits from new control devices, or neglecting to consider the cost advantages of installing new machines instead of "retrofitting" old equipment, or including in its sums compliance costs for firms that were already in substantial compliance with the new standard on their own initiative and so would incur no new costs. *AFL-CIO v. Marshall, supra,* 617 F.2d at 660. And OSHA can revise any gloomy forecast that estimated costs will imperil an industry by allowing for the industry's demonstrated ability to pass through costs to consumers. *Id.,* 617 F.2d at 661.[110]

*   *   *

B. The Circularity Problem

Despite these useful guidelines, the cases have left one very serious and elusive ambiguity at the heart of the feasibility question, an ambiguity highlighted —or exacerbated—by a new twist OSHA has placed in the "Means of Compliance" section of the new lead standard. The problem is essentially that we cannot know if a standard is feasible until we know exactly what it expects of employers, and to the extent that we find OSHA's rules to be less demanding than they would otherwise appear, we may find it easier to accept them as feasible. There should be a clear line between construing the precise terms of the demands OSHA places on industry and assessing the feasibility of those demands. But some of the cases have blurred this line. They have justified their deference to OSHA's feasibility findings, at least in part, by discovering flexibility in OSHA's regulatory scheme that renders the standards less stringent than they may first appear. These apparently very reasonable decisions, however, do not address a serious analytic problem in the standards.

For example, two cases, in upholding OSHA stan-

[109]In the cotton dust decision Judge Bazelon also noted that an agency will inevitably have difficulty achieving precise cost estimates when it must depend on often recalcitrant industry sources for much of its data. *AFL-CIO v. Marshall, supra. . . ,* 617 F.2d at 661.

[110]The cotton dust decision does bar OSHA from one means of revising outside estimates: OSHA cannot act in circular fashion by subtracting from the total cost of the standard the costs of compliance of those firms which are likely to go out of business precisely because they cannot afford these costs. *Id.,* 617 F.2d at 671 n.219.

dards as feasible, have pointed to the availability of variance proceedings as means by which companies can gain relief from standards which prove too demanding. *American Iron & Steel Institute v. OSHA, supra*, 577 F.2d at 835; *Society of Plastics Industries, Inc. v. OSHA, supra*, 509 F.2d at 1310.[112] And at least one court has noted that even if industry cannot find the technology to meet the PEL, OSHA has simply required companies to install new engineering controls and work practices "to the extent feasible," and allowed them to use respirators to meet the PEL when feasible engineering and work practice controls prove insufficient to meet the PEL by themselves. *Id.*

The reference these courts make to variances glosses over important details in the OSHAct. The statute speaks of variances in two sections. In 29 U.S.C. §655(d)(1976) the statute describes what are in effect "permanent variances." These variances from a standard, however, are only available to employers whose workplaces "are as safe and healthful as those which would prevail if [they] complied with the standard." *Id.* These variances are therefore useless to the employer who claims that he can find no practical way of meeting the health and safety demands of an OSHA standard, assuming, of course, that OSHA does not find respirators as "safe and healthful" as engineering and work practice controls. Indeed, these variances seem generally irrelevant to the whole scheme of the lead standard. Section 655(d) seems to contemplate a standard that requires employers to use particular means of compliance to meet a health or safety goal; the section would allow employers to find an alternate route to the same goal. The lead standard, by contrast, essentially leaves the choice of particular engineering controls or work practices to the employer, and simply specifies the goal—the PEL.[113]

The other type of statutory variance, essentially a "temporary variance,"[114] derives from 29 U.S.C. §655(b)(6)(A) (1976). If an employer cannot comply with a standard by its effective date "because of unavail-

ability of professional or technical personnel or of materials and equipment . . . or because necessary construction or alteration of facilities cannot be completed by the effective date," *id.*, he may obtain temporary relief from compliance. This section appears to grant some relief at least to the employer who is technologically incapable of meeting a standard.[115] The problem, of course, is that such a variance is *very* temporary. Relief is generally for no more than one year, though some employers may obtain renewals that could effectively extend the variance to three years.[116] These variances may suffice for many employers claiming infeasibility. But OSHA's predictions about feasibility may turn out to be so faulty as to make even three years' relief insufficient. In any event, the temporary variance hardly seems designed to relieve an entire industry—as opposed to an isolated employer—of a standard which the industry appears unable *ever* to meet.

The Second Circuit's reassuring reference to the provision in the vinyl chloride standard allowing employers to rely on respirators when "feasible" engineering and work practice controls prove inadequate to meet the PEL, *Society of Plastics Industries, Inc. v. OSHA, supra* 509 F.2d at 1310 also glosses over problems in determining OSHA's burden of proving feasibility. Assuming, as we probably can in the present case, that the PEL is feasible so long as the employer can rely on respirators when the best possible engineering and work practice controls prove insufficient,[117] a court should have no trouble find-

---

[112]The Second Circuit misstated the issue somewhat by speaking of "amendments" to standards available under 29 U.S.C. §655(b)(6)(A) (1976), which is in fact the temporary variance provision. *Society of Plastics Industries, Inc. v. OSHA.* 509 F.2d 1301, 1310 (2d Cir.), *cert. denied.* 421 U.S. 992 (1975).

[113]OSHA has argued in its brief that a permanent variance from the lead standard may be available to at least one industry—telecommunications. OSHA brief at 246. We need not address that argument now, since this industry is one of those for which we remand the question of feasibility. . . .

[114]The parties—even including OSHA—have enhanced the confusion over the variance issue by referring to what

can only be temporary variances as "abatement proceedings"—a term which appears nowhere in the statute. We also note one minor and very special form of variance under the statute: OSHA, without any express restrictions, can grant a variance from any standard to permit an employer to participate in an experiment designed by OSHA or HEW to test new safety or health controls. 29 U.S.C. §655(b)(6)(C) (1976).

[115]OSHA construes the legislative history of the statute as barring claims of economic hardship as grounds for temporary variances.

[116]LIA argues that the maximum statutory length of a temporary variance, including renewals, is two years. Though the statutory language may leave some doubt, we believe LIA misreads it. Under 29 U.S.C. §655(b)(6)(A) (1976) a variance can be as long as one year when necessary, and may be renewed twice. The statute does say that no *interim* order of renewal can last more than 180 days, but the 180-day limit does not appear to apply to a non-interim renewal granted after a hearing.

[117]LIA nowhere seriously contends that respirators cannot reduce the lead level in the air workers breathe below 50 $\mu g/m^3$, or that the industries cannot afford these respirators.

ing an OSHA standard feasible if, in somewhat circular fashion, the employer need use only those non-respirator controls which are feasible. But the Second Circuit never considered the potential circularity in this scheme. If employers need only use those engineering and work practice controls which prove feasible, what meaning is there in the statutory requirement of feasibility? How can such a standard ever be infeasible?

The cases have apparently assumed that there *is* some meaning to the feasibility test—that it does not suffer a fatal logical flaw. In this assumption, as we shall see, they are doubtless right, but their failure to address the problem has left us somewhat rudderless in approaching the great difficulty of the feasibility question in the present case. Similarly, the cases are ultimately right in their assumption that procedures in the OSHA regulatory scheme giving employers relief from impractical standards reduce OSHA's burden of proof on feasibility. But the cases have never carefully considered the range and significance of these procedures.[118] We find it necessary to examine both these matters, not just because this rulemaking is so complex, but also because the industry parties here have directed our attention to language in the lead standard different from that of all previous standards which, the industry argues, demonstrates that the lead standard is significantly more stringent than all earlier standards, and so harder to prove feasible. We believe the industry is right in focusing on this language, but we conclude, after what we hope will be clarifying analysis, that the difference is ultimately in emphasis, not substance.

C. Resolving the Circularity

1. *Construing the earlier standards.* Every earlier OSHA standard restricting toxic substances that has created a hierarchy among preferred means of compliance has by its terms built the very concept of feasibility into the standard. The cotton dust standard is typical. It requires employers to meet the PEL solely through engineering and work practice controls "except to the extent that the employer establishes that such

controls are not feasible." Where "feasible engineering and work practice controls" cannot achieve the PEL, the employer may and must use respirators to make up the difference, though he must continue to use engineering and work practice controls to reduce exposure as well as he can.

The feasibility test for a standard that only expects feasible improvements from employers may appear circular. But reasonable construction of such a standard avoids circularity. The cases have apparently treated these standards as creating a general *presumption* of feasibility for an industry. A company could not simply refuse to pursue engineering or work practice controls by asserting their infeasibility. Rather, it would have to attempt to install controls to the limits of contemporary technical knowledge and of its own financial resources. Judicial review of feasibility would have some meaning, because the court would have to find substantial evidence to justify this presumption—evidence that the technical knowledge to meet the PEL without relying on respirators was available or would likely be available when deadlines arrive, and that enough firms could afford to so meet the PEL that the market structure of the industry would survive. Thus the court would do a preliminary test of feasibility on any pre-enforcement challenge to the rulemaking.

But the court always reserves the power to test feasibility bargain later—in reviewing denial of a temporary variance or, where an employer found such a variance insufficient, in judicial review of an enforcement proceeding under 29 U.S.C. §659(1976). In the temporary variance proceeding the employer would of course argue that the standard was infeasible for his company in particular—at least at the effective date of the standard. But in an enforcement proceeding the employer could also expect the agency and court to entertain the affirmative defense that the standard had proved *generally* infeasible—even if the court had earlier found otherwise on pre-enforcement review. In any of these proceedings, of course, the employer would bear the burden of proof.[119]

---

[118]The Third Circuit has indeed analyzed OSHA *enforcement* proceedings in some detail. *Atlantic & Gulf Stevedores v. OSHA.* 534 F.2d 541 (3d Cir. 1976). But it offered that analysis in considering an appeal from an enforcement proceeding. The enforcement proceeding has received no judicial attention in the context of a general pre-enforcement review of an entire standard, where the court would be able to consider the effect of the future availability of a feasibility defense in such a proceeding on the *initial* feasibility decision.

[119]Because the financial incapacity of a single firm to meet the PEL cannot normally prove a standard infeasible, presumably an employer could only raise a defense of *technological* infeasibility in an enforcement proceeding. Indeed, one OSHA standard, the asbestos standard, generally allows respirators only where engineering and work practice controls prove "technically" infeasible. 29 C.F.R. §1910.100(d)(ii)(1979). On the other hand, an employer could attempt to defend against an enforcement citation by showing, through his particular

Thus, since the presumption of feasibility remains rebuttable, in pre-enforcement review the court would not expect OSHA to prove the standard *certainly* feasible for *all* firms at *all* times in *all* jobs. But it would have to justify the presumption, and the attendant shift in burden, with reasonable technological and economic evidence and analysis. Assuming this construction of the earlier standards is correct, the question is whether any of it changes under the lead standard.

2. *The lead standard.* The key change is in the following provision in the lead standard:

(e) *Methods of compliance—*(1) *Engineering and work practice controls.* The employer shall implement engineering and work practice controls to reduce and maintain employee exposure to lead in accordance with the implementation schedule in Table I below. Failure to achieve exposure levels without regard to respirators is sufficient to establish a violation of this provision.[120]

This section does not, like the earlier standards, incorporate feasibility into the rules for means of compliance. Rather, it appears to declare the standard and timetable feasible without regard to respirators, and thus announces any failure to meet the PEL by engineering and work practice controls alone a violation of the standard.

* * *

circumstances, that a standard is *generally* economically infeasible. *See Atlantic & Gulf Stevedores v. OSHA, supra* notes 118, 534 F.2d at 555.

The dissent takes the view that in shifting the burden of proving infeasibility to the employer in such later proceedings, the agency has acted inconsistently with the recent plurality opinion in *Industrial Union Dept't, AFL-CIO v. American Petroleum Institute, supra.* . . . We note, however, that Justice Stevens' opinion in that case spoke of OSHA's evidentiary burden only in the context of the §3(8) requirement that the agency find "significant risk of harm,". . . and was not dealing with the agency's responsibilities under §6(b)(5), the source of the feasibility requirement. In any event, we think it consistent with *American Petroleum Institute* that OSHA bear the initial burden of proving the general feasibility of the standard for the industry as a whole at the rulemaking stage, shifting to the employer in later proceedings the task of overcoming OSHA's initial finding.

[120]On the other hand, the lead standard provisions explaining when employers may use respirators are virtually identical to the parallel provisions in the earlier standards, requiring respirators whenever engineering and work practice controls cannot reduce air-lead levels to the PEL.

VII. THE FEASIBILITY OF THE LEAD STANDARD
A. OSHA's General Approach

* * *

In its Preamble explanation of the standard OSHA thoroughly described its approach to analyzing feasibility. Relying on ample judicial authority for examining feasibility with respect to particular industries and the general capacity of each to comply, e.g., *Industrial Union Dep't, AFL-CIO v. Hodgson, supra* 499 F.2d at 478 the Secretary outlined the criteria it used in determining such feasibility: (1) The general innovativeness of the industry; (2) the financial and technical resources available to the industry; (3) the degree of change needed and its stage of development; (4) the certainty of the product market; (5) the size and complexity of the plant or process requiring change; and (6) the experience of recent technological change in similar industries.

In measuring technological feasibility for the great majority of affected industries, OSHA found that the exposure levels in these industries were sufficiently low that very simple engineering and work practice controls could meet the PEL. For the five major industries OSHA acknowledged the need for major technological change, attributing much of the problem to the fact that these industries had generally never attempted even basic controls. OSHA thus adopted a strategy of "reasonable 'planning horizons.' " It left the choice of the best particular techniques to industry, while setting reasonable timetables to allow firms to make and carry out that choice.

* * *

In measuring economic feasibility for the many low-exposure industries, OSHA states that it inferred from its consultants' reports an estimate of $118.4 million in capital costs and $84.5 million in annual costs, which would mean costs per exposed employee of $188 and $134 respectively. OSHA asserts that it received no evidence whatever that these industries could not afford these costs.

OSHA, of course, does not pretend to have achieved strictly accurate estimates for the five major industries, where its need to force experimental technology and to leave the choice of means to industry makes such a goal quixotic. But it nevertheless constructed estimates out of the various industry and consultants' reports.

Revising available estimates by these means, OSHA attempted to predict the costs of the standard to each major industry in terms of capital expenses—

the costs of new engineering controls and hygiene facilities—and annual expenses—the costs of monitoring, medical surveillance, record-keeping, medical removal, meeting respirator and hygiene standards, protecting and maintaining equipment, and depreciation and interest expenses.

The difficulty of measuring all these complex elements of feasibility, and perhaps also the difficulty of obtaining accurate information from the industry itself, left OSHA with a very limited data base for many specific feasibility questions. And LIA's attack on the sufficiency of that data base amounts to its strongest attack on any part of the lead standard, especially since LIA blames OSHA itself for not developing enough information on feasibility. On the less important lead industries, LIA argues that OSHA relied on a document—the Short Report—which the agency itself conceded to be unreliable. On the more important industries, LIA contends that OSHA marshalled insufficient information to counter what it claims to be clear statements in the DBA report that even the 100 $\mu g/m^3$ standard might be infeasible—though LIA does not effectively answer OSHA's analysis of consistent double-counting in DBA's economic findings.

But most important, on all the industries LIA argues that since OSHA never gave notice that the final PEL would be 50 $\mu g/m^3$ rather than the proposed 100 $\mu g/m^3$, and since OSHA never released its consultants' post-hearing report on the feasibility of this lower PEL, virtually all the evidence developed at the rulemaking concerned the higher PEL. In our earlier discussion of alleged procedural flaws in the rulemaking we addressed the strict procedural legality of OSHA's setting the final PEL at 50 $\mu g/m^3$ when the rulemaking was chiefly directed at the higher proposed PEL. Here, our concern is to apply the substantial evidence test to determine whether evidence chiefly directed to the higher proposed PEL can logically, if indirectly, support the feasibility of a lower final PEL. LIA, of course, argues that OSHA cannot prove the feasibility of the final PEL by this means. But we believe there is nothing *inherently* invalid about such means of proof. The question is always one of reasonable inferences from the available facts about a particular industry. We thus see no point in deciding, as a matter of substantial evidence, whether research directed to a higher standard can *generally* prove the feasibility of a lower standard.

Moreover, regardless of whether OSHA directly invited evidence on the feasibility of the 50 $\mu g/m^3$

standard, the rulemaking did in fact produce some evidence on at least the technological feasibility of the lower PEL. Thus, industrial hygiene engineer Maier Schneider expressly testified that proper engineering and work practice controls would reduce lead exposure in lead processing and handling even below 50 $\mu g/m^3$, and Dr. Melvin First, on whom OSHA reasonably relied for a comprehensive analysis of proper dust control throughout lead industries, very clearly suggests that once industry sets its mind to adopt such a strategy it could achieve incredibly low levels of air-lead exposure. As for economic feasibility, OSHA concedes that its cost estimates generally derived from data on the likely expenses of the 100 $\mu g/m^3$ standard. However, given the speculative technology on which it could ask industry to rely, and the difficulty in obtaining accurate cost data from industry, it was not necessarily wrong for OSHA to conclude that attempting to extend the highly speculative "guesstimates" on the costs of the 100 $\mu g/m^3$ standard to account for the incremental costs of the lower standard would produce very diminishing analytic returns.[132] Rather, where OSHA found that industry and consultants' estimates for the higher PEL were overstated because of double-counting, and where their economic analyses failed to account for the ability of healthy industry to absorb great costs or pass them through to the market, there is nothing inherently invalid about OSHA projecting its determination of the feasibility of a 100 $\mu g/m^3$ PEL to a well-phased-in 50 $\mu g/m^3$ PEL.[133]

*   *   *

[132]Although it does not rely very heavily on this second theory, OSHA also had evidence that reductions in the PEL below 100 $\mu g/m^3$ were likely to increase compliance costs in a linear rather than an exponential fashion.

[133]LIA argues that the statutory requirement that OSHA act on the "best available evidence," 29 U.S.C. §655(b)(5) (1976), means that OSHA had to direct the rulemaking to the 50 $\mu g/m^3$ standard. But the "best available evidence" rule, as construed by the recent Supreme Court benzene decision and our cotton dust decision, requires OSHA to act immediately to protect workers as best it can, without waiting for scientific certainty, *Industrial Union Dep't, AFL-CIO v. American Petroleum Institute, supra. . .*, at 48 U.S. L. Week at 5035 (plurality opinion); *AFL-CIO v. Marshall, supra. . .*, 617 F.2d at 658. We decline to read this phrase as setting an unprecedented evidentiary burden on the agency—to show not only that its evidence was substantial, but also that its evidence was the best it could possibly have presented.

◇ NOTES

1. OSHA proposed at first to lower the standard from 200 to 100 $\mu g/m^3$. However, after the hearings on the proposed standard the agency issued a 50 $\mu g/m^3$ standard. Did OSHA provide industry an ample opportunity to object to this lower standard?

2. Important parts of the standard, and the court's review, addressed medical removal of workers. This concept is addressed in more detail in Section 8 on page 171.

3. In standards other than lead, e.g., vinyl chloride and asbestos, employers must first use engineering controls and then, if necessary, respirators. The lead standard requires that compliance be achieved without resort to respirators.

4. In its discussion of the "circularity problem," the court suggests that the general infeasibility of the standard for the industry as a whole could be argued by a company contesting a citation for noncompliance. This defense is related to a defense at common law of "impossibility of performance," which can be raised by a party in violation of a legal duty in response to an action to enforce compliance.

◇

5. Emergency Temporary Standards

Asbestos Information Association/North America v. OSHA

727 F. Supp 415 (5th Cir. 1984)

E. GRADY JOLLY, Circuit Judge.

The Asbestos Information Association (AIA), an organization of American and Canadian manufacturers of asbestos products, asks this court to determine whether the Occupational Safety and Health Association (OSHA) properly by-passed normal notice-and-comment rulemaking procedures in favor of creating an Emergency Temporary Standard (ETS) lowering workers' permissible exposure level (PEL) to ambient asbestos fibers from 2.0 fibers per cubic centimeter (f/cc) to 0.5 f/cc. We hold that OSHA did not invoke its ETS powers properly.

I

Congress passed the Occupational Safety and Health Act (the Act) in 1970, codified at 29 U.S.C. §§651–678, to assure safe and healthful working conditions for the nation's work force and to preserve the nation's human resources. Toward that goal, the Act allows the Secretary of Labor (the Secretary), after public notice and opportunity for comment by interested persons, to promulgate rules and standards for occupational safety and health. The Act also allows the Secretary to by-pass these normal procedures in favor of promulgating an ETS to take effect immediately upon publication in the Federal Register if he determines that "employees are exposed to grave danger from exposure to substances or agents determined to be toxic or physically harmful or from new hazards," and also determines "that such emergency standard is necessary to protect employees from such danger." 29 U.S.C. §655(c)(1). The ETS statute further provides that the ETS as published shall serve as a proposed rule, and that the secretary shall act on the rule no later than six months after publication.[1]

[1]The full text of the statute says:
    (1) The Secretary shall provide, without regard to the requirements of chapter 5 of Title 5, for an emergency temporary standard to take immediate effect upon publication in the Federal Register if he determines (A) that employees are exposed to grave danger from exposure to substances or agents determined to be

* * *

On November 4, 1983, acting pursuant to its ETS enabling statute, OSHA published in the Federal Register an ETS lowering the time-weighted average PEL for ambient asbestos fibers from 2.0 f/cc that are 5 microns or more in length[3] to 0.5 f/cc. In the November 4 publication, the Secretary also included a statement of reasons to support his action as he is required by law to do. 29 U.S.C. §655(c). *See also Dry Color Manufacturers' Association v. Department of Labor,* 486 F.2d 98 (3d Cir. 1973) (applying §655(c) to ETS promulgations).

The ETS allows "any practical combination" of engineering controls, work practices and personal protective equipment to meet the lower PEL.

* * *

Subsequently, on November 17, 1983, the AIA petitioned this court for an emergency stay pending judicial review of OSHA's action, arguing that its members would suffer irreparable harm if the stay were not granted, and arguing its likely ultimate success on the merits. After reviewing the arguments of both the AIA and OSHA, analyzing them according to well established legal criteria[4] for determining when a court should grant equitable interim relief,

toxic or physically harmful or from new hazards, and (B) that such emergency standard is necessary to protect employees from such danger.

(2) Such a standard shall be effective until superseded by a standard promulgated in accordance with the procedures prescribed in paragraph (3) of this subsection.

(3) Upon publication of such standard in the Federal Register the Secretary shall commence a proceeding in accordance with subsection (b) of this section, and the standard as published shall also serve as a proposed rule for the proceeding. The Secretary shall promulgate a standard under this paragraph no later than six months after publication of the emergency standard as provided in paragraph (2) of this subsection. 29 U.S.C. §655(c).
[3]As much as 98% of ambient asbestos fibers may be less than 5 microns long, and, given the sophistication of measuring devices, too small to regulate.
[4]To obtain equitable relief pending further judicial action on the merits, an applicant must establish (1) a substantial likelihood of success on the merits; (2) danger of irreparable harm if the court denies interim relief; (3) that other parties will not be harmed substantially if the court grants interim relief; and (4) that interim relief will not harm the public interest. *Virginia Petroleum Jobbers Assn'n v.*

and balancing the equities involved,[5] this court granted the stay but expedited full hearing on the merits. We now hold that OSHA did not properly act pursuant to its ETS enabling statute and that the standard in question, therefore, should not become effective absent notice-and-comment rule-making.

II

OSHA has regulated asbestos since 1971. Its first asbestos PEL was 12.0 f/cc. In 1972 OSHA reduced this standard to 5 f/cc,[6] and in 1976 OSHA again reduced the standard to the currently effective 2 f/cc. In 1975 OSHA proposed to reduce the standard to 0.5 f/cc, but founded its proposal on a policy to set PEL's for carcinogens as low as technologically and economically feasible. The Agency did not act quickly and, in 1980, the Supreme Court rejected the proposition that such a general policy may serve as the basis for any rule, and held that OSHA must make an actual finding that the workplace is unsafe before it promulgates a standard. *Industrial Union Department v. American Petroleum Institute,* 448 U.S. 601, 100 S. Ct. 3844, 65 L. Ed. 2d 1010 (1980). OSHA bases its 1983 promulgation of an ETS lowering the PEL to 0.5 f/cc, however, on specific data compiled and analyzed by OSHA that lead it to conclude that a "grave danger" exists, necessitating immediate action.

No new data or discovery leads OSHA to invoke its extraordinary ETS powers and lower the asbestos PEL. Rather, OSHA bases its conclusion that a grave danger exists on quantitative risk assessments, which are mathematical extrapolations, of the likelihood

Federal Power Commission, 104 U.S. App. D.C. 106, 259 F.2d 921, 925 (1958). In *Taylor Diving & Salvage Co. v. Department of Labor,* 537 F.2d 819, 821 n.8 (5th Cir. 1976) this court applied *Virginia Petroleum Jobbers'* four criteria to application for stay pending review of an OSHA ETS.
[5]Subsequent interpretations of the four legal criteria de-emphasize the likelihood of success criteria and emphasize balancing the equities of the situation. *United States v. Baylor University Medical Center,* 711 F.2d 38 (5th Cir. 1983).
[6]OSHA adopted the 12.0 f/cc PEL by adopting a national consensus standard existing in 1971, the year the Occupational Safety and Health Act became effective. Section 6 of the Act explicitly allowed OSHA to adopt national consensus standards, and directed the agency to establish standards "as soon as practicable." 20 U.S.C. §655(a). On December 7, 1971, OSHA lowered the 12.0 f/cc standard to 5.0 f/cc by using, for the first time, its ETS powers. The action was not challenged, and in June of 1972, it became a permanent standard through normal notice-and-comment procedures.

of contracting an asbestos-related disease at various levels of exposure to asbestos particles.

* * *

OSHA calculated the likelihood of developing lung cancer, mesothelioma,[7] and gastrointestinal cancer due to contact with ambient asbestos fibers at different exposure levels. By applying its calculations to an estimated working population exposed to asbestos, OSHA claims that 210 lives eventually can be saved from cancer by lowering the PEL to 0.5 f/cc for six months. These figures include deaths that will occur at OSHA's estimated current actual exposure levels and include employees working in environments where the density of ambient asbestos particles is 20 f/cc, ten times the current PEL. Even if, however, OSHA removes from the computation those employees who do not enjoy the benefit of the current 2.0 f/cc PEL because it is not enforced in their work place, and counts only those employees who are exposed to ambient asbestos between the levels of 2.0 f/cc and 0.5 f/cc, OSHA estimates it can save 80 lives by lowering the PEL for six months.[8]

* * *

III

We note at the outset of our analysis that immediately after its November publication, OSHA commenced regular notice-and-comment rule-making

---

[7]Mesothelioma is an incurable cancer infecting the mesothelium, a layer of fat cells lining the membranes enclosing the heart, abdominal cavity, and thoracic cavity and lungs.

[8]OSHA's own data, however, indicates that the actual number of asbestos-related cancer deaths prevented by a 0.5 f/cc standard would be approximately 40 for six months. Approximately 71% of the benefits of OSHA's ETS accrue in the drywall construction industry where demolition and other activities generate large amounts of ambient asbestos particles. OSHA specifically estimated it could save 57 lives in that industry by lowering the PEL for six months, assuming the entire industry currently complies with the 2 f/cc standard. Its estimate of employee exposure, however, indicates that of all 51,621 employees estimated to be working in that industry, 38,666 or approximately 75% currently are exposed to only 0.2 f/cc. Consequently, these workers would not benefit by lowering the standard to 0.5 f/cc. Reducing OSHA's calculations by 75% indicates that approximately 14 drywall construction workers will benefit from the ETS. Stated differently, approximately 43 of the 57 workers already are exposed to levels below that which the ETS would permit.

to decide whether to impose a new permanent PEL for asbestos. The statute requires that the Secretary promulgate a permanent standard no later than six months after publication of the ETS. 29 U.S.C. §655(c) (3). At oral argument OSHA's counsel stated that OSHA could complete full notice-and-comment rulemaking within one year, presumably even without the impetus of the ETS requirement that it do so within six months.[13] Consequently, the practical effects of our decision on the regulations enforced in the workplace will endure only a short time. We are, however, concerned not only with practical implications, but also with the legal issue of the extent of the Secretary's power to determine when an emergency situation exists and his power to act in such a situation.

The standard under which we review OSHA's new PEL is whether the Agency's action is "supported by substantial evidence in the record considered as a whole." 29 U.S.C. §655(f).

* * *

Even though we must apply the substantial evidence test, OSHA urges us to apply it less rigorously in reviewing an informally promulgated ETS than we would in reviewing a standard imposed after formal notice-and-comment rulemaking procedures. Indeed, the anomaly of being required to make a searching review of the evidence, but being provided only with a record of a volume and technical complexity that would tax the competency of any court, forces us to concede the Agency's contention to a degree. The record fills nine large boxes, and contains years' worth of accumulated asbestos reports and studies from all over the world. It also includes mathematical and statistical computations, and letters and memoranda to, from and between government agencies.

* * *

In reviewing the ETS, we also must remain aware that the plain wording of the statute limits us to assessing the harm likely to accrue, or the grave danger that the ETS may alleviate, during the six-month period that is the life of the standard. OSHA

---

[13]At oral argument OSHA intimated that even with the ETS requirement that it promulgate a permanent standard within six months, the complexity of the data in this case might make impossible such rapid action. We note, however, that the statute says the agency "shall promulgate a standard no later than six months after publication," 29 U.S.C. §655(c)(3). The statute does not contemplate the agency's allowing the new rule to lapse.

urges us to assess the harm likely to accrue over at least a year, even though the ETS expires six months from its promulgation. At oral argument OSHA said that even if the ETS lapsed before OSHA promulgated a permanent regulation, the benefits of the ETS likely would continue because employers will have expended the resources to comply with the new lower standard and would have no incentive to revert to old practices. These post hoc rationalizations cannot be accepted as basis for our review; first, because the ETS statute does not contemplate the Secretary's allowing an ETS to lapse before he promulgates a permanent standard, and second, because to assume that employers will not revert to less exacting standards is pure speculation. The opposite is equally plausible, especially given that OSHA allows compliance with the ETS through methods as simple as wetting floors or wearing respirators.

In its November 4 publication OSHA partially justified its decision to issue an ETS on the fact that notice-and-comment rulemaking often takes several years to complete, excluding possible subsequent postponements of the effective date caused by court-ordered stays pending judicial review. OSHA apparently would have us assess benefits in this light. We cannot do so. As noted earlier, OSHA concedes that it can complete rulemaking within one year. Additionally, as its legislative history makes clear, the ETS statute is not to be used merely as an iterim relief measure, but treated as an extraordinary power to be used only in "limited situations" in which a grave danger exists, and then, to be "delicately exercised." *Public Citizen Health Research Group v. Auchter,* 702 F.2d at 1150 (D.C. Cir. 1983). *See also Taylor Diving & Salvage v. Department of Labor,* 537 F.2d 819, 820-21 (5th Cir. 1976); *Florida Peach Growers,* 489 F.2d at 129 [5th Cir. 1974]; *Dry Color Manufacturers Ass'n,* 486 F.2d at 104 n.9a (3d Cir. 1973). The Agency cannot use its ETS powers as a stop-gap measure. This would allow it to displace its clear obligations to promulgate rules after public notice and opportunity for comment in any case, not just in those in which an ETS is necessary to avert grave danger. *See* 29 U.S.C. §655(b).

IV

A

The AIA urges us to hold that OSHA must have new information before it promulgates an ETS.[16] An "emergency" cannot exist, it argues, when the Agency has

[16]The AIA argues that OSHA's successful invocation of its ETS powers are distinguishable from its unsuccessful uses on the basis of the existence of new information. We do

known for years that asbestos constitutes a serious health risk and, in fact, has had all the data it uses to support its November 4 action at hand, but nevertheless failed to act on it. Although new information may be a sound basis for an ETS, we decline to hold that OSHA cannot issue an ETS in its absence. As OSHA admits, the Agency's failure to act may be evidence that a situation is not a true emergency, but we agree with OSHA that failure to act does not conclusively establish that a situation is not an emergency.

The ETS statute itself, allowing the Secretary to promulgate an ETS in response to "grave danger . . . or . . . new hazards," precludes our imposing a "new information" requirement on OSHA. Additionally, to impose such a requirement would imprudently circumscribe the Secretary's ability to act in response to serious situations. If exposure to 2.0 f/cc of asbestos fibers creates a grave danger, to hold that because OSHA did not act previously it cannot do so now only compounds the consequences of the Agency's failure to act.

OSHA should, of course, offer some explanation of its timing in promulgating an ETS, especially

not wholly agree. Certainly, OSHA has used its ETS powers successfully to address a situation that recently came to light. In 1974 OSHA issued an ETS for vinyl chloride within weeks after learning that workers' deaths were attributable to exposure to that substance. No one contested the Agency's action. *See also Society of Plastics Industries, Inc. v. OSHA,* 509 F.2d 1301 (2d Cir. 1975) (upholding final rule). Similarly, in 1977 OSHA issued an ETS for 1,2 Dibromo-3-Chloropropane (DBCP), after having become aware in just a few months that exposure to even small amounts of the chemical caused sterility. *See* 42 Fed. Reg. 45,536 (1977). In 1971, however, OSHA used its ETS powers to lower the asbestos PEL from 12 f/cc to 5 f/cc, having concluded only that asbestos presented a "grave danger." The 1971 action was not challenged.

Other invocations of the ETS power have failed. None, however, has failed solely because the Agency did not act pursuant to newly acquired information. *See American Petroleum Institute v. OSHA,* 581 F.2d 493, 503 (5th Cir. 1978) *aff'd sub nom Industrial Union Dept. v. American Petroleum Institute,* 448 U.S. 607, 100 S. Ct. 2844, 65 L. Ed. 2d 1010 (1980) (benzene ETS failed for lack of substantial evidence); *Taylor Diving & Salvage Co., Inc. v. Dept. of Labor,* 537 F.2d 819, 821 (5th Cir. 1976) (temporary stay granted because petitioners showed likelihood of success on the merits and irreparable harm); *Florida Peach Growers Ass'n Inc. v. Dept. of Labor,* 489 F.2d 120, 129 [5th Cir., 1974] (organophosphorous pesticides ETS failed for lack of substantial evidence); *Dry Color Mfrs Ass'n. Inc. v. Dept. of Labor,* 486 F.2d (3d Cir. 1973) (fourteen carcinogens ETS failed for failure to adequately state reasons).

when, as here, for years it has known of the serious health risk the regulated substances poses, and has possessed, albeit in unrefined form, the substantive data forming the basis for the ETS.[17] In this case OSHA says it acted in response to new awareness of the danger of asbestos and in response to extrapolated data that did not become available until July of 1983, four months before it promulgated the ETS. We are not prepared to say that such heightened awareness cannot justify the Secretary's action.

Additionally, even if adequately explained, an ETS must, on balance, produce a benefit the costs of which are not unreasonable. The protection afforded to workers should outweigh the economic consequences to the regulated industry. *American Petroleum Institute v. OSHA*, 581 F.2d 493, 502-03 (5th Cir. 1978) *Aff'd sub nom Industrial Union Department v. American Petroleum Institute*, 448 U.S. 607, 100 S. Ct. 2844, 65 L. Ed. 2d 1010 (1980); *Florida Peach Growers*, 489 F.2d at 130 [5th Cir. 1974].[18] OSHA conducted a benefits analysis prior to promulgating the ETS, and concluded that the cost of compliance with the lower PEL is reasonable compared to total industry sales volume.

* * *

The AIA does not complain of the cost of compliance with the new PEL, however, as much as of the anticipated ripple effects that OSHA's action will have on the asbestos products market. The AIA argues that asbestos users will substitute other products for asbestos because of the alarm the ETS causes throughout the industry in labeling the situation as an emergency. Indeed, such considerations are not insignificant. As this court has noted, "It is essential that employees be protected against exposure to highly toxic materials, but this should be done with-

out eliminating the [asbestos industry] and the associated jobs." *Florida Peach Growers*, 489 F.2d at 130 [5th Cir. 1974].[20] The industry, however, already will have felt any ripple effects precipitated by OSHA's declaration of an emergency, and our holding today cannot undo whatever harm has been done, especially in light of the fact that the ETS now is a proposed permanent standard. The AIA, moreover, fails to convince us that the ETS seriously jeopardizes the asbestos industry, or even that the harm due to lost sales will be significant.

B

The ETS statute requires that the Secretary issue an ETS only after he finds substantial evidence indicating both that a "grave danger" exists and that an emergency standard is "necessary" to protect workers form such danger. Thus, the gravity and necessity requirements lie at the center of proper invocation of the ETS powers. No one doubts that asbestos is a gravely dangerous product. The gravity we are concerned with, however, is not of the product itself, but of six months exposure to it at 0.5 f/cc, as compared with six months exposure at 2.0 f/cc. Our inquiry, then, is a narrow one, and requires us to evaluate both the nature of the consequences of exposure, and also the number of workers likely to suffer those consequences.

According to the Secretary, the consequences of exposure to significant amounts of asbestos are likely to be fatal. Victims of lung cancer, mesothelioma, and gastrointestinal cancer have poor survival rates. Additionally, workers exposed to significant amounts of asbestos run a risk of developing asbestosis, a serious condition caused by the accumulation of asbestos fibers in the lungs.

* * *

OSHA claims that by permanently lowering the present 2.0 f/cc PEL to 0.5 f/cc, it will save sixty-four lives per one thousand workers over a working lifetime of forty-five years. *See* 48 Fed. Reg. at 51,100. Over six months, this works out to eighty lives out of

---

[17]OSHA contemplated a risk assessment in 1981 that concluded that between 89 and 260 deaths per year would occur at the 2.0 f/cc PEL. At oral argument, counsel for OSHA said the Agency did not act then because it considered the data to be too unrefined.

[18]Although in this case the agency conducted a formal cost-benefit analysis, we do not imply that the Occupational Safety and Health Act required the agency to do so before it promulgates an ETS. Indeed, in true "emergency" situations, that the agency would have time to conduct such an analysis is unlikely. The *American Petroleum Institute* and *Florida Peach Growers* cases require only that in reviewing whether the agency's action was reasonable under the circumstances, we analyze the anticipated benefit of the ETS in light of its probable consequences.

[20]OSHA itself noted that "the nature of the action itself, and the accompanying enforcement program will undoubtedly boost the incentives to comply with all protective provisions of the asbestos standard." Exactly what OSHA intended by this remark is unclear from the context of the publication. Certainly, for OSHA to use its ETS powers expressly to alarm the industry is illegitimate, and would count against the Agency in a judicial challenge.

an estimated worker population of 375,399.[21] As the Supreme Court has noted, the determination of what constitutes a risk worthy of Agency action is a policy consideration that belongs, in the first instance, to the Agency. *Industrial Workers Union,* 448 U.S. at 656, n.62, 100 S. Ct. at 2871, 65 L. Ed. 2d at—. "Some risks are plainly acceptable and others are plainly unacceptable." *Id.* at 655, 100 S. Ct. at 2870, 65 L. Ed. 2d at—[1603]. The Secretary determined that eighty lives at risk is a grave danger. We are not prepared to say it is not.

* * *

OSHA has made the *number* of deaths avoided—at least 80—the basis for its rulemaking. Yet it is apparent from an examination of the record that the actual number of lives saved is uncertain, and is likely to be substantially less than 80.[22] Both the gravity of the risk as defined by OSHA and the necessity of an ETS to protect against it are therefore questionable.

* * *

C

Even assuming that OSHA's projected benefits would accrue from the ETS, however, we hold that OSHA's action must fail for another reason. The Agency has not proved that the ETS, OSHA's most dramatic weapon in its enforcement arsenal, is "necessary" to achieve the projected benefits.

As OSHA concedes, the probable practical effect of the ETS, which allows compliance through "any feasible combination of engineering controls, work practices, and personal protective equipment and devices," would be that employers would require employees to wear respirators. Current regulations already require employers to outfit workers with respirators that can provide up to one hundred-fold protection. . . . Yet OSHA did not include in its calculations the effect of enforcing the current standard by requiring employers in the drywall construction and demolition industry to furnish these respirators. Counsel for OSHA informed the court at oral argu-

ment that the Secretary considers the regulation requiring construction and demolition workers to wear respirators to be unenforceable absent actual monitoring to show that ambient asbestos particles are so far above the permissible limit that respirators are necessary to bring the employees' exposure within the PEL of 2.0 f/cc. Fear of a successful judicial challenge to enforcement of OSHA's permanent standard regarding the respirator use hardly justifies resort to the most dramatic weapon in OSHA's enforcement arsenal.[24] Thus, lacking a satisfactory explanation why the ETS is a necessary means to achieve the added saving obtainable by application of the current regulations, we must assume that OSHA's claimed benefit should be discounted by some additional, uncertain amount.

OSHA also attempts to justify the ETS by emphasizing that the ETS does more to protect worker's health than simply lowering the asbestos fiber PEL. An ETS, however, is not necessary to achieve these ancillary benefits. The ETS requires employers to educate employees concerning the risks of asbestos exposure and the proper steps necessary to minimize exposure. While education is a worthy objective, OSHA could achieve it without invoking its extraordinary ETS power. Indeed, current regulations provide for worker training and education. Similarly, OSHA supports its action by arguing that it plans to increase enforcement efforts, with the aim of encouraging greater compliance with the new standard than it estimates currently exist under the present standard. Increasing enforcement is another worthy objective; but it likewise cannot justify use of the ETS power, especially when, as in this case, much of the claimed benefit could be obtained simply by enforcing the current standard.

In sum, although asbestos doubtless may present a grave danger to workers, the record considered as a whole does not substantially support OSHA's conclusion that an ETS lowering worker PEL from 2.0 f/cc to 0.5 f/cc is necessary to alleviate a grave risk of worker deaths during its six-month term. This court, in ruling on a challenge to an ETS, has "reject [ed] any suggestion that deaths must occur before health and safety standards may be adopted," *Florida Peach Growers Association, supra* at 132, and we make no such suggestion here. Additionally, we do not decide whether the record would support a conclusion that

---

[21] In the November 4 publication, OSHA stated that 210 lives will be saved over six months. The Agency, however, concedes that this figure is inflated because it includes those lives that OSHA could save by enforcing its current 2.0 f/cc standard. *See supra* note 8 and accompanying text.
[22] *See supra* note 8.

[24] Indeed, Occupational Safety and Health Review Commission decisions do not support the Secretary's interpretation. *See Anaconda Aluminum Co.* (1981), 9 OSHC 1460.

some threat to workers' health of a magnitude substantially less than 80 deaths during the ETS periods constitutes a grave danger necessitating an ETS of 0.5 f/cc. Gravity of danger is a policy decision committed to OSHA, not to the courts. We hold only that an ETS that lacks support in the record for the basis OSHA has articulated must be declared invalid.

V

OSHA may, of course, continue its plan to increase enforcement of the current PEL. If danger is imminent, it should complete notice-and-comment rulemaking as quickly as possible to determine if a new standard lowering the PEL to 0.5 f/cc, or even lower, is appropriate. Our finding here that substantial evidence does not exist to support a six-month ETS should not be construed as a prediction that a lower asbestos PEL would fail under judicial scrutiny if OSHA promulgates it pursuant to proper notice-and-comment rulemaking procedures.

We determine the Emergency Temporary Standard to be invalid because the record, considered as a whole, does not indicate that the risk the ETS seeks to eliminate is "grave," as OSHA itself has defined it, or that the ETS is "necessary," as those terms are used in the ETS statute.

ENFORCEMENT OF EMERGENCY TEMPORARY STANDARD STAYED.

◇ NOTES

1. Did the court substitute its judgment for that of the agency?

2. In this case, the court says it is not prepared to say that OSHA erred in concluding that 80 lives saved constituted a grave danger averted. Is the word "grave" to be measured in seriousness of disease or the number of workers affected? The Supreme Court's benzene decision talks of "significant risk of material impairment," where significance is measured in numbers for the permanent 6(b) OSHA standards. Is the court of appeals hinting at a significant risk requirement for emergency temporary standards?

3. Reread Section 6(c) of the Act. Does the ETS lapse after six months? If so, could OSHA renew the ETS for subsequent six month periods?

4. The court hints that because respirators could be used while OSHA sets a lower PEL to 0.5 f/cc, the ETS is not necessary. Using this reasoning, why would engineering controls be preferable over respirators for a permanent standard?

5. Is the court's deference to OSHA on the "gravity" of the danger consistent with its refusal to accept OSHA's determination of "necessity"?

6. The court states that "an ETS must, on balance, produce a benefit the costs of which are not unreasonable. The protection offered should outweigh the economic consequences to the regulated industry." The Court then cites the Supreme Court's benzene decision. Is this assertion consistent with the Court's cotton dust decision?

7. In 1986 OSHA ultimately revised the permanent asbestos standard from 2 to 0.2 fibers per cubic centimeter.

6.   Short-Term Exposure Limits

Short-term exposures to higher levels of carcinogens are generally considered more hazardous than longer exposures to lower levels. OSHA issued a new standard for exposure to ethylene oxide (EtO) in 1984, but excluded a short-term exposure limit (STEL) that had originally been prepared, in deference to objections from the Office of Management and Budget. Ralph Nader's Health Research Group sued the Secretary of Labor in 1986 over OSHA's continuing failure to issue the STEL.

Public Citizen Health Research Group v. Brock

823 F.2d 626 (D.C. Cir. 1987)

Before ROBINSON, Circuit Judge, and WRIGHT and MCGOWAN, Senior Circuit Judges.

Per Curiam. Petitioners Public Citizen Health Research Group, *et al.*, in Nos. 84-1252 and 85-1014 (hereinafter petitioners) allege that the Occupational Safety and Health Administration has contemptuously and unreasonably delayed promulgation of a "Short-Term Exposure Limit" (STEL) for the toxin ethylene oxide, despite this court's specific order in *Public Citizen Health Research Group v. Tyson*, 796 F.2d 1479 (D.C. Circuit 1986). This allegation places the court in a delicate position. Although the courts must never forget that our constitutional system gives the Executive Branch a certain degree of breathing space in its implementation of the law, we cannot countenance maneuvering that merely maintains a facade of good faith compliance with the law while actually achieving a result forbidden by court order. We understand that technical questions of health regulations are not easily untangled. We understand that an agency's limited resources may make impossible the rapid development of regulation on several fronts at once. And we understand that the agency before us has far greater medical and public health knowledge than do the lawyers who comprise this tribunal. But we also understand, because we have seen it happen time and time again, that action Congress has ordered for the protection of the public health all too easily becomes hostage to bureaucratic recalcitrance, factional infighting, and special interest politics. At some point, we must lean forward from the bench to let an agency know, in no uncertain terms, that enough is enough.

At issue here, then, is whether that point has been reached. We conclude that it has, but that the court's proper role within the constitutional system counsels caution in fashioning a remedy.

BACKGROUND

The history of OSHA's attempts to regulate ethylene oxide (EtO) is one of hesitation and lack of resolve. In January 1982, OSHA first issued an advance notice of proposed rulemaking for EtO in response to growing evidence of its toxicity. In 1983, this court found OSHA's delays in promulgating a final rule to be unjustifiable, and ordered the Administration to complete its rulemaking proceedings "within a year." *Public Citizen Health Research Group v. Auchter*, 702 F.2d 1150, 1154 n.12 (D.C. Cir. 1983) (*per curiam*). OSHA subsequently published a proposed rule (1983), that included both a "Permissible Exposure Limit" (PEL) and a "Short-Term Exposure Limit" (STEL).

After extensive public hearings, OSHA was ready to issue a final rule on June 14, 1984. In compliance with Executive Order No. 12291, OSHA sent the final rule to the Office of Management and Budget (OMB) for approval. But approval was not to be had. OMB balked at OSHA's inclusion of the short-term exposure limit, objecting primarily on the ground of cost-effectiveness. OSHA dutifully issued a final rule that had been sanitized of all mention of short-term exposure limits.

Almost immediately, petitioners challenged both the level of OSHA's ethylene oxide PEL and the agency's failure to include a STEL in the final regulation. Last July 25th, this court affirmed OSHA's PEL regulations, but determined that OSHA's decision to forego a STEL did not have adequate support in the rulemaking record. *Public Citizen Health Research Group v. Tyson*, 796 F.2d 1479 (D.C. Cir. 1986). Our instruction to the agency on the need for a STEL was fairly simple:

> On remand, we expect the agency to ventilate the issues on [the STEL] point thoroughly and either adopt a STEL or explain why empirical or expert

evidence on exposure patterns makes a STEL irrelevant to controlling long-term average exposures.

796 F.2d at 1507. In contention presently is whether OSHA's failure to issue even a notice of proposed regulation in the nine months between issuance of our mandate and filing of the instant motion constitutes contempt of court, unreasonable delay under the Administrative Procedure Act, or both.

DISCUSSION

This is a troubling case. We are mindful that OSHA's rulemaking determinations are "essentially legislative and rooted in inferences from complex scientific and factual data," *United Steelworkers of America v. Marshall*, 647 F.2d 1189, 1206 (D.C. Cir. 1980), *cert. denied sub nom. Lead Industries Ass'n v. Donovan*, 453 U.S. 913 (1981). They are thus entitled to great deference from the court. *Public Citizen v. Auchter*, 702 F.2d at 1156. At the same time we cannot help but note that OSHA's EtO regulations, first proposed in 1982, are not final in 1987, despite repeated orders and exhortations from this court. In fact, OSHA informs us that the final STEL regulations will not issue until March 1988, even assuming that the rulemaking process suddenly changes what has been its essential character and proceeds according to schedule. With lives hanging in the balance, six years is a very long time.

Petitioners contend that the *Tyson* opinion does not support OSHA's recent decision to undertake a full-blown rulemaking proceeding on the STEL issue. And even assuming rulemaking is permissible, they submit that OSHA's failure to issue a notice of proposed rulemaking in the months that have passed since this court's *Tyson* remand constitutes both contemptuous failure to comply with that order and unreasonable delay under the Administrative Procedure Act. OSHA responds that, in order to comply with *Tyson*, it decided a rulemaking was necessary and quickly contracted with a private firm to collect data on the STEL question. Any delay in the process, it says, stems from practical difficulties encountered by the contractor and from the very nature of the rulemaking process.

A contempt citation under these circumstances would be a draconian and disproportionate remedy. OSHA decided in good faith that the record required supplementation on the issue of the public health necessity of a STEL when a strict PEL is already in place. Viewed fairly, our mandate in *Tyson* does not

preclude such supplementation. To the contrary, we specifically instructed OSHA "to ventilate [the STEL] point thoroughly." 796 F.2d at 1507. We see nothing in OSHA's record supplementation decision, therefore, that may properly be labeled "contemptuous" of our order. Even if *Tyson* did not *require* supplementation, it certainly did not *preclude* it.

Granted, OSHA's failure to issue a notice of proposed rulemaking simultaneously with hiring a contractor to supplement the record is more difficult to defend. After five years of struggle to promulgate a final EtO regulation, this court had hoped that OSHA would act with greater alacrity on our specific order to either promulgate a STEL or explain why a STEL was unnecessary. We are more than a little perturbed that OSHA chose to construe the *Tyson* opinion to allow close to two years for compliance. Nevertheless, we see bureaucratic inefficiency rather than bad faith as the source of OSHA's laggard pace in this regard.

Consequently, a finding of "unreasonable delay" under 5 U.S.C. §706(1)(1982)[1] would be more closely tailored than a contempt order to correct any deficiency in performance here. After all, once we have admitted the permissibility of record supplementation under *Tyson*, the timetable for STEL regulation inevitably stretches out beyond petitioners' wishes. But that is not to say *any* timetable, however, dilatory, is reasonable. When lives are at stake, as they assuredly are here, OSHA must press forward with energy and perseverance in adopting regulatory protections. *See Public Citizen v. Auchter*, 702 F.2d at 1157-58.[2]

Deciding whether a particular set of agency actions constitutes "unreasonable delay" is a tricky proposition. Our sense of concern notwithstanding, we are loath to rush in to manage the details of OSHA's ethylene oxide rulemaking procedure. OSHA not only possesses enormous technical expertise we lack, but must juggle competing rulemaking demands on its limited scientific and legal staff. *See e.g., Farmworker Justice Fund, Inc. v. Brock*, 811 F.2d 613, 633 (D.C. Cir. 1987), *vacated as moot*, 817 F.2d 890 (D.C. Cir. 1987)

---

[1]That provision of the Administrative Procedure Act gives the court power to "compel agency action unlawfully withheld or unreasonably delayed." *See Telecommunications Research & Action Center v. FCC*, 750 F.2d 70 (D.C. Cir. 1984).
[2]As we noted in *Tyson*, OSHA's PEL regulation, standing by itself, leaves exposed workers at significant risk. OSHA itself estimates that under the present one part-per-million PEL, ethylene oxide exposure will be responsible for 12-23 excess cancer deaths per 10,000 workers. *See Tyson*, 796 F.2d at 1502-03.

(ordering OSHA to promulgate farmworker field sanitation standard within 30 days of mandate); *United Steelworkers of America v. Pendergrass*, No. 83-3554 (3d Cir., May 29, 1987) (ordering OSHA to issue a "hazard communication" standard within 60 days). This court should intervene to *override* agency priorities and timetables only in the most egregious of cases.

Fortunately, OSHA has presented to us a specific timetable in this case. OSHA represents that a final rule will issue in March 1988. Although we are disappointed with this target data, we cannot find any specific aspect of the proposed rule-making schedule that is impermissibly slow, in light of the complexity of the health questions involved and OSHA's limited resources.

The five-year history of the EtO regulatory process convinces us, however, that the proposed timetable treads at the very lip of the abyss of unreasonable delay. In light of the fact that OSHA's timetable representations have suffered over the years from a persistent excess of optimism, we share petitioners' concerns as to the probable completion date of the rulemaking. Consequently, we find that any delay whatever beyond the proposed schedule is unreasonable.

Having made this determination, we must fashion an order that balances two competing concerns. On the one hand, we should avoid if possible any direct judicial meddling with the details of OSHA's rule-making schedule. On the other, the public health, as defined by Congress, requires that we set a clear end point to the regulatory snarl that is the EtO short-term exposure limit rule-making.

As a clarification of our mandate in *Tyson*, therefore, we order OSHA to adhere to the schedule set out in this response to petitioners' contempt motion. OSHA's final decision on the EtO short-term exposure limit regulation is to issue no later than March 1988. Failure to comply with this timetable may well expose OSHA to liability for contempt. Moreover, given OSHA's apparent reluctance to keep petitioners informed as to the progress of STEL rulemaking, we order OSHA to submit to the court a concise progress report every 90 days from the order's date of issuance until the final rule is in place. With these two commands, we hope and expect that the court's Sisyphean efforts to force agency action on this matter will finally be at an end.

So ordered.

◇ NOTES

1. In the case, the court orders OSHA either to act or explain why a STEL is not needed. It derives its authority under the APA, not the OSHAct (see footnote 1 of the case).

2. Is contempt of court a realistic remedy against a recalcitrant federal agency?

3. OSHA posted the STEL for ethylene oxide in the Federal Register on March 30, 1988 and thereby met the requirements of the court order. The STEL limits exposure to 5 ppm over a 15-minute period.

7. Protection of Workers from Exposure to Noise: A Special Case or a Troubling Precedent?

In the Carter Administration, OSHA had planned to revise the 6(a) standard for noise exposure by reducing the PEL from 90 db over eight hours to 85 db. Significant numbers of workers were undoubtedly affected, but substantial costs would have been imposed on industry. The new standard was not issued during the Reagan Administration. Instead a compromise was offered, providing mandatory automatic testing for workers exposed above 85 db. In its first review of an OSHA standard the Fourth

Circuit Court of Appeals vacated the testing requirement. The Court held that the amended standard was invalid because it required employers "to take actions in regard to hazards existing outside the workplace."

Forging Industry Association v. Secretary of Labor

748 F.2d 210 (4th Cir. 1984)

CHAPMAN, Circuit Judge.

This case is before the court pursuant to Section 6(f) of the Occupational Safety and Health Act of 1970 ("Act").[1] The Forging Industry Association ("FIA") petitions this court to review the Secretary of Labor's promulgation of a "hearing conservation amendment" ("amendment") to its occupational noise exposure standard, 29 C.F.R. §1910.95 ("standard"). Finding that the Department of Labor's Occupational Safety and Health Administration (OSHA) exceeded its authority in adopting the amendment, we vacate the amendment and remand.

I

An occupational noise exposure standard has existed since OSHA's inception in 1971. The current standard, which is found at 29 C.F.R. §1910.95, was originally promulgated under the Walsh-Healey Public Contracts Act, 41 U.S.C. §35 et seq. for the purpose of protecting employees from workplace exposure to damaging levels of noise. The Walsh-Healey standard was adopted by OSHA pursuant to Section 6(a) of the Occupational Safety and Health Act, which allowed the Secretary to promulgate any established Federal standard within two years of the effective date of the Act without regard to established rulemaking procedure.

The standard establishes a permissible workplace limit of 90 decibels (db)[2] calculated using an 8-hour time-weighted average.[3] If the 90 db exposure limit is exceeded, the employer must reduce noise to or below this level by using feasible engineering or administrative controls.[4] If such controls are infeasible, employers may use hearing protectors, such as ear muffs or plugs, to reduce employee noise exposure to permissible limits. Prior to amendment, the standard also contained a generally phrased requirement that employers administer "a continuing effective hearing conservation program" in workplaces where sound levels exceeded the permissible exposure level.

When studies revealed that many employees suffered significant hearing impairment at noise levels below the 90 db threshold, OSHA began the process of collecting and evaluating the information necessary to issue a comprehensive new regulation with a reduced permissible exposure level of 85 db. As an interim measure, OSHA adopted a hearing conservation amendment to replace the general conservation program requirement.

Despite its interim nature the requirements of the amendment are substantial. The amendment requires

[1]Section 655(f) provides that:

Any person who may be adversely affected by a standard issued under this section may at any time prior to the sixtieth day after such standard is promulgated file a petition challenging the validity of such standard with the United States court of appeals for the circuit wherein such person resides or has his principal place of business, for a judicial review of such standard.

[2]Decibels are a measure of sound loudness. The entire spectrum of audible sound pressure can be compressed into a logarithmic scale of 0 to 140 db. Hertz ("Hz"), in contrast, measure the frequency of sound. Frequency is determined by the number of times that a complete cycle of compressions and expansions occurs in a second. The audible range of frequencies for people with good hearing is 20 Hz to 20,000 Hz.

[3]The time-weighted average (TWA) combines noise level and duration of exposure to measure the accumulation of noise levels experienced by an employee over a workshift. OSHA computes the relationship between noise level and exposure time by using a 5 db "exchange rate." This means that for each 5 db increase in noise level, the exposure time must be cut in half. For example, an employee who works for 4 hours in continuous noise of 95 db would be exposed to an 8 hour TWA of 90 db, as would an employee exposed to 100 db for 2 hours.

[4]Engineering controls involve modification of plant, equipment, processes, or materials to reduce noise; for example, adding a muffler to a vehicle. Administrative controls involve modifications of work assignments to reduce employee exposure to noise; for example, rotating employees so that they work in noisy areas for a short time.

employers to determine which employees are exposed to or above an "action level" of 85 db measured as an 8-hour time-weighted average. Such employees must be notified of the amount of sound they are exposed to and provided with an audiometric test to determine their hearing level. At least annually thereafter, the employer must provide the exposed employee with an additional test to determine whether the employee has suffered an average loss of hearing of 10 db, known as a standard threshold shift, or "STS."[5] If there has been an STS, the employer must take follow-up measures to prevent the employee from reaching the material impairment stage. These measures include fitting the employee with hearing protectors, providing training, and requiring the employee to use the protectors. The protectors must reduce the employee's exposure to an 8-hour TWA of 85 db or below.

In addition, the employer must institute a training program on audiometric testing, hearing protectors, and effects of noise on hearing for all employees who are exposed to noise at or above an 8-hour TWA of 85 db. *Id.* at 1910.95(k). The employer must also retain records of employee exposure measurements and audiometric tests.

The provisions of the amendment apply to all employees covered by the Act, except those in construction, agriculture, and oil and gas well drilling and services. OSHA estimates the annual cost of compliance for the amendment at $254,321,000.00. *Final Regulatory Analysis of the Hearing Conservation Amendment,* U.S. Department of Labor, Occupational Safety and Health Administration, Office of Regulatory Analysis (January 1981), part IV.

II

An initial inquiry that must be made in determining the validity of any regulation adopted by a federal agency is whether the regulation is within the scope

[5]Hearing loss is measured by an audiometer. Audiometers produce pure tones at specific frequencies (e.g., 250, 500, 1000, 2000, 3000, 4000, 6000, and 8000 Hz) and at specific sound levels. The record of a given individual's hearing sensitivity is called an audiogram. An audiogram shows hearing threshold level measured in decibels as a function of frequency in hertz. It indicates how intense or loud a sound at a given frequency must be before it can be perceived. Thus under the amendment, follow up measures are required whenever the quietest sound an employee can hear at 2000, 3000, and 4000 Hz is an average 10 db louder than it was when the baseline audiometric text was performed.

of the agency's statutory authority. *Citizens to Preserve Overton Park, Inc. v. Volpe,* 401 U.S. 402, 415 (1971). Examining the language of the Occupational Safety and Health Act and the Supreme Court decisions interpreting it, we find it clear that Congress only authorized the Secretary to adopt those standards which relate to health and safety *at the workplace.*

The Act in its statement of findings and declaration of purpose and policy refers repeatedly to "working conditions." The Act defines term "occupational safety and health standard" as "a standard which requires conditions . . . reasonably necessary or appropriate to provide safe or healthful *employment* and *places of employment.*" (emphasis added).

In addition, the Supreme Court stated in the first OSHA case it considered that: "The Act created a new statutory duty to *avoid maintaining unsafe or unhealthy working conditions.*" *Atlas Roofing Co. v. OSAHRC,* 430 U.S. 422, 445 (1977) (emphasis added). In later cases, the Court further defined the scope of the Secretary's authority under the Act. "[B]efore he can promulgate *any* permanent health or safety standard [emphasis in the original], the secretary is required to make a threshold finding that *a place of employment* is unsafe." *Industrial Union Department v. American Petroleum Institute,* 448 U.S. 607, 642 (emphasis added). Congress placed pre-eminent value on assuring employees "a safe and healthy *working environment.*" *American Textile Mfrs. Institute v. Donovan,* 452 U.S. 490, 540 (1981) (emphasis added).

Most importantly, given the vast number of factors outside of the workplace that can potentially affect an employee's health or safety (e.g. social and recreational activities, alcohol or drug use, defective consumer products), interpreting the Act to extend to hazards existing outside the workplace would place under OSHA's control areas already subject to regulation by other federal agencies (the Alcohol, Drug Abuse and Mental Health Administration, the Consumer Product Safety Commission, the Environmental Protection Agency, the Food and Drug Administration and the National Highway Traffic Safety Administration to name but a few).

In light of the foregoing, we follow the approach of the Eleventh Circuit which is that "the conditions to be regulated [by OSHA] must fairly be considered *working* conditions, the safety and health hazards to be remedied *occupational,* and the injuries to be avoided *work-related.*" *Frank Diehl Farms v. Secretary of Labor,* 696 F.2d 1325, 1332 (11th Cir. 1983) (emphasis in the original) (OSH Act does not extend to hazards associated with housing provided to seasonal farm employees unless such housing is a condition of employment).

A standard is invalid if it requires an employer to take actions in regard to hazards existing outside the workplace. It is clear from the language of the hearing conservation amendment, as well as the record before this court, that under the amendment employers may be subjected to requirements and penalties may be imposed as a result of non-workplace hazards. The amendment's requirements are triggered whenever an employee suffers a standard threshold shift loss in hearing. It is obvious that such a hearing loss can result from non-occupational noise exposure just as easily as it can from occupational exposure.[6] Airplanes, hunting rifles, loud music and a myriad of other sources produce noise potentially as damaging as any at the workplace. Yet the amendment makes no distinction between hearing loss caused by workplace sources and loss caused by non-workplace sources. The rule-making record clearly provides that once a hearing loss is found, the amendment requires the *same* actions by the employer "whether or not the [loss] is work-related," and that the subject rule contains no requirement that there be "a determination of work relatedness."[7]

Thus the hearing conservation amendment clearly imposes responsibilities on employers based on non-work-related hazards. Under the amendment, an employer whose workers are unaffected by workplace noise may be subject to numerous requirements

simply because its workers choose to hunt, listen to loud music or ride motorcycles during their non-working hours. Hearing loss caused by such activities is regrettable but it is not a problem that Congress delegated to OSHA to remedy. The amendment is therefore vacated and remanded to OSHA for the creation of a valid standard.

VACATED AND REMANDED

DISSENTING OPINION

SPROUSE, Circuit Judge, dissenting.

I respectfully dissent. Doubtless the majority is correct in asserting that the Occupational Safety and Health Act only authorizes regulation of unsafe conditions of the workplace, but I believe that the majority affords insufficient deference to the Secretary's conclusion that the hearing loss regulated by the Hearing Conservation Amendment is employment-related.

The Hearing Conservation Amendment proposes to regulate by detecting industrially-produced hearing loss and then arresting further deterioration by requiring the use of protective devices, employee training, and the posting of warning signs. The majority opinion obviated discussion of the details of the regulation by holding that the entire regulation exceeded OSHA's authority. I, therefore, confine this discussion to the majority decision. In my opinion there is more than substantial evidence to support the factual findings which form the basis of the Secretary's conclusion that the Hearing Conservation Amendment is "reasonably necessary or appropriate to provide safe or healthful employment"; *Industrial Union Dep't v. American Petroleum Institute*, 448 U.S. 607, 630-35, 642 (1980), and is therefore a

---

[6]As the Department of Labor itself acknowledges:

There are few places you can go now to escape it. In any urban area, large or small, it's the roar of traffic, the thud of a pile driver, the staccato of a pneumatic drill, the shriek of a fire engine, the blast of a motorcycle, the blare of a rock and roll group, the whine of a jet overhead. Noise has become an inescapable component of modern, mechanized life.

Noise. The Environmental Problem. A Guide to OSHA Standards, U.S. Department of Labor, Occupational Safety and Health Administration (OSHA 2067) (reprinted from the July-August issue of Safety Standards magazine).

[7]In a letter to this Court following oral argument, the Secretary refers to Section 8(g)8(ii) of the amendment in support of his contention that the requirements of the hearing conservation amendment are *not* triggered by non-occupational noise. Section 8(g)8(ii) provides (emphasis added)

Unless a physician determines that the standard threshold shift is not work related or aggravated by occupational noise exposure, the employer shall ensure that the following steps are taken when a standard threshold shift occurs.

This clause is of little effect. An earlier draft of the amendment required that the professional reviewing employee audiograms determine whether any significant threshold shifts detected were caused by occupational noise exposure. In deleting this requirement in the final versions, OSHA emphasizes:

Commenters stated that in some cases it is very difficult, even for an audiologist or otolaryngologist, to determine the cause, or work relatedness, of a significant threshold shift because of the similarity between an occupational and nonoccupational audiometric hearing loss configuration on the audiogram.

Thus it is apparent that OSHA itself places little confidence in a physician's ability to identify those workers whose hearing loss is not work related or aggravated by occupational noise exposure.

valid exercise of his authority to promulgate occupational safety and health standards.

OSHA made factual determinations of the extent of the danger posed by workplace noise before issuing this interim regulation. It found that there are 2.2 million workers in American production industries exposed to an eight-hour TWA between 85-90 db. Ten to fifteen percent of the workers exposed to an eight-hour TWA of 85 db will suffer material hearing impairment, as will twenty-one to twenty-nine percent of employees exposed to an eight-hour TWA of 90 db. These figures represent a composite of studies done by the Environmental Protection Agency, the National Institute for Occupational Safety and Health, the International Organization for Standardization, and Dr. W. Baughn of the General Motors Corporation. On the basis of these figures, OSHA identified the risk of hearing loss to these 2.2 million workers as a serious public health problem requiring regulation.

Besides determining from several objective sources the magnitude of the risk to workers, OSHA also considered two scientific studies of the projected benefits of the Hearing Conservation Amendment. Bolt, Beranek, and Newman, Inc., a consulting firm under contract to OSHA, concluded that over a twenty-year exposure period the amendment would save a maximum of 324,000 workers from occupationally-caused material hearing impairment. A closely related study of the Center for Policy Alternatives, under contract to the Environmental Protection Agency, concluded that if noise exposure were reduced from 90 db to 85 db, over a forty-year exposure period, 580,000 employees would be spared occupationally-caused material hearing impairment. OSHA carefully reviewed both studies and relied on them in preparing its own benefit analysis. After making what it considered to be appropriate adjustments for changes in the size of the exposed work force, for the OSHA definition of material impairment, and for the protective ability of various devices, OSHA estimated that the total number of workers spared material impairment would be 212,000 in the tenth year, 477,000 in the twentieth year, 696,000 in the thirtieth year, 799,000 in the fortieth year, and 898,000 in the seventieth year.

Regulation of the industrial cause and effect of hearing loss obviously is not as simple as regulation of most mechanical hazards. Dangers inherent in the operation of moving machinery such as a table saw are easily perceived, and their causal relation to mutilated human limbs or eyes are readily understood and frustrated. Prevention of more subtle hazards requires more sophisticated solutions. OSHA resorted to scientific institutions to define the problem relating to industrially-caused hearing loss and relied on that information in designing its proposal The court must apply the substantial evidence test deferentially, particularly when the Secretary's factual findings are based upon complex scientific and factual data or involve speculative projections. *Baltimore Gas & Electric Co. v. Natural Resources Defense Council, Inc.*, 462 U.S. 87, 103 S. Ct. 2246, 2256 (1983); *United Steelworkers of America v. Marshall*, 647 F.2d 1189, 1206-07 (D.C. Cir. 1980), *cert. denied sub nom. National Ass'n of Recycling Industries, Inc. v. Secretary of Labor*, 453 U.S. 913 (1981). In these circumstances, the Agency's finding must only be within a "zone of reasonableness." *United Steelworkers*, 647 F.2d at 1207.

The Secretary's Hearing Conservation Amendment is clearly within the zone of reasonableness. In the first place, the amendment covers only those industries with a noise level that has been scientifically demonstrated to be a high risk factor to hearing health. Secondly, a threshold test is administered in those high risk industries, and hearing loss is measured after continued exposure to that high risk noise level.

To be sure, some hearing loss occurs as a part of the aging process and can vary according to non-occupational noises to which employees are exposed. The Hearing Conservation Amendment, however, is concerned with occupational noses—a hazard of the workplace. The hazard is identified as sustained noise of great intensity—85 db and above. Non-occupational noise of that intensity sustained over a period of eight hours each day is hard to imagine.

The amendment provides that non-occupationally caused hearing loss be excluded form its regulation. Assuming, however, that some loss caused by aging or smaller amounts of noise sustained for shorter periods also aggravates the hearing loss incurred by an individual employed in a high noise-producing industry, that is scant reason to characterize the primary risk factor as non-occupational. Breathing automobile exhaust and general air pollution, for example, is not healing to a wounded lung. That hardly justifies failure to regulate noxious workplace fumes that inflicted the primary wound. Nor would there be logic to characterizing regulation of the fumes as non-occupational because the condition inflicted is aggravated by outside irritants.

OSHA found that the amendment's cost to the regulated industries would average $41.00 annually per employee, and that it was economically feasible. Believing that the facts found by OSHA are supported by substantial evidence, and that the Secretary acted within his statutory authority and adequately explained the logic and policies underlying his regulation, *United Steelworkers,* 647 F.2d at 1207 I would affirm.

◇ NOTES

1. If carried to its logical extension, would this decision invalidate any OSHA standard that attempted to regulate toxic substances also found in the general environment or in the home?

2. Notice that the blame for nonoccupational noise exposure is ascribed to lifestyle factors such as "listening to loud music" or "riding a motorcycle." Does the fact that toxic pollution in the environment is not the worker's "fault" distinguish toxic substances from noise? What about toxic substances in consumer products?

3. Note that OSHA did not appeal this decision to the Supreme Court. The Justice Department, rather than the federal agencies, makes the decision to appeal.

◇

8. Medical Removal Protection

In order to prevent adverse health effects from continuing exposure, it is sometimes necessary to order the medical removal (MR) of exposed workers from the workplace. To fulfill the OSHAct's mandate that employers bear the burden of ensuring occupational safety and health, OSHA has included medical removal protection (MRP) provisions in some standards. With an MRP provision in place, the employer must guarantee that the removed employee suffer no loss of wages or seniority during the period of removal. The propriety of this approach was addressed by the D.C. Circuit in the lead case.

United Steelworkers of America, AFL-CIO-CLC v. Marshall and Bingham

647 F.2d 1189 (D.C. Cir. (1980))

WRIGHT, Chief Judge.

* * *

One of the most vigorously contested issues in the case is the substantive validity of the provision for Medical Removal Protection (MRP). OSHA regards MRP as a *sine qua non* of the lead standard, insisting that without it employees, fearing they will lose jobs if they demonstrate high blood-lead levels, will refuse to participate in medical surveillance. The industry regards MRP as an illegal and unwarranted extension of OSHA's authority—a cash subsidy to encourage employees to cooperate with medical surveillance when the statute expressly *requires* employees to comply with all provisions of the standard. 29 U.S.C. §654(b) (1976). The essential question is one of statutory interpretation. LIA [the Lead Industries Association] contends that MRP lies outside OSHA's statutory grant of authority, contradicts the will of

Congress as manifested by the OSHAct's legislative history, and violates an express prohibition on OSHA's power within the statute.[55] We reject LIA's contentions.

1. *The MRP program.* MRP supplements the medical surveillance provisions of the lead standard. All employers are required to measure the air-lead content of all workplaces to determine whether employees suffer exposure to lead above the "action level" of 30 μg/m^3, or above the PEL. For any workers exposed to air-lead above the action level for more than 30 days a year, the employer must offer a program of biological monitoring, including exhaustive blood measurements, other laboratory tests, and detailed medical examinations. If these tests and examinations reveal that a worker has a high blood-lead level or some ailment attributable to lead exposure, the employer must remove the worker from the high-exposure workplace. The standard sets out the criteria for and terms of this removal in great detail.

The standard will ultimately require employers to remove any worker who is exposed to air-lead at or above the action level[56] and whose blood-lead level averages 50 μg/100g or more on three consecutive blood tests (or on all tests conducted over a six-months period if the worker's last three tests took place over a period of less than six months), unless the worker's blood-lead level on the last of these tests is at or below 40 μg/100g. However, since a great number of employees currently have blood-lead levels well above these figures, OSHA found it infeasible to enforce these removal criteria immediately. OSHA therefore has phased in the removal standard gradually, with three preliminary stages requiring removal at progressively lower blood-lead and air-lead levels.[57] For each stage of the removal standard OSHA has also designated the reduced blood-

lead level which employees must achieve before they can return to the jobs from which they were removed.[58] Finally, the employer must also remove an employee when a "final medical determination"[59] is made that the lead exposure at the employee's workplace threatens his health, and the employer cannot return the employee to the workplace until a final medical determination is made that exposure in the workplace no longer places the employee's health at risk.[60]

The rule on MRP benefits, Section 1910.1025(k)(2), is the true center of the controversy. An employer enjoys the discretion to place a removed worker in a low-exposure job in the same plant, or at a job at a non-lead facility, or even to simply retain him at his high-exposure job for a smaller number of hours per week to reduce his time-weighted average exposure level below the action level. 52975/2. If these actions fail, however, the employer must lay the worker off. *Id.* And most important, whatever form removal takes, the employer must maintain the worker's earnings and seniority rights during removal for a period up to 18 months,[61] and must restore him to

---

[55]LIA also contends that MRP fails the statutory test of economic feasibility. Since the cost of MRP cannot be disentangled form the industries' technological capacity to lower air-lead exposure, we address this question later as part of our general inquiry into the feasibility of the standard.

[56]To reduce lead exposure below the action level, the employer must achieve an air-lead level of less than 30 μg/m^3 *without relying on respirators.*

[57]During the first year of the standard the employer must remove any worker exposed to air-lead at or above 100 μg/m^3 whenever a periodic and follow-up blood test reveals the worker's blood-lead is at or above 80 μg/100g. During the second year the employer must remove a worker exposed to air-lead at or above 50 μg/m^3 when

these blood tests measure at or above 70 μg/100g. During the third and fourth years the employer must remove any worker exposed to lead above the action level whose blood-lead is at least 60 μg/100g.

[58]An employee removed because of a blood-lead level of 80 μg/100g or more must show a level at or below 60 μg/100g on two consecutive tests before return. For a worker removed for blood-lead at or above 70 μg/100g, the return figure is 50 μg/100g. For a worker removed because of blood-lead at or above 60 μg/100g or at an average level of at least 50 μg/100g, the return figure is 40 μg/100g.

[59]"Final medical determination" means the outcome of the multiple physician review scheme or alternate medical determination scheme. . . .

[60]With two special exceptions, the employer may choose to return a worker to his former job while the final medical determination scheme to decide whether the worker can be returned is running its course, so long as that choice is consistent with the opinion of at least one physician who has reviewed the worker's health status. Section 1910.1025(k)(1)(v).

[61]However, if an employee's blood-lead level fails to decline to a safe level within 18 months, the employer may have to maintain his financial benefits beyond that period until a final medical determination is made that the worker can never return to his former job, or until the worker is in fact returned pursuant to a final medical determination that such return is medically sound.

all the rights of his original job status when he becomes medically eligible for return.[62]

2. *General authority under the OSH Act.* The fact of the statute and the legislative history both demonstrate unmistakably that OSHA's statutory mandate is, as a general matter, broad enough to include such a regulation as MRP.[63] A number of terms of the statute give OSHA almost unlimited discretion to devise means to achieve the congressionally mandated goal. *See Industrial Union Dep't, AFL-CIO v. Hodgson, supra,* 499 F.2d at 474. Thus OSHA is to ensure worker safety and health "by developing innovative methods, techniques, and approaches for dealing with occupational safety and health problems[.]" 29 U.S.C. 651(b)(5)(1976). The definition of an "occupational health and safety standard" speaks of "a standard which requires . . . the adoption or use of one or more practices, means, methods, operations, or processes, reasonably necessary or appropriate to provide safe or healthful employment and places of employment." *Id.* 652(8). Moreover, an OSHA standard should, "[w]here appropriate, . . . prescribe suitable . . . control . . . procedures" to prevent hazards. *Id.* 655(b)(7). Finally, "[t]he Secretary . . . shall . . . prescribe such rules and regulations as he may deem necessary to carry out [his] responsibilities" under the statute. *Id.* 657(g)(2).

* * *

The scheme of the statute, manifest in both the express language and the legislative history, also appears to permit OSHA to charge to employers the cost of any new means it devises to protect workers. Thus Congress directed all enforcement authority for OSHA standards against the employer, 29 U.S.C. §658, 659(a)(1976), and charged employers with the expense of providing medical examinations to employees, *id.* §655(b)(7), and of keeping all records necessary for enforcement, *id.* §657(c). The employer's general financial responsibility under the Act is also implicit in the requirement that

> [e]ach employer shall promptly notify any employee who has been or is being exposed to toxic material . . . and shall inform any employee who is being thus exposed *of the corrective action being taken.*

Id. §657(c)(3) (emphasis added).

The report of the Senate subcommittee from which the statute emerged stressed the need to place the cost of standards on employers, noting that

> many employers—particularly smaller ones—simply cannot make the necessary investment in health and safety, and survive competitively, unless all are compelled to do so. The competitive disadvantage of the more conscientious employer is especially evident where there is a long period between exposure to a hazard and manifestation of an illness. In such instances a particular employer has no economic incentive to invest in current precautions, not even in the reduction of workmen's compensation costs, because he will seldom have to pay for the consequences of his own neglect.

S. Rep. No. 91-1282, 91st Cong., 2d Sess. 4 (1970), *reprinted in Legislative History, supra,* at 144.[65] Senator Yarborough, chairman of the parent Committee on Labor and Public Welfare, told the Senate:

> We need a Federal statute, not to try to federalize things, but to equalize the cost in one industry vis-a-vis another. We know the costs would be put into consumer goods but that is the price we should pay for the 80 million workers in America.

---

[62]The employer may remove an employee at air-lead or blood-lead levels less dangerous than those which trigger mandatory removal, but must then maintain the employee's earnings and seniority rights as fully as the standard demands for required removal.

[63]In the cotton dust case we held that a medical removal provision was within OSHA's statutory power. *AFL-CIO v. Marshall, supra* . . . 617 F.2d at 674-75. That provision, however, was a very modest one compared to that in the lead standard, since it only required removal of workers who were incapable for medical reasons of wearing respirators, and even then only when there was a low-exposure job available. The wage-guarantee rule in that provision thus posed a far smaller economic threat to employers than does the lead MRP, and we made clear in the cotton dust case that we were reserving the question of the statutory validity of the more stringent types of removal program. *Id.* 617 F.2d at 674-75 n.238. In the asbestos case we described as "salutary" a removal program identical to that in the cotton dust case. *Industrial Union Dep't, AFL-CIO v. Hodgson, supra* note 31, 499 F.2d at 485, but the program had not actually been at issue in the appeal there.

[65]The report added that
> employers are . . . bound by this general and common duty to bring no adverse effects to the life and health of their employees throughout the course of their employment. Employers have primary control of the work environment and should insure that it is safe and healthful.

Legislative History, supra, at 444. Senator Eagleton, a member of the subcommittee, added that "[t]he costs that will be incurred by employers in meeting the standards of health and safety to be established under this bill are . . . reasonable and necessary costs of doing business." *Id.* at 1150.

LIA points to the statutory requirement that employees as well as employers comply with all OSHA regulations. 29 U.S.C. §654(b) (1976). But the subcommittee report made clear that this requirement is essentially an exhortation to employees to cooperate in the standards and is not meant

> to diminish in anyway [*sic*] the employer's compliance responsibilities or his responsibility to assure compliance by his own employees. Final responsibility for compliance with the requirements of this act remains with the employer.

S. Rep. No. 91-1282, *supra,* at 10-11, *reprinted in Legislative History, supra,* at 150-151; *see Atlantic & Gulf Stevedores, Inc. v. OSHRC,* 534 F.2d 541, 553 [1070] (3d Cir. 1976).[66]

MRP thus appears to lie well within the general range of OSHA's powers. LIA nevertheless focuses on two signposts of congressional intent to show that the drafters of the statute specifically meant to deny OSHA power to create a program like MRP. First, LIA points out that the OSH Act was passed only a year after the Federal Coal Mine Health and Safety Act of 1969, which required job removal with earnings protection for miners whose chest x-rays revealed they had pneumoconiosis. 30 U.S.C. §843(b)(2)-(3) (1976). Thus, argues LIA, in 1970 Congress was well aware of the concept of medical removal protection, and so its failure either to require MRP in the OSH Act or at least to expressly delegate to OSHA the power to create it proves that Congress intended that OSHA not include MRP in any standards.[67] Thus

LIA relies on the principle of *expresio unius est exclusio alterius. Marshall v. Gibson's Products, Inc. of Plano,* 584 F.2d 668, 675-676 (5th Cir. 1978).[68]

We agree that this rule of statutory construction may provide some evidence of congressional intent, but the evidence is hardly decisive, or even strong:

> This maxim is increasingly considered unreliable . . . for it stands on the faulty premise that all possible alternative or supplemental provisions were necessarily considered and rejected by the legislative draftsmen. . . .

Nat'l Petroleum Refiners Ass'n v. FTC, 482 F.2d 672. 676 (D.C. Cir. 1973). *cert. denied,* 415 U.S. 951 (1974). Moreover, the rule can only apply sensibly when we compare very similar statutes, and we find a crucial difference between the two in question here. In the OSH Act Congress invested a new agency with extremely broad jurisdiction to prevent all types of safety and health hazards throughout American industry. In the Coal Act, on the other hand, Congress culminated decades of intense concern for a single industry by creating a sharply focused statute, with 30 pages of safety and health regulations that even surpass an OSHA standard in their detail. 30 U.S.C. §§811-878 (1976). Congress may well have avoided all mention of medical removal protection in the OSH Act simply because it thought that mandating such a specific program was inappropriate in a statute so much broader and so much more dependent on agency implementation than the Coal Act.

LIA's second special argument on legislative intent is that the Congress that passed the OSH Act deliberately rejected an amendment, sponsored by Representative Daniels, that would have created a form of medical removal protection. And indeed, Section 19(a)(5) of the House Bill, H.R. 16785, 91st Cong., 2d Sess. (1970), *Legislative History, supra,* at 842, became

---

[66]In *Budd Co. v. OSHA,* 513 F.2d 201, 203-205 (3d Cir. 1975), the Third Circuit upheld an OSHA decision that employers did not have to pay for the protective footwear which OSHA regulations required employees to wear. But the court stressed the special character of protective devices which the employee would wear off-the-job as well as on-the-job and made clear it was expressing no opinion on the proper party to be charged for other devices and methods. Moreover, the court there failed to address the relevant parts of the legislative history of the OSH Act.

[67]LIA also points out that neither the Coal Act nor the Federal Mine Safety and Health Act of 1977, 30 U.S.C. 811(a)(7) (Supp. 4, 1977), protects *seniority* rights as well as earnings, and that whereas OSHA claims plenary author-

ity to institute MRP for any occupational disease, the Mine Safety Administration, in implementing the Coal Act, could provide earnings protection only for pneumoconiosis, even though the chest x-rays it gave miners could reveal other occupational diseases.

[68]In *Gibson's Products, Inc.,* the Fifth Circuit pointed to the express power to obtain inspection injunctions granted other agencies by statute to show that OSHA, which was not given this power by express statute, did not hold it by implication. But the court there relied just as heavily on provisions within the OSH Act showing that Congress had carefully delineated the limited circumstances in which courts had jurisdiction to issue injunctions against employers. *Marshall v. Gibson's Products, Inc.,* 584 F.2d 668 (5th Cir. 1978).

controversial and even infamous as the "strike-with-pay" provision. LIA notes that Section 19(a)(5) had a purpose similar to that of MRP:

> There is still a real danger that an employee may be economically coerced into self-exposure in order to earn his livelihood so the bill allows an employee to absent himself from that specific danger for the period of its duration without loss of pay. . . .

H.R. Rep. No. 91-1291, 91st Cong., 2d Sess. 30 (1970), *reprinted in Legislative History, supra,* at 860. LIA also points out that Senator Williams, in advocating the Senate bill, felt compelled to reassure his colleagues that it contained no "strike-with-pay" clause and so did not "rais[e] a possibility for endless disputes over whether employees were entitled to walk off the job with full pay[.]" *Legislative History, supra,* at 416.

LIA's argument, however, does not survive close reading of Section 19(a)(5), which reveals that the "strike-with-pay" clause differs markedly from the MRP regulation. The rejected clause reads as follows:

> The Secretary of Health, Education, and Welfare shall publish within six months of enactment of this Act and thereafter as needed but at least annually a [l]ist of all known or potentially toxic substances and the concentrations at which such toxicity is known to occur; and shall determine following a request by any employer or authorized representative of any group of employees whether any substance normally found in the working place has potentially toxic or harmful effects in such concentration as used or found; and shall submit such determination both to employers and affected employees as soon as possible. Within sixty days of such determination by the Secretary of Health, Education, and Welfare of potential toxicity of any substance, an employer shall not require any employee to be exposed to such substance designated above in toxic or greater concentrations unless it is accompanied by information, made available to employees, by label or other appropriate means, of the known hazards or toxic or long-term ill effects, the nature of the substance, and the signs, symptoms, emergency treatment, and proper conditions and precautions of safe use, and personal protective equipment is supplied which allows established work procedures to be performed with such equipment, or unless such exposed employee may absent himself from such risk of harm for the period necessary to

avoid such danger without loss of regular compensation for such period.

H.R. Rep. No. 1291, *supra,* at 12, *Legislative History, supra,* at 842. Two features of this clause merit emphasis. First, the grounds for removal derive from *ad hoc,* informal action by HEW. By comparison, the grounds for removal in MRP result from OSHA rulemaking. Second, and perhaps more important, under the "strike-with-pay" clause the employee himself can apparently make the individual judgment that the grounds for removal apply, and so he can effectively remove himself. Under MRP removal is determined solely by OSHA's objective criteria and may indeed occur against the worker's will.[69] Thus, the "strike-with-pay" clause would probably invite controversy and abuse in a way that MRP would not, so the reasons for which Congress rejected the former may well not apply to the latter.

[69]For this reason the Supreme Court's recent decision in *Whirlpool Corp. v. Marshall,*—U.S.—. 48 U.S. L. Week 4189 (Feb. 26, 1980), does not support LIA's argument that MRP violates congressional intent. In *Whirlpool* the Court upheld an OSHA regulation forbidding an employer to discriminate against an employee who walks off the job or refuses to perform an assigned task because he reasonably apprehends death or serious injury from a hazard. 29 C.F.R. §1977.12 (1979). In holding that the regulation lay within OSHA's statutory power the Court rejected the petitioner's argument that Congress' rejection of the Daniels "strike-with-pay" provision bespoke a legislative animus against the regulation in question.—U.S. at—. 48 U.S. L. Week at 4192-4194. Specifically, the Court stated that the Daniels bill was not concerned with the "highly perilous and fast-moving situations" with which the challenged regulation was concerned, and that the congressional debate over the Daniels bill chiefly focused on a worker's right to continued compensation after he left the job, whereas the challenged regulation made no provision for continued salary during a work stoppage.

The Court's analysis of the Daniels bill might appear to threaten the statutory basis for MRP, since MRP is not concerned with sudden, imminent threats to worker safety, while it does, of course, guarantee a worker's wage when he is removed for safety or health reasons. But a careful reading of the Court's opinion only reinforces our sense of the important difference between MRP and the Daniels bill. The Court stated:

> It is also important to emphasize that what primarily troubled Congress about the Daniels bill's "strike-with-pay" provision was its requirement that employees be paid their regular salary *after having properly invoked their right to refuse to work under the section.* It is instructive that virtually every time the issue of *an employee's right to absent himself* from hazardous work was discussed in

3. *The Section 4(b)(4) prohibition.* LIA's most serious statutory argument against MRP is that MRP violates the express prohibition contained in Section 4(b)(4) of the OSH Act, 29 U.S.C. §653(b)(4) (1976):

> Nothing in this chapter shall be construed to supersede or in any manner affect any workmen's compensation law or to enlarge or diminish or affect in any other manner the common law or statutory rights, duties, or liabilities of employers and employees under any law with respect to injuries, diseases, or death of employees arising out of, or in the course of, employment.

LIA argues vigorously that MRP contravenes this prohibition in both purpose and effect. As a matter of purpose, it cites evidence that OSHA designed MRP to remedy the inadequacies of state workmen's compensation laws. But this evidence shows little more than that LIA was able to put its ideas into the mouths of OSHA witnesses during cross-examination, and in any event we do not construe Section 4(b)(4) as being concerned with agency motive. LIA's argument on the effect of MRP does, however, deserve serious attention.

Under workmen's compensation law, a worker suffering disablement of a designated type can recover, depending on the state, up to two thirds of his lost wages. These laws presume that if a disabled worker could recover all his lost wages, he would have no incentive to return to work. LIA contends that under MRP a worker would never seek worker's compensation, nor would he ever become medically or financially eligible for it. A worker removed from such a disability would be guaranteed all the earnings rights of his high-exposure job; suffering no loss of wages, he would be entitled to no wage replacement under worker's compensation. Moreover, under MRP most workers would be removed because of high blood-lead levels—before they exhibit clinical symptoms of disablement. Thus they would enjoy their guaranteed salary before they became disabled in the eyes of workmen's compensation law. LIA concludes that for workers vulnerable to lead disease MRP would "supersede" and "affect" worker's compensation law to the point of wholly replacing it.

LIA acknowledges that under the "credit" provision of MRP, the employer could reduce his payments under MRP by the amount the removed employee receives under worker's compensation. But LIA contends that this credit is meaningless since no worker harmed by lead will ever receive worker's compensation in the first place.

To resolve this issue we first recognize that Section 4(b)(4) is vague and ambiguous on its face. We must seek then the best, not a perfect, reading.[70] Thus LIA's argument under Section 4(b)(4) really shows that no literal reading of it is possible, since any health standard that reduces the number of workers who become disabled will of course "affect" and even "supersede" worker's compensation by ensuring that those workers never seek or obtain workmen's compensation benefits.

---

[70]Unfortunately, the legislative history of the OSH Act tells us essentially nothing about §4(b) (4). Section 27 of the Act (omitted from the 1976 codification), created a National Commission to Study State Workmen's Compensation Laws, and the Senate subcommittee report discussion of this commission is the only allusion the legislative history makes to worker's compensation laws. The report lamented the failure of state law to cover enough workers to apply to enough forms of disablement, and to keep compensation apace with inflation. It explains the Commission's purpose as being to study the feasibility of federal legislation to remedy these flaws, but emphasizes "that by authorizing this study it is not impliedly recommending federalization of the existing workmen's compensation system," and leaves the whole issue to "an informed decision by Congress . . . in the future." *Id.* at 165. LIA construes this language as denying OSHA the power to draw regulations displacing workmen's compensation law. But as we explain in text below, there is a great difference between a regulation that has the effect of reducing the number of claims made under state law and one that actually alters the terms of workmen's compensation law. OSHA has not "federalized" workmen's compensation law. It has left the legal scheme of that law wholly intact.

the legislative debates, it was in the context of the employee's right to continue to receive his usual compensation . . . But the regulation at issue here does not require employers to pay workers *who refuse to perform their assigned tasks* in the face of imminent danger. . . .
—U.S. at—48 U.S. L. Week at 4193-4194 (emphasis added; footnotes omitted). In characterizing the Daniels bill here the Court may have stressed the issue of compensation, but it was clearly speaking only in the context of a worker's self-initiated decision to leave the job after making a subjective decision that the workplace was dangerous. The Court did not address, and of course had no reason to address, the question of requiring compensation of workers who are removed from a job—perhaps against their will—according to objective regulatory criteria. The *Whirlpool* decision thus does not bear on MRP.

Applied to those workers who, thanks to MRP, never become disabled, the argument certainly proves too much. Nevertheless, LIA's argument, if sharpened, remains plausible, since it can still sensibly apply to a small and special group of workers; those who are removed under MRP pursuant to a final medical determination that they are already disabled, and for whom no low-exposure job of equal salary is available.

We must first reject OSHA's attempt to rebut LIA by asserting that, whereas worker's compensation is "compensatory" in nature, MRP is "preventive." At best, this assertion argues that MRP differs from worker's compensation in purpose, whereas we have said that Section 4(b)(4) is concerned with effects. At worst, OSHA's assertion is a *non sequitur*.

Nevertheless, though we must take seriously LIA's argument with respect to the special group of workers we have described, we do not think it proves MRP to violate Section 4(b)(4). First, this special group of workers is probably very small, and will become progressively smaller. As the new PEL lowers lead exposure throughout industry, and as older workers with high accumulations of lead retire, fewer and fewer workers will require removal because of manifest disablement from lead. Moreover, workers remaining in this group will still have an incentive to file worker's compensation claims. First, worker's compensation laws universally reimburse workers for the medical expenses of their disablement, while MRP does not. Second, MRP benefits can only last slightly more than 18 months, whereas worker's compensation may replace lost wages for longer periods or even indefinitely. 82 Am. Jur. 2d *Workmen's Compensation* §382 (1976).[71]

The question remains, then, what *does* Section 4(b)(4) mean, if it *does not* mean that OSHA is barred from creating medical removal protection? We see two plausible meanings. First, as courts have already held, Section 4(b)(4) bars workers from asserting a private cause of action against employers under OSHA standards. *Jeter v. St. Regis Paper Co.*, 507 F.2d 973 (5th Cir. 1975); *Byrd v. Fieldcrest Mills, Inc.*, 496 F.2d 1323 (4th Cir. 1974). Second, when a worker actually asserts a claim under workmen's compensation law or some other state law, Section 4(b)(4) intends that neither the worker nor the party against whom the claim is made can assert that any OSHA regulation or the OSH act itself *preempts* any element of the state law. For example, where OSHA protects a worker against a form of disablement not compensable under state law, the worker cannot obtain state relief for that disablement. Conversely, where state law covers a wider range of disablements than OSHA aims to prevent, an employer cannot escape liability under state law for disablement not covered by OSHA. In short, OSHA cannot *legally* preempt state compensation law, even if it *practically* preempts it in some situations.

We conclude that though MRP may indeed have a great practical effect on workmen's compensation claims, it leaves the state schemes wholly intact as a *legal* matter, and so does not violate Section 4(b)(4).

4. *Interference with national labor policy.* As its final statutory argument LIA contends that MRP violates the policies of federal labor legislation. Specifically, LIA asserts that MRP sets in concrete a program that is a traditional and mandatory subject of collective bargaining under federal law, and thus violates the congressional principle that the substantive provisions of labor-management relations be left to the bargaining process. *Local 24, Int'l Brhd of Teamsters v. Oliver,* 358 U.S. 283, 295 (1959). We do not doubt that MRP will have a noticeable affect on future collective bargaining, but such an effect hardly proves that OSHA has misconstrued legislative intent.

Earnings protection is no doubt a mandatory subject of collective bargaining, but so is any issue directly related to worker safety. *NLRB v. Gulf Power Co.*, 384 F.2d 822, 824-825 (5th Cir. 1967). In passing a massive worker health and safety statute, Congress certainly knew it was laying a basis for agency regulations that would replace or obviate worker safety provisions of many collective bargaining agreements. Congress may well have viewed collective bargaining agreements along with state worker's compensation laws as part of the status quo that had failed to provide workers sufficient protection. *See Legislative History, supra,* at 164.

[71]OSHA has also compiled evidence of the many collective bargaining agreements that include some form of earnings protection for disabled workers, and concludes that these programs have not interfered with the operation of workmen's compensation. It notes that under these agreements employers maintain wages pending disposition of worker's compensation claims, receive credits or paybacks once compensation awards are made, and sometimes supplement the compensation awards up to 100% of the worker's lost wages. We find this evidence inconclusive, however, since OSHA has not explained whether these agreements would allow workers to receive long-term wage replacements even if they never file compensation claims.

*   *   *

5. *Reasonableness of MRP.* Having concluded that a program of earnings protection for removed workers lies within OSHA's statutory power, we need finally to inquire whether MRP is a reasonable exercise of that power. Like any OSHA program, MRP must pass the statutory tests of judicial review: OSHA must demonstrate substantial evidence to support any conclusions of determinable fact that underlie the program and, where the new provision cannot rely on factual certainty, must carefully explain the bases of its "legislative" decision to create it. *Industrial Union Dep't, AFL-CIO v. Hodgson, supra* 499 F.2d at 475-476. This test essentially reinforces the principle that where a statute empowers an agency to make rules necessary to carry out the provisions of the statute, the court will uphold such a rule if it is " 'reasonably related to the purposes of the enabling legislation.' " *Mourning v. Family Publications Service, Inc.,* 411 U.S. 356, 369 (1973), *quoting Thorpe v. Housing Authority of the City of Durham,* 393 U.S. 268, 280-281 (1969).

Here, we have no trouble upholding the agency's decision as reasonable and well grounded in the evidence. OSHA has very precisely articulated the bases for MRP, set forth factual findings where relevant, made all inferences and speculations explicit, and explained why it rejected counter-evidence and proposed alternate programs.... We need only review the major points briefly here.

OSHA offered two primary bases for MRP. First, the standard's PEL leaves little margin of safety for workers, since even at an air-lead level of $50 \mu g/m^3$ as many as 29.3 percent of all exposed workers will at any time have a blood-level over $40 \mu g/100g$, the level which OSHA sets as its goal for *all* workers. However, lead disease is highly reversible in its early stages, since a person removed to a low-exposure workplace can excrete accumulated lead. OSHA thus concluded that removal was a preventive device crucial to the standard.

OSHA found, however, that unless workers were guaranteed all their wage and seniority rights upon removal, they would resist cooperating with the medical surveillance program that determined the need for removal, since they reasonably might fear being fired or sent to lower-paying jobs if they revealed dangerously high blood-lead levels. The record showed that workers often consumed self-prescribed chelating agents,[73] and lied to physicians about their sub-

jective symptoms, all because they held job security more dear than their health. (citing evidence). OSHA also found existing earnings protection programs in private bargaining agreements too few and too limited. Moreover, exercising its statutory authority to rely on experience gained under other health and safety laws in fashioning a standard, 29 U.S.C. §655(b)(5) (1976), OSHA compiled evidence that even under a congressionally-mandated earnings protection program like the one that is part of the Federal Coal Mine Health and Safety Act, *see* text and notes at notes 67-68 *supra,* workers—perhaps because they were not guaranteed the seniority rights and pay increases of their high-exposure jobs—frequently refused to cooperate in medical review. Finally, OSHA found evidence that seniority rights were often the linchpin of any earnings protection, and that required seniority protection could easily be coordinated with collective bargaining provisions on seniority.

OSHA also supported each particular criterion and term of the removal program with evidence and explanation. For example, it offered both medical testimony and worker health statistics to demonstrate that workers needed removal protection for at least 18 months to allow the reversible phase of lead disease to subside and reduce their blood lead to safe levels. And beyond justifying each element of the final MRP, OSHA soundly explained why it rejected a number of proposed MRP schemes.

For example, OSHA rejected the suggestion that it directly *force*, rather than encourage, workers to participate in medical surveillance, since such force might violate their freedom of religion and right to privacy, and in any event could not prevent workers from hiding symptoms and using chelating agents. OSHA also struck down a proposal that instead of requiring removal with earnings protection, OSHA simply require employers to provide workers information on their blood-lead levels and then let the workers decide whether to remove themselves; OSHA found no evidence that workers, given the choice, would always or even often choose health over job security. And last, OSHA rejected the idea that workers with high blood-lead levels use respirators rather

---

[73]Chelating agents are drugs which bind themselves to lead when they enter the body and then flush themselves— along with the lead—out of the body. OSHA compiled exhaustive evidence that chelation produces such ill

effects as anxiety, nausea, hypertension, and anemia. It found that though "therapeutic" chelation—to relieve manifest symptoms of lead disease—and "diagnostic" chelation—to help determine lead disease—might be acceptable, "prophylactic" chelation, which physicians and employees themselves practice simply to lower blood-lead levels, is unacceptable under modern medical principles.

than be removed, pointing to all the evidence it had compiled for other purposes on the inherent flaws of respirator protection.

We therefore conclude that MRP is a reasonable exercise of legitimate authority.[74]

*   *   *

[74]*Amici* representing public interest law organizations and California state labor agencies have argued that MRP is not only legally *valid* under the OSH Act, but is legally *required* by Title VII of the Civil Rights Act of 1964, 42 U.S.C. §2000e *et seg.* (1976 & Supp. 11 1978). They argue that without MRP employers will discriminate against fertile women—to whom lead exposure poses an even greater threat than it does to other workers—by excluding them from all lead-exposed jobs at the outset. A review of an OSHA proceeding, however, is not the place to address hypothetical Title VII questions, and in any event we think fertile women can find statutory protection from such discrimination in the OSH Act's own requirement that OSHA Standards ensure then "*no* employee will suffer material impairment of health. . . . 29 U.S.C. §655(b)(5) (1976) (emphasis added).

◇ NOTES

1. Note especially the discussion of MRP and its relationship to the workers' right to refuse hazardous work and workers' compensation. These subjects are addressed in chapters 8 and 9.

2. Note, too, the discussion of Section 4(b)(4) of the OSHAct.

3. Finally, after reading chapter 6 on the National Labor Relations Act, reread the section of the case on interference with national labor policy.

4. OSHA also included an MRP provision in the cotton dust standard. In 1981, the Supreme Court vacated the provision and remanded the issue to OSHA for further proceedings to determine whether MRP was necessary to protect workers. OSHA issued an advance notice of proposed rulemaking (ANPR) on the provision in 1982 and received comments from the public. After consideration of new evidence, in 1985 OSHA again issued an MRP provision in the cotton dust standard.

5. When OSHA promulgated a new formaldehyde standard at 1.0 ppm, it refrained from requiring MRP. The D.C. Circuit Court of Appeals rejected OSHA's reasons for not establishing an MRP program and remanded the standard to OSHA for further explanation on June 9, 1989.

◇

E. EVOLUTION OF THE GENERAL DUTY OBLIGATION

While its application has fluctuated over time, the importance of the Act's general duty obligation cannot be overstated. It imposes a general responsibility upon employers to take affirmative action to ensure safe working conditions. It has a preventive focus and can significantly empower workplace health and safety professionals if they are risk averse. The genery duty clause has been interpreted broadly by the courts.

Crisis in the Workplace: Occupational Disease and Injury

N.A. Ashford

The legislative history of the OSHAct does not reveal a clear congressional intent regarding the general duty provision. The original bills, which contained broad general duty requirements and no penalties for violations, were redrafted and amended until in the final Act specific requirements and penalty provisions were adopted.[73] The House-Senate Conference Committee found that the concept of a general duty is firmly rooted in the common law. The Senate committee on Labor and Public Welfare noted that "under principles of common law individuals are obliged to refrain from actions which cause harm to others. Courts often refer to this as a general duty of care to others.... The committee believes that employers are equally bound by this general and common duty to bring no adverse effects to the life and health of their employees.... [The Act] merely restates that each employer shall furnish this degree of care." Further, there is existing precedent for the general duty concept in federal and state laws which compel employers to provide a safe and healthful place of employment. Three federal statutes and the laws of over 36 states recognize this concept.

* * *

Early Review Commission cases established the principle that employer liability under the general duty provision does not require the occurrence of an accident or employee injury. Liability is conditioned only on whether the employer has exercised reasonable diligence in providing his employees with a place of employment free from recognized hazards. The question of liability has turned upon the definition of the concept of "recognized hazards" in any particular case. Congressman William Steiger, in discussion on the Conference Report, stated that the "recognized hazards" language of the general duty clause applies to hazards which are of "the type that can be detected on the basis of the human senses. Hazards which require technical or testing devices to detect them are not intended to be within the scope of the general duty requirement."

MIT Press, Cambridge, 1976.
[73]Marjorie E. Gross, *Loyola University Law Journal*, Vol. 3 (1972), p. 246; see also H. Rep., No. 91-1291, 91st Cong., 2d Session (1970), p. 21.

On the other hand, OSHA's position in its *Compliance Operations Manual* is that "A hazard is 'recognized' if it is a condition that is (a) of common knowledge or general recognition in the particular industry in which it occurs, and (b) detectable (1) by means of the senses (sight, smell, touch and hearing) or (2) is of such wide general recognition as a hazard in the industry that *even if it is not detectable by means of the senses, there are generally known and accepted tests for its existence which should make its presence known to the employer.*"

The Review Commission has rejected the Steiger position in favor of OSHA's interpretation on the basis of the relevant legislative history, including remarks by Congressman Dominick Daniels. In *American Smelting and Refining Co.,* the Commission found that "serious hazards that are not obvious and can be detected by instrumentation are within the scope of Section 5(a)(1) of the Occupational Safety and Health Act, which is not limited to 'recognized hazards' of types that are detectable by basic human senses."[79]

The Eighth Circuit Court of Appeals specifically upheld the Commission's interpretation of "recognized hazard" in its review of *American Smelting*.[80] The court stated that a narrow construction of the term would endanger the health and safety of workers: "Where hazards are recognized but not detectable by the senses, common sense and prudence demand that instrumentation be utilized." In an earlier case, *Vy Lactos Laboratories, Inc.,* the same circuit court had held that "recognized hazards" include an employer's actual knowledge as well as the generally recognized knowledge of the industry.[82]

* * *

Given that the employer is under an obligation to provide a workplace free of "recognized hazards," the question remains whether the employer is liable for conditions created by his employees when they are not acting in accordance with his instructions. While the OSHAct specifically states that each employee must comply with occupational safety and

[79]*American Smelting and Refining Co.*, Occupational Safety and Health Review Commission (hereinafter OSHRC) Docket No. 10 (August 17, 1973), as reported by the Bureau of National Affairs, in Vol. 1 of *Occupational Safety and Health Cases* (OSHC), p. 1256.
[80]501 F.2d 504 (8th Cir. 1974).
[82]494 F.2d 460 (8th Cir. 1974).

health standards and all rules, regulations, and orders issued pursuant to the Act, no mechanism to force compliance by employees exists under the Act. Under Section 5(b) ultimate responsibility lies with the employer. The report of the Senate Committee makes this point clear: "The Committee does not intend the employee-duty provided in Section 5(b) to diminish in any way the employer's compliance responsibility or his responsibility to assure compliance by his own employees. Final responsibility for compliance with the requirements of this Act remains with the employer."[87]

A number of Review Commission cases have concluded that an employer is *not* liable for a violation of the general duty clause if the condition in question results from an unauthorized or disobedient action which could not have been prevented by the exercise of reasonable employee supervision. This position has its basis in the common-law tort concepts which hold that the employer cannot be held negligent in such a situation. In *Richmond Block, Inc.*, the employer had instructed all employees to "lock out" the flow of electric power to a cement mixer prior to entering the mixer for cleaning purposes. The employee responsible for cleaning the mixer had been so instructed twice in the ten days preceding the accident, but on the day in question, he failed to follow instructions and was killed when a coworker accidently started the mixer. The Review Commission affirmed the trial judge's vacating of the citation because the fatality was caused by "the employee's disregard of a simple safety rule and procedure, clearly and recently communicated to the employee, which he apparently understood."[89]

A similar result was reached by the Seventh Circuit in *Republic Creosoting Co., Division of Reilly Tar and Chemical Corp.* In this case, an employee disregarded specific instructions and cut the steel band encasing a package of railroad ties. As a result, five ties fell on the employee, fatally injuring him. The Seventh Circuit affirmed the Review Commission's vacating of the citation, stating that "a reasonably diligent employer would not have foreseen that the employee would injure himself."[90]

In several cases where there has been found evidence of inadequate supervision, a violation of the general duty clause has been cited. In *John B. Kelly, Inc.*, an employee assigned to remove masonry from a freestanding wall was told to brace the wall before starting the work. The employee was observed by his supervisor on at least two occasions working without any bracing and was instructed to erect the bracing. The employee never complied, and the wall later collapsed, killing the employee. The Review Commission affirmed the trial judge's finding of a violation because the supervisor "should have remained on the scene in order to insure that the employee, who had twice contravened his instructions, properly carried them out."[91]

In *REA Express, Inc.*, the Second Circuit Court of Appeals affirmed the Review Commission's finding of a violation of the general duty clause where the employer's supervisory personnel permitted untrained employees to attempt repairs of high-voltage equipment on a wet concrete floor without using protective equipment.[92]

*   *   *

The question of employer and employee duties has perhaps been best summarized by an OSHA hearing judge who stated that "despite the mutuality of responsibility which the Act imposes on employer and employee, it is clearly apparent that because the employer controls the work environment, the standard of care he owes to his employees under the general duty cluse of the Act is high, but not so high . . . that the employer becomes the virtual insurer of the conduct of his employees, and thus absolutely liable for all their acts of commission and omission."[95]

---

[87]S. Rep., No. 1282, p. 91.
[89]*OSHR*, BNA, Vol. 2 (1972), p. 119; *Richmond Block, Inc.*, OSHRC Docket No. 82 (January 11, 1974), 1 OSHC 1505. See also *Arnold Hansen d.b.a. Hansen Brothers Logging*, OSHRC Docket No. 141 (October 13, 1972), 1 OSHC 1060.

[90]501 F.2d 1200 (7th Cir. 1974).
[91]*John B. Kelly, Inc.*, OSHRC Docket No. 154 (August 3, 1973), 1 OSHC 1301.
[92]495 F.2d 822 (2d Cir. 1974).
[95]*Richmond Block, Inc.* [supra].

What is the role of the general duty clause in abating a hazard for which an OSHA standard already exists? In the following case, the D.C. Circuit rejects the *dictum* of a much earlier decision on this issue, and affirms the potential power of the general duty clause in protecting worker health.

International Union, United Automobile, Aerospace and Agricultural Implement Workers of America v. General Dynamics Land Systems Div.

815 F.2d 626 (D.C. Cir. 1987)

BUCKLEY, Circuit Judge.

The Occupational Safety and Health Administration ("OSHA") cited General Dynamics Land Systems Division ("General Dynamics or the "Company") for workplace violations of a statute and of administrative safety standards. An administrative law judge ("ALJ") found that as General Dynamics conformed with the safety standards, it could not be found to have violated the statutory requirement. The Occupational Safety and Health Review Commission (the "Commission") adopted the ALJ's decision as its own. Petitioners urge us to vacate the order insofar as it holds that a specific OSHA safety standard preempts enforcement of a general statutory duty to assure workplace safety. We vacate the challenged portion of the order and hold that, in the circumstances of this case, an employer's compliance with OSHA's standards will not discharge his statutory obligation to provide employees with safeguards against recognized hazards.

I. BACKGROUND

General Dynamics manufactures M-1 Abrams tanks in a Department of Defense facility called the Detroit Arsenal Tank Plant ("Plant"). Abrams tanks have internal hydraulic systems that sometimes leak during assembly. For several months prior to November 1983, General Dynamics employees had used a solvent called 1,1,2 trichloro 1,2,2 trifluoroethane ("solvent" or "freon") to clean up resulting oil spills. The solvent evaporates quickly. While it is less toxic than other commercial solvents, in its gaseous state it is heavier than air and may cause serious illness or death. It tends to accumulate in assembly-line pits and tank hulls, displacing oxygen and creating a risk of asphyxiation. In high concentrations it may also cause cardiac arrythmia and eventual arrest.

* * *

The incidents leading to this litigation began shortly after the company commenced production of the M-1 tank in March 1982. Brief for Secretary at 7. A Plant employee was overcome by fumes in August 1982 after entering an assembly-line pit containing freon vapors. *Id.* at 11. After an inspection, OSHA issued a citation authorized by section 9 of the Occupational Safety and Health Act of 1970 (the "Act"), 29 U.S.C. §658 (1982). The citation charged General Dynamics with violations of section 5(a) of the Act, 29 U.S.C. §654(a) (1982), which provides:

Each employer—

(1) shall furnish to each of his employees employment and a place of employment which are free from recognized hazards that are causing or are likely to cause death or serious physical harm to his employees; (2) shall comply with occupational safety and health standards promulgated under this chapter.

Specifically, General Dynamics was charged with a violation of its statutory duty to provide safe working conditions under subsection 5(a)(1) ("general duty clause") and to observe safety standards issued pursuant to subsection 5(a)(2) ("safety standards" or "specific standards"). General Dynamics reached a settlement with OSHA in June 1983 in which the Company affirmed that it had implemented amended confined-space procedures and agreed that "in the future [it] will in good faith continue to comply[ ] with the provisions of the Act, *and* applicable standards promulgated pursuant thereto." Joint Appendix ("J.A.") at 964 (emphasis added).

In March 1983, another Plant employee became dizzy and weak after driving a fully assembled tank. Brief for Secretary at 12. As a result of this incident, General Dynamics posted a safety bulletin at the Plant that read in part:

[Trichloro-trifluoroethane] vapors being 6½ times heavier than air will readily displace oxygen in pits and enclosed spaces such as the inside of tracked vehicles. Ventilation must be utilized when this solvent is used in enclosed spaces, otherwise its use must be limited to one pint quantities.

Again on July 9, 1983, a General Dynamics employee from another plant, Frederick Spearing, became seriously ill at the Plant's test track while sitting inside the driver's compartment of a tank. He remained in that position while two employees poured several gallons of solvent onto the floor of the tank. Moreover, to clear an opening for the solvent to enter, employees rotated the tank's turret in such a manner that it blocked the only exit from the compartment where Mr. Spearing was sitting. He lost consciousness in the presence of a supervisor. The relevant General Dynamics injury report ascribed the injury to "inhalation of fumes from Genesolv D. (trichloro-trifluoroethane)." Supervisors Report of Accident (July 11, 1983).

In yet a third incident on September 21, 1983, Plant employee Charles Paling also suffered from exposure to solvent fumes. After entering a tank and pouring approximately two gallons of solvent to clean a hydraulic leak, Paling exited and used a portable device to ventilate the tank. This was in apparent compliance with the ventilation requirement in the safety bulletin that was issued following the incident of March 1983. Paling then reentered the tank. Another employee later discovered him inside the driver's compartment shaking and foaming from the mouth.

As it had done before, the UAW complained to OSHA about the Paling incident. While OSHA conducted an investigation, a General Dynamics employee at a different plant died from exposure to solvent fumes. Shortly thereafter, on November 29, 1983, OSHA cited General Dynamics for violations of section 5(a)(1) of the Act ("statutory charge"), and of OSHA's specific standard for governing an employee's exposure to the solvent, 29 C.F.R. §1910.1000(a)(2) & (e). OSHA sought penalties totaling $18,000.

The statutory charge stated that General Dynamics had violated section 5(a)(1) because it "did not furnish employment and a place of employment which were free from recognized hazards" associated with exposure to solvent vapors in confined spaces requiring special entry procedures. General Dynamics contested this citation, and the matter was referred to the ALJ. After hearing evidence for eighteen days, he determined, as a matter of law, that "the circumstances of this case are governed solely by the regulations at 29 C.F.R. §1910.1000(a)-(d) which set forth the limits of employee exposure to trichloro-trifluoroethane." Subsection (a) provides that employers must not expose employees to a time-weighted average of more than 1,000 parts per million of trichloro-trifluoroethane vapors during any eight-hour work shift of a forty-hour work week. 29 C.F.R. §1910.1000(a) (the "freon standard"). The freon standard calculates the concentration of solvent vapors by volume at a given temperature and pressure.

The ALJ reasoned that "a ruling upon consideration of all the evidence will not be made as the allegations of the [section 5(a)(1)] violation are inappropriate and must be vacated because the alleged hazard is *addressed by* a specific standard" (emphasis added). In support of this conclusion he cited (1) the legislative history of the Act and its general policy aims, (2) a rule of statutory construction that the "specific takes precedence over the general," and (3) OSHA's preemption regulation in 29 C.F.R. §1910.5(c)(1) as interpreted by the Commission. The ALJ also held that "the Secretary failed to meet his burden in proving the violation" of the freon standard, and that the evidence offered "is not convincing that [the employee's] exposure was in excess of the permissible level according to the standard."

The ALJ's decision and order became a final order of the Commission by virtue of the failure of any commissioner to direct internal review within thirty days. *See* 29 U.S.C. §661(j) (1982). The UAW filed a petition for review in this court. The Secretary first proceeded in the United States Court of Appeals for the Eighth Circuit, but his petition was transferred to this court and consolidated with that of the UAW. General Dynamics responds to both petitions, and the Commission responds only to the Secretary's petition.

Petitioners do not challenge the ALJ's conclusion that General Dynamics did not violate the freon standard. They both argue, however, that the ALJ erred in vacating that part of the citation charging a violation of the general duty clause. They contend that OSHA's standard for trichloro-trifluoroethane defines a narrow set of unsafe workplace conditions. The relationship between these sets is the crux of the dispute. Petitioners argue that the hazards cited at the Plant extend beyond those addressed by the freon standard, and that an employer's compliance with that standard is not a substitute for compliance with the general duty imposed by section 5(a)(1). They also maintain that the court should defer to the Secretary's construction of OSHA regulations. Finally, the Secretary contends that the challenged portion of the final order represents an unexplained departure from Commission precedent.

General Dynamics responds that the hazards at issue are within the scope of the freon standard, that its compliance with that standard relieves it of any responsibility for the alleged violations under the general duty clause, that OSHA's regulations so require, and that the ALJ's interpretation of these regulations is not unreasonable and deserves deference. The Company also argues that the final order does not ignore Commission precedent.

We conclude that the Commission erred in its construction of OSHA's regulations because it misunderstood the structure of the Act that authorizes the regulations. We also conclude that the Commission acted arbitrarily and capriciously in refusing to follow its own precedent.

II. DISCUSSION

A. The Nature of the Inquiry

The issue before us is whether, as a matter of law, the Company's compliance with the freon standard relieved it of responsibility for alleged violations of the statute's general duty clause. This raises two fundamental principles. The first concerns the power of an agency to make law. Without such power OSHA could not prescribe a regulation that preempts an otherwise applicable statutory provision. The second concerns the relationship between such agency-made law and statutory law. Treaties aside, the Supremacy Clause provides that a federal statute must always be superior to all other forms of law, including regulations. U.S. Const. art. VI.

These principles are implicated in the issue before us because OSHA's freon standard could preempt section 5(a)(1) only if Congress, the repository of all legislative power, U.S. Const. art. I, §1, and supreme lawmaker, U.S. Const. art. VI, permitted it. Because Congress may delegate its legislative power, *see Chevron U.S.A. Inc. v. Natural Resources Defense Council, Inc.*, 467 U.S. 837, 843-44, 865-66 (1984), the question in this case is whether OSHA has been delegated the power to issue regulations preempting an employer's duty under section 5(a)(1).

Neither petitioners nor General Dynamics approach this case as one directly involving a delegation issue. The Secretary relies entirely on his own regulation at 29 C.F.R. §1910.5(f), and dictum in a decision of this court, *National Realty & Constr. Co. v. OSHRC*, 489 F.2d 1257, 1261 n.9 (D.C. Cir. 1973), for the proposition that section 5(a)(1) "does not apply with respect to any 'condition . . . ' covered by the [OSHA] standard." The UAW agrees that "to preempt a general duty clause citation a specific standard must address the particular hazard cited under the general duty clause."

* * *

Rather than questioning the Commission's apparent premise that a specific standard will be given preemptive effect even when an employer knows it to be inadequate, petitioners rely instead on their argument that the hazard alleged in the statutory charge is distinguishable from that to which the specific standard is addressed. Nevertheless, as all the parties accept that premise, we believe it necessary to place it in proper focus.

We cannot address the relationship between section 5(a)(1) and the regulations issued under section 6 without turning at the outset to the statutory basis for those regulations. Neither can we decide whether the Secretary or the Commission deserves our deference without reference to what Congress has said on the matter.

We consider first whether the ALJ's analysis of the legislative history and policy of the Act, and his invocation of a rule of statutory construction, identify any congressional delegation of authority to preempt a duty under section 5(a)(1) with a specific standard. We then examine the effect of OSHA's preemption regulations. Subsequently we consider concretely the role of section 5(a)(1) in this case as it relates to section 5(a)(2). We end with the Secretary's argument that the final order is in violation of Commission precedent.

B. Preemptive Effect of the OSHA Standard

* * *

Section 5(a)(1) clearly and unambiguously imposes on an employer a general duty to provide for the safety of his employees that is distinct and separate from the employer's duty, under section 5(a)(2), to comply with administrative safety standards promulgated under section 6 of the Act. Therefore, unless we find this interpretation in conflict with another source of statutory laws, we are bound to reject any agency interpretation that is inconsistent with what we find to be the plain meaning of section 5(a).

Neither the commission nor General Dynamics has cited such a source, and we have found none.

* * *

We conclude that the Act does not empower the Secretary, and hence OSHA, to absolve employers who observe specific standards from duties otherwise imposed on them by the general duty clause. To the degree that the final order makes such a claim, it is in error.

C. Preemption Regulations

The final order also relies on an OSHA preemption regulation, 29 C.F.R. §1910.5(c)(1), which General Dynamics here supplements with another, namely 29 C.F.R. §1910.5(f). Section 1910.5(c)(1) provides:

> if a particular standard is specifically applicable to a condition, practice, means, method, operation, or process, it shall prevail over any different general standard which might otherwise be

applicable to the same condition, practice, means, method, operation, or process. . . .

Section 1910.5(f) provides:

an employer who is in compliance with any standard in this part shall be deemed to be in compliance with the requirement of section 5(a)(1) of the Act, but only to the extent of the condition, practice, means, method, operation, or process covered by the standard.

First we note that the application of section 1910.5(c)(1) to this case is less than obvious. The regulation does not refer to section 5(a)(1), and its reference to an otherwise applicable general standard has no necessary relationship to a statutory duty to protect employees from a recognized hazard. Although the Commission has applied the principle incorporated in section 1910.5(c)(1) in support of a decision setting aside a citation for a violation of the general duty clause, *Brisk Waterproofing Co.*, 73 OSAHRC 30/E1, 1 BNA OSHC 1263, 1973-74 CCH OSHD ¶15,392 (No. 1046, 1973), the language used in the case would more appropriately apply to section 1910.5(f), the regulation cited by General Dynamics in justifying its claim of preemption. While we believe that section 1910.5(f) may be read merely to prohibit a double penalty for an employer who has violated a safety standard, that is to say, one penalty under the safety standard and the other under the general duty clause, we now consider whether the Company's broader reading of this regulation is authorized by Congress.

Both preemption regulations were promulgated pursuant to OSHA's authority under section 6 of the Act, 29 U.S.C. §655 (1982). Section 6 authorizes the Secretary to promulgate safety standards. It nowhere suggests that the Secretary may promulgate standards that displace the general duty imposed by section 5(a)(1). When we compare the lack of statutory support for the construction of sections 1910.5(c)(1) and (f) that the Commission and General Dynamics reach, and the clear and unambiguous language of the general duty clause, there is no contest. On the facts in this case, section 5(a)(1) can no more be denied legal effect on the basis of OSHA's preemption regulations than it can on the basis of its specific standard.

D. The Role of Section 5(a)(1)

Any apparent conflict between section 5(a)(2) or the preemption regulations on the one hand, and the general duty clause on the other, is resolved when one focuses on the words "recognized hazard" in section 5(a)(1). As the commission has pointed out in *Con. Agra, Inc., McMillan Co. Division* (1983):

In order to establish a section 5(a)(1) violation, the Secretary must prove: (1) the employer failed to render its workplace free of a hazard, (2) *the hazard was recognized* either *by the cited employer* or generally within the employer's industry, (3) the hazard was causing or likely to cause death or serious physical harm, and (4) there was a feasible means by which the employer could have eliminated or materially reduced the hazard. (emphasis added).

This analysis emphasizes the fact that the duty to protect employees is imposed on the employer, and the hazards against which he has the obligation to protect necessarily include those of which he has specific knowledge. Therefore if (as is alleged in this case) an employer knows a particular safety standard is inadequate to protect his workers against the specific hazard it is intended to address, or that the conditions in his place of employment are such that the safety standard will not adequately deal with the hazards to which his employees are exposed, he has a duty under section 5(a)(1) to take whatever measures may be required by the Act, over and above those mandated by the safety standard, to safeguard his workers. In sum, if an employer knows that a specific standard will not protect his workers against a particular hazard, his duty under section 5(a)(1) will not be discharged no matter how faithfully he observes that standard. Scienter is the key.

By the same token, absent such knowledge, an employer may rely on his compliance with a safety standard to absolve him from liability for any injury actually suffered by employees as a consequence of a hazard the standard was intended to address, and he will be deemed to have met his obligation under the general duty clause with respect thereto. In other words, compliance with a safety standard will not relieve an employer of his duty under section 5(a)(1); rather, it satisfies that duty. It is in this sense that it may be said that an OSHA standard preempts obligations under the general duty clause.

Thus a decision in the case at hand depends on the following factual determinations: Is the hazard alleged in the statutory charge accurately described? If so, is it adequately addressed by the freon standard? And if it is not, did General Dynamics have knowledge of the fact and take appropriate measures to mitigate the hazard? We do not have the answers because, having decided that "the allegations of the violation [of the general duty clause] are inappropriate and must be vacated because the alleged hazard is addressed by a specific standard," the ALJ declined to address the basic questions. Although General Dynamics recognized that certain uses of the solvent inside tracked vehicles could cause serious harm to employees, the ALJ has not determined whether it believed that its observance of the freon standard was sufficient to protect them from that harm.

One thing we can conclude is that the hazard alleged in the statutory charge is distinguishable from that addressed by the freon standard. The hazard is described as follows:

> Employees working in the Heavy Repair, Test and Adjust, Marriage, M-1 Hull Line, and Engine Test areas were required to spray or pour varying quantities of 1,1,2 trichloro 1,2,2 trifluoroethane into the turret, driver's, and engine compartments of M-1 tanks and immediately enter these compartments to perform clean-up and other routine tasks thereby exposing themselves to the hazard of asphyxiation and/or chemical poisoning. A confined space entry procedure, specific for these operations, had not been implemented when toxic compounds were introduced into the vehicles.

The citation also lists the minimum requirements of the "confined space entry procedure."

The occupational hazard described by that language is clearly distinct from the one addressed by 29 C.F.R. §1910.1000(a). While the regulation places a limit on the permissible level of time-weighted exposure to freon vapors over an eight-hour work shift, the hazard described in the citation is both broader and more specific than that described in the regulation. The citation refers to the danger, in confined spaces requiring special entry procedures, of short-term exposure to toxic vapors in concentrations that may displace oxygen. Given the facts alleged, the ALJ should have determined whether the description of the hazard was accurate; and if so, whether General Dynamics had knowledge of the hazard.

* * *

III. CONCLUSION

We hold that that part of the Commission's final order vacating the statutory charge (1) is not in accordance with law because it impermissibly construes regulations in a manner unauthorized by the clear and unambiguous language of the Act, and (2) is arbitrary and capricious because it inadequately explains the commission's failure to follow the holding of *Con Agra*. We therefore grant both petitions for review, vacate that part of the Commission's final order that vacates the section 5(a)(1) portion of the citation, and remand the cases to the Commission with instructions to address the merits of the section 5(a)(1) citation.

It is so ordered.

◇ NOTES

1. The standard under discussion in this case addressed freon's toxic effects, not its ability to displace oxygen in confined places. The court makes broad statements about the reach of the general duty obligation, that the employer has a duty under Section 5(a)(1) if "the safety standard will not adequately deal with the hazards." Does this mean that if a 6(a) standard inadequately protects workers because it was based on "average" risk of harm, the employer could be in violation of 5(a)(1) even if he complied with the 6(a) PEL?

2. On December 9, 1987 the Supreme Court denied certiorari, leaving the D.C. Circuit decision in place.

◇

DISCUSSION QUESTIONS FOR THE CHAPTER

1. You are an administrative law judge hearing the following case.

A construction worker was electrocuted when he fell from a ladder onto a live power line and then fell another 85 feet to his death. The worker was not using fall protection equipment (i.e., a safety belt) at the time of the accident. OSHA issued the construction company a citation for a serious violation of an OSHA regulation that holds the employer responsible for requiring the use of appropriate personal protective equipment in all hazardous operations. The company contested the citation, stating that it had a safety rule requiring the use of safety belts and that the fatality was due to "unforeseeable employee misconduct."

What information would you want before making a decision? Would you decide differently if the company had a rigorous safety program than if it had a program "on the books" only? What if OSHA did not have the personal protection regulation?

2. In an attempt to streamline its standard setting process, OSHA has established new permissible exposure limits (PELs) for 376 air contaminants *en masse* under one rulemaking procedure. Is this a good idea? Why or why not? On what basis might labor and industry challenge this action? How might the court rule?

3. Suppose Congress becomes interested in amending the OSHAct of 1970. What revisions or modifications would you suggest and why?

4. In regulating toxic materials, OSHA must use *the best available evidence* to set a standard that most adequately assures *to the extent feasible* that *no employee* will suffer *material impairment to health or functional capacity.* On the basis of your understanding of the Act and relevant court decisions, explain what the *italicized* phrases mean.

5. If OSHA's generic cancer standard were in effect and OSHA made lists of Class I and Class II carcinogens as required, could OSHA use those lists for enforcement purposes under the OSHAct even if it promulgated no standards for the listed carcinogens? If so, how?

CHAPTER 4

The Toxic Substances

Control Act of 1976

A. INTRODUCTION

Six years after the passage of the OSHAct, and after concerted lobbying by the labor movement and environmentalists, Congress passed the Toxic Substances Control Act (TSCA) of 1976. The Act permits "cradle to grave" regulation of chemical hazards and confers far-reaching authority on the Environmental Protection Agency (EPA) to require testing, control, recordkeeping, and reporting for individual chemicals and chemical mixtures. The Act is not as limited as the OSHAct; many of its provisions reflect lessons learned during the early administration of the OSHAct. TSCA also reflects the advancements in toxicology and epidemiology that occurred in the six years between the two statutes.

In a nutshell, the Act accomplishes four things. First, it allows EPA to regulate toxic substances in the broadest possible way, from outright banning of chemical substances to requiring labeling of chemicals. Secondly, in an unprecedented way, it authorizes EPA to require industry to test both old and new chemicals. Thirdly, it permits EPA to exercise regulatory control over the introduction of new chemicals into commerce long before capital investment in production is made and workers and consumers are exposed. Finally, it contains wide-reaching recordkeeping and reporting requirements that go far beyond existing obligations found in many other environmental or workplace statutes.

TSCA is a very powerful Act, but unlike the OSHAct it cannot be fully appreciated by reading it from beginning to end. TSCA is cumbersome, lengthy, and complex. This chapter offers a more natural order for reading and thinking about its various sections. The reader will note that TSCA specifically addresses both occupational health and environmental concerns. The following discussion describes the major features of the Act, which is printed in its entirety in Appendix B.

1. Purposes

The three major purposes of the Act are described in Section 2. They are (1) to encourage or require industry to develop adequate data on the health and environmental effects of chemicals, (2) to regulate chemicals that pose unreasonable risk of injury to health or the environment and to take action against imminent hazards, and (3) not to unnecessarily impede technologic innovation. The third purpose is subservient to the second.

2. Definitions

Section 3 of TSCA contains salient definitions and distinctions. Definitions are important both for what they include and what they exclude. The definition of "chemical substance" in TSCA *excludes* chemicals regulated under other acts, such as pesticides, nuclear material, food additives, drugs, cosmetics, alcohol, and tobacco. (See Section 3(2)(B).) It also excludes chemical mixtures. Where the Act intends to control mixtures, it uses the word "mixtures." For example, no premanufacturing notification is generally required for mixtures of existing chemicals even if the mixture represents a new combination of those chemicals, while a premanufacturing notification *is* required for *new chemical substances*. (See Section 5 of the Act.)

The Act regulates both manufacturers and processors. Manufacturers include those who import, produce, or manufacture chemicals. (See Section 3(7).)

Section 3(9) distinguishes "new" chemical substances from old chemical substances. New chemical substances are those not listed in the original inventory of chemicals required to be reported in the early days of the Act. (See Section 8(b).) New chemicals and old chemicals are subject to different regulatory provisions.

The term "health and safety study" includes studies of occupational exposure to chemicals and mixtures as well as studies of health and environmental effects generally. (See Section 3(6).)

"Standards for the development of test data" are specified with particularity in the Act. (See Section 3(12).) These standards are prescriptions for the types of health and environmental data to be developed and specifications for any test protocol or methodology to be used in the development of these data.

Although the term "unreasonable risk" is a crucial term in the Act, it is not defined in Section 3.

3. Authority to Regulate

The authority to regulate chemicals is found in several sections of the Act. Section 6 is the major regulatory section of TSCA. Subsection 6(a) states that if the EPA Administrator finds that there is a "reasonable basis to conclude that . . . a chemical substance or mixture . . . presents or will present an unreasonable risk of injury to health or the environment," the Administrator shall regulate "using the least budensome requirements" as are necessary to address that risk. Unlike OSHA, EPA may, in appropriate cases, explicitly prohibit the manufacture, processing, and/or distribution of a chemical. To effectuate regulation under 6(a) of the Act, EPA promulgates rules. In doing so, the Administrator must publish a rationale that explains his or her consideration of a variety of factors. These include 1) human health effects and the magnitude of human exposure, 2) environmental effects, 3) the benefits of the regulated substances and availability of substitutes therefore, and 4) the economic consequences of the rule, including its effects on technological innovation. (See Section 6(c)(1).) While Section 6 does not define "unreasonable risk," EPA clearly must consider and balance the above-mentioned factors to determine whether a particular chemical poses a risk that is reasonable or unreasonable. (For a discussion of unreasonable risk, see *EDF v. EPA* in Section C of this chapter.)

(a) Rulemaking under Section 6

EPA rulemaking under Section 6(a) of TSCA resembles OSHA rulemaking under Section 6(b)(5) of the OSHAct. A notice of proposed rulemaking (NPR) must be followed by an opportunity for interested parties to submit comments and by an informal hearing (if one is requested) at which a verbatim record is kept and limited cross-examination is provided. (See Section 6(c)(2) and (3).)

EPA also has another alternative. Under Section 6(d) of TSCA, the agency may declare a Section 6(a) rule to be effective *immediately* upon its publication in the Federal Register if it determines that the chemical substance or mixture is *"likely* to result in an unreasonable risk of *serious* or *widespread* injury to health or the environment" before a hearing can be held on the proposed rule (emphasis added). Thus far, this looks quite similar to OSHA's authority to issue emergency temporary standards under Section 6(c) of the OSHAct. However, if the proposed rule would *prohibit* the manufacture, processing, or distribution of the chemical or mixture, EPA cannot make the rule immediately effective unless it has first obtained relief in court against the risk under Section 7 of the Act (discussed below). When it *does* make a 6(a) rule immediately effective, however, this action is not subject to further judicial review. The manufacturer's only recourse at this stage is to request a hearing before EPA, which the agency must provide within 5 days if so requested. Within 10 days of the conclusion of the hearing, EPA must either revoke the rule or promulgate a final 6(a) rule, which is then subject to judicial review.

(b) Imminent hazards

Imminent hazards are further regulated under Section 7 of TSCA. When EPA determines that a chemical or mixture presents an imminent hazard, Section 7 specifies that EPA must either 1) commence a civil action in U.S. District Court to seize the substance and/or obtain other relief against a manufacturer, processor or disposer *or* 2) issue an immediately effective rule under Section 6(d). EPA may take court action against an imminent hazard regardless of whether the substance is already regulated under Section 4, 5, or 6 of TSCA. Consistent with Section 6(d), Section 7(f) defines an "imminently hazardous chemical substance or mixture" as "a chemical substance or mixture which presents an imminent and unreasonable risk of *serious* or *widespread* injury to health or the environment" (emphasis added). Regulation of imminent hazards under Section 7 is similar to the regulation of "imminent danger" under Section 13 of the OSHAct. Both rely on the immediacy of the harm. (See section C of this chapter for a discussion of the reach of TSCA's imminent hazard authority in *United States v. Commonwealth Edison Co.*)

(c) Regulating carcinogens, mutagens, and teratogens

Additional regulatory requirements and opportunities are triggered by other sections of TSCA. Congress was particularly concerned with carcinogens, mutagens, and teratogens. Under Section 4(f), if EPA finds that "there may be a reasonable basis to conclude that a chemical substance or mixture presents or will present a significant risk of serious or widespread harm from cancer, gene mutations, or birth defects," it must, within 180 days, either take appropriate regulatory action or explain by publication in the Federal Register the basis for its finding that "such risk is not unreasonable." This "act or explain" provision of TSCA is a safeguard against administrative foot-dragging. (See the provision of Section 4(f) (as applied to formaldehyde) in section D of this chapter.)

(d) Premanufacturing notification

Section 5 provides for premanufacturing notification of new chemical substances and for their regulation. Manufacturers are required to submit a premanufacturing notification (PMN) of their intent to manufacture a new chemical substance or to use a chemical substance for "a significant new use." The manufacturers must submit the PMN 90 days before the intended commencement of production, and the EPA Administrator may delay the commencement for yet another 90 days. The PMN must contain all manufacturing data and health and safety information known by the manufacturer. If EPA takes no action on the PMN, the new chemical becomes part of its chemical "inventory" and is thereafter treated as an "old" or "existing" chemical. If the Administrator fails to take any action on the PMN, the Administrator must explain the reasons for his or her inaction in the Federal Register.

Note that Section 5 is not a premarket clearance provision such as those found in the Food, Drug, and Cosmetic Act or the Federal Insecticide, Fungicide, and Rodenticide Act. Unless the Administrator requires the manufacturer to submit additional data, the chemical may be allowed into commerce. This distinguishes a premanufacturing notice requirement from a premarket clearance provision. However, if (1) the data are not sufficient to enable the Administrator to make a determination as to whether or not the chemical substance "may present an unreasonable risk," or (2) the chemical is or will be produced in substantial quantities and may either enter the environment in substantial quantities *or* may lead to significant or substantial human exposure, the Administrator may limit or prohibit the manufacture of the chemical pending the development of information. (See Section 5(e).) If necessary, the Administrator may effectuate this control by seeking an injunction in the district court.

If, in reviewing the PMN, EPA determines that "there is a reasonable basis to conclude" that the substance "presents or will present an unreasonable risk" before a rule promulgated under Section 6 can protect against such risk, the agency is required to take some action, short of prohibiting the manufacture of the substance. EPA *must* issue a proposed rule under Section 6(a) to regulate the substance, effective upon its publication in the Federal Register. (See Section 5(f).) The agency *may*, at its discretion, also issue a proposed order to *prohibit* the manufacture of the substance, effective on the expiration of the notification period, or it may apply to the district court for an injunction to prohibit its manufacture, processing, or distribution into commerce.

(e) Significant new use rule

At the time EPA allows the manufacturing of a new chemical to begin, the agency may also issue a significant new use rule (SNUR) under Section 5(a)(2). SNURs restrict the use of the chemical to the specific uses for which the chemical was cleared and put the maufacturer on notice that any new use must be accompanied by a new PMN. A SNUR may also be issued at any later time, even though a chemical is on the inventory, because the PMN implicitly is approved only for the uses articulated in the PMN application.

(f) PMN exemption

Under Section 5(h) the Administrator may provide exemptions to the PMN requirements for certain chemical substances or for research purposes.

4.   Authority to Require Testing

Section 4 compels the Administrator to require the testing of a chemical substance or mixture, whether new or old, if 1) there are insufficient data to make an unreasonable risk determination and testing is necessary and 2) the chemical substance or mixture "may present an unreasonable risk" *or* the chemical will be produced in substantial quantities *and* either may enter the environment in substantial quantities or may lead to significant or substantial human exposure. The Administrator is expected to issue test rules and protocols for the required testing. (See section C in this chapter for a discussion of *NRDC v. Costle* and *NRDC v. EPA,* which address testing rules.) Exemptions from testing rules are allowed under certain conditions.

Section 4(e) establishes the Inter-Agency Testing Committee (ITC) to recommend to EPA which substances should be tested and which tests are necessary. The ITC may construct a prioritized list of as many as 50 chemicals on which EPA must act (by either initiating rulemaking or publishing in the Federal Register its reasons for not requiring testing) within 12 months. TSCA requires that the ITC give priority attention to carcinogens, mutagens, and teratogens.

The testing authority found in TSCA is notably absent in the OSHAct.

5.   Recordkeeping and Reporting Requirements

Beyond testing rules, TSCA imposes substantial additional requirements on chemical manufacturers and processors to develop and report health effects data.

Section 8 allows EPA to promulgate rules that require chemical manufacturers, processors, and distributors to maintain records and make reports on chemicals and mixtures. The type of information required under this section includes the chemical's trade name, chemical identity, and molecular structure; its categories of use; the amount of the substance or mixture being manufactured in total and for each category of use; a description of the byproducts resulting from its manufacture, processing, use, or disposal; all existing data regarding environmental and health effects; *the number of people exposed and an estimate of the number who will be exposed on the job;* and the manner of its disposal. Under Section 8(a), EPA has promulgated regulations requiring general reporting on some 350 chemicals.

In addition to the general reports required for specific chemicals listed in the

Section 8(a) regulations, EPA has promulgated rules for the submission of health and safety studies for over 400 substances under Section 8(d). Section 8(c) requires manufacturers, processors, and distributors to retain records of "significant adverse reactions to [employee] health" for 30 years, and Section 8(e) imposes a statutory duty on manufacturers, processors, and distributors to report immediately to the EPA "information which reasonably supports the conclusion that such substance or mixture presents a substantial risk of injury to health or the environment . . ." (See Chapter 7 for a fuller discussion of TSCA's recordkeeping and reporting requirements.)

6. Relationship to Other Acts

TSCA's relationship to other federal environmental laws is described in two sections of the Act. Section 6(c)(1) addresses the Administrator's duties in the event that a risk under consideration "could be eliminated or reduced to a sufficient extent by actions taken under another Federal law (or laws) administered in whole, or in part" by EPA. (These include, among others, the Clean Air Act, the Clean Water Act, the Resource Conservation and Recovery Act, the Comprehensive Environmental Response, Compensation, and Liability Act—"Superfund"—and the Federal Insecticide, Fungicide, and Rodenticide Act.) In that case, in order to regulate the risk under TSCA, the Administrator must make a finding that exercising authority under TSCA rather than under other laws is "in the public interest." In making this determination, EPA must consider relative costs, risks, and administrative efficiency. (See also Section 9(b), which requires the Administrator to coordinate actions under TSCA with other EPA-administered authorities.) In addition, Section 9(a) requires EPA to formally refer regulation of an unreasonable risk to other agencies, through a detailed report transmitted to the other agency and also published in the Federal Register, if:

> (1) the Administrator "has reasonable basis to conclude that . . . a chemical substance or mixture . . . presents or will present an unreasonable risk" and
> (2) that risk "may be prevented or reduced to a sufficient extent under a federal law not administered by the Administrator."

Possible recipients or "referral agencies" for the report include OSHA and the Consumer Product Safety Commission (CPSC). The referral agency must respond to EPA within 90 days, by either 1) issuing (and publishing in the Federal Register) a statement disagreeing with EPA's determination of the risk (meaning it will take no action) or 2) initiating regulatory action. Failure to respond leaves regulatory authority with EPA.

7. Citizen Suits and Citizen Petitions

Section 20 provides that "any person" may commence a civil action in United States District Court against any person or governmental agency who is in violation of TSCA or its rules or regulations to restrain such violation. This section also authorizes suits against the Administrator to compel the performance of nondiscretionary duties mandated by the Act. In *Welch v. Schneider National Bulk Carriers,* 676 F. Supp. 571 (D.N.J. 1987), the District Court for New Jersey held that "no [express or] implied private right of action eixsts under Section 20 by which [the] plaintiff can seek redress [i.e., monetary damages for his or her personal injuries.]"

Section 21 provides for citizen petitions to the Administrator for the issuance of regulations or orders under Sections 4, 5(e), 6, or 8. If the Administrator denies the petition, the petitioner may ask the United States District Court for a *de novo* proceeding (a factual hearing) to obtain the relief requested.

Court costs and attorneys' fees may be awarded to substantially prevailing parties in actions under Sections 20 and 21.

8. Judicial Review

Rules promulgated pursuant to the authority to control a chemical risk (Sections 6(a), 6(e), 5(a)(2) and 5(b)(4)), the authority to issue testing rules (Section 4(a)) and the authority to require recordkeeping and reporting (Section 8) are subject to judicial review. Petitions for judicial review must be filed, not later than 60 days after the promulgation of such rules, with the U.S. Court of Appeals for the D.C. Circuit or for the circuit in which the petitioner resides or has a principal place of business. Except for SNURs (Section 5(a)(2)), the standard of judicial review is "substantial evidence in the rulemaking record . . . taken as a whole" (Section 19(c)(B)(i)), as it is for occupational safety and health standards under the OSHAct. For SNURs, the standard of review is the familiar "arbitrary and capricious" standard specified by the Administrative Procedure Act for informal rulemaking.

B. THE IMPLEMENTATION OF TSCA

TSCA became effective on January 1, 1977, at the beginning of the Carter administration. Whereas OSHA was placed in the Department of Labor, an agency with little prior experience with toxic substances, the Office of Toxic Substances administering TSCA was established in the Environmental Protection Agency, an agency experienced in toxic substances, risk assessment, and enforcement of health standards. TSCA had the full support and interest of environmental groups and labor unions, which, along with industry, participated actively in guiding the early development of TSCA through membership in the Administrator's Toxic Substances Advisory Committee (ATSAC).

Even with this promising history, TSCA has not enjoyed the success that OSHA has had against greater odds. Very little was accomplished during the Carter years, or during the Reagan and Bush years that followed. ATSAC was discontinued when the first Reagan administration began. Riddled with scandal, EPA neglected the enforcement of all environmental statutes, but TSCA was especially compromised. Congressional oversight hearings in 1981 and 1988 were critical of the lack of accomplishments under TSCA, as were a report by the Conservation Foundation[a] and three reports by the General Accounting Office (GAO).[b,c,d]

[a]*Conservation Foundation Letter,* 1983, pp. 2–7.

[b]General Accounting Office. 1984a. EPA's Efforts to Identify and Control Harmful Chemicals in Use. GAO/RCED-84-100.

[c]General Accounting Office. 1984b. Assessment of New Chemical Regulation Under the Toxic Substances Control Act. GAO/RCED-84-84.

[d]General Accounting Office. 1990. EPA's Chemical Testing Program Has Made Little Progress. GAO/RCED-90-112.

In oversight hearings (appropriately titled "What Ever Happened to the Toxic Substances Control Act?") Jacqueline Warren of the Natural Resources Defense Council stated: "The record of accomplishment by EPA [under TSCA] is so sparse that it has become a public embarrassment."[e] Karen Florini of the Environmental Defense Fund testified: "TSCA unfortunately has not been well utilized. There has been too much emphasis on information gathering and not enough on action. To date, the saga is primarily one of missed opportunities."[f] Both Warren and Florini argued for replacement of an "unreasonable risk" standard with "significant risk."[g] Eric Frumin, of the Amalgamated Clothing and Textile Workers Union—AFL-CIO, testified in stronger terms: "The promise of the Toxic Substances Control Act has been broken, torn apart on the shoals of agency indifference, OMB interference, and the unbridled treachery of the chemical industry."[h] At these same hearings, Charles Elkins, Director of the EPA Office of Toxic Substances, testified: "It is clear to me that the current level of accomplishment of the Existing Chemicals Program is inadequate."[i] Only the chemical industry was complimentary of EPA efforts to implement TSCA, testifying that "EPA has made considerable and appropriate progress in implementing the act."[j] The lack of accomplishment cannot be explained entirely by a lack of leadership or will, especially during the Carter administration. Before exploring the possible reasons, we examine below the accomplishments of the first 12 years of the Act.

1. Regulation

(a) Section 6 regulations

To date, EPA has exercised its Section 6 authority in promulgating PCB regulations (which are specifically mandated by Section 6(e)); in restricting ozone-depleting CFC's in aerosol propellants (1978); in promulgating dioxin-disposal regulations (1980) that have since been transferred to RCRA (1988); in promulgating asbestos-in-schools regulations (1987); and most recently in banning almost all asbestos-containing products in stages over the next seven years (1989). Of 41 carcinogens identified in 1987 by the U.S. Office of Technology Assessment as warranting TSCA attention, only asbestos has been regulated under TSCA.[k]

Rather than regulate nitrates used in metal working fluids, EPA chose to warn

[e]J. Warren. Testimony in "What Ever Happened to the Toxic Substance Control Act?" Hearing before a Subcommittee of the Committee on Government Operations, The House of Representatives October 3, 1988, p. 88.

[f]K. Florini. Testimony in "What Ever Happened to the Toxic Substance Control Act?" Hearing before a Subcommittee of the Committee on Government Operations, The House of Representatives October 3, 1988, p. 33.

[g]Florini, p. 114.

[h]E. Frumin. Testimony in "What Ever Happened to the Toxic Substance Control Act?" Hearing before a Subcommittee of the Committee on Government Operations, The House of Representatives October 3, 1988, p. 33.

[i]C. Elkins. Testimony in "What Ever Happened to the Toxic Substance Control Act?" Hearing before a Subcommittee of the Committee on Government Operations, The House of Representatives October 3, 1988, p. 220.

[j]J. Condray. Testimony in "What Ever Happened to the Toxic Substance Control Act?" Hearing before a Subcommittee of the Committee on Government Operations, The House of Representatives October 3, 1988, p. 4.

[k]House of Representatives, "What Ever Happened to the Toxic Substance Control Act?" Hearing before a Subcommittee of the Committee on Government Operations October 3, 1988, p. 445.

manufacturers and workers about the carcinogenic risk through a "chemical advisory" designed to encourage voluntary changes in production and use. As of September 30, 1987, 10 chemical advisories on 11 chemicals had been issued.[l]

(b) Imminent hazard regulation (Section 7)

Section 7 was not used until 1985, and then only with regard to violation of PCB regulations.

(c) Section 4(f) actions

As of January 1984, EPA had designated only two chemicals for priority consideration: 4,4'-methylenedianiline (MDA) and 1,3-butadiene, because of their possible carcinogenic risk.[m] Subsequently formaldehyde (1984) and methylene chloride (1985) were also designated 4(f) chemicals, the former with a complex and checkered history. (See Section D on page 215.) All were ultimately referred to OSHA for regulatory consideration.

In 1985, EPA refused to designate perchloroethylene as a 4(f) chemical, arguing that evidence for its carcinogenicity was not "new" but existed prior to January 1, 1979, when Section 4(f) came into effect.[n] To date, no other chemicals have received priority consideration under Section 4(f), even though there are many suspect carcinogens that could qualify for such action.

(d) PMN activity

From July 1979 through September 1987, EPA received approximately 11,000 new chemical notices.[o] For about 80 percent, no EPA action was deemed necessary. Over half of the submitted PMNs had no toxicity data. Of the chemicals that cleared the PMN process as of the fall of 1988, approximately 4800 were actually manufactured or imported for commercial use.[p] PMN actions as of September 30, 1987, are summarized in Table 4-1.

On July 27, 1988, EPA promulgated a generic SNUR (54 Fed. Reg. 31298) to expedite the simultaneous issuance of restrictions applying to all manufacturers of a particular new chemical. The rule also codifies the criteria for subjecting a new chemical to a SNUR.[q] As of September 30, 1987, 11 SNURs had been issued on 19 chemicals.[r]

In sum, regulation under TSCA included controls on four existing chemicals, restrictions on four new chemicals, and 11 SNURs. This is not an impressive record for 12 years of implementation.

2. Testing

As of May 1, 1989, the Interagency Testing Committee (ITC) had submitted 24 semiannual reports of chemicals recommended for priority testing under Section 4(e). In the first 23 reports, ITC recommended 77 chemicals and 20 groups of

---

[l]*Chemicals-in-Progress Bulletin*, EPA Office of Toxic Substances, 1988, 9(4):10.

[m]GAO, 1984a, p. 13.

[n]J. Moore. Testimony in "What Ever Happened to the Toxic Substance Control Act?" Hearing before a Subcommittee of the Committee on Government Operations, The House of Representatives October 3, 1988, p. 230.

[o]*Chemicals-in-Progress Bulletin.*

[p]*Chemicals-in-Progress Bulletin.*

[q]*Chemicals-in-Progress Bulletin* 1988, 10(3):7.

[r]*Chemicals-in-Progress Bulletin* 1988, 9(4):10.

Table 4-1. New chemical actions mid-1979–September 30, 1987

Actions	Aggregated Total to Date
Total new chemical substance submissions received (PMNs; applications for exemptions)	10,842
Valid Premanufacture notifications (PMNs) received	9132
PMNs requiring no further action	7166
Voluntary testing in response to EPA concerns	149
Voluntary control actions by submitters	45
PMNs withdrawn in face of regulatory action	183
PMNs subject to control, pending data	349
PMNs resulting in prohibition or restrictions	4
Final Action Taken Granted	1473
Withdrawn	147
Denied	12

Source: EPA 1988a. *Environmental Progress and Challenges: EPA's Update;* Office of Policy Planning and Evaluation, Washington DC, EPA-230-07-88-33.

chemicals for priority consideration within 12 months. In addition, 29 chemicals and one group of chemicals were recommended without being designated for response within 12 months.[s] In the 24th report, the ITC added no new chemicals to this list. There are now some 68,000 chemicals on the chemical inventory. Of those, approximately 15,000 are believed not to be currently in commerce and 19,000 are polymers, generally of low toxicity. This leaves approximately 34,000 that deserve active attention. As of 1990, the ITC has designated 386 chemicals for testing.[t] This represents a little over 1 percent of the chemicals of current concern. Moreover, by 1990, EPA had compiled complete test data for only six chemicals (GAO 1990, p. 3). As of 1984, only 11 percent of industrial chemicals with consumption exceeding over 1 million pounds per year had been tested adequately by anyone for the purpose of assessing their health risks; 82 percent of those chemicals had been subjected to no reported toxicity testing whatsoever.[u]

In 1983, after years of delay, EPA promulgated test guidelines that established methodologies for the broad range of human health effects, environmental effects and chemical fate studies that may be performed under Section 4. In 1986, EPA promulgated regulations that established procedures and timetables for negotiating testing consent orders in order to speed up the testing process. These orders have the same force and effect under TSCA as test rules.[v] As of October 1988 EPA had issued over 165 actions on ITC chemicals. These included 34 final rulemakings, 19 negotiated testing

[s]*Chemicals-in-Progress Bulletin,* 1989, 10(2):4.
[t]GAO 1990, p. 3.
[u]Frumin, pp. 13–14.
[v]Condray, p. 20.

agreements, 3 consent orders, 53 notices of proposed rulemaking, 9 advanced notices of proposed rulemaking, and 47 decisions not to test.[w]

In fiscal years 1987 and 1988, a total of 32 test rule modification requests were received. Twenty-seven of these were for reporting deadline extensions and five were for technical modifications in test protocols. Twenty-five of the extension requests (all seeking extensions of less than six months) were approved after EPA review, but not always for the full duration requested. Under current TSCA Section 4 procedural rules, requests for extensions of less than six months (in the aggregate) that are reasonably justified by technical or administrative difficulties may be, and regularly are, granted by EPA without soliciting public comment.[x] EPA is considering amending its current regulations so that it can grant 12-month extensions for chemical testing conducted under Section 4 of TSCA without requiring public comment.[y]

3.   Reasons for Slow Implementation

Aside from political interference at EPA that began in 1980 under Administrator Anne Gorsuch, other reasons exist to explain the slow progress made under TSCA. Rather than viewing the heart of the statute as regulation, EPA in the Carter administration viewed the statute as requiring a prioritization of risks and a methodical progression from risk assessment (based on testing) to risk management (regulation). Delays in deciding protocols for demonstrating a variety of health effects resulted in very few risk assessments and very little follow-up regulatory action. Further, Sections 4, 5, and 6 were administered by three separate deputy assistant administrators, with little coordination among them. Finally, Section 6 actions were driven by a desire to stake out a special path for TSCA. Although most chemicals have a serious, if not predominant, occupational exposure, EPA wanted to establish TSCA as an "environmental" statute. As a result the regulations focused on nonoccupational problems: CFCs, PCBs, asbestos in schools, and dioxin disposal. EPA referred six substances to OSHA for regulatory action and declined to look for other occupationally focused hazards. Unfortunately, there are very few chemicals that do not have a serious occupational component as part of their exposure profiles. EPA's shunning of occupational hazards left it very little to regulate.

Failure to connect new chemical regulation with existing chemical regulation also resulted in significant missed opportunities. Upon notification that a new, safer chemical may be substituted for an existing more dangerous one in some or all uses, EPA could subject the latter to enhanced regulatory scrutiny. In this way, TSCA could encourage the replacement of existing chemicals through innovations in product and process design.

EPA's Office of Toxic Substances now has authority for implementing Section 313 of Title III under the Superfund Amendment and Reauthorization Act (SARA), and has shifted its attention still further from implementing TSCA. Information submitted by industry under SARA's Toxic Release Inventory could be utilized to establish testing priorities under Section 4 of TSCA, which authorizes testing of chemicals which are produced in "substantial quantities" and which "may enter the environment" or "may lead to significant or substantial human exposure."

[w]EPA 1988b. Testimony and documents submitted for the record in "What Ever Happened to the Toxic Substance Control Act?" Hearing before a Subcommittee of the Committee on Government Operations, The House of Representatives, p. 461.
[x]EPA 1988b, p. 500.
[y]*Chemicals-in-Progress Bulletin* 1989, 10(1):5.

◇ NOTES

1. Enforcement of TSCA violations is coordinated by EPA's National Enforcement Investigation Center, Office of Compliance Monitoring, in Denver.

2. The federal District Court for Rhode Island has held that, because EPA has the explicit authority to enter facilities to make inspections under TSCA, the agency has the implied authority to seek an administrative search warrant to carry out the purpose of that law. *Boliden Metech Inc v. United States,* 695 F. Supp 77 (D.C. R.I. 1988).

3. On April 19, 1988, EPA reported that BASF Corporation agreed to pay a $1.3 million penalty for violating the PMN and import provisions of TSCA. Such "mega-fines" are likely to have significant deterrent effects.

◇

C. JUDICIAL INTERPRETATIONS OF THE ACT

Although the judicial history of TSCA is not nearly as rich as that of the OSHAct, several important opinions have emerged. Several deal with concepts closely related to those interpreted by the courts under the OSHAct. On the whole, these cases illustrate the courts' recognition of the need to control the proliferation of hazardous substances, the importance of TSCA, and the responsibility of EPA to fulfill its statutory mandate expeditiously.

1. Unreasonable Risk

In 1980, the D.C. Circuit Court of Appeals was asked to address the meaning of "unreasonable risk" under Section 6 of the Act. Note especially the court's reliance on OSHA cases in applying the substantial evidence test.

Environmental Defense Fund v. Environmental Protection Agency

636 F.2d 1267 (D.C. Cir. 1980)

EDWARDS, Circuit Judge.

In this case the Environmental Defense Fund (EDF) petitions for review of regulations, issued by the U.S. Environmental Protection Agency (EPA), that implement Section 6(e) of the Toxic Substances Control Act (TSCA). That section of the Act provides broad rules governing the disposal, marking, manufacture, processing, distribution, and use of a class of chemicals called polychlorinated biphenyls (PCBs).

EDF seeks review of three aspects of the regulations. First, it challenges the determination by EPA that certain commercial uses of PCBs are "totally enclosed," a designation that exempts those uses from regulation under the Act. Second, it claims that the EPA acted contrary to law when it limited the applicability of the regulations to materials containing concentrations of PCBs greater than fifty parts per million (ppm). Third, EDF challenges the decision by EPA to authorize the continued use of eleven non-totally enclosed uses of PCBs.

From our examination of the record, we find that there is no substantial evidence to support the EPA determination to classify certain PCB uses as "totally enclosed." We also find that there is no substantial evidence in the record to support the EPA decision to exclude from regulation all materials containing concentrations of PCBs below fifty ppm. Accordingly, on these first two points, we hold unlawful and set aside the challenged regulations, and remand to EPA for further proceedings consistent with this opinion.

We find, however, that there is substantial evidence in the record to support the EPA determination to allow continued use of the eleven non-totally enclosed uses. Accordingly, on this third point, we uphold the EPA regulations.

I. BACKGROUND

A. Polychlorinated Biphenyls

Polychlorinated biphenyls (PCBs) have been manufactured and used commercially for fifty years for their chemical stability, fire resistance, and electrical resistance properties. They are frequently used in electrical transformers and capacitors. However, PCBs are extremely toxic to humans and wildlife. The extent of their toxicity is made clear in the EPA Support Document accompanying the final regulations, in which the EPA Office of Toxic Substances identified several adverse effects resulting from human and wildlife exposure to PCBs.

Epidemiological data and experiments on laboratory animals indicate that exposure to PCBs pose carcinogenic and other risks to humans. Experimental animals developed tumors after eating diets that included concentrations of PCBs as low as 100 parts per million (ppm). Experiments on monkeys indicate that diets with PCB concentrations of less than ten ppm reduce fertility and cause still births and birth defects. Other data show that PCBs may adversely affect enzyme production, thereby interfering with the treatment of diseases in humans.

EPA has found that PCBs will adversely affect wildlife as well as humans. Concentrations below one ppb (part per billion) are believed to impair reproductivity of aquatic invertebrates and fish. Some birds suffered "severe reproductive failure" when fed diets containing concentrations of only ten ppb of PCBs. Because PCBs collect in waterways and bioaccumulate in fish, fish-eating mammals run a special risk of adverse effects. Such mammals may have "significantly higher concentrations of PCBs in their tissues than the aquatic forms they feed on."

EPA estimates that by 1975 up to 400 million pounds of PCBs had entered the environment. Approximately twenty-five to thirty percent of this amount is considered "free," meaning that it is a direct source of contamination for wildlife and humans.

The rest, "mostly in the form of industrial waste and discarded end use products, is believed to be in landfill sites and thus constitutes a potential source of new free PCBs." Other significant sources of PCBs include atmospheric fallout and spills associated with the use or transportation of PCBs.

EPA concluded in the Support Document that "the additional release of PCBs" into the environment would result in widespread distribution of the PCBs and "will eventually expose large populations of wildlife and man to PCBs." EPA concluded further that:

> As a practical matter, it is not possible to determine a "safe" level of exposure to these chemicals. Because PCBs are already widely distributed throughout the biosphere, they currently pose a significant risk to the health of man as well as that of numerous other living things. As a consequence, any further increase in levels of PCBs in the biosphere is deemed undesirable by EPA. Because "PCBs released anywhere into the environment will eventually enter the biosphere . . . EPA has determined that any such release of PCBs must be considered 'significant.' "

* * *

B. Congressional Response

Responding to the dangers associated with the use of PCBs and other toxic chemicals, Congress in 1976 enacted the Toxic Substances Control Act (TSCA), Pub. L. No. 94-469, 90 Stat. 2003 (1976). Although the Act is generally designed to cover the regulation of all chemical substances, Section 6(e) refers solely to the disposal, manufacture, processing, distribution, and use of PCBs. No other section of the Act addresses the regulation of a single class of chemicals.

* * *

As enacted, Section 6(e) of the Act 10 sets forth a detailed scheme to dispose of PCBs, to phase out the manufacture, processing, and distribution of PCBs, and to limit the use of PCBs.

* * *

III. USE AUTHORIZATIONS

A. Criteria for the "Unreasonable Risk" Determination

The Act permits the Administrator to authorize "by rule" non-totally enclosed uses of PCBs if he finds that such uses "will not present an unreasonable risk of injury to health or the environment." 15 U.S.C. §2605(e)(2)(B). Using the criteria set forth in subsection 6(c)(1), the Administrator found that eleven non-totally enclosed uses did not present an unreasonable risk. On the basis of these findings, EPA authorized the continued use of the eleven non-totally enclosed uses here in dispute. In attacking these use authorizations, EDF claims that the Administrator employed the wrong criteria in making his determinations concerning "unreasonable risk."[20] In particular, EDF argues that Congress intended to preclude the Administrator from using the subsection 6(c)(1) criteria in promulgating the PCB use authorization rules.

The basis for EDF's argument is found in subsection 6(e)(4), which requires the Administrator to promulgate rules in accordance with the procedural provisions in subsections 6(c)(2), (3), and (4); no reference to subsections 6(c)(1) is found in subsection 6(e)(4). EDF claims that this omission evidences a Congressional intent to preclude EPA from using the 6(c)(1) criteria in making "unreasonable risk" determinations pursuant to subsection 6(e)(2)(B). Because the Administrator used those criteria, EDF argues that the unreasonable risk determinations were "fatally flawed . . . [placing] disproportionate weight . . . on the adverse economic impact of a ban, [and seriously undermining] the Congressional objective of bringing about the development and use of substitutes in existing PCB activities. . . ."

Without more, however, we find that the omission of a reference to subsection 6(c)(1) in 6(e)(4) does not imply that congress meant to *prevent* EPA from considering the challenged criteria in making unreasonable risk determinations under 6(e)(2)(B). There is nothing in the wording of the statute or the legislative history that affirmatively supports the position of EDF.

The structure of section 6(e) also indicates that EDF's position is incorrect. Because section 6(e) uniquely applies to a single class of chemicals, some provision was needed to establish *procedural* guidelines for the issuance of rules regulating the use of PCBs. Reference to subsections 6(c)(2), (3), and (4) fulfilled that need, whereas reference to subsection 6(c)(1), a *substantive* provision, was simply unnecessary. Because there is no compelling evidence to support the position advanced by EDF, and because of the deference that we must accord an agency's reasonable interpretation of its statute, we conclude that the statute does not preclude the EPA from using the subsection 6(c)(1) criteria in making the unreasonable risk determinations under 6(e)(2)(B).

Moreover, because the expression "unreasonable risk of injury to health or the environment" is left undefined in section 6(e), the Administrator was required to give some meaning to it. Since the 6(c)(1) criteria obviously pertain to factors of "unreasonable risk," it was entirely appropriate for EPA to consider such criteria in ascribing a meaning to the use authorization provision in 6(e)(2)(B). EDF has shown nothing to indicate otherwise. In fact, EDF does not really contest use of the first three criteria in 6(c)(1)—i.e., the effects on health and on the environment, and the availability of substitutes. Rather, EDF's primary focus is on the fourth criterion in 6(c)(1), relating to the economic consequences of the authorization. Yet, EDF's objections to the "economic consequences" criterion cannot stand in the face of section 2(c) of the Act, which expressly requires the Administrator to consider such factors.[23]

Furthermore, the particular economic factors that EPA took into account were plainly reasonable. The Administrator did not simply propose to consider the effect of the ban on industry, but also the effects on "the national economy, small business, technological innovation, the environment, and public health." This formulation, which considers a broad range of benefits and costs of the ban and use authorization, is entirely consistent with the section 2(c) requirement that the Administrator consider the economic and social impact on his actions.

Because the 6(c)(1) criteria fulfill an express mandate of the statute and reflect a reasonable interpretation of an ambiguous phrase, we conclude that the Administrator did not err in choosing those criteria to make the unreasonable risk determinations under 6(e)(2)(B).[25]

[20]The dispute over the proper criteria arises because §6(e) offers no definition of the expression "unreasonable risk of injury to health or the environment."

[23]Section 2(c) of the Act requires the Administrator to "consider the environmental, economic, and social impact of *any action* the Administrator takes or proposes to take under this chapter." 15 U.S.C. §2601(c) (1976) (emphasis added).

[25]Because we find that the Administrator's decision to use the §6(c)(1) criteria for the unreasonable risk determination in §6(e)(2)(B) was reasonable, those are the criteria against which we review the findings of unreasonable risk.

B. Application of the Criteria

EDF's final attack on the use authorizations is that the Administrator did not properly apply his own criteria in making the unreasonable risk determinations. Here, too, we reject EDF's position.

The standard of judicial review for rules promulgated under section 6(e) is expressly set forth in subsection 19(c)(1)(B)(i): "the court shall hold unlawful and set aside such rule if the court finds that the rule is not supported by substantial evidence in the rulemaking record." 15 U.S.C. §2618(c)(1)(B)(i). Evidence includes "any matter in the rulemaking record."

The substantial evidence standard[27] mandated by the Act is generally considered to be more rigorous than the arbitrary and capricious standard normally applied to informal rulemaking.[28] Under the substantial evidence standard, a reviewing court must give careful scrutiny to agency findings and, at the same time, accord appropriate deference to administrative decisions that are based on agency experience and expertise. Because administrative decisions often involve judgments based on incomplete or even conflicting scientific data, the agency "may have to fill gaps in knowledge with policy considerations." *AFL-CIO v. Marshall,* 617 F.2d 636, 651 (D.C. Cir. 1979). Consequently, reviewing courts "must examine both factual evidence and the agency's policy considerations set forth in the record." *Id.* The court's role in reviewing regulations is to ensure public accountability "by requiring the agency to identify relevant factual evidence, to explain the logic and the policies underlying any legislative choice, to state candidly any assumptions on which it relies, and to present its reasons for rejecting significant contrary evidence and argument." *United Steelworkers of America v. Marshall.* With these general guidelines in mind, we review EPA's PCB use authorizations.

In an attempt to reduce the costs of compliance and the risks associated with exposure to PCBs, the Administrator created two categories for transformers—PCB-contaminated transformers (containing PCB concentrations between fifty and 500 ppm) and PCB transformers (containing PCB concentrations greater than 500 ppm).

Because the Administrator found that proper protective clothing and good management practices should reduce PCB exposure to "very low levels," the regulations permit routine servicing of PCB transformers and electromagnets. Additionally, the Administrator heard uncontradicted evidence that a prohibition of routine servicing would significantly increase the chances of catastrophic transformer failure, presenting "far greater risks to health and the environment than that associated with the minimal PCB exposure during routine servicing." However, the Administrator found that the rebuilding of PCB transformers and electromagnets (i.e. non-routine servicing) risks greater exposure to PCBs due to leaks, spillage, or volatilization of the dielectric. The Administrator also found that a prohibition against "any servicing (including rebuilding) of PCB transformers that involves removing the coils from the casing . . . will cost about $12 million the first year and steadily less each year thereafter." Applying these findings to the 6(c)(1) criteria, the Administrator ruled that continued routine servicing, without rebuilding, and not involving the removal of coils, would present no unreasonable risk of injury.[30] The uncontradicted evidence and the explication of policy considerations in the present record is sufficient to uphold the use authorization for routine servicing of PCB transformers and electromagents.

Because they contain lower concentrations of PCBs, PCB-contaminated transformers present correspondingly smaller risks associated with exposure. Accordingly, the Administrator found that routine servicing of PCB-contaminated transformers presents no unreasonable risk of injury. Furthermore, because ninety-nine percent of all large transformers are PCB-contaminated transformers, a prohibition on rebuilding could cause "extremely high" costs. Bal-

[27]The Supreme Court has said that "[s]ubstantial evidence is more than a mere scintilla. It means such relevant evidence as a reasonable mind might accept as adequate to support a conclusion." *Consolidated Edison Co. v. NLRB,* 305 U.S. 197, 229 (1938).

[28]*But see Pacific Legal Foundation v. Dep't of Transportation,* 593 F.2d 1338, 1343 n. 35 (D.C. Cir. 1979) (court agrees with "emerging consensus of the Courts of Appeals that the distinction between the arbitrary and capricious standard and substantial evidence review is largely semantic").

[30]EDF's repeated attempt to equate risk of exposure to PCBs and unreasonable risk is misguided. Although the EPA has determined that no level of exposure can be considered safe, and that therefore, any exposure should be considered "significant," that determination does not imply that the exposure is unreasonable. EPA expressly noted that its definition of insignificant "is not a determination that any exposure to PCB's present an unreasonable risk."

ancing these factors, the Administrator concluded that there was no unreasonable risk associated with the rebuilding or other extensive servicing of PCB-contaminated transformers.

Through these two classifications the Administrator has sought to encourage users to convert to PCB-contaminated transformers by draining their PCB transformers and refilling them with some other dielectric fluid. Only after converting the transformers can users rebuild their transformers, thereby reducing operating costs. Thus, the Administrator has created an incentive to dispose of PCBs without imposing extraordinary costs on industry. These policy considerations and findings reflect the criteria outlined in subsection 6(c)(1). Because the Administrator has carefully articulated his policy judgments, and because there is substantial evidence in the record to support his findings, we uphold the use authorization for PCB-contaminated transformers.

The Administrator's authorization of the use and servicing of railroad transformers also reflects a proper balancing of the subsection 6(c)(1) criteria. Because of the strenuous conditions under which they operate, railroad transformers often leak PCBs onto railroad beds, risking exposure to the environment and to workers and other persons near rail lines. On the other hand, nearly 1,000 of these transformers are currently used in railroad engines. A flat prohibition of their use would produce "an unacceptably severe curtailment of railroad service."

In order to balance the social and economic impact of a prohibition against the risks to health and the environment, the Administrator sought a solution that would permit continued use while promoting conversion to non-PCB dielectric fluid. In reaching his solution, the Administrator considered the ninety million dollars in costs associated with immediate conversion to non-PCB dielectrics and the undetermined safety risks associated with fire and explosion in using non-PCB dielectrics. Consequently, EPA issued regulations requiring railroads to reduce the concentration of PCBs in railroad transformers to six percent (60,000 ppm) by 1982, and to 1,000 ppm by 1984. This timetable, EPA believes, will give it sufficient time to evaluate the risks associated with use of non-PCB fluids, and will also substantially reduce the costs of compliance.

It is clear that the Administrator has properly applied the 6(c)(1) criteria in making the unreasonable risk determinations. Where scientific knowledge is incomplete, EPA has set forth specific policy considerations explaining the final regulations. Finding substantial evidence in the record to support the Administrator's findings, we uphold the authorizations for railroad transformers.

IV. THE FIFTY PPM REGULATORY CUTOFF

As a part of the regulatory scheme for PCBs under section 6(e), EPA limited application of the Disposal and Ban Regulations to materials containing concentrations of at least fifty ppm of PCBs. With one exception, materials with lower concentrations remain unregulated under the TSCA regulations. EDF contends that this limitation contravenes the statutory command in subsections 6(e)(2)(A) and 6(e)(3)(A) to regulate "any polychlorinated biphenyl." While we do not adopt all of EDF's reasoning, we find that, under the applicable standard for judicial review, there is no substantial evidence in the record to support the Administrator's decision to establish a regulatory cutoff at fifty ppm.

*   *   *

In the proposed Ban Regulations, EPA listed four reasons for setting the regulatory cutoff at fifty ppm. First, EPA believed that a fifty ppm limit would "exclude from the municipal sludges and other mixtures containing low (less than 50 ppm) levels of PCB's whose presence is due to ambient levels of PCB present in the air or water." As EPA develops in its brief, Congress did not design section 6(e) to regulate ambient sources of PCBs. Second, EPA believed that some industrial chemical processes inevitably produce traces of PCBs, and that careful controls could reduce the concentrations of PCBs only to fifty ppm. Third, EPA felt that it was impractical to regulate the "diffuse and extremely numerous PCB sources" with concentrations below fifty ppm. EPA believed that the proposed cutoff would ensure maximum effectiveness of the regulation by focusing "Agency attention under TSCA upon the most significant and controllable sources of PCB exposure." Fourth, the agency believed that other statutes were available to regulate low concentrations of PCBs, particularly municipal sludges and dredge soils.

In the final Ban Regulations, EPA adopted the proposed fifty ppm regulatory cutoff. Although industry favored a cutoff of 500 ppm in order to reduce the costs of complying with the regulations, EPA found that industry could comply with the more stringent standard. Furthermore, lowering the cutoff from 500 to fifty ppm would "result in substantially increased health and environmental protection."

A cutoff below fifty ppm, on the other hand, would "provide an additional degree of environmental protection but would have a grossly disproportionate effect on the economic impact and would have a serious technological impact on the organic chemicals industry." While it did not have firm data, EPA believed that for some chemical processes, it was technically impossible to eliminate the inadvertent production of PCBs. EPA also feared that because of limited disposal facilities, a lower cutoff would increase disposal requirements and interfere with the disposal of high concentration wastes. In short, EPA believed that the fifty ppm cutoff "provides adequate protection for human wealth and the environment while defining a program that EPA can effectively implement."

Both EPA and EDF claim that the statutory language and legislative history support their positions on the regulatory cutoff. The statutory language is simple: "no person may . . . use any polychlorinated biphenyl in any manner other than in a totally enclosed manner." 15 U.S.C. §2605(e)(2)(A). Similarly, the prohibitions on manufacture, processing, and distribution refer to "any polychlorinated biphenyl." *See id.* §2605(e)(3)(A). Taken literally, this language might require EPA to regulate every molecule of PCB. We are reluctant, however, to impose such an extreme interpretation absent support in the legislative history.

The legislative history reveals that Congress was aware of existing environmental contamination by PCBs—the so-called ambient sources of contamination. For example, during the Senate debate of the amendment adding section 6(e) to the TSCA bill, Senator Nelson, who introduced the amendment, read into the record reports showing the widespread environmental contamination by PCBs.

*   *   *

EPA concluded, we believe correctly, that despite Congress' recognition that existing contamination of PCBs in the environment posed continuing risks to humans and wildlife, Congress did not design section 6(e) to regulate ambient sources of PCBs.

*   *   *

Partly in order to incorporate congressional intent, the Administrator chose a regulatory cutoff at a level that he felt would exclude the ambient sources from regulation. We are troubled by this regulation, however, since the purpose of section 6(e) is to prevent the "introduction of additional PCB's into the environment." The selection of a cutoff undermines the congressional intent to regulate non-ambient sources of PCBs if non-ambient sources of contamination remain unregulated. It is equally troubling that the Administrator apparently is not aware of the amount of PCBs excluded from regulation by the fifty ppm or other possible cutoffs. Particularly because the Administrator has found that any exposure to PCBs may have adverse effects, the Administrator's flat exclusion of some industrial sources of contamination must undergo careful scrutiny. While some cutoff may be appropriate, we note that the Administrator did not explain why the regulation could not be designed expressly to exclude ambient sources, thus directly fulfilling congressional intent, rather than achieve that goal indirectly with a cutoff, thereby partly contravening congressional intent. Thus, a desire to exclude ambient sources of contamination, without more, cannot support the regulatory cutoff.

EPA also seeks to justify the regulatory cutoff on the basis of the serious impact a lower cutoff would have on industries that inadvertently produce PCBs during the manufacturing process. As EPA readily concedes, however, the inadvertent commercial production of PCBs is to be regulated under the Act. By providing a blanket exemption for concentrations below fifty ppm, the Administrator has circumvented the authorizations and exemptions requirements provided in the statute. EPA made no finding that the cutoff would involve no unreasonable risk to health or the environment. As the EPA noted in its Support Document for the final Ban Regulations, justifying a fifty rather than a 500 ppm cutoff, "the authorization and exemption processes are the most effective way to deal with any difficulties. The authorization and exemption processes allow the Agency to tailor the compliance requirements and to be informed as to which companies are having problems and how they are disposing of their waste streams." We agree with EPA. Consequently, the burdens faced by industries cannot be the basis for the fifty ppm cutoff.

*   *   *

The record in the present case is replete with findings and data that PCBs are toxic to wildlife in concentrations well below fifty ppm. Furthermore, the record shows that PCBs bioaccumulate in animals, concentrating as they move up the foodchain. Most importantly, EPA expressly found that any exposure of PCBs to the environment of humans could cause adverse effects. These findings leave us unable to say that the Administrator could rationally conclude that the benefits of regulating concentrations below fifty ppm are of no value. Consequently, we conclude that the *de minimis* exception to the Act is not available to justify the fifty ppm cutoff.

We reemphasize that the Administrator has other, more appropriate means providing him with flexibility to avoid disproportionate impact on industries or on health and the environment. Those tools are the authorization and exemption provisions in subsections 6(e)(2) (B) and 6(e)(3) (B). The standards enunciated therein, requiring findings of no "unreasonable risk of injury to health and the environment" and, in the case of exemptions, good faith efforts to find substitutes, reflect a plain congressional intention that cannot be ignored. For if there is an unreasonable risk of injury, as there may be given EPA's findings, surely Congress did not intend to permit the continued use, manufacture, processing or distribution of PCBs in concentrations below fifty ppm. EPA's *ad hoc* consideration of economic impact and disposal requirements, leading to a conclusion that the fifty ppm cutoff "provides adequate protection for human health and the environment," is neither as rigorous nor as strict as the statutorily required unreasonable risk determination based on the subsection 6(c) (1) criteria.[47] Thus, we remand this part of the record to EPA for further proceedings.

V. TOTALLY ENCLOSED USES

EDF also petitions for review of the Administrator's decision to list several uses, including non-railroad transformers, capacitors, and electromagnets, as to-

[47]So that there is no misunderstanding, we are not striking down the 50 ppm cutoff because EPA has failed to justify that particular level instead of a slightly lower level. "In reviewing a numerical standard, we must ask whether the agency's numbers are within a 'zone of reasonableness,' not whether its numbers are precisely right." *Hercules, Inc. v. EPA*, 598 F.2d 91, 106-07 [12 ERC 1376] (D.C. Cir. 1978). EPA has failed to adduce substantial evidence that 50 ppm is within the zone of reasonableness.

tally enclosed uses and therefore exempt from the regulations promulgated under section 6(e). Because we find no substantial evidence in the record to support the Administrator's classifications, we remand this part of the record for further proceedings.

* * *

VI. CONCLUSION

On the basis of the foregoing, we find that there is substantial evidence in the record to support the use authorizations; therefore, we uphold those regulations. However, because we find no substantial evidence in the record to support either the fifty ppm cutoff or the EPA classification of certain PCB uses as totally enclosed, these latter two regulations cannot be upheld. Consequently, we set aside the regulations dealing with the fifty ppm cutoff and the classification of certain PCB uses as totally enclosed, and remand those portions of the record for further proceedings consistent with this opinion.

We feel constrained to add one final note to emphasize our concern in this case. Human beings have finally come to recognize that they must eliminate or control life threatening chemicals, such as PCBs, if the miracle of life is to continue and if earth is to remain a living planet. This is precisely what Congress sought to do when it enacted section 6(e) of the Toxic Substances Control Act. Yet, we find that forty-six months after the effective date of an act designed to either totally ban or closely control the use of PCBs, 99% of the PCBs that were in use when the Act was passed are still in use in the United States.[53] With information such as this in hand, timid souls have good reason to question the prospects for our continued survival, and cynics have just cause to sneer at the effectiveness of governmental regulation.

The EPA regulations can hardly be viewed as a bold step forward in the battle against life threatening chemicals. There is no substantial evidence in the record to support certain of the EPA regulatory enactments, and portions of the regulations are plainly contrary to law. Thus, the effort by EPA has, in

[53]A recent House committee report on the proposed Toxic Substances Control Act Amendment of 1980, H.R. 7126, lamented that the "EPA definition [of totally enclosed uses] exempted from the [§6(e)] ban approximately 99 percent of all PCB's found in the United States." H.R. Rep. No. 968, 96th Cong., 2d Sess. 6 (1980).

certain respects, fallen far short of the mark set by the congressional mandate found in section 6(e) of the Toxic Substances Control Act.

On remand, we trust that EPA will act with a sense of urgency to find effective solutions to enforce the Act. We are not so naive as the assume or suggest that hasty responses will ensure effective regulations.

However, we are well able to see, from the plain text of the Act, that the deadlines for the enactment of regulations to enforce section 6(e) have passed. We therefore believe that EPA should act with expedition to complete the important task assigned to it by Congress.

So ordered.

2. Imminent Hazard

In 1985, EPA used the imminent hazard provisions of Section 7 to clean up PCB spills even though Section 6 regulations had been issued three years earlier. The U.S. District Court for the Northern District of Illinois affirmed the use of the provision and defined its reach in the following case.

United States v. Commonwealth Edison Co.

620 F. Supp. 1404 (N.D. IL. 1985)

MORAN, District Judge.

This case for injunctive and declaratory relief concerns Commonwealth Edison's responsibility for cleaning up highly toxic polychlorinated biphenyls (PCBs) spilled from its electrical equipment, often in residential areas. Edison, impliedly at least, recognizes a responsibility to clean up spills, with a dispute over the extent of that responsibility. It argues, however, that the extent of its responsibility cannot be determined through the medium of this lawsuit as the issues are presently framed. This court disagrees.

*   *   *

III

The government has pleaded two claims with respect to each of the seven incidents of PCB contamination covered in the complaint. In the odd-numbered counts the government claims that Edison violated the PCB regulations promulgated by the EPA pursuant to 15 U.S.C. §2605. In the even-numbered counts the government claims that the dielectric fluid containing PCBs spilled from Edison's transformers and capacitors is an "imminently hazardous chemical substance" under 15 U.S.C. §2606.

The government seeks an order compelling Edison to submit a list of sites where there have been known or suspected discharges of PCB-contaminated dielectric fluid, to initiate a program of sampling and laboratory analysis to evaluate the extent of PCB contamination at each spill site, and to provide detailed results of each analysis to the owner and occupant of the affected property. It also seeks to compel Edison to clean up each of the sites contaminated by PCBs, using methods designed to minimize human exposure and to limit the spread of PCBs into the environment. Finally, the government seeks an order declaring that "Edison has a duty to remove all PCB . . . contamination resulting from any future discharge of dielectric fluid from Edison's PCB capacitors, transformers and other PCB electrical equipment."

The TSCA makes it unlawful to fail or refuse to comply with the requirements of §2605 or EPA regulations promulgated thereunder, 15 U.S.C. §2614(1). It gives district courts jurisdiction over civil actions to, *inter alia*, (1) restrain any violation of §2614, (2) restrain action prohibited by §2605 or applicable regulations, and (3) compel any action required by or under the TSCA, 42 U.S.C. §2616. The parties agree that §2616 gives this court jurisdiction over counts 1, 3, 5, 7, 9, 11 and 13.

The more difficult question is whether the government can proceed under §2606. That section permits the EPA administrator to bring an action in district court

for relief . . . against any person who manufactures, processes, distributes in commerce, or uses, or disposes of, an imminently hazardous chemical substance or mixture or any other article containing such a substance or mixture.

42 U.S.C. §2606(a)(1)(B). The statute defines "imminently hazardous chemical substance or mixture" as

> a chemical substance or mixture which presents an imminent and unreasonable risk of serious or widespread injury to health or the environment. Such a risk to health or the environment shall be considered imminent if it is shown that the manufacture, processing, distribution in commerce, use, or disposal of the chemical susbtance or mixture, or that any combination of such activities, is likely to result in such injury to health or the environment before a final rule under section 2606(f) of this title can protect against such risk.

15 U.S.C. §2607(f).

The EPA may commence a civil action under §2606 "notwithstanding the existence of a regulation or the pendency of any administrative or judicial proceeding." 15 U.S.C. §2606(a) (1). Section 2606(b) authorizes the district court to grant temporary or permanent relief "necessary to protect health or the environment from the unreasonable risk associated with the chemical substance," §2606(b).

Edison's position is that dielectric fluid containing PCBs cannot be considered "imminently hazardous" as a matter of law and thus §2606 is inapplicable. It points out that "imminently hazardous" is defined with reference to whether a chemical substance "is likely to result in . . . injury to health or the environment before a final rule under §2605 . . . can protect against such risk." Section 2606(f). Because the EPA promulgated regulations three years ago that permit the continued use of dielectric fluid containing PCBs in transformers and capacitors, Edison argues that a final rule already exists that protects against the risk of PCB contamination from leaks and spillages. But that position rests upon a *non sequitur.* That use of PCBs in nonruptured enclosures may not be "imminently hazardous," does not mean that the continuing presence of PCBs in the environment as a result of uncontrolled spills, however inevitable such spills may be, is equally tolerable.

The government's case under §2606 is a narrow one. It seeks to establish that PCBs spilled in residential areas constitute an unusually great health risk and to compel Edison to clean up those areas contaminated by malfunctioning electrical equipment. Obviously, the government would not come into federal court to compel Edison to disclose where PCB spills have occurred and to clean up those sites

if existing regulations forced Edison to keep public records of spills, to clean up spill sites and to inform unsuspecting residents of the extent of the PCB contamination.

Section 2606's definition of "imminently hazardous chemical substance" explicitly distinguishes between "use" and "disposal." While the regulations are directed primarily at the "use" of PCBs in electrical equipment, the government's action is directed at a certain form of "disposal"—spills from rupturing electrical equipment. The existence of regulations governing the disposal of PCBs do not preempt an action under §2606. First, the disposal regulations do not focus on the specific problem the government seeks to redress. Second, the government may bring an action against an imminent hazard notwithstanding the existence of a regulation. *See* §2606(a).

The imminence of the hazard posed by the use or disposal of a chemical is defined in part with reference to the likelihood that the injury will occur before a regulation can be promulgated to protect against the harm, §2606(f). This does not mean that the EPA administrator can bring an action under §2606 only when a regulation is to be promulgated. Rather, §2606 is best viewed as a complement to §2616. Section 2616 permits the government to bring an action for violation of existing regulations. Section 2606 permits the government to close regulatory loopholes by taking action against applications of toxic chemicals whose health and environmental risks are not sufficiently minimized by the regulatory scheme. Requiring the EPA to couple new regulations with every §2606 action would both complicate enforcement efforts and lead to an ever-growing regulatory scheme of impenetrable complexity.

By bringing an action under §2606 rather than §2616, the government presumptively establishes that the regulatory scheme is incomplete. The primary focus of a §2606 case should be upon the seriousness of the real or threatened hazard posed by a toxic chemical or the particular application of a toxic chemical. The requirement that the government establish that the health or environmental threat is both (1) imminent and (2) commensurate to that posed by other substances covered by the TSCA ("unreasonable risk"), protects regulated companies from the arbitrary or frivolous enforcement efforts that Edison fears.

*   *   *

[Motion for summary judgment in favor of Commonwealth Edison denied.]

3. Testing Rules

Three years after TSCA was passed, EPA had not issued testing requirements even though it had designated 18 chemicals for priority action. The Natural Resources Defense Council sued EPA to compel action.

Natural Resources Defense Council v. Costle

14 Env' Rep. Cases 1858 (S.D. N.Y. 1980)

PIERCE, District Judge.

Plaintiff Natural Resources Defense Council, Inc. ("NRDC"), a non-profit membership corporation organized under the laws of the State of New York, brought this suit against the United States Environmental Protection Agency ("EPA"), and Douglas Costle, as Administrator of the EPA, seeking an order requiring the EPA to adhere to the provisions of the Toxic Substances Control Act, 15 U.S.C. §§2601, *et seq.* ("the Act"). The Chemical Manufacturers Association ("CMA"), the Synthetic Organic Chemical Manufacturers Association ("SOCMA"), the American Petroleum Institute ("API"), Koppers Co., and Reilly Tar and Chemical Corporation ("Reilly") were given permission to intervene on behalf of the defendants pursuant to Rule 24(a) of the Fed. R. Civ. P. predicated upon their property interest as manufacturers, or associations representing manufacturers, of the chemicals involved in this litigation.

Defendants moved for summary judgment pursuant to Rule 56 Fed. R. Civ. P.; plaintiff filed a cross-motion for partial summary judgment seeking a declaratory judgment that defendants have failed to comply with certain nondiscretionary duties under the Act. Oral arguments on the motions were heard on October 10, 1979. For the following reasons, defendant's motion for summary judgment is denied; plaintiff's motion for partial summary judgment is hereby granted. The following shall constitute the Court's findings of fact and conclusions of law pursuant to Rule 56 Fed. R. Civ.

THE STATUTE

The Act, effective as of January 1, 1977, was enacted by the Congress in order to bring about the institution of comprehensive procedures for the testing and control of chemicals believed to present unreasonable risks of injury to health or to the environment. It was "designed to fill a number of [then existing] regulatory gaps. The legislature perceived a need for this legislation because "[t]he last century has witnessed the ever-accelerating growth of the chemical industry . . . [and] too frequently we have discovered that certain of these chemicals present lethal health and environmental dangers."

The core regulatory sections of the Act are 15 U.S.C. §§2603, 2604, and 2605. Section 2603(a) (1)(A) provides for the testing of chemical substances and mixtures found by the EPA Administrator: (1) to present an unreasonable risk of injury to health or the environment, (2) to have insufficient data available to determine the effects of the particular substance, or (3) to need testing to develop data on risk to health and the environment. Section 2604 requires manufacturers to notify EPA of new chemicals and significant new uses of old chemicals before manufacture begins. Section 2605 outlines the remedies available to EPA with respect to chemicals which are found to present unreasonable risks to health or the environment.[2] Plaintiff alleges that defendants have failed to comply with the testing requirements of section 2603.

Section 2603(a) provides that if the Administrator makes one of the legislatively prescribed findings, then "the Administrator shall by rule require that testing be conducted on such substances or mixtures . . . ." Pursuant to this testing mandate, section 2603(e)(1)(A) establishes the Interagency Testing Committee (the "Committee") to recommend chemicals for priority consideration by the EPA Administrator.

*   *   *

Section 2603(e)(1)(A) requires that the Committee make its recommendations to the EPA Administrator of chemical substances for priority consideration based upon "all relevant factors" including eight specifically enumerated factors concerned with the extent of human and environmental exposure. The recommended chemicals are to be listed by the committee in the order of recommended priority for rulemaking action by the EPA. In establishing

[2]These remedies include prohibition or limitation on the manufacture of the chemical.

this list, the Committee is required to give priority attention to those chemicals known to cause or suspected of causing or contributing to "cancer, gene mutations, or birth defects." 15 U.S.C. §2603(e) (1) (A).

* * *

Section 2603(e)(1)(B) provides that the Committee shall publish this list in the Federal Register, and submit it to the EPA Administrator "[a]s soon as practicable, but not later than nine months after January 1, 1977. . . ." Section 2603(e)(1)(B) then states the deadline for EPA action as:

"Within the 12-month period beginning on the date of the first inclusion on the [Committee's] list of a chemical substance or mixture designated by the committee . . . the Administrator shall with respect to such chemical substance or mixture either initiate a rulemaking proceeding under subsection (a) or, *if such a proceeding is not initiated within such period, published* [*sic*] *in the Federal Register the Administrator's reason for not initiating such proceeding.*" (Emphasis added.)

In July, 1977, the Committee in compliance with its statutory mandate published a "Preliminary List of Chemical Substances for Further Evaluation," which identified approximately 330 substances or categories of substances along with background information on the methods used by the Committee in making its selections.

* * *

On October 5, 1977, the Committee transmitted to the EPA its "Initial Report" which designated for consideration by the EPA within the following twelve months, ten substances and categories of substances for the promulgation of regulations under Section 4(a), along with specific testing regulations. The Committee determined that each of the ten chemicals was of an equally high testing priority.

On April 10, 1978, the Committee transmitted its "Second Report" to the EPA, designating eight more chemicals and categories of chemicals for certain recommended health and environmental effects testing, and describing them as having the same priority as the original ten.

On October 10, 1978, the EPA responded to the Committee's Initial Report." In its response, consisting of approximately one page plus appendix, the EPA essentially stated that it had been unable to evaluate the recommended chemicals.

In a letter dated March 2, 1979, plaintiff NRDC informed EPA that it considered EPA's first response inadequate, and that it believed that the EPA was in violation of the Act. In April, 1979, Dr. Warren Muir answered the letter, and provided the NRDC with a more detailed explanation of why EPA was not able to initiate testing procedures at that time. The Muir letter essentially explained the current status and duties of the EPA with respect to the Act. The letter made reference to difficulties the EPA was experiencing, including a shortage in trained personnel and office space.

On May 8, 1979, the EPA published its response to the ITC's "Second Report." In summary, the EPA stated:

As discussed further, below, the EPA has not yet completed full evaluation of the recommendations, which must be done before the Agency can determine whether to propose testing rules for them. In addition, the test standards that must be identified in proposing section 4 test rules have not yet been properly proposed for public review and comment. For these reasons the Agency is not at this time proposing section 4 test rules for the chemicals and categories included by the ITC in its second report." . . .

Plaintiff contends that the reasons published by the EPA in its responses are not sufficient to comply with the mandates of the Act. Defendant and intervenors rely upon a statement in the Act's legislative history that Congress did not intend for the EPA to devote "an inordinate amount of resources to justify the failure to require testing." They assert that the EPA is in compliance with the statutory mandate. The Court disagrees and holds that the reasons published by the EPA are insufficient as a matter of law to comply with the Congressional intent and the mandate of the Act.

* * *

The Muir letter, . . . sent in response to NRDC's notice of intention to sue, which gave further details on the EPA manpower and financial difficulties, was never published in the Federal Register as an official supplement to the EPA's "reasons." In this letter Dr. Muir also predicted that the first test rule would be issued by the EPA by the end of December, 1979. As of the date of this opinion the Court has not received any notice that a test rule has in fact been issued. Furthermore, at oral arguments herein, the attorney representing EPA predicted that the first quarter of 1980 would likely be the time period in which the *first* rule would be issued; he was not able to provide the Court with any further predictions beyond that date.

EPA contends that having published its "reasons" in the Federal Register, it is under no further deadlines with respect to the aforesaid 18 chemical substances. The EPA argues that the twelve month section 2603 deadline was no more than a deadline for a progress report which "serves the obvious purpose of focusing for Congress and the citizens what the agency's capabilities are with respect to this program and what it is going to take in the future to require action with greater speed, if that is what is required or with lesser speed if that is what is required and warranted."

Upon oral argument of the motions, counsel for the EPA reemphasized the funding problems experienced by the Agency, but, upon inquiry by the Court, was unable to provide any evidence that the Agency had sought additional funds from the Congress to enable it to carry out its new responsibilities under the Act.

This Court cannot agree with EPA's interpretation of what Congress intended when it imposed a twelve month deadline with respect to this critical health and environmental legislation. If Congress intended to require what is essentially a progress report from the EPA when it directed the EPA to state its "reasons," if it found that it could not comply with the twelve month deadline for the initiation of a rulemaking proceeding, it would likely have tracked the procedure it adopted with respect to Section 2629 of the Act.

Section 2629 requires the Administrator to prepare and submit to the President and the Congress an annual "comprehensive report on the administration of this chapter during the preceding fiscal year" and, to include Section 2603 of the Act, by February 29, 1980. Plaintiff shall have two weeks from the date of such submission to submit comments regarding defendants' proposed plan.

SO ORDERED.

In response to this suit, EPA developed its controversial "negotiated testing program" in which industry and EPA would agree to testing without the issuance of a formal testing rule for a particular chemical. In 1984, NRDC again brought suit to force promulgation of formal testing rules.

NRDC v. EPA

595 F. Supp. 1255 (S.D.N.Y. 1984)

DUFFY, District Judge.

BACKGROUND

Plaintiffs brought this action seeking an order requiring the EPA'S compliance with provisions of the Toxic Substances Control Act, 15 U.S.C. §§2601 *et seq.* ("TSCA" or "the Act"). TSCA was enacted in response to what Congress perceived as the unreasonable risks associated with the increasing marketing of chemical products whose potential toxicity was as yet untested.

*   *   *

TSCA, 15 U.S.C. §2603 provides for EPA issuance of rules requiring testing of chemicals which may present unreasonable risks of injury to human health or the environment. The testing required by such rules is to be carried out and financed by the manufacturers and/or processors of the chemical substances. *See* TSCA, 15 U.S.C. §2603(b) (3). In section 2603(e) of the Act, Congress established an expert panel of government scientists, known as the "Interagency Testing Committee" ("ITC"). ITC is directed to select and recommend to EPA a list of those chemicals whose potential risks to health and the environment are determined to warrant "priority consideration by the agency for the promulgation of a rule. . . ." 15 U.S.C. §2603(e)(1)(A). Thereafter the EPA is required within twelve months of the date on which the substances are first designated to "either initiate a rulemaking proceeding under subsection (a) . . . or if such a proceeding is not initiated within such period, publish in the Federal Register the [EPA] Administrator's reason for not initiating such a proceeding." 15 U.S.C. §2603(e)(1)(B).

A test rule "shall" be promulgated if EPA finds that:

(A)(i) the manufacture, distribution in commerce, processing, use, or disposal of a chemical substance or mixture . . . may present an unreasonable risk of injury to health or the environment,

(ii) there are insufficient data and experience upon which the effects of such manufacture . . . on health or the environment can reasonably be determined or predicted, and

(iii) testing of such substance or mixture with respect to such effects is necessary to develop such data; or

(B) (i) a chemical substance or mixture is or will be produced in substantial quantities, and (I) it enters or may reasonably be anticipated to enter the environment in substantial quantities or (II) there is or may be significant or substantial human exposure to such substance or mixture,

(ii) there are insufficient data and experience upon which the effects of such manufacture . . . on health or the environment can reasonably be determined or predicted, and

(iii) testing of such substance or mixture with respect to such effects is necessary to develop such data.

15 U.S.C. §2603(a)(1).

The final test rule must identify, *inter alia,* the chemical(s) to be done,[1] the test standards or protocols, and the deadlines for test completion and submission of data. To formulate a final rule, EPA is required first to publish proposed rules with these characteristics, soliciting public commentary. *See* Administrative Procedure Act ("APA"), 5 U.S.C. §553 (notice and comment provision); *see also* 15 U.S.C. §2603(b) (5).

The central issue in this case is whether EPA's implementation of TSCA with respect to sixteen ITC-designated chemicals satisfies this statutory mandate to either initiate rulemaking or to publish EPA's reasons for not initiating such proceedings.

*   *   *

[1] The health and environmental effects for which standards for the development of test data may be prescribed include carcinogenesis, mutagenesis, teratogenesis, behavioral disorders, cumulative or synergistic effects, and any other effect which may present an unreasonable risk of injury to health or the environment. 15 U.S.C. §2603(b) (2) (A).

DISCUSSION

I. NEGOTIATED VOLUNTARY TESTING AGREEMENTS

A. Description

In late 1981 and early 1982, EPA announced in the Federal Register that it would consider accepting voluntary testing programs in certain circumstances. These programs would be negotiated and conducted by manufacturers or processors of ITC designated chemical substances in lieu of initiating a rulemaking proceeding. EPA asserts that it adopted this new policy based upon the belief that such agreements would provide the required health and environmental effect test data in an expeditious manner.

According to EPA, after ITC designates priority lists of chemical substances in the Federal Register, EPA conducts two public meetings scheduled at ten and sixteen weeks after the ITC designation. At the second meeting, EPA announces its preliminary decision whether to require testing. If EPA determines that no testing is necessary, it publishes a Federal Register notice describing its reasoning. If it determines that testing is necessary, EPA begins work on a test rule and "simultaneously [invites] industry initiation of negotiations for the purpose of developing a negotiated testing program."

By week twenty-four, EPA requires that its scientists and the industry representatives have reached preliminary agreement on a testing program. Between weeks sixteen and twenty-four EPA apparently holds informal meetings with industry to attempt to reach this negotiated test agreement. EPA reviews the manufacturers' study plans which include test standards and schedules for submission of test data. Acceptable proposals are published and the ensuing comments are reviewed before publication of EPA's final decision to adopt a negotiated voluntary testing agreement.

Plaintiff's first four claims are based on ITC's designation, between November 1980 and May 1982, of four chemicals warranting priority rulemaking consideration and review by EPA. In the two to four years since their designations, EPA has not initiated rulemaking proceedings, rather, within twelve months of each of the designations EPA accepted or tentatively accepted voluntary testing programs negotiated with industry.

B. Jurisdictional Defenses

Plaintiffs brought this action pursuant to 15 U.S.C. §2619(a)(2) which provides that "any person may commence a civil action . . . against the [EPA] Administrator to compel the Administrator to perform any duty under this Act whch is not discretionary." EPA contends that it has no mandatory duty to issue test rules and therefore its discretionary acts are not subject to review by a citizen-suit civil action.

"While it is intended that a recommendation of the [ITC] be given great weight by the Administrator, it should be emphasized that the decision to require testing rests with the Administrator." It is the province of EPA to evaluate the criteria of section 2603(a)(1)(A) or (a)(1)(B) to determine whether testing is appropriate. Section 2603(e)(1)(B) provides the agency with a choice of either initiating a rulemaking proceeding or publishing its reasons for not doing so. It is evident, however, that the Administrator's duty to choose either to initiate rulemaking proceedings or to publish its reasons for not doing so is a mandatory choice that it must make.[4] Thus, plaintiffs may invoke section 2619(a)(2) to review whether EPA carried out this nondiscretionary act.[5]

C. Negotiated Testing Programs and TSCA

At the outset, EPA notes that with respect to each of the four chemicals in Claims One through Four of plaintiff's complaint there have been no formal findings that the threshold requirements for initiating rulemaking proceedings have been met. For example, there has been no express finding that there is unsufficient data on these chemicals presenting an unreasonable risk of injury to public health or to the environment. *See* section 2603(a). EPA asserts that it

is under no obligation to make such determinations. In addition, EPA contends that even if the standards were applied to plaintiffs' first four claims, the facts demonstrate that "testing . . . is [not] necessary to develop [data revealing] the effects . . . of such substance . . . on health or the environment. . . ." 15 U.S.C. §2603(a)-(1)(A)(ii), (iii). Because EPA has arranged to obtain testing data by the negotiated voluntary agreements, defendant maintains that testing is no longer necessary to develop such data.

Although it does not appear that publication of findings concerning the threshold determinations is obligatory, I find for the reasons that follow that EPA's program of negotiating voluntary testing agreements subverts the statutory scheme.

Both in the legislative history of TSCA and on the face of the statute, Congress has evinced its intention that chemicals on which there is insufficient data will be tested pursuant to formal rulemaking.

* * *

EPA's assertion that it has not made the requisite threshold findings necessitating the initiation of rulemaking proceedings is specious. Section 2603 was promulgated to mandate the testing of potentially dangerous chemicals on which there was insufficient data existing at the present time. EPA's negotiation and acceptance of voluntary testing agreements by the manufacturers obviously reflects EPA's belief that additional data concerning the chemicals in question needs to be developed. The absence of a formal finding of testing necessity cannot hide EPA's evidence *de facto* findings of such a necessity.[6] It is the substance not the form of agency action that is relevant upon review.

Furthermore, I can find no support for EPA's decision to utilize negotiated testing agreements instead of the statutorily-prescribed initiation of rulemaking proceedings either on the face of the statute or based on some vague assertion of agency discretion.

---

[4]Use of the word "shall" in the statute is indicative of mandatory obligations. *See Association of American Railroads v. Costle*, 562 F.2d 1310, 1312 (D.C. Cir. 1977).

[5]Defendants also maintain that plaintiffs' only, or at least preliminary, recourse is the petition provisions of section 2620(a) which states that "[a]ny person may petition the Administrator to initiate a proceeding for the issuance, amendment, or repeal of a rule under section 2603. . . ." I agree, however, with Judge Pierce's conclusion in *NRDC v. Costle*:

> Subsection (b)(5) of this statute provides that "the remedies under this section shall be in addition to, and not in lieu of, other remedies provided by law." The Act apparently contains two provisions for citizen review of EPA compliance. In the Court's view, plaintiff is not required to rely exclusively upon utilization of a citizen's petition.

10 Envtl. L. Rep. (Envtl. L. Inst.) 20274, 20277 (S.D.N.Y. Feb. 4, 1980). The petition provision is broader permitting some review of both discretionary and nondiscretionary agency acts; the citizen-suit provision only applies to mandatory agency acts.

[6]For example, EPA states "[a]ssuming data is not already available, the only reason testing would *not* be 'necessary' to obtain it [sic] is if there is reason to believe that the information *will be* developed through means other than tests required by rule." EPA's Reply Memorandum at 9 n.* (emphasis supplied in part). However, TSCA section 2603(a)(1) requires EPA to evaluate the *present* availability of information and need for testing. *See also* 15 U.S.C. §2603(e)(1)(A)(vii). Defendants' contrary interpretation would obviate the necessity of ever initiating formal rulemaking.

"The agency charged with implementing the statute is not free to evade the unambiguous directions of the law merely for administrative convenience." *Brown v. Harris*, 491 F. Supp. 845, 849 (N.D. Cal. 1980) (citing *Manhattan General Equipment Co. v. The Commissioner of Internal Revenue*, 297 U.S. 129, 134 (1936). "It is not an agency's prerogative to alter a statutory scheme even if its assertion is as good or better than the congressional one." *Mid-Louisiana Gas Co. v. Federal Energy Regulatory Commission*, 664 F.2d 530, 535 (5th Cir. 1981). In fact, in the more than seven years since TSCA's enactment, though seventy-three chemicals have been designated by ITC for priority rulemaking consideration, EPA has yet to finalize a single test rule. Congress could not have intended (or envisioned) this result.

In addition to violating the test rule promulgation process, EPA's failure to utilize the rulemaking process circumvents several other statutory provisions. *See, e.g.,* 15 U.S.C. §2603(d) (prompt notice for public review of test rule-generated data); *id.* §2604(a) & (b) (submission to EPA of test data prior to new manufacture of chemicals subject to test rule); *id.* §2607(a)(3)(A) (record-keeping and reporting requirements of small quantity manufacturers of test rule chemicals); *id.* §2611(b) (notice to importing foreign government of test rule data availability); *id.* §2615(a)(1) (sanctions for noncompliance with testing program); *id.* §2617(a)-(2)(A) (test rule's preemption of state and local programs); *id.* §2618(a)(1)(A) (authorized judicial review of testing rules), and *id.* §2619(a)(1) (authorized citizen suits to compel compliance with test rules process).

The intervenor Chemical Manufacturers Association's ("CMA") argument that "most" of these provisions are "essentially" satisfied by other mechanisms misses the point. Congress detailed a complex and comprehensive statute for the testing of potentially toxic chemical substances. It further specified several important additional safeguards triggered by rulemaking results. It is not now the prerogative of EPA to substitute for this intricate framework a number of haphazard and informal purported equivalents.

Furthermore, despite CMA's assertion, several of these statutory provisions have no equivalents whatsoever. *See, e.g.,* 15 U.S.C. §2607(a)(3)(A) (record-keeping and reporting requirements for small manufacturers). And many of the putative substitutes appear far less than adequate. For example, it is not clear whether the substance of negotiated testing agreements are judicially reviewable, whereas section 2618 provides for review of final test rules in the Circuit Court of Appeals. Moreover, even if negotiated agreements were reviewable under the Administrative Procedure Act ("APA") as final agency action, the standard of "arbitrary and capricious" is a far less searching inquiry than section 2618 (c) (1) (B)(i)'s "substantial evidence" standard. Similarly, should EPA be permitted to use voluntary testing programs in lieu of following the rulemaking process, the burden of notifying foreign governments of the availability of test data concerning chemicals they import would fall on the foreign nations themselves rather than the manufacturers and EPA as set forth in the statute.

Finally, I note that most of the supposed benefits of negotiated testing agreements claimed by EPA and the intervenors can be preserved even if EPA complies with the statute's mandate. In support of these agreements, EPA has asserted that negotiated voluntary testing expeditiously provides the desired data in much the same way a test rule would—i.e. the negotiated agreement contains test standards, deadlines, interim review and so forth. Therefore, as plaintiff suggests, it should be a straightforward task to incorporate existing voluntary programs into statutory test rules.

> Plaintiffs have no objection in principle to EPA's negotiation of testing agreements with affected industries. Nor are [they] challenging here the substance or the scientific basis of the negotiated agreements. The gravamen of plaintiffs' complaints is that federal defendants' failure to incorporate these negotiated testing schemes into binding rules has subverted TSCA's statutory foundation.

Plaintiffs' Memorandum of Law at 23.

I agree that the negotiated programs without rulemaking cannot be sanctioned under TSCA, though negotiation to determine appropriate test protocols as well as other relevant criteria certainly is not only permissible but indeed preferable to blind, often impractical, bureaucratic blundering. Accordingly, partial summary judgment is granted in favor of plaintiffs on their first four claims. The parties are ordered to meet to attempt to stipulate to a reasonable timetable for the incorporation of the existing voluntary programs into statutory test rules. In the absence of agreement within thirty (30) days from the date hereof, the parties should submit their respective contentions to this court for a determination of any unresolved issues.

*   *   *

◇ NOTES

1. In 1986, after EPA's continued failure to issue testing rules, EPA, NRDC, and the Chemical Manufacturer's Association entered into a consent agreement for an interim final testing rule. 51 Fed. Reg. 23,713 (1986); 40 C.F.R. §§790.20-28. This agreement changed EPA's procedures for reaching a consensus on negotiated testing by providing for participation by environmental groups in the effort and by requiring an open negotiation process leading to a final formal rule with sanctions for noncompliance. Nineteen Negotiated Test Agreements have been reached and three consent orders have been issued as of October 1988.

2. In 1987, nine years after the Interagency Testing Committee (ITC) listed mesityl oxide as a priority chemical, EPA finally issued a testing rule under Section 4(a)(1). Shell contested the necessity of testing because manufacturing conditions had changed and "only 100 workers" would be affected by exposure to the chemical. In *Shell Chemical v. EPA,* 826 F. 2d 295 (1987), the Fifth Circuit Court of Appeals remanded the rule to EPA for reconsideration.

3. Why might industry prefer negotiated testing? Is there a stigma attached to the issuance of a formal testing rule for a chemical that one produces?

◇

D.   THE RELUCTANCE TO REGULATE: THE REGULATORY HISTORY OF FORMALDEHYDE

The manner in which EPA responded to evidence of the carcinogenicity of formaldehyde provides a stark example of the limitations of regulatory tools in unwilling political hands.

A Hard Look at Federal Regulation of Formaldehyde: A Departure from Reasoned Decisionmaking

N.A. Ashford, C.W. Ryan, and C.C. Caldart

Formaldehyde, one of the most widely used chemicals in modern industry, has recently become one of the most controversial as well. A plethora of lawsuits, Congressional hearings, and scholarly analyses have centered on formaldehyde, and more particularly on federal agency responses to new data indicating that it may be a carcinogen (cancer-causing substance). These developments were sparked by an October 1979 report from the Chemical Industry Institute of Toxicology (CIIT) that formaldehyde causes cancer in rats.

Source: From 7 *Harvard Environmental Law Review* 297 (1983). © 1983 by the President and Fellows of Harvard College. Citations and references in the original article are omitted here.

*   *   *

IV. EPA'S DECISION NOT TO DESIGNATE FORMALDEHYDE A SECTION 4(f) CHEMICAL

A. Background

1. Chronology of Events. EPA received the preliminary results of the CIIT bioassay in November 1979. In response, the agency participated in efforts by the Interagency Regulatory Liaison Group (IRLG) to estimate formaldehyde exposures. A year later, the agency's Office of Toxic Substances (OTS) received the Federal Panel on Formaldehyde report on the chemical's carcinogenicity, and, shortly thereafter, IRLG's exposure data. Based on the available carcinogenicity and exposure information, OTS determined that formaldehyde might be a candidate for action under section 4(f) of TSCA. In January 1981,

OTS began to prepare a Priority Review Level 1 (PRL-1) document on formaldehyde. A PRL-1 designation was an internal EPA mechanism for identifying items of highest priority within the agency. In March 1981, after estimating the potential formaldehyde cancer risk to the general population, EPA's Deputy Assistant Administrator for Toxic Substances, together with other EPA officials, determined that there "may be a reasonable basis to conclude" that formaldehyde poses a significant cancer risk and that the threshold requirement of section 4(f) had thus been met. Accordingly, they drafted a Federal Register notice indicating that section 4(f) had been invoked, that additional information would be required before the agency could formulate a proper response, and that an expedited investigation to obtain the necessary information would begin.

When newly confirmed EPA Administrator Anne Gorsuch and Deputy Administrator John Hernandez assumed office in May 1981, the draft notice awaited them. Also awaiting them was a letter from John Byington, a lawyer representing the Formaldehyde Institute, who disputed the work of the EPA staff. Byington argued that section 4(f) had not been "triggered" and that the available data was insufficient to support *any* immediate action. He urged that further studies be undertaken outside of official agency auspices.

Hernandez delayed further EPA action on formaldehyde and met with representatives of the Formaldehyde Institute to review their position. Thereafter, he determined that the available scientific information was insufficient to support a section 4(f) determination and called for further study. During July and August 1981, he held a series of three closed meetings with EPA staff, industry representatives, and a few selected scientists from other institutions. In testimony at subsequent congressional hearings, Hernandez characterized these meetings as "exclusively scientific," and described their purpose as having been to "shed some light" on the "scientific" issues.

During the following months, EPA acted on several fronts. In a September 11 memorandum, Don Clay, the newly appointed OTS director, indicated to John Todhunter, the newly designated Assistant Administrator for Pesticides and Toxic Substances, that consideration of formaldehyde under section 4(f) was inappropriate. Clay based his position on several factors, including his belief that section 4(f) should be reserved for "a crash effort to remedy a

very serious hazard to public health." Between August and November 1981, the OTS staff revised the original agency document on formaldehyde, reclassifying it from "Priority Review" status to a "Technical Document" status. During October and November 1981, Administrator Gorsuch repeatedly stated that the agency had yet to take a position on formaldehyde.

EPA formally announced its position on February 12, 1982, by releasing a memorandum from Todhunter that embraced Clay's earlier position on formaldehyde. Unlike Clay, Todhunter supported this position with an analysis of the "scientific" issues inherent in formaldehyde risk assessment. He also outlined a plan for further formaldehyde study. EPA characterized the Todhunter memorandum as "concluding present agency action on the subject."

Todhunter's analysis advanced theories that conflicted with prevalent scientific opinion in the field, and that departed from prior cancer policies of EPA and other regulatory agencies. The memorandum accordingly created considerable controversy, within both the scientific community and other agencies. In response, congressional hearings were held in May 1982 to explore the scientific basis for EPA's formaldehyde determination. Following those hearings, both EPA and industry have proceeded with further formaldehyde research.

* * *

3. The Statutory Mandate. As noted, the courts have not yet interpreted section 4(f) of TSCA, nor is there any legislative history specifically pertinent to this section. The plain language of the statute, however, lends itself to common-sense interpretation. Section 4(f) requires the agency to take action whenever the available information indicates

> that there may be a reasonable basis to conclude that a chemical substance or mixture presents or will present a significant risk of serious or widespread harm to human beings from cancer, gene mutation, or birth defects.

Congress apparently designed section 4(f) as a mechanism for early identification and regulation of those chemicals that were of particular concern because they are likely carcinogens, mutagens, or teratogens. EPA's current interpretation of that section, however, will frustrate this scheme.

a. The Possibility of Harm. As noted, section 4(f) requires agency action if there "*may* be a reasonable basis" to conclude that a risk of harm exists. In both common usage and judicial interpretation, the term "may" indicates the possibility of occurrence. Under the plain language of section 4(f), then, EPA cannot delay its threshold determination until a risk has become certain or probable, but rather must take action upon learning of a credible possibility of such risk.

Presumably, EPA had made that threshold determination in early 1981, when it prepared its preliminary Federal Register notice on formaldehyde. In subsequently reversing that determination, the agency offered a markedly different interpretation of the section 4(f) threshold. In essence, EPA redefined the word "may" in a manner inconsistent with both common usage and TSCA's statutory framework.

Todhunter's memorandum is particularly noteworthy in this respect. In summarizing the formaldehyde data, Todhunter noted that "there *may* be human exposure situations . . . which *may not* present carcinogenic risk which is of significance." He thus stated the required statutory finding in the negative. The logical converse of this statement—that there may be human exposure situations that *do* present significant carcinogenic risk—is precisely the finding that requires EPA to proceed under section 4(f). The agency's failure to do so is simply a misinterpretation of statutory language.

EPA documents indicate that the agency continues to endorse this interpretation of section 4(f). EPA's Office of General Counsel has prepared a "Primer on TSCA 4(f)," which details the agency's interpretation of its duties under that section. The Primer concludes that Congress must have intended "may" in this section to refer to a "reasonably high probability" of occurrence. Given the agency's limited resources, the Primer argues, section 4(f) must be viewed as having been reserved for only a small number of highly probable risks. In addition to being inconsistent with the plain language of section 4(f), this interpretation ignores that section's place in the overall scheme of TSCA. Section 7 of the Act specifically provides for expedited action on "imminent hazards," and section 6 details the actions to be taken against "unreasonable risks." If EPA's interpretation were correct, section 4(f) would merely duplicate these other sections. As one commentator has noted, the agency's interpretation "effectively writes the [4(f)] provision out of TSCA."

b. The Nature and Extent of the Possible Harm. The agency's assessment of the kind of risk that it is to consider under section 4(f) may also be inaccurate. The Clay memorandum, for example, argues that section 4(f) "should be reserved only for [carcinogens] of the most serious concern" and for "those situations that require a crash effort to remedy a very serious hazard to public health."

Once again, the statute itself provides relatively clear guidelines, which point to a contrary interpretation. Congress dealt with both short-term and long-term risk in 4(f), which addresses chemical substances that either "present" or "will present" a significant risk of harm. The Act does not define the phrase "significant risk," but the context suggests that "significance" pertains to the likelihood of occurrence. By providing that the risk that *may* exist must be *significant,* the Act seems to require only the *possibility* of a *probable* occurrence. Evidence indicating the possibility of a significant risk thus triggers the threshold determination that compels EPA to assess the risk more precisely.

In its risk assessment, the agency must consider *both* "serious" and "widespread" harm. By significantly distinguishing between these two categories of harm in section 4(f), Congress clearly indicated that either one will trigger a threshold determination. One element focuses on the extent to which the chemical may pose a risk of *serious* harm. Here the concern is not so much the number of people who may be affected, but how severely they may be affected. A low incidence of a debilitating cancer, then, would suffice. An alternate element is the extent to which the chemical may pose a risk of *widespread* harm. Here a higher incidence is required, but the harm need not be as severe.

c. The Nature of the Required Action. EPA's formaldehyde decision also reveals some confusion within the agency as to what it must do once it makes a section 4(f) threshold determination. Although Clay's reference to "a crash effort to remedy a very serious hazard to public health" may occasionally describe section 4(f) actions, it more properly applies to actions under the "imminent hazard" provisions of section 7.

After making a threshold determination of possible significant risk, EPA must decide, within a prescribed time period, whether regulatory action is appropriate. If the agency determines that the risk is not unreasonable, it must subject this finding to public scrutiny by publishing it in the Federal Register. If, on the other hand, the agency does *not* conclude that the potential risk is not unreasonable, it must "initiate appropriate action ... to prevent or reduce to a sufficient extent such risk."

Depending on the severity of the risk, various actions may be appropriate. The agency is free to exercise its reasoned discretion in selecting which of TSCA's other regulatory authorities to call into play, so long as it designs its action to reduce the risk sufficiently. In some cases, a chemical labeling requirement may be adequate. In others, the agency may have to limit the use of the chemical, or ban it outright. While we express no opinion here as to the appropriate regulatory response to formaldehyde under TSCA, it appears that section 4(f) requires something more of EPA than the agency's actions to date.

The response to EPA's inaction was a Congressional hearing and a major lawsuit by the Natural Resources Defense Council. The chronology of regulatory action is depicted in Table 4-2. The lawsuit was dismissed after EPA designated formaldehyde a 4(f) chemical for two major exposed groups: workers in the apparel industry and

Table 4-2. History of regulatory action regarding formaldehyde

Year	Action
1979	CIIT study implicates formaldehyde as an animal carcinogen.
1980	Federal Panel on Formaldehyde concludes that it is prudent to regard formaldehyde as a carcinogenic risk to humans.
1981	NYU study corroborates CIIT findings.
1981	UAW petitions OSHA for an Emergency Temporary Standard (ETS). OSHA denies request on 1/19/82.
1981	EPA Office of Toxic Substances determines that formaldehyde might be a candidate for action under section 4(f) of TSCA. EPA ultimately concludes that the available scientific information is insufficient to trigger section 4(f) and drops it as a priority for consideration.
1982	CPSC bans use of urea-formaldehyde foam insulation (UFFI) in residences and schools. Overturned in 1983 when Fifth Circuit Court of Appeals finds that CPSC failed to support its ban with substantial evidence.
1984	EPA designates formaldehyde a 4(f) chemical and performs a risk assessment. EPA passes the record on informally to OSHA in March 1986.
1984	District court in Washington, D.C. remands UAW petition for an ETS to OSHA for reconsideration or for initiation of permanent rulemaking.
1985	HUD issues regulations covering formaldehyde emissions from pressed wood products in manufactured homes and requires that plywood and particle board emit 0.2 ppm and 0.3 ppm respectively.
1985	OSHA again denies UAW petition for ETS, but indicates continuing consideration of need for a permanent standard.
1985	OSHA issues notice of proposed rulemaking in which it indicates that it will reduce the permissible exposure limit to either 1.5 or 1.0 ppm.
1987	An EPA study indicates that formaldehyde is "a probable human carcinogen," and the agency indicates that it will consider regulatory action.
1987	OSHA promulgates final formaldehyde standard, limiting workplace exposure level to 1.0 ppm.

Source: Adapted from K. Rest and N. Ashford. 1985. Regulations and Technological Options: The Case of Occupational exposure to Formaldehyde. *Harvard Journal of Law and Technology* 1985(I):63–96.

mobile home residents. EPA did not regulate formaldehyde in either context, but instead informally transferred the issue to OSHA in March 1986. OSHA established a workplace standard in 1987, but the unions have contested its adequacy (see Chapter 3). As this text goes to publication, EPA has yet to deal with formaldehyde in building materials and components of cabinets found in both mobile and conventional homes, although the agency could make a Section 9(a) referral to the Consumer Product Safety Commission (CPSC). The formaldehyde experience is illustrative of many of the reasons why TSCA has failed to serve as a centerpiece for federal chemical regulation or as a driving force for workplace and consumer protection.

DISCUSSION QUESTIONS FOR THE CHAPTER

1. In what ways is the regulation of "unreasonable risk" under Section 6(a) of TSCA different from the regulation of "significant risk of material impairment" under the OSHAct?.

2. EPA made three formal referrals to OSHA under Section 9(a) in the period 1985–1986. Since then, EPA has instead used its consulting and coordinating function under Section 9(d). For formaldehyde, MBOCA, and toluendiamines, EPA simply passed on its investigating records to OSHA without making an "unreasonable risk" determination. EPA stated that it believed OSHA would regulate those chemicals "to a sufficient extent." Because no formal Section 9(a) referral was made, OSHA did not have to respond to EPA or to submit a report in the Federal Register. Does EPA have the authority to do this?[z]

3. What is the difference between the citizen actions allowed under Sections 20 and 21 of TSCA?

4. Section 28 provides that EPA may make grants to the states for state plans that complement, but do not preempt, federal authority. In what ways would state plans under TSCA be expected to differ from OSHA state plans?

5. In *NRDC v. Costle,* what remedies exist against EPA if it fails to adhere to the court's order?

6. OSHA is the primary federal agency designated to protect the health and safety of workers through its standard setting and enforcement activities. Do you think that TSCA can be an additional and useful tool in this regard? Explain how workers or their representatives could use TSCA to improve workplace health and safety.

7. If your company was developing a new chemical substance that would require premanufacturing notification to the EPA under TSCA, what kinds of data would you plan to provide the agency? What determination must EPA make in order to permit the commencement of manufacturing? Can EPA totally *prohibit* the manufacture of chemical substances? If yes, on what basis?

[z]See C. Ruggerio. 1989. Referral of Toxic Chemical Regulation Under the Toxic Substances Control Act: EPA's Administrative Dumping Ground. *Boston College Environmental Affairs Law Review* 17(1):75–122.

CHAPTER 5

Economic Issues in Occupational

Health and Safety

A. INTRODUCTION

An understanding of economic aspects of occupational health and safety is important because (1) economic forces explain, in part, the behavior of employers and workers regarding workplace hazards, (2) imperfections in private economic arrangements—that is, in the job market—help explain the need for government intervention, and (3) economic mechanisms and policies can be introduced by government to improve worker health and safety.

It is obvious, even on a superficial level, that economic factors play a significant role in controlling occupational hazards. Job-related injuries and illnesses impose economic costs in many forms. These include medical and rehabilitation expenses, lost wages, and productivity losses from lost work time. The safety equipment, engineering controls, worker training, and other resources that are required to prevent occupational illness and injury also have associated economic costs.

The relative importance of the costs of occupational hazards and the costs of their prevention and control in the context of worker health and safety depends on one's perspective. *Firms* may resist either market opportunities or regulatory obligations, the net effect of which increases their costs and thereby reduces their profits. A *regulator's* ability to impose a standard, according to judicial interpretations of the Occupational Safety and Health Act, is limited by the economic feasibility (as well as the technological feasibility) of that standard, where economic feasibility has been interpreted to

The authors are indebted to Robert F. Stone for assistance with this chapter.

mean that the standard must not threaten the regulated industry's long-run profitability and competitiveness, even though individual firms in the industry may be forced out of business.[a] According to the cost-benefit criterion developed by *economists,* the imposition of a standard is justified if the reduction in injury and illness costs (the benefits) occasioned by the standard exceeds the costs of complying with the standard (or, to word it differently and more accurately, the standard should be set at such a level that the costs of additional control would then exceed the marginal benefits).

In this chapter, the reader is encouraged to assume the role of a "decisionmaker" confronted with the task of controlling workplace hazards—or imposing or accepting workplace risks—and to consider the following questions: What are the true costs and benefits of promoting occupational health and safety and who bears those costs? Are the true costs and benefits accurately perceived, and if not, what are the consequences? Are any other factors, aside from the usual costs and benefits, pertinent in choosing the level of workplace health and safety?

Assuming the decisionmaker to be a firm or worker, the answer to these questions will help determine whether the marketplace is the appropriate arena for making occupational health and safety decisions or whether, instead, government intervention is needed. Assuming the decisionmaker to be a government policymaker, the answer to these questions will help determine the appropriate methods for evaluating the social desirability of alternative government programs to promote occupational safety. Following a description of the market for health and safety, this chapter considers two topics in detail: (1) the basis for government intervention in decisions regarding occupational health and safety, and the various forms that such government intervention can take and (2) the logic and limitations of using cost-benefit analysis to fashion government policy concerning occupational health and safety.

B. THE MARKET FOR HEALTH AND SAFETY

In our society, the preferred mechanism for conducting economic and social activities, and for making economic and social decisions, is the private market—and usually with good reason. A well-functioning market system possesses two important properties. First, under suitable conditions, a market system—a decentralized network of private transactions through which information about individual preferences is imparted—is economically (allocatively) efficient in the following sense: resources are allocated where they are most highly-valued; the appropriate mix of goods and services, embodying the desired bundle of characteristics (for example, size, color, and style for clothing), is produced; and all possible mutually beneficial exchanges will take place, so that further improvements in the welfare of any member of society cannot be attained without making at least one other member worse off. Second, consistent with libertarian values, marketplace transactions are entirely voluntary; only if the interested parties are able to negotiate to mutual advantage will a market exchange occur.

For some economic activities, of course, no marketplace exists. For instance, in the case of environmental hazards, the party releasing a pollutant and those parties at risk are generally economic strangers. There is no market mechanism through which polluters and citizens can negotiate to reduce the environmental risk, at least not prior to an environmental release. Occupational hazards, however, take place within the context of a job (labor) market that permits and sometimes requires (as discussed in

---

[a]*Industrial Union Department, AFL-CIO v. Hodgson,* 499 F.2d 467 (D.C. Cir. 1974).

Chapter 6) the employer and the employee to bargain over workplace health and safety conditions to their mutual advantage. Our natural starting point then is to consider whether the unfettered market (which includes the courts to enforce the agreed-upon terms of real or implied market transactions) promotes the socially desired degree of occupational health and safety, and if not, why not.

1. The Idealized Job Market

Filling a specific job vacancy in a market system represents the intersection of employer demand and worker supply, where each job embodies a particular combination of workplace characteristics, such as salary and other worker benefits, job content (affecting worker job satisfaction), and job-related risks to worker health and safety. According to economic theory, a job market operating under ideal conditions will reduce workplace hazards to their socially optimal level.

Employers compete among themselves in the job market to attract workers, who will obviously choose the most desirable job in terms of the overall "bundle" of workplace characteristics provided. Therefore, in order to hire workers to perform relatively unpleasant tasks, employers must offer some offsetting benefit, such as a higher wage rate. Similarly, to induce workers to accept hazardous jobs, employers must offer to compensate the risk. Risk compensation is typically evaluated in terms of a wage premium (or wage differential) for risk, but it may equally well take the form of insurance benefits, employer assumption of liability for worker injuries and illness, or some combination of the above. The magnitude of the compensation required to attract applicants for hazardous jobs depends on the amount of extra risk involved and the intensity of worker preferences for risk avoidance. If workers are relatively risk-averse, they will demand more compensation to accept any specified level of job hazard. However, since workers compete with each other for jobs, the risk premium and other risk benefits an individual worker can command are limited to the extent that other workers require less risk compensation to accept hazardous work.

To this point, job hazards have been treated as fixed. In reality, job hazards and other working conditions are variables under the control of the employer. Safety equipment, engineering controls, and related employer investments can reduce the level of workplace risk and allow an employer to lower, by a commensurate amount, the risk compensation required to attract workers. An employer will find it profitable to adjust the level of workplace safety such that the cost of the last unit of job risk avoided just equals the associated savings in worker risk compensation. Further reductions in workplace hazards would be undesirable to the employer, and to workers as well, to the extent that workers would prefer that additional employer expenditures on workplace safety be distributed to them in the form of increased risk compensation (higher wages).

The equilibrium level of occupational risk, as derived through job market transactions, is depicted in Figure 5-1. The labor supply curve as a function of job risk is upward-sloping, indicating that workers want additional compensation to perform more hazardous work.[b] The downward-sloping curve reflects the costs confronting employers of lowering workplace hazards at different levels of risk. Additional reductions in workplace risks (moving to the left in the figure) become increasingly more

[b]Note that the labor supply curve for hazardous work is a mirror image of the labor demand curve for safety, which has the familiar negative slope. In fact, Figure 1 is just a mirror image of the graph of the demand and supply curves for job safety.

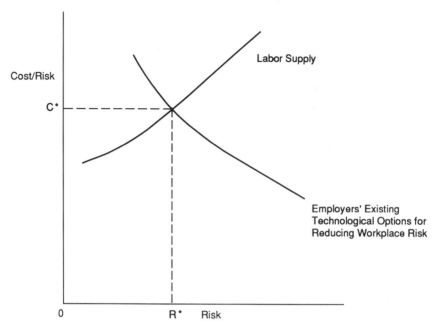

Figure 5-1. Equilibrium level of occupational risk as determined by costs of reducing risks by employers and valuation of reducing risks by workers.

expensive. The intersection of the two curves indicates the market-derived level of job risk, R*, and the market-derived level of risk compensation (per unit of risk), C*.

Thus, according to economic theory, the operation of the private market provides an allocatively efficient level of occupational risk and is capable of providing the socially optimal level of occupational risk as well. However, as we will examine later in more detail, four conditions must be satisfied for the marketplace to function as postulated by economic theory. First, workers must be fully informed about their job opportunities, including the degree of risk present in alternative work environments. Second, those parties engaging in marketplace transactions concerning the level of job risk must bear all of the costs and derive all of the benefits of their actions (that is, the job market must not create externalities). Third, the job market must be perfectly competitive so that individual employers or workers (unions) are not able to influence the wage level or the wage premium for risk. Fourth, since market outcomes vary depending on the preexisting distribution of wealth and other social parameters concerned with justice and equity, for the market-derived level of occupational risk to be *socially* optimal (not just allocatively efficient), the moral and distributional setting, which transcends or precedes the market, must be socially acceptable.[c]

2.   Performance of the Job Market: The U.S. Experience

Until the beginning of the twentieth century, the magnitude of occupational risk in the United States was determined almost entirely by the unconstrained workings of the private job market. Unfortunately, its performance fell far short of the accomplishments ascribed to economic theory's ideally functioning market.

[c]See the later discussion on cost-benefit analysis, where the failure to meet this condition renders the use of economic criteria questionable as a decision rule.

Rapid industrialization in this country was accompanied by a noticeable increase in workplace injuries. Despite extremely hazardous working conditions in many industries, employers, with rare exception, failed to introduce even the most rudimentary health and safety measures.[d] Nor were workers compensated for the job risks they encountered. Wages for hazardous work were largely undifferentiated from wages for less hazardous but otherwise equivalent work, and employers generally offered no wage-loss, medical, or impairment benefits to workers sustaining job-related injuries (United States Office of Technology Assessment 1985, pp. 207–208).

A worker could receive compensation for job-related injuries through the courts, but only by showing that the employer had been negligent and was therefore liable for the resultant damages at common law. However, negligence was often hard to prove, both because most of the relevant evidence was in the possession of the employer and because the conduct of the particular employer did not differ substantially from the accepted (albeit unsafe) practice within the industry. Further, employers could invoke three powerful defenses to worker claims: contributory negligence, the fellow-servant doctrine, and assumption of risk. Contributory negligence allowed the employer to escape liability if the worker's own negligence, however minor, contributed to the injury. The fellow-servant rule prevented the worker from collecting for damages if a fellow worker's negligence contributed to the injury. Assumption of risk denied recovery for damages if the worker was aware of the occupational risks and voluntarily accepted them. Initially, the courts interpreted these defenses so liberally that workers were rarely able to obtain a successful judgment. At the turn of the century, however, more courts began to find for the worker.

Given the historic failure of the private job market to control or to provide compensation for workplace hazards and given industry interest in avoiding increased court awards for workers, states began introducing workers' compensation legislation. Workers' compensation programs are essentially a type of no-fault insurance, providing wage-loss and medical coverage to workers suffering job-related injuries.[e] The cost of the workers' compensation system is borne by employers in the form of the insurance premiums they pay. In return for employers' acceptance of liability for the workers' injuries—through workers' compensation—regardless of fault, workers' compensation statutes bar workers from suing their employers through common law, except in cases of gross negligence or intentional harm. Although originally limited to workplace injuries, workers' compensation has, over time, been expanded somewhat to cover occupational diseases as well.

Other than workers' compensation, itself a quasi-market insurance system, the job market operated with limited government intervention from 1910 to 1970, when the Occupational Safety and Health Act was passed. (Some federal health and safety regulations existed prior to 1970, but they applied to only a few categories of workers— e.g., those working under federal contracts and miners. State occupational health and safety regulations provided more pervasive coverage, but most tended to be weak and poorly enforced.) How did the private job market perform?

After a downward trend for some decades, the occupational injury rate began rising in the mid-1950s. By 1970, annual deaths from workplace injuries were estimated to exceed 14,000, with another 2.2 million workers suffering disabling injuries (United States National Safety Council 1972). Data on occupational diseases were far

[d]Evidence concerning work conditions in America at the turn of the century is provided in Chapter 1.
[e]A more detailed discussion of workers' compensation programs is provided in Chapter 9.

less reliable, but even more disturbing. The most commonly cited figures were 390,000 cases of occupational illness annually, resulting in 100,000 fatalities each year, but the actual number of cases and deaths from occupational disease might reasonably have been half or twice those amounts (Ashford 1976, p. 93). Regardless, the magnitude of occupational risks resulting from marketplace decisions was unnecessarily high and, by almost any standard, far exceeded the social optimum. Furthermore, despite workers' compensation and tentative evidence that workers received a (modest) wage premium for the risks they assumed, worker benefits were not nearly commensurate with workplace hazards, especially those involving risks of occupational disease (Ashford and Stone 1988, pp. 17–29).

C. MARKET IMPERFECTIONS AS THE BASIS FOR GOVERNMENT INTERVENTION

1. Imperfections in the Job Market

What accounts for the job market's disappointing performance? Recall that the efficient and socially optimal functioning of the private market relies, according to economic theory, on four conditions: full information, lack of externalities, perfect competition, and a just and equitable social setting. The failure of the job market to produce socially desirable outcomes reflects the violation of those conditions. Let us examine the imperfections in the job market one by one.

(a) Imperfect information

The economist's model of an idealized market system assumes that the participants are fully informed of their options, or able to acquire needed information without cost. Applied to the job market, this means that workers and employers must know the consequences of any decisions they might make, including the associated workplace risks, in order to negotiate to best advantage. Workers unaware of job hazards would not seek compensation for the risks they bear, and as a result, employers would have insufficient incentives to invest in safer working conditions. Similarly, if employers overestimate the cost of eliminating workplace hazards, then the level of occupational risk to which workers are exposed would exceed the social optimum.

As a practical matter, the attainment of full risk information consists of a three-stage process: the information must exist or be created, it must be obtained by the affected parties, and they must be able to understand the information they receive. In the job market, each stage of the information process is flawed.

Regarding the first stage, market-generated information relating working conditions and exposures to health and safety outcomes is relatively poor. Individual employers or workers would incur substantial costs in generating and analyzing data about occupational risks, but could recapture only a minute fraction of the benefits such knowledge would provide. A private enterprise involved in creating and selling occupational risk information would encounter similar problems in appropriating benefits because of the "public good" nature of information. Once a worker purchased risk information, it would be difficult to prevent that worker from revealing the information to other workers.

The second stage of the information process, acquiring information about occupational risks, also functions poorly. Employers that discover their own or industry-wide workplace hazards have little economic incentive to share that knowledge with their workers, who could be expected to demand additional compensation from them in light of the risk information. Similarly, firms that develop cost-effective methods of improving workplace health and safety have little incentive to share such (unpatentable) information with their competitors.

The third stage of the information process concerns the question of whether employers and workers are able to understand the risk information they receive. Both experimental and market-related evidence suggests that individuals have considerable difficulty in processing information about low-probability, high-consequence events—such as occupational risks—in a rational manner. For example, individuals systematically underestimate less-publicized risks and risks over which they exercise some control (e.g., some 80 percent of drivers believe they drive more safely than the average driver), and for life-threatening risks, they are inexperienced in valuing, *a priori*, undesirable outcomes (Slovic, Fischhoff, and Lichtenstein 1981, pp. 589-598). In addition to these problems of risk perception, most individuals are unable to comprehend or rationally act on risk information when presented in the technical language preferred by most risk analysts—a 1/100,000 versus a 1/10,000 annual risk of worker death, for example.

Despite these defects in the information process, the private job market may function tolerably in response to many occupational *safety* hazards. Some working conditions are inherently unsafe and commonly recognized as such. For example, workers can reason that activities involving explosive materials are dangerous. Furthermore, workers develop some, albeit limited, knowledge of the workplace safety hazards confronting them from their own and their coworkers' on-the-job injury experience.

It is in the area of occupational *disease* that informational deficiencies most seriously compromise the performance of the private job market. Whereas the relationship between an occupational accident and the resultant injury is both obvious and immediate, the connection between the work environment and job-related disease generally is not. Most diseases have multiple potential causes and may be the result of synergistic effects, making it virtually impossible to determine whether an individual's disease is job-related rather than an "ordinary disease of life" resulting from lifestyle, genetic, physiological, or other nonoccupational environmental factors. This problem is compounded by the fact that there is frequently a long latency period—sometimes 20 years or more—between exposure to the occupational health hazard and the manifestation of the consequent disease. As a result, workers usually cannot rely on their knowledge or "intuition" to draw a connection between workplace conditions and a chronic disease, as would be the case for an acute injury. For example, would a worker attribute a heart attack to cigarette smoking, excess weight, hypertension, or genetic predisposition, or to occupational factors such as exposure to carbon disulfide, on-the-job noise, or other job-related stress? Placed within the job market's decision calculus, workers could hardly be expected to know that, by accepting a particular employment position or by agreeing to perform a particular task, they are increasing significantly the risk of heart attack 20 years later.

It is true that, on rare occasions, the cause of a disease is unique or nearly so. Examples of such "signature" diseases are mesothelioma and angiosarcoma, which are caused by exposure to asbestos and vinyl chloride, respectively. Even in these cases, however, workers may be unaware that they were exposed to the hazardous substance in question. In other cases, the hazardous substance may be sufficiently new on the market so that the deleterious health consequences have not yet been identified.

An additional area of imperfect information in the job market concerns the productivity losses associated with occupational disease. For example, most employers do not realize the extent to which respiratory disorders of occupational origin increase the rate of worker absenteeism or that occupational noise stress adversely affects the quality of workmanship and magnifies the risk of workplace accidents (Ashford 1976, pp. 337–338). As a result, employers usually underinvest in technologies to reduce such occupational hazards.

(b) Externalities

Externalities arise when the actions of economic agents impose costs or bestow benefits on other parties that are direct (not those caused indirectly by price adjustments) and that are not recognized in market transactions. The classic example of an externality is water pollution, generated by a firm's production process, which diminishes the welfare of individuals downstream, but for which they are not compensated by the firm. The presence of externalities undermines the efficiency of the market because, given the resulting divergence between social and private costs, the market imparts inaccurate signals to economic agents. In the water pollution example, the firm is able to treat water as a costless resource, but its private market production decisions impose social costs on the individuals downstream. As a consequence, the firm will produce beyond the socially optimal level and overuse the water resource.

Workplace health and safety hazards involve significant externalities since many of the costs of occupational injury and illness are borne not by those whose market decisions determine the level of occupational risk, but by the rest of society. Four distinct types of externalities affect occupational hazards: externalities within the workers' compensation system, externalities associated with limitations in workers' compensation coverage, supplier-induced externalities, and intrafirm intertemporal externalities.

The way in which workers' compensation is financed causes externalities within the system. Only the largest firms, representing about 15 percent of employees, are self-insured. The remaining firms are either class-rated, based on industry-wide safety and health performance, or experience-rated, which is a class rating adjusted by the individual firm's safety and health performance. For these firms, payments into workers' compensation are only loosely related to the costs their employees' injuries and diseases impose on the system. Such risk spreading mechanisms therefore dampen the incentives of these individual firms, representing some 85 percent of employees, to improve job health and safety.

The largest source of externalities from occupational hazards, however, arises from the injury and illness costs not covered by workers' compensation. In principle, workers' compensation benefits are supposed to reimburse medical expenses and to replace approximately two-thirds of lost income for ill or injured workers. In one sense, these benefits would appear to exceed their nominal value, from the workers' viewpoint, because of their tax-exempt status and the security they provide (although these features do not affect employer incentives to reduce workplace hazards). In another sense, these workers' compensation benefits are overstated because income benefits do not include adjustments for cost-of-living or normal promotions, because of constraints on maximum benefit levels and duration, and because death benefits are extremely limited. But these qualifications are dwarfed by the fact that, in practice, a worker suffering from an occupational disease is rarely awarded any workers' compensation at all because the worker's disease is usually not recognized as job related.

In order to qualify for workers' compensation, workers must demonstrate by a "preponderance of evidence" (that it is more likely than not) that they were exposed to a workplace hazard and that such exposure caused (or significantly contributed to) their disease (i.e., "arose out of and in the course of employment").[f] Because of the informational deficiencies, identified earlier, concerning the relationship between occupational hazards and disease, it is often impossible to meet this test, especially for diseases with multiple etiologies. Even if a causal relationship between a hazardous exposure and a disease is known, the worker's history of exposure may be unknown, since exposure frequently occurs a decade or more before disease symptoms appear. (Furthermore, workers' compensation statutes in some states exclude "ordinary diseases of life" from benefits entitlement, impose a restrictive statute of limitations, or require that workers be employed at the firm where the disease occurred for a minimum period of time—on average, eight years—even if the usual gestation period for the disease is shorter)[g] (Viscusi 1988, pp. 168–169). The net effect is that fewer than 5 percent of (severely disabled) workers who believe their disease is job-related receive workers' compensation benefits (and many workers are themselves unaware their disease is or may be due to exposure to occupational hazards). (United States Department of Labor 1980, p. 3, and Viscusi 1988, p. 169).

As indicated above, workers' compensation benefits cover many, but not all, of the costs of occupational injury and almost none of the costs of occupational disease. Only a portion of the residual costs (those not covered by workers' compensation) is borne by the incapacitated worker or the employer. Victims of occupational illness and injury and their families receive life insurance benefits, health care, rehabilitation, retraining, and direct income maintenance, which are predominantly paid for by society through Social Security and social welfare programs. In addition, the entire medical care system is heavily subsidized by government so that part of the cost of treating an injured or ill worker is paid for by the rest of society (Nichols and Zeckhauser 1977, pp. 44–45). Finally, there are intangible, but real psychic losses that befall the family and friends of an ill or injured worker. Since much of the residual cost of occupational injury and illness is not internalized, the job market negotiations of employers and workers produce an inefficiently high level of occupational risk.

The third type of externality associated with occupational risks—supplier-induced externalities—occurs when (1) a product used in the workplace, such as a piece of equipment or a chemical substance, causes worker injury or disease, and (2) the manufacturer of the product does not bear the resultant social costs. In principle, such externalities need not arise: A worker with a supplier-related injury or disease may file a lawsuit against the manufacturer to recover damages (since the workers' compensation exclusion usually applies only to a worker's employer, not to third parties). To be successful, the worker must show, again by a preponderence of evidence, that the product was defective, that the defect caused the injury or illness, and that the manu-

---

[f]For a more detailed discussion, see Chapter 9.
[g]See Chapter 9. However, in many states, some of these strict requirements have been removed or relaxed in recent years.

facturer was responsible (under negligence or strict liability principles) for the defect.[h] In the case of a supplier-related *disease,* however, the same informational deficiencies that make it difficult for a worker to obtain workers' compensation benefits also make it difficult to recover damages through the tort system. Most diseases have multiple etiologies, so the worker's illness cannot be readily attributed to a single hazardous product. Also, because of the disease's long latency period, it usually cannot be ascertained with any degree of certainty whether or to what extent the worker was exposed to the hazardous product in question or, when the hazardous product is a generic chemical or material, the identity of the manufacturer of the specific substance to which the worker was exposed.

In practice, only in the relatively rare case of a signature disease is the worker likely to prevail in a third-party tort suit. Even then, the worker may not be able to collect. During the long latency period between exposure and manifestation of the disease, the manufacturer may have gone out of business. If still in business, the firm may not have retained sufficient funds to pay damages, particularly if the number of worker claims is large.[i] As a result, the monetary and psychic costs of diseases caused by hazardous products in the workplace are borne almost entirely by the worker, the worker's family, and the rest of society rather than by the manufacturer of the product. These externalities result in socially excessive workplace health hazards.

Even if the preceding three types of externalities did not exist, a fourth type—related to management objectives—could affect the level of occupational risk. Most American workers are employed by corporations, a form of business in which ownership and management do not ordinarily coincide. In many American corporations today, both top and middle management believe they must demonstrate their effectiveness immediately or risk being replaced. Since, for many stockholders, the most visible evidence of management performance is corporate profit, management objectives are frequently dominated by attempts to maximize short-run corporate profits. The problem with such management behavior is that profitable corporate investments will be rejected whenever the investment returns are not immediate. The reason is that only the costs of the investment will appear in the short-run, creating a detrimental effect on short-run corporate profitability. The implication for occupational hazards is that management, operating myopically, will tolerate uneconomic workplace risks if the costs to the firm do not appear in the short-run. In the case of most occupational health hazards, this situation arises because of the long latency periods between exposure and manifestation of disease. In effect, present corporate management will attempt to externalize the costs of occupational health hazards to future corporate management. The obvious consequence of such intrafirm, intertemporal externalities in the private job market is an excessive amount of workplace disease.

(c) Imperfect competition

The economist's idealized market system is predicated on a model of perfect competition. That is, the market for each commodity is assumed to contain such a large number of buyers and such a large number of sellers that no individual economic agent is able,

[h]See Chapter 9 for a discussion of worker tort remedies against third parties.
[i]A prominent example is asbestos manufacturers, who are confronted with over 100,000 potential lawsuits representing a potential liability in the tens of billions of dollars. As a result, at least one of them, the Manville Corporation, has sought protection under Chapter 11 of the federal bankruptcy laws (Viscusi 1988, p. 179).

through his or her actions, to influence the price of the commodity. Each buyer and seller therefore treats prices as given and makes maximizing decisions by comparing marginal gains or losses against the corresponding market price. As a result, the market transmits accurate information about tastes and technology in the form of prices, and prices act as signals in guiding the market decisions of economic agents. The allocative efficiency of the decentralized market system relies on these conditions.

Presumably because of the large number of buyers and sellers of labor services, the job market is frequently considered representative of the economist's conception of a perfectly competitive market. In reality, the job market is not one market, but many markets differentiated by occupation, geography, and other factors. To a greater or lesser degree, these job markets violate the conditions necessary for perfect competition, and wages, the market price for labor services, generally do not adequately reflect the market value of labor services to employers.

Some job markets contain only a few employers. An extreme example, but hardly a unique one, would be a small Appalachian town whose inhabitants are geographically immobile and whose sole employer is a coal mining company. Another example would be the job market for a specific occupation, such as astronaut, in which the government is the only employer (although, at least in this case, potential astronauts can seek employment in related professional markets). In such job markets, each employer has an appreciable influence on market-clearing wages, and if the sole employer, determines wages. The profit-maximizing strategy of these employers results in workers being paid less than their value to the firm. The fewer employment alternatives workers have (and, in economics jargon, "the more inelastic is the supply of labor"), the larger will be the disparity between the workers' wages and their job value. Wages, including risk premiums, in these markets will fail to impart accurate information, so that employers, comparing the costs of reducing workplace hazards to the risk premiums they pay, will underinvest—in relation to the social optimum—in worker health and safety.

A more pervasive source of imperfect competition in job markets is related to the fact that workers are not indifferent between continuing to perform at their present job and being fired, since contrary to the model of perfect competition, they cannot costlessly secure a similar job at the same wage with another employer. Intrafirm promotion and training opportunities, pension rights, and wages often depend on job tenure, making the opportunity cost of voluntary job changes prohibitively high. Employers derive market power from the fact that a portion of the compensation they provide to their current workers is not transferable to other jobs. In addition, job loss can be one of the most devastating psychological events in a person's life. Because jobs typically embody an indivisible bundle of features, many workers are willing to tolerate an undesirably high level of workplace risk rather than jeopardize their job security. As a consequence, the market-derived level of workplace hazard will be socially excessive.

(d) Market-transmitted injustices or inequities

Market transactions do not take place in a vacuum. They occur in a social setting with a preexisting distribution of wealth and a specified set of individual rights and obligations. Market transactions reflect the preexisting distribution of wealth and are constrained by individual rights and obligations. Therefore, market outcomes will vary according to the prevailing social conditions.

The economist's idealized market system allocates resources efficiently; in an effi-

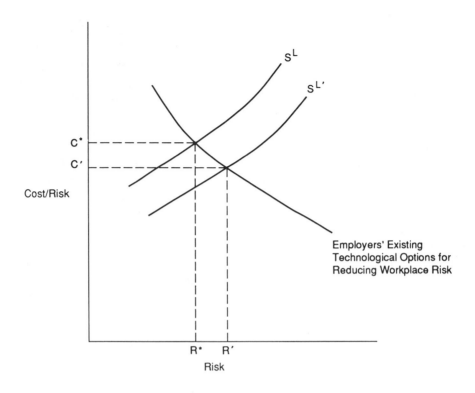

Figure 5-2. The wealth effect on worker decisions to accept hazardous employment.

cient market, it is impossible to *reallocate* resources in a way that makes someone better off without making someone else worse off. Economists refer to such an allocation of resources as "Pareto optimal" (or, more accurately, "Pareto efficient").[j] However, as indicated above, the Pareto optimal allocation of resources generated by the market depends on the preexisting distribution of wealth. The wealth effect on worker decisions to accept hazardous employment can be seen in Figure 5-2. Curve S_L' represents the supply of labor as a function of job risk for the sector of society that possesses the fewest financial resources. Curve S_L represents the supply of labor as a function of job risk for the remainder of society (with at least some minimum endowment of wealth). Since poor people, on average, are willing to accept hazardous employment for lower compensation than wealthier individuals would demand, Curve S_L' is below Curve S_L for every given level of risk. In equilibrium, the poor will perform more hazardous work (R' versus R*) for less compensation (C' versus C*) than do wealthier individuals.

If the initial endowment of wealth were distributed in a socially undesirable manner, then the resulting market outcome would, in all likelihood, not be socially optimal. The poor would sustain too many injuries and illnesses. A socially preferred outcome could, as a rule, be achieved by some more equitable, nonmarket reallocation of resources, even though there would be associated efficiency losses.[k]

[j]Pareto optimality is named after the Italian economist and sociologist, Vilfredo Pareto, who first enunciated it as a criterion for measuring improvements in social welfare.
[k]In economic terms, unless a market outcome is socially optimal as well as Pareto optimal, then non-Pareto optimal allocations exist that are socially preferred (Mishan 1981, pp. 345–353).

In addition, some individual actions are circumscribed by rights and duties that take precedence over market opportunities. Market transactions in such circumstances may be socially unacceptable on ethical grounds, regardless of their voluntary nature. For example, we do not permit individuals to sell themselves into slavery or to bring their children into factories to assist them in earning wages for the family. Similarly, the right to vote is a privilege and criminal penalties are sanctions that cannot be transferred to others, through the market or otherwise.

The preceding considerations about social justice and equity, applied to the job market, challenge not only the optimality but potentially the legitimacy of the market-determined level of workplace risk. For example, a socially unacceptable distribution of wealth could create a large class of impoverished individuals whose financial desperation makes them willing to perform highly dangerous work. In that case, reducing workplace risk towards the level that would obtain under a more equitable distribution of wealth would be socially preferable to the market-derived level of risk. Furthermore, permitting financially desperate individuals to exchange their health and safety for money would be morally unacceptable if the source of the financial desperation were itself deemed unjust. This would be so even if the financially desperate individuals consented to and benefited from the transactions; these factors are superseded by the morality of the economic motivation (in the same sense that "voluntary" blackmail transactions are, nonetheless, morally unacceptable).[1]

2. Types of Government Intervention

The preceding discussion demonstrates that the conditions necessary for the private market to produce socially optimal outcomes are violated in the job market. The various defects in the unfettered job market—imperfect information, externalities, imperfect competition, and potential inequities—result in a socially excessive level of workplace risk and create a need for government intervention to correct the market failure.[m]

For some individuals, government intervention to promote social health and safety objectives is perceived as taking on only one form: the imposition, by the applicable regulatory authority, of mandatory health and safety standards. In reality, the range of policy instruments through which government intervention can be effected is considerably more extensive than mandatory standards. Other possible policy instruments include research, education and training, economic charges, financial and tax incentives, insurance (including workers' compensation), and liability rules.

The various policy instruments are not necessarily mutually exclusive or independent of one another; they may interact in complex ways to promote workplace health and safety. Government research to identify occupational hazards provides a useful example. For many workplace chemicals and materials, the current state of scientific and medical knowledge is inadequate to establish the association between exposure

[1]Note, however, that we seem to have violated this ethical posture in exempting migrant farm workers from minimum wage requirements and allowing their children to work along with them on the farms.

[m]However, the role of government intervention is not necessarily restricted to correcting perceived market failures. As Paul Weaver argues, "The real purpose of government regulation is not to correct the deficiencies of markets, but to transcend markets altogether—which is to say, government regulation is not economic policy but social policy." (Weaver 1978, p. 56).

and disease. Such lack of knowledge was previously determined to be a major cause of market failure, both directly as imperfect information and indirectly as a contributor to job market externalities (since recovery from workers' compensation or third-party tort suits is precluded for diseases of unknown origin). In addition, however, the formulation of any rational government policy, involving specific workplace chemicals and materials, depends on information about whether, and at what levels, exposure constitutes a risk to workers. Therefore, research to identify occupational hazards would seem to be a prerequisite for any other type of government intervention.

Government policies to reduce workplace hazards might be categorized according to the market defects they attempt to remedy. Consider, for instance, *imperfect information* in the job market. Government-conducted or government-sponsored research in the fields of toxicology, epidemiology, and occupational medicine is only one of several policy tools whose purpose is to address imperfect information. Another is government research to improve the knowledge base concerning the prevention of work-related injury and disease. Several other policy instruments attempt to rectify information-dissemination deficiencies in the job market through right-to-know legislation and regulation.[n] Other examples of information-enhancing interventions include government-supported training and education of occupational physicians and nurses, industrial hygienists, safety engineers, business managers, and other professionals concerned with occupational health and safety. Faculty development materials for vocational education teachers, high school science teachers, unions, and others involved with training workers and future workers, serve to instruct them—the educators—in the recognition and control of workplace hazards. Providing workers with the information they need through direct worker training and education in hazard recognition and control and notifying workers of their legal procedural rights to workplace health and safety are also ways of disseminating information. Another type of intervention includes government-conducted or government-funded demonstration projects to inform firms of available hazard-reducing technologies.

A diverse set of policy instruments—in addition to the indirect contribution of improved information—might be deployed to remove job-market *externalities* created by workplace hazards. In theory, the most effective method for internalizing the costs of workplace hazards would be to charge employers an "injury tax" for each work-related injury (Smith 1974 and 1976) and an "illness tax" for each work-related disease that their workers sustain. In this case the "tax"—actually, the economic charge—would be set to equal the costs borne by society (including uncompensated costs imposed on workers) as a result of the specific type of injury or disease. The advantage of this approach is that the charge is directly related to the undesirable outcome (the injury or disease itself). In practice, the problem is that occupational diseases, which account for the vast majority of workplace externalities, often manifest themselves only after lengthy latency periods. Therefore, in addition to the obstacles in distinguishing job-related diseases from nonoccupational ones, there would be considerable difficulty in determining which prior employer was responsible for the disease. A more practical economic charge might be based on the amount of hazardous material present in the workplace, or better yet, on the level of worker exposure to hazardous substances; but these alternatives would create significant inspection and monitoring problems of their own.

Other ways of removing externalities caused by workplace hazards include modi-

---

[n]See Chapter 7 for a detailed discussion of information transfer laws.

fying the workers' compensation system and the tort system so that employers and suppliers responsible for workplace injuries and disease pay a larger share of the associated costs. For example, the workers' compensation system might increase the degree of experience-rating and institute employer deductibles or copayments, and both workers' compensation and the tort system might remove institutional restrictions or relax evidentiary requirements to establish the work-relatedness of the individual's illness. An additional policy instrument is the imposition of financial responsibility requirements, which ensure that suppliers of dangerous machinery or materials possess the financial assets to compensate workers for the damages their products may cause. Finally, rather than removing externalities, the government may provide employers with positive tax incentives or financial assistance to stimulate investment in occupational health and safety—thereby neutralizing employer disincentives occasioned by the externalities. These policy instruments include investment tax credits, accelerated depreciation, government loan programs, and direct subsidies or grants to firms making investments to reduce occupational hazards.

Government policies to address imperfect competition in the job market tend to be the identical policies that remedy social inequities transmitted through the job market. For example, government programs to train the unemployed or to retrain workers have the dual effect of improving job mobility—that is, increasing competition among employers in the job market—and enhancing the equity of (primarily) disadvantaged citizens. Similarly, programs to redistribute wealth to the poorest members of society also provide them the economic resources to acquire job skills or to move to areas where job opportunities are more favorable—that is, these programs expand their occupational and geographic mobility.

In contrast to mandatory health and safety standards, the aforementioned policy instruments attempt to correct, to the extent possible, the functioning of the job market rather than to supplant the market process. Maintaining market-like private incentives may offer significant efficiency gains in a decisionmaking arena as potentially complex and dynamic as the workplace. Whereas a well-functioning job market can provide the socially preferred level of workplace health and safety at lowest cost, mandatory standards (e.g., specification standards for safety equipment) may lock in uniform technological choices that, for many employers, are neither statically nor dynamically efficient.

Nevertheless, workplace health and safety standards may, in some cases, preserve much of the flexibility of the market. For example, the imposition of a hazard communication standard, which requires that dangerous substances in the workplace be clearly labeled (and that workers receive training concerning them), attempts to ensure that workers have access to job-risk information (and can understand it), but entrusts the determination of occupational health and safety outcomes to the workings of the market—a market no longer impaired by information-dissemination defects. (Other regulations limited to improving the development and dissemination of hazard and hazard-control information might include government requirements that each firm conduct a health and safety audit and that each firm have a worker-elected health and safety steward.) Similarly, performance standards specifying the maximum allowable level of worker exposure to designated hazardous substances allow the firm to select the low-cost means of compliance. Another mechanism is a legally imposed "general duty" that employers maintain a safe and healthful workplace, which leaves the technological details to the discretion of the firm.

Even so-called design or engineering standards, which specify the precise characteristics that the workplace or workplace equipment must meet, may contain flexible elements. The regulatory authority may make fine distinctions in applying design standards, for instance on the basis of firm size, new versus existing plants, or the nature of the production process. Similarly, the regulatory authority can target inspection rates and the size of penalties for regulatory violations according to policy-relevant attributes (e.g., the expected magnitude of health consequences associated with a specific violation). In addition, the firm may be permitted compliance exemptions on the basis of desirable market performance. An obvious example is the innovation waiver, which encourages firms to develop new technologies that are either more effective than the design standard, less costly, or both (Ashford, Ayers, and Stone 1985, pp. 443-462).

Finally, in some situations, the inflexibility of the design standard may be a desirable feature. For example, if market imperfections cannot be remedied by other forms of government intervention, then mandating a specific workplace technology might be socially preferable to outcomes influenced by private decisions, when such decisions are based on distorted market signals. The need for absolute control over a workplace hazard is even more obvious in the case of a highly toxic substance, because the cost (in the loss of life or disability) of faulty market decisions by individual firms and workers is just too large and unjust to be permitted.

D.   COST-BENEFIT ANALYSIS

The various types of government intervention identified above are, separately and in combination, capable of reducing the socially excessive level of workplace risk produced by the defective functioning of the private job market, but they will normally impose social costs as well. Government decisionmakers have the responsibility of identifying which—if any—public policies concerning workplace health and safety will, on balance, make society "better off" in some meaningful sense, and of selecting from among them the policy or mix of policies that will provide the greatest social improvement. The key underlying issue is what analytic technique is appropriate for evaluating social states before and after the enactment of alternative government policies.

During the past two decades, cost-benefit analysis (encompassing closely related economic approaches) has become the dominant method used by policymakers to evaluate government intervention in the areas of health and safety. As conceived in theory, cost-benefit analysis (1) enumerates *all* possible consequences, both positive and negative, that might arise in response to the implementation of a candidate government policy; (2) estimates the probability of each consequence occurring; (3) estimates the benefit or loss to society should each occur, *expressed in monetary terms;* (4) computes the *expected* social benefit or loss from each possible consequence by multiplying the amount of the associated benefit or loss by its probability of occurrence; and (5) computes the net *expected* social benefit or loss associated with the government policy by summing over the various possible consequences. The reference point for these calculations is the social state in the absence of the government policy, termed the "baseline."

The mechanics of cost-benefit analysis can be seen in Figure 5-3, which presents a relatively disaggregated matrix of the various positive and negative consequences of a government policy for a variety of actors. The consequences are separated into economic, health and safety, environmental, and other effects, and those affected are

Group \ Effect	Economic	Health/Safety	Environmental
Producers	$C_\$$		
Workers	$C_\$$	$B_{H/S}$	
Consumers	$C_\$$	$B_{H/S}$	
Others	$C_\$$	$B_{H/S}$	$B_{Env't}$

$C_\$$ refers to the monetary costs of reducing hazards that impair health and safety.

$B_{H/S}$ refers to benefits of reducing hazards that impair health or safety.

$B_{Env't}$ refers to environmental benefits resulting from efforts to improve health and safety.

Figure 5-3. Matrix of policy consequences for different actors.

organized into policy-relevent groups of actors, such as firms, workers, and consumers. At this point, the consequences are represented in their natural units: Economic effects are expressed in monetary units, health and safety effects are expressed in mortality and morbidity terms, and so on.[o]

Further, the consequences are described fully in terms of the times during which they occur. What cost-benefit analysis does is translate all of these consequences into "equivalent" monetary units, discount each to present value,[p] and aggregate them into a single dollar value intended to express the net social effect of the government policy. The matrix is essentially collapsed into a single monetary value.[q]

[o]Note that a single cell in the matrix could contain both policy benefits and policy costs. Employers, for example, incur the costs of complying with the policy, but also derive economic benefits in the form of productivity improvements and reduced risk premiums paid to workers. However, in order not to confuse monetary benefits with health or safety benefits, we describe the sum of the monetary costs and benefits as net costs $C_\$$.

[p]Even if inflationary effects are ignored, the value of a dollar—the bundle of goods it represents—changes over time. One reason is that a dollar today can be invested to earn interest over time. A dollar in a later period is worth less than a dollar today since the future dollar will not have accrued interest from earlier periods. In order to make the costs and benefits from different periods commensurate, as required by cost-benefit analysis, future costs and benefits are adjusted downward, using a specified discount rate, to derive their present value. Note that, since most government policies involve an investment of resources in early periods that generate benefits in later periods, the major effect of discounting is to reduce the magnitude of the benefits (the larger the discount rate the greater the reduction) and thereby to make government policies in general, and particularly those whose benefits are not realized until later periods, less attractive.

[q]The cost-benefit calculation can be expressed in simple mathematical terms by the following equation:

$$V = \sum_{i=1}^{n} \sum_{j=1}^{m} \frac{(B_{ij} - C_{ij})}{(1 + r)^i}$$

where B_{ij} and C_{ij} are the jth type of policy benefit and cost, respectively, in the ith year after the policy is introduced, and B and C are expressed in monetary units; r is the appropriate discount rate; and V is the (discounted) present value of the policy.

When there is only one policy option, cost-benefit analysis dictates that the option should be implemented only if its anticipated net social effect is positive. In general, however, numerous policies or sets of policies are possible, where each policy can be differentiated according to the various features—type of policy instrument, policy level or stringency, firms covered, and so on—that comprise it. In this situation, according to the cost-benefit criterion, the policy with the largest expected net social benefit, when compared to the baseline, should be implemented.

As a decisionmaking tool, cost-benefit analysis offers several compelling advantages. First, cost-benefit analysis clarifies choices among alternatives by evaluating consequences in a systematic and rational manner. Second, it professes to foster an open and fair policymaking process by making explicit the estimates of costs and benefits and the assumptions on which those estimates are based. Third, by expressing all of the gains and losses in monetary terms, discounted to their present value, cost-benefit analysis permits the total impact of a policy to be summarized using a common metric and represented by a single dollar amount. (In contrast, *cost-effectiveness* analysis—which attempts to find the minimum cost of achieving a target level of reduced injury and disease—and *health-effectiveness* analysis—which attempts to find the maximum reduction in injury and disease obtainable for a given cost—cannot be conducted using a single unit of measurement. As a result, the health and safety target or the budget constraint will be arbitrary—that is, it will have no uniquely defensible normative significance, although it may be socially endorsed through legislation—and the policies selected by these procedures may not be socially optimal.)[r]

As a practical matter, however, cost-benefit analysis possesses several serious limitations. The remainder of this chapter will be devoted to an exploration of these limitations, but three introductory comments are in order. First, cost-benefit analysis utilizes economic methods and economic terminology that are inappropriate or inaccurate measures of some types of policy effects. Second, examination of the shortcomings of cost-benefit analysis will frequently parallel the previous examination of defects in the private job market. This should hardly come as a surprise, since job market imperfections affect the market-derived estimates of the costs and benefits of government intervention. Third, the ensuing dissection of cost-benefit analysis is not intended to suggest a wholesale rejection of the technique, but to caution against the uncritical application of an imperfect methodology and the unqualified acceptance of its results.

1. Problems in Estimating Public Policy Benefits

The benefits of a specific government policy concerning occupational health and safety are generally the reduced social costs associated with a decrease in the number (or severity) of job-related injuries and illnesses, where the decrease is brought about by the policy in question. Prominent examples of policy benefits include reductions in medical expenses, productivity losses, physical disability, pain and suffering, and loss of life. Estimation of the policy benefits in cost-benefit analysis is a formidable task

---

[r]Cost-effectiveness or health-effectiveness is expressed as a benefit-to-cost ratio, $\frac{B}{C}$, which has the desirable feature of not requiring a monetization of health benefits. The dimensions of the ratio $\frac{B}{C}$ are, for example, disease prevention per dollar expended. However, problems of comparability still arise among health and safety benefits. For example, how does a policymaker trade off injuries of various types and fatalities?

because it is difficult to predict the reduced risk of injury and disease and to monetize the associated benefits.

There are many problems in trying to determine the effects of a government policy on the incidence of job-related injuries and disease. The baseline occupational risks may not be scientifically established. In most cases, the relationship between exposure and disease is simply not known. Estimating the effects of the policy on worker exposure levels may also be rather uncertain, depending as it does on assumptions about firm and worker behavior as well as on technical production relationships. (Even in the case of a design standard, the firm has the choice of not complying. The policy analyst must estimate the likelihood of this response not simply as a function of the expected regulatory fine, but also by taking into account the influence of regulatory "compulsion," the fact that noncompliance is a violation of the law and may involve loss of reputation because of the social stigma attached and perhaps criminal charges as well.)

Additionally, many of the benefits of government policy, such as reductions in physical disability, pain and suffering, and loss of life, have no clearly defined economic value (as compared to the market prices established for labor and medical services). The traditional methods of monetizing these benefits—surveys and market studies—have been, to a large extent, unsuccessful. Interviews and questionnaires asking individuals what they would be willing to pay for a stated reduction in risk have inherent limitations since answers to hypothetical questions have been shown to be poor indicators of a person's behavior.[s] Imputing the value of risk reduction from an individual's market behavior is also a seriously flawed approach (Fischer 1979). Individual actions are normally undertaken for a variety of reasons, and it is difficult to isolate what portion is motivated by a desire to reduce the risk of bodily impairment, pain and suffering, or a premature death. Furthermore, consumers are rarely well informed about the risks confronting them and have a well-documented history of being unable to process the risk information at their disposal in an expected manner.[t] As a result, the assumption of economic efficiency underlying attempts to value risks from consumer market decisions is untenable in practice.[u]

Where policy analysts have most frequently turned to derive the value of a reduction in risk is the job market itself. Recall that, according to economic theory, the risk-compensating wage premium (wage differential) represents the workers' valuation of job risk. But, the same job market imperfections that produce a socially excessive level of workplace risk and create a need for government intervention also undermine

---

[s]This is particularly so for decisions involving job hazards, when the mood and motive of actual choice are difficult to simulate. In drawing the subject's attention to a particular risk, a willingness-to-pay question may elicit a response that measures the immediate anxiety accompanying any contemplation of mortality or bodily harm rather than a representative value. The likelihood of an unrealistic answer is increased because the respondents have little incentive to determine or to reveal their true preferences (Ashford and Stone 1988, p. 18).

[t]The literature documenting risk perception problems is huge. (See, for example Tversky and Kahneman 1974, pp. 1124–1131; Machina 1987, pp. 141–147; and Fischhoff 1977, pp. 187–189).

[u]For example, attempts to impute the value of noise reductions from residential property values, using traffic and aircraft noise as explanatory variables, suffer from the obvious problem that most homeowners are not aware of life-threatening risks associated with noise stress. Therefore these risks will not be captured by residential property values. What such studies measure, assuming the various econometric difficulties can be overcome, is only the imputed value of noise-related annoyance and perhaps risk of hearing damage, but not the broader and more serious health consequences.

the usefulness of the risk premium as a measure of the worker's risk valuation. For example, job-related diseases that the worker does not know about will not be reflected in the wage premium for risk. Moreover, workers may have difficulty in understanding risk information. In theory, they are just as likely to overreact as underreact to hazard information, but in practice, worker risk perception appears to be dominated by an "it-can't-happen-to-me" syndrome. This results in known risks being understated and therefore undervalued.[v] Another job market defect, externalities, causes the observed wage premium for risk to measure only the *worker's* valuation of an incremental risk, but not the value family members, friends, and other interested parties attach to the risk. Furthermore, models of the risk-compensating wage differential assume a perfectly competitive job market; violation of this assumption means that the resulting estimates will "misinterpret" the true wage premium for risk. This is a particularly serious problem, since there may be no way to adjust the estimates to correct for the misspecification.

A further complication in monetizing policy benefits arises because most benefits (and some costs) occur years after the government policy is implemented. Since the value of money changes over time, these later-year benefits and costs must be discounted in order to translate them into a single, common monetary unit. In addition, however, changes in individual incomes and tastes over time can influence the value of future policy benefits and costs. Health benefits, in particular, have become more valuable relative to other goods over time and as the standard of living has improved.[w] Unfortunately, most cost-benefit studies perform discounting, but do not adjust benefits for the temporal appreciation in the value of health amenities. As a consequence, the effective discount rate for health benefits, used in these studies, will exceed its true value, and discounted benefits will therefore be undervalued. The net result is that socially desirable government policies may be rejected by the cost-benefit criterion if the excessively large discount rate causes calculated future benefits, discounted to present value, to be exceeded by discounted costs.[x]

[v]See Ashford (1976 p. 357), which also explores other attitudinal factors that cause workers to understate the risks they know.

[w]Recall that the present value of a benefit in year i was previously expressed as:

$$\frac{B_i}{(1 + r)^i}$$

However, if the benefit is a health amenity that increases in value at an annual rate of (e), then the proper mathematical expression for the benefit's present value is:

$$\frac{B_i (1 + e)^i}{(1 + r)^i}$$

For small values of (r) and (e), that expression is approximately equivalent to:

$$\frac{B_i}{(1 + r - e)^i}$$

Thus the "effective" discount rate is approximately $(r-e)$. In principle, if the social valuation of health benefits increases at a relatively rapid rate, the effective discount rate for benefits could even be negative.

[x]Using a benefit-cost ratio $\frac{B}{C}$ similarly suffers from an excessive discounting of benefits and yields an inappropriately low ratio.

2.   Problems in Estimating Public Policy Costs

Although the costs imposed by a government policy seem rather easy to identify and to express in economic terms, they are usually no more certain or reliable than the benefits. One reason is that policy analysts rarely have access to detailed, independent information about actual—and potential—production relationships and associated costs in an industry. Instead, they must depend to a large extent on industry-provided data to develop estimates of the costs to industry of complying with the public policy. Since higher compliance costs make a policy less attractive, industries adversely affected by the policy may choose to inflate their reported compliance costs.

In addition, compliance cost estimates often fail to take three significant factors into account: 1) economies of scale, which reflect the fact that an increase in the production of compliance technology often reduces unit costs; 2) the ability of industry to learn over time to comply more cost-effectively—what the management scientists refer to as the learning curve; and 3) compliance costs based on present technological capabilities ignore the crucial role played by technological innovation in reducing those costs.

To see the estimation bias caused by failing to predict correctly industry's techno-logical response—in the form of input substitution, process redesign, or product reformulation—consider Figure 5-4. Assume a firm's activities create a level of risk at point A prior to the introduction of a government policy, in this case a performance standard specifying the maximum level of allowable workplace risk (represented by the dotted line). The cheapest way for the firm to comply with the new standard, using existing technology, is to move to point B. Suppose, however, that in response to the regulation, the firm develops a process innovation that permits the firm to comply by

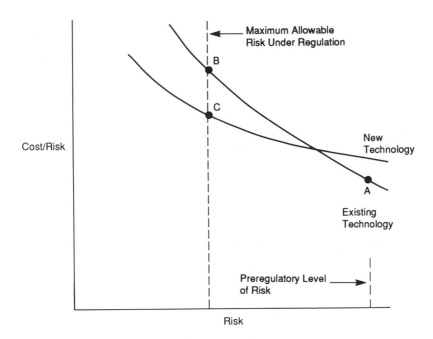

Figure 5-4. Comparison of costs of new and existing technologies for reducing workplace risks.

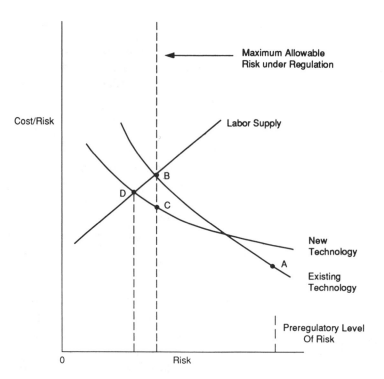

Figure 5-5. Comparison of equilibrium levels of risks as determined by new and existing technologies.

moving to point C. Ignoring the firm's innovative activity therefore results, in this example, in an overestimation of compliance costs by the amount BC.[y]

Furthermore, ignoring the possibility of a firm's innovative response may lead to the selection of an inferior level of job health and safety. Figure 5-5, which is simply a reproduction of Figure 5-4 with the supply curve for labor as a function of job risk added (as a conservative measure of the social cost of occupational risk), demonstrates the nature of the potential bias. Taking into account the effect of the new technology should lead not to equilibrium point C, but to the socially preferred point D, which represents an even larger reduction in workplace risk—and at lower cost than using existing technology. Moving from A to D extablishes a new "dynamic equilibrium," representing win-win situations for both management and labor.

Other errors of omission and faulty attribution plague attempts to estimate policy costs. A typical problem is firm activity that, although causally related to compliance efforts, is not necessary for compliance. The classic reason for this phenomenon is indivisibilities in investment decisions; it is usually less costly to make multiple changes in production simultaneously rather than individually. Hence, government intervention provides the stimulus to make other production improvements, not justified in

[y] The firm's technological response to a government policy is predictable, within bounds, in a given industrial context. The most important factor influencing the technological response appears to be the "stringency" of the policy, where stringency signifies either 1) that a significant reduction in workplace hazards is required, 2) that compliance using existing technology is costly, or 3) that compliance requires a significant technological response (Ashford, Ayers, and Stone 1985, pp. 421–443).

isolation, as well as the changes needed to comply with the new policy. In addition, costs ascribed to public policy are frequently offset by joint productivity improvements. For example, sealing reactor vessels during manufacture to limit worker exposure to hazardous substances offers the simultaneous advantage of increasing the production yield, since less material is lost during processing. Similarly, removing hazardous substances from the production process to reduce worker exposure also reduces the firm's costs of preventing environmental release and of disposing of the resultant hazardous residues. Finally, government intervention sometimes provides "leveraged" benefits by inducing firms to deal preventively with other, unregulated workplace hazards (Ashford 1981, pp. 131–132). These concomitant, but indirect, effects of public policy are often omitted in cost-benefit calculations.

3.   Problems of Equity and Ethics

Policy decisions regarding the level of workplace risk rely fundamentally on considerations of equity and concern for individual justice. Yet, traditional cost-benefit analysis is unable to address these issues or, relatedly, to assess whether implementation of a government policy truly constitutes an improvement in social welfare.

The economist's normative standard is Pareto optimality—achieving a situation in which it is impossible to make someone better off without making someone else worse off (because all mutually beneficial exchanges will already have occurred). The cost-benefit criterion closely resembles the test of Pareto optimality. If the net effects of a government policy are positive, then those who gain as a result of the policy *could* pay off those who lose and still have some benefits left over for themselves. *Potentially,* no one loses and at least some gain. But a potential Pareto improvement (referred to by economists as the Kaldor-Hicks criterion) is not the same as an actual Pareto improvement *unless* the redistribution of benefits to the losers actually takes place. The uncompensated distributional effects of a government policy vitiate the normative content of cost-benefit analysis as used to measure the policy's effects on social welfare.

For some types of government policy, of course, the distributional consequences are negligible; in such instances, the cost-benefit criterion may serve as a reasonable indicator of real Pareto improvements. However, many workers risk their health and safety, and frequently their lives, to perform tasks for which, because of market imperfections, they are inadequately compensated. Furthermore, the workers at risk are not a random cross-section of the general population; they represent a disproportionate share of the poor and unskilled, that sector of society with the smallest preexisting endowment of wealth. Thus, for government policies involving workplace hazards, the distributional effects are likely to be substantial, in which case cost-benefit analysis will be an unsatisfactory measure of Pareto improvements.

A related point is that, despite its appearance as an ethically neutral technique, cost-benefit analysis is highly value-laden. The observed market values used to represent the benefits and costs of public policy reflect the existing distribution of wealth and income, regardless of how inequitable or socially undesirable that distribution is. Therefore, determining the value of benefits and costs by market criteria is itself value-laden.

In addition, certain policy effects are not intrinsically economic, but concern more fundamental social attributes, such as individual rights and justice. The procedure of quantifying and monetizing these attributes, as part of a cost-benefit analysis, is more likely to obfuscate and misconstrue their essential qualities than to clarify their values. (For example, what monetary value should society place on protecting all workers equally, although it may be less costly and more cost-effective to protect workers in new plants more than workers in old plants?)

Whether, and to what extent, policies concerning workplace health and safety involve individual rights is assuredly a controversial matter. The central issue seems to revolve around whether workers voluntarily accept job risks (for which they are compensated). Clearly, "voluntary" and "involuntary" are not absolutes; these terms must be evaluated within a specific social context and relative to some operational standard. Nevertheless, voluntary transactions would seem to require some acceptable degree of information and the absence of coercion or desperation. Therefore, to the extent that workers have no desirable job opportunities because of their social background and education, are denied pertinent information about job hazards or are unable to assess risks rationally, and are unable to respond to acquired and understood risk information because of geographic or occupational immobility or other employment impediments, these workers cannot be said to have voluntary choices.[z]

Also, when policy effects are long term, involving intergenerational consequences, the preferences of future generations are not counted. Discounting their benefits simply reflects the way in which we have voted to represent our concern for them. In such situations, particularly when the rights of future generations to a minimum quality of life are at stake, the validity of the cost-benefit procedure might properly be called into question. (For example, would we be willing to discount our grandchildren's or their grandchildren's right to vote?)

4.   Other Misuses and Abuses

In addition to the inherent limitations of cost-benefit analysis already discussed, the use of cost-benefit analysis, in practice, has raised a host of other interrelated problems, among them suboptimization, quantification bias, "bottom-line" myopia, and the politicization of cost-benefit analysis as a decisionmaking tool.

Suboptimization has been defined as discovering the best way to do things that might better be left undone (Boulding 1974, p. 136). Cost-benefit analysis could fall prey to suboptimization for either of two reasons. First, the cost-benefit analysis might unwittingly neglect superior policy options. For instance, policies concerned with the design of personal protective equipment to reduce worker exposure to hazardous materials may ignore the possibility of redesigning the production process so as to remove the workplace hazard entirely. Second, cost-benefit analysis is an "instrumental technique" (Tribe 1973, p. 635), concerned not with selecting ultimate societal ends, but merely with helping to choose the best means to achieve those ends. A policy satisfying the cost-benefit criterion might still be socially undesirable because the policy objective itself is suboptimal. For example, a policy to limit worker risks in disposing of radioactive wastes from nuclear power plants overlooks the larger question of whether nuclear power production is a socially desirable activity. If not, then the best policy to limit worker exposure to radioactive wastes is not to create them in the first place.

Quantification bias refers to the tendency of many policy analysts to identify policy

[z]Using Caldart's (1986, p. 83) terminology their decisions are not autonomous ones, because autonomy requires the availability of a large variety of employment or sustenance options.

effects that are difficult to quantify or to express in monetary terms—such as distributional impacts or violations of individual rights—but to omit these effects entirely in the ensuing cost-benefit calculations. What this does, in essence, is to impose a value of zero on those policy effects for which no objective economic value is available, even though those effects may be of larger social significance than all of the monetized consequences combined. Another manifestation of the quantification bias is the valuation of certain policy effects on the basis of their readily available, but inappropriate, economic dimensions. The most obvious example is the long-time use of the sum of a person's discounted future earnings (the human capital approach), now discredited, to value lives saved due to government policy. Despite its superficial appeal, the human capital approach bears no relation to the social valuation of a reduction in life-threatening risks and, in fact, is consistent with the morally offensive proposition that the death of a no-longer-productive retired or handicapped person confers a net benefit onto society.

"Bottom-line" myopia concerns the double-edged property of cost-benefit analysis to be able to express the total (expected) effect of a policy in a single dollar amount. The summary cost-benefit value makes alternative government policies commensurate and facilitates comparisons among them, but it does so at a price. Collapsing the various consequences of a policy into a single, bottom-line value does not reveal the myriad assumptions and data on which that value is based; it compresses them and removes them from view. As a practical matter, policymakers and the public are apt to accept the summary value of cost-benefit analysis as gospel without considering the plausibility of the underlying assumptions and the accuracy of the supporting data, particularly when such information is not readily available. Similarly, the uncertainty surrounding the estimates of the costs and the benefits, the sensitivity of the expected policy impacts to specific assumptions and data, and the "confidence interval" that represents the range within which the true effect of the policy will probably fall are all matters to be considered in making policy decisions, but they are suppressed by the reduction of the cost-benefit details to a single bottom-line value. For example, even minute changes in the (often controversial) choice of discount rate may have a profound effect on the estimated impact of a policy. But unless this information accompanies the bottom-line value, the results of the cost-benefit analysis, for policy-making purposes, are liable to be misleading and misunderstood. In that case, rather than promoting rational, candid policymaking, cost-benefit analysis exposes the policymaking process to potential abuse.

The politicization of cost-benefit analysis is one of the ways in which the policymaking process can be abused. Political groups, in the hope of furthering ideological or special interests, may endeavor to influence either the cost-benefit estimates of policy impacts or how those estimates are used in policymaking. In the former case, policy analysts with a vested interest in the outcome might attempt to "construct" the cost-benefit analysis so as to arrive at a predetermined outcome. Note that unless the underlying assumptions and data are revealed and subject to public scrutiny, the manipulation of the estimation procedure is liable to go unnoticed. In the latter case, the abuse of the policymaking process is normally accomplished by imposing cost-benefit decision rules on the policymaker—thereby improperly circumventing government checks and balances, for example. Using authority to oversee regulatory impact analyses given to it through Presidential Executive Orders, the Office of Management and Budget

(OMB) has been able to reject regulations prior to publication, and to advise the agencies in their preparation of cost-benefit studies.[aa]

Using the cost-benefit criterion as a *decision rule* for policies whose estimated impacts have been influenced by politically motivated "guidance" subverts rather than promotes the democratic process. Democratic policymaking is threatened with replacement by political tactics whose purpose is to reorient legislative mandates to serve narrow interests. The policymaker's ability to make decisions is compromised and his or her accountability eroded. In short, as a decision rule, cost-benefit analysis no longer clarifies and facilitates the democratic process, but becomes a substitute for it.

5. Conclusions

Despite the various theoretical and practical shortcomings of cost-benefit analysis, its basic objective and approach—to help evaluate government policy by enumerating all of its consequences—are arguably desirable. Many of the admitted limitations of cost-benefit analysis are, in fact, unavoidable by-products of any such systematic approach to decisionmaking. For example, the imprecision of the cost-benefit estimates of policy impacts involving workplace hazards simply mirrors the technical uncertainties and the social complexities surrounding the problem. Obviously, more accurate estimates of impacts could be achieved by improved scientific methods and knowledge, but the same could be said for any policy evaluation technique, not just cost-benefit analysis.

The fact that an objective, unambiguously correct assessment of policy impacts cannot be guaranteed reinforces the importance of the process by which a particular assessment is performed. It is in this area—the evaluative process—that cost-benefit analysis is most vulnerable. The process of conducting a cost-benefit analysis forces the policy analyst to make explicit assumptions and data choices. However, in practice, collapsing the impact of policy choices into a single, bottom-line value has tended to conceal the underlying assumptions and supportive data from public examination.

One way to remedy this problem is for policy analysts to acknowledge the limitations of their craft and to provide policymakers and the public not merely the bottom-line expected value, but also a meaningful critique of the policy evaluation exercise, including uncertainties, confidence intervals, and the sensitivity of the results to specific assumptions and data choices. Even if this were to be done, the specter of political misuse and abuse of cost-benefit analysis would remain a viable threat.

A possible method for defusing the threat is to have the policy analyst calculate the various policy consequences, as was done in Figure 5-3, *without* translating the various economic, health, and other effects into a single dollar metric, without discounting them to present value, and without summing the benefits and costs

---

[aa]For example, OMB Circular A-94, introduced during the Nixon administration, directed most federal agencies to apply a 10 percent real rate of discount when calculating the present value of the costs and benefits of federal projects. The Office of Management and Budget continued to require use of a 10 percent real discount rate for regulatory impact analyses during the Reagan administration. The effects of such a large discount rate on future benefits is dramatic. For instance, after discounting at 10 percent per annum, 17 lives saved 30 years from now (perhaps from a reduction in occupational diseases with a long latency period) are valued at less than one life saved currently. For intergenerational hazards, such as damage to stratospheric ozone, the methodological consequences of applying a 10 percent discount rate are even more incredible. Saving a life today would be worth more than saving the entire current population of the world (some five billion people) 235 years from now.

accruing to actors in order to come up with a net benefit or a benefit-to-cost ratio. The consequences, when presented in disaggregated form, permit decisionmakers to examine the real policy trade-offs, guided by the social expression of preferences *provided in the law.*[bb] For example, the Clean Air Act provides special protection to certain social subgroups—such as children from exposure to lead; and the setting of standards under Section 6(b)(5) of the Occupational Health and Safety Act does not require the use of cost-benefit analysis. In these instances, Congress has already performed the social balancing; the collapsing of consequences by cost-benefit analysis may upset that balance. Of course, in some cases, a particular legislative mandate may be consistent with collapsing the various consequences into a single value, as is done in cost-benefit analysis; but in these cases, at least the application of the cost-benefit procedure has not been prematurely imposed.

Thus, "trade-off" analysis avoids unnecessarily obscuring the differences between noncommensurables, such as economic commodities, risks to life, and individual rights, or between those who benefit and those who suffer from the public policy (Ashford and Ayers 1985, pp. 882–884). Not only does this type of analysis expose to public scrutiny the policy analyst's disaggregated estimates, and the assumptions and data on which they are based, it also *forces the policymakers to comply with legislative mandates* and to make explicit their value judgments and trade-offs, thereby preventing them from abdicating responsibility for their decisions. In this way, instead of compromising Congressional intent, economic analysis can contribute to furthering legislative goals in occupational health and safety, environmental preservation, economic growth, and technological advance.

REFERENCES

Ashford, N. A. 1976. *Crisis in the Workplace: Occupational Disease and Injury.* Cambridge: MIT Press.

Ashford, N. A. 1981. Alternatives to Cost-Benefit Analysis in Regulatory Decisions. *Annals New York Academy of Science* 129–137.

Ashford, N. A. and C. Ayers. 1988. Policy Issues for Consideration in Transferring Technology to Developing Countries. *Ecology Law Quarterly.* 12:4: 871–906.

Ashford, N. A., C. Ayers, and R. F. Stone. 1985. Using Regulation to Change the Market for Innovation. *Harvard Environmental Law Review* 9:2:419–465.

Ashford, N. A. and R. F. Stone. 1988. Cost-Benefit Analysis in Environmental Decision-Making: Theoretical Considerations and Applications to Protection of Stratospheric Ozone, research supported by the Office of Policy Analysis and Review in the Office of Air and Radiation, U.S. Environmental Protection Agency.

Boulding, K. E. 1974. Fun and Games with the Gross National Product: The Role of Misleading Indicators in Social Policy, in *Environment and Society* (Roelofs et al., eds.). Englewood Cliffs: Prentice-Hall.

Caldart, C. C. 1986. Promises and Pitfalls of Workplace Right-to-Know. *Seminars in Occupational Medicine* 1:81–90.

Fischer, G. W. 1979. Willingness to Pay for Probabilistic Improvements in Functional Health Status: A Psychological Perspective," in *Health: What Is It Worth? Measures of Health Benefits* (Mushkin et al., eds.). New York: Pergamon Press.

[bb]The level of disaggregation in such a "trade-off" matrix is not uniquely correct; that depends on the particular social problem in question and what the associated policy-relevant variables are.

Fischoff, B. 1977. Cost Benefit Analysis and the Art of Motorcycle Maintenance. *Policy Sciences* 8:177–202.

Machina, M. J. 1987. Choice Under Uncertainty: Problems Solved and Unsolved. *Economic Perspectives* 1:1: 121–154.

Mishan, E. J. 1981. *Introduction to Normative Economics.* New York: Oxford University Press.

Nichols, A.L. and R. Zeckhauser. 1977. Government Comes to the Workplace: An Assessment of OSHA. *The Public Interest* 49:36–69.

Schrader-Frechette, K. S. 1985. *Science Policy, Ethics, and Economic Methodology.* Dordrecht: D. Reidel Publishing Company.

Slovic, P., B. Fischhoff, and S. Lichtenstein. 1981. Informing the Public about the Risks from Ionizing Radiation. *Health Physics* 41:4: 589–598.

Smith, R. S. 1974. The Feasibility of an "Injury Tax" Approach to Occupational Safety. *Law and Contemporary Problems* 38:4:730–744.

Smith, R. S. 1976. *The Occupational Safety and Health Act: Its Goals and Its Achievements.* Washington, D.C.: American Enterprise Institute.

Tribe, L. H. 1973. Technology Assessment and the Fourth Discontinuity. *Southern California Law Review* 46:3: 617–660.

Tversky, A. and D. Kahneman. 1974. Judgment Under Uncertainty: Heuristics and Biases. *Science* 185:458–468.

United States Department of Labor. 1989. *An Interim Report to the Congress on Occupational Diseases.* Washington, DC, 1980.

United States National Safety Council. 1972. *Accident Facts.*

United States Office of Technology Assessment. 1985. *Preventing Illness and Injury in the Workplace.* Washington, DC: U.S. Congress (OTA-H-256).

Viscusi, W. K. 1988. Liability for Occupational Accidents and Illnesses (Litan et al., eds.). *Liability: Perspectives and Policy.* Washington, DC: The Brookings Institution.

Weaver, P. H. 1978. Regulation, Social Policy, and Class Conflict. *The Public Interest,* 50:45–63.

Regulation of

Labor-Management

Relations Under the National

Labor Relations Act

The history of the labor movement in the United States has already been touched on in Chapter 1. The purpose of this chapter is to provide a general understanding of the legal framework governing union organizing and collective bargaining, especially as it pertains to issues of health, safety, and workplace technology. We look first at the overall system of rights and responsibilities that governs the American labor union and at the role played by the union in today's workplace. We will then explore the extent to which unions are permitted under the current system to participate with management in decisions affecting health, safety, and technological change. Throughout, we focus on the National Labor Relations Act (NLRA).

The NLRA covers the bulk of the workers in the private manufacturing sector. However, it does *not* cover:

- Agricultural employees (who are covered under some state labor laws)
- "Domestic" employees (servants, butlers, maids, and others employed in the home)
- Independent contractors (who are not considered "employees")
- Supervisors and other "managerial" employees (who are considered part of "management")
- Employees covered by the federal Railway Labor Act (which governs most private sector employees in the transportation industry, including airline pilots)
- Government employees

This chapter does not discuss the Railway Labor Act or the network of federal and state laws that governs public employees. Although there are many similarities—especially conceptual ones—between these laws and the NLRA, there are important differences as well. Before generalizing about the collective bargaining rights available to workers covered by these laws, one should consult the specific laws themselves.

A.   THE ROLE OF UNIONS IN THE UNITED STATES

Although one may think that the labor movement as an institution has been around since colonial times, this is not the case. The right to unionize came relatively late to the United States legal system. Although individual workers always had the right to bargain over wages, early efforts of workers to organize and bargain *collectively* were thwarted by the courts. Initially, many courts applied a criminal conspiracy concept to collective worker action. This concept fell into disfavor in the mid 1800s, after the Supreme Judicial Court of Massachusetts dismissed a criminal conspiracy indictment against seven union organizers. Courts then began to issue injunctions prohibiting strikes and other trade union activities under the Sherman Antitrust Act of 1890, which forbade combinations of capital in restraint of trade. A 1918 decision of the U.S. Supreme Court made union organizing even more difficult. In *Hitchman Coal and Coke Co. v. Mitchell*, 245 U.S. 229 (1918), the Supreme Court upheld the constitutionality of the so-called "yellow-dog" contract, thus allowing employers to condition employment on a worker's agreement not to join a union.

Outside the courtroom, however, public feeling in favor of the labor movement was strengthening. As criticism of the judiciary's antilabor stance grew, Congress was able to pass several important pieces of prolabor legislation. The Norris-LaGuardia Act of 1932 limited the injunction powers of the federal courts and made peaceful picketing and strikes virtually unenjoinable. The National Labor Relations Act (also known as the Wagner Act), passed in 1935, guaranteed workers the right to organize without employer interference, and imposed on employers the duty to bargain in good faith with the employees' representatives. The Walsh-Healy Act of 1936 set minimum standards of wages and hours for workers under federal contracts. Finally, the 1938 Fair Labor Standards Act set a minimum wage and forbade the employment of children in industries engaged in interstate commerce or producing goods for interstate commerce.

After World War II, the prolabor tide in Congress began to turn. The year 1947 saw the passage of the Taft Hartley Act, which introduced the concept of unfair labor practices by unions. Later, in response to charges of illegal and unethical practices by certain labor unions, Congress passed the Landrum-Griffin Labor-Management Reporting Disclosure Act of 1959. This Act opened the internal operations of unions to public scrutiny and broadened the class of union activities that would be treated as unfair labor practices.

Perhaps not surprisingly, this change in Congressional attitude has been followed by a gradual decline in the power and influence of unions. As discussed in Chapter 1, unionization has decreased substantially over the past 30 years. Indeed, some commentators have argued that the United States is witnessing a shift from unionized labor-management relations to a system of nonunion industrial relations.

The Transformation of American Industrial Relations

T. Kochan, H. Katz, and R. McKersie

At the same time that collective bargaining was developing and maturing, an alternate nonunion system of industrial relations was steadily emerging in the same companies and industries—first, slowly and quietly and then, by the 1970s, more rapidly and visibly. Between 1960 and 1980 the growth and diffusion of this nonunion human resource management system took place across a broad enough array

of industries and firms to cause major changes to be introduced in unionized relationships during the 1980s.

* * *

Perhaps the best way to make our point is to briefly summarize the magnitude and scope of the decline in unionization that has occurred in the private sector of the U.S. economy since 1960 . . . [Figure 6-1] summarizes its distribution across broad industry classifications.

It is not possible to provide a single precise measure of growth and decline in unionization since World War II because no single historical data series is available to track union membership changes over

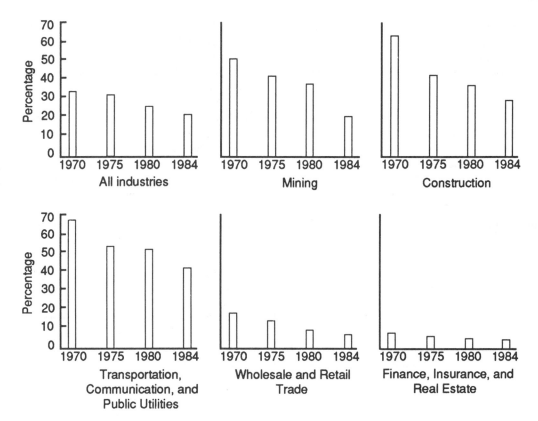

Figure 6-1. Union membership as a percentage of wage and salary workers in selected industries, 1970–1984. Source: Richard B. Freeman and James L. Medoff, "New Estimates of Private Sector Unionization in the United States," Industrial and Labor Relations 32 (January 1969): 143–74; and Larry T. Adams, "Changing Employment Patterns of Organized Workers," Monthly Labor Review, 107 (February 1985): 25–31.

this period. Piecing together several different data sources does, however, provide a reasonably clear pattern. Estimates obtained by the Bureau of Labor Statistics (BLS) from surveys of national unions indicate that the peak levels of union membership (measured as a percentage of the nonagricultural labor force organized) occurred in the mid-1950s, at around 35 percent. In 1960 approximately 31 percent of the nonagricultural labor force was unionized. Estimates shown in figure 6.1 suggest that this percentage declined slowly during the 1960s and then more rapidly in the 1970s and 1980s. By 1984 the BLS estimated union membership among employed wage and salaried workers had fallen to approximately 19 percent. The 1984 estimate most comparable to the earlier BLS measures obtained from surveys of national unions puts the figure slightly lower, at about 18 percent. The declines were particularly strong between 1980 and 1984 since the most highly unionized industries were hit hardest by the effects of the 1981–82 recession, the declining competitiveness of U.S. manufactured goods on world markets, and the economic and organizational restructuring underway within these industries.

While these overall numbers demonstrate the seriousness of the decline in unionization, they mask an even more important characteristic of the current pattern of unionization that was identified in the various case studies we conducted. Over the two decades of the 1960s and the 1970s, a significant nonunion sector emerged in virtually every situation that we examined. For some industries such as petroleum refining, where union membership had declined from approximately 90 percent of all blue-collar workers in the early 1960s to approximately 60 percent in the 1980s, the change is not surprising, given the tradition of independent unions, decentralized bargaining, and elaborate personnel policies that bear the imprint of dominant founders. The decline has also been dramatic, however, in three industries that have historically been viewed as strongholds of unionism: mining, construction, and trucking.

In underground bituminous coal mining, the nonunion sector has grown from virtually zero to approximately 20 percent of the industry. Major companies have opened nonunion mines, often in the heartland of United Mineworkers territory, such as Kentucky and West Virginia. In construction, the penetration of nonunion activity has occurred even in large commercial construction. During the 1950s and 1960s, open-shop construction companies moved into residential and light commercial construction but unionized companies continued to dominate the important, heavy construction sector. However, starting in the 1970s, the situation changed; increasingly, major office buildings and large industrial sites are being put up on a nonunion basis. Overall, between 1975 and 1984 nonunion activity has increased from approximately 50 percent to 75 percent of all construction work. Over-the-road trucking presents a similar picture. Since deregulation in 1980, the industry has experienced a sharp transformation. Approximately 20 percent of the large, unionized carriers went into bankruptcy and were replaced by smaller, independent, and generally nonunion companies.

At the company level, data from a number of highly visible firms further illustrate this general trend. In the early sixties, General Electric had thirty to forty nonunion plants. By the mid-eighties, this number roughly doubled as a result of the nonunion status of the new plants that went into operation in the 1960s and the 1970s. The drop in percentage of workers organized has not been as great as these figures suggest since the newer plants are smaller (often dedicated to a particular component) and usually have about five hundred to one thousand employees. Companies such as Monsanto and 3M have seen a drop in the representation rate for production and maintenance workers from approximately 80 percent to 40 percent or less. Similarly, some of the major companies in the auto parts field, such as Dana, Bendix, and Eaton, have developed sizable nonunion sectors. The most complete transformation has occurred at Eaton, where as of the early 1980s, the only unionized operation to remain is at its main facility in Cleveland. The other union plants have been closed down and all the new plants that have been opened have remained nonunion.

In short, the decline in union membership is both steep and broad.

Much of the legal framework for this emerging "nonunion system of industrial relations" is embodied in the laws discussed elsewhere in this text, such as the laws providing protection against discrimination and unjustifiable discharge discussed in Chapter 8. This raises a question: if these laws and other mechanisms truly are supplanting the NLRA as the dominant system of labor-management relations in this country, why is it still necessary to focus considerable attention on the NLRA and its collective bargaining process? There are at least two reasons. The first is that collective bargaining—even with all its inherent difficulties and contradictions—offers a substantial opportunity for labor and management to work together toward a creative redesign of workplace technologies. The second is that it is likely that the trend toward the development of a system of stronger protections for workers *outside* of the collective bargaining system has been made possible by the political and economic pressure created by unions and collective bargaining. Indeed, union membership has always been low in some industries. Union membership in the textile industry, for example, has long stood at around 1 percent. Nonetheless, union activities have always been important in establishing the legal precedents that have improved worker protection throughout the industry. Further, the unionized shops have set a standard for labor-management relations to which the nonunion firms have tended to adhere in order to avoid unionization. In short, if unions were allowed to whither and die, the worker protections within the nonunion sector might well die with them.

B. THE ESSENTIAL RIGHTS AND OBLIGATIONS

In general, the NLRA is concerned with the dynamic between labor (especially the labor union) and management, as well as between the labor union and the individual employee. (See the appendix at the end of this chapter for excerpts from the NLRA.) We deal here with the first of these dynamics. Labor's basic rights under the Act are set forth in Section 7, which authorizes union organizing and collective bargaining. Section 7 also provides that employees may engage in other concerted activities for mutual aid or protection. Under this section, employees have the right to *form a union* that is *independent* from management. Further, existing unions have the right to attempt to *unionize* a workplace that is presently without a union. In either case, the goal of the organizing effort is for the union to be recognized by the National Labor Relations Board as the *bargaining representative* for the workers—or for a particular category of workers—within a given workplace. Once it is so recognized, a union may begin the process of collective bargaining with the employer on behalf of those workers.

1. Unfair Labor Practices

The fundamental Section 7 rights are protected and limited by the provisions of Section 8. Section 8 describes a number of "unfair labor practices" that are prohibited by the Act. For example, Section 8(a)(1) makes it an unfair labor practice for an employer to "interfere with, restrain, or coerce employees in the exercise of the rights guaranteed in Section 7." That is, management may not act unfairly in seeking to prevent an independent union from being formed or recognized as the bargaining agent. Similarly, Section 8(b)(1) prohibits unions from taking action to "restrain or coerce" the exercise of Section 7 rights. That is, a union may not act unfairly in seeking to promote its recognition as the bargaining agent.

2. The Organizing Process

To understand the meaning of these protections and limitations, one needs to understand the labor organizing process under the NLRA. As noted, the goal of the union organizing process is recognition of the union as the bargaining representative for a definable group of workers within the workplace. This group is the prospective "bargaining unit." The end-point of the process is a National Labor Relations Board (NLRB)-certified election in which the affected workers indicate, via secret ballot, whether they choose to be represented by the union. Before the election, the employer and the union each wages a campaign to persuade the workers. In essence, the union argues that the workers would be better off with union representation; the employer argues that the workers have been and will be better off without it.

Section 8 of the NLRA is designed to protect the integrity of this election process and of the organizing effort that precedes it. Although both sides are permitted to wage a vigorous, hard-hitting campaign, neither may use false promises, threats, or other forms of misstatement or intimidation in an attempt to influence the workers. To do so is to commit an unfair labor practice and to risk sanction from the NLRB. This does not mean that unfair labor practices are rare occurrences. As discussed in Chapter 8, the relative leniency of the sanctions levied against employers for committing unfair labor practices, and the long delays that accompany the NLRB enforcement process, have stripped Section 8 of much of its deterrent effect. A partial solution to this for the disaffected union is what is known as a "Gissel bargaining order." If unfair labor practices by management undermine the fairness of an election, the NLRB may nonetheless order the employer to bargain with the union as if it had won the election. See *NLRB v. Gissel Packing Co.*, 395 U.S. 575 (1989). One commentator, however, has argued that the Gissel bargaining order is not a particularly useful device in practice.

Promises to Keep: Securing Workers' Rights to Self-organization under the NLRA

P. Weiler

The Gissel Bargaining Order.—If either the union or the employer violates the legal standards governing the representation campaign, the Board may delay or even rerun the election to ensure that the outcome is untainted by illegal intimidation. If the employer has committed unfair labor practices so serious that they render a fair and free election

Source: From 96 *Harv. L. Rev.* 1769, 1793–1795 (1983). Copyright © 1983 by the Harvard Law Review Association.

impossible, the Board may order the employer to bargain with the union even though the union has not won a secret ballot election. Approved by the Supreme Court in *NLRB v. Gissel Packing Co.*, the bargaining order is specifically designed to repair the harm done to the group rights promised by section 7 of the NLRA. At the same time, the prospect of such an order should be a significant deterrent to antiunion conduct, because an employer that violated the law in order to avoid unionization could end up with a union anyway, whereas an employer that campaigned vigorously within the law would stand a fair chance of winning the election and thereby excluding the union from its plant. *Gissel*-type orders, then, appear to promise the happiest of results: they would repair the damage done to the statutory rights of the employees while also providing a tangible disincentive to the type of behavior that produced the damage.

In practice, this promise has proved—and must inevitably prove—illusory. The vigorous debate about matters of high principle that still rages over the legitimacy of using bargaining orders in place of secret ballot elections is really beside the point. A bargaining order is an effort by an outside agency to construct a lasting collective bargaining relationship between a trade union and an employer whose antiunion behavior has been so egregious that the Board is prepared to bypass the normal secret ballot election. What can the union do with the bargaining order? Although the order requires the employer to sit down at the negotiating table and go through the motions of trying to reach an agreement, the governing principle of freedom of contract under the NLRA means that the employer is not required to consent to any significant changes in working conditions. The Board cannot direct the employer to make a reasonable contract offer.

If a decent employment package is to be extracted from a recalcitrant employer, it must come through the efforts of the workers themselves—that is, through the threat of strike action. Here lies the catch-22 of *Gissel*. The bargaining order has been issued because the employer's behavior is thought to have so thoroughly cowed the employees that they cannot express their true desires about collective bargaining even within the secrecy of the voting booth. But all the order can do is license the union to bring negotiations to the point at which its leadership must ask those same employees to put their jobs on the line by going on strike.

Again, timing is crucial. If a bargaining order is granted within a few weeks (or even months) of the organizing drive, while the attraction of collective bargaining remains strong among the employees, it might still be effective. But as time passes, employee interest wanes. Normal turnover will deprive the union of some key supporters, and to many of the replacements the union will seem a remote outsider that caused some trouble in the distant past. It is highly unlikely, therefore, that an order issued by the NLRB after protracted legal proceedings will actually produce a viable and enduring collective bargaining relationship.

◇ NOTE

1. Some unfair labor practices are more or less blatant. For example, if the employer fires all the union organizers in its workforce or threatens to decrease wages if the union is designated as the bargaining representative, a Section 8 violation will be relatively easy to prove. Similarly, if the union guarantees that wages and benefits will go up if the workers elect to unionize, it will be guilty of a Section 8 violation. Quite often, however, tactics are far more subtle. Assume, for example, that a union begins a concentrated organizing effort at a plant that employs 500 workers, and that plant is just one of seven owned and operated by the employer throughout the country. A year later, the company decides to do a "test automation implementation" at the plant. The employer announces a plan to install machines to do the majority of the work previously done by the workers. Along with this, the employer plans to discharge 350 workers, and replace them with 50 technical people who will run the machines. Does this action violate Section 8(a)(1) of the NLRA? The key is the employer's *purpose* is going to an automated plant in this fashion. If the primary purpose was to discourage union activity at its plants, it likely is an unfair labor practice. See *Textile Workers v. Darlington Mfg. Co.*, 380 U.S. 263 (1965).

C. THE COLLECTIVE BARGAINING PROCESS

Once a union has been recognized by the NLRB as the bargaining representative for a particular bargaining unit, it may then begin the process of collective bargaining with the employer.

1. General Overview

Past and Future Tendencies in American Labor Organizations

J. T. Dunlop

THE METHOD OF COLLECTIVE BARGAINING

The American labor movement regards collective bargaining as its major business. George Meany [former head of the AFL–CIO] has defined collective bargaining in the following perceptive way: "On its philosophical side, collective bargaining is a means of assuring justice and fair treatment. In the economic realm it is a means of prodding management to increase efficiency and output, and of placing upon trade unions great responsibility to limit their demands to practical realities. A failure to recognize the unique role of collective bargaining is a failure to understand the distinctive new nature of American private enterprise as it has evolved over the past seventy-five years."[10]

Collective bargaining, in common parlance, is used in this country to refer to at least three separate forms of labor-management activity: (a) periodic negotiations for a new collective agreement that may take place yearly or every two or three years depending on the duration of the agreement setting compensation and other terms and conditions of employment; (b) the day-to-day administration of the provisions of an existing agreement, including the vital steps in the grievance procedure and arbitration provisions; and (c) in some relationships, informal joint consultations and conferences that may take place outside agreement negotiations to explore common problems, to improve productivity, and to review broad questions of common interest in the industry or locality.

Source: Reprinted by permission of *Daedalus,* Journal of the American Academy of Arts and Sciences, "A New America?" Winter 1978, Vol. 107, No. 1, pp. 82–85, Cambridge, Massachusetts.
[10]George Meany, What Labor Means by "More," Fortune Magazine 92, 93, 172, 174, 177 (March 1955).

In the industrial-relations system of the United States the no-strike, no-lockout clause in [collective agreements] developed concomitantly with the provision for binding arbitration over disputes arising during the term of the agreement, or more narrowly, over disputes as to the meaning and application of the agreement. The British, in contrast, generally have not made the distinction between periodic negotiations and administration, between issues of interests and rights; and accordingly almost any issue may be raised at any time on the shop floor and pressed to a work stoppage.

The collective-bargaining arrangements in this country are peculiarly the product of American unions, workers, and managers in the setting of our institutions; more recently and to a small degree, they have been constricted by legislation. The U.S. labor movement did not arise with the Wagner Act, nor were the fundamental features of collective bargaining created by government. Collective bargaining is a highly decentralized and diversified accommodation to union structure, managerial independence, market and locality forces, and the size of our country. Moreover, collective-bargaining institutions have changed gradually with developments in these features. There are probably more than 200,000 separate collective agreements, and negotiations are designed to address the problems of each pair of parties, including their employees (members). As Thomas R. Donohue [Secretary of the AFL-CIO] has said, "Each of these agreements is of paramount importance to the workers involved—no matter whether it is Firestone Rubber or a small electrical shop, whether it's New York City or Kalispell, Montana."[11] The individual parties are not entirely free, of course,

[11]Thomas R. Donohue, The Future of Collective Bargaining, International Conference on Trends in Industrial and Labour Relations, Montreal, Canada, May 26, 1976, AFL-CIO Free Trade Union News 6 (September 1976).

to consider alone their preferences. The collective bargaining processes compel them to pay attention sooner or later to the practical realities of market competition, to industrial relations developments in closely related negotiations, and to legal constraints.

For an appreciation of the decentralized quality of American collective bargaining it would be well to compare the texts of agreements in a variety of sectors—airplane pilots, seamen, railroad engineers, basic steel workers, construction pipefitters, insurance agents, TVA employees. The methods of wage payment; fringe benefits; rules on hiring, transfers, promotions, and layoffs; grievance procedures; and arbitration—each one reflects the problems of technology and markets posed to workers and managements and their organizational histories. No government agency could ever promulgate rules of the workplace with such diversity or adjust them so well to changing conditions. Within some sectors collective bargaining is relatively centralized on a national basis, at least on key questions such a compensation, as is the case in the basic steel industry, railroads, and motor freight trucking. In other industries, negotiations may be conducted on a regional or locality basis as in maritime occupations, construction, and paper, whereas in still others agreement making is decentralized to the establishment level.

American collective bargaining is distinguished from European and Japanese industrial relations systems in that there is generally a single line of responsibility on the union side from the national level to the immediate workplace, local unions, and workers. In this country one agreement between management and the local or international union, or both, sets the full terms and conditions of employment. Elsewhere the works council, enterprise organization, or shop steward constitutes a more or less independent authority. Negotiations take place in several tiers at varying levels with wage drift and separate provisions as each level negotiates additional items.

Each agreement in the United States typically specifies a grievance procedure, or steps in which representatives of the two sides seek to resolve disputes or differences over the meaning or the application of the agreement that arise during its term. Agreements also tend to specify an arbitration process, including the selection of arbitrators or umpires; they may also provide that certain issues may be resolved by the controlled resort to strike or lockout, as in the case of production standards in the automobile industry. In this way the grievance procedure provides a way for responsible officers to review the operation of the agreement, to clarify the application of new provisions, and to consider questions of interpretation that had not been anticipated. The industrial jurisprudence that develops is shaped by the particular parties rather than by government agents; management supervision and workers and stewards participate in resolving most issues at the work level. In this process our industrial relations system has developed a large cadre of private arbitrators with wide understanding of plant-level issues.

One of the most significant—and often overlooked—effects of collective bargaining in the United States has been its influence on management generally as well as on unorganized or nonunion establishments, particularly of large size. Fifty years ago there were much sharper differences in wages, benefits, and conditions of work than now prevail in American industry between workplaces that are under collective bargaining and those that are not. In part to keep unions out and in part because multiplant companies may have to deal with employees in both unionized and unorganized plants, these differences have been narrowed, even with regard to procedures for discharge, discipline, or layoff. "The challenge that unions presented to management has, if viewed broadly, created superior and better-balanced management, even though some exceptions must be recognized. . . . If one single statement were sought to describe the effect of unions on policy making, it would be: they have encouraged investigation and reflection."[12]

The American collective bargaining arrangements purport to perform a wide range of vital functions in the society—it sets the vital rules of the workplace relating to hiring, discharge, transfer, layoff, retirement, and promotion; it establishes procedures to review these decisions under collective agreements with resort to third-party neutrals; it provides means for plant-level employees and supervisors to participate in many decisions; it provides means of communications by means of the enterprise hierarchy; it constitutes a means to resolve many industrial conflicts; and it permits periodic review of rules and bargaining arrangements in the light of new technological or market conditions. These are primarily activities of local and national unions.

[12]Sumner H. Slichter, James J. Healy, and E. Robert Livernash, The Impact of Collective Bargaining on Management 951, 952 (1960).

Although the basic features of collective bargaining have continued through the past several decades, several new features are to be noted. The scope of bargaining, in the sense of the subjects bargained about and incorporated into agreements, has grown. Health and welfare plans and pension arrangements have expanded to cover both more people and more situations. Some negotiations improve the retirement benefits of those already retired. Supplemental unemployment benefits and enhanced job security in numerous ways have been specified. Special training provisions and some modifications of the units in which seniority and rights are exercised have been developed to agree with the requirements of law respecting equal employment opportunities. The steelworkers and the basic steel industry imaginatively created an experimental agreement to avoid the costs of instability for both sides that grow out of a normal expiration date. A legal services benefit has been negotiated in a few situations. Collective agreements have grown more detailed and much longer. The greater complexity and technical nature of issues has introduced more experts and specialists on each side in some phases of the bargaining process. The negotiations process is continuing to concern itself with the greater intrusion of government into the employment relationship by means of statutes and regulations concerned with health and safety, pensions, affirmative-action programs, and the like.

But collective bargaining is not a cure-all of social and economic problems. It does not apply to a majority of workplaces, although its influence is widespread. It is not a principal means to eliminate poverty, to change the distribution of income, to reform the production process, to eliminate discrimination, or to reform the health-care delivery system, although it may make major contributions to all these purposes. Collective bargaining should not be demeaned because it is not an all-purpose social tool, but no one should prescribe for[workers]or the workplaces of America without knowledge and experience with collective bargaining and consultation with the parties.

2.   The Duty to Engage in Collective Bargaining

In practice, the concept of collective bargaining would have little meaning without a clear legal *duty* to bargain. Under the NLRA, both labor and management have a duty to "meet at reasonable times" and "confer in good faith" over so-called mandatory subjects of bargaining. This duty arises from Sections 8(a)(5) and 8(b)(3) of the Act, which specify that it is an unfair labor practice for either the employer or the union to "refuse to bargain collectively." The subjects over which they must bargain are defined (without amplification) in Section 8(d): "wages, hours, and other terms and conditions of employment." In addition, the parties are permitted to bargain over any *other* subjects (called "permissive" subjects of bargaining), so long as to do so would not violate some other law. The distinction between mandatory and permissive subjects of bargaining was enunciated by the Supreme Court in *NLRB v. Wooster Division of Borg-Warner Corp.*, 356 U.S. 312, 319 (1958). The provisions of Section 8, noted the court,

> . . . establish the obligation of the employer and representative of its employees to bargain with each other in good faith with respect to "wages, hours, and other terms and conditions of employment. . . . " The duty is limited to those subjects, and within that area neither party is legally obligated to yield. . . . As to other matters, however, each party is free to bargain or not to bargain, and to agree or not to agree.

Given the relative brevity of the Congressional language, it has fallen largely to the NLRB (and to the courts in reviewing NLRB decisions) to determine the range of subjects that falls within the reach of Section 8. The Board has taken up this task on a case-by-case basis. Over the years, many of the basic and integral aspects of the employment relationship have been designated mandatory subjects of bargaining. These include holidays, vacations, bonuses, discharges, pensions, profit sharing, work loads, work standards, insurance benefits, the union shop, subcontracting, shop rules, work schedules, rest periods, and merit increases. See generally, C. Morris *et al.,* The Developing Labor Law, BNA, Washington, D.C. (2d ed. 1971), p. 380. This is not an exhaustive list of all mandatory subjects of bargaining, but it is indicative of the kind of activity or decision that is likely to come within the ambit of the statutory definition.

The issue of whether the employer should provide an on-site daycare center for young children of employees would not be considered a mandatory subject of bargaining. It would, however, be a permissive subject. In general, permissive subjects of bargaining are all those that are *not* considered part of "wages, hours, or other terms and conditions of employment," so long as they are not otherwise illegal. A bargaining subject can be illegal because it violates some provision of the NLRA itself (such as a proposal to restrain trade in violation of Section 8(e)), or because it violates the Constitution or some other federal statute. For example, a male-dominated union could not lawfully meet with management to bargain over ways to discriminate against the company's women employees, because this would violate Title VII of the Civil Rights Act (see Chapter 8). If the subject does not involve some inherent illegality, however, the parties are free to bargain and to attempt to reach an accord. But neither party may use the failure to reach an accord on a permissive subject—or the failure of the other side to bargain over that subject—as a reason to refuse to bargain over a mandatory subject. Such a refusal would violate Section 8.

Although the NLRA requires the parties to engage in collective bargaining, it does not require them to come to agreement. The NLRA sets up a system within which bargaining can occur. It does not mandate any particular result at the bargaining table. Indeed, the parties are not required to reach *any* contractual result at all. As noted by the Fifth Circuit Court of Appeals in *NLRB v. Gulf Power Co.,* 384 F.2d 822, 823 (5th Cir. 1967), discussed below, labor and management must only make "a genuine effort to reach agreement if it is possible to do so. The Act does not compel agreement." (Citations omitted.)

◇ NOTES

1. In addition to the requirement to meet and confer in good faith, the NLRA's duty to bargain over mandatory subjects of bargaining incorporates three other important requirements:

(a) The requirement that each side meet with the specific representative chosen by the other side (so long as that person is authorized by the NLRB);

(b) The requirement that neither side take any *unilateral* action on matters regarding which it is required to bargain; and

(c) The requirement that management, upon proper request, supply to the union "relevant and necessary" information that enables the union to bargain intelligently and effectively regarding mandatory subjects of bargaining.

The failure to comply with any of these requirements constitutes an unfair labor practice.

2. It is an unfair labor practice for the employer to *raise* wages without first bargaining with the union on this issue. Why, as a matter of policy, is this consistent with the philosophy underlying Section 8? Why might management want to raise wages unilaterally?

3.   Bargaining over Safety and Health

The issue of what safety and health practices will be followed at the workplace is a mandatory subject of bargaining.

National Labor Relations Board v. Gulf Power Company

384 F.2d 822 (5th Cir. 1967)

AINSWORTH, Circuit Judge.

The National Labor Relations Board has petitioned this court for enforcement of its January 5, 1966 order against Gulf Power Company, wherein it found that the Company had engaged in unfair labor practices in violation of Section 8(a)(5) and (1) of the Act (29 U.S.C. §151 et seq.) for refusal to bargain with the Union. The issue we must decide is whether safety rules and safe work practices are mandatory subjects of collective bargaining under the provisions of Section 8(d) of the Act relating to the mutual obligation of the employer and the Union "to meet at reasonable times and confer in good faith with respect to wages, hours and *other terms and conditions of employment. . . .*" (Emphasis supplied.)

For more than twenty years the Union has been the exclusive bargaining agent for Gulf Power's employees in an appropriate unit. Gulf Power Company is a public utility engaged in the generation, transmission and sale of electric power, with principal offices at Pensacola, Florida, and is engaged in commerce within the meaning of the Act. A collective bargaining contract was entered into between the Company and the Union on December 27, 1962, effective until August 15, 1964, and from year to year thereafter unless either party should give notice of termination or revision. Article VI of said contract entitled "Safety" contained several provisions relative to safe employment practices. The Company also embodied other safety provisions in a 60-page "Safe Work Practices" handbook, prepared by it and

issued to all employees, which provided also for discipline or discharge of employees who violated its provisions. The most recent edition of the handbook was published and distributed on January 2, 1963, and was promulgated unilaterally as were other handbooks previously issued. On April 30, 1963, the Company unilaterally revised its safety rules. On June 3, 1963, the Union wrote the Company that it protested the Company's action in having on April 30, 1963 unilaterally placed a revised set of safety rules in operation without having bargained on this subject with the Union. The Company replied that is was required by law to provide reasonably safe and healthful places of employment for its employees, to provide compensation for workman injured in the course of their employment; that this responsibility rested exclusively with it; and that it could not share its obligations with the Union and could not agree that the Union "take any part in the formulation of policy in respect to safety procedures or other matters for which it bears no responsibility under the law." The Union accordingly filed an unfair labor practice charge against the Company which it withdrew shortly thereafter. On June 15, 1964, the Company and the Union notified each other of intention to revise the contract between them. In the Union's request, "General Asking" No. 8, it sought a meeting to negotiate safety rules during 1964 terms of agreement.

A number of bargaining sessions were held at which the Union stated that it was concerned about safety and wanted some type or program of cooperation with the Company on safety. The Company replied that it would be glad to discuss safety with the Union but was of the view that it could not delegate its responsibility for promulgating and regularly revising safety rules. After six bargaining sessions the Company steadfastly declined to bargain with reference to the promulgation and revision of safe work practice rules, maintaining that this was exclusively the legal and moral nondelegable responsibility of the Company. The present complaint of unfair labor practice was accordingly filed. The Board held a hearing after which it concluded that the Company's action in declining to bargain with the Union with respect to safe work practices and/or safety rules violated Section 8(a)(5) and (1) of the Act. We agree and accordingly order enforcement of the order. It is inescapable that in a public utility electric generating and transmission company the workers, through their chosen representative, should have the right to bargain with the Company in reference to safe work practices. This is not to say that there are not areas where the Company's obligations to the public are paramount and which may not properly be the subject of an agreement with the Union after a bargaining session is held. Nonetheless, the parties by their own actions in the past have clearly indicated that they consider safety rules and practices a bargainable issue by including several important provisions in Article VI of the contract already referred to.

Mr. Justice Stewart in his concurring opinion in Fibreboard Paper Products Corp. v. N.L.R.B., 379 U.S. 203, 85 S. Ct. 398, 13 L. Ed. 2d 233 (1964), stated:

> What one's hours are to be, what amount of work is expected during those hours, what periods of relief are available, what safety practices are observed, would all seem conditions of one's employment. (379 U.S. at 222, 85 S. Ct. at 409.)

Cf. Westinghouse Electric Corporation v. N.L.R.B., 4 Cir., 1966, 369 F.2d 891; N.L.R.B. v. Washington Aluminum Company, 370 U.S. 9, 82 S. Ct. 1099, 8 L. Ed. 2d 298 (1962).

We hold, therefore, in agreement with the Board, that the phrase "other terms and conditions of employment" contained in Section 8(d) of the Act is sufficiently broad to include safety rules and practices which are undoubtedly conditions of employment, and that Section 8(d) requires good faith bargaining as a mutual obligation of the employer and the Union in connection with such matters.[3]

By its failure to bargain with the Union in connection with safety rules and practices, the Company violated the provisions of the Act. The Company's obligation is, therefore, to bargain in good faith as required by Section 8(d) of the Act. The duty to bargain in good faith requires the parties to confer and negotiate in a genuine effort to reach an agreement if it is possible to do so. N.L.R.B. v. Mayes, 5 Cir., 383 F.2d 242 August 22, 1967, No. 23,710. The Act does not compel agreement. National Labor Relations Bd. v. American Nat. Ins. Co., 343 U.S. 395, 72 S. Ct. 824, 96 L. Ed. 1027 (1952). The Board held, and we agree with its holding, that the Company refused to bargain with the Union relative to safety rules and practices; that the Company's contention that in all matters pertaining to safety it was immune from bargaining since safety was a prerogative of management was without merit; and thus that the Company had engaged in unfair labor practices within the meaning of the Act.

Enforced.

[3]Pertinent thereto is the reference by the Board in the margin of its opinion (fn. 1) which reads as follows:

> See Bulletin No. 1201 of the United States Department of Labor, December 1956, entitled "Collective Bargaining Clauses, Labor-Management Safety, Production, and Industry Stabilization Committees," which shows that of a total of 1954 major agreements under study, covering a total of more than 7,000,000 workers, 281 contracts, covering 1,773,800 workers, provided for joint safety committees of management and union members. In the public utility field, the study shows that of a total of 61 major agreements, covering 176,900 employees, 14 agreements, covering 61,600 employees, contained such provisions.

Although it would be overstatement to call it a loosening of the floodgates, the *Gulf Power* decision certainly opened the door for broader labor-management negotiation over matters of safety and health. The general trend toward increased attention to these issues at the bargaining table, and the kinds of contractual provisions that have ensued, are explored in the following analysis.

Labor Union Involvement in Occupational Safety and Health, 1957–1987

J. Robinson

It is . . . difficult to measure the extent of collective bargaining over job safety and health issues, much less [to assess] the trends in bargaining. Two complementary data sources were analyzed for this study. The Bureau of National Affairs (BNA), a private Washington-based organization, has maintained a file of collectively bargained contracts and monitored the topics they cover since the 1940s. The file contains over 5000 contracts and . . . date[d] with new contract changes. Contracts are deleted from the file when the bargaining relationship is terminated, as in the case of a plant closure. A sample of approximately 400 contracts, chosen to cover a cross section of industries and unions, is analyzed periodically. Seven reports, spanning the period between 1957 and 1987, were obtained for this study.[1] The second data source is the series of analyses performed by the Bureau of Labor Statistics (BLS) between 1970 and 1980 of union contracts covering 1000 or more workers, supplemented with occasional earlier studies.[2] This series of reports was terminated in 1981 due to cutbacks in . . . [BLS] funding. Tables [6-1] and [6-2] present the percentage of surveyed contracts in the BNA and BLS files . . . [with] various types of health and safety-related clauses.

There has been a clear growth in collective bargaining activity over health and safety since the 1950s. While 69% of BNA manufacturing . . . [contracts] and 38% of BNA nonmanufacturing contracts made some mention of safety in 1957, 89% of manufacturing [contracts] and 77% of nonmanufacturing contracts made some mention in 1987 (Table [6-1]). These figures must be treated with caution, [however,] since they are based upon contracts representing fairly large numbers of workers. It is to be expected that contracts covering small worker groups would be less elaborate and leave more to be adjudicated informally between the employer and the employees. Nevertheless, the upward trend in union activity is clear, with the major increase occurring during the 1970s in tandem with the general growth of public interest in occupational safety and health.

* * *

The most common type of [safety] clause . . . in collectively bargained contracts is the general statement of management's responsibility for providing a safe work environment. In the BNA data, general duty clauses were found in 66% of manufacturing contracts and 43% of nonmanufacturing contracts in 1987, up from 43% and 12%, respectively, in 1957 (Table [6-1]). The general duty clause may or may not

Source: From 13(3) *Journal of Health Politics, Policy and Law* 453–468 (1988).
[1]Bureau of National Affairs, Basic Patterns in Union Contracts, Wash. D.C., BNA (selected editions, 1957–1987).
[2]U.S. Bureau of Labor Statistics, Characteristics of Major Collective Bargaining Agreements, Wash. D.C., U.S. Govt. Printing Office (selected editions, 1971–1981).

Table 6-1. The percentage of union contracts covering selected issues in workplace health and safety: the Bureau of National Affairs sample

	1957	1961	1966	1971	1979	1983	1987
Any Safety Clause							
Manufacturing	69	71	69	71	87	87	89
Nonmanufacturing	38	48	48	52	73	72	77
General Duty Clause							
Manufacturing	43	39	43	48	58	64	66
Nonmanufacturing	12	20	19	28	36	38	43
Joint Safety Committee							
Manufacturing	31	34	35	38	55	57	62
Nonmanufacturing	12	14	18	19	24	26	27
Ongoing Physical Exams	12	14	10	17	22	23	22

make some reference to cooperation with the union or consistency with governmental [safety] laws and standards. . . . While in some cases these statements are just pro forma acknowledgements of the issue, carrying no real content, in other instances they play very significant roles. They bring all safety-related issues under the purview of the contract and thus, in principle, under . . . [the purview] of the grievance and arbitration system that enforces the contract. Workers and their union representatives have the right to raise safety issues through these latter channels, where some amount of equity if ensured, rather than being open to summary rejection by lower level management personnel.

While the general duty clause guarantees workers and unions the right to use the grievance and arbitration system, it does no ensure a favorable outcome from that process. Thus, many contracts include clauses specifying particular safeguards that must be employed or harmful products or processes that must be eliminated. These clauses cover environmental conditions such as lighting, temperature, noise, radiation, fire, and exposure to noxious gases and dusts. The 1974 agreement between U.S. Steel and the United Steelworkers' union, for example, specified precise engineering controls that were to be installed to limit carcinogenic emissions from coke ovens at the Clairton, Pennsylvania, plant.[3] In the BLS sample, 16.4% of manufacturing contracts and 6.3% of nonmanufacturing contracts mandated some kind of environmental protections by 1980 (Table [6-2]). Far more common are clauses that require management to supply safety clothes and equipment such as gloves, steel-toed shoes, and respiratory devices. By 1980, 59% of manufacturing contracts and 50% of nonmanufacturing contracts required management to supply employees with safety equipment of one kind or another (Table [6-2]). Some contract clauses specify minimum crew sizes for particular hazardous operations or prohibit work in isolation.

* * *

A major component of the occupational safety and health problem is the lack of adequate information. [See Chapter 7.] Scientists are in many cases unsure of the biological effects of agents used in production, and even those effects known are often not conveyed to the workers on the shop floor. Some contract clauses require management to provide to the union and the workers any safety information it possesses. In its 1974–75 special survey, the Bureau of Labor

[3]L. Bacow, *Bargaining for Job Safety and Health* (1980).

Table 6-2. The percentage of union contracts covering selected issues in workplace health and safety: the Bureau of Labor Statistics sample

	1970	1971	1972	1973	1974	1975	1976	1978	1980
Environmental Protections									
Manufacturing	NA	11.7	12.6	10.0	11.5	14.0	14.0	15.1	16.4
Nonmanufacturing	NA	7.9	4.6	4.7	5.7	6.7	6.0	5.9	6.3
Safety Equipment									
Manufacturing	47.6	52.3	51.7	51.8	52.8	56.3	56.5	56.5	58.7
Nonmanufacturing	35.0	35.3	42.9	43.8	47.0	48.0	46.1	51.0	49.5
Joint Safety Committee									
Manufacturing	39.6	39.2	36.5	39.0	38.2	40.6	44.7	47.6	55.0
Nonmanufacturing	16.7	18.2	15.8	15.3	16.6	16.7	16.3	18.1	19.9
Right to Refuse Hazardous Work									
Manufacturing	NA	NA	NA	NA	14.8	16.7	16.6	17.3	26.3
Nonmanufacturing	NA	NA	NA	NA	25.0	26.1	24.7	26.5	24.9
Hazard Pay									
Manufacturing	NA	9.4	10.7	11.6	11.0	9.6	10.2	10.0	9.8
Nonmanufacturing	NA	30.3	29.4	28.8	28.7	31.0	29.8	29.8	31.2

NA = not available.

Statistics found that 16% of contracts had such clauses,[4] Chown . . . provides examples of clauses guaranteeing to the union information on toxic materials being used, the results of workplace monitoring performed by the company, reports on lost-time accidents, and copies of Workers' Compensation data.[5]

Access to existing information is, however, only a first step. Innovative unions bargain for programs that increase the amount of knowledge about the particular hazards found in particular work settings and the general understanding of the health effects of various toxins. Some contracts require management to monitor exposure levels or provide monitoring equipment to the joint health and safety committee. Others guarantee regular medical surveillance and health checkups to catch problems in their early stages. Particular types of safety and health training, sometimes with active union participation, are guaranteed in some contracts. The United Auto Workers Union has developed an epidemiological program to evaluate patterns of mortality among specific subgroups of the membership.

A particularly interesting contract clause concerns periodic employee health screening. Many companies routinely require health examinations before employment to reduce the possibility that a prospective employee will become an immediate burden on the employer's health insurance plan and/or to guarantee that the prospective employee is physically able to perform the required tasks. Of more immediate interest to this analysis, however, are the physical exams that take place at periodic intervals after hiring, since these exams can in principle detect the effects of work-related changes in health status. These examinations have become a topic of intense interest in policy circles, since they . . . increasingly [include] tests that measure biological markers of exposure, such as chromosomal changes that may eventually result in cancer. [See Chapter 8.] While unions are worried that results from these tests may be used as the basis for discharge or other forms of discrimination against workers who test positive, they are increasingly bargaining to have these tests performed. In the BNA sample, 22% of all contracts required ongoing physical exams in 1987, up from 12% in 1957 (it was not possible to separate manu-

facturing from nonmanufacturing industries for this item).

* * *

Some unions have found it desirable to bargain for wage compensation for hazardous exposures. During the 1970s, hazard pay clauses were found in approximately 10% of large manufacturing contracts and in 30% of large nonmanufacturing contracts, mostly in the construction and transportation industries (Table [6-2]). No trend was observed over the course of the 1970s in the prevalence of hazard pay clauses. Other contract clauses focus on the financial impacts of job-related accidents and illnesses. Of particular importance are rate retention, job transfer, and seniority rights for workers unable to do their former jobs due to accident or illness but able to do some other kind of work. Other clauses guarantee retention of benefits and accrual of seniority for workers unable to do any work for a certain period of time due to accident or illness related to the job.

An important category of safety-related contract clauses guarantees the rights of workers and their union representatives in the pursuit of safety and health objectives. The employer may be required to pay workers for time spent accompanying OSHA inspectors on tours of the plant or while engaged in other safety activities. Although the 1970 Occupational Safety and Health Act prohibits retaliation against workers for calling in an OSHA inspector or otherwise participating in health and safety activities, some unions find it wise to guarantee nondiscrimination against such workers in their contracts in order to be able to defend workers through the grievance system.

The bottom line in many disagreements over working conditions is whether or not employees will continue working in the presence of particular hazardous conditions. Some contracts explicitly protect individual or collective refusals to perform hazardous tasks. Once again, these rights are in part guaranteed by OSHA and the National Labor Relations Act [See Chapter 8.], but inclusion in the contract adds another avenue through which they may be pursued. Data from the BLS file indicates that approximately one fourth of large contracts had right to refuse clauses by 1980 (Table [6-2]).

An important point on safety clauses in union-management contracts concerns the enforcement mechanism. While the union's power to have a par-

[4]U.S. Bureau of Labor Statistics, Major Collective Bargaining Agreements: Safety and Health Provisions, Bulletin 1425-16, Wash. D.C., U.S. Govt. Printing Office (1976).
[5]Chown, P., Workplace Health and Safety: A Guide to Collective Bargaining. Berkeley, Univ of Cal. Labor Occupational Health Program (1980).

ticular clause included in the contract depends ultimately on its ability to wage a successful strike, the enforcement of those clauses actually in the contract is via the grievance mechanism, and ultimately in the arbitration procedure that constitutes the final level of the grievance mechanism in most contracts. While it would be very difficult to obtain an appreciation of the extent and nature of safety-related grievances, two studies survey arbitration rulings related to hazardous working conditions.[6] The most common rulings fall into three categories: protection of individual refusals to perform hazardous tasks, protection of collective work stoppages related to hazards, and the defense of workers disciplined by the employer for not following safety rules and procedures.

While in the first two types of grievances the desired effect of the union's activities is to force management to provide a safer workplace, in the third case the union apparently is defending workers' rights to act in an unsafe manner. This points out in a particularly graphic manner the extent to which occupational safety and health in the collective bargaining arena is first and foremost a matter of the allocation of rights and power, rather than a singleminded attempt to reduce hazard levels. Primary emphasis is placed on establishing who will decide what is hazardous and what response should be made, and the procedure by which those decisions are to be made. The union's position is consistently one of supporting the participation of the workers and their representatives in those decisions,

even if in some cases this leads to the defense of workers engaged in unsafe behavior.

The primary mechanism for union participation in day to day workplace health and safety matters is the joint union-management health and safety committee. These committees meet regularly, at least in principle, to discuss safety problems and propose remedies. They give to the union a continuous platform from which to deliver its demands, and provide the firm with an ongoing means for canvassing the opinions of its employees, as filtered by the union. Joint committees may be established by a contract clause but need not be.

Contract clauses mandating joint union-management safety committees have become very common, with a doubling between 1957 and 1987 from 31% to 62% in manufacturing contracts and from 12% to 27% in nonmanufacturing contracts (Table [6-1]). Safety committees were mandated in 14% of the contracts surveyed by the BLS in 1950, 7% of the California contracts surveyed by the California Department of Industrial Relations in 1951, and 22% of the contracts surveyed by the BLS in 1954–55. In the BLS sample of large contracts, safety committee clauses grew over the course of the 1970s from 40% to 55% in manufacturing and from 17% to 20% in nonmanufacturing (Table [6-2]). The influence of OSHA on management's willingness to establish joint committees was evidenced in a 1975 survey of approximately 50 firms in New York State whose employees were represented by the International Association of Machinists. Half of the 42 existing committees had been established in the four years since the passage of the Occupational Safety and Health Act.[7]

---

[6]C. Olson, Trade Unions, Wages, and Occupational Injuries: An Empirical Analysis, Ph.D. dissertation, Dept. of Economics, Univ. of Wisconsin (1979); J. Gross and P. Greenfield, Arbitral Value Judgments in Health and Safety Disputes: Management Rights Over Workers' Rights, 34 Buffalo Law Rev. 645 (1985).

[7]T. Kochan, L. Dyer, and D. Lipsky, The Effectiveness of Union Management Safety and Health Committees, (1977).

◇ NOTES

1. The role of the labor-management health and safety committee is discussed in section D, below.

2. As the foregoing analysis suggests, inclusion in the collective bargaining agreement of provisions that require compliance with other laws (such as OSHA standards) are more than mere redundancy. Placement of these requirements in the collective bargaining agreement accomplishes at least two things. It not only emphasizes (both to the employer *and* the individual employees) the importance of health and safety, it also provides another means of enforcement.

3. Lawrence Bacow of MIT has argued that the OSHAct can never be an efficient means of delivering the optimum level of occupational health and safety, and he has suggested that much more can be accomplished through the collective bargaining process. See L. Bacow, Bargaining for Job Safety and Health (1980). To the extent that he may be correct in this assessment, how much does a realistic projection of continued bargaining activity in this area depend on the continued vitality of the OSHAct? Note that the greatest increase in health and safety bargaining came in the 1970s, when OSHAct activity was at its strongest.

4. Conversely, the available evidence also indicates that enforcement of the OSHAct has been much more pronounced in those industries where there is a strong union interest in occupational health and safety. See D. Weil, Government and Labor at the Workplace: The Role of Labor Unions in the Implementation of Federal Health and Safety Policy, Ph.D. dissertation in Public Policy, Harvard University (1987), to be published in part as Weil (1991).

4. The Timing of the Duty to Bargain: "Decision" vs. "Effects" Bargaining

Obviously, a wide variety of management actions will have an effect on wages, hours, or conditions of employment. Does the NLRA give the union a right to bargain over the *decision* to take the action—that is, to bargain over whether the action should be taken, and, if so, what form it should take—or does the Act merely give the union the right to bargain over how to deal with the *effects* of the action on the workforce? If the action itself is concerned directly with something that is a mandatory subject of bargaining, so-called decision bargaining clearly is required. Thus, for example, if the employer is thinking of lowering wages or increasing hours, the issue must be submitted to bargaining before a final decision is reached. However, if the contemplated action is only *indirectly* concerned with a mandatory subject of bargaining, the answer is not always clear. The *First National Maintenance* decision, excerpted below, is the Supreme Court's major statement to date on this issue.

First National Maintenance Corporation v. National Labor Relations Board

452 U.S. 66 (1981)

Justice BLACKMUN delivered the opinion of the Court.

Must an employer, under its duty to bargain in good faith "with respect to wages, hours, and other terms and conditions of employment," §§8 (d) and 8 (a)(5) of the National Labor Relations Act (Act), as amended. 49 Stat. 452. 29 U.S.C. §§158 (d) and 158 (a)(5), negotiate with the certified representative of its employees over its decision to close a part of its business? In this case, the National Labor Relations Board (Board)

imposed such a duty on petitioner with respect to its decision to terminate a contract with a customer, and the United States Court of Appeals, although differing over the appropriate rationale, enforced its order.

I

Petitioner, First National Maintenence Corporation (FNM), is a New York corporation engaged in the business of providing housekeeping, cleaning, maintenance, and related services for commercial customers in the New York City area. It supplies each of its customers, at the customer's premises, contracted-for labor force and supervision in return for reimbursement of its labor costs (gross salaries, FICA and FUTA taxes, and insurance) and payment of a set fee. It contracts for and hires personnel separately for each customer, and it does not transfer employees between locations.[1]

During the spring of 1977, petitioner was performing maintenance work for the Greenpark Care Center, a nursing home in Brooklyn. Its written agreement dated April 28, 1976, with Greenpark specified that Greenpark "shall furnish all tools, equiptment [sic], materials, and supplies," and would pay petitioner weekly "the sum of five hundred dollars plus the gross weekly payroll and fringe benefits." Its weekly fee, however, had been reduced to $250 effective November 1, 1976. The contract prohibited Greenpark from hiring any of petitioner's employees during the term of the contract and for 90 days thereafter. Petitioner employed approximately 35 workers in its Greenpark operation.

Petitioner's business relationship with Greenpark, seemingly, was not very remunerative or smooth. In March 1977, Greenpark gave petitioner the 30 days' written notice of cancellation specified by the contract, because of "lack of efficiency." This cancellation did not become effective, for FNM's work continued after the expiration of that 30-day period. Petitioner, however, became aware that it was losing money at Greenpark. On June 30, by telephone, it asked that its weekly fee be restored at the $500 figure and, on July 6, it informed Greenpark in writing that it would discontinue its operations there on August 1 unless the increase were granted. By telegram on July 25, petitioner gave final notice of termination.

While FNM was experiencing these difficulties, District 1199, National Union of Hospital and Health Care Employees, Retail, Wholesale and Department Store Union, AFL-CIO (union), was conducting an organization campaign among petitioner's Greenpark employees. On March 31, 1977, at a Board-conducted election, a majority of the employees selected the union as their bargaining agent. On July 12, the union's vice president, Edward Wecker, wrote petitioner, notifying it of the certification and of the union's right to bargain, and stating: "We look forward to meeting with you or your representative for that purpose. Please advise when it will be convenient." Petitioner neither responded nor sought to consult with the union.

On July 28, petitioner notified its Greenpark employees that they would be discharged three days later. Wecker immediately telephoned petitioner's secretary-treasurer, Leonard Marsh, to request a delay for the purpose of bargaining. Marsh refused the offer to bargain and told Wecker that the termination of the Greenpark operation was purely a matter of money, and final, and that the 30 days' notice provision of the Greenpark contract made staying on beyond August 1 prohibitively expensive. Wecker discussed the matter with Greenpark's management that same day, but was unable to obtain a waiver of the notice provision. Greenpark also was unwilling itself to hire the FNM employees because of the contract's 90-day limitation on hiring. With nothing but perfunctory further discussion, petitioner on July 31 discontinued its Greenpark operation and discharged the employees.

The union filed an unfair labor practice charge against petitioner, alleging violations of the Act's §§8 (a)(1) and (5). After a hearing held upon the Regional Director's complaint, the Administrative Law Judge made findings in the union's favor. Relying on *Ozark Trailers, Inc.,* 161 N.L.R.B. 561 (1966), he ruled that petitioner had failed to satisfy its duty to bargain concerning both the decision to terminate the Greenpark contract and the effect of that change upon the unit employees. The judge reasoned:

> That the discharge of a man is a change in his conditions of employment hardly needs comment. In these obvious facts, the law is clear. When an employer's work complement is represented by a union and he wishes to alter the hiring arrangements, be his reason lack of money or a mere desire to become richer, the law is no less clear that he must first talk to the union about it. . . . If Wecker had been given an opportunity to talk, something might have been worked out to transfer these people to other parts of [petitioner's] business. . . . Entirely apart from whether open discussion between the parties—with the Union speaking on behalf of the employees as was its right—might have persuaded [petitioner] to find a way of continuing this part of its operations, there was always the possibility that Marsh might have persuaded Greenpark to use these same

---

[1]The record does not show the precise dimension of petitioner's business. See 242 N.L.R.B. 462, 464 (1979). One of the owners testified that petitioner at that time had "between two and four" other nursing homes as customers. *Ibid.* The Administrative Law Judge hypothesized, however: "This is a large Company. For all I know, the 35 men at this particular home were only a small part of its total business in the New York area." *Id,* at 465.

employees to continue doing its maintenance work, either as direct employees or as later hires by a replacement contractor. 242 N.L.R.B. 462, 465 (1979).[5]

The Administrative Law Judge recommended an order requiring petitioner to bargain in good faith with the union about its decision to terminate its Greenpark service operation and its consequent discharge of the employees as well as the effects of the termination. He recommended, also, that petitioner be ordered to pay the discharged employees backpay from the date of discharge until the parties bargained to agreement, or the bargaining reached an impasse, or the union failed timely to request bargaining, or the union failed to bargain in good faith.

The National Labor Relations Board adopted the Administrative Law Judge's findings without further analysis, and additionally required petitioner, if it agreed to resume its Greenpark operations, to offer the terminated employees reinstatement to their former jobs or substantial equivalents; conversely, if agreement was not reached, petitioner was ordered to offer the employees equivalent positions, to be made available by discharge of subsequently hired employees, if necessary, at its other operations.

The United States Court of Appeals for the Second Circuit, with one judge dissenting in part, enforced the Board's order, although it adopted an analysis different from that espoused by the Board. 627 F.2d 596 (1980). The Court of Appeals reasoned that no *per se* rule could be formulated to govern an employer's decision to close part of its business. Rather, the court said, §8 (d) creates a *presumption* in favor of mandatory bargaining over such a decision, a presumption that is rebuttable "by showing that the purposes of the statute would not be furthered by imposition of a duty to bargain," for example, by demonstrating that "bargaining over the decision would be futile," or that the decision was due to "emergency financial circumstances," or that the "custom of the industry, shown by the absence of such an obligation from typical collective bargaining agreements, is not to bargain over such decisions." *Id.*, at 601–602.

The Court of Appeals' decision in this case appears to be at odds with decisions of other Courts of Appeals, some of which decline to require bargaining over any management decision involving "a major commitment of capital investment" or a "basic operational change" in the scope or direction of an enterprise, and some of which indicate that bargaining is not mandated unless a violation of §8 (a)(3) (a partial closing motivated by antiunion animus) is involved. The Court of Appeals for the Fifth Circuit has imposed a duty to bargain over partial closing decisions. See *NLRB* v. *Winn-Dixie Stores, Inc.*, 361 F.2d 512, cert. denied, 385 U.S. 935 (1966). The Board itself has not been fully consistent in its rulings applicable to this type of management decision.

Because of the importance of the issue and the continuing disagreement between and among the Board and the Courts of Appeals, we granted certiorari. 449 U.S. 1076 (1981).

II

A fundamental aim of the National Labor Relations Act is the establishment and maintenance of industrial peace to preserve the flow of interstate commerce. *NLRB* v. *Jones & Laughlin Steel Corp.*, 301 U.S. 1 (1937). Central to achievement of this purpose is the promotion of collective bargaining as a method of defusing and channeling conflict between labor and management. §1 of the Act, as amended, 29 U.S. C. §151. Congress ensured that collective bargaining would go forward by creating the Board and giving it the power to condemn as unfair labor practices certain conduct by unions and employers that it deemed deleterious to the process, including the refusal "to bargain collectively." §§3 and 8, 29 U.S. C. §§153 and 158.

Although parties are free to bargain about any legal subject, Congress has limited the mandate or duty to bargain to matters of "wages, hours, and other terms and conditions of employment." A unilateral change as to a subject within this category violates the statutory duty to bargain and is subject to the Board's remedial order. *NLRB* v. *Katz*, 369 U.S.

---

[5]The judge further found that petitioner's "regular and usual" method of operation involved "taking on, finishing, or discontinuing this or that particular job," 242 N.L.R.B., at 466, and that "[t]here was no capital involved when it decided to terminate the Greenpark job. The closing of this one spot in no sense altered the nature of its business, nor did it substantially affect its total size." *Ibid.* The Administrative Law Judge therefore found inapplicable the Board's ruling in *Brockway Motor Trucks, Division of Mack Trucks, Inc.*, 230 N.L.R.B. 1002, 1003 (1977), enf. denied, 582 F.2d 720 (CA3 1978), that an employer's decision to close part of its business is not a mandatory subject of bargaining if it involves such a " 'significant investment or withdrawal of capital' as to 'affect the scope and ultimate direction of an enterprise,' " quoting from *General Motors Corp., GMC Truck & Coach Div.*, 191 N.L.R.B. 951, 952 (1971).

736 (1962). Conversely, both employer and union may bargain to impasse over these matters and use the economic weapons at their disposal to attempt to secure their respective aims. *NLRB v. American National Ins. Co.*, 343 U.S. 395 (1952). Congress deliberately left the words "wages, hours, and other terms and conditions of employment" without further definition, for it did not intend to deprive the Board of the power further to define those terms in light of specific industrial practices.

Nonetheless, in establishing what issues must be submitted to the process of bargaining, Congress had no expectation that the elected union representative would become an equal partner in the running of the business enterprise in which the union's members are employed. Despite the deliberate open-endedness of the statutory language, there is an undeniable limit to the subjects about which bargaining must take place:

> Section 8 (a) of the Act, of course, does not immutably fix a list of subjects for mandatory bargaining. . . . But it does establish a limitation against which proposed topics must be measured. In general terms, the limitation includes only issues that settle an aspect of the relationship between the employer and the employees.

Chemical & Alkali Workers v. *Pittsburgh Plate Glass Co.*, 404 U.S. 157, 178 (1971).

See also *Ford Motor Co.* v. *NLRB*, 441 U.S. 488 (1979); *Fibreboard Paper Products Corp.* v. *NLRB*, 379 U.S. 203 (1964); *Teamsters* v. *Oliver*, 358 U.S. 283 (1959).

Some management decisions, such as choice of advertising and promotion, product type and design, and financing arrangements, have only an indirect and attenuated impact on the employment relationship. See *Fibreboard*, 379 U.S., at 223 (Stewart, J., concurring). Other management decisions, such as the order of succession of layoffs and recalls, production quotas, and work rules, are almost exclusively "an aspect of the relationship" between employer and employee. *Chemical Workers*, 404 U.S., at 178. The present case concerns a third type of management decision, one that had a direct impact on employment, since jobs were inexorably eliminated by the termination, but had as its focus only the economic profitability of the contract with Greenpark, a concern under these facts wholly apart from the employment relationship. This decision, involving a change in the scope and direction of the enterprise, is akin to the decision whether to be in business at all, "not in [itself] primarily about conditions of employment, though the effect of the decision may

be necessarily to terminate employment." *Fibreboard*, 379 U.S., at 223 (Stewart, J., concurring). Cf. *Textile Workers* v. *Darlington Co.*, 380 U.S. 263, 268 (1965) ("an employer has the absolute right to terminate his entire business for any reason he pleases"). At the same time, this decision touches on a matter of central and pressing concern to the union and its member employees: the possibility of continued employment and the retention of the employees' very jobs. See *Brockway Motor Trucks* v. *NLRB*, 582 F.2d 720, 735–736 (CA3 1978); *Ozark Trailers, Inc.*, 161 N.L.R.B. 561, 566–568 (1966).

Petitioner contends it had no duty to bargain about its decision to terminate its operations at Greenpark. This contention requires that we determine whether the decision itself should be considered part of petitioner's retained freedom to manage its affairs unrelated to employment.[15] The aim of labeling a matter a mandatory subject of bargaining, rather than simply permitting, but not requiring, bargaining, is to "promote the fundamental purpose of the Act by bringing a problem of vital concern to labor and management within the framework established by Congress as most conducive to industrial peace," *Fibreboard*, 379 U.S., at 211. The concept of mandatory bargaining is premised on the belief that collective discussions backed by the parties' economic weapons will result in decisions that are better for both management and labor and for society as a whole. Ford Motor Co., 441 U.S., at 500-501; *Borg-Warner*, 356 U.S., at 350 (condemning employer's proposal of "ballot" clause as weakening the collective-bargaining process). This will be true, however, only if the subject proposed for discussion is amenable to resolution through the bargaining process. Management must be free from the constraints of the bargaining process[17] to the extent

---

[15]There is no doubt that petitioner was under a duty to bargain about the results or effects of its decision to stop the work at Greenpark, or that it violated that duty. Petitioner consented to enforcement of the Board's order concerning bargaining over the effects of the closing and has reached agreement with the union on severance pay.

[17]The employer has no obligation to abandon its intentions or to agree with union proposals. On proper subjects, it must meet with the union, provide information necessary to the union's understanding of the problem, and in good faith consider any proposals the union advances. In concluding to reject a union's position as to a mandatory subject, however, it must face the union's possible use of strike power. See generally Fleming, The Obligation to Bargain in Good Faith, 47 Va. L. Rev. 988 (1961).

essential for the running of a profitable business. It also must have some degree of certainty beforehand as to when it may proceed to reach decisions without fear of later evaluations labeling its conduct an unfair labor practice. Congress did not explicitly state what issues of mutual concern to union and management it intended to exclude from mandatory bargaining. Nonetheless, in view of an employer's need for unencumbered decisionmaking, bargaining over management decisions that have a substantial impact on the continued availability of employment should be required only if the benefit, for labor-management relations and the collective-bargaining process, outweighs the burden placed on the conduct of the business.

The Court in *Fibreboard* implicitly engaged in this analysis with regard to a decision to subcontract for maintenance work previously done by unit employees. Holding the employer's decision a subject of mandatory bargaining, the Court relied not only on the "literal meaning" of the statutory words, but also reasoned:

> The Company's decision to contract out the maintenance work did not alter the Company's basic operation. The maintenance work still had to be performed in the plant. No capital investment was contemplated; the Company merely replaced existing employees with those of an independent contractor to do the same work under similar conditions of employment. Therefore, to require the employer to bargain about the matter would not significantly abridge his freedom to manage the business. 379 U.S., at 213.

The Court also emphasized that a desire to reduce labor costs, which it considered a matter "peculiarly suitable for resolution within the collective bargaining framework," *id.*, at 214, was at the base of the employer's decision to subcontract:

> It was induced to contract out the work by assurances from independent contractors that economies could be derived by reducing the work force, decreasing fringe benefits, and eliminating overtime payments. These have long been regarded as matters peculiarly suitable for resolution within the collective bargaining framework, and industrial experience demonstrates that collective negotiation has been highly successful in achieving peaceful accommodation of the conflicting interests. *Id.*, at 213–214.

The prevalence of bargaining over "contracting out" as a matter of industrial practice generally was taken as further proof of the "amenability of such subjects to the collective bargaining process." *Id.*, at 211.

With this approach in mind, we turn to the specific issue at hand: an economically motivated decision to shut down part of a business.

III

A

Both union and management regard control of the decision to shut down an operation with the utmost seriousness. As has been noted, however, the Act is not intended to serve either party's individual interest, but to foster in a neutral manner a system in which the conflict between these interests may be resolved. It seems particularly important, therefore, to consider whether requiring bargaining over this sort of decision will advance the neutral purposes of the Act.

A union's interest in participating in the decision to close a particular facility or part of an employer's operations springs from its legitimate concern over job security. The Court has observed: 'The words of [§8 (d)] . . . plainly cover termination of employment which . . . necessarily results' from closing an operation. *Fibreboard*, 379 U.S., at 210. The union's practical purpose in participating, however, will be largely uniform: it will seek to delay or halt the closing. No doubt it will be impelled, in seeking these ends, to offer concessions, information, and alternatives that might be helpful to management or forestall or prevent the termination of jobs.[19] It is unlikely, however, that requiring bargaining over the decision itself, as well as its effects, will augment this flow of information and suggestions. There is no dispute that the union must be given a significant opportunity to

---

[19] We are aware of past instances where unions have aided employers in saving failing businesses by lending technical assistance, reducing wages and benefits or increasing production, and even loaning part of earned wages to forestall closures. See S. Slichter, J. Healy, & E. Livernash, The Impact of Collective Bargaining on Management 845–851 (1960); C. Golden & H. Rutenberg, The Dynamics of Industrial Democracy 263–291 (1942). See also *United Steel Workers of America, Local No. 1330, v. United States Steel Corp.*, 492 F. Supp. 1 (ND Ohio), aff'd in part and vacated in part, 631 F.2d 1264 (CA6 1980) (union sought to purchase failing plant); 104 LRR 239 (1980) (employee ownership plan instituted to save company); *id.*, at 267–268 (union accepted pay cuts to reduce plant's financial problems). These have come about without the intervention of the Board enforcing a statutory requirement to bargain.

bargain about these matters of job security as part of the "effects" bargaining mandated by §8 (a)(5). See, *e.g.*, *NLRB* v. *Royal Plating & Polishing Co.*, 350 F.2d 191, 196 (CA3 1965); *NLRB* v. *Adams Dairy, Inc.*, 350 F.2d 108 (CA8 1965), cert. denied, 382 U.S. 1011 (1966). And, under §8 (a)(5), bargaining over the effects of a decision must be conducted in a meaningful manner and at a meaningful time, and the Board may impose sanctions to insure its adequacy. A union, by pursuing such bargaining rights, may achieve valuable concessions from an employer engaged in a partial closing. It also may secure in contract negotiations provisions implementing rights to notice, information, and fair bargaining. See BNA, Basic Patterns in Union Contracts 62–64 (9th ed., 1979).

Moreover, the union's legitimate interest in fair dealing is protected by §8(a)(3), which prohibits partial closings motivated by antiunion animus, when done to gain an unfair advantage. *Textile Workers* v. *Darlington Co.*, 380 U.S. 263 (1965). Under §8(a)(3) the Board may inquire into the motivations behind a partial closing. An employer may not simply shut down part of its business and mask its desire to weaken and circumvent the union by labeling its decision "purely economic."

Thus, although the union has a natural concern that a partial closing decision not be hastily or unnecessarily entered into, it has some control over the effects of the decision and indirectly may ensure that the decision itself is deliberately considered. It also has direct protection against a partial closing decision that is motivated by an intent to harm a union.

Management's interest in whether it should discuss a decision of this kind is much more complex and varies with the particular circumstances. If labor costs are an important factor in a failing operation and the decision to close, management will have an incentive to confer voluntarily with the union to seek concessions that may make continuing the business profitable. Cf. U.S. News & World Report, Feb. 9, 1981, p. 74; BNA, Labor Relations Yearbook—1979, p. 5 (UAW agreement with Chrysler Corp. to make concessions on wages and fringe benefits). At other times, management may have great need for speed, flexibility, and secrecy in meeting business opportunities and exigencies. It may face significant tax or securities consequences that hinge on confidentiality, the timing of a plant closing, or a reorganization of the corporate structure. The publicity incident to the normal process of bargaining may injure the possibility of a successful transition or increase the economic damage to the business. The employer also may have no feasible alternative to the closing, and even good-faith bargaining over it may both be futile and cause the employer additional loss.

There is an important difference, also, between permitted bargaining and mandated bargaining. Labeling this type of decision mandatory could afford a union a powerful tool for achieving delay, a power that might be used to thwart management's intentions in a manner unrelated to any feasible solution the union might propose. See Comment, "Partial Terminations"—A Choice Between Bargaining Equality and Economic Efficiency, 14 UCLA L. Rev. 1089, 1103–1105 (1967). In addition, many of the cases before the Board have involved, as this one did, not simply a refusal to bargain over the decision, but a refusal to bargain at all, often coupled with other unfair labor practices.

*   *   *

In these cases, the employer's action gave the Board reason to order remedial relief apart from access to the decisionmaking process. It is not clear that a union would be equally dissatisfied if an employer performed all its bargaining obligations apart from the additional remedy sought here.

While evidence of current labor practice is only an indication of what is feasible through collective bargaining, and not a binding guide, see *Chemical Workers*, 404 U.S., at 176, that evidence supports the apparent imbalance weighing against mandatory bargaining. We note that provisions giving unions a right to participate in the decisionmaking process concerning alteration of the scope of an enterprise appear to be relatively rare. Provisions concerning notice and "effects" bargaining are more prevalent. See II BNA, Collective Bargaining Negotiations and Contracts §65:201–233 (1981); U.S. Dept of Labor, Bureau of Labor Statistics, Bull. 2065, Characteristics of Major Collective Bargaining Agreements, Jan. 1, 1978, pp.96, 100, 101, 102–103 (1980) (charting provisions giving interplant transfer and relocation allowances; advance notice of layoffs, shutdowns, and technological changes; and wage-employment guarantees; no separate tables on decision-bargaining, presumably due to rarity). See also U.S. Dept. of Labor, Bureau of Labor Statistics, Bull. No. 1425–10. Major Collective Bargaining Agreements, Plant Movement, Transfer, and Relocation Allowances (July 1969).

Further, the presumption analysis adopted by the Court of Appeals seems ill-suited to advance harmonious relations between employer and employee. An employer would have difficulty determining beforehand whether it was faced with a situation requiring bargaining or one that involved economic necessity sufficiently compelling to obviate the duty to bargain. If it should decide to risk not bargaining, it might be faced ultimately with harsh remedies forcing it to pay large amounts of backpay to employees who likely would have been discharged regardless of bargaining, or even to consider reopening a failing operation. See, *e.g.*, *Electrical Products Div. of Midland-Ross Corp.*, 239 N.L.R.B. 323 (1978), enf'd, 617 F. 2d 977 (CA3 1980), cert. denied, 449 U.S. 871 (1981). Cf. *Lever Brothers Co.* v. *International Chemical Workers Union*, 554 F.2d 115 (CA4 1976) (enjoining plant closure and transfer to permit negotiations). Also, labor costs may not be a crucial circumstance in a particular economically based partial termination. See, *e.g.*, *NLRB* v. *International Harvester Co.*, 618 F.2d 85 (CA9 1980) (change in marketing structure); *NLRB* v. *Thompson Transport Co.*, 406 F.2d 698 (CA10 1969) (loss of major customer). And in those cases, the Board's traditional remedies may well be futile. See *ABC Trans-National Transport, Inc.* v. *NLRB*, 642 F.2d 675 (CA3 1981) (although employer violated its "duty" to bargain about freight terminal closing, court refused to enforce order to bargain). If the employer intended to try to fulfill a court's direction to bargain, it would have difficulty determining exactly at what stage of its deliberations the duty to bargain would arise and what amount of bargaining would suffice before it could implement its decision. Compare *Burns Ford, Inc.*, 182 N.L.R.B. 753 (1970) (one week's notice of layoffs sufficient), and *Hartmann Luggage Co.*, 145 N.L.R.B. 1572 (1964) (entering into executory subcontracting agreement before notifying union not a violation since contract not yet final), with *Royal Plating & Polishing Co.*, 148 N.L.R.B. 545, 555 (1964), enf. denied, 350 F.2d 191 (CA3 1965) (two weeks' notice before final closing of plant inadequate). If an employer engaged in some discussion, but did not yield to the union's demands, the Board might conclude that the employer had engaged in "surface bargaining," a violation of its good faith. See *NLRB* v. *Reed & Prince Mfg. Co.*, 205 F.2d 131 (CA1), cert. denied, 346 U.S. 887 (1953). A union, too, would have difficulty determining the limits of its prerogatives, whether and when it could use its economic powers to try to alter an employer's decision, or whether, in doing so, it would trigger sanctions from the Board. See, *e.g.*, *International Offset Corp.*, 210 N.L.R.B. 854 (1974) (union's failure to realize that shutdown was

imminent, in view of successive advertisements, sales of equipment, and layoffs, held a waiver of right to bargain); *Shell Oil Co.*, 149 N.L.R.B. 305 (1964) (union waived its right to bargain by failing to request meetings when employer announced intent to transfer a few days before implementation).

We conclude that the harm likely to be done to an employer's need to operate freely in deciding whether to shut down part of its business purely for economic reasons outweighs the incremental benefit that might be gained through the union's participation in making the decision,[22] and we hold that the decision itself is *not* part of §8 (d)'s "terms and conditions," . . . over which Congress has mandated bargaining.

B

In order to illustrate the limits of our holding, we turn again to the specific facts of this case. First, we note that when petitioner decided to terminate its Greenpark contract, it had no intention to replace the discharged employees or to move that operation elsewhere. Petitioner's sole purpose was to reduce its economic loss, and the union made no claim of antiunion animus. In addition, petitioner's dispute with Greenpark was solely over the size of the management fee Greenpark was willing to pay. The union had no control or authority over that fee. The most that the union could have offered would have been advice and concessions that Greenpark, the third party upon whom rested the success or failure of the contract, had no duty even to consider. These facts in particular distinguish this case from the subcontracting issue presented in *Fibreboard*. Further, the union was not selected as the bargaining representative or certified until well after petitioner's eco-

[22] In this opinion we of course intimate no view as to other types of management decisions, such as plant relocations, sales, other kinds of subcontracting, automation, etc., which are to be considered on their particular facts. See, *e.g.*, *International Ladies' Garment Workers Union* v. *NLRB*, 150 US. App. D.C. 71, 463 F.2d 907 (1972) (plant relocation predominantly due to labor costs); *Weltronic Co.* v. *NLRB*, 419 F.2d 1120 (CA6 1969) (decision to move plant three miles), cert. denied, 398 U.S. 938 (1970); *Dan Dee West Virginia Corp.*, 180 N.L.R.B. 534 (1970) (decision to change method of distribution, under which employee-drivers became independent contractors); *Young Motor Truck Service, Inc.*, 156 N.L.R.B. 661 (1966) (decision to sell major portion of business). See also Schwarz, Plant Relocation or Partial Termination—The Duty to Decision-Bargain, 39 Ford. L. Rev. 81, 100–102 (1970).

nomic difficulties at Greenpark had begun. We thus are not faced with an employer's abrogation of ongoing negotiations or an existing bargaining agreement. Finally, while petitioner's business enterprise did not involve the investment of large amounts of capital in single locations, we do not believe that the absence of "significant investment or withdrawal of capital," *General Motors Corp., GMC Truck & Coach Div.*, 191 N.L.R.B., at 952, is crucial. The decision to halt work at this specific location represented a significant change in petitioner's operations, a change not unlike opening a new line of business or going out of business entirely.

The judgment of the Court of Appeals, accordingly, is reversed, and the case is remanded to that court for further proceedings consistent with this opinion.

[Dissenting opinion of BRENNAN, J., in which MARSHALL, J. joined, omitted].

◇ NOTES

1. How does one go about applying the general standard the court gives us for determining whether decision bargaining is required in a given situation? How does the court characterize the "purpose" of the NLRA?

2. The court concludes—apparently as a matter of law—that the decision to terminate a portion of the company's business represents a "significant change in [the company's] operations." How important is this to the conclusion that decision bargaining is not required in such a situation? Is the degree of the effect of the decision in question on the "direction" of the company a factor that influences the "balancing" to be done under the court's articulated standard?

3. What about a decision by an employer to close down part of the business— perhaps to close down a plant—in the middle of a union organizing battle? The decision by First National Maintenance to cancel the contract in question came in the midst of such an effort. Thus, the court is careful to note that the decision appeared to be motivated by "economic" considerations only. Had the union instead been able to establish that the decision had been motivated by what the court calls "antiunion animus," the employer's action presumably would have been an unfair labor practice and would have been assailable on those grounds.

4. The court is also careful to note that it "intimate[s] no view as to other types of management decisions, such as plant relocations, sales, other kinds of subcontracting, automation, etc., which are to be considered on their particular facts." (See footnote 22.)

5. How does the court characterize the "interest" of the union in plant closing situations? Do you think that this is an accurate characterization? Whether or not it is, could one convince the court that there are different, more complicated interests at stake with other kinds of management decisions and that these decisions would thus benefit more from union participation at the decision stage? Would *this* factor influence the "balancing" to be done under the *First National Maintenance* standard?

Thomas Kohler of Boston College Law School has argued that, as effects bargaining will always be available, the creative and competent union will still have ample opportunity to influence the ultimate decision even if no "decision" bargaining is required.

Distinctions Without Differences: Effects Bargaining in Light of First National Maintenance

T. Kohler

The goals of decision and effects bargaining are essentially identical: to afford the affected employees' bargaining representative notice sufficiently in advance of the implementation of an operational change to permit the union the opportunity, through bargaining, to preserve jobs and otherwise protect the interests of employees. Further, their mechanical features are alike. Within their respective spheres, the scope of bargaining is equally broad, and in both, the union has the right to secure information under the employer's control which the union needs in order to bargain intelligently. Finally, and most critically, the duties attach at virtually the same time, i.e., sufficiently in advance of the implementation of a change as to permit the union a "meaningful opportunity" to bargain. In the final analysis, the differences between the two duties seem more of degree than kind. The emphasis in decision bargaining is on an exploration of alternatives which the employer may find attractive enough to forego the contemplated change. In effects bargaining, the focus is on ways to ameliorate the impact of the change's execution, particularly through discussions concerning opportunities for continued employment for affected workers at the employer's other facilities. The duties then, are actually variations on a theme.

In theory, the cases envision decision and effects bargaining occurring in sequential, isolated phases. However, because "effects are so inextricably interwoven with the decision itself," there seems no way to prevent bargaining over the former from having an impact on the latter. Because the *First National Maintenance* opinion has upheld the duty to effects-bargain, its impact on the ability of unions to protect employee interests may be less than first expected. The consideration of a hypothetical partial closing situation may aid in illuminating this point An imaginary business enterprise, the "Mas Corporation,"

operates three plants at which it produces components used in the manufacture of automobiles. In order to reduce its costs, Mas has decided to consolidate its operations and to close its plant I, the eldest and least profitable of its three facilities. Prior to *First National Maintenance,* Mas' duties, at least under the Board's general rule, would be as follows: upon reaching its decision to close plant I, Mas would have to notify the union representing the employees at that facility, Local One, of its contemplated plan. As seen, such notice would have to be given before Mas was irrevocably committed to effectuating the plan, and sufficiently in advance of the closing to permit Local One an opportunity to study the situation and to formulate suggestions as to possible alternatives. At the union's request, of course, Mas would be under an obligation to supply it with documentation concerning the plant's lack of profitability, and other financial information the union might require to formulate proposals concerning severance pay, benefit continuation, and, particularly, the opportunities for affected employees to transfer to Mas' other plants. Bargaining would then proceed, during which all aspects of the plan, including alternatives to its effectuation and ways to ameliorate its impact, would be discussed. Only by consideration of the costs of the plan's implementation could the amount to be saved by the termination be ascertained.

After *First National Maintenance*, Mas, of course, need not discuss or consider any union suggestions as to alternatives to the planned closing, but must bargain about the effects of its decision. The Court's opinion indicates that this bargaining must occur at a "meaningful time," which, precedent indicates, is roughly contemporaneous with the time at which the duty to decision-bargain attaches. It appears that Mas would be under no duty to supply any information concerning its decision, such as documentation concerning the plant's profitability. Mas would, however, be obligated to supply Local One with other information pertaining to transfers, and financial information by which the union could formulate economic demands concerning severance pay, insurance continuation, and the like. Through such requests, the union

Source: From 5(3) *Indus. Rel. L.J.* 402, 421–423 (1983). © 1983 by *Industrial Relations Law Journal.*

would probably be able to obtain most of the information to which it would have had access prior to *First National Maintenance*. Most importantly, because effects bargaining occurs prior to the implementation of a change, Local One, through the magnitude of its termination demands, might convince Mas that the continued operation of plant I, possibly under altered conditions, e.g., reductions in rates of pay or changes in contract terms concerning manning requirements, would prove cheaper in the long run than closing the facility. Finally, although the union could only insist

on bargaining about effects, it might nevertheless propose alternatives to the employer's plan and, while not insisting to impasse over them, create sufficient pressure by its "effects" demands to convince the employer to agree to discuss the permissive topic of "decision." The point of this exercise is that, as a practical matter, because the duties to decision- and effects-bargain attach at the same time, their substance and results will be largely similar. The dynamics of bargaining will not permit the line between decision and effects to stand unblurred.

Although Kohler doubtless is correct that much substantive bargaining is possible within the context of effects bargaining, it is arguable that he underestimates the *psychological* difference between the two types of bargaining. It is one thing to approach the bargaining table knowing that federal law dictates that the decision in question cannot be finalized until you have your say, but quite another to come to the table knowing that you will be able to discuss the wisdom of the decision only if the employer can be convinced to listen. Further, there is a clear difference between the two types of bargaining with regard to the nature and scope of the information that the employer will be required to provide to the union. Significantly, effects bargaining will not permit the union to compel the employer to provide information regarding the alternatives it considered—but rejected—before coming to its decision. This may be especially important to union efforts to bargain over proposed technological changes in the workplace.

5.   Collective Bargaining over Issues of Technological Change

Where do proposed changes in workplace technology fit into the *First National Maintenance* framework? Under what circumstances is management obligated to decision-bargain with labor over whether to introduce technological change into the workplace? As the following excerpt suggests, the answer likely will turn on the court's view of the importance of the role that labor can play if it participates in the decision.

Changes and Opportunities in the Environment for Technology Bargaining

N. Ashford and C. Ayers

II. TECHNOLOGY BARGAINING

For the purposes of this Article, "technology bargaining" refers to collective bargaining over the introduction of technological change into the workplace. Technological change can take numerous forms, some

Source: From 62(5) *Notre Dame L. Rev.* 810, 818–819, 842–855 (1987). Reprinted with permission. © 1987 by *Notre Dame Law Review*, University of Notre Dame.

of the most common forms are the following: Increased mechanization, the introduction of automation into the office or plant, the use of robotics, and the introduction of new raw materials or new power sources. The introduction of technological change may or may not negatively impact the work environment.

A. Interests Involved

Predictably, management and labor hold divergent views regarding the introduction of technological change. Management typically favors technological change since management benefits from product and process improvements, increased productivity and reduced costs as a result of the change. Additionally, management favors the unilateral discretion to make decisions regarding the change, typically arguing that such discretion fosters unified decision making, reducing costs and delays, that it allocates resources flexibly, maximizing profit and economic efficiency, and that it is consistent with management's fiduciary responsibilities to the corporate ownership.

On the other hand, labor is typically ambivalent towards technological change, even where it could improve the work environment.[41] Paramount among labor's concerns are the issues surrounding health, safety and job security. Technological innovation in the workplace often results in the introduction of hazardous machinery and equipment, the introduction of toxic substances, job lay offs, transfers of operations, the reclassification of workers—frequently downward—which results in pay losses, and the like. Labor's position is that an employee has, in addition to health and safety considerations, a substantial personal investment in his or her job to protect, since an employee typically has developed a level of skill that may or may not be salable to another employer, accumulated seniority and pension rights. From that position, labor argues that because technological change affects conditions of employment so significantly, unions should have an extensive opportunity to influence decisions regarding technological change at the decision making stage rather than merely bargaining over the effects of such change.

*   *   *

[41]Although labor is typically threatened by technological change, this observation does not suggest that labor unions uniformly oppose the introduction of technological change. Union views vary substantially over time, between and within unions, and according to the nature of the changes introduced. Much technological change takes the form of incremental improvements and thus does not change the workplace abruptly or substantially. Incremental change is much less threatening than change which results in replacing an entire process at one time. In addition, many unions have supported technological change as a means of ensuring better working conditions and higher wages. For example, the United Mine Workers actively promoted mechanization to improve the working conditions of miners even at the risk of diminishing the number of jobs available to miners. Even the most positive technological change can have a negative impact on the lives of workers.

C. Critical Evaluation of the Law Bearing on Technology Bargaining

*   *   *

The Court in *First National Maintenance* expressed concern that certain aspects of mandatory decision bargaining would endanger the employer's financial position. The Court cited an employer's need for "speed, flexibility, and secrecy in meeting business opportunities and exigencies." Most decisions concerning operational changes, such as the introduction of technological change, are not made in an emergency situation requiring "speed, flexibility, and secrecy." The Court also cited the risk of economic damage to the business posed by publicity incident to the bargaining process. Risks incident to publicity, however, already exist since labor must receive notice and an opportunity to bargain effectively prior to implementation of the decision. The employer can diminish costs if it plans ahead to incorporate labor's participation into its decision making process. The NLRA established a process of negotiation and agreement through which the relationship between labor and management could be developed. Under the collective bargaining process, the employer is relieved of direct regulation of labor-management relations, and is allowed to retain final control over economic decisions. There are costs to the employer as well as to labor that are incident to the bargaining process. Congress, however, determined that the societal benefits of mandatory negotiations outweigh those costs to individual parties.

*   *   *

[F]ew cases deal with the duty to bargain over technological change per se. The vast majority of cases stemming from the introduction of technological change focus on individual problems that arise from the impact of the change, *e.g.*, the various effects on the bargaining unit such as lay offs or reclassification, transfers of operations, partial closing, health and safety problems. As noted above, the complexity of the impact of technological change coupled with the piecemeal fashion in which the Board and courts have approached the problem has resulted in a confused set of decisions. Those decisions strongly favor management by limiting the duty to bargain to the *effects* of the technological change while according management unlimited discretion to make decisions regarding the introduction of the technology.

An important question is whether the presence of a health or safety issue should be treated any differently in the case of technological change. The impact of technological change on the health and safety of the work environment can generally take one of six basic forms:

1. More hazardous technology introduced with worker displacement;
2. More hazardous technology introduced with no worker displacement;
3. Safer technology introduced with worker displacement;
4. Safer technology introduced with no worker displacement;
5. Technology introduced with worker displacement and no effect on health or safety of the worker;
6. Technology introduced with no worker displacement and no effect on health or safety of the worker.

Board and court decisions mandate that an employer has a duty to bargain over the effects of technological change on labor that result in any one of those situations; however, after *First National Maintenance* an employer would appear to have a duty to bargain over the *decision* to introduce that change only in those situations in which health and safety issues are raised. Furthermore, a strong argument can be made that *Gulf Power* mandates a duty to bargain over decisions concerning health and safety, although the case itself is not absolutely clear on that point. However, reading *First National Maintenance* and *Gulf Power* together, decision bargaining over health and safety issues seems to be required, since most adverse effects on health and safety can only meaningfully be addressed and mitigated if the decision to adopt technology or change production process is reversed or altered. These scenarios will be evaluated in light of these decisions.

The first scenario of the six listed above is the most compelling, both from a legal and a policy point of view. The introduction of technology into the workplace that results both in job losses and in greater hazards to the remaining workers clearly exerts tremendous impact on the terms and conditions of employment for those workers involved. The holding in *Gulf Power* that health and safety issues are mandatory subjects of bargaining, coupled with the high priority accorded health and safety in the legislative history of the NLRA, strongly supports the proposition that considerations of health and safety override management prerogatives, and, therefore, should be mandatory subjects of bargaining *at the decision making*

stage. Board and court decisions, however, do not generally mandate a duty to decision bargain where issues concerning displacement are involved. Nevertheless, in this first situation, where the health and safety issues appear to give rise to a duty to decision bargain, the job displacement issues presumably would also be included as subjects of bargaining at the decision making stage.

Examples of the situation where new technology could result in both increased hazards *and* job displacement may help to clarify the point that this enormous impact on workers compels a duty to decision bargain. One example of this situation can occur in the chemical industry when a change is made from a batch process to a closed chemical process under extremely high pressure. This change in process not only results in a decreased need for workers, but it also may pose greater dangers in the workplace in the form of an increased potential for explosions and toxic leaks. This situation also can occur with the introduction of high-speed cutting machinery which is used in various industries. This machinery is not only more efficient and requires less labor, but it can be extremely dangerous as well.

This first situation can arise in an office context as well as in the factory setting. The use of computers and word processors is a prime example. Computers and word processors are being used with ever increasing frequency in offices throughout the country. As a result, clerical workers are required to be better trained and more highly skilled, and certain lower-level clerical positions are being eliminated altogether. As time goes on, increased automation in the office may result in nearly fully automated offices thereby eliminating the need for clerical staff. In addition to the job displacement problems caused by the use of office automation, the video display terminals (VDTs) found on computers and word processors are proving to be health hazards as well. Although some of the studies concerning the hazards of VDTs are as yet inconclusive, those studies do indicate that VDTs result in such problems as severe eye strain, back strain and exposure to low-level radiation emitted by VDTs, which some responsible scientists have postulated can produce birth defects and miscarriages in addition to cataracts and skin rashes. Although the studies may as yet be inconclusive and the injuries caused by VDTs may not be the same acute, immediate injuries often associated with hazards in the factory, they are of no less significance, and thus should be accorded a high degree of priority in the collective bargaining process.

The job displacement problems in the examples given above affect the lives of workers just as substantially as the health and safety problems. The issues surrounding job displacement, however, generally appear to be mandatory subjects of decision bargaining only where health and safety issues are also involved, assuming that one is able to successfully argue that precedent dictates that considerations of health and safety are mandatory subjects of decision bargaining.

If one is successful in making that argument, the outcome of the second situation, in which more hazardous technology is introduced with no displacement, will be the same as the outcome in the first situation since considerations of health and safety establish the duty to decision bargain in both situations. An example of the second situation in the factory setting would be the simple substitution of a more toxic substance in the manufacturing process for another substance. The union would have a substantial interest in protecting the health and safety of its members by participating in that decision. An example in the office context would again be the introduction of automation; however, the focus would be on the professional workforce rather than the clerical staff. The focus shifts in this second scenario because the introduction of automation frequently results in less job displacement among the professional staff than the clerical staff. Since the professional staff will be utilizing the automation, however, they will be exposed to the same hazards as members of the clerical staff who also use it.

In the third and fourth situations, in which safer technology is introduced rather than more hazardous technology, one cannot rely as clearly on the high priority accorded health and safety issues as one could in the first two situations. Nevertheless, even though safer technology is introduced, health and safety could well have been at issue during the decision making process, in which case management should be obligated to bargain with labor at the time. An example of a situation where safer technology is introduced with worker displacement and where health and safety are at issue in the decision is the introduction of robotics to perform hazardous work previously performed by human workers. Although the use of robots to do hazardous work is admittedly a benefit to workers in one respect, it can have a negative effect on their lives as well. The widespread use of robots can result in lay offs and seniority problems and changes in the composition of the bargaining unit by making certain blue-collar

jobs obsolete and requiring increased technical expertise. Compensation issues would be raised where the use of robots cuts down on the need for overtime and thus directly affects wages. These are but a few of the myriad ways in which the use of robots can affect employees, positively and negatively. One of the purposes of collective bargaining is to provide workers with an opportunity to participate in decisions which profoundly affect their lives by proposing alternative ideas to management. Certainly, this is a situation in which workers would wish to propose alternatives, and, perhaps, to make concessions to avoid complete automation of the process by using robots. Workers may often be even more concerned with job security than with the question of their own personal health and safety. Workers have had a long tradition of engaging in hazardous occupations (consider, *e.g.*, coal miners) in order to support their families at the risk of their own lives. That is not to say that management should not accord the highest priority to making the workplace as safe as possible for workers; rather, where various considerations surrounding health and safety must be weighed against those of job security, workers should be given a voice in those decisions. Certainly, if labor is to pay for safer working conditions by losing jobs, it ought to have any opportunity to suggest job savings *and* safer technological alternatives.

A question surrounding the introduction of *"safer"* technology, regardless of whether it results in worker displacement or not, is *whether the technology really is safer or whether the hazards are latent and as yet unknown.* Management does not always possess complete and perfect information. Today many labor unions have become very sophisticated in their efforts to protect the health and safety of workers from hazards inherent in many technological advances. Often acting in concert with unions in other industrial markets abroad, American unions are expending substantial time, money and effort on researching and studying the effects on workers of exposure to various toxic substances, VDTs and the like. To assume that management alone possesses all of the information necessary to make the most well-informed decisions concerning the health and safety of workers may be paternalistic and short-sighted. The use of personal protection equipment, such as hearing protectors or rubber suits, while ostensibly adopted to protect the worker, also can stress the worker. Unions consistently argue for engineering controls on technology rather than enclosing the worker in a hot suit or burdening the worker with devices which impair his movement. Even where management intends in good faith to introduce safer technology, labor's participation in the decision making process contributes to a fully informed decision.

The discussion thus far has focused upon health and safety as a consideration in the decision making process. Even if health and safety issues are not involved in the decision to introduce technological change, consideration should be given to mandating a duty to bargain over the decision. Technological change can have an even larger effect on job security than health and safety. Health and safety affects individual workers, and can possibly affect their later offspring. If new technology results in transfers of operations, partial plant closing or an entire plant closing in a small community, those changes affect the lives of workers, their families and the entire community. The effect can be devastating in economic, social and personal terms.

Board and court decisions accord high priority to issues of health and safety; however, technology can affect workers just as significantly in many other ways. The union can represent the interests and concerns of the workers involved, and thereby mitigate the impact most effectively through participation at the decision making stage. As one commentator ably articulates:

> Without full collective bargaining—no matter how enlightened or benevolent management may be—working men and women simply don't participate in the basic decisions which govern their jobs, their income, and their lives. Collective bargaining is essential to meet the challenge of technological change with a minimum of social and human dislocation; . . . to assure workers of dignity on the job and protection against arbitrary action by management; and to work for general economic, social and political conditions which enhance human welfare, human dignity and human freedom.[190]

D. Employee Participation in Management Decision Making

Numerous Board and court decisions outlined above, particularly the *First National Maintenance* decision, and recent Board decisions since *First National Maintenance* deferred to management's perceived need for unfettered decision making. Implicit in the Court's arguments in *First National Maintenance* is the assumption that labor can generally only hinder rather than

advance the decision making process. The introduction of technological change into the workplace, however, is rarely done under compelling conditions requiring speed, flexibility or secrecy. Diffusion of existing technologies (usually tried in other plants) into a particular workplace, rather than technological innovation, is the norm. In addition, the employer does not always necessarily have complete and perfect information concerning alternatives so as to know whether bargaining would be futile. Although the Board and court decisions have attempted to diminish labor's potential contribution to the decision making process, discussed below are various experiences in employee participation in management which stand in stark contrast to advocates of management's rights who argue that labor's participation is unworkable, or at best a hindrance.

Some labor unions have increasingly assumed a more collaborative role with management in the decision making process. Unions have been participating in joint committees with management for some time.[193] Many of those joint committees still play only a consultative role with management, exercising no real control. However, some unions, departing from this consultative role, are assuming a more collaborative role in the introduction of technology into the workplace. Substantial participation in management decision making is occurring currently in both the United States and abroad.[194] In-depth documentation of the successes and failures of these activities may be desirable to undertake at a future time in order to fully explore the opportunities presented by technology bargaining. Extensive documentation of labor participation in management

[190]Roberts, *The Impact of Technology on Union Organizing and Collective Bargaining*, in *Labor and Technology: Union Response to Changing Environments* 26 (1982).

[193]See e.g., L. Bacow . . . , [*Bargaining for Job Safety and Health* (1980)]; Recent Initiatives in Labor-Management Cooperation, Report No. CP76003, National Center for Productivity and Quality of Working Life (Feb. 1976); More Awareness of Health, Safety Issues Comes from Joint Committees, Survey Finds, O.S.H. Rep. (BNA), at 159 (July 12, 1984); Auto Supplier Says Joint Committee Sharply Increases Productivity, Quality, O.S.H. Rep. (BNA), at 450 (Nov. 8, 1984).

[194]For discussions of employee participation activities abroad, see Chamot and Dymmel, *supra* note 186; Chamot, Technological Change and Unions: An International Perspective, a publication of the Dept. of Professional Employees, AFL-CIO, Aug. 1982; Work and Health in the 1980's: Experiences of Direct Workers' Participation in Occupational Health (Bagnara, Misiti, and Wintersberger eds. 1985) (proceedings of the Conference on Direct Workers' Participation in Matters of Work, Safety and Health, held in Nov. 1982 at the Institute of Psychology of the National Research Council of Italy in Rome).

decision making is beyond the scope of this Article. However, this Article will discuss some representative examples of labor participation in the introduction of technological change.

The United Automobile Workers (UAW) is one union in particular which has taken great strides in acquiring some degree of participation in the decision making process surrounding the introduction of new technology. The president of the UAW currently sits on the Chrysler Corporation Board of Directors.[195] Certainly this is a striking, albeit atypical, example of a step toward labor participation in major decisions. The UAW has taken an aggressive role in ameliorating the impact of technological change on the work force by promoting joint management-union efforts. Since 1979, as part of many UAW collective bargaining agreements, management has had a duty to notify and consult with the union over any decision to introduce new technology well in advance of implementation. The agreements also provide that a National Joint Committee on New Technology, composed of five UAW members and five company members must be established to discuss at the corporate level any development of new technology, to discuss anticipated problems concerning the introduction of new technology, and to review and decide cases appealed from the plant level concerning disputes over the introduction of new technology. In 1982, the UAW reached an agreement with both the General Motors Corporation and Ford Motor Corporation to establish extensive training and retraining programs to mitigate the impact of technological change. In 1984, the UAW reached agreement with General Motors concerning a notable job security feature, the Job Opportunity Bank-Security (JOB Security) Program. Under the JOB Security Program, jobs that become unnecessary because of the introduction of new technology are put into the bank and slowly phased out. The workers continue to be employed, but they are placed into other jobs as those jobs open up or new jobs are created. Eventually the bank shrinks as those workers are placed elsewhere. Additionally, for more than a decade, the UAW along with both General Motors and Ford has jointly sponsored Quality of Work Life (QWL) activities which have been designed to ensure that new technology is introduced so as to enhance the quality of

work life in a plant. The UAW's position is that technology that advances workers' skills and enhances job content should be selected over that which downgrades or eliminates job skills whenever possible.

The UAW is involved in several particularly noteworthy joint labor-management efforts with General Motors. The UAW is currently involved in a pilot project with General Motors called the "Saturn Project," a program to manufacture small cars in the United States. The Saturn Project is a management-labor project in which management and the union are jointly studying the feasibility of introducing a new small car. The union now continues to participate in the design, engineering and production of the project, which involves many innovative concepts.

The second significant joint labor-management effort between the UAW and General Motors is a $100 million New Venture Fund. The program is an ongoing activity to develop and mutually direct ventures into new, non-traditional business areas in an effort to expand employment opportunities for UAW workers. The program is financed entirely by corporate funds, but provides for full input by the union. The program will be administered by a negotiated Growth and Opportunity Committee, made up of equal numbers of UAW and corporate representatives. The Committee will be assisted by a full-time staff assigned by General Motors and the UAW known as the New Business Venture Development Group. The Group will review and study the feasibility of proposals for entry into new business ventures, make recommendations to the Committee, and develop employee participation in the new ventures. The Committee will be required to report periodically to the International Union and to the General Motors Executive Committee concerning the identification of viable opportunities for employment growth. The negotiated agreement that established the program specified that the Committee should pay particular attention to developing new ventures in communities that have been affected by the loss of UAW-GM employment opportunities through plant closings.

An additional UAW-GM cooperative effort is the launch of a comprehensive, five-year scientific study concerning the effects on workers of exposure to chemicals in machining. The jointly administered Occupational Health Advisory Board has commissioned the Harvard School of Public Health to conduct the study. The advisory board will work with the Harvard team during the period of the study, the board will advise the company and the union concerning appropriate measures or changes in technology.

---

[195]Walton, New Perspectives on the World of Work: Social Choice in the Development of Advanced Information Technology, 35 Hum. Rel. 1073, 1078 (1982).

A second union that has become involved with management in the decision making process is the Communications Workers of America (CWA). In its collective bargaining agreements, CWA typically includes provisions that require management to notify the union at least six months prior to the introduction of technological change. These agreements also contain provisions for training and retraining programs. Although CWA has been active in QWL committees, CWA has not generally been as aggressively involved with management in the decision making process as the UAW has been successful in doing. CWA's involvement varies from case-to-case. One notable case involves the AT&T hotel billing and information systems (HOBIS) office in Tempe, Arizona, near Phoenix. The company wanted to redesign a new system in that office, its goal, being a successful, self-managed office. The company approached CWA about participating in the total system design. The joint management-labor effort resulted in an innovative autonomous work group where the workers police themselves without supervisors. The Tempe office has proved to be the most profitable of all AT&T's HOBIS units throughout the country.

A third union that has taken strides toward becoming involved in decisions to introduce technological change is the International Association of Machinists and Aerospace Workers (IAM). IAM has prepared model contract language calling for advance notification of technological change along with a meaningful opportunity to negotiate adjustments in the plan:

> In the event of management's introducing new technology it is imperative that the union firmly establish the right to advance notice, the right to certain kinds of information and the obligation to bargain over necessary adjustments through clear and specific contract language. By being required to give advance notice of plans to introduce technological change, the Union will have time to negotiate all of the necessary adjustment program.[207]

In 1983 IAM negotiated a collective bargaining contract with the Boeing Company which provides that Boeing must notify the union at least one year in advance of any technological changes, such brief-

ings to include any anticipated schedules of introduction of new technology. The union in turn is required to protect the confidentiality of any sensitive or proprietary information disclosed during the discussions. The IAM-Boeing contract also provides for an extensive training program to mitigate the effects of technological change.

The advance notification provisions exemplified by UAW, CWA and IAM collective bargaining contracts are becoming increasingly common, and represent a significant step toward a cooperative labor-management approach to the introduction of technological change. Although advance notification may take different forms in different contracts, it offers the potential for labor to influence design features and plans of proposed technological change when decisions are being made, thereby offering greater potential for solving problems and fulfilling needs of both management and labor.

Labor unions have become quite sophisticated in their efforts to make positive contributions to the direction of technological development, and, as a result, have become an information resource in matters concerning technological change and development. For example, during 1985 the Subcommittee on Employment and Housing of the House Committee on Government Operations solicited the aid of several labor unions in developing legislation on employee protection at hazardous waste cleanup sites. Also during 1985, a job health research agency, the Workplace Health Fund, was instituted through the efforts of various labor unions with assistance from groups representing business, churches, the environment, and various communities, The Fund is designed to develop and mobilize support for occupational disease research and education programs to be conducted by the labor community. It is also designed to provide job health services to workers. These are only two examples of research efforts undertaken by labor unions through which organized labor is making positive contributions to the direction and implementation of technological development. Many more extensive research programs are currently being conducted by labor. Deutsch suggests that research support is vitally important in order to identify creative approaches to the introduction of technology that benefit both management *and* labor.

The foregoing discussion concerning current involvement of organized labor in management decision making and in the development of technology should illustrate that the Board and some courts are not adequately considering current industrial activi-

[207]Nulty, Case Studies of Local IAM Experience with the Introduction of New Technologies, in Labor and Technology: Union Response to Changing Environments 132 (1982).

ties when they argue that labor has nothing mean-ingful to offer to management in its decision making process. To requote the Supreme Court's majority opinion in *Fibreboard:* "Industrial experience is not only reflective of the interests of labor and manage-ment in the subject matter but is also indicative of

the amenability of such subjects to the collective bargaining process."[215]

[215]Fibreboard Paper Prods. Corp. v. NLRB, 379 U.S. 203, 211 (1964).

◇ NOTES

1. The authors rest much of their legal argument for mandatory decision bar-gaining on the fact that health and safety are mandatory subjects of bargaining. How strong is this argument? Is there a difference between bargaining over the question of whether a potentially dangerous new technology *should be introduced,* and bargaining over, say, the development of health and safety rules to protect workers from the effects of that technology *once it is introduced?* Which of these is closer to the situation facing the Fifth Circuit Court of Appeals in *Gulf Power?*

2. The stronger argument for mandatory decision bargaining here is that there is much to be gained by it. Look again at the "test" applied by the Supreme Court in *First National Maintenance* in deciding whether decision bargaining is required. How well do the benefits of technology bargaining fare under this test? Note the importance of stressing, as the foregoing article does, the sophisticated role that labor could play.

3. There appears to be a trend toward regulatory policies that encourage or require companies to modify production processes to reduce reliance on toxic chemicals. See, e.g., the Massachusetts Toxics Use Reduction Law, M.G.L. ch. 21i (1989). Although the driving force behind these policies is to minimize the release of toxic chemicals into the environment, these policies may also serve to prompt greater cooperation between management and labor on issues of product and process reformulation.

6.   Arbitration Clauses and the Right to Strike

Traditionally, unions flexed their political and economic muscle by going out on strike—by refusing to come back to work until the employer agreed to take certain actions (or to bargain over the need for such actions). This is much less the case today. Although the NLRA guarantees the right to strike, the system of "industrial self-government" that has developed under the Act has tended to limit the role of the strike as a means of effecting change in the workplace.

(a) Types of strikes

The Act recognizes two types of lawful strikes: the "unfair labor practice" strike, and the "economic" strike. The distinction lies with the *purpose* of the strike. Employees who strike to protest an unfair labor practice committed by their employer are called unfair labor practice strikers. Absent serious misconduct during the strike, unfair labor practice strikers may not be discharged or permanently replaced. When the strike is over, the employer must give them their old jobs back, even if replacement employees have been hired in the interim. Economic strikers, on the other hand, are those who go on strike to try to obtain some economic concession, such as higher pay, shorter hours, or better working conditions. These strikers may not be discharged, but they may be *replaced* by new *bona fide* permanent replacements. Thus, although they must be given priority for any new openings that arise in the future, economic strikers may find themselves out of work for an extended period of time after the strike is over. For this reason, the staging of an economic strike poses a greater risk for the union. Nonetheless, the NLRA gives employees a clear right to stage both unfair labor practice and economic strikes.

(b) The no-strike clause

However, employees may bargain away their right to strike under the Act. Typically, this is done by agreeing to include a no-strike clause in the collective bargaining agreement. Such a clause may apply only to a few issues, or it may be a broad covenant not to stage a strike under any circumstances. In either case, the clause is lawful. Further, as the Supreme Court made clear in the following decision, no-strike clauses generally will be enforced by the courts.

Boys Markets, Inc. v. Retail Clerks Union, Local 770

398 U.S. 235 (1970)

Mr. Justice BRENNAN delivered the opinion of the Court.

In this case we re-examine the holding of *Sinclair Refining Co.* v. *Atkinson*, 370 U.S. 195, 50 LRRM 2420 (1962), that the anti-injunction provisions of the Norris-LaGuardia Act preclude a federal district court from enjoining a strike in breach of a no-strike obligation under a collective bargaining agreement even though that agreement contains provisions enforceable under Section 301(a) of the Labor Management Relations Act for binding arbitration of the grievance dispute concerning which the strike was called. The Court of Appeals for the Ninth Circuit, considering itself bound by *Sinclair,* reversed the grant by the District Court for the Central District of California of petitioner's prayer for injunctive relief.... Having concluded that *Sinclair* was erroneously decided and that subsequent events have undermined its continuing validity, we overrule that decision and reverse the judgment of the Court of Appeals.

I

In February 1969, at the time of the incidents that produced this litigation, petitioner and respondent were parties to a collective bargaining agreement which provided *inter alia,* that all controversies concerning its interpretation or application should be resolved by adjustment and arbitration procedures set forth therein and that, during the life of the contract, there should be "no cessation or stoppage of work, lock-out picketing or boycotts...." The dispute arose when petitioner's frozen foods supervisor and certain members of his crew who were not members of the bargaining unit began to rearrange merchandise in the frozen food cases of one of petitioner's supermarkets. A union representative insisted that the food cases be stripped of all merchandise and be restocked by union personnel. When petitioner did not accede to the union's demand, a strike was called and the union began to picket petitioner's establishment. Thereupon petitioner demanded that the union cease the work stoppage and picketing and sought to invoke the grievance and arbitration procedures specified in the contract.

The following day, since the strike had not been terminated, petitioner filed a complaint in California Superior Court seeking a temporary restraining order, a preliminary and permanent injunction, and specific performance of the contractual arbitration provision.... Shortly thereafter, the union removed the case to the federal district court and there made a motion to quash the state court's temporary restraining order.... Concluding that the dispute was subject to arbitration under the collective bargaining agreement and that the strike was in violation of the contract, the District Court ordered the parties to arbitrate the underlying dispute and simultaneously enjoined the strike, all picketing in the vicinity of petitioner's supermarket, and any attempts by the union to induce the employees to strike or to refuse to perform their services.

* * *

IV

We have . . . determined that the dissenting opinion in *Sinclair* states the correct principles concerning the accommodation necessary between the seemingly absolute terms of the Norris-LaGuardia Act and the policy considerations underlying §301(a), 370 U.S., at 215. Although we need not repeat all that was there said, a few points should be emphasized at this time.

The literal terms of §4 of the Norris-LaGuardia Act must be accommodated to the subsequently enacted provisions of §301(a) of the Labor-Management Relations Act and the purposes of arbitration. Statutory interpretation requires more than concentration upon isolated words; rather, consideration must be given to the total corpus of pertinent law and the policies which inspired ostensibly inconsistent provisions. See *Richards* v. *United States*, 369, U.S. 1, 11, 82 S. Ct. 585, 592, 7 L. Ed. 2d 492 (1962); *Mastro Plastics Corp.* v. *NLRB*, 350 U.S. 270, 285, 76 S. Ct. 349, 359, 100 L. Ed. 309 (1956); *United States* v. *Hutcheson*, 312 U.S. 219, 235, 61 S. Ct. 463, 467, 85 L. Ed. 788 (1941).

The Norris-LaGuardia Act was responsive to a situation totally different from that which exists today. In the early part of this century, the federal courts generally were regarded as allies of management in its attempt to prevent the organization and strengthening of labor unions; and in this industrial struggle the injunction became a potent weapon which was wielded against the activities fo labor groups. The result was a large number of sweeping decrees, often issued *ex parte*, drawn on an *ad hoc* basis without regard to any systematic elaboration of national labor policy. See *Drivers' Union* v. *Lake Valley Co.*, 311 U.S. 91, 102 (1940).

In 1932 Congress attempted to bring some order out of the industrial chaos that had developed and to correct the abuses which had resulted from the interjection of the federal judiciary into union-management disputes on the behalf of management. See Declaration of Public Policy, Norris-LaGuardia Act, §2, 47 Stat. 70 (1932). Congress, therefore, determined initially to limit severely the power of the federal courts to issue injunctions "in any case involving or growing out of any labor dispute. . . . " 47 Stat. 70. Even as initially enacted, however, the prohibition against federal injunctions was by no means

absolute. See Norris-LaGuardia Act, §§7, 8, 9, 47 Stat. 70 (1932). Shortly thereafter Congress passed the Wagner Act, designed to curb various management activities which tended to discourage employee participation in collective action.

As labor organizations grew in strength and developed toward maturity, congressional emphasis shifted from protection of the nascent labor movement to the encouragement of collective bargaining and to administrative techniques for the peaceful resolution of industrial disputes. This shift in emphasis was accomplished, however, without extensive revision of many of the older enactments, including the anti-injunction section of the Norris-LaGuardia Act. Thus it became the task of the courts to accommodate, to reconcile the older statutes with the more recent ones.

A leading example of this accommodation process is *Brotherhood of R. R. Trainmen* v. *Chicago River & Ind. R. R.*, 353 U.S. 30 (1957). There we were confronted with a peaceful strike which violated the statutory duty to arbitrate imposed by the Railway Labor Act. The Court concluded that a strike in violation of a statutory arbitration duty was not the type of situation to which the Norris-LaGuardia Act was responsive, that an important federal policy was involved in the peaceful settlement of disputes through the statutorily-mandated arbitration procedure, that this important policy was imperiled if equitable remedies were not available to implement it, and hence that Norris-LaGuardia's policy of nonintervention by the federal courts should yield to the overriding interest in the successful implementation of the arbitration process.

The principles elaborated in *Chicago River* are equally applicable to the present case. To be sure, *Chicago River* involved arbitration procedures established by statute. However, we have frequently noted, in such cases as *Lincoln Mills*, the *Steelworkers Trilogy*, and *Lucas Flour*, the importance which Congress has attached generally to the voluntary settlement of labor disputes without resort to self-help and more particularly to arbitration as a means to this end. Indeed, it has been stated that *Lincoln Mills*, in its exposition of §301(a), "went a long way towards making arbitration the central institution in the administration of collective bargaining contracts."

The *Sinclair* decision, however, seriously undermined the effectiveness of the arbitration technique as a method peacefully to resolve industrial disputes without resort to strikes, lockouts, and similar devices. Clearly employers will be wary of assuming obligations to arbitrate specifically enforceable against them when no similarly efficacious remedy is available to enforce the concomitant undertaking of the union to refrain from striking. On the other hand, the central purpose of the Norris-LaGuardia Act to foster the growth and viability of labor organizations is hardly retarded—if anything, this goal is advanced—by a remedial device which merely enforces the obligation that the union freely undertook under a specifically enforceable agreement to submit disputes to arbitration. We conclude, therefore, that the unavailability of equitable relief in the arbitration context presents a serious impediment to the congressional policy favoring the voluntary establishment of a mechanism for the peaceful resolution of labor disputes, that the core purpose of the Norris-LaGuardia Act is not sacrificed by the limited use of equitable remedies to further this important policy, and consequently that the Norris-LaGuardia Act does not bar the granting of injunctive relief in the circumstances of the instant case.

V

Our holding in the present case is a narrow one. We do not undermine the vitality of the Norris-LaGuardia Act. We deal only with the situation in which a collective bargaining contract contains a mandatory grievance adjustment or arbitration procedure. Nor does it follow from what we have said that injunctive relief is appropriate as a matter of course in every case of a strike over an arbitrable grievance. The dissenting opinion in *Sinclair* suggested the following principles for the guidance of the district courts in determining whether to grant injunctive relief—principles which we now adopt: "A District Court entertaining an action under §301 may not grant injunctive relief against concerted activity unless and until it decides that the case is one in which an injunction would be appropriate despite the Norris-LaGuardia Act. When a strike is sought to be enjoined because it is over a grievance which both parties are contractually bound to arbitrate, the District Court may issue no injunctive order until it first holds that the contract *does* have the effect; and the employer should be ordered to arbitrate, as a condition of his obtaining an injunction against the strike. Beyond this, the District Court must, of course, consider whether issuance of an injunction would be warranted under ordinary principles of equity—whether breaches are occurring and will continue, or have been threatened and will be committed; whether they have caused or will cause irreparable injury to the employer; and whether the employer will suffer more from the denial of an injunction than will the union from its issuance." 370 U.S., at 228. (Emphasis in original.)

In the present case there is no dispute that the grievance in question was subject to adjustment and arbitration under the collective bargaining agreement and that the petitioner was ready to proceed with arbitration at the time an injunction against the strike was sought and obtained. The District Court also concluded that, by reason of respondent's violations of its no-strike obligation, petitioner "has suffered irreparable injury and will continue to suffer irreparable injury." Since we now overrule *Sinclair,* the holding of the Court of Appeals in reliance on *Sinclair* must be reversed. Accordingly, we reverse the judgment of the Court of Appeals and remand the case with directions to enter a judgment affirming the order of the District Court.

[Concurring opinion of STEWART, J., and dissenting opinion of BLACK, J., omitted. MARSHALL, J. took no part in the decision of the case.]

A key point in the court's willingness to enforce the no-strike clause here was the existence of an arbitration clause covering the issue in question. That is, the agreement by the employer to enter into binding arbitration over the issue was seen as a necessary *quid pro quo* for the union's agreement not to strike over that issue. By implication, then, arbitration was deemed to be an adequate substitute for the strike. Indeed, it is fair to say that arbitration has become the primary means of formal dispute resolution in the workplace.

(c) Arbitration

The NLRA does not compel either side to submit to mandatory arbitration. However, as part of the collective bargaining agreement, the parties can agree that some or all disputes arising under the agreement will be settled by submitting the matter to binding arbitration. That is, both the employer and the union can agree in advance that any disagreement about contract interpretation will be resolved through arbitration rather than through the courts or the NLRB.

The arbitration procedure is rather straightforward. When a dispute arises, an independent arbitrator or a panel of arbitrators is selected as specified in the collective bargaining agreement. Typically, the agreement requires that the arbitrator be selected from a specified "pool" of arbitrators and be acceptable to both parties. The arbitrator then convenes a hearing, and both parties present evidence and arguments supporting their points of view. The hearing is quasi-judicial in nature. Initially, labor arbitration hearings were less formal than full court trials and often were done without lawyers. More recently, however, the proceedings have become considerably more formal, with lawyers representing each side. Either way, the end result is the arbitrator's binding decision as to the proper interpretation of the collective bargaining agreement regarding the dispute at hand. If necessary, the arbitrator may "specifically enforce" the agreement by issuing an order compelling one of the parties to take some action—or refrain from taking some action—to come into compliance with the agreement.

This means of labor dispute resolution has received considerable support from the courts. In general, when there is a question as to whether a particular issue is covered by an arbitration clause in a collective bargaining agreement, the courts employ a strong presumption *in favor* of arbitrability. As noted by a plurality of the Supreme Court in *United Steelworkers of America v. Warrior and Gulf Navigation Co.*, 363 U.S. 574, 584-585 (1960), "In the absence of any express provision excluding a particular grievance from arbitration, we think only the most forceful evidence of a purpose to exclude the claim from arbitration can prevail."

◇ NOTES

1. The employee seeking to walk off the job to protest unsafe working conditions often will have greater legal protections than those available to the economic, or even the unfair labor practice, striker. As discussed in detail in Chapter 8, there is a limited *right to refuse hazardous work* under both the NLRA and the OSH Act.

2. Management has a tool analogous to the strike at its disposal. A "lockout" is in some sense the opposite of a strike. When the collective bargaining expires, the employer may put pressure on the union by simply "locking out" the employees and refusing to allow them to come to work. Through the collective bargaining agreement, the union may be able to negotiate a promise of no lockout from the employer. The delay of the 1990 major league baseball season is a tribute to the continuing vitality of the lockout.

3. Unless the collective bargaining agreement specifies otherwise, courts generally interpret an agreement to *arbitrate* a certain issue or set of issues as an implicit promise *not to strike* over that issue or set of issues. To the extent that there is a presumption in favor of arbitration, then, there is a corresponding presumption *against* strikes. As noted by the plurality opinion in *Warrior & Gulf*, cited above, "[arbitration], rather than strike, is the terminal point of disagreement."

4. Can one make the argument that the courts have moved from a view of the NLRA as the protector of the unions as "underdogs" to a view of the NLRA as a "neutral" law that merely provides a vehicle for labor negotiations? Could this help to explain the decisions in *Boys Markets* and *First National Maintenance*?

5. Cases settled by arbitration have little precedential value beyond the particular parties and contract subject to the dispute.

6. As an alternative to arbitration, the parties can agree to submit a dispute to mediation. This is a form of negotiation that is facilitated by a professional mediator.

D.   LABOR-MANAGEMENT HEALTH AND SAFETY COMMITTEES

Joint health and safety committees can be an important vehicle for improving working conditions. See, for example, Robinson, J.C., "Labor Union Involvement in Occupational Safety and Health, 1957–1987," 13 *Journal of Health Politics, Policy and Law* 453 (1988); Ruttenberg, R. "The Role of Labor-Management Committees in Safeguarding Worker Safety and Health," U.S. Department of Labor, Report BLMR 121 (1988); and Witt, M., and Early, S. "The Worker as Safety Inspector," *Working Papers*, September-October (1980): 21. Such committees either can be created by collective bargaining, or can exist and function "outside the collective bargaining contract" so that they do not become embroiled in labor-management disputes unrelated to health and safety. They can also be established in nonunion workplaces. However, if management appoints the worker members, or is the committee's sole financial supporter, or sets the committee's agenda, the committee might be regarded as a "management dominated labor organization" in violation of the NLRA.

Contract clauses mandating joint safety committees have doubled between 1957 and 1987, from 31 percent to 62 percent in manufacturing contracts and from 12 percent to 27 percent in nonmanufacturing contracts. See Robinson, *supra.* The breakdown by industry for 1983 and 1986 is shown in Table 6-3, which is taken from Ruttenberg, *supra.,* at p. 3.

Table 6-3. Collective-bargaining contracts with provisions for labor–management safety committees (frequency expressed as percentage of industry contracts)

	1983	1986
All Industries	45	49
Manufacturing	57	62
Apparel	11	22
Chemicals	75	75
Electrical Machinery	30	40
Fabricated Metals	68	68
Food	52	52
Furniture	50	33
Leather	50	53
Lumber	14	57
Machinery	56	68

(continued)

Table 6-3. continued

	1983	1986
Paper	50	43
Petroleum	71	71
Primary Metals	84	88
Printing	13	13
Rubber	100	100
Stone, Clay & Glass	62	62
Textiles	20	40
Transportation Equipment	77	83
Nonmanufacturing	26	27
Communications	40	50
Construction	10	10
Insurance & Finance	—	—
Maritime	38	50
Mining	83	83
Retail	4	4
Services	22	22
Transportation	32	32
Utilities	50	50

Source: Bureau of National Affairs.

The roles for joint committees have expanded as well. Some contracts direct joint committees to make recommendations for management consideration; in others, the committees actually have responsibility for correcting and preventing hazardous situations. Committee functions may include submission of monthly reports covering suggestions and recommendations as to safety appliances, equipment, clothing, rules, and practices; study of safety performance and recommendations for plantwide programs and standards; investigation of serious accidents, injuries, working conditions, and practices; conferring with the safety director; making inspections; undertaking investigations; offering suggestions; and shutting down unsafe machinery. Although they often contribute to safer working conditions by helping to inspect and maintain existing production technology, the joint committees rarely are a catalyst for technology and job redesign. This situation may well be due to the fact that they are largely advisory in nature and are not endowed with real decisionmaking powers.

◇ NOTES

1. Legislation in many European countries mandates the creation of joint health and safety committees or "works' council committees." See Witt and Early, *supra.;* Ashford, *Crisis in the Workplace,* pp. 500–519 (1976). Given recent evidence that union involvement in health and safety leverages the effectiveness of OSHA, similar legislation might be warranted in the United States. See Weil, "Government and Labor at the Workplace." *Proceedings of the 1987 Industrial Relations Research Association*. 332–335 (1987).

2. An employer's rationale for not giving too much decisionmaking power to health and safety committees probably is obvious. Unions may balk at the idea as well, however, out of a concern that it would increase their liability for health and safety problems at the worksite. (This concern should be lessened by a recent decision of the U.S. Supreme Court. See the discussion of union liability in Chapter 9.)

E. THE DISSENTING EMPLOYEE

Although the bulk of labor law is concerned with the relationship between labor and management, the relationship between employees and their union is also important. Certain protections are available under the NLRA for employees who disagree with their union and for employees who may not want to join a union at all.

Nonetheless, union representation involves a majoritarian form of government, and the "first principle" of union representation is that, on collective bargaining issues, a majority vote of the union members is binding. Thus, for example, if the union membership votes to accept safety and health concessions that will have an adverse impact on certain workers, those workers generally will have no right to challenge this result under the NLRA. Instead, their recourse is to attempt to change the political climate *within the union,* so as to place a greater emphasis on safety and health.

1. Impermissible Discrimination and the Duty of Fair Representation

There are two important exceptions to this general principle. The first flows from the rule that the collective bargaining agreement is not permitted to violate the law. For example, if the *purpose* of the safety and health concessions were to make life more difficult for a group of minority employees who worked in the affected area, the vote might well violate Title VII of the federal Civil Rights Act, which prohibits employers from discriminating against workers on the basis of race.

The second limitation on the "majority rule" principle of the NLRA is the doctrine of "fair representation." Under this doctrine, which grew out of cases involving racial discrimination within labor unions, the union has an affirmative duty to "fairly represent" *all* the employees in the bargaining unit. See *Vaca v. Sipes,* 386 U.S. 171 (1967). In the example above, this doctrine would be another basis for setting aside the contract provision if it were racially motivated. Indeed, the concept of fair representation extends well beyond racial, religious, or sex-based discrimination. Importantly, it provides the union dissident with at least some protection from union retaliation. For example, the union would violate its duty of fair representation if it failed to process an employee's grievance with management, or refused to put the employee on a union hiring list, because that employee had advocated a change in union leadership.

Some courts had held that the duty of fair representation also protected employees from union "malpractice" in processing grievances. If a worker could prove that the union did an inadequate job of processing a grievance and that he or she would have prevailed in the grievance had the union provided adequate representation, the worker sometimes could pursue a cause of action against the union for a breach of the duty of fair representation on that basis alone. That is, the employee did not need to show that the union's failure was due to discrimination or other hostility; it was enough to show that the union was negligent. See *Ruzick. v. General Motors*, 707 F.2d 259 (6th Cir. 1983), cert. denied, 464 U.S. 982 (1983); *De Arroyo v. Sindicato De Trabajadores Packinghouse*, 425 F.2d 281 (1st Cir. 1970), cert. denied, 400 U.S. 877 (1970). However, in *United Steelworkers of America v. Rawson*, __U.S.__, 110 S. Ct. 1904 (1990), discussed in Chapter 9, the Supreme Court has indicated that "mere negligence" does not violate the duty of fair representation.

2. Must Employees Join the Union?

What about employees who do not want to join a union at all? Can they be forced to become union members as a condition of employment? The answer is found in Sections 7 and 8 of the Act. Section 7 specifies that employment must be extended to nonunion members "under the same terms and conditions" as union members. Thus, this section prohibits a "closed shop"—an arrangement under which only union members are employed at a given workplace and the union has absolute control over who can join the union. However, Section 8(a)(3) does allow the union to enter into an agreement with the employer for the creation of what is commonly termed a "union shop." Under this arrangement, as with the closed shop, union membership is a prerequisite for employment. But the union cannot deny union membership to an otherwise qualified person whom the employer wishes to hire. The purpose of the union shop is to protect the union from "free riders." This is the name commonly given to nonunion employees within the bargaining unit who reap the benefits of the union's collective bargaining activities without bearing any of the costs through union dues.

◇ NOTES

1. A common variation of the union shop is the "union security agreement." Under this arrangement, new employees do not have to become members of the union, but they are required to pay the union a "security fee," equivalent to what their union dues would have been if they were members. This, too, is permissible under the NLRA.

2. Dissenting employees do have some right to limit how the union uses their security fees. If such an employee objects, a union may not use the employees' fees for purposes unrelated to organizing and collective bargaining, such as for making political campaign contributions. The union must refund to the dissenting employee that portion of the fee that would otherwise be used for such purposes. See *International Association of Machinists v. Street*, 367 U.S. 740 (1961); *Abood v. Detroit Board of Education*, 431 U.S. 209 (1977); *Ellis v. Brotherhood of Railways, Airline and Steamship Clerks*, 466 U.S. 435 (1984); *Chicago Teachers Union Local No. 1 v. Hudson*, 475 U.S. 292 (1986); and *Beck v. C.W.A.*__U.S.__, 108 S. Ct. 2897 (1988).

3. States are free to *prohibit* the establishment of a union shop—or a union security agreement—by passing so-called right-to-work laws. As of 1990, nineteen states, mostly in the south and west, had done so.

F. THE ROLE OF THE NLRB

Finally, we turn to the National Labor Relations Board, whose function it is to administer the Act. The following explanation of the organization, function, and authority of the NLRB comes from the Board's own publication.

A Guide to Basic Law and Procedures

NLRB

The rights of employees declared by Congress in the National Labor Relations Act are not self-enforcing. To ensure that employees may exercise these rights, and to protect them and the public from unfair labor practices, Congress established the NLRB to administer and enforce the Act.

The NLRB includes the Board, which is composed of five members with their respective staffs, the General Counsel and staff, and the Regional, Subregional, and Resident Offices. The General Counsel has final and independent authority on behalf of the Board, in respect to the investigation of charges and issuance of complaints. Members of the Board are appointed by the President, with consent of the Senate, for 5-year terms. The General Counsel is also appointed by the President, with consent of the Senate, for a 4-year term. Offices of the Board and the General Counsel are in Washington, D.C. To assist in administering and enforcing the law, the NLRB has established 33 Regional and a number of other field offices. These offices, located in major cities in various States and Puerto Rico, are under the general supervision of the General Counsel.

The Agency has two main functions: to conduct representation elections and certify the results, and to prevent employers and unions from engaging in unfair labor practices. In both kinds of cases the processes of the NLRB are begun only when requested. Requests for such action must be made in writing on

forms provided by the NLRB and filed with the proper Regional office. The form used to request an election is called a "petition," and the form for unfair labor practices is called a "charge." The filing of a petition or a charge sets in motion the machinery of the NLRB under the Act. Before discussing the machinery established by the Act, it would be well to understand the nature and extent of the authority of the NLRB.

The NLRB gets its authority from Congress by way of the National Labor Relations Act. The power to Congress to regulate labor-management relations is limited by the commerce clause of the United States Constitution. Although it can declare generally what the rights of employees are or should be, Congress can make its declaration of rights effective only in respect to enterprises whose operations "affect commerce" and labor disputes that "affect commerce." The NLRB, therefore, can direct elections and certify the results only in the case of an employer whose operations affect commerce. Similarly, it can act to prevent unfair labor practices only in cases involving labor disputes that affect, or would affect, commerce.

"Commerce" includes trade, traffic, transportation, or communication within the District of Columbia or any Territory of the United States; or between any State or Territory and any other State, Territory, or the District of Columbia; or between two points in the same State, but through any other State, Territory, the District of Columbia, or a foreign country. Examples of enterprises engaged in commerce are:

Source: From U.S. Government Printing Office 39–44 (1987)

* A manufacturing company in California that sells and ships its product to buyers in Oregon.

+ A company in Georgia that buys supplies in Louisiana.

+ A trucking company that transports goods from one point in New York State through Pennsylvania to another point in New York State.

+ A radio station in Minnesota that has listeners in Wisconsin.

Although a company may not have any direct dealings with enterprises in any other State, its operations may nevertheless affect commerce. The operations of a Massachusetts manufacturing company that sells all of its goods to Massachusetts wholesalers affect commerce if the wholesalers ship to buyers in other States. The effects of a labor dispute involving the Massachusetts manufacturing concern would be felt in other States and the labor dispute would, therefore, "affect" commerce. Using this test, it can be seen that the operations of almost any employer can be said to affect commerce. As a result, the authority of the NLRB could extend to all but purely local enterprises.

The scope of the commerce clause is limited however, by the First Amendment's prohibition against Congress' enacting laws restricting the free exercise of religion. Because of this potential conflict, and because Congress has not clearly expressed an intention that the Act cover lay faculty in church-operated schools, the Supreme Court has held that the Board may not assert jurisdiction over faculty members in such institutions.

Although the National Labor Relations Board could exercise its powers to enforce the Act in all cases involving enterprises whose operations affect commerce, the Board does not act in all such cases. In its discretion it limits the exercise of its power to cases involving enterprises whose effect on commerce is substantial. The Board's requirements for exercising its power or jurisdiction are called "jurisdictional standards." These standards are based on the yearly amount of business done by the enterprise, or on the yearly amount of its sales or of its purchases. They are stated in terms of total dollar volume of business and are different for different kinds of enterprises. The Board's standards in effect on July 1, 1976 are as follows:

1. *Nonretail business:* Direct sales of goods to consumers in other States, or indirect sales through others (called outflow), of at least $50,000 a year; or direct purchases of goods from suppliers in other States, or indirect purchases through others (called inflow), or at least $50,000 a year.

2. *Office buildings:* Total annual revenue of $100,000 of which $25,000 or more is derived from organizations which meet any of the standards except the indirect outflow and indirect inflow standards established for nonretail enterprises.

3. *Retail enterprises:* At least $500,000 total annual volume of business.

4. *Public utilities:* At least $250,000 total annual volume of business, or $50,000 direct or indirect outflow or inflow.

5. *Newspapers:* At least $200,000 total annual volume of business.

6. *Radio, telegraph, television, and telephone enterprises:* At least $100,000 total annual volume of business.

7. *Hotels, motels, and residential apartment houses:* At least $500,000 total annual volume of business.

8. *Privately operated health care institutions:* At least $250,000 total annual volume of business for hospitals; at least $100,000 for nursing homes, visiting nurses associations, and related facilities; at least $250,000 for all other types of private health care institutions defined in the 1974 amendments to the Act. The statutory definition includes: "any hospital, convalescent hospital, health maintenance organization, health clinic, nursing home, extended care facility or other institution devoted to the care of the sick, infirm, or aged person." Public hospitals are excluded from NLRB jurisdiction by Section 2(2) of the Act.

9. *Transportation enterprise, links and channels of interstate commerce:* At least $50,000 total annual income from furnishing interstate passenger and freight transportation services; also performing services valued at $50,000 or more for businesses which meet any of the jurisdictional standards except the indirect outflow and indirect inflow of standards established for nonretail enterprises.

10. *Transit systems:* At least $250,000 total annual volume of business.

11. *Taxicab companies:* At least $500,000 total annual volume of business.

12. *Associations:* These are regarded as a single employer in that the annual business of all association members is totaled to determine whether any of the standards apply.

13. *Enterprises in the Territories and the District of Columbia:* The jurisdictional standards apply in

the Territories; all businesses in the District of Columbia come under NLRB jurisdiction.

14. National defense: Jurisdiction is asserted over all enterprises affecting commerce when their operations have a substantial impact on national defense, whether or not the enterprises satisfy any other standard.

15. Private universities and colleges: At least $1 million gross annual revenue from all sources (excluding contributions not available for operating expenses because of limitations imposed by the grantor).

16. Symphony orchestras: At least $1 million gross annual revenue from all sources (excluding contributions not available for operating expenses because of limitations imposed by the grantor).

17. Law firms and legal assistance programs: At least $250,000 gross annual revenues.

18. Employers that provide social services: At least $250,000 gross annual revenues.

Through enactment of the 1970 Postal Reorganization Act, jurisdiction of the NLRB was extended to the United States Postal Service, effective July 1, 1971.

In addition to the above-listed standards, the Board asserts jurisdiction over gambling casinos where these enterprises are legally operated, when their total annual revenue from gambling is at least $500,000.

Ordinarily if an enterprise does the total annual volume of business listed in the standard, it will necessarily be engaged in activities that "affect" commerce. The Board must find, however, based on evidence, that the enterprise does in fact "affect" commerce.

The Board has established the policy that where an employer whose operations "affect" commerce refuses to supply the Board with information concerning total annual business, etc., the Board may dispense with this requirement and exercise jurisdiction.

Finally, Section 14(c)(1) authorizes the Board, in its discretion, to decline to exercise jurisdiction over any class or category of employers where a labor dispute involving such employees is not sufficiently substantial to warrant the exercise of jurisdiction, provided that it cannot refuse to exercise jurisdiction over any labor dispute over which it would have asserted jurisdiction under the standards it had in effect on August 1, 1959. In accordance with this provision the Board has determined that it will not exercise jurisdiction over racetracks, owners, breeders, and trainers of racehorses, and real estate brokers.

Appendix to Chapter 6

SECTIONS 7 AND 8 OF THE NATIONAL LABOR RELATIONS ACT, 29 U.S.C. §§157 and 158

Rights of Employees

Sec. 7. Employees shall have the right to self-organization, to form, join, or assist labor organizations, to bargain collectively through representatives of their own choosing, and to engage in other concerted activities for the purpose of collective bargaining or other mutual aid or protection, and shall also have the right to refrain from any or all of such activities except to the extent that such right may be affected by an agreement requiring membership in a labor organization as a condition of employment as authorized in section 8(a)(3).

Unfair Labor Practices

Sec. 8. (a) It shall be an unfair labor practice for an employer—

(1) to interfere with, restrain, or coerce employees in the exercise of the rights guaranteed in section 7;

(2) to dominate or interfere with the formation or administration of any labor organization or contribute financial or other support to it: Provided, That subject to rules and regulations made and published by the Board pursuant to section 6, an employer shall not be prohibited from permitting employees to confer with him during working hours without loss of time or pay;

(3) by discrimination in regard to hire or tenure of employment or any term or condition of employment to encourage or discourage membership in any labor organization: Provided, That nothing in this Act, or in any other statute of the United States, shall preclude an employer from making an agreement with a labor organization (not established, maintained, or assisted by any action defined in section 8(a) of this Act as an unfair labor practice) to require as a condition of employment membership therein on or after the thirtieth day following the beginning of such employment or the effective date of such agreement, whichever is the later, (i) if such labor organization is the representative of the employees as provided in section 9(a), in the appropriate collective-bargaining unit covered by such agreement when made, [and has at the time the agreement was made or within the preceding twelve months received from the Board a notice of compliance with Section 9(f), (g), (h)], and (ii) unless following an election held as provided in section 9(e) within one year preceding the effective date of such agreement, the Board shall have certified that at least a majority of the employees eligible to vote in such election have voted to rescind the authority of such labor organization to make such an agreement: Provided further, That no employer shall justify any discrimination against an employee for nonmembership in a labor organization (A) if he has reasonable grounds for believing that such membership was not available to the employee on the same terms and conditions generally applicable to other members, or (B) if he has reasonable grounds for believing that membership was denied or terminated for reasons other than the failure of the employee to tender the periodic dues and the initiation fees uniformly required as a condition of acquiring or retaining membership;

(4) to discharge or otherwise discriminate against an employee because he has filed charges or given testimony under this Act;

(5) to refuse to bargain collectively with the representatives of his employees, subject to the provisions of section 9(a).

(b) It shall be an unfair labor practice for a labor organization or its agents—

(1) to restrain or coerce (A) employees in the exercise of the rights guaranteed in section 7: Provided, That this paragraph shall not impair the right of a labor organization to prescribe its own rules with respect to the acquisition or retention of membership therein; or (B) an employer in the selection of his representatives for the purposes of collective bargaining or the adjustment of grievances;

(2) to cause or attempt to cause an employer to discriminate against an employee in violation of subsection (a)(3) or to discriminate against an employee with respect to whom membership in such organization has been denied or terminated on some ground other than his failure to tender the periodic dues and the initiation fees uniformly required as a condition of acquiring or retaining membership;

(3) to refuse to bargain collectively with an employer, provided it is the representative of his employees subject to the provisions of section 9(a);

(4)(i) to engage in, or to induce or encourage [the employees of any employer] any individual employed by any person engaged in commerce or in an industry affecting commerce to engage in, a strike or a [concerted] refusal in the course of [their] his employment to use, manufacture, process, transport, or otherwise handle or work on any goods, articles, materials, or commodities or to perform any services[,]; or (ii) to threaten, coerce, or restrain any person engaged in commerce or in an industry affecting commerce, where in either case an object thereof is—

(A) forcing or requiring any employer or self-employed person to join any labor or employer organization [or any employer or other person to cease using, selling, handling, transporting, or otherwise dealing in the products of any other producer, processor, or manufacturer, or to cease doing business with any other person;] or to enter into any agreement which is prohibited by section 8(e);

(B) forcing or requiring any [any employer or other] person to cease using, selling, handling, transporting, or otherwise dealing in the products of any other producer, processor, or manufacturer, or to cease doing business with any other person, or forcing or requiring any other employer to recognize or bargain with a labor organization as the representative of his employees unless such labor organization has been certified as the representative of such employees under the provisions of section 9[;]; Provided, That nothing contained in this clause (B) shall be construed to make unlawful, where not otherwise unlawful, any primary strike or primary picketing;

(C) forcing or requiring any employer to recognize or bargain with a particular labor organization as the representative of his employees if another labor organization has been certified as the representative of such employees under the provisions of section 9;

(D) forcing or requiring any employer to assign particular work to employees in a particular labor organization or in a particular trade, craft, or class rather than to employees in another labor organization or in another trade, craft, or class, unless such employer is failing to conform to an order or certification of the Board determining the bargaining representative for employees performing such work:

Provided, That nothing contained in this subsection (b) shall be construed to make unlawful a refusal by any person to enter upon the premises of any employer (other than his own employer), if the employees of such employer are engaged in a strike ratified or approved by a representative of such employees whom such employer is required to recognize under this Act: Provided further, That for the purposes of this paragraph (4) only, nothing contained in such paragraph shall be construed to prohibit publicity, other than picketing, for the purpose of truthfully advising the public, including consumers and members of a labor organization, that a product or products are produced by an employer with whom the labor organization has a primary dispute and are distributed by another employer, as long as such publicity does not have an effect of inducing any individual employed by any person other than the primary employer in the course of his employment to refuse to pick up, deliver, or transport any goods, or not to perform any services, at the establishment of the employer engaged in such distribution:

(5) to require of employees covered by an agreement authorized under subsection

(a)(3) the payment, as a condition precedent to becoming a member of such organization, of a fee in an amount which the Board finds excessive or discriminatory under all the circumstances. In making such a finding, the Board shall consider, among other relevant factors, the practices and customs of labor organizations in the particular industry, and the wages currently paid to the employees affected; [and]

(6) to cause or attempt to cause an employer to pay or deliver or agree to pay or deliver any money or other thing of value, in the nature of an exaction, for services which are not performed or not to be performed[.]; and

(7) to picket or cause to be picketed, or threaten to picket or cause to be picketed, any employer where an object thereof is forcing or requiring an employer to recognize or bargain with a labor organization as the representatives of his employees, or forcing or requiring the employees of an employer to accept or select such labor organization as their collective bargaining representative, unless such labor organization is currently certified as the representative of such employees:

(A) where the employer has lawfully recognized in accordance with this Act any other labor organization and a question concerning representation may not appropriately be raised under section 9(c) of this Act.

(B) where within the preceding twelve months a valid election under section 9(c) of this Act has been conducted, or

(C) where such picketing has been conducted without a petition under section 9(c) being filed within a reasonable period of time not to exceed thirty days from the commencement of such picketing: Provided, That when such a petition has been filed the Board shall forthwith, without regard to the provisions of section 9(c)(1) or the absence of a showing of a substantial interest on the part of the labor organization, direct an election in such unit as the Board finds to be appropriate and shall certify the results thereof: Provided further, That nothing in this subparagraph (C) shall be construed to prohibit any picketing or other publicity for the purpose of truthfully advising the public (including consumers) that an employer does not employ members of, or have a contract with, a labor organization, unless an effect of such picketing is to induce any individual employed by any other person in the course of his employment, not to pick up, deliver or transport any goods or not to perform any services.

Nothing in this paragraph (7) shall be construed to permit any act which would otherwise be an unfair labor practice under this section 8(b).

(c) The expressing of any views, argument, or opinion, or the dissemination thereof, whether in written, printed, graphic, or visual form, shall not constitute or be evidence of an unfair labor practice under any of the provisions of this Act, if such expression contains no threat of reprisal or force or promise of benefit.

(d) For the purposes of this section, to bargain collectively is the performance of the mutual obligation of the employer and the representative of the employees to meet at reasonable times and confer in good faith with respect to wages, hours, and other terms and conditions of employment, or the negotiation of an agreement, or any question arising thereunder, and the execution of a written contract incorporating any agreement reached if requested by either party, but such obligation does not compel either party to agree to a proposal or require the making of a concession: Provided, That where there is in effect a collective-bargaining contract covering employees in an industry affecting commerce, the duty to bargain collectively shall also mean that no party to such contract shall terminate or modify such contract, unless the party desiring such termination or modification—

(1) serves a written notice upon the other party to the contract of the proposed termination or modification sixty days prior to the expiration date thereof, or in the event such contract contains no expiration date, sixty days prior to the time it is proposed to make such termination or modification;

(2) offers to meet and confer with the other party for the purpose of negotiating a new contract or a contract containing the proposed modifications;

(3) notifies the Federal Mediation and Conciliation Service within thirty days after such notice of the existence of a dispute, and simultaneously therewith notifies any State or Territorial agency established to mediate and conciliate disputes within the State or Territory where the dispute occurred, provided no agreement has been reached by that time; and

(4) continues in full force and effect, without resorting to strike or lock-out, all the terms and conditions of the existing contract for a period of sixty days after such notice is given or until the expiration date of such contract, whichever occurs later: The duties imposed upon employers, employees, and labor organizations by paragraphs (2), (3), and (4) shall become inapplicable upon an intervening certification of the Board, under which the labor organization or individual, which is a party to the contract, has been superseded as or ceased to be the representative of the employees subject to the provisions of section 9(a), and the duties so imposed shall not be construed as requiring either party to discuss or agree to any modification of the terms and conditions contained in a contract for a fixed period, if such modification is to become effective before such terms and conditions can be reopened under the provisions of the contract. Any employee who engages in a strike within [the sixty-day] any notice period specified in this subsection, or who engages in any strike within the appropriate period specified in subsection (g) of this section, shall lose his status as an employee of the employer engaged in the particular labor dispute, for the purposes of sections 8, 9, and 10 of this Act, but such loss of status for such employee shall terminate if and when he is reemployed by such employer. Whenever the collective bargaining involves employees of a health care institution, the provisions of this section 8(d) shall be modified as follows:

(A) The notice of section 8(d)(1) shall be ninety days; the notice of section 8(d)(3) shall be sixty days; and the contract period of section 8(d)(4) shall be ninety days.

(B) Where the bargaining is for an initial agreement following certification or recognition, at least thirty days' notice of the existence of a dispute shall be given by the labor organization to the agencies set forth in section 8(d)(3).

(C) After notice is given to the Federal Mediation and Conciliation Service under either clause (A) or (B) of this sentence, the Service shall promptly communicate with the parties and use its best efforts, by mediation and conciliation, to bring them to agreement. The parties shall participate fully and promptly in such meetings as may be undertaken by the Service for the purpose of aiding in a settlement of the dispute.

(e) It shall be an unfair labor practice for any labor organization and any employer to enter into any contract or agreement, express or implied, whereby such employer ceases or refrains or agrees to cease or refrain from handling, using, selling transporting or otherwise dealing in any of the products of any other employer, or to cease doing business with any other person, and any contract or agreement entered into heretofore or hereafter containing such an agreement shall be to such extent unenforcible and void: Provided, That nothing in this subsection (e) shall apply to an agreement between a labor organization and an employer in the construction industry relating to the contracting or subcontracting of work to be done at the site of the construction, alteration, painting, or repair of a building, structure, or other work: Provided further, That for the purposes of this subsection (e) and section 8(b)(4)(B) the terms "any employer", "any person engaged in commerce or an industry affecting commerce", and "any person" when used in relation to the terms "any other producer, processor, or manufacturer", "any other employer", or "any other person" shall not include persons in the relation of a jobber, manufacturer, contractor, or subcontractor working on the goods or premises of the jobber or manufacturer or performing parts of an integrated process of production in the apparel and clothing industry: Provided further, That nothing in this Act shall prohibit the enforcement of any agreement which is within the foregoing exception.

(f) It shall not be an unfair labor practice under subsections (a) and (b) of this section for an employer engaged primarily in the building and construction industry to make an agreement covering employees engaged (or who, upon their employment, will be engaged) in the building and construction industry with a labor organization of which building and construction employees are members (not established, maintained, or assisted by any action defined in section 8(a) of this Act as an unfair labor practice) because (1) the majority status of such labor organization has not been established under the provisions of section 9 of this Act prior to the making of such agreement, or (2) such agreement requires as a condition of employment, membership in such labor organization after the seventh day following the beginning of such employment or the effective date of the agreement, whichever is later, or (3) such agreement requires the employer to notify such labor organization of opportunities for employment with such employer, or gives such labor organization an opportunity to refer qualified applicants for such employment, or (4) such agreement specifies minimum training or experience qualifications for employment or provides for priority in opportunities for employment based upon length of service with such employer, in the industry or in the particular geographical area: Provided, That nothing in this subsection shall set aside the final proviso to section 8(a)(3) of this Act: Provided further, That any agreement which would be invalid, but for clause (1) of this subsection, shall not be a bar to a petition filed pursuant to section 9(c) or 9(e).

(g) A labor organization before engaging in any strike, picketing, or other concerted refusal to work at any health care institution shall, not less than ten days prior to such action, notify the institution in writing and the Federal Mediation and Conciliation Service of that intention, except that in the case of bargaining for an initial agreement following certification or recognition the notice required by this subsection shall not be given until the expiration of the period specified in clause (B) of the last sentence of section 8(d) of this Act. The notice shall state the date and time that such action will commence. The notice, once given, may be extended by the written agreement of both parties.

CHAPTER 7

Toxics Information Transfer in

the Workplace

A. INTRODUCTION

This chapter deals with a topic that is having a profound effect on working conditions: the generation and communication of information regarding the use of, exposure to, and effects of toxic substances in the workplace. More specifically, it addresses the ways in which the law either promotes or retards the development and dissemination of such information.

Until recently, employers retained the right to determine when and even if information regarding the nature and identity of workplace chemicals should be made available to those who worked with them. To a large extent, information of this nature was considered "proprietary" and thus unavailable for public scrutiny. This changed dramatically in the 1980s, both within the workplace and within the community at large. With the growing acknowledgement that chemical exposures are often not detectable by the human senses—and that, even when they are detectable, the dangers of such exposures may not be apparent—legislatures, courts, and administrative agencies acted to create legal mechanisms, commonly called *right to know laws,* to transfer to the exposed or potentially exposed worker or community resident what we will call *toxics information.* We use this term to designate the broad range of data relevant to the creation, measurement, evaluation, and avoidance of chemical, biologic, or radioactive exposures. Most right to know laws deal only with chemical exposures, and few provide for the transfer of all relevant toxics information about a particular workplace or industrial process.

Although broad worker and community right to know programs are now mandated by federal statute and regulation, they evolved only after many years of community organizing campaigns at the state and local levels. Until the passage of the community right to know sections of the federal Superfund Amendment and Reauthorization Act (SARA) in 1986, and the promulgation of the revised OSHA Hazard Communication Standard for workers the following year, state and local right to know laws were the dominant force in providing toxics information to workers and community residents.

Industry's response to the emergence of right to know has been far from uniform. Some employers began implementing toxics disclosure policies far in advance of current regulation, and some have spoken in favor of mandatory information disclosure. In testimony before OSHA, for example, the Chemical Manufacturers Association indicated that it "strongly believes that the substantive provisions of the Hazard Communication Standard are sound as a matter of science and policy." 17 *Occ Safety and Health Reporter* 525 (1987). Other industry representatives resisted right to know laws vigorously, however, and mounted intensive political and legal campaigns to prevent their passage and implementation.

During the debate preceding the passage of the Massachusetts right to know statute in 1983, for example, a major industry lobbying group, the Associated Industries of Massachusetts (AIM), distributed a political flier in which it characterized an early version of the law as "the single most dangerous piece of legislation to business and local government to be filed in years," and declared that it "must be defeated." The group's principal arguments against right to know were the costs of implementation and, in a somewhat more creative vein, the alleged threat to public safety. This latter argument was based on the supposition that social miscreants would use the law to locate and exploit "drugs, explosives and other sensitive material." "Thus," argued AIM, "the right-to-know bill is in reality a right-to-steal bill—compelling government and business to do important R&D work for the criminal and extremist elements of our society." After the passage of a somewhat modified piece of compromise legislation, however, AIM worked to educate Massachusetts industry with regard to the requirements of the law, and played an important role in helping smaller companies comply with these requirements.

The Massachusetts story is not unique. Indeed, although right to know remains unpopular in some industry circles, it appears that American business has finally accepted right to know as a fact of life. Arguments will remain over questions of implementation and degree, but the basic concept likely is here to stay.

B.   THE RIGHT TO KNOW

Promises and Pitfalls of Workplace Right-to-Know

C. Caldart

The people of Manville, N.J. have paid a high price for their piece of the American dream. The hardworking immigrants who settled this town struggled, and often succeeded, in earning the

Source: From 1(1) *Seminars in Occupational Medicine* 81-86, 89 (1986). New York: Thieme Medical Publishers, Inc., 1986. Reprinted by permission.

rewards our society promises. They bought houses, raised children and sent them to college. For many, that dream has ended with the cruel reality of oxygen tanks, slow descents into weakness and incurable cancers. And they never made that choice for themselves. "The people I know, they feel bitter," [one asbestos worker] says. "When we were hired, they should have said, 'Hey, there's a risk involved. Do you want to take it with the risk?' And you'd have had the option to say no."[1]

* * *

A FRAMEWORK FOR UNDERSTANDING RIGHT-TO-KNOW LAWS[4]

Although the popular phrase "right-to-know" is a useful generic designation, it is an inadequate description of the legal rights and obligations governing the transfer of workplace information on toxic substances. What is meant by the right-to-know? Who has the right? How is it exercised? What information is within its scope? These questions touch on a relatively complicated area of the law, and their resolution demands attention to the specific provisions of a number of different statutes, ordinances, and regulations.

* * *

Fortunately, however, this area of the law is amenable to a rather simple conceptual framework.

In essence, the right-to-know embodies a democratization of the workplace. It is the mandatory sharing of information between management and labor. Through a variety of laws, manufacturers and employers are directed to disclose information regarding toxic (and potentially toxic) substances in the workplace to a number of parties: to workers, to unions in their capacity as worker representatives, to members of the surrounding community, and to governmental agencies charged with the protection of the public health. It is this broad range of laws that defines a workers' right-to-know.

Network of Rights and Obligations

One cannot have a meaningful right to information unless someone else has a corresponding duty to provide that information. Thus, a worker's right-to-know is secured by requiring a manufacturer or employer to disclose. The disclosure requirement can take a variety of forms, and the practical scope of that requirement may depend on the nature of the form chosen. For example, a duty to disclose only information that has been requested probably will provide a narrower flow of information than a duty to disclose all information regardless of whether it has been requested. The various rights and obligations in the area of toxics information transfer can be divided into three distinct categories.

The duty to generate and/or retain information refers to the obligation to compile a record of certain workplace events or activities and/or maintain such a record for a specified period of time if it has been compiled. An employer may, for example, be required to monitor its workers regularly for evidence of toxic exposures (biologic monitoring) and to keep written records of the results of such monitoring.

The right of access (and the corresponding duty to disclose upon request) refers to the right of a worker, union, community member, or agency to request and secure access to information held by a manufacturer or employer. Drawing on the previous example, such a right of access would provide workers with a means of obtaining copies of biologic monitoring records pertaining to their exposure to toxic substances.

Finally, the *duty to inform* refers to an employer's or manufacturer's obligation to disclose, without request, information pertaining to chemicals, bioengineered organisms, or radiation used in the workplace. An employer may, for example, have a duty—independent of any worker's exercise of a right to access—to inform a worker whenever biologic monitoring reveals that his exposure to a toxic substance has produced bodily concentrations of that substance above a specified level.

Information Subject to Transfer

Just as they vary according to the category of right or obligation they create, right-to-know laws vary according to the nature of the data to which they pertain. Here again, three categories are appropriate. The data that may be subject to transfer obligations within the workplace may be classified as scientific information, technologic information, or legal information.

Scientific information refers to data concerning the nature and consequences of workplace exposures. These data, in turn, can be divided into three subcategories.[9] Ingredients information is that which provides the worker with the identity of the substances to which he is exposed. Depending on the circumstances, this may involve only the generic

---

[1]Jubak J: They are the first. *Environ Action* 1983; Feb:9-14.
[4]Major portions of this section are excerpted from Ashford NA, Caldart CC: The "right to know": Toxics information transfer in the workplace, in Breslow L, Fielding JE, Lave LB (eds): *Annual Review of Public Health*, vol. 6. Palo Alto, Annual Reviews, 1985.

[9]O'Reilly JT: *Unions' Right to Company Information.* Philadelphia, University of Pennsylvania, Industrial Relations and Public Policy Series No. 21, 1980.

classifications of the various chemicals involved, or may include the specific chemical identities of all relevant chemicals and the specific contents of all chemical mixtures. Exposure information, on the other hand, encompasses all data regarding the amount, frequency, duration, and route of workplace exposures. This information may be of a general nature (such as the results of ambient air monitoring performed at a central workplace location) or may take individualized form (such as the results of personal environmental or biologic monitoring of a specific worker). Finally, there is effects information. This is information on the known or potential health effects of workplace exposures. These may be general data regarding the effects of a particular substance (such as material safety data sheets or epidemiologic studies) or individualized data regarding the health of particular workers (such as worker medical records compiled as a result of medical surveillance).

Technologic information covers data pertaining to present and potential workplace technology, and includes all information relevant to the question of how—from a technical or mechanical standpoint—workplace exposures are created and controlled. For a given workplace, this would include not only data regarding process technology already in use, but also data regarding alternative processes, possible chemical substitutes, and control technologies that would reduce exposures to known or suspected toxins.

Legal information encompasses all data pertaining to the existence and nature of workers' legal rights in the workplace. Conceptually, this covers a broad range of subjects: specific statutory and regulatory rights under TSCA, the OSHAct, and a number of other health and safety laws; rights of compensation for workplace injuries under worker compensation and tort law; rights to be free from retaliatory employment practices under certain state and federal laws; and other rights secured by the other federal labor laws. This article, however, is concerned primarily with legal information pertaining to the rights and duties governing toxics information transfer in the workplace, as this is the information that workers must have if those rights and duties are to be meaningful. Such information is essential, both if workers are to exercise their own rights of access and if they are to encourage employer and manufacturer compliance with the various duties of information transfer and generation.

THE CONCEPTUAL BASIS OF RIGHT-TO-KNOW

As is true for most laws that grow out of political compromise, the underlying rationale for the promulgation of workplace right-to-know laws will tend to vary according to the differing perspectives of those who have supported them. As one who supports the right-to-know as a logical and necessary component of United States labor policy, I can do no more than to offer what I believe to be the most important moral, economic, and political justifications for this position.

The Moral Rationale

The moral basis for the promulgation of right-to-know laws is similar to that which underlies the concept of informed consent in medicine. It is the principle of personal autonomy. This does not mean just self-determination, but the freedom and capacity to develop into autonomous people. Undoubtedly, many would disagree about the extent to which a worker should be autonomous within the workplace. At some point, this becomes, in the broad sense, a political as well as a moral question. Nonetheless, whatever one's political or social views on the issue, most people would agree that the opportunity to participate in decisions affecting one's own health and physical integrity is an essential feature of the development of personal autonomy.

The worker who may be exposed to toxic substances in the course of employment faces such decisions. "Should I remain in the workplace?" "If so, under what conditions—what risk will I accept?" "Should I take actions to try to bring about a reduction of the danger?" Without the information provided through right-to-know, the worker usually will have little basis on which to sort through these choices. Indeed, without right-to-know, the worker (and, in some cases, the employer) often will not be aware of the danger and will thus not even realize that there are decisions of this nature to be made.

The argument for personal autonomy becomes more complicated—but all the more compelling—when one acknowledges that many of the exposures that workers will encounter are known or putative reproductive toxins. The worker facing a decision involving such exposures must thus address concerns that go beyond his own physical welfare (and beyond the ancillary effects that changes in his welfare may have on family and loved ones). This worker must also be concerned with two types of possible reproductive outcomes: adverse effects on the physical or mental integrity of future offspring and adverse effects on the ability to have children. Regardless of how the concept of privacy is derived and defined, these concerns are, or ought to be, highly private. When the worker is denied access to information that is relevant to decisions touching directly on these concerns, however, the worker's need for privacy in these matters is trivialized. The locus of decision-making is elsewhere, removed from the private domain of the worker. To the extent that the notion of privacy is central to the demarcation of legitimate autonomy, the worker's exercise of personal autonomy is that much further compromised.

There is more at stake here than just the interests of the worker or the worker's offspring. To be sure, the worker has an interest in securing a fuller sense of personal autonomy, both in making health decisions and in making decisions that affect obligations to future offspring. Equally important, however, is the employer's interest in fulfilling an obligation to respect the autonomy of the worker. To the extent that it provides the worker with information that may be relevant to the difficult choices that may have to be made regarding workplace exposures, right-to-know serves the moral interest of both the worker and the employer. The great promise of these laws, however, is that they will do more than merely provide information to one who has a conceptual right to make a particular decision. The promise of right-to-know is that it will help create a situation in which the worker is truly capable of making that decision.

One of the ways in which one acquires the capacity to make certain decisions is through the exercise of actually making those kinds of decisions. If one is not in the habit of making certain kinds of decisions, one is far less likely to be able to make them if called upon to do so. Traditionally, major decisions affecting the nature and level of chemical exposures in the workplace have been made by the employer. The message to the worker has been a paternalistic one: "Leave the decisions to us." By denying the worker even the access to the information with which informed decisions could be made, the system has tended to reinforce a perception that the worker lacks the capacity to make those decisions. Right-to-know seeks to remove this level of paternalism by creating a situation in which worker access to such information is the status quo.

If the great promise of right-to-know is its capacity to promote personal autonomy and integrity, the great danger is that it may be seen as a solution to the issue of worker autonomy. The nature of this danger can be seen through an examination of the difference between right-to-know and informed consent. Although the concept of autonomy is essential to both situations, the ethical implications of worker autonomy in the employer-employee setting differ considerably from those of patient autonomy in the physician-patient setting. Consider the following argument for informed consent offered by Glenn Graber of the University of Tennessee:

> There is a morally significant difference between (a) choosing *for oneself* to face the pain and (b) having this decision made *on one's behalf* by another; and the moral quality of the first of these alternatives (which amounts to a sort of courage) is denied to the patient if the physician makes the decision.

Although a moral good can come out of the commission of a courageous act, this characterization of the nature of autonomous decisionmaking is far more appropriate to the medical setting than it is to the workplace. The key distinction is the origin of the pain to be faced in each situation. Although the nature of the physician-patient relationship clearly can affect the availability of medical techniques or technology to reduce the pain to be faced by the patient, the primary source of that pain (absent some form of medical malpractice) is the disease or trauma giving rise to the patient's medical condition and not the relationship itself. Informed consent furthers patient autonomy if it enables the patient to face that condition in a forthright and realistic fashion.

In the workplace, however, the pain grows directly out of the employment relationship. The disease emanates from the workplace exposures generated by the technology that the employer puts in place and with which the employee "consents" to work. If it is to contribute to worker autonomy in any real sense, right-to-know must facilitate enough of a change in the employer-employee relationship to enable the worker to effect changes in workplace technology (or changes in the way in which that technology is employed in the workplace). To the extent that the transfer of information through right-to-know laws provides workers with the potential to make such changes, workers will be better able to make their own decisions regarding the nature of the workplace within which they want to work. But if all right-to-know does is provide a mechanism whereby workers are able to summon the courage to risk occupational disease or reproductive disaster, it will be an empty exercise.

The Economic Rationale

The argument for right-to-know is not simply a moral one. The economic justification for right-to-know is grounded firmly in the tenets of the free market. From the standpoint of overall economic efficiency, toxics information transfer laws act as market-correcting mechanisms. Unlike most environmental or occupational health regulations, however, right-to-know laws do not explicitly seek to internalize market externalities. Rather, they seek to improve the quality of the information on which decisions within the market are based. Classic market economic theory is premised on the assumption that all parties have free access to "perfect" information regarding their options within the marketplace. In the modern workplace, the disparity between reality and this ideal has often been great. Especially with regard to workplace exposure to toxic substances, the burden of inadequate information transfer has often fallen most heavily on the health of the worker.

Right-to-know seeks to ameliorate this situation in two ways. Most obviously, it provides workers with access to existing information that was heretofore often denied them. Often, it will not provide all of the information that a rational decisionmaker would like. Indeed, many times the information provided may raise more questions than it answers. However, this is all a part of the information process and will serve to narrow the information gap between employers and employees. Perhaps just as important, however, right-to-know laws serve to facilitate the creation of information regarding the effects of toxins in the workplace. To the extent that they prompt disclosure of the nature and extent of chemical, biologic, and radiologic exposures within a given workplace, they will permit the correlation of these exposures with later human effects, if any, experienced in that workplace. Accordingly, they will contribute to the store of baseline epidemiologic data regarding toxic substances, and thus provide new information relevant to worker health.

In both ways, right-to-know laws provide the worker with additional data regarding the safety of the workplace. In theory, the worker will be better able to make a rational, informed decision regarding the desirability of remaining in that work situation. Over-

all, assuming that the additional transaction costs imposed by right-to-know laws are not too high,[22] more efficient choices will thus be made regarding the allocation of resources within the labor market. However, this theory will be for naught if the worker lacks the bargaining power to carry through on an informed decision. For this reason, the critical test of right-to-know laws will be their ability to facilitate the exercise of a worker's political and legal power.

The Political Rationale

It is perhaps trite to note that information can empower its holder; nonetheless, empowerment is what right-to-know is all about. In practice, right-to-know will succeed to the extent that it enables and encourages workers to make more aggressive use of the mechanisms that are open to them under the political and regulatory systems. Were each worker truly free to choose from a variety of workplaces—representing a range of health risk levels—he could simply select, on the basis of information obtained through right-to-know, the particular workplace that corresponded to his own personal set of risk criteria. In reality, this situation often does not prevail. Whether because the relevant toxic risks are industry-wide or because market imperfections routinely impede the operation of the labor market, many workers are not free to "shop" for a safer workplace, but rather are left to try to fashion solutions to toxic substance risk problems within their own workplace.

Conceptually, right-to-know could assist in the exercise of a number of the political and legal tools that are available to the workers who participate in this process. Workers have, for example, a right to refuse hazardous work under both the OSHAct and the NLRA. Although the extent of the applicability of the right to refuse hazardous work to long-term health or reproductive hazards has not yet been clearly defined, generally the right clearly extends to exposure to toxic substances. With the use of information obtained through right-to-know, workers could, in appropriate circumstances, build a case

[22]The costs to the employer (and to the relevant administrative agency) of implementing a right-to-know program are often consequential; as always, cost-efficiency should be an important criterion as we design regulatory programs in this area. It is important to acknowledge, however, that it is the function of right-to-know laws to "internalize" the costs created by an inadequate supply of information in the workplace. Before right-to-know, these costs were borne most heavily by the worker.

to support the argument that exposure to a particular workplace toxin created a health hazard. Their exercise of a right to refuse hazardous work could thus set the stage for corrective action.

Right-to-know should also be useful to worker participation in the "command and control" regulatory process. Access to information through right-to-know should make it easier for workers to discover violations of OSHA and TSCA standards, to discover situations that may merit general duty clause citations under the OSHAct, and to target workplace chemicals—and chemical uses—that might properly be the subject of new rulemaking procedures. Existing legal mechanisms, such as the right to ask for an OSHA inspection or the right to petition for standard-setting or chemical testing under TSCA, could then be used more effectively.

Compensatory remedies embodied in the worker compensation and tort systems should also be buttressed by the data available through right-to-know. Recording and recordkeeping requirements should facilitate proof of the nature, duration, and extent of particular chemical exposures, and thus provide a firmer baseline for the evaluation of potential claims and lawsuits.

More aggressive use of legal mechanisms such as these will not go unnoticed, and may, in certain workplaces, bring reprisals from the employer. Fortunately, there is a growing trend within the law to provide workers with protection from such reprisal. TSCA, the OSHAct, the NLRA, and a number of other federal health, safety, and environmental statutes specifically prohibit discrimination against employees who take actions to carry out the purposes of these statutes, and provide administrative and judicial remedies against such discrimination. Further, many states have recognized common law protections for workers who are subjected to retalia-

tion for exercising a legal right or for making a good faith effort to carry out some "articulated public purpose." Thus, workers have been held to have a common law cause of action for retaliatory discharge when they are fired for filing a worker compensation claim or for reporting violations of workplace sanitation standards.

Most assuredly, these statutory and common law projections are not iron-clad. Not only are they all subject to the delay and uncertainty that ordinarily plague the administrative and judicial system, but the availability and scope of the common law remedies differ considerably among the various states. Nonetheless, the courts appear increasingly willing to apply these protections vigorously, and the business community appears increasingly resigned to their emergence as an integral part of the labor/management relationship. Thus, there is cause for cautious optimism about the extent to which workers will actually be able to put right-to-know to use in the exercise of the legal tools at their disposal.

Ultimately, an aggressive use of these mechanisms will have an effect on the overall "political balance" within the ongoing bargaining process between management and labor. To the extent that such use tends to make it more expensive for an employer to maintain an unsafe workplace, it provides an economic incentive for employers to modify workplace technology to reduce risks. Further, it tends to give credence to worker opinion on the dangers of workplace technology, and may encourage employers to involve workers in the decisionmaking process pertaining to the design and implementation of that technology. In either case, it has the effect of strengthening the voice of the worker. If the transfer of toxics information can assist this process, the moral and economic promises of right-to-know may begin to be fulfilled.

C. EMPLOYEE AND WORKPLACE MONITORING

Among the many types of information covered by right to know laws, perhaps the most sensitive is employee-specific or workplace-specific data on toxic substance exposure or effects. Here, the process of information transfer begins with the transfer of physical data from the employee or workplace to a third party, usually the employer, a government agency, or a union. This is done through the monitoring of the worker, the workplace, or both. Monitoring the workplace to determine the ambient levels of toxic substances in the air is generally called *environmental monitoring*. Testing individual workers to determine their level of exposure to specific toxic substances or the effects of those substances on their bodies is called *human monitoring*. The major categories of monitoring and the nature of the information they produce are discussed below.

Monitoring the Worker for Exposure and Disease: Scientific, Legal and Ethical Considerations in the Use of Biomarkers

N. Ashford, C. Spadafor, D. Hattis, and C. Caldart

The discussion of terms and definitions that follows —medical surveillance, genetic monitoring, biological monitoring, and environmental monitoring—is consistent with the meaning and usage of the same in regulations promulgated by OSHA.

MEDICAL SURVEILLANCE

Medical surveillance is designed in an occupational setting to detect adverse health *effects* (or health status) resulting from hazardous exposures in the workplace. Medical surveillance tests are generally diagnostic tools used in routine medical practice. They most commonly include tests such as chest X-rays, pulmonary function tests, routine blood analyses, serum liver function tests, serum kidney function tests, and routine urinalyses.

Medical surveillance testing serves to obtain certain types of information, such as the identification of workers who are suffering from an occupational injury or illness, the epidemiological data on occupational disease, and the general or specific data on categories or types of workers. These data are intended to aid in testing workers by monitoring specific organ systems that may be affected by exposure to workplace hazards. Testing may be instituted by an employer on its own initiative, in response to OSHA requirements, or at the request of employees or their unions.

Medical surveillance is most useful in three situations: (1) if compliance with the permissible exposure limits established by OSHA will not adequately ensure worker health; (2) if air measurement cannot sufficiently monitor worker exposure (e.g., if a significant route of entry is not inhalation); and (3) if high-risk groups are exposed. Medical removal may also be appropriate in these three situations.

Source: Excerpts from *Monitoring the Worker for Exposure and Disease: Scientific, Legal and Ethical Considerations in the use of Biomarkers.* Baltimore, MD: Johns Hopkins University Press, 1990. 6–14, xiv.

The *Assessment of Toxic Agents* implies that in some circumstances medical surveillance can prevent occupationally related disease [Berlin, Yodaiken and Henman, eds., *Assessment of Toxic Agents*, XII-XIII]. If a disease is *reversible* or *arrestible*, medical surveillance may be preventive insofar as it serves as a warning signal prompting timely action to avoid future exposures and continuing or progressive adverse health effects.

Pursuant to statute, OSHA can require employee medical surveillance. Medical surveillance provisions in OSHA standards traditionally include both the general diagnostic tools as well as some laboratory tests that could be classified under "biological monitoring." For this reason, it has been suggested that the OSHA provisions should be recharacterized as "Medical Surveillance and Monitoring." The required medical surveillance provisions vary for the twenty-four OSHA health standards promulgated since 1972, but generally they are routine diagnostic tests used in medical practice. All OSHA health standards contain required medical surveillance provisions that articulate the tests that physicians are required to perform. However, the ethylene oxide standard relaxes such requirements. In that standard, the Agency departed from giving the specific medical guidance as it had in previous standards, opening up the possibility of unchecked discretion in testing. This new flexible approach providing only very broad, general guidelines for testing gives reason for concern because many examining physicians are not specially trained and certified in occupational health. Realizing this lack of specialized medical expertise, OSHA should propose and promulgate standards that provide detailed medical guidelines, sufficient for *all* examining physicians, which are essential to promoting good health in the workplace.

In January 1981, OSHA and NIOSH published the *Occupational Health Guidelines for Chemical Hazards.* This volume includes a variety of information on approximately 450 substances for which OSHA adopted consensus standards under section 6(a) of the OSHAct. The information for each substance usually includes chemical, toxicological and health hazard data, as well as recommendations for industrial hygiene and medical surveillance practices.

[As discussed in Chapter 3, OSHA revised its set of section 6(a) standards in 1989. With this revision, OSHA has established permissible exposure limits (PELs) for a total of 614 substances. The *Occupational Health Guidelines* have not been revised to include these new substances. However, OSHA has issued a proposed *generic* medical surveillance standard that would be applicable. See 53 *Fed. Reg.* 37, 595 (Sept. 27, 1988).]

GENETIC MONITORING

One particular type of medical surveillance that has received much attention is genetic monitoring. This monitoring includes the periodic testing of employees working with or possibly exposed to certain substances (such as known or potential carcinogens) that cause changes in chromosomes or deoxyribonucleic acid (DNA) and is classified in two categories. The first, *cytogenetic monitoring*, involves monitoring for changes in chromosomes. A sample of blood is usually collected, and the chromosomes are visualized after inducing cell division in specific types of white cells. The second, *noncytogenetic monitoring*, includes monitoring for DNA reaction products (adducts), small changes at specific places in DNA, or the presence of mutagens in body fluids. Generally, both types of genetic monitoring are conducted in an attempt to determine if environmental exposures of a specific population (e.g., workers in the same job category) to particular substances cause changes in genetic material in statistically significant numbers above background levels.

BIOLOGICAL MONITORING

The third practice is biological monitoring. Biological monitoring is defined in the *Assessment of Toxic Agents* as "the measurement and assessment of workplace agents or their metabolites either in tissues, secreta, excreta, expired air or any combination of these to evaluate *exposure and health risk* compared to an appropriate reference." More broadly defined, biological monitoring is a collection of activities designed to determine both whether a person has

been exposed to and whether his or her body fluids or organs contain unusual amounts of a particular substance or its metabolites. Using this broader definition, biological monitoring would include what we and others call "genetic monitoring," discussed in the previous subsection. [We make] a distinction between monitoring for "internal dose" and monitoring for "biologically effective dose." In this work, we treat the former as biological monitoring; the latter we term genetic monitoring.

Ideally, the best indicator of biologically relevant exposure or risk would be a direct measure of the chemical or its metabolite *at the target site*, not the broad concept of toxin "uptake." According to one researcher,

[a] direct measure [at the site of action] is not usually feasible because the sites of action are frequently located in tissues not accessible for sampling (e.g., brain acetylcholinesterase activity). The concentration of the pollutant or its metabolites in another body compartment (blood, urine) or the amount bound to another molecule may be used for this purpose if one has demonstrated that the latter parameter is in equilibrium with the amount at the site of action.[44]

In practice, biological monitoring is used to measure the actual total "uptake" (intake × fractional absorption) through several pathways simultaneously (inhalation, ingestion, dermal absorption). Biological monitoring differs from medical surveillance in that the former measures uptake (and sometimes also the persistence of the measured substance in the body) and the latter is used to determine the effects of exposure. For example, medical surveillance (e.g., chest X-ray) would be an appropriate part of a pulmonary evaluation for a worker exposed to silica dust, as the X-ray would most likely show any major *effects* due to such exposure. An X-ray, however, would not yield useful biological monitoring information in this situation because the X-ray would not assess the worker's level of uptake of silica dust from an X-ray.

Biological monitoring measures primarily the levels of an agent and/or its metabolites in biological specimens. For some chemicals, it may also measure

[44]Robert R. Lauwerys, *Industrial Chemical Exposure: Guidelines for Biological Monitoring* (Davis: Biomedical Publications, 1983) 5.

agent-specific reversible nonadverse effects (e.g., blood zinc protoporphyrin levels for exposure to lead) but the definition of biological monitoring does not include the measurement of nonspecific effect parameters (e.g., decreased nerve conduction velocity or decreased FEV_1). The analysis most commonly uses urine, breath and blood specimens, but sometimes utilizes hair, nails, tears, breast milk, or perspiration, as well.

Some observers believe that the information obtained from biological monitoring can be used in conjunction with environmental monitoring results to determine whether ambient data predict the true exposure of workers and thereby evaluate occupational hygiene control methods. Others, however, believe that ambient environmental measurements cannot accurately be correlated with biological measurements because of individual pharmacokinetic and metabolic variability. One researcher stated that "biological measurements reflect *uptake* and not *exposure* . . . there are numerous instances in which significant uptake of toxic materials have [sic] occurred in spite of low air-levels of the contaminant in question." Obviously the use of particular biological monitoring tests for evaluating specific exposures needs to be evaluated on a case-by-case basis in the light of both (1) the degree of interindividual variability, and (2) the dynamics of the marker that is being measured in the workers in relation to the dynamics of their exposures.

Biological monitoring should not substitute for environmental monitoring. Rather, "environmental and biological monitoring are ways of investigating different problems and should be seen as complementary and not mutually exclusive procedures."[50] The general substitution of biological monitoring in favor of environmental monitoring to determine compliance and control consistent with the OSHA standards is not appropriate for the following reasons:

♦ First, it is not clear whether OSHA has the authority to require workers to submit to biological monitoring procedures for determining compliance.

♦ Second, a biological standard may provide an incentive for employers to intervene in altering specific parameters in their workers. For example, chelation therapy may be used in industrial settings to decrease worker blood lead levels.

♦ Third, biological standards may reinforce a "blame the worker" attitude among employers

with regard to specific employees, rather than focusing attention on the workplace.

♦ Finally, in some cases, biological standards may involve greater risk of health damage due to possible delays between dangerous air exposure and the monitored biological response.

One cannot always define biochemical tests as either medical surveillance or biological monitoring. Certain tests not only are indicators of metabolic effects (medical surveillance) but also can be quantitatively linked to effects of exposure. Examples of these effect tests include zinc protoporphyrin (ZPP) and delta aminolaevulinic acid dehydrase (ALA-D), both of which are quantitatively related to lead exposure.

ENVIRONMENTAL MONITORING

Environmental monitoring measures the concentration of harmful agents *in the workplace,* while the other types of monitoring involve tests performed *on the workers.* It is used primarily to assess whether there is actual or impending overexposure in the workplace and whether there is a related increased health risk from that exposure. The multiple components of a comprehensive environmental monitoring program are listed below. The program should:

1. signal an unhealthy environment;

2. effectively monitor the atmosphere;

3. detect unexpected noxious contaminants;

4. maintain surveillance on the efficacy of their containment;

5. help determine the effectiveness of alteration of the environment (e.g., workplace renovation, removal from exposure);

6. test for compliance with workplace standards; and

7. help evaluate the effectiveness of corrective or possibly therapeutic measures.

Environmental monitoring includes both ambient monitoring and personal monitoring. Personal monitoring is "a term designating the determination of the inhaled dose of an airborne toxic material or of an air-mediated hazardous physical force by the continuous collection of samples in the breathing or auditory zone, or other appropriate exposed body area, over a finite period of exposure time." Personal monitors placed in the breathing zone (e.g., on shirt collars) are considered to provide a representative dose of inhaled air that transports any airborne hazardous agents. Ambient monitoring is defined as

[50]D. Gompertz, "Assessment of Risk by Biological Monitoring," *Br. J. Indus. Med.* (1981) 38(2):201.

"the measurement and assessment of agents at the workplace and . . . [the evaluation of] ambient exposure and health risk compared to an appropriate reference." Ambient (work area) monitoring is useful if the hazard is a specific one for which a permissible exposure limit is known (e.g., a consensus guideline or legal standard). An advantage is that it does not use the worker as a sampling device. In this work, we do not treat environmental monitoring in any detail, but mention it only in reference to the other monitoring practices.

SENSITIVITY SCREENING

A fourth kind of human monitoring, sensitivity screening, is practiced on an employee only once, usually as part of a preemployment or preplacement exam. Such examinations are very common. According to the National Occupational Exposure Survey, conducted in 1981-3, medical screening of potential new workers was mandatory in 22 percent of "small" manufacturing plants (less than 100 workers), 57 percent of medium-sized plants (100-499 workers), and 78 percent of larger plants.

Sensitivity screening seeks to determine whether an *individual* possesses certain inherited or acquired traits that may predispose him or her to an increased risk of disease if exposed to particular substances. If that trait is inherited, the term *genetic screening* applies. (One example of this is an inherited deficiency in alpha-1 antitripsin, which is associated with increased susceptibility for emphysema, particularly among smokers.) Laboratory tests on body fluids (usually blood) identify these traits. However, a number of other common components of medical examinations can also be thought of as more general forms of sensitivity screening, such as back X-rays (intended to detect susceptibility to back injuries) or inquiries about past histories of allergic responses or cardiovascular symptoms.

*   *   *

Most of the literature, until recently, has lent itself to a characterization based upon the *type of monitoring* conducted, e.g., medical surveillance, biological monitoring, genetic screening, and genetic or sensitivity monitoring. However, a new focus has emerged on *biological markers*, i.e., the actual substance or biochemical marker being monitored. The National Academy of Sciences offered the following definitions:

Biological markers are indicators signaling events in biological systems or samples. It is useful to classify biological markers into three types, those of *exposure, effect,* and *susceptibility,* and to describe the events particular to each type. A biological marker of *effect* may be an indicator of an endogenous component of the biological system, a measure of the functional capacity of the system, or an altered state of the system that is recognized as impairment or disease. A biological marker of *susceptibility* is defined as an indicator that the helath of the system is especially sensitive to the challenge of exposure to a xenobiotic compound (a compound originating outside the organism). A biological marker of *exposure* may be the identification of an exogenous substance within the system, the interactive product between a xenobiotic compound and endogenous components, or other event in the biological system related to the exposure. Of utmost importance is the correlation of biological markers of exposure with health impairment or potential health impairment.

It must be emphasized that there is a continuum between markers of exposure and markers of health status, with certain events being relatable to both types of markers. The terms biological monitoring and health monitoring are also in common use, and their distinguishing features are subject to debate. In essence, biological markers can be used for both biological monitoring and health monitoring.

Once exposure has occurred, a continuum of biological events may be detected. These events may serve as markers of the initial exposure, internal dose, biologically effective dose (dose at the site of toxic action, dose at the receptor site, or dose to target macromolecules), altered structure/function with no subsequent pathology, or potential or actual health impairment. Even before exposure occurs, there may be biological differences between humans that cause some individuals to be more susceptible to environmentally induced disease. Biological markers, therefore, are tools that can be used to clarify the relationship, if any, between exposure to a xenobiotic compound and health impairment.[11]

[11]Committee on Biological Markers of the National Research Council, "Biological Markers in Environmental Health Research," *Environ. Health Perspec.* (1987) 74:3–9.

D.   FEDERAL INFORMATION TRANSFER LAWS

At the federal level, workplace right to know is secured primarily under the three statutes discussed previously: the OSHAct, TSCA, and the NLRA. In addition, there are strong *community* right to know provisions in the federal Emergency Planning and Community Right-to-Know Act.

1.   The OSHAct

(a) The Authority to require the generation of information

Monitoring the Worker for Exposure and Disease: Scientific, Legal and Ethical Considerations in the Use of Biomarkers (1990)

N. Ashford, C. Spadafor, D. Hattis, and C. Caldart

The OSHAct grants both OSHA and NIOSH the authority to promulgate regulations that require employers to conduct human monitoring. The authority granted to NIOSH for monitoring is broader in scope than that vested in OSHA, but financial limitations on NIOSH in exercising that authority give OSHA the greater practical grant of authority.

While employers are mandated to make medical testing programs available (generally consisting of medical surveillance alone or coupled with biological monitoring in some circumstances), participation by employees in all OSHA health standards testing programs is not required *as a matter of law.*

The implication that employee participation in medical testing is voluntary for health standards provisions is found in Section 6(b)(7) of the OSHAct which states that medical examinations "shall be made available" to certain employees. There is no mention that employees must participate, only that employers must make such tests available.

Certain standards such as the asbestos, vinyl chloride and carcinogen standards imply the voluntary nature of the employees' participation as well as articulate the mandatory nature of the employer's duties. The OSHA standard for coke oven emissions furthers the notion of voluntary employee participation by providing guidelines for an employer on what to do when an employee refuses to be tested. The regulation states that the employer is to

inform the employee who refuses to be examined of any possible health consequences of such refusal. In addition, the employer is to obtain a statement signed by the employee that the employee understands the risk taken by refusing to be examined. The OSHA lead standard specifically states that participation in the medical surveillance program is not mandatory for the employee. The agency did, however, initially consider mandating that all employees participate in the medical surveillance program. It finally decided not to institute a mandatory program because the program could interfere with the voluntaries and meaningfulness of worker participation. This participation was believed to be essential to the success of the medical surveillance provisions and consistent with employees' personal privacy and religious concerns.

The policy decision to permit voluntary participation in medical surveillance programs has not been significantly opposed by either employee or employer representatives. Employees' rights and privacy interests are believed to be protected if the examinations are made available. Further, there are concerns that the results of mandatory examinations could be used as a basis for adverse personnel action against them. Employers find acceptable the voluntariness of participation because mandatory provisions could prove difficult to enforce.

OSHA

Some types of *medical surveillance* provisions are contained in all OSHA health standards promulgated under section 6(b) of the OSHAct. The agency's

Source: Excerpts from *Monitoring the Worker for Exposure and Disease: Scientific, Legal, and Ethical Considerations in the Use of Biomarkers.* Baltimore, MD: Johns Hopkins University Press, 1990. 15–24.

authority to promulgate such provisions is found in section 6(b)(7) of the Act, which specifies that, where appropriate, section 6(b) standards (standards issued, modified or revoked through rulemaking procedures) "*shall* prescribe the type and frequency of medical examinations or other tests which *shall* be made available, by the employer or at his cost, to employees exposed to [the regulated hazard] in order to most effectively determine whether the health of such employees is adversely affected by such exposure." Further, if such examinations are "in the nature of research," NIOSH may provide the funding. These provisions give clear authorization for OSHA to require medical surveillance to determine the health effects of hazards regulated under section 6(b), even if such surveillance is considered "research."

In addition, OSHA may order *biological monitoring* under each of three sections of the OSHAct—Sections 8(c)(1), 8(c)(3) and 6. Section 8(c)(1) provides general authority:

> Each employer shall make, keep and preserve . . . such records regarding his activities relating to this chapter as the Secretary, in cooperation with the Secretary of Health, Education, and Welfare, may prescribe by regulation as necessary or appropriate . . . *for developing information* regarding the causes and prevention of occupational accidents and illnesses. In order to carry out the provisions of this paragraph such regulations may include provisions requiring *employers to conduct periodic inspections.*

Section 8(c)(3) contains a more specific mandate. It provides that OSHA "*shall* issue regulations requiring *employers to maintain accurate records of employee exposures* . . . which are required to be monitored or measured under Section 6." This section also requires employers to "promptly notify" employees if they have been or are being exposed to any hazard in violation of "an applicable occupational safety and health standard promulgated under section 6." The Senate deliberated over this provision while considering the OSHAct. Senator Peter H. Dominick (R-Colo.), who led an unsuccessful effort to pass a substitute Nixon Administration bill, proposed an amendment that would have eliminated the provision. In objecting to the language of section 8(c)(3), Senator Dominick noted that it "indirectly requires excessive employer monitoring of his entire operation," and thus requires the employer "to be his own policeman, judge and jury." The language of this section is broad as enacted. The fact that Congress chose to include it in the Act, rather than accept the substitute bill deleting this language, supports the proposition that the OSHAct requires employers to conduct biological or environmental monitoring or both for *any* exposure regulated under section 6.

Section 6 itself, however, casts some doubt on this interpretation. On the one hand, section 6(a) makes no mention of human monitoring, but merely requires OSHA to adopt previously existing health standards. Section 6(b), on the other hand, specifically discussed both biological monitoring and medical surveillance. Section 6(b)(7) mandates that, where "appropriate," a section 6(b) standard "*shall* provide for monitoring or measuring employee exposure . . . in such manner as may be necessary for the protection of employees."

This last provision raises an interesting issue. Unquestionably, section 8(c)(3) applies both to section 6(a) standards (national consensus standards and established federal standards adopted without rulemaking) and to section 6(b) standards, as both are occupational health standards that require "employee exposures to be measured" within the meaning of section 8(c)(3). If section 8(c) grants broad power to order biological monitoring, the question is raised as to why section 6(b) *also* grants such power. The solution may lie in the "accuracy" limitation of section 8(c).

Section 8(c)(3) requires only that employers maintain "accurate" records of employee exposures. For many exposures, accurate measurements may be possible without biological monitoring, i.e., by using environmental monitoring. Arguably, section 8 would not require biological monitoring in those situations. Under section 6(b), however, OSHA could go further and order biological monitoring if it were "necessary for the protection of the employees." This interpretation finds an implicit Congressional attempt to balance the need for reliable information against the cost and inconvenience of a physically invasive monitoring procedure. Accordingly, potentially invasive monitoring would be subject to the specificity and increased scrutiny of the section 6(b) standard-setting procedure.

Currently, only the OSHA lead standard requires routine biological monitoring. The recently promulgated benzene standard requires biological monitoring for urinary phenol in emergency situations. Concerning the consensus health standards adopted under section 6(a) of the Act, OSHA recommends biological monitoring for certain substances such as carbon monoxide, fluoride (inorganic), and pesticides, including endrin and parathion.

Under the *NIOSH/OSHA Occupational Health Guidelines*, OSHA issued biological monitoring guidelines for some consensus health standards adopted under section 6(a). These guidelines "provide a basis for promulgation of *new* occupational health regulations." As these guidelines are not regulations, they were not subject to judicial review. On September 27, 1988 OSHA issued an advance notice of proposed rulemaking on a generic standard for exposure monitoring (53 Fed. Reg. 37,591). No specific mention of biological monitoring is made, but OSHA might well include it as well as environmental monitoring.

Biological monitoring results must be preserved as a part of an employee's *exposure* record under the OSHA rule [discussed later in this chapter] governing access to employee exposure and medical records. The OSHA rule defines this record as including any information concerning "biological monitoring results which directly assess the absorption of a substance or agent by body systems (e.g., the level of a chemical in the blood, urine, breath, hair, fingernails, etc.) but not including results which assess the biological *effect* of a substance or agent." This OSHA definition of biological monitoring clearly distinguishes the results of these tests from the results of medical surveillance tests. OSHA further distinguished between exposure records and medical records by providing that an employee medical record must contain any information concerning the health status of the employee including "the results of medical examinations (preemployment, preassignment, periodic, or episodic) and laboratory tests (including X-ray examinations . . .)."

THE "OSHA LOG": RECORDKEEPING AND REPORTING OCCUPATIONAL INJURIES AND ILLNESSES

Results from medical surveillance and biological monitoring testing must also be included in the OSHA Log that is maintained both to develop information related to the causes and prevention of occupational illnesses and to collect, compile, and analyze occupational health standards. The specific regulation, entitled "Recording and Reporting Occupational Injuries and Illnesses," requires employers with more than ten employees to (1) maintain a log and summary of all recordable occupational *injuries and illnesses* in each establishment and (2) enter each recordable injury and illness in the log and summary as soon as practicable, but no later than six working days after receiving the information that the incident occurred. The Logs are to be retained by the employer for five years after the end of the year of recording and are to be made available by the employer upon request, to any employee, former employee or representative.

The regulation requires that "recordable occupational injuries or illnesses" be entered in the Log. There are three categories under this classification:

- fatalities, regardless of the time between the injury and death, or of the length of the illness;
- lost workday cases other than fatalities, that result in lost workdays;
- nonfatal cases without lost workdays which result in transfer to another job or termination of employment, or require medical treatment (other than first aid) or involve: loss of consciousness or restriction of work or motion. This category also includes any diagnosed occupational illnesses which are reported to the employer but are not classified as fatalities or lost workday cases.

Medical surveillance and biological monitoring test results apply most logically to the third category or "nonfatal cases." Yet, the implementation of this reporting requirement may be difficult and produce inconsistent recordings because it is not clear what constitutes a reportable item under this category when applied to the testing results.

* * *

THE AUTHORITY TO REQUIRE MEDICAL SURVEILLANCE ON CONSENSUS STANDARDS

Section 6(a) of the OSHAct does not mention human monitoring. The *NIOSH/OSHA Occupational Health Guidelines*, however, contains extensive medical surveillance guidelines for approximately 450 substances. Because the guidelines provide a basis for new regulations and do not constitute required practices, the legality of adding medical surveillance requirements to existing section 6(a) standards was not an issue at the time the guidelines were established. With a proposed rule on medical surveillance programs for section 6(a) standards under consideration [see page 307], the issue may be raised.

OSHA HAS NOT AUTHORIZED GENETIC SCREENING FOR ANY PROMULGATED HEALTH STANDARDS

The particular kind of medical surveillance involving genetic testing of workers has caused some concern. Some standards that OSHA promulgated under section 6(b) . . . require medical examinations to include a personal history of the employee, his family or both, including "genetic and environmental factors." OSHA issued a clarification of these standards, emphatically denying that the standards require genetic testing of any employee.

* * *

NIOSH

In its capacity as a research agency, NIOSH has broad power to order human monitoring. Inherent authority to do so is granted in sections 20(a)(1), 20(a)(4), and 20(a)(7) of the OSHAct, all of which include mandates to NIOSH to conduct various studies pertaining to occupational health. Section 20(a)(5) gives specific authority to order both biological monitoring and medical surveillance. It states that NIOSH may

> prescribe regulations requiring employers to measure, record, and make reports on the *exposure* of employees to substances or physical agents which [NIOSH] reasonably believes may

endanger the health or safety of employees . . . [and] establish such programs of *medical examinations* and tests as may be necessary for determining the incidence of occupational illnesses and the *susceptibility* of employees to such illnesses.

This section envisions collecting information for extensive epidemiological studies and is not limited to hazards already regulated under section 6. Therefore, its potential scope is much broader than that pertaining to OSHA. Section 20(a)(5), however, also directs NIOSH to "furnish full financial or other assistance" to "any employer who is required to measure and record *exposure* of employees . . . under this subsection," to defray "additional expense" the employer incurs in fulfilling those requirements. Budgetary limitations thus place a decided constraint on NIOSH's ability to impose biological monitoring requirements. The reimbursement provision does not appear to apply to medical surveillance for the measuring and recording of *effects*.

(b) The OSHA hazard communication standard

The primary federal right to know law is the OSHA *Hazard Communication Standard.* As a result of a challenge brought by unions, states, and citizen groups in the Third Circuit Court of Appeals, the standard currently provides for greater disclosure of information to workers than it did when first promulgated by OSHA in 1983. Those challenging the standard argued that it was actually a Section 8 "regulation" rather than a Section 6 "standard" under the OSHAct. Accordingly, they argued that it could not be used to preempt *state* right to know laws. The Third Circuit disagreed. Noting that the Hazard Communication Standard "is aimed at eliminating the specific hazard that employees handling hazardous substances will be more likely to suffer impairment to their health if they are ignorant of the contents of those substances," the court held that workplace right to know requirements could be promulgated as a section 6(b)(5) standard. Nonetheless, the court agreed with the challengers that the standard, as written, did not satisfy the criteria of that section.

United Steelworkers of America, AFL-CIO-CLC v. Auchter

763 F.2d 728 (3d Cir. 1985)

GIBBONS, Circuit Judge.

This case involves consolidated petitions for judicial review of the Hazard Communication Standard promulgated by the Secretary of Labor on the authority of the Occupational Safety and Health Act of 1970 (OSH Act), Pub. L. 91-596, 84 Stat. 1590, 29 U.S.C. §651 *et seq.* (1982).

* * *

I. EVOLUTION OF THE STANDARD

Section 6 of the OSH Act directs the Secretary of Labor to promulgate occupational safety and health standards to further the purpose of the Act "to assure so far as possible every working man and woman in the Nation safe and healthful working conditions. . . . " 29 U.S.C. §§651(b) and 655 (b)(1) (1982). Any standard promulgated by the Secretary

> shall prescribe the use of labels or other appropriate forms of warning as are necessary to insure that employees are apprised of all hazards to which they are exposed, relevant symptoms and appropriate emergency treatment, and proper conditions and precautions of safe use or exposure.

29 U.S.C. §655(b)(7) (1982).

In 1974, the National Institute for Occupational Safety and Health (NIOSH), an agency created by section 22 of the OSH Act, 29 U.S.C. §671 (1982), recommended that the Secretary promulgate a standard requiring employers to inform employees of potentially hazardous materials in the workplace. 47 Fed. Reg. 12095 (1982). Later that year the Secretary appointed an advisory committee to develop standards for implementation of the statutory provision requiring labels or other appropriate forms of warning. That advisory committee issued its report on June 6, 1975, recommending a classification of hazards, the use of warning devices such as labels and placards, disclosure of chemical data, and employee training programs. *Id.* at 12096.

The 1975 Committee report did not result in prompt action by the Secretary. In 1976 a House of Representatives subcommittee held oversight hearings during which several committee members expressed concern over the Secretary's failure to promulgate a comprehensive Hazard Communication Standard. Control of Toxic Substances in the Workplace: Hearings Before the Subcomm. on Manpower and Housing of the House Comm. on Government Operations, 94th Cong. 2d Sess. 87, 89-90 (1976). Seventeen months later, the full House Committee on Government Operations issued a Report which criticized the agency for "miserly use of its delegated powers to deal with disease and death-dealing toxic substances." House Comm. on Government Operations, Failure to Meet Commitments Made in the Occupational Safety and Health Act, H.R. Rep. No. 710, 95th Cong., 1st Sess. 13 (1977). The Committee concluded that:

> The Department of Labor should exercise its power under the Occupational Safety and Health Act to insure that employers and workers can and will know what kinds of toxic dangers are present in the Nation's workplaces. OSHA should require chemical formulators to identify any regulated substance in products they sell.

Id. at 15.

Eventually, on January 16, 1981 the agency published a notice of proposed rulemaking entitled "Hazards Identification." 46 Fed. Reg. 4412-53. The standard proposed would be applicable to employers in Division D, Standard Industrial Classification Codes 20-39, which include only employers in the manufacturing sector. *Id.* at 4426. This classification of employers is made by type of activity for the purpose of promoting uniformity and comparability in the presentation of statistical data. Executive Office of the President, Office of Management and Budget, Standard Industrial Classification Manual 9 (1972). This initial proposal was withdrawn by the Secretary on February 12, 1981 for further consideration of regulatory alternatives. 46 Fed. Reg. 12214. The notice of proposed rulemaking which resulted in the rule challenged in the instant proceedings, entitled "Hazard Communication," was published on March 19, 1982. 47 Fed. Reg. 12091. Like the January 16, 1981 proposal, it was limited to employers in the manufacturing sector. The most significant difference from the rule proposed in 1981 was the inclusion in the March 19, 1982 proposal of a trade secret exception to the requirement that the chemical identities of all hazardous chemicals be disclosed. *Compare* 46 Fed. Reg. 4426 (1981) *with* 47 Fed. Reg. 12105 (1981).

The standard was published in its final form on November 25, 1983. 48 Fed. Reg. 53279. It requires that chemical manufacturers and importers "evaluate chemicals produced in their workplaces or imported by them to determine if they are hazardous." 29 C.F.R. §1910.1200(d)(1) (1984). It refers to several compilations of toxic materials. These lists establish a floor of toxic substances which chemical manufacturers or importers must treat as hazardous. 29 C.F.R. §1910.1200(d)(3) (1984). Chemicals not included in the designated compilations must be evaluated for hazardousness by reference to "available scientific evidence." 29 C.F.R. §1910.1200(d)(2) (1984). A manufacturer or importer of chemicals found to be hazardous must "ensure that each container . . . leaving the workplace is labeled" with the chemical identity, with appropriate hazard warnings, and with the name and address of the source. 29 C.F.R. §1910.1200(f)(1) (1984). Manufacturers or importers must also prepare a "material safety data sheet" (MSDS) containing the chemical common names of each hazardous ingredient, and information necessary for safe use of the product. 29 C.F.R. §1910.1200(g) (1984). The MSDS must be provided to each employer in the manufacturing sector (Standard Industrial Classification Codes 20-39) purchasing a hazardous chemical. That employer must in turn make the MSDS available for employee inspection, 29 C.F.R. §1910.1200(g)(8) (1984) and "shall provide employees with information and training on hazardous chemicals in their work area. . . ." 29 C.F.R. §1910.1200(h) (1984).

The rule allows an exception from the labeling and MSDS ingredient disclosure requirements when a chemical manufacturer or importer claims that the chemical identity is a trade secret. 29 C.F.R. §1910.1200(i) (1984). In such a case, the manufacturer or importer must provide a MSDS disclosing the hazardous properties of the chemical and suggesting appropriate precautions. In the case of a medical emergency, the manufacturer or importer must disclose the chemical identity to a treating physician or nurse, and may later require such a health professional to sign a confidentiality agreement. 29 C.F.R. §1910.1200(i)(2) (1984). Absent a medical emergency, the manufacturer or importer may be required to disclose the chemical identity to a health professional who makes a written request detailing the occupational need for the information, and who is willing to sign a confidentiality agreement containing a liquidated damages clause. 29 C.F.R. §1910.1200(i)(3) & (4). In no case is the manufacturer required to disclose the precise formula, as opposed to the identity of chemicals in the compound.

* * *

B. Limitation of Coverage to the Manufacturing Sector

None of the petitioners still before us contend that the manufacturing sector should be not covered by a hazard communication standard. The Secretary's decision to provide coverage only for employees in the manufacturing sector is based on a finding that this sector, which includes 32% of total employment, accounts for more than 50% of the reported cases of illness due to chemical exposure. 48 Fed. Reg. 53285 (1983). From this datum the Secretary determined that employees in the manufacturing sector have the greatest risk of experiencing health effects due to chemical exposure. *Id.* Agricultural employees have a higher chemical source incidence rate than manufacturing employees. The Secretary discounted this datum, however, because 80% of the reported chemical source cases among agricultural workers involved skin illnesses from handling plants, which would not be regulated by the proposed Hazard Communication Standard. *Id.* Moreover the Secretary concluded that the Environmental Protection

Administration, has, under the Federal Insecticide, Fungicide and Rodenticide Act, exercised jurisdiction over regulation of field use of pesticides. Excluding agricultural employees, there is substantial evidence in the record that the manufacturing sector has the highest incidence rate of chemical exposures which the Agency has authority to regulate.

Several petitioners, while conceding that the finding about incidence rate of illnesses in the manufacturing sector is supported by substantial evidence, contend that the Secretary's exclusion of other sectors such as service, construction, and agriculture, is unsupported by reasons that are consistent with the purposes of the statute. They urge that while the incidence rate for employees in the manufacturing sector is high overall, some employees in specific nonmanufacturing categories, such as hospital workers, are exposed to a greater number of toxic substances than are typical workers in the manufacturing sector. Moreover some workers in specific non-covered industries have higher reported rates of chemical source illness and injury than do workers in many covered industries. The Standard Industrial Classification breakdown, they contend, is not relevant to the statute, since that classification is made for a myriad of statistical purposes, mainly economic, having little to do with exposure to hazards. The result of the standard is that spray painters in the manufacturing sector, for example, must be provided with MSDS's and with information and training on hazardous chemicals in the products they use, while spray painters in the construction industry using the same products are not so protected.

In explaining the limited coverage, the Secretary reasoned:

> It should be emphasized that the Agency does not believe that employees in other industries are not exposed to hazardous chemicals, or that they should not be informed of those hazards. OSHA has merely exercised its discretion to establish rulemaking priorities, and chosen to first regulate those industries with the greatest demonstrated need.

45 Fed. Reg. 53286. Rejecting arguments of participants in the rulemaking proceeding that other workers, such as painters in the construction industry, are exposed to the same hazards as are workers in the manufacturing sector, the Secretary reasoned:

> As stated previously, OSHA acknowledges that exposures to hazardous chemicals are occurring in other industries as well. A limited coverage of them is included in the final standard since all

containers leaving the workplace of chemical manufacturers, importers, or distributors will be labeled, regardless of their intended destination. This will alert downstream users to the presence of hazardous chemicals, and the availability of material safety data sheets. The Agency contends that the focus of this standard should remain on the manufacturing sector since that is where the greatest number of chemical source injuries and illnesses are occurring. This focus will also serve to ensure that hazard information is being generated for chemicals produced or imported into this country, and this increased availability will benefit all industry sectors.

Id. The Secretary's reasoning does not address the petitioners' contention that reliance on the Standard Industrial Classification is inappropriate because it ignores the high level of exposure in specific job settings outside the manufacturing sector.

The Secretary maintains that section 6(g) of the Act affords him unreviewable discretion to determine what industries shall be covered by a standard. That section provides in relevant part:

In determining the priority for establishing standards under this section, the Secretary shall give due regard to the urgency of the need for mandatory safety and health standards for particular industries, trades, crafts, occupations, businesses, workplaces or work environments.

29 U.S.C. §655(g) (1982). We reject the Secretary's contention that his priority-setting authority under section 6(g) vitiates judicial review of his determination that only the manufacturing industry need be covered by the Hazard Communication Standard. Section 6(g) must be read in conjunction with section 6(f), which provides for judicial review of standards. Indeed the language "due regard to the urgency of the need for mandatory safety standards for particular industries" suggests to us a statutory standard by which to measure the exercise of the Secretary's priority-setting discretion. In *United Steelworkers of America v. Marshall*, 647 F.2d 1189, 1309-10 [8 OSHC 1810, 1900-01] (D.C. Cir. 1980), *cert. denied*, 453 U.S. 913 [11 OSHC 1264] (1981) the court reviewed under section 6(f) the Secretary's decision to exempt the construction industry from a standard limiting exposure to lead. Although the court did not explicitly address section 6(g), it implicitly rejected the contention that the Secretary's priority-setting authority is unreviewable. We do so explicitly.

Our difficulty with the Secretary's reliance on section 6(g) arises from the Secretary's failure to explain why coverage of workers outside the manufacturing sector would have seriously impeded the rulemaking process. Section 6(g) clearly permits the Secretary to set priorities for the use of the agency's resources, and to promulgate standards sequentially. Once a standard has been promulgated, however, the Secretary may exclude a particular industry only if he informs the reviewing court, not merely that the sector selected for coverage presents greater hazards, but also why it is not feasible for the same standard to be applied in other sectors where workers are exposed to similar hazards. *See United Steelworkers*, 647 F.2d at 1309-10 [8 OSHC at 1900-01]. The explanation for the Secretary's action quoted above is deficient in the latter respect. Thus the Secretary has failed to carry the burden of persuading us that section 6(g) justifies limitation of coverage to the manufacturing sector.

We are also unpersuaded by the Secretary's contention that the communication rule will "trickle down" to uncovered workers because containers will be labeled. Section 6(b)(5) requires that:

The Secretary, in promulgating standards dealing with toxic materials or other harmful physical agents . . . shall set the standard which most adequately assures, to the extent feasible, on the basis of the best available evidence, that no employee will suffer material impairment of health or functional capacity even if such employee has regular exposure to the hazard dealt with by such standard for the period of his working life.

29 U.S.C. §655(b)(5) (1982). There is record evidence that workers in sectors other than manufacturing are exposed to the hazards associated with use of toxic materials or other harmful physical agents. The Secretary has given no statement of reasons why it would not be feasible to require that those workers be given the same MSDS's and training as must be given to workers in the manufacturing sector. Section 6(c)(7) provides that:

Any standard . . . shall prescribe the use of labels or other appropriate forms of warning as are necessary to insure that employees are apprised of all hazards to which they are exposed, relevant symptoms and appropriate emergency treatment, and proper conditions and precautions of safe use or exposure.

29 U.S.C. §655(c)(7) (1981). The Secretary has given reasons why the labeling, MSDS, and instruction requirements comply with section 6(c)(7) for employ-

ees in the manufacturing sector, but no explanation why the same information is not needed for workers in other sectors exposed to industrial hazards. Such a statement of reasons is required by section 6(f). *Synthetic Organic*, 503 F.2d at 1160 [2 OSHC at 1163].

We hold, therefore, that the petitions for review of those petitioners who object to the limitation of the Hazard Communication Standard must be granted. That standard may continue to operate in the manufacturing sector, but the Secretary's explanation for excluding other sectors does not withstand the scrutiny mandated by section 6(f). Thus the Secretary will be directed to reconsider the application of the standard to employees in other sectors and to order its application to other sectors unless he can state reasons why such application would not be feasible. 29 U.S.C. §655(b)(5).

*   *   *

1. The Definition of Trade Secret. Trade secret protection may arise from two sources: state law or a federal statute. *Ruckelshaus v. Monsanto Co.* 467 U.S. 986 (1984); *Chevron Chemical Co. v. Costle*, 641 F.2d 104, 115 (3d Cir.), *cert. denied*, 452 U.S. 961 (1981). The OSH Act does not create substantive trade secret protection. It deals with that subject in section 15, which provides:

> All information reported to or otherwise obtained by the Secretary or his representative in connection with any inspection or proceeding under this chapter which contains or which might reveal a trade secret referred to in section 1905 of Title 18 shall be considered confidential for the purpose of that section, except that such information may be disclosed to other officers or employees concerned with carrying out this chapter or when relevant in any proceeding under this chapter. In any such proceeding the Secretary, the Commission, or the court shall issue orders as may be appropriate to protect the confidentiality of trade secrets.

29 U.S.C. §664 (1982). The cross-reference to the Trade Secrets Act, 18 U.S.C. §1905 (1982) confirms that Congress intended section 15 to protect against agency disclosure or misuse of data submitted to it under an expectation that the agency would treat that data as a trade secret to the extent that applicable state law did so. *See Ruckelshaus v. Monsanto, supra*, 52 U.S.L.W. at 4892. Section 18 cannot be read as authorizing the creation of trade secret protection going beyond that afforded under state law. Moreover that section deals only with what the agency and

its employees may disclose, not with what disclosures the agency may compel in the interest of safety in the workplace.

The Secretary contends that while section 15 deals with disclosures by the agency and its employees, it embodies a recognition by Congress of the significance in the economy of state law trade secret protection, and that it was quite appropriate, in drafting the Hazard Communication Standard, to balance that recognition against the competing congressional concern over safety in the workplace. *See* 48 Fed. Reg. 53312 (1983). Aside from section 15, however, the Secretary points to no provision in the OSH Act authorizing such balancing. Indeed trade secret protection is not even mentioned in section 6, which directs that "[t]he Secretary . . . shall set the standard which most adequately assures, to the extent feasible, on the basis of the best available evidence, that no employee will suffer material impairment of health . . . ." 29 U.S.C. §655(b)(5). The quoted provision indicates that Congress struck a balance in favor of safety in the workplace at the expense of competing interests. In a slightly different context, the Supreme Court held that in section 6(b)(5), "Congress itself defined the basic relationship between costs and benefits, by placing the 'benefit' of worker health above all other considerations save those making attainment of this 'benefit' unachievable," *American Textile Manufacturers Institute v. Donovan*, 452 U.S. 490, 509 [9 OSHC 1913, 1920] (1981).

The Secretary originally proposed that while traditional trade secrets such as chemical formula and process information could be withheld, chemical identity information must be disclosed. 46 Fed. Reg. at 4426 (1981). When representatives of the chemical industry commented on the importance of trade secrets to the economic health of that industry, *see* 48 Fed. Reg. 53312-14 (1983), the Secretary adopted an entirely new approach. First, it defined a trade secret as:

> [A]ny confidential formula, pattern, process, devise, information or compilation of information (including chemical name or other unique chemical identifier) that is used in an employer's business, and that gives the employer an opportunity to obtain an advantage over competitors who do not know or use it.

29 C.F.R. §1910.1200(c) (1984). The Secretary contends that this definition was adopted from section 757 of the Restatement of Torts, 48 Fed. Reg. 53314. In fact, however, inclusion of the parenthetical "(including chemical name or other unique chemical iden-

tifier)" not found in section 757, enlarges considerably the Restatement definition. The Secretary's comments indicate that the definition provides trade secret protection for chemical identity which is determinable by reverse engineering. *Id.* This type of information has not traditionally been afforded trade secret protection under state law. *See Kewanee Oil Co. v. Bicron Corp.*, 416 U.S. 470, 476 (1974). The Restatement provides that information that may be properly acquired or duplicated without great difficulty should not be considered a trade secret. Restatement of Torts, §757, comment b. The Secretary justified the enlargement of the Restatement definition, reasoning:

> Many products can be reverse engineered if sophisticated analytical techniques are applied, yet cannot be if less advanced technology is used. The determination of what is "practical" in terms of reverse engineering capability rests with this degree of analysis, rather than with a definitive finding of "ability" to be reversed engineered or not. Furthermore, the definition of a trade secret says that the competitor or potential competitor does not know *or* use the information. Thus, even though a competitor could theoretically "reverse engineer" and discover the components of a product, if this information is not in fact used, it remains a *bona fide* trade secret.

48 Fed. Reg. 53314 (emphasis in original). Plainly the Secretary has provided greater protection for chemical manufacturers and importers than that afforded in those states utilizing the Restatement of Torts trade secret definition. Even the Restatement definition, moreover, goes beyond the protection afforded to trade secrets in other regulatory contexts. *See Public Citizen Health Research Group v. Food and Drug Administration*, 704 F.2d 1280, 1287 (D.C. Cir. 1983). (Food and Drug Administration's adoption of Restatement of Torts trade secret definition inconsistent with Freedom of Information Act.)

Section 15 deals only with disclosure by the agency or its employees, and section 6(b)(5) does not permit the Secretary to balance employee safety against competing economic concerns. No other statutory provision has been called to our attention which would justify enlarging trade secret protection beyond that afforded by state law. Indeed it seems plain that state law cannot prevent the implementation of section 6 safety standards that are otherwise feasible.

* * *

We agree that there is no legal justification for affording broader trade secret protection in the Hazard Communication Standard than state law affords. No petitioner urges that the Secretary's original proposal, which would have protected formula and process information but required disclosure of hazardous ingredients, is inadequate. That proposal was consistent with "[t]he general policy of OSHA . . . that the interests of employee safety and health are best served by full disclosure of chemical identity information." 48 Fed. Reg. 53312 (1983). The petition for review will therefore be granted and the proceedings remanded to the Secretary for reconsideration of the definition of trade secrets, which definition shall not include chemical identity information that is readily discoverable through reverse engineering.

2. The Access Rule. In addition to objections to the definition of trade secrets, several petitioners challenge the provisions of the Hazard Communication Standard relating to access to the allegedly confidential information. The petitioners contend that once a manufacturer raises a claim of trade secret, the Standard places overly stringent procedural barriers to the discovery of information relevant to the assessment of health hazards. These objections go to: a) the requirement that the request be in writing with a statement of need; b) the limitation of access to "health professionals"; and c) the requirement that a confidentiality agreement with a liquidated damages clause be signed. 29 C.F.R. §1910.1200(i).

a) The Written Request Requirement. The Secretary justifies the requirement that a request for trade secret information be in writing with supporting documentation as a means of facilitating dispute resolution:

> Then if the matter is to be referred to OSHA to settle any dispute between the requesting party and the employer protecting a trade secret, the Agency will be able to base a decision upon a review of these written materials. Should the matter not be resolved to the satisfaction of all parties, it may result in a citation and referral to the Occupational Safety and Health Review Commission (OSHRC) for judicial review.

48 Fed. Re. at 53315. The tendered justification is both reasonable and consistent with the purposes of the Act.

b) The Restriction of Access to Health Professionals. The restriction limiting access to trade secrets to health professionals is more troublesome. Commentators on the proposed standard were generally in agreement that trade secret chemical identity information must be disclosed at least to a treating physician. *Id.* at 53316. The Secretary concluded that other health professionals, particularly those engaged in the prevention of disease or injury, had a legitimate need for chemical identity information. *Id.* at 53318. It declined, however, to authorize direct employee access to specific chemical identities of hazardous substances for which a trade secret is claimed. *Id.* The Secretary advances three rationales to justify restricting employee access to secret chemical identity information.

One reason is that "by and large professional training would be required" for any purpose that would amount to a "need to know" confidential information. *Id.* The United Steelworkers of America, AFL-CIO-CLC, a petitioner, points to a number of instances in the record, however, where non-health professional workers used chemical identity information to improve workplace safety. They urge that employees and local union safety officers, although not health professionals, have often received training in health and safety, and thus know how to use the basic literature on chemical hazards, and know how to obtain technical assistance. Steelworkers' brief at 40.

A second reason advanced by the Secretary for allowing access only to health professionals, and not directly to employees, is that "providing access to trade secret chemical identities only to health professionals on a confidential basis will protect those employees adequately." 48 Fed. Reg. at 53318. The Steelworkers point to record evidence, however, that it is quite difficult for many workers to obtain the services of a health professional, at least prior to the need for treatment. Steelworkers' brief at 41-42. There

is no substantial evidence in the record that significant numbers of unorganized workers will be able to obtain the services of a health professional prior to the time that treatment becomes necessary. Even for organized workers, the record evidence suggests that few local unions retain health professionals.

The Secretary's final justification for limiting access to health professionals involves the risk of disclosure.

> This is not to say that "downstream" employees are more likely to disclose trade secrets or violate confidentiality agreements than health professionals, but it is an unmistakable fact that the more people who have access to confidential information, the more difficult it is to preserve its secrecy or to locate the source of a leak if one occurs.

48 Fed. Reg. at 53318. The Secretary correctly notes that the chance of a leak increases as the number of people having access to information increases. The issue posed by the petitioners, however, is not the number of persons obtaining access, but the type of persons. There is no record evidence supporting the Secretary's apparent conclusion that employees who are not health professionals will be more likely to breach a confidentiality agreement than would the same number of health professionals. We conclude that the restriction in the Hazard Communication Standard of access to trade secret information to health professionals is not supported by substantial evidence in the record, and is inconsistent with the mandate of section 6(b)(5) that OSHA promulgate the standard that "most adequately assures, to the extent feasible, on the basis of the best available evidence, that no employee will suffer material impairment of health. . . ." 29 U.S.C. §655(b)(5).

* * *

(i) Purpose of the Hazard Communication Standard

In response to this decision, OSHA promulgated a revised Hazard Communication Standard in 1987 that expanded coverage to all workers within OSHA's jurisdiction. Like its 1983 predecessor, the standard is intended "to ensure that the hazards of all chemicals produced or imported are evaluated, and that information concerning their hazards is transmitted to employers and employees." 29 CFR 1910.1200 (a). This is in keeping with OSHA's view that "when employees have access to, and understand, the nature of the chemical hazards they are exposed to during the course of their employment, they are better able to participate in the employers' protective programs, and take steps to protect themselves." Preamble to OSHA's Final Rule to Expand Scope of Hazard Communication Standard, 52 FR 31852, August 24, 1987.

(ii) Definitions

The standard defines an employee as "a worker who may be exposed to hazardous chemicals under normal operating conditions or in foreseeable emergencies." Thus, coverage is determined by exposure, not by industrial sector, SIC code, or job description. A hazardous chemical is any chemical that is a physical or health hazard. A health hazard means any "chemical for which there is statistically significant evidence based on at least one study conducted according to established scientific principles that acute or chronic health effects may occur in exposed employees." These include chemicals that are carcinogenic, toxic, irritating, corrosive, sensitizing, or have target organ effects.

(iii) Requirements

Chemical manufacturers and importers are required to evaluate the chemicals they produce or import to determine if they are hazardous. The standard provides criteria to be used in making this determination. The hazard determination requirement is performance-oriented, but it imposes on manufacturers and importers a duty to conduct a thorough evaluation. They must examine all relevant data and produce a scientifically defensible evaluation. Although the manufacturers and importers are free to determine for themselves the relevance of existing data and the scope of their evaluation, they must describe in writing the evaluation procedures used to make the hazard determination.

Chemical manufacturers and importers are required to obtain or develop a Material Safety Data Sheet (MSDS) for each hazardous chemical they produce or import and provide these MSDSs to downstream users (distributors and employers). If a MSDS is not provided, the employer is responsible for obtaining it as soon as possible. Each MSDS must be in English and contain at least the following information: the chemical and common name, subject to trade secret restrictions; its physical and chemical characteristics; its physical and health hazards; its route of entry; any OSHA PEL, ACGIH TLV, or other exposure limit; handling precautions; applicable control measures; emergency and first aid procedures; the date of MSDS preparation; and the name, address, and telephone number of the chemical manufacturer, importer, employer, or other responsible party. The MSDS must also indicate if the chemical is listed as a carcinogen by the National Toxicology Program or has been found to be a potential carcinogen by the International Agency for Research on Cancer (IARC) or by OSHA.

The standard requires that manufacturers, importers, or distributors label each container of hazardous chemicals and that employers ensure that every container of hazardous chemicals in the workplace is labeled. The label must identify the chemical(s), include appropriate hazard warnings, and list the name and address of the manufacturer, importer, or other responsible party.

Employers are not required to evaluate chemicals used in their workplaces under the OSHAct or the Hazard Communication Standard. They can rely on the information provided by the manufacturer or importer on the MSDS. They are under no obligation to amend inadequate or incorrect information or request that the manufacturer or importer do so. They are, however, required to develop, implement, and maintain a written hazard communication program. This program must describe how the employer plans to deal with the labeling, MSDS, and training sections of the standard. It must also include a list of the hazardous chemicals known to be present in the workplace and a description of the method the employer will use to inform workers of the hazards of nonroutine tasks.

(iv) Employee information and training

The employee information and training section of the standard specifies that employers must inform workers about the standard itself, tell them where hazardous chemicals are present in their work area, and inform them of the location and availability of the employer's written hazard communication program. Additionally, workers must be trained in methods to detect the presence or release of hazardous chemicals; the physical and health hazards of the chemicals; appropriate protective measures; and details of the employer's hazard communication program.

(v) Trade secrets

Trade secret information is protected from disclosure except in specific emergency and nonemergency situations described in the standard. Health professionals, employees, and their designated representatives can obtain access to trade secret information for certain occupational health needs. These include: the need to assess the hazards of the chemicals to which employees will be exposed; the need to conduct or assess environmental sampling to determine exposure levels; and the need to assess engineering controls or other protective measures. In requesting access to trade secret information in such nonemergency situations, the requestor must explain in detail why the disclosure is essential and why alternative information would not suffice. The requestor must also sign a written confidentiality agreement, promising not to disclose the trade secret information under any circumstances, except to OSHA. The agreement may provide for appropriate legal remedies in the event of a breach.

(vi) "Tailored" provisions

Because the revised standard has broad coverage, it includes several "tailoring" provisions that were not included in the original version. For example, laboratories need only ensure that labels on incoming chemicals are not removed, that MSDSs are received and accessible, and that employees are apprised of the hazards of the chemicals pursuant to the information and training section of the standard. They are not required to develop a written hazard communication program or a list of hazardous chemicals found in the lab. These abbreviated requirements reflect OSHA's belief that laboratories are different than typical industrial workplaces. OSHA has recently promulgated a generic laboratory standard, Final Rule on Occupational Exposures to Hazardous Chemicals in Laboratories (55 FR 3300, January 31, 1990). However, the labeling requirements remain the same as described above, with the addition of clarifying remarks. There are similar "tailored" provisions for establishments where employees only handle chemicals in sealed containers, such as cargo handling, warehousing, and retail sales.

(c) The OSHA medical access rule

Beyond the Hazard Communication Standard, the OSH Act has given rise to a number of other pathways for the disclosure of toxics information in the workplace. As noted in the discussion on personal confidentiality later in this chapter, both OSHA and NIOSH have rather broad access to employee medical and exposure records. Further, the various OSHA Section 6(b)(5) standards for toxic substances commonly include chemical-specific monitoring and disclosure requirements. From a worker perspective, the most powerful of the other disclosure tools is the *Medical Access Rule*.

Monitoring the Worker for Exposure and Disease: Scientific, Legal and Ethical Considerations in the Use of Biomarkers

N. Ashford, C. Spadafor, D. Hattis, and C. Caldart

A private employer may not limit or deny an employee access to his or her own medical or exposure records. The current OSHA regulation [29 C.F.R. §1910.20], promulgated in 1980 and revised in 1988, grants employees a general right of access to medical and exposure records kept by their employer. Furthermore, it requires the employer to preserve and maintain these records for an extended period of time. As defined in the regulation, both "medical" and "exposure" records may include the results of biological monitoring. The former, however, are generally defined as those pertaining to "the health status of an employee," while the latter are defined as those pertaining to employee exposure to "a toxic substance or harmful physical agent."

The employer's duty to make these records available is a broad one. The regulation provides that upon any employee request for access to a medical or exposure record, "the employee *shall* assure that access is provided in a reasonable time, place, and manner." If the employer "cannot reasonably provide access to the record within fifteen (15) working days," the employer is required to notify the employee of "the reason for the delay and the earliest date when the record can be made available." Although this may appear at first glance to be rather open-ended, OSHA has made it clear that it expects "the vast majority" of requests "to be satisfied within fifteen days."

Source: Excerpts from *Monitoring the Worker for Exposure and Disease: Scientific, Legal and Ethical Considerations in the Use of Biomarkers.* Baltimore, MD: Johns Hopkins University Press, 1990. 132–137.

Because the regulation defines "access" as including the right to make copies of records, the employer appears to have an affirmative duty to maintain such procedures as are necessary to ensure that, in most circumstances, the employee will have a copy of the records in his or her possession within fifteen days after the request. The employer cannot escape this duty by contracting with others to maintain the records. Although the regulation does not specifically require a physician, health maintenance organization, or other health care provider to permit employee access to records, it does require the employer to "assure that the preservation and access requirements of this section are compiled with *regardless of the manner in which records are made or maintained.*" Thus, any employer contract with a third party must provide for the disclosure of those records.

An employee's right of access to *medical* records is limited to records pertaining specifically to that employee. The regulation allows physicians some discretion as well in limiting employee access. The physician is permitted to "recommend" to the employee requesting access that the employee: (1) review and discuss the records with the physician; (2) accept a summary rather than the records themselves; or (3) allow the records to be released instead to another physician. Further, where information in a record pertains to a "specific diagnosis of a terminal illness or a psychiatric condition," the physician is authorized to direct that such information be provided only to the employee's designated representative. Although these provisions were apparently intended to respect the physician-patient relationship and do not limit the employee's ultimate right of access, they could be abused. In situations in which the physician feels loyalty to the employer rather than the employee, the physician could use these provisions to discourage the employee from seeking access to his or her records.

Similar constraints do not apply to employee access to *exposure* records. In the first instance, the employee is assured access to records of his or her own exposure to toxic substances. To the extent that those records fail to detail the amount and nature of the substances to which the employee has been exposed, however, the employee is also granted access to the exposure records of other employees "with past or present job duties or working conditions related to or similar to those of the employee." In addition, the employee enjoys access to all general exposure information pertaining to the employee's workplace or working conditions and to any workplace or working condition to which he or she is to be transferred.

Employers are only required to preserve records of employee exposure to substances defined as "toxic." The regulation defines this term with appropriate breadth, and includes as "toxic" all chemicals identified as potential human toxicants in the "Registry of Toxic Effects of Chemical Substances" compiled by NIOSH. During the Reagan administration, OSHA had proposed to narrow this definition significantly, and to include only those chemicals that have *already* been shown to be toxic in humans or toxic at specified significant levels in animals. As the agency noted at the time, the proposed redefinition would have resulted in "a greater than 90 percent decrease in the number of chemicals specified."

Not only would the proposed definition have excluded certain human and animal mutagens and certain chemicals that display metagenic potential in short-term, *in vitro* tests, but it would also have discouraged epidemiologic research on chemicals not already known to be toxic. One major purpose of occupational epidemiology is to determine the toxic effects of substances not presently known to be toxic to humans, so that employers can take the steps necessary to reduce workplace exposure. The OSHAct was intended to facilitate that process. The value of epidemiologic research in this area, however, depends in large part on the availability of reliable data regarding employee exposure to substances not already proven toxic. Fortunately, the revised OSHA access rule continues to apply to the compilation of data on a wide variety of substances.

One criticism of the OSHA regulation is that it does not require the employer to compile medical

or exposure information but merely requires employee access to such information if it is compiled. The scope of the regulation, however, should not be underestimated. The term "record" is meant to be "all-encompassing," and the access requirement appears to extend to all information gathered on employee health or exposure, no matter how it is measured or recorded. Thus, if an employer embarks upon any program of human monitoring, no matter how conducted, he or she must provide the subjects access to the results. Any recordkeeping required under TSCA, for example, presumably is available to workers under the OSHA rule. It is conceivable, however, that the access requirement may serve as a disincentive for an employer to monitor employee exposure or health voluntarily where it is not clearly in the employer's interest to do so. The access rule does not prevent employers from denying employees the benefit of having health or exposure data for a number of substances simply by failing to record such data if not otherwise required to do so by law.

The trade secret interest of employers places a limitation on employee access. Section 15 of the OSHAct requires OSHA to be sensitive to employer trade secrets in its collection and use of occupational safety and health information. OSHA originally read this section as requiring the agency to *balance* employee safety and health against competing economic interests of the employer, and proposed during the Reagan administration to significantly expand the protections given to trade secrets under the access rule. This interpretation did not survive court review. Responding to a challenge to the definition of "trade secret" in OSHA's Hazard Communication standard, the Third Circuit Court of Appeals held in a 1985 decision that neither Section 15 nor any other part of the OSHAct permits OSHA to provide trade secret protection beyond that which is afforded by state law. Further, the court noted that Section 15 "deals only with what the agency and its employees may disclose, not with what disclosures the agency may compel in the interest of safety in the workplace."

The trade secret provisions in the revised OSHA access rule appear to have been drafted with the Third Circuit's admonition in mind. Nonetheless, these provisions do provide for somewhat greater protection of trade secrets than had been afforded under the original rule. The revised regulation permits the employer to deny access to "trade secret data which discloses manufacturing processes . . . or . . . the percentage of a chemical substance in a mixture," provided that the employer:

1. notifies the party requesting access of the denial; and

2. if relevant, provides alternative information sufficient to permit identification of when and where exposure occurred.

To this extent, the revised rule follows the language of its predecessor. The original rule, however, ensured employee access to the precise identities of chemicals and physical agents, whether or not the employer claimed this information as a trade secret. This access is especially critical for chemical exposures. Within each "generic" class of chemicals there are a variety of specific chemical compounds, each of which may have its own particular effect on human health. The health effects can vary widely within a given family of chemicals. Accordingly, the medical and scientific literature on chemical properties and toxicity is indexed by specific chemical name, not by generic chemical class. To discern any meaningful correlation between a chemical exposure and a known or potential health effect, an employee must know the precise chemical identity of that exposure. Furthermore, in the case of biological monitoring, the identity of the toxic substance or its metabolite is itself the information monitored.

The revised rule places barriers on the disclosure of such information. The rule permits the employer to delete from an exposure record "the specific chemical identity, including the chemical name and other specific identification of a toxic substance," so long as the employer

1. notifies the requesting party of the deletion;

2. can support the claim that the information is a trade secret;

3. gives the requesting party "all other available information on the properties and effects" of the toxic substance in question; and

4. provides the specific chemical identity of the substance to a treating physician or nurse if such information is "necessary for emergency for first-aid treatment."

Were the trade secret protections to end here, they would defeat much of the public health purpose of granting employee access to exposure information, and would make it difficult to establish necessary baseline information on the relationship of various substances to various health outcomes. Sensibly,

however, the revised rule establishes additional procedures which should make specific chemical identity information available to interested employees in most circumstances.

If the employee is not satisfied with exposure records from which specific chemical identities have been deleted, the employee may file a written request with the employer for such information. This request must contain:

1. a description, in "reasonable detail," of one or more "occupational health needs" for which the information is being sought;

2. an explanation, "in detail," as to why certain enumerated kinds of other information would not be adequate to meet the occupational health needs in question;

3. a description of "the procedures to be used to maintain the confidentiality of the disclosed information; and

4. an indication of the employee's willingness to be bound by a written confidentiality agreement.

The "occupational health needs" for which disclosure may be sought under this section are rather broadly defined, and include "studies to determine the health effects of exposure." Further, the rule includes provisions designed to discourage employers from using the mantle of trade secrecy as a ruse for denying legitimate requests.

The employer must respond to the request for chemical identity information within thirty days. If the request is denied, the employer must provide a written response which includes "evidence to support the claim that the chemical identity is a trade secret," states "the specific reason why the request is being denied," and explains "in detail" how the occupational health needs in question can be met without the requested information. Upon such a denial, the requesting party may refer the matter to OSHA for resolution. If OSHA determines that the claim of trade secrecy is not *bona fide,* or that the information can be disclosed to the requesting party for a legitimate public health need without jeopardizing the employer's proprietary interest, "the employer will be subject to citation by OSHA." Finally, if the employer demonstrates that a confidentiality agreement would not be sufficient to protect a *bona fide* trade secret, OSHA "may issue such orders or impose such additional information as may be appropriate to assure [sic] that the occupational health needs are met without an undue risk of harm to the employer."

The revised rule gives employers somewhat greater freedom in fashioning the "confidentiality agreements" that may be required as a condition of the disclosure of specific chemical identities. The original rule also permitted such agreements, but the explanatory comments to the rule made it clear that they could not be used "as a pretext for more onerous requirements such as the posting of penalty bonds, liquidated or punitive damages clauses, or other preconditions." The revised rule softens this approach, and specifically permits the "reasonable pre-estimate of likely damages" in the event of a breach of confidentiality. However, the prohibition against the use of penalty bonds continues in effect.

This would appear to strike an appropriate balance between legitimate competing interests. By specifying that any liquidated damage provision must be "reasonable," this section should help insure that such provisions are used as a legitimate means of guarding against disclosure, and not as a means of discouraging employees from following through on their request for information. Further, by specifically disallowing the penalty bond— which presumably would have to be purchased by the requesting party prior to the receipt of the information—the rule helps keep the costs of such requests to a minimum.

2.   The Toxic Substances Control Act

The Environmental Protection Agency also has a role in setting right to know policies at the federal level. The Toxic Substances Control Act, described in detail in Chapter 4, vests EPA with broad authority to stimulate the flow of information regarding the production, use, and toxicity of chemical substances.

Monitoring the Worker for Exposure and Disease: Scientific, Legal and Ethical Considerations in the Use of Biomarkers

N. Ashford, C. Spadafor, D. Hattis, and C. Caldart

The Toxic Substances Control Act (TSCA) imposes substantial requirements on chemical manufacturers and processors to develop health effects data. Section 4 of TSCA requires chemical testing, section 5 requires premarket manufacturing notification, and section 8 requires the reporting and retention of certain information. TSCA imposes no specific medical surveillance or biological monitoring requirements. However, human monitoring may be used to meet the more general TSCA reporting and recordkeeping requirements discussed below, and the data resulting from such monitoring presumably are subject to an employer's recording and retention obligations under the OSHAct. Further, EPA has proposed agency-wide generic guidelines for exposure assessment used in risk assessment under TSCA and other statutes [53 Fed. Reg. 48,830,

Source: Excerpts from *Monitoring the Worker for Exposure and Disease: Scientific, Legal and Ethical Considerations in the Use of Biomarkers.* Baltimore, MD: Johns Hopkins University Press, 1990, 24–26.

et seq. (December 2, 1988)], and these guidelines cover the gathering of exposure data through biological monitoring.

Beginning in 1982, EPA has promulgated a series of "chemical information rules" under section 8(a) of TSCA [40 C.F.R. §712]. Presently, these rules require the submission of certain data, including information on occupational health effects and exposure, on some 350 chemicals. Section 8(a)(2) allows the agency to require the reporting and maintenance of such data "insofar as known...or...reasonably ascertainable." EPA would thus appear to be authorized to require human monitoring under section 8(a)(2), as a way of securing information that is "reasonably ascertainable." However, the chemical information rules require only the reporting of "information that is readily obtainable by management and supervisory personnel." Nonetheless, if monitoring is undertaken, the results must be reported.

To supplement this general reporting requirement, the agency issued "specific chemical reporting requirements" under section 8(a) in 1988 [40 C.F.R. §704]. The focus of these requirements is a Comprehensive Assessment Information Rule (CAIR), which sets forth ten categories of information that, depending on the circumstances, may be required to be reported. One of these categories—"worker exposure" —encompasses data gathered through human monitoring. Although initially applicable only to nineteen chemicals, the CAIR is intended by EPA to "standardize certain section 8(a) rules" by establishing uniform questions, reporting forms, and reporting and recordkeeping requirements for a detailed list of TSCA-related data.

Beyond the section 8(a) regulations, EPA has promulgated a rule under section 8(d) of TSCA requiring submission of health and safety studies for over 400 chemical substances and mixtures [40 C.F.R. §716]. As defined in the regulation, a health and safety study includes "[a]ny data that bear on the effects of a chemical substance on health." Examples are "[m]onitoring data, when they have been aggregated and analyzed to measure the exposure of humans . . . to a chemical substance or mixture." Although Section 8(d) requires that all such data as are "known" or "reasonably ascertainable" be reported, the regulation itself is somewhat less inclusive.

Section 8(c) of TSCA mandates that records of any "significant adverse reactions to [employee] health" alleged to have been caused by a chemical be maintained for thirty years. The rule implementing this section [40 C.F.R. §717] defines significant adverse reactions as those "that may indicate a substantial impairment of normal activities, or long-lasting or irreversible damage to health." Under the rule, human monitoring data, especially if derived from a succession of tests, clearly appears reportable. Genetic monitoring of employees, if some basis links the results with increased risk of cancer, also seems to fall within the rule.

Finally, section 8(e) imposes a statutory duty to report "immediately . . . information which supports the conclusion that [a] substance or mixture presents a substantial risk of injury to health." In a policy statement issued in 1978, EPA interpreted "immediately" in this context to require receipt by the agency within fifteen working days after the reporter obtains the information. Substantial risk is defined exclusive of economic considerations. Evidence can be provided by either designed, controlled studies or undesigned, uncontrolled studies, including "medical and health surveys" or evidence of effects in workers. EPA has distinguished this reporting requirement from the "significant adverse reactions" requirement of section 8(c) by noting that "[a] report of substantial risk of injury, unlike an allegation of a significant adverse reaction, is accompanied by information which reasonably supports the seriousness of the effect of the probability of its occurrence." Human monitoring results indicating a substantial risk of injury would thus seem reportable to EPA. Either medical surveillance or biological monitoring data would seem to qualify.

* * *

[Finally,] in addition to requiring the creation of information that becomes subject to the OSHA access rule, the Toxic Substances Control Act gives workers certain rights to obtain monitoring information that has been submitted to EPA. Section 14(b) of TSCA gives EPA the authority to disclose from health and safety studies all data pertaining to chemical identities, except for the proportion of chemicals in a mixture. In addition, EPA may disclose information, otherwise classified as a trade secret, "if the Administration determines it necessary to protect . . . against an unreasonable risk of injury to health." Monitoring data thus seem subject to full disclosure.

EPA has been decidedly low-key in exercising its authority to require the disclosure of information under TSCA. The agency could choose to use this authority to act as a coordinator and "clearinghouse" for information—by facilitating its transfer to health clinics, unions, state agencies, trade associations, and other parties for use in the interest of worker health—and it could identify and exploit the ways in which TSCA reporting requirements might be used in conjunction with other state and federal laws to promote wider dissemination of workplace toxics information. Thus far, however, the agency has largely deferred to OSHA on these issues. Indeed, it appears much more likely that the community right to know provisions of the 1986 amendments to the federal "Superfund" law, discussed in section 4 below, will be the focal point for EPA's activities in this area.

3.   The National Labor Relations Act

Beyond the specific avenues to toxic substance data available under the more recent federal safety and health statutes, employees are often able to obtain access to these data through the collective bargaining process. As discussed in Chapter 6, the National Labor Relations Act obligates employers to supply the authorized representatives of their employees with such information as enables them to bargain knowledgeably over mandatory subjects of bargaining. Because workplace safety and health are mandatory bargaining topics, information regarding the origin, nature, and extent of workplace exposures to chemicals, radiation, and bioengineered organisms are all within the ambit of this general disclosure obligation. The following two cases highlight the scope, and the limitations, of this route of access. The first of these—an Administrative Law Judge decision that was subsequently upheld by the National Labor Relations Board, 257 N.L.R.B. No. 101 (1981)—confirms that an employer's obligation to supply information pursuant to the collective bargaining process exists independently of any obligations under the OSHAct.

Winona Industries, Inc. and International Chemical Workers Union, Local 893, AFL-CIO

Case-18-Ca 6509 Before the National Labor Relations Board Division of Judges

ALMIRA ABBOT STEVENSON, Administrative Law Judge.

The issue is whether or not the Respondent [Winona Industries] refused to bargain in good faith in violation of Section 8(a)(5) and (1) of the National Labor Relations Act by refusing to grant an industrial hygienist of the Charging Party Union access to the Respondent's plant for the purpose of evaluating health and safety conditions in the plant. For the reasons given below, I conclude that the Respondent violated the Act as alleged.

*   *   *

The Respondent is engaged at its plant in Winnona, Minnesota in the manufacture of speaker cabinets and furniture. Robert Whitcomb is labor-relations manager and quality assurance manager; Dennis Lubinski is manufacturing manager. The Union was certified exclusive bargaining representative of an appropriate unit of all full time and regular part-time production and maintenance employees on August 17, 1973. Since then the approximately 185 employees in the unit have been covered by successive collective-bargaining agreements, the most recent effective from May 1, 1977 until April 30, 1980. At the time of the hearing, the parties were in negotiations for a subsequent agreement. Terry Edwards is

an International representative and Denise Stark is president of the Union.

This dispute began with the posting of a notice to all employees by Safety Director Wiegel on July 16, 1979 to the effect that wearing tank-top shirts was contrary to OSHA requirements.[6] Upon being advised by OSHA that it had no such requirement, Union President Stark so informed Manufacturing Manager Lubinski; Lubinski responded that the rule was due to "industrial dermatitis" problems. Stark, who did not know whether the rule was a dress code based on a management "moral code of ethics" or designed to deal with a problem in the factory which caused dermatitis, filed a grievance dated July 20, 1979 requesting access to the plant by Stan Eller, an industrial hygienist employed by the International Union, to test for materials harmful to the skin.

The following week, Industrial-Relations Manager Whitcomb telephoned International Representative Terry Edwards to complain that the employees were not complying with the posting and to tell Edwards that the problem was a possible exposure of employees to formaldehyde in the air. Concerned about this information, which indicated that there might be health problems in the plant, Edwards passed the information along to Union President Stark, advising compliance and informing her that a meeting

[6]A proscription against tank tops had been in existence since 1973 but enforcement had been lax.

would be held on the grievance August 7. At the August 7 meeting, Labor-Relations Manager Whitcomb presented a written denial of the grievance on the ground that the Company had the right under ARTs. VII and XXIII of the contract to establish rules pertaining to health and safety and reasonable rules governing employment and working conditions; Whitcomb informed the Union orally that Stan Eller would not be allowed in the plant. However, International Representative Edwards, who has no special training, was permitted to tour the plant for the purpose of investigating the grievance.

There were three more meetings between management and Union on this matter.[7] On September 28, both parties expressed uncertainty over the existence of any health problem, and the Union explained that its request was for access by an industrial hygienist of the International Union, who was a professional qualified in areas with which the Union was concerned, possible health and safety problems from formaldehyde in the air. The Union also requested a complete list of chemicals used in the plant by generic names. The Respondent took the requests under advisement.

On October 4, Edwards modified the Union's request by announcing that it would accept access by any industrial hygienist on the staff of the International Union and would not insist on designating Stan Eller. The Company suggested that in view of the then current OSHA inspection, it was not a good time for any Union investigation and the Union agreed. International Representative Edwards followed up this meeting with a letter dated October 9 to Whitcomb in which he explained that the International Union had three industrial hygienists on its staff, any of whose access would be acceptable, and also explained that the purpose for the Union's access request was,

> that the Company had made us aware of a possible health and safety problem per your publication of the new dress code in the plant posted July 16, 79. The health and safety of our members today is enough of a reason to make this request, but with the present contract expiring April 30, 1980 we need the information gained from this tour to bargain intelligently over health and safety issues.

[7]Facts with respect to these meetings are based on an amalgamation of accounts given by International Representative Edwards and the Respondent's negotiators, Robert Whitcomb and Leonard Grannes, in light of the probabilities.

The third meeting was held October 25. Edwards again requested a tour by an industrial hygienist, and suggested that a proper time was after the closing conference with OSHA and before any corrections were made. Management raised some questions to which the Union responded by assuring the Company that the industrial hygienist would not talk to employees long enough to interfere with production, would take no air samples, would be accompanied by the President of the Union and an employee-member knowledgeable in each area; and any problem discovered would be handled through the Company before OSHA was contacted. The Union offered to supply names of the companies where industrial-hygienist inspections had been conducted, and the Company representatives agreed to talk to top management about access to the plant.

In response to the Union's request for a list of chemicals used in the plant, the Respondent provided material data safety sheets of two of its suppliers which, however, were, to the Union's knowledge, non-specific and incomplete. The Union did not provide the Respondent with the names of other companies which had permitted industrial-hygienist inspections. The tank-top issue was referred to the joint labor-management health and safety committee where, according to Union President Stark, it "is being taken care of," although it still is not completely settled.

During its early October inspection, OSHA investigators tested for formaldehyde fumes by taking air samples; its report stated that minimal OSHA standards were met.

* * *

On November 6, 1979, Labor-Relations Manager Whitcomb addressed a letter to International Representative Edwards dealing with the request for plant access by an industrial hygienist of the Union, in which he stated, in part, as follows:

> It is our decision that to accept your request at this time would serve no useful purpose since OSHA has inspected the plant thoroughly and your people are provided the opportunity to participate in a closing meeting with the members of the local unit.
>
> Discussions with OSHA during their visit indicate that there are no problems of a major nature which would result in large capital expenditures.
>
> As we have indicated to you previously we are willing at some future date to discuss a visit by these people provided we can reach agreement

on the conditions under which the visit would be conducted.

* * *

I cannot agree with the Respondent that it was not obligated to grant access by the Union's industrial hygienist because the information sought had been furnished or was available. The supplier's material data safety sheets provided were non-specific and incomplete. Information obtainable by the employee-members of the health-and-safety committee and Union officers is circumscribed by their lack of the technical training and experience possessed by an industrial hygienist who knows what to look for and how to assess possible impact on the health of the employees. And although the investigation reports of OSHA may be available, the evidence shows that the Union's industrial hygienists follow procedures which would reveal data supplementing and expanding that obtained by OSHA investigators. Moreover, the Board has said, with regard to laws governing plant safety similar to that of OSHA:

> Such laws, like the minimum wage and a variety of governmental regulations, merely establish certain minimum requirements in their respective fields as conditions of doing business and are not intended to preempt their fields of regulations to such an extent as to exclude therefrom the concept of collective bargaining.[11]

Finally, Labor-Relations Manager Whitcomb's letter of November 6, 1979 must be construed as a refusal of the Union's request to permit access. There is no merit to the Respondent's contention that all it was required to do was bargain and that it met this obligation by its meetings and discussions with the Union. Although the matters discussed at those meetings—selection of the industrial hygienist, time and method of inspection, non-interference with production, disposition of information acquired—were appropriate subjects for bargaining, the Respondent's obligation to grant the Union's request for access in circumstances such as those present here is basic and unqualified and not subject to ultimate refusal after negotiations. Nor can I find merit in the contention that the Respondent was merely awaiting the names of other companies whose plants had been inspected by an industrial hygienist; Whitcomb's letter made no reference to the Union's failure to provide the companies' names. The letter's denial of the request "at this time" and expressed willingness to "discuss" the request "at some future date" subject to further bargaining, after 3 months and 4 meetings during which the Union modified its request and offered assurances in response to all questions raised by the Respondent, amounts to a refusal inconsistent with the requirements of the Act.

I conclude that by effectively denying the Union's request for an in-plant inspection by an industrial hygienist designated by the Union for investigation of health and safety conditions, which was relevant to the Union's discharge of its bargaining obligation, the Respondent refused to bargain in good faith and violated Section 8(a)(5) and (1) of the Act.

---

[11]Gulf Power Company. [384 F.2d 822 (5th Cir. 1967)]. The Respondent had directed my attention to McCulloch Corporation, 132 NLRB 201, in which the Board upheld an employer's refusal to furnish information which was readily available in a publically distributed handbook. That case, however, is distinguishable as the Union here wishes to obtain relevant information not contained in the material available from the Employer. As the information sought is obtainable only from an in-plant inspection by a qualified industrial hygienist, and granting access would not be burdensome to the Employer, Cincinnati Steel Castings Company, 86 NLRB 592, is also distinguishable. The Respondent's contention that it had no bargaining obligation to the Union with respect to its request for access by an industrial hygienist is equally without merit. (1) Notification of the Federal Mediation and Concilia-tion Service and other requirements of Section 8(d) of the Act are not applicable, as the Union was not proposing to amend or modify the contract. (2) Contrary to the Respondent, the Union did, in my opinion, pursue its request through the grievance procedure. In any event, neither the grievance-arbitration provisions, the "Entire Agreement" clause, the management-rights clause, nor the joint labor-management health and safety committee provision constitutes a waiver by the Union of its statutory right to the information requested. Proctor and Gamble Manufacturing Company, 237 NLRB 747, 751, enforced 603 F.2d 1310 (C.A. 8); Globe Stores, Inc. et al., 227 NLRB 1251; Worcester Polytechnic Institute, 213 NLRB 305. The Respondent does not contend that the information sought is confidential.

The second case—which has its origins in labor disputes involving three separate companies—explores the extent to which an employer's concern for the confidentiality of its employees will operate as a limitation on the employer's obligation to supply otherwise relevant information. The following description by Michelle Mentzer aptly summarizes the facts giving rise to the three disputes.

Union's Right to Information About Occupational Health Hazards Under the National Labor Relations Act

M. Mentzer

In 1977, in response to the discovery of sterility among union members working with the pesticide dibromo chloropropane (DBCP) at an Occidental Chemical plant in California, the Oil, Chemical & Atomic Workers (OCAW) initiated a campaign to gain disclosure of hazard information at many of its organized plants. The International Union sent 560 locals a form letter requesting company disclosure. One hundred ten of the locals chose to send the letter on their own stationery to their company representatives. More than half of the 110 companies furnished the information in some form or another. OCAW filed section 8(a)(5) charges against two of the companies that had refused—the Minnesota Mining & Manufacturing Co. and Colgate-Palmolive Co.

The letter requested the following data:

1. morbidity and mortality statistics on all past and present employees;

2. the generic name (chemical name, as opposed to trade name or code number) of all substances used and produced at the plant;

3. results of clinical and laboratory studies of any employee undertaken by the company, including the results of toxicological investigations

Source: From *Indus. Rel. L. J.* Vol 5 (2) 255-256 (1983).
Mr. Mentzer is an attorney with the Farm Worker Division of Evergreen Legal Services, Granger, WA.

regarding agents to which employees may be exposed;

4. certain health information derived from insurance programs covering employees, as well as information concerning occupational illness and accident data related to workers' compensation claims;

5. a listing of contaminants monitored by the company, along with a sample protocol;

6. a description of the company's hearing conservation program, including noise level surveys;

7. radiation sources in the plant, and a listing of radiation incidents requiring notification of state and federal agencies; and

8. an indication of plant work areas which exceed proposed National Institute for Occupational Health heat standards and an outline of the company's control program to prevent heat disease.

The *Borden Chemical* case arose out of a narrower request from the International Chemical Workers Union (ICW), which had been in contact with OCAW about the initiative. In that case, the International's Vice President made a written request for a list of "all materials and chemicals" which union members handled by trade or code names as well as by generic names. The ICW request was limited to item (2) of the OCAW request.

The NLRB issued its decision in 1982. Citing the general proposition that employers are required to bargain over health and safety conditions when requested to do so, the Board adopted a broad policy favoring union access. "Few matters can be of greater legitimate concern to individuals in the workplace, and thus to the bargaining agent representing them, than exposure to conditions potentially threatening their wealth, well-being, or their very lives," the Board noted.

The NLRB did not, however, grant the unlimited right of access. The union's right of access is constrained by the individual employee's right of personal privacy. Furthermore, the Board acknowledged an employer's interest in protecting trade secrets. While ordering the employer in each of the three cases to disclose the chemical identities of substances to which the employer did not assert a trade secret defense, the Board indicated that employers are entitled to take reasonable steps to safeguard "legitimate" trade secret information. The Board did not delineate a specific mechanism for achieving the balance between union access and trade secret disclosure. Instead, it ordered the parties to attempt to resolve the issue through collective bargaining.

Both sides appealed the decision to the District of Columbia Court of Appeals. In 1983, finding that the Board had correctly applied the guidelines laid down by the Supreme Court in its earlier *Detroit Edison* case, the D.C. Circuit enforced the order in full.

Oil, Chemical & Atomic Workers Local Union No. 6-418, AFL-CIO, v. National Labor Relations Board

711 F.2d 348 (D.C. Cir 1983)

EDWARDS, Circuit Judge.

* * *

II. DISCUSSION

A. General Principles

The disposition of these petitions for review does not require the development of any novel legal principles. An employer's duty to bargain in good faith with a labor organization representing its employees has long been acknowledged to include a duty to supply a union with "requested information that will enable [the union] to negotiate effectively and to perform properly its other duties as bargaining representative." This fundamental obligation to furnish relevant information is "rooted in recognition that union access to such information can often prevent conflicts which hamper collective bargaining," and it undoubtedly extends to data requested in order properly to administer and police a collective bargaining agreement as well as to requests advanced to facilitate the negotiation of such contracts. In either instance, the employer's duty is predicated on the need of the union for information that will promote "intelligent representation of the employees."[24]

[24]Westinghouse Elec. Corp., 239 N.L.R.B. 106, 107 (1978) (footnote omitted), *modified on other grounds sub nom.* International Union of Elec., Radio & Mach. Workers v. NLRB, 648 F.2d 18 (D.C. Cir. 1980); *see* Soule Glass &

* * *

B. Relevance

The right to relevant information, the previous section makes clear, validates requests for data reasonably necessary to enable unions effectively to administer and police collective bargaining agreements or intelligently to seek their modification. Employee health and safety indisputably are mandatory subjects of collective bargaining,[34] about which the unions in each of these cases have insisted on negotiating. The resulting contractual health and safety provisions, of course, are not set in stone and may appropriately be modified in future negotiations. In the meantime, it is "the duty of union representatives . . . to see that . . . [the employers meet their] obligations under those clauses." *International Union of Electrical, Radio & Machine Workers v. NLRB*, 648 F.2d 18, 25 (D.C. Cir. 1980).

These observations notwithstanding, the employers here argue that the unions' requests could properly be denied because (1) they were overbroad, (2)

Glazing Co. v. NLRB, 652 F.2d 1055, 1092 (1st Cir. 1981); Local 13, Detroit Newspaper Printing & Graphic Communications Union v. NLRB, 598 F.2d at 271.

[34]United Steelworkers of America v. Marshall, 647 F.2d 1189, 1236 (D.C. Cir. 1980), *cert. denied*, 453 U.S. 913, 101 S.Ct. 3148, 69 L.Ed.2d 997 (1981); NLRB v. Miller Brewing Co., 408 F.2d 12, 14 (9th Cir. 1969); NLRB v. Gulf Power Co., 384 F.2d 822, 824-25 (5th Cir. 1967); Winona Indus., 257 N.L.R.B. 695, 697 n. 9 (1981); San Isabel Elec. Servs., 225 N.L.R.B. 1073, 1078 n. 6 (1976).

the information covered by the requests is not presumptively relevant, and (3) the unions failed to document the relevance of the requested information under the facts of these cases. The first contention is easily countered, for

> the mere fact that a Union's request encompasses information which the employer is not legally obligated to provide does not automatically excuse him from complying with the Union's request to the extent that it also encompasses information which he would be required to provide if it were the sole subject of the demand.

Fawcett Printing Corp., 201 N.L.R.B. 964, 975 (1973) (footnote omitted). The companies' second claim need not be resolved to decide these cases; the NLRB does not purport to rely on the presumptive relevance theory, and our disposition of the employers' third argument renders unnecessary a full explication of the scope of the presumption.

That third contention, on which the outcome of this case turns, is actually an agglomeration of several tangentially related arguments. Thus, the employers claim, alone or in combination, that (1) the unions' requests were not motivated by specific problems in their respective plants, (2) the mere presence of hazardous substances or conditions within the workplace does not demonstrate the relevance of the requested information, (3) the disclosure of a list of hazardous substances or conditions would not aid the unions in performing their bargaining obligations, and (4) the companies' extensive health and safety programs ensure their employees a safe and healthy working environment. It is clear, however, that a labor organization's right to relevant information is not dependent upon the existence of some particular controversy or the need to dispose of some recognized problem. *J.I. Case Co. v. NLRB*, 253 F.2d 149, 155 (7th Cir. 1958); *Westinghouse Electric Corp.*, 239 N.L.R.B. 106, 109 (1978), *modified on other grounds sub nom, International Union of Electrical, Radio & Machine Workers v. NLRB*, 648 F.2d 18 (D.C. Cir. 1980). And the proposition that a union must rely on an employer's good intentions concerning the vital question of the health and safety of represented employees seems patently fallacious.

In cases like those now before us, where the employees admittedly are exposed to a variety of potential hazards and have expressed growing and legitimate concerns over their health and safety, where the unions explained the rationales underlying their requests in considerable detail, and where the pertinent collective bargaining agreements obligate both management and the unions to take specified actions to safeguard employees' health and safety, the relevance of a wide range of information concerning the various elements of the working environment and employees' health experiences cannot be gainsaid.[35] Under these circumstances, at least, the goals of occupational health and safety are inadequately served if employers do not fully share with unions available information on working conditions and employees' medical histories. Requiring the release of exposure and medical data in such cases will facilitate the identification of workplace hazards, promote meaningful bargaining calculated to remove or reduce those hazards, and enable unions effectively to police the performance of employers' contractual obligations as well as to carry out their own responsibilities under the respective collective bargaining agreements. As a result, we hold that the NLRB correctly assessed the relevance of the information requested in each of the cases under review.

(C) Trade Secrets

In defense of their refusal to disclose, the employers contend that some of the data encompassed within the unions' requests constitute trade secrets. As the NLRB found, however, this trade secret defense applies to only a small portion of the requested information and thus could not justify the companies' total noncompliance with the unions' requests.[36] And, while *Detroit Edison* makes clear that a union's interest in relevant information will not always outweigh an employer's legitimate and substantial interests in maintaining the confidentiality of such information, *Detroit Edison Co. v. NLRB*, 440 U.S. at 318, 99 S. Ct. at 1132, that case certainly affords no support for the proposition that an employer is absolutely privileged from revealing relevant proprietary or trade secret information. Even a cursory examination of the NLRB's decisions in these cases reveals the Board's faithfulness to the accommodative philosophy espoused by the *Detroit Edison* Court.

In each of the cases under consideration, the NLRB concluded, explicitly or implicitly, that the General Counsel had not established that the refusal to disclose trade secret information violated the

---

[35]Nothing in this opinion should be read to suggest that the presence of each of the factors identified herein is essential to the establishment of relevance in cases presenting similar issues.

[36]*See* Fawcett Printing Corp., 201 N.L.R.B. at 975.

NLRA. The Board held, however, that the employers' blanket denials of relevance, which it rejected, had precluded meaningful bargaining over the conditions under which this proprietary information *might* be disclosed; accordingly, the employers were ordered to bargain in good faith on this point. This position, it seems clear, fully respects the legitimacy of the companies' interests in the confidentiality of their trade secrets, for, as we read the Board's orders, they express no view on whether trade secret information ultimately need be disclosed. If conditions can be devised to accommodate both the employers' confidentiality interests and the unions' interest in obtaining relevant information, a refusal to disclose under those conditions likely will be found to violate the Act. But if no such conditions can be created, the Board might be forced to sanction the companies' refusal to disclose trade secret information.[37]

The NLRB's decision to defer, and hopefully to eliminate the need for, an administrative balancing of the parties' competing interests by allowing the parties to attempt an accommodation at the bargaining table is vehemently attacked by the unions. This decision, the unions argue, amounts to a refusal to determine whether the employers breached their bargaining obligations by withholding trade secret information. Although such a refusal would violate section 10(c) of the NLRA,[38] our reading of the decisions comports with the one advanced by counsel for the Board. Accordingly, we conclude that the Board *did decide* that the companies had not been shown, on these records, to have violated the Act by failing unconditionally to disclose their trade secrets. The employers had, to be sure, failed to satsify their bargaining obligations concerning such material. But the NLRB's determination that these violations did not necessitate a full-scale attempt to balance the competing interests on the sparse records before it represents a legitimate and appropriate exercise of the Board's remedial discretion.

Where, as here, requested information is both relevant to a union's representational responsibilities and subject to an employer's legitimate confidentiality interest, the NLRB's reliance, in the first instance, on a longstanding bargaining relationship for the development of an accommodation of these interests does not contravene the Board's statutory

obligation to resolve unfair labor practice charges. The unions acknowledge that the NLRB may properly leave to the parties the determination of conditions governing disclosure of information to which it has found a union entitled. *See International Union of Electrical, Radio & Machine Workers v. NLRB*, 648 F.2d at 26; *Ingalls Shipbuilding Corp.*, 143 N.L.R.B. 712, 718 (1963). But they nevertheless contend that, before ordering parties to bargain over the conditions under which confidential information might appropriately be disclosed, the Board must declare unequivocally that the union is entitled to that information—regardless whether conditions sufficiently protective of the employer's legitimate interests can be devised. This contention plainly lacks validity in light of the decision in *Detroit Edison*. These cases may, of course, ultimately come before the Board again for a balancing of the parties' interests. But they will do so on records that will allow a more effective balancing than would those now before us; accordingly, we reject the unions' challenge to the NLRB's orders.

D. Medical Records

3M and Colgate both argue that they legitimately withheld certain of the requested information because it involved disclosure of individually identified medical data that would violate employees' rights to privacy and confidentiality. The NLRB recognized that employers may have a legitimate and substantial interest in protecting the confidentiality of employees' medical records,[39] but found that neither Local 6-418 nor Local 5-114 had sought acccess to individually identified medical records.[40] While we have previously suggested, in the context of discrimination complaints, that the mere deletion of names may not sufficiently ensure employees' confidentiality,

---

[37]We express no view concerning how the balance between the parties' interests should be struck in the event that no satisfactory conditions can be developed.

[38]29 U.S.C. §160(c) (1976); *see* International Union, UAAAIW v. NLRB, 427 F.2d 1330, 1331-32 (6th Cir. 1970).

[39]See *Johns-Manville Sales Corp.*, 252 N.L.R.B. 368, 368 (1980); cf. *Detroit Edison Co. v. NLRB*, 440 U.S. at 318-19, 99 S.Ct. at 1132-33 (recognizing employer's interest in protecting confidentiality of data bearing on employees' basic competence); *United States v. Westinghouse Elec. Corp.*, 638 F.2d 570, 577-78 (3d Cir. 1980) (employees' medical records fall within zone of privacy entitled to protection, but may be subject to disclosure on showing of proper governmental interest).

[40]There is no merit to the contention that the propriety of a union's continuous request for information must be determined solely from the wording of the initial communication of that request. See *Ohio Power Co.*, 216 N.L.R.B. 987, 990 n. 9 (1975), enforced mem., 531 F.2d 1381 (6th Cir. 1976).

International Union of Electrical, Radio & Machine Workers v. NLRB, 648 F.2d at 27, the Board's orders in these cases permit the deletion of *any* information that could reasonably be used to identify specific employees. We therefore reject the medical confidentiality defenses asserted by the companies.

E. Burdensomeness

The claims of burdensomeness advanced by 3M and Colgate also may be disposed of quickly. While the excessive breadth of a union's request "may in particular circumstances relieve the employer from the obligation to provide some or all of information to which the Union would be entitled on proper demand,"[41] the cost and burden of compliance ordinarily will not justify an initial, categorical refusal to supply relevant data. *See International Union of Electrical, Radio & Machine Workers v. NLRB*, 648 F.2d at 26. Issues focused on these factors typically are considered instead at the compliance stage of a disclosure proceeding, where the parties may bargain over the allocation of the costs of producing the requested information. *Safeway Stores*, 252 N.L.R.B. 1323, 1324 (1980), *enforced*, 691 F.2d 953 (10th Cir.

[41] *Fawcett Printing Corp.*, 201 N.L.R.B. at 975 (footnote omitted).

1982).[42] The NLRB adhered to these principles in holding that the alleged burdens of compliance did not excuse the companies' refusals to disclose, and we have found no reason to upset its conclusions.

III. CONCLUSION

We have not attempted to discuss every argument advanced by all of the parties to this proceeding, but we have carefully considered and rejected those not specifically addressed herein. For the reasons set forth above, we enforce in full the Board's orders in each of these cases.

[42] See also *Food Employer Council*, 197 N.L.R.B. 651, 651 (1972) (footnote omitted):

> If there are substantial costs involved in compiling the information in the precise form and at the intervals requested by the Union, the parties must bargain in good faith as to who shall bear such costs, and, if no agreement can be reached, the Union is entitled in any event to access to records from which it can reasonably compile the information.

Notwithstanding Colgate's suggestion to the contrary, we do not read this statement, or the corresponding declarations in the NLRB's opinions in the cases presently under review, to contain any implication that an employer's *confidentiality* interests must give way, in all cases, to a union's right of access to relevant information.

◇   **NOTES**

1. Who has the burden of proof if there is a dispute as to whether the disclosure of particular information actually would compromise trade secrecy? Does the employer have the obligation to prove the legitimacy of its trade secret claims?

2. Where legitimate trade secrets are involved, one practical solution may be for the employer to disclose the information to an appropriate third party (such as an industrial toxicologist) trusted by the union. That person would be free to discuss any potential health concerns with the union, but would be obligated not to disclose trade secret information.

3. Individual employees may have a limited right of access to medical and exposure records under federal labor law. Logically, the right to refuse hazardous work inherent in Section 7 of the NLRA and Section 502 of the Labor Management Relations Act (discussed in Chapter 8) carries with it the right of access to the information necessary to determine whether a particular condition is hazardous. In the case of toxic substance exposure, this may mean a right of access to all information relevant to the health effects of the exposure and may include access to both medical and exposure records. At present, however, there is no systematic mechanism for enforcing this right.

4. The Emergency Planning and Community Right to Know Act

In 1986, Congress amended the federal hazardous waste clean-up law (commonly referred to as the Superfund statute) with the Superfund Amendment and Reauthorization Act of 1986 (known as SARA). Beyond clean-up, Congress took in SARA what may prove to be a significant step toward reducing the likelihood of new hazardous substance contamination in the future. Title III of SARA (called the Emergency Planning and Community Right to Know Act, now codified at 42 U.S.C. §§11001, *et seq.*) enacted a comprehensive federal community right to know program, implemented by the states under guidelines promulgated by EPA. The central feature of this federal program is broad public dissemination of information pertaining to the nature and identity of chemicals used at commercial facilities.

Although Title III is not a workplace right to know law *per se,* it does provide an alternative means through which many employees can learn about toxic substance use, not only in their own workplaces, but in other workplaces in which they may wish to work.

The implementation of Title III began with the creation of state and local bodies to implement this community right to know program. Section 301 of the Act required the governor of each state to appoint a "state emergency response commission (SERC)," to be staffed by "persons who have technical expertise in the emergency response field." In practice, these state commissions have tended to include representatives from the various environmental and public health and safety agencies in the state. Each state commission, in turn, was required to divide the state into various "local emergency planning districts" and to appoint a "local emergency planning committee" (LEPC) for each of these districts. These state and local entities are responsible for receiving, coordinating, maintaining, and providing access to the various types of information required to be disclosed under the Act.

Title III established four principal requirements for reporting information about hazardous chemicals. Section 304 requires all facilities that manufacture, use, or store certain "extremely hazardous substances" in excess of certain quantities to provide "emergency" notification to the SERC and the LEPL of any unexpected release of one of these substances. Section 311 requires facilities covered by the OSHA Hazard Communication Standard to prepare and submit to the LEPC and the local fire department material safety data sheets for chemicals covered by the OSHA standard. Under Section 312, many of these same firms are required to prepare and submit to the LEPC an "emergency and hazardous substance inventory form" that describes the amount and location of certain hazardous chemicals on their premises. Finally, Section 313 requires firms in the manufacturing sector to provide to EPA an annual reporting of certain routine releases of hazardous substances. In addition, Section 303 requires certain commercial facilities to cooperate with their respective LEPCs in preparing "emergency response plans" for dealing with major accidents involving hazardous chemicals. The applicability of these provisions to any particular facility depends on the amount of the designated chemicals that it uses or stores during any given year.

Taken as a whole, these requirements constitute a broad federal declaration that firms that choose to rely heavily on hazardous chemicals in their production processes may not treat information regarding their use of those chemicals as their private domain. Indeed, except for trade secrecy protections that generally parallel those available under the OSHA Hazard Communication Standard, there are no restrictions on the disclosure of Title III information to the general public. Instead, Section 324 of the Act mandates that most of the information subject to Title III reporting requirements "be made available to the general public" upon request, and requires that each local emergency planning committee publicize this fact in a local newspaper.

5. High Risk Notification

To date, right to know efforts have relied on employers to transfer generic information to workers about hazardous substances. The more complicated issue of individual risk has been avoided. Recent federal initiatives have attempted to extend the government's role in the transfer of risk information. The proposed High Risk Occupational Disease Notification and Prevention Act would require the federal government to individually notify workers determined to be at risk of serious occupational disease.

The recent interest in worker notification was sparked in October 1984 by Ralph Nader and Sidney Wolfe, Director of the Public Citizen Health Research Group. In a press release, Nader and Wolfe accused the National Institute for Occupational Safety and Health (NIOSH) of failing to notify between 200,000 and 250,000 workers, who had been identified in 66 NIOSH retrospective cohort studies, that they were at increased risk of contracting cancer and other diseases due to their workplace exposures. Three months later, Public Citizen released a list of 253 worksites where NIOSH epidemiologists had previously determined over 200,000 workers to be at risk of cancer, lung disease, heart disease, and other serious illnesses.

In February 1985, Representatives Gaydos and Hawkins introduced the first worker notification bill into Congress, H. R. 1309. One year later, Senators Metzenbaum and Stafford introduced a similar bill, S. 2050. The legislation sought to "establish a system for identifying, notifying, and preventing illness and death among workers who are at increased risk or high risk of occupational disease" (H.R. 1309). Successive versions of the proposed legislation have varied, but all have expanded the notification requirement well beyond subjects of NIOSH epidemiologic studies.

The legislation would require the federal government to notify and counsel individual workers determined to be members of a population at risk; certify health facilities for training and advising local community-based physicians regarding recognition, diagnosis, and treatment of occupational disease; expand federal research efforts; and establish a set of protections to prohibit discrimination on the basis of identification and notification. Employers would be required to provide, at no cost, medical monitoring to current employees who are notified that they are members of a population at risk and, in some cases, provide them with medical removal protection (MRP) if an employee's physician determines that the employee shows evidence of the disease in question, or has other symptoms or conditions that increase his or her risk of acquiring the disease.

Under the proposed legislation, a substance or process is considered an occupational health hazard if there is statistically significant evidence (based on clinical or epidemiologic studies conducted according to sound scientific principles) that chronic health effects have occurred in persons exposed to the substance or process under consideration. A "population at risk" means a class or category of employees exposed to the hazard under working conditions comparable to the clinical or epidemiologic data cited above, that is, at the same concentrations or for the same duration of exposure.

Not unexpectedly, industry was generally opposed to early versions of the bill. From mining to farming to manufacturing, the arguments were consistent. Industry found the bills unnecessary, redundant, and duplicative of other federal legislation, notably the OSHA Hazard Communication Standard, and argued that the basis for notification was vague, arbitrary, unreasonable, and unscientific. Industry representatives warned that the bill would be excessively costly. Costs would arise from the notification process itself, but more significantly from the unnecessary medical monitoring, inappropriate clinical intervention, and unjustified liability claims both within and outside the workers' compensation system.

NIOSH and OSHA also voiced opposition to the legislation. NIOSH argued that worker notification was already an integral part of the agency's efforts, and that it was considering how best to notify subjects of past and future cohort studies. Moreover, the agency director testified that the bill would hinder the agency's other research and training mandates. OSHA argued that existing OSHA regulations, primarily the Medical Access Rule and the Hazard Communication Standard, were already accomplishing the goals of the proposed legislation. The agency also expressed concern that the bills could inappropriately suggest liability in workers' compensation and tort claims.

Prominent in early hearings on the house and Senate bills was the testimony of the Department of Justice (DOJ). DOJ concluded that the bill did not adequately protect the federal government or federal employees from liability arising out of any act or omission performed under several sections of the legislation and suggested that the legislation would increase litigation substantially. At the same time, claimed DOJ, the bills would represent an unacceptable intrusion into state workers' compensation systems.

But there has been sustained support for the bills, as well. Labor has offered detailed counter-arguments regarding necessity and duplication. Acknowledging that the proposed notification criteria would result in some false notification, the AFL-CIO argued that "our inability to identify with certainty those individuals who will become ill or die because of an occupational exposure means that we must treat all members of the population as if they are equally at risk and will inevitably suffer without aid." AFL-CIO Department of Occupational Safety, Health and Social Security. 1985. Testimony on H.R. 1309, "The High Risk Occupational Disease Notification and Prevention Act of 1985," before the House Education and Labor Subcommittee on Safety and Health. The United Auto Workers saw the need for worker notification as an issue of civil rights in the workplace. Professional and public health organizations such as the American Public Health Association, the American Cancer Society, and the American Lung Association also testified in favor of the bills. In addition, the Senate heard compelling testimony from a panel of victims, some of whom were subjects of previous NIOSH studies.

Although the legislation has yet to become law, supporters of the worker notification concept appear prepared for a protracted battle. They argue that it is a necessary supplement to federal risk communication and right to know efforts, which will provide understandable information to individuals actually at risk of occupational disease and suggest a medical monitoring protocol that may decrease their risk. This information, they argue, will further empower workers, who need knowledge of risk to bargain more effectively for improved workplace health and safety.

E.   STATE AND LOCAL INFORMATION TRANSFER LAWS

As noted, the first right to know laws were promulgated at the state and local level, and today's federal workplace and community information transfer requirements owe much to the example set by these early efforts. The importance of state and local laws, however, extends well beyond their historical significance. In many instances, states or communities have passed right to know laws that provide for more disclosure of information than that required by federal law. For example, the OSHA Hazard Communication Standard, and the great majority of the federal community right to know provisions of Title III of SARA, apply only to commercial facilities. Public facilities—such as public schools and public utilities—are not covered. Many state and local information transfer laws, however, *do* extend to these public facilities. Further, some state and local laws require more from commercial facilities than do the relevant federal regulations. What is the validity of these state and local laws in the face of the broad federal workplace and community right to know programs? This is an issue of federal preemption.

As discussed in Chapter 2, the question of whether a state law is preempted by a federal statute or regulation is resolved by examining the intent of Congress. In the right to know area, Congress has spoken directly to the preemption issue. In general, state and local *community* right to know laws that require greater disclosure than that required under Title III of SARA are *not* preempted by federal law. OSHA has made it clear that its workplace Hazard Communication Standard is intended to have no impact on community right to know laws. Further, Section 321 of SARA contains an explicit declaration that most state and local laws providing for disclosure of information to community residents are not preempted, and the legislative history of this section indicates that Title III was intended to set a *minimum* level of disclosure in all states, rather than a maximum level above which states could not go. The only exception to this expansive policy is Section 321(b), which requires that all state and local laws passed after August 1, 1985 have MSDS requirements that are identical to those contained in Title III.

Federal preemption *is* a factor, however, with state and local *workplace* right to know laws. Once a standard has been promulgated for a particular hazard under Section 6 of the OSHAct, states are precluded from regulating the same hazard unless they do so under a state plan approved by OSHA under Section 18 of the Act. When OSHA first promulgated the Hazard Communication Standard in 1983, the agency expressed its intent to preempt *all* state and local workplace right to know laws, in order "to reduce the regulatory burden posed by multiple state laws." 48 Fed. Reg. 53,284. In the *Steelworkers* case, discussed above in section D, the Third Circuit Court of Appeals rejected this approach and held that Section 6 of the OSHAct did not permit the agency to use the Hazard Communication Standard to provide regulatory relief for industry. OSHA's purpose, said the court, is to protect worker safety and health, not to protect employers from the conflicting regulatory burdens that state and local laws might impose. "Congress," the court noted, "struck a balance in favor of safety in the workplace at the expense of competing interests." 763 F.2d at 734. In a subsequent case, the Third Circuit also rejected a claim by the New Jersey State Chamber of Commerce that New Jersey's stringent state right to know law was preempted in its entirety by the OSHA standard. Nonetheless, the court did permit OSHA to preempt state right to know laws to the extent that they pertain to the protection of employee safety and health within the industries covered by the Hazard Communication Standard. This case, excerpted below, sets forth the analytical test to be applied in determining whether a particular state or local law is preempted by the OSHA standard.

New Jersey State Chamber of Commerce v. Hughey

774 F.2d 587 (3d Cir. 1985)

GIBBONS, Circuit Judge.

This is an appeal from a final summary judgment in consolidated actions challenging the constitutionality of the New Jersey Worker and Community Right to Know Act, N.J.Stat.Ann. §34:5A-1 to -31 (West 1984), which requires the disclosure of substances that may pose environmental hazards. The plaintiffs[1] contend that the statute is preempted in its entirety by the federal Occupational Safety and Health Act of 1970, Pub. L. No. 91-596, 34 Stat. 1590 (1970), *codified at,* 29 U.S.C. §§651-678 (1982) (OSH Act).

* * *

I. THE RIGHT TO KNOW ACT

The legislative findings and declaration of purpose are included in the statute, which provides in relevant part:

[T]he proliferation of hazardous substances in the environment poses a growing threat to the public health . . . [and] individuals have an inherent right to know the full range of risks they face so that they can make reasoned decisions and

take informed action concerning their employment and their living conditions.

. . . [L]ocal health, fire, police, safety and other government officials require detained [sic] information about the identity . . . of hazardous substances . . . in order to adequately plan for, and respond to, emergencies. . . .

. . . [T]he toxic contamination of the air, water, and land in this State has caused a high degree of concern among its residents . . . [which is] needlessly aggravated by the unfamiliarity of these substances to residents.

The Legislature therefore determines that it is in the public interest to establish a comprehensive program for the disclosure of information about hazardous substances in the workplace and the community, and to provide a procedure whereby residents of this State may gain access to this information.

N.J. Stat. Ann. §34:5A-2 (West 1984). Thus the New Jersey Legislature intended the Right to Know Act to make information about toxic chemicals available, (1) to all New Jersey residents who might be exposed to such chemicals, in the workplace or elsewhere, and (2) to public safety officers who might need such information to prevent, or respond to, emergencies.

The Right to Know Act directs the New Jersey Department of Environmental Protection (DEP) to develop an environmental hazardous substance list, which must contain:

substances used, manufactured, stored, packaged, repackaged, or disposed of or released into the environment of the State which, in the department's determination, may be linked to the incidence of cancer; genetic mutations; physiological malfunctions, including malfunctions in

---

[1]The plaintiffs include in the New Jersey State Chamber of Commerce, three chemical and business associations, eight pharmaceutical and chemical companies, the Fragrance Materials Association of the United States, the Flavor and Extract Manufacturers' Association, and thirteen corporations which manufacture, compound, mix, or blend fragrances. The original defendants are the New Jersey commissioners of Environmental Protection, Health, and Labor, and the State. The New Jersey Public Advocate and twenty-nine labor unions and environmental organizations are intervening defendants.

reproduction; and other diseases; or which, by virtue of their physical properties, may pose a threat to the public health and safety.

N.J. Stat. Ann. §34:5A-4(a). The DEP promulgated such a list pursuant to the Act. N.J. Admin. Code tit. 7, §1G-2.1 (1984). DEP is further directed to develop an environmental survey "designed to enable employers to report information about environmental hazardous substances at their facilities." N.J. Stat. Ann. §34:5A-4(b).

A separate section of the Act directs the Department of Health to develop a workplace hazardous substance list which must include: (1) all substances regulated by the Occupational Safety and Health Administration (OSHA) under 29 C.F.R. §1910, subpart Z; (2) all environmental hazardous substances; and (3) all substances that "pose[] a threat to the health or safety of an employee." N.J. Stat. Ann. §34:5A-5(a). The workplace hazardous substance list, therefore, by definition includes all substances on the environmental hazardous substance list, and may include some additional substances. The list as promulgated is codified at N.J. Admin. Code, tit. 8, §59 App. A (1984). The Department of Health must develop a "hazardous substance fact sheet" for every item on the workplace hazardous substance list. N.J. Stat. Ann. §34:5A-5(d). The Department of Health is further directed to designate a "special health hazard substance list" indicting those substances that are so dangerous that employers will not be permitted to make a trade secret claim as to them. N.J. Stat. Ann. §34:5A-5(b). Finally, the Department of Health is directed to develop a workplace survey to facilitate reporting of workplace hazardous substances by employers. N.J. Stat. Ann. §34:5A-5(c).

Employers must provide completed surveys to county health departments, local fire and police departments, and to the Department of Health (workplace surveys) or the DEP (environmental surveys). N.J. Stat. Ann. §34:5A-7. Upon receipt of an employer's completed workplace survey, the Department of Health must provide the employer with a hazardous substance fact sheet (prepared by the Department, pursuant to section 5(d) of the Act) for each hazardous substance reported by that employer. N.J. Stat. Ann. §34:5A-8(a). The Department of Health must keep the workplace surveys on file, and make them available to any person who submits a written request. N.J. Stat. Ann. §34:5A-10. Likewise, DEP must maintain a current file of environmental surveys and make them available to any person who makes a written request. N.J. Stat. Ann. §34:5A-9. The Act provides that "any person" may obtain access to envi-

ronmental and workplace surveys; this information is not limited to employees.

Several sections of the Act are designed to ensure that information about hazardous substances is communicated to employees in particular. Employers must maintain a central file of workplace and environmental surveys, and notify employees of the file's availability. N.J. Stat. Ann. §34:5A-12. Each employer must also implement an employee education and training program designed to inform employees of the risk of hazardous chemicals, and to train them in precautions for safe handling of hazardous substances in the workplace. N.J. Stat. Ann. §34:5A-13. The Act also requires employers to label containers of hazardous substances. N.J. Stat. Ann. §34:5A-14.

The Act allows employers to assert a trade secret claim against disclosure of confidential information "which is not patented, which is known only to an employer and certain other individuals, and which is used in the fabrication and production of an article of trade or service, and which gives the employer possessing it a competitive advantage over businesses who do not possess it. . . ." N.J. Stat. Ann. §§34:5A-3, 34:5A-15. Employers may not, however, refuse to disclose information about substances that are on the "special health hazard substance list." N.J. Stat. Ann. §34:5A-5(b).

II. PREEMPTION

Whether a state law is preempted by a federal statute is a question of congressional intent. Preemption may be found if, on the face of the federal statute, Congress expressly stated an intent to preempt a state law. *Fidelity Federal Savings and Loan Association v. de la Cuesta,* 458 U.S. 141, 152-53, 102 S. Ct. 3014, 3022-23, 73 L. Ed. 2d 664 (1982). Even if there is no such express preemption, an intent to preempt state law will be implied if: (1) it is impossible to comply with both the state and the federal law, or (2) "the state law stands as an obstacle to the accomplishment of the full purposes and objectives of Congress." *Silkwood v. Kerr-McGee Corporation,* 464 U.S. 238, 248, 104 S. Ct. 615, 621, 78 L. Ed. 2d 443 (1984). Administrative regulations promulgated pursuant to congressional authorization have the same preemptive effect as federal statutes. *Fidelity Federal,* 458 U.S. at 153-54, 102 S. Ct. at 3022-23.

A. Express Preemption

The Chamber of Commerce and the Fragrance Materials Association contend that the entire Right to Know Act is expressly preempted by the Occupational Safety and Health Act of 1970 (OSH Act), 29 U.S.C. §§651-678 (1982), and by the Hazard Communication Standard, 29 C.F.R. §1900.1200 (1984), promulgated pursuant to the OSH Act. Our decision in *United Steelworkers of America v. Auchter,* 763 F.2d 728, 738 (3d Cir. 1985), compels a holding that the OSH Act and the Hazard Communication Standard preempt the Right to Know Act "with respect to disclosure to employees in the manufacturing sector." *Id.* at 736. We are left with the question not resolved by *Steelworkers*: to what extent are the other provisions of the New Jersey Right to Know Act preempted by the Hazard Communication Standard?

As we held in *Steelworkers,* section 18 of the OSH Act expressly preempts state law pertaining to issues that are addressed by OSHA standards unless a state plan has been approved by the Secretary. 763 F.2d at 733-34. Section 18(a) of the OSH Act expressly gives the states authority to regulate matters that are not governed by a standard.

> Nothing in this chapter shall prevent any State agency or court from asserting jurisdiction under State law over any occupational safety or health issue with respect to which no standard is in effect.

29 U.S.C. §667(a). The statute provides that states may develop and enforce "standards relating to any occupational safety or health issue with respect to which a Federal standard has been promulgated" if the state submits a plan to OSHA, and OSHA approves that plan. 29 U.S.C. §667(b).

The Fragrance Materials Association contends that these provisions should be read expansively to preempt expressly all state statutes that "relate to" a safety or health issue for which a standard has been promulgated. The Association further contends that all provisions of the Right to Know Act relate to issues that are regulated by the Hazard Communication Standard, and that the New Jersey statute is thus preempted in its entirety.

We reject this broad reading of the preemptive affect of OSHA standards. The Fragrance Materials Association cites a number of cases in support of its contention that Congress' language in the OSH Act "evidences a sweeping preemptive intent." The cases cited, however, all involve interpretations of the preemption language of the Employee Retirement Income Security Act of 1974 (ERISA), 29 U.S.C. §§1001-1461 (1982), which is much more explicit than the language of the OSH Act. (The ERISA provisions "shall supersede any and all State laws insofar as they may now or hereafter relate to any employee benefit plan. . . . " 29 U.S.C. §1144(a)).

The preemption language of the OSH Act must be read with the overall purpose of the Act "to assure so far as possible every working man and woman in the Nation safe and healthful working conditions. . . . " 29 U.S.C. §651(b). In light of this purpose and the plain language of section 18 of the OSH Act, it seems clear that the state laws are expressly preempted only to the extent that a federal standard regulating the same issue is already in effect. It would thwart the overriding congressional intent to promote worker safety if federal standards preempted state laws governing issues that are not federally regulated.

Because OSHA standards by definition govern occupational safety and health issues, they do not preempt state laws that regulate other concerns. The Secretary has authority to promulgate standards only as to occupational safety and health and those standards cannot have a preemptive effect beyond that field. Indeed the Secretary argued in *Steelworkers* that the Hazards Communication Standard should not preempt state laws addressing "general environmental problems originating in the workplace, but whose effects are outside it. . . . " Secretary's Br. at 88, *Steelworkers*. The contention of the various New Jersey business interests that Congress intended the OSH Act to preempt all the environmental protection provisions of the New Jersey law is unpersuasive. We hold that the New Jersey Right to Know Act is expressly preempted by the OSH Act and the Hazard Communication Standard only insofar as the New Jersey Act pertains to protection of employee health and safety in the manufacturing sector.

B. Implied Preemption

Other portions of the New Jersey Act might still be preempted, however, under the doctrine of implied preemption, if it is impossible to comply with both the New Jersey law and the Hazard Communication Standard, or if enforcement of the New Jersey Act will thwart congressional purpose. It should be noted from the outset that the mere fact that a state law provision increases the regulatory burden on employers does not make the state law provision contrary to congressional intent. Although OSHA declared that the Federal Hazard Communication Standard was designed "to reduce the regulatory burden posed by multiple state laws," 48 Fed. Reg. 53284, the legislative history of the OSH Act does not indicate that Congress was significantly concerned with reducing the regulatory burden. *See Steelworkers,* 763 F.2d at 734.

Each section of the Right to Know Act must be examined to determine whether it must be held to be preempted by federal law. We may hold a section preempted only if it is clear that Congress would have intended to preempt such a provision. *See Commonwealth Edison Co. v. Montana*, 453 U.S. 609, 634, 101 S. Ct. 2946, 2962, 69 L. Ed. 2d 884 (1981). As this court recently noted:

> Preemption must either be explicit, or compelled due to an unavoidable conflict between the state law and the federal law. Consideration of whether a state provision violates the supremacy clause starts with the basic assumption that Congress did not intend to displace state law.

Penn Terra Ltd. v. Department of Environmental Resources, 733 F.2d 267, 272-73 (3d Cir. 1984) (citations omitted). A section should be held to preempt therefore, only to the extent that: 1) it is expressly preempted because it regulates communication of hazards to employees in the manufacturing sector; 2) it is impossible to comply both with state law and with federal law; or 3) the provision serves as an obstacle to the accomplishment of Congress' object in the OSH Act to promote safe and healthful working conditions.

C. Preemption of the Right to Know Act

Applying these legal standards for preemption, we have examined each section of the Right to Know Act. Listed in the margin are those sections which in our view unquestionably present no issue of express or implied preemption.[4] We will discuss only those sections as to which some plausible preemption contention may be made.

[4]
34:5A-1	Title
34:5A-2	Legislative findings and declarations
34:5A-3	Definitions
34:5A-4(c)	Spanish translation
34:5A-4(d)	Deadlines
34:5A-5(d)	Department of Health must develop hazardous substance fact sheets
34:5A-5(e)	Spanish translation
34:5A-5(f)	Deadlines
34:5A-6	Distribution of workplace and environmental hazard surveys to employers
34:5A-7(b)	Completion and distribution of environmental hazardous substance surveys
34:5A-8(a)	Department of Health must provide each employer with a fact sheet for each hazardous substance reported
34:5A-8(b)	Employers exempt from Act
34:5A-9	DEP must maintain file of environmental surveys; may require employers to update or clarify surveys: surveys will be made available to the public

1. State Hazardous Substance Lists. Sections 34:5A-4(a) and 5(a) direct the New Jersey DEP and Department of Health to develop environmental and workplace hazardous substance lists. Sections 34:5A-4(b) and 5(c) direct those departments to develop environmental and workplace hazard surveys for employers to report hazardous substances. Section 34:5A-5(d) requires the Department of Health to develop hazardous substance fact sheets for each hazardous substance. In these provisions, New Jersey has opted for a different hazard identification procedure than adopted in the federal Hazard Communication Standard, which depends primarily upon identification of hazards by the original manufacturer or importer of the substance. 29 C.F.R. §1910.1200(d). *See Steelworkers*, 763 F.2d at 732. New Jersey's undertaking to develop its own list of hazardous substances through governmental agencies, and to make surveys of environmental and workplace hazards in all sectors of the economy, will in no way inhibit the implementation of the federal standard. The fed-

34:5A-11	Spanish translation
34:5A-14(c)	Department of Health may certify containers labeled in compliance with any federal act as complying with 34:5A-14
34:5A-14(d)	Substances for which chemical name need not be included on label
34:5A-15	Trade secret claims
34:5A-17	Employees may not be discharged or penalized for exercising rights under the Act
34:5A-18 to 20	Establishment, powers and duties of Right to Know Advisory Council
34:5A-21	Implementation of Act
34:5A-22	County health departments must maintain files of workplace and environmental surveys and make them available to the public
34:5A-23	Civil actions for violations of Act
34:5A-24	Absence of substance from lists does not affect employer's duty to warn, or take other precautions
34:5A-25	Role of fire and police departments
34:5A-27	State law preempts municipal and county ordinances passed after May 11, 1983
34:5A-28	State agencies will evaluate Act in two years and present report to legislature
34:5A-29	Agencies may enter employer's facilities to assure compliance
34:5A-30	Agencies may adopt necessary rules and regulations
34:5A-31	Remedies for failure to comply

eral and state lists may comfortably coexist. Thus sections 34:5A-4 and 34:5A-5 are not preempted.

2. Employer Completion of Workplace and Environmental Surveys.

Section 34:5A-7(a) requires employers to complete workplace hazard surveys and to provide the completed surveys to the State Department of Health, the County Department of Health, and local fire and police departments. Section 34:5A-7(b) imposes parallel requirements for completion and distribution of environmental hazard surveys. Section 34:5A-7(b) is not preempted, since it requires reporting of environmental hazards to agencies concerned with public health and safety, a matter not governed by OSHA standards. Section 34:5A-7(a) is not preempted in areas of the economy other than the manufacturing sector, so long as no federal standard is in effect in those areas.

Section 34:5A-7(a) is preempted in the manufacturing sector, SIC Codes 20-89. The requirement that workplace surveys be furnished not only to state agencies concerned with the protection of employees, but also to state and local agencies concerned with the protection of the public at large suggests that section 34:5A-7(a) may have a broader purpose than the federal Hazard Communication Standard. Nevertheless any hazardous substance listed as a workplace hazard pursuant to section 34:5A-5(a), and not listed as an environmental hazard pursuant to section 34:5A-4(a), is deemed to be one posing a specific threat to workers. It appears therefore that the primary purpose of section 34:5A-7(a) is the promotion of occupational health and safety through hazard communication. The federal Hazard Communication Standard expressly preempts section 34:5A-7(a) in the manufacturing sector. Because section 34:5A-7(a) may not operate in the manufacturing sector, the provision in section 34:5A-10 directing the Department of Health to maintain a file of workplace surveys is preempted in the same extent. Moreover, the obligation of employers in the manufacturing sector to keep a central file of workplace surveys, imposed in section 34:5A-12, is also preempted. Likewise, section 34:5A-16 is preempted to the extent that it guarantees employees in the manufacturing sector access to workplace surveys and hazardous substance fact sheets for items on the workplace hazardous substance list. Employers in the manufacturing sector must, however, pursuant to section 34:5A-12, keep and make available the environmental surveys required by section 34:5A-7(b). This provision of the Act is not preempted.

3. Funding.

Section 34:5A-26 provides that the Right to Know Act will be funded by assessments against all New Jersey employers. The Fragrance Materials Association argues that this indicates that the Act is aimed solely at occupational safety and health, rather than at community health and safety in general, and thus that all provisions of the Act should be preempted by the OSH Act. That contention is without merit. The Act discloses a broad community health and safety purpose. The funding provision is a logical means for funding the accomplishment of that purpose. It does not interfere with compliance with the OSH Act or impose obstacles to the accomplishment of the OSH Act's purposes.

4. Container Labeling.

Section 34:5A-14 requires New Jersey employers to label containers of hazardous substances with the chemical name and Chemical Abstracts Service number. The federal Hazard Communication Standard requires chemical manufacturers and importers to label each container leaving the workplace with the identity of the hazardous chemical, which may be indicated by either a chemical or a common name. 29 C.F.R. §1910.1200(f), (d). The labeling requirements of the federal Hazard Communication Standard expressly preempt section 34:5A-14 to the extent that they require labeling of workplace hazards in the manufacturing sector. Section 34:5A-14 continues to operate, however, outside SIC Codes 20-89.

The Hazard Communication Standard does not expressly preempt section 34:5A-14 to the extent that it requires that containers of environmental hazardous substances be labeled. Plaintiffs contend that the labeling provisions of the New Jersey Act are impliedly preempted because they are an obstacle to the accomplishment of congressional purpose in the OSH Act. Section 34:5A-14 requires that within two years containers and pipelines be labeled with chemical names and chemical Abstracts Service Registry numbers of the five predominant substances passing through them. The plaintiffs contend that compliance with these labeling requirements, which are intended to furnish information to firefighters, police officers, and members of the community at large, will, if imposed on employers in the manufacturing sector, stand as an obstacle to the accomplishment of the purposes of the federal Hazard Communication Standard. Their theory, set forth in several affidavits filed in support of their motion for summary judgment, is that the environmental hazard labeling, in conflicting with the labeling required by the federal standard, will lead to confusion of workers. The district court, because it determined that the preempted provisions of the Act were unseverable, did not decide whether the plaintiffs were entitled to a summary judgment that the environmental hazard labeling features of the Right to Know Act were impliedly preempted by the federal Hazard Communication Standard. Our examination of the summary judgment record suggests that there are disputed issues of material fact which preclude a summary judgment for the plaintiffs or the defendants on that issue. If New Jersey's imposition of the environmental hazard labeling requirements in the manufacturing sector, not for the purpose of protecting workers, but in the interest of firefighters, police officers, and the general public, does in fact stand as an obstacle to the accomplishment of the purposes of the federal standard, the New Jersey law must yield. Otherwise it may operate. On this record we cannot determine that question as a matter of law.

D. Severability. If any provisions of the Right to Know Act are preempted by the OSH Act, we must look to New Jersey law to determine whether the preempted provisions may be severed from the rest, or whether the statute as a whole must fall.

* * *

Legislative intent as to severability may also be gleaned from the structure of the statute. In *State v. Lanza* the court declared that an invalid provision should be deemed severable from a valid one "unless the two are so intimately connected and mutually dependent as reasonably to sustain the hypothesis that the Legislature would not have adopted the one without the other." 143 A.2d at 577. This is largely a question of whether the statute will continue to make sense after the challenged portion is excised. A regulatory statute may be permitted to stand if, "stripped of those provisions which are invalid, [it] remains a comprehensive and cohesive regulatory ordinance with appropriate sanctions for its enforcement." *Oxford Consumer Discount Co. v. Stefanelli*, 246 A.2d at 467 (quoting *Cranberry Lake Quarry Co. v. Johnson*, 95 N.J. Super. 495, 517, 231 A.2d 837, 848 (App. Div.), *cert. denied*, 50 N.J. 300, 234 A.2d 407 (1967)). In sum, the New Jersey courts use a common-sense approach to severability, holding that an invalid provision is severable if that is in keeping with legislative intent; legislative intent is ascertained by looking to the broad purpose of the statute, the degree too which the valid and invalid provisions are intertwined with one another, and the extent to which the statute remains comprehensive and logical after the invalid provisions are excised.

The intention of the New Jersey legislature will be best advanced by severing preempted portions of the Right to Know Act and leaving the remainder operative. The legislature's intent to protect all New Jersey residents by informing them of hazardous substances in manifest. *See* N.J. Stat. Ann. 34:5A-2. Removal of the preempted provisions makes the New Jersey Act somewhat less comprehensive; it does not, however, make any drastic changes that would appear to be contrary to legislative purpose, and it does not broaden the scope of the regulatory scheme by eliminating exceptions to the regulation. The preemptive effect of the OSH Act and the Hazard Communication Standard is narrow. The Right to Know Act remains a "comprehensive and cohesive regulatory ordinance with appropriate sanctions for its enforcement," *Oxford*, 246 A.2d at 467, which should be enforced to further the intent of the New Jersey legislature.

◇ NOTES

1. On remand, the district court held that the environmental hazard warning requirement was not preempted. On the appeal of *that* decision, the Third Circuit agreed. See *New Jersey State Chamber of Commerce v. Hughey*, 868 F.2d 621 (3d Cir. 1989).

2. With its extension of the Hazard Communication Standard to *all* OSHA-covered workplaces in the 1987 revisions, OSHA indicated that the preemptive effect of the standard was to be extended to these workplaces as well. Thus, subject to the limitations set forth in the *Hughey* case, it is prudent to assume that the OSHA standard preempts state and local right to know laws in all commercial workplaces—and not just in the manufacturing sector. The viability of such laws in public workplaces is not affected by the OSHAct.

3. The Third Circuit applied a similar analysis to the Pennsylvania right to know law in *Manufacturers' Association of Tri-County v. Knepper,* 801 F.2d 130 (3d Cir. 1986).

4. The Sixth Circuit has held that the preemptive effect extends to municipal right to know laws as well. See *Ohio Manufacturers' Association v. City of Akron,* 801 F.2d 824 (6th Cir. 1986).

5. The First and Second Circuits have addressed the issue of whether the Hazard Communication Standard, together with the OSHA asbestos standard, preempt stricter state laws governing workplace asbestos removal. Using analyses that differ somewhat from that applied by the Third Circuit, both courts concluded that those parts of the regulations that pertained to medical surveillance of asbestos abatement workers, or to the use of respirators by those workers, were preempted. However, the bulk of the regulations, dealing with licensure, certification, training, notice, and work practice requirements, were allowed to stand. In the words of the Second Circuit, to the extent that a state or local government can demonstrate that its regulations have a "legitimate and substantial purpose apart from promotion of occupational health and safety," there is no preemption by the OSHAct. *Environmental Encapsulating Corp. v. New York City,* 855 F.2d 48, 57 (2d Cir. 1988). The First Circuit, on the other hand, looked to the *effect* of the state regulation. "If the effect is to protect the public, the state regulation is not preempted. If the effect is solely to protect workers, the OSHA standard prevails and the state regulation fails. If the effect is to protect the public by regulating workers and work places, the regulation stands because its ultimate effect is protection of the public." *Associated Industries of Massachusetts v. Snow,* 898 F.2d 274 (1st Cir. 1990).

◇

F. POTENTIAL BARRIERS TO DISCLOSURE

As we have seen, laws requiring the transfer of information within the workplace often face two potential constraints: employee confidentiality and employer trade secrecy. Both of these interests, especially the personal privacy of individual employees, find some protection in the federal constitution. Conversely, it would appear that the constitution also places limits on the steps that the government may take to protect these interests from disclosure under a right to know law. Beyond the limitations imposed by the constitution, important practical and policy questions remain.

1. Employee Confidentiality

An employee's interest in personal privacy arises at two important stages in the process of toxics information transfer. The first occurs when the employer—acting at the direction of an agency or on its own—seeks to compel the employee to submit to human monitoring. The second occurs when someone other than the employee (an agency, a union, the employer, or some other third party) seeks access to the information generated through human monitoring procedures.

Monitoring the Worker for Exposure and Disease: Scientific, Legal and Ethical Considerations in the Use of Biomarkers

N. Ashford, C. Spadafor, D. Hattis, and C. Caldart

In the abstract sense, an employee may always refuse to be the subject of human monitoring. OSHA, NIOSH, and the employer have no legal authority to compel employees to cooperate. Refusal to participate, however, may well mean the loss of a job. Thus, the relevant inquiry is the extent to which the employer may condition employment on such cooperation. For example, may an employer require a prospective employee to submit to genetic or biological screening as a precondition to employment? May he or she require a current employee to submit to periodic biological monitoring or medical surveillance?

* * *

At the outset, a distinction must be made between human monitoring that OSHA, NIOSH, or the EPA requires and monitoring that the employer implements on his or her own initiative. In the first case, in which a federal agency requires monitoring, the worker will have a valid objection only if (a) the requirement exceeds the agency's statutory authority, or (b) the requirement violates the constitution.

* * *

Congress was mindful of constitutional considerations in drafting the OSHAct. For example, it specifically acknowledged the need for a balancing of interests where an employee asserted a religious objection to a monitoring procedure. Human monitoring can also impinge on the worker's constitutional right to privacy. In the case of human monitoring, the privacy right may be articulated in two ways: the right to *physical integrity*, and the right to *withhold information* likely to prove detrimental to one's self-interest.

If an employee does not wish to comply with a monitoring procedure required by agency regulation, imposing that procedure as a condition of employment may invade that employee's constitutional right to physical integrity. It may, depending on the nature of the procedure, infringe upon the right to be free

Source: Excerpts from *Monitoring the Worker for Exposure and Disease: Scientific, Legal and Ethical Considerations in the Use of Biomarkers.* Baltimore, MD. Johns Hopkins University Press, 1990. 125–131.

from unwelcome physical intrusions and/or on the right to make decisions regarding one's own body. While these rights are obviously related, they are derived from separate constitutional doctrines. The former is grounded in the Fourth Amendment's proscription against unreasonable search and seizure, while the latter is closely associated with the rights of personal privacy commonly identified with the Ninth and Tenth amendments. Although protected by the Constitution, these privacy interests are not inviolate. The Supreme Court has followed a general policy of balancing the privacy interests of the individual with the public health interests of society. In some situations, the former will be deemed to outweigh the latter, but in others intrusion will be permitted in the name of public health.

The public health significance of human monitoring, when properly used, is difficult to deny. Gathering information through human monitoring to develop standards for the protection of worker health, or for the enforcement or evaluation of existing standards, serves an important public health purpose. Furthermore, although the Constitution places limits on governmental paternalism, the fact that this public health interest parallels the affected worker's own interest in a healthy workplace may make monitoring of this nature a less onerous invasion of privacy than it would otherwise be. To the extent that monitoring serves a legitimate public health purpose, a limited intrusion of physical privacy appears constitutionally permissible. The less the accuracy, reliability, or predictability of a particular intrusion, however, the weaker the case for violating physical privacy.

The scope of permissible intrusion depends on the nature of the monitoring. The insertion of a urethral tube, for example, involves a greater invasion of personal privacy than does the taking of a blood sample. Some monitoring procedures also involve greater risk than others. A program of periodic lung X-rays, for instance, poses a greater risk than a program of periodic lung function tests. At some point, the degree of risk or intrusiveness may become sufficiently compelling to outweigh the public health interests. Indeed, some forms of human monitoring may simply be too risky or too intrusive to be constitutionally permissible. Furthermore, the worker may well have a right to insist on an alternate, less intrusive procedure that adequately fulfills public health purposes. To survive constitutional challenge, then, a regulation requiring human monitoring should not only be reasonably related to an important public health goal, but should also impose the least intrusive method necessary to achieve that goal.

The employee's privacy interest in refusing to participate in a program of agency-directed monitoring is heightened where the resulting information may be used as a basis for termination, transfer, or demotion. For example, the worker who suffers reduced lung capacity as a result of workplace particulate exposure may fear that a program of medical surveillance will reveal this condition to the employer and thus induce removal. In this sense, participation in a monitoring program can be tantamount to self-incrimination. Presumably, the fact that a compulsory monitoring program may cause employees to lose their jobs in this manner will have an effect on the nature of the "balance" to be struck under the Fourth Amendment. If the public health need for the monitoring in question is not clear and compelling, or if the program is drawn more broadly than necessary, the underlying regulation may not survive constitutional challenge.

The Supreme Court held in a 1989 decision that Federal Railroad Administration regulations requiring compulsory drug and alcohol testing of railroad workers *are* permissible under the Fourth Amendment, even though such testing could result in job loss or criminal sanctions for individual employees [*Skinner v. Railway Labor Executives Association*,—U.S.—, 109 S.Ct. 1402 (1989)]. However, it would require a substantial conceptual leap to extend the reasoning of this case to the more general issue of compulsory health and exposure monitoring. In upholding the railroad worker testing requirements, the Court acknowledged that Fourth Amendment considerations were at stake. In permitting this invasion of worker privacy, the Court was careful to point out that: (a) the monitoring in question was designed to reveal the presence of substances associated with criminal activity (drinking and drug-taking on the job); (b) the targeted activity had an undeniable relationship to public safety; and (c) the testing was required only after the occurrence of events which raised public safety concerns (major train accidents and the violation of certain safety rules). Industrywide programs of health and exposure monitoring do not fit easily into this framework. Simply put, a testing program that results in the discharge of employees in a safety-sensitive profession whose illegal on-the-job behavior has put innocent third parties at grave risk is one thing. A program that results in the discharge of employees merely because they have had the misfortune of being exposed to toxic chemicals, and thus might themselves become sick, is quite another.

In developing human monitoring requirements, an agency should seriously consider the constitutional dimensions of the issue. To avoid a challenge on a "self-incrimination" basis, OSHA might consider including mandatory MRP programs as part of its monitoring requirements. Properly used, an MRP program would safeguard the employment rights of employees whose health was damaged or threatened by workplace exposure and would help ensure employee cooperation.

* * *

If an employer institutes a human monitoring program in the absence of agency directive, he or she is still subject to applicable restrictions under state common law, state statute, and federal labor law. Common law requires that human monitoring be implemented in a "reasonable" fashion. Determining reasonableness involves balancing the benefits gained by monitoring against the risk, discomfort and intrusiveness of the monitoring procedure. In a given jurisdiction, the balance might be affected by a state statute defining a right of personal privacy. Further, employees may be able to impose limitations on human monitoring through the collective bargaining process authorized by federal labor law.

* * *

EMPLOYEES' RIGHT TO CONFIDENTIALITY: ACCESS TO EMPLOYEE RECORDS BY AGENCIES, UNIONS, AND EMPLOYERS

Of all of the issues raised by human monitoring, employee confidentiality may have received the most attention. An employee's right to maintain the confidentiality of information regarding his or her body and health places a significant limitation on the ways in which others can use that information. As programs of human monitoring are developed, mechanisms must be found that maximize both the employee's interest in privacy and society's interest in promoting general workplace health and safety. In the final analysis, this may be more a technological challenge than a legal or ethical one.

In a broad sense, private citizens do have a right to protect the confidentiality of their personal health information. With regard to *governmental* invasions of privacy, this right is created by the Bill of Rights and is one component of the constitutional right of personal privacy discussed above. With regard to *private* intrusions, the right is grounded in statute and common law. In the medical setting, it grows out of the confidential nature of the physician-patient relationship, although rights of confidentiality exist outside this relationship as well. In essence, the recognition of a right of privacy reflects an ongoing societal belief in the need to protect the integrity of the individual.

Although the right to privacy is not absolute, courts nonetheless remain sensitive to society's concern for individual confidentiality. They generally look for a reasonable middle ground when faced with legitimate interests on both sides of the confidentiality question. They prefer an approach that permits both the use of health information for a socially useful purpose and the protection of the privacy of the individual. The key is the development of technology that will make that approach more readily available.

Developing such technology will be especially important in protecting the confidentiality of information generated by human monitoring. Both medical and exposure records contain health information of a confidential nature, and the employee has a legitimate interest in limiting its disclosure. At the same time, public agencies, unions and employers have a legitimate interest in using this information. From a technical point of view, the solution will lie in mechanisms that allow third parties to use meaningfully health information that is not tied by name or other common identifier (such as a social security number) to any one individual, and that allow data relevant to toxic substances exposure to be separated from other health information. When disclosure is limited to relevant medical and exposure information that cannot be traced to any worker by name, any detrimental effect of disclosure will be held to a minimum.

Proper development of the necessary technology, however, will not follow from piecemeal solutions devised by reviewing courts. Rather, what is needed is a concerted, comprehensive, multi-disciplinary approach. OSHA, NIOSH, and the EPA could pursue such technology either as an agency research and development project or through cooperation with private industry. This technology, once developed, could then be made available to employers at the cost of installation and equipment. No system will resolve completely the conflict between confidentiality and disclosure and the potential for abuse will always be present. However, methods of record-keeping that permit the effective use of relevant health information without requiring the disclosure of other personal data would eliminate much of that conflict. Presently, the conflict remains substantial, especially regarding medical records prepared under methods that are ill-adapted to protective disclosure.

AGENCY ACCESS

In addition to the limitations it places on the government's authority to compel an employee to submit to monitoring procedures, the Constitution imposes limitations on federal agency *access* to monitoring information once it has been compiled. The Supreme Court has outlined a number of important issues in this general area in *Whalen v. Roe* [429 U.S. 589 (1977)], a case involving a New York law that required physicians to provide the state with the names of persons receiving prescriptions for certain controlled drugs. The court upheld the statute against privacy claims raised by both patients and physicians because the law was narrowly drawn to apply only to a limited class of arguably dangerous drugs and New York had a legitimate public health interest in controlling the dissemination and use of those drugs. In doing so, the Court indicated the broad framework within which questions of constitutional privacy rights must be decided. On the one hand, the right to confidentiality clearly can be limited when such limitation is necessary to meet a legitimate public health purpose:

> [D]isclosures of private medical information to doctors, to hospital personnel, to insurance companies, and to *public health agencies* are often an essential part of modern medical practice even where the disclosure may reflect unfavorably on the patient. Requiring such disclosures to representatives of the State having responsibility for the health of the community does not automatically amount to an impermissible invasion of privacy.

On the other hand, the Court also indicated that the disclosure of confidential information should be no broader than necessary to meet the desired public health purpose. "The right to collect and use such data for public purposes is typically accompanied by a concomitant statutory or regulatory duty to avoid *unwarranted* disclosures." As with the protection of physical privacy, the Constitution demands a careful balancing of the individual's right to confidentiality and the legitimate interests of society.

Both OSHA and NIOSH have sought to achieve this balance. They have, however, taken markedly different paths toward this end. OSHA access to medical records is secured by a regulation designed to protect employee confidentiality. In general, records obtained under this regulation must be secured through a specific, written access order, must be used only for the purposes indicated on the order, and must be destroyed or returned after OSHA has completed such use. A significant flaw in this regulation is the fact that it applies only to "personally identifiable employee *medical* information." By its terms, it in inapplicable to "*exposure* records, including biological monitoring records." Instead, OSHA access to exposure records is governed by the OSHA access rule discussed above. A provision of this rule grants OSHA "prompt" entry to such records, without any mechanism for limiting the use to which the agency may put such information. To the extent that these records contain constitutionally protected health data, this lack of privacy protection may violate the doctrine enunciated in *Whalen v. Roe.*

NIOSH, on the other hand, has not promulgated access regulations. Instead, it has sought access on a case-by-case basis by using its subpoena power. Employers have resisted the subpoena on the basis of the employees' constitutional rights to privacy. As a result, decisions of various federal courts have developed limitations on NIOSH access.

In general, the courts have applied the *Whalen* doctrine and have conditioned access by NIOSH upon the development of procedures designed to limit the intrusion on individual worker privacy. The Court of Appeals for the Sixth Circuit, for example, noted that there should be "no public disclosure of the medical information" beyond the agency itself [*General Motors Corp. v. NIOSH*, 636 F.2d 163 (6th Cir. 1980)]. Significantly, the court also recognized that the conflict between confidentiality and public health grows more out of practical than philosophical considerations. It "[did] not believe that the parties' interests . . . were mutually exclusive. With proper security administration, [NIOSH] should be able to complete [its health studies] without jeopardizing the constitutional rights of the individuals involved." This recognition of practical constraints underscores the need for a creative approach to medical and exposure information storage and transfer.

UNION ACCESS

Although a union is usually presumed to be acting on behalf of its members, at times the union's assertion of access to medical or exposure records will conflict with an employee's interest in keeping those results confidential. All employees have an interest in ensuring that the union's right of access is not unchecked, but rather is limited to legitimate purposes.

Where the union seeks medical and exposure information from an employer as a part of the collective bargaining process, employee privacy interests will be balanced against the union's interest in securing disclosure. In two cases before the NLRB, the *employer* asserted the physician-patient privilege as a defense to a union's request for access to medical records. In striking the balance between disclosure and privacy, the NLRB ordered the employer to provide access "to the extent that such data do not include medical records from which identifying data have not been removed." But the NLRB did indicate in both of these decisions that the union's interest in securing health and safety information could be satisfied without disclosure of personal identifiers. It thus did not foreclose the possibility that a broader right of access might be appropriate if the union establishes a legitimate need for such data. In upholding the Board's decision, however, the District of Columbia Circuit Court of Appeals noted that "employers may have a legitimate and substantial interest in protecting the confidentiality of employees' medical records . . . [and] the mere deletion of names may not sufficiently ensure employees' confidentiality" [*O.C.A.W. v. NLRB*, at page 333 *supra*]. The precise nature of the appropriate balance, then, is not altogether clear.

Nor is it clear whether more exacting employee protection might be required when the employee, rather than the employer, asserts the right to confidentiality. The additional protection, if any, available to dissenting union employees has not yet been delineated. As a matter of policy, it seems that the rationale for protecting personal privacy is as compelling in the case of union access as it is in the case of agency access and that the union's interest in collective bargaining should be accommodated in a manner that respects the confidentiality of the individual employee.

Unions also have a right of access to employee records under the OSHA access rule. Where a union has been recognized as the bargaining representative for a certain group of employees under federal labor law, it is authorized to secure access to their exposure records under the OSHA rule as their "designated representative." In general, the union has the same right of access to exposure records as do the employees themselves. The union is not required to secure individual employee authorization before it is entitled to such access. Noting that "there are occupational health benefits to be gained by permitting [unrestricted] union access to records," OSHA in 1988 turned down a request by employers that the rule be revised to require individual authorization as a prerequisite to union access. In response to employer complaints that union access requests were often burdensome, however, the agency revised the rule to require designated representatives to "specify with reasonable particularity" the records requested and the "occupational health need" for disclosure. The OSHA rule also grants designated representatives a right of access to employee medical records. In recognition of the more confidential nature of medical information, however, the rule requires "specific written consent" from each affected employee before such access may be secured.

EMPLOYER ACCESS

Perhaps the most obvious threat to employee confidentiality is that posed by employer access. Of all parties seeking access to employee health information, the employer has the most directive economic incentive to use that information in ways detrimental to the employee. On the other hand, human monitoring data is essential for those employers who strive in good faith to eliminate workplace hazards.

By accepting or seeking employment, the employee implicitly consents to certain limitations on his or her expectations of confidentiality. To the extent that individual employee information—even information of a personal nature—is relevant to a legitimate employer interest, the employer may condition employment on its disclosure. In general, employers are entitled to information that bears upon the employee's ability to perform his or her job, and to information that indicates the levels of toxic substance exposure in the workplace. Although not all information gathered through health and exposure monitoring meets these criteria, much of it does. In the absence of protective statutes, employers will have a right of access to such data. Health and personal information that is not reasonably related to the work environment, however, arguably is pro-

tected either by the physician-patient relationship or by tort concepts of personal privacy. In theory, it should be unavailable to the employer.

In practice, however, this may not be the case. Testimony taken by OSHA in 1980, for example, indicates that many employers routinely gain access to an employee's complete medical file. According to OSHA, the following statement by a member of the United Auto Workers was representative of the testimony received:

> I have been in medical . . . trying to talk to the company doctor. A member of [management] would come down and just, you know, hi, doc, and then go through the records, the medical records, and pull a particular individual's medical record and without even consulting the doctor first or a nurse or anybody as far as that goes, just directly [go] to the cabinet and pull an individual's record. . . . They will just go directly down and pull the file themselves. So there is no confidentiality.

There is little reason to believe that this does not remain the practice in many workplaces.

OSHA has acknowledged that this is a "serious problem" but has thus far declined to take any specific remedial action. It did not discuss the issue when it revised the employee access rule in 1988. Where the information in question was gathered pursuant to a monitoring program that was required or expressly permitted by OSHA regulation, the agency may well have a constitutional obligation to protect employee privacy. Indeed, even where the information *is* related to a legitimate employer interest, the doctrine of *Whalen v. Roe* may require the agency to protect the employee's confidentiality in such a situation.

As a practical matter, much of the problem might be alleviated if human monitoring and the maintenance of medical and exposure records were undertaken by a third-party health care provider, such as a health maintenance organization. Much of the abuse inherent in employer access to employee health information arises from simple proximity. The employer is often the keeper of the information to which claims of confidentiality attach. If this information were held by a third party, such as a health maintenance organization, that party would be in a better *practical* position to ensure that all those with legitimate rights of access—the employee, the agency, the union and the employer—exercise those rights in full compliance with the law.

2.   Trade Secrets

The Right to Know: Toxics Information Transfer in the Workplace

N.A. Ashford and C.C. Caldart

Under our current system of commercial laws, a chemical manufacturer or other industrial employer is entitled to take reasonable steps to prevent the disclosure of information regarding its manufacturing processes that gives it a true commercial advantage over its competitors. This information is said to comprise that company's "trade secrets." Quite often, the need for the dissemination of data regarding workplace exposures is characterized as being in conflict with an employer's interest in maintaining its trade secrets. In many cases, however, this characterization may be misleading.

Much of the data subject to information transfer laws simply does not rise to the level of trade secrecy. The *identity* of a chemical substance, for example, will be a true trade secret only if its disclosure reveals a technological process or functional chemical characteristic that is not known to other companies. In addition, such use of the chemical must represent a clear advance in the field, must provide the owner with a distinct competitive advantage, and must not be available to others through investigation or other fair means. Certainly, there will be times when these conditions are met. Often, however, disclosure of a specific chemical identity will reveal no information that is not either already known to competitors or discoverable through reverse engineering. Further, disclosure of data regarding the *effects* of chemical exposure need only rarely reveal a legitimate trade secret.

Even where a claim of trade secrecy would be otherwise valid, disclosure under information transfer laws can often be done in such a way as to

Source: Reproduced with permission from the *Annual Review of Public Health,* Vol. 6. © 1985 by Annual Reviews, Inc., pp. 383-401.

maintain a firm's competitive advantage. Trade secrecy is violated only where the protected information is given to a party who will use or disclose the secret to the owner's detriment. In many cases, a trade secret can be released to parties entitled to access under information transfer laws without its ultimate disclosure to third party competitors. A few years ago, for example, Polaroid Corporation released an expansive list of specific chemical identities—many of which the company considered to be trade secrets—to EPA under Section 8 of TSCA, subject to the agency's agreement to keep the information on a separate computer system that guarded against outside disclosure. Arrangements of this nature could well be adopted by other agencies, or by unions, to facilitate the release of arguably sensitive technological information.

Nonetheless, claims of trade secrecy—both justifiable and unjustifiable—will continue to pose a potential barrier to information transfer. Some laws contain specific provisions designed to protect disclosure of trade secrets. Under ordinary circumstances, for example, information obtained by EPA under TSCA is explicitly exempted from disclosure to the extent that it reveals either a chemical manufacturing *process* or the precise *percentage* composition of a chemical mixture. Further, regardless of the existence or scope of specific statutory protections, all information-transfer regulations are subject to the constitutional prohibition against the appropriation of private "property"—which may include trade secrets—without due process. In the inevitable close cases, then, courts and agencies will face the difficult, but necessary, task of weighing the legitimate trade secret interests of employers against the legitimate health interests of workers. Although the enunciation of a general rule is perhaps not appropriate, sound law and public policy would appear to require a tipping of the balance in favor of health.

The courts *have* begun this balancing process and have indicated that the question of whether trade secrets should be afforded protection under right to know laws is largely a question of federal and state legislative policy, not a question of federal constitutional law. The landmark U.S. Supreme Court case in this area did not arise under a workplace right to know law, but under a provision of an environmental protection statute, the Federal Insecticide, Fungicide, and Rodenticide Act (FIFRA). The section in question provides public access to health and safety data submitted to EPA by a pesticide manufacturer as part of the federal permitting process. FIFRA was originally adopted in 1949 but was amended substantially in 1972 and again in 1978. This time frame provides the basis for the Court's opinion.

Ruckelshaus v. Monsanto Co.

467 U.S. 986 (1984)

Justice BLACKMUN delivered the opinion of the court.

In this case, we are asked to review a United States District Court's determination that several provisions of the Federal Insecticide, Fungicide, and Rodenticide Act (FIFRA), 61 Stat. 163, as amended, 7 U.S.C. §136 *et seq.*, are unconstitutional. The provisions at issue authorize the Environmental Protection Agency (EPA) to use data submitted by an applicant for registration of a pesticide in evaluating the application of a subsequent applicant, and to disclose publicly some of the submitted data.

* * *

II

Appellee Monsanto Company (Monsanto) is an inventor, developer, and producer of various kinds of chemical products, including pesticides. Monsanto, headquartered in St. Louis County, Mo., sells in both domestic and foreign markets. It is one of a relatively small group of companies that invent and develop new active ingredients for pesticides and conduct most of the research and testing with respect to those ingredients.

These active ingredients are sometimes referred to as "manufacturing-use products" because they are not generally sold directly to users of pesticides. Rather, they must first be combined with "inert ingredients"—chemicals that dissolve, dilute, or stabilize the active components. The results of this process are sometimes called "end-use products," and the firms that produce end-use products are called "formulators." See the opinion of the District Court in this case, *Monsanto Co. v. Acting Administrator, United States Environmental Protection Agency*, 564 F. Supp. 552, 554 (ED Mo. 1983). A firm that produces an active ingredient may use it for incorporation into its own end-use products, may sell it to formulators, or may do both. Monsanto produces both active ingredients and end-use products. *Ibid.*

The District Court found that development of a potential commercial pesticide candidate typically requires the expenditure of $5 million to $15 million annually for several years. The development process may take between 14 and 22 years, and it is usually that long before a company can expect any return on its investment. *Id.*, at 555. For every manufacturing-use pesticide the average company finally markets, it will have screened and tested 20,000 others. Monsanto has a significantly better-than-average success rate; it successfully markets one out of every 10,000 chemicals tested. *Ibid.*

Monsanto, like any other applicant for registration of a pesticide, must present research and test data supporting its application. The District Court found that Monsanto had incurred costs in excess of $23.6 million in developing the health, safety, and environmental data submitted by it under FIFRA. *Id.*, at 560. The information submitted with an application usually has value to Monsanto beyond its instrumentality in gaining that particular application. Monsanto uses this information to develop additional end-use products and to expand the uses of its registered products. The information would also be valuable to Monsanto's competitors. For that reason, Monsanto has instituted stringent security measures to ensure the secrecy of the data. *Ibid.*

It is this health, safety, and environmental data that Monsanto sought to protect by bringing this suit. The District Court found that much of this data "contains or relates to trade secrets as defined by the Restatement of Torts and Confidential, commercial information." *Id.*, at 562.

* * *

III

In deciding this case, we are faced with four questions: (1) Does Monsanto have a property interest protected by the Fifth Amendment's Taking Clause in the health, safety, and environmental data it has submitted to EPA? (2) If so, does EPA's use of the data to evaluate the applications of others or EPA's disclosure of the data to qualified members of the public effect a taking of that property interest? (3) If there is a taking, is it a taking for a public use? (4) If there is a taking for a public use, does the statute adequately provide for just compensation?

This Court never has squarely addressed the applicability of the protections of the Taking Clause of the Fifth Amendment to commercial data of the kind involved in this case. In answering the question now, we are mindful of the basic axiom that " '[p]roperty interests . . . are not created by the Constitution. Rather, they are created and their dimensions are defined by existing rules or understandings that stem from an independent source such as state law.' " *Webb's Fabulous Pharmacies, Inc.* v. *Beckwith,* 449 U.S. 155, 161 (1980), quoting *Board of Regents* v. *Roth,* 408 U.S. 564, 577 (1972). Monsanto asserts that the health, safety, and environmental data it has submitted to EPA are property under Missouri law, which recognizes trade secrets, as defined in §757, Comment *b,* of the Restatement of Torts, as property. See *Reddi-Whip, Inc.* v. *Lemay Value Co.,* 354 S.W. 2d 913, 917 (Mo. App. 1962); *Harrington* v. *National Outdoor Advertising Co.,* 355 Mo. 524, 532, 196 SW 786, 791 (1946); *Luckett* v. *Orange Julep Co.* 271 Mo. 289, 302-304, 196 SW 740, 743 (1917) The Restatement defines a trade secret as "any formula, pattern, device or compilation of information which is used in one's business, and which gives him an opportunity to obtain an advantage over competitors who do not know or use it." §757, Comment *b.* And the parties have stipulated that much of the information, research, and test data that Monsanto has submitted under FIFRA to EPA "contains or relates to trade secrets as defined by the Restatement of Torts." App. 36.

Because of the intangible nature of a trade secret, the extent of the property right therein is defined by the extent to which the owner of the secret protects his interest from disclosure to others. See *Harrington, supra; Reddi-Wip, supra;* Restatement of Torts, *supra;* see also *Kewanee Oil Co.* v. *Bicron Corp.,* 416 U.S. 470, 474-476 (1974). Information that is public knowledge or that is generally known in an industry cannot be a trade secret. Restatement of Torts, *supra.* If an individual discloses his trade secret to others who are under no obligation to protect the confidentiality of the information, or otherwise publicly discloses the secret, his property right is extinguished. See *Harrington, supra;* R. Milgrim, Trade Secrets. §1.01[2] (1983).

*   *   *

We therefore hold that to the extent that Monsanto has an interest in its health, safety, and environmental data cognizable as a trade-secret property right under Missouri law, that property right is protected by the Taking Clause of the Fifth Amendment.

Having determined that Monsanto has a property interest in the data it has submitted to EPA, we confront the difficult question whether a "taking" will occur when EPA discloses that data or considers the data in evaluating another application for registration. The question of what constitutes a "taking" is one with which this Court has wrestled on many occasions. It has never been the rule that only governmental acquisition or destruction of the property of an individual constitutes a taking, for

courts have held that the deprivation of the former owner rather than the accretion of a right or interest to the sovereign constitutes the taking. Governmental action short of acquisition of title or occupancy has been held, it its effects are so complete as to deprive the owner of all or most of his interest in the subject matter, to amount to a taking. *United States v. General Motors Corp.,* 323 U.S., at 378.

See also *Pruneyard Shopping Center* v. *Robins,* 447 U.S. 74 (1980); *Pennsylvania Coal Co.* v. *Mahon,* 260 U.S. 393, 415 (1922).

As has been admitted on numerous occasions, "this Court has generally 'been unable to develop any "set formula" for determining when "justice and fairness" require that economic injuries caused by public action' " must be deemed a compensable taking. *Kaiser Aetna* v. *United States,* 444 U.S. 164, 175 (1979), quoting *Penn Central Transportation Co.* v. *New York City,* 438 U.S. 104, 124 (1978); accord, *Hodel* v. *Virginia Surface Mining and Recl. Assn.,* 452 U.S. 264, 295 [16 ERC 1027] (1981). The inquiry into whether a taking has occurred is essentially an "ad hoc, factual" inquiry. *Kaiser Aetna,* 444 U.S., at 175. The Court, however, has identified several factors that should be taken into account when determining whether a governmental action has gone beyond "regulation" and effects a "taking." Among those factors are: "the character of the governmental action, its economic impact, and its interference with reasonable investment-backed expectations." *PruneYard Shopping Center* v. *Robins,* 447 U.S., at 83; see *Kaiser Aetna,* 444 U.S., at 175; *Penn Central,* 438 U.S., at 124. It is to the last of these three factors that we now direct our attention, for we find that the force of this factor is so overwhelming, at least with respect to certain of the data submitted by Monsanto to EPA, that it disposes of the taking question regarding that data.

A "reasonable investment-backed expectation" must be more than a "unilateral expectation or an abstract need." *Webb's Fabulous Pharmacies*, 449 U.S., at 161. We find that with respect to any health, safety, and environmental data that Monsanto submitted to EPA after the effective date of the 1978 FIFRA amendments—that is, on or after October 1, 1978 —Monsanto could not have had a reasonable, investment-backed expectation that E.P.A. would keep the data confidential beyond the limits prescribed in the amended statute itself.

Monsanto was on notice of the manner in which EPA was authorized to use and disclose any data turned over to it by an application for registration.

Monsanto argues that the statute's requirement that a submitter give up its property interest in the data constitutes placing an unconstitutional condition on the right to a valuable government benefit. But Monsanto has not challenged the ability of the Federal Government to regulate the marketing and use of pesticides. Nor could Monsanto successfully make such a challenge, for such restrictions are the burdens we all must bear in exchange for " 'the advantage of living and doing business in a civilized community.' " *Andrus v. Allard*, 444 U.S. 51, 67 [13 ERC 2057] (1979), quoting *Pennsylvania Coal Co. v. Mahon*, 260 U.S. 393, 422 (1922) (Brandeis, J., dissenting); see *Day-Brite Lighting, Inc. v. Missouri*, 342 U.S. 421, 424 (1952). This is particularly true in an area, such as pesticide sale and use, that has long been the source of public concern and the subject of government regulation. That Monsanto is willing to bear this burden in exchange for the ability to market pesticides in this country is evidenced by the fact that it has continued to expand its research and development and to submit data to EPA despite the enactment of the 1978 amendments to FIFRA.[11] 564 F. Supp., at 561.

Thus, as long as Monsanto is aware of the conditions under which the data are submitted, and the

[11]Because the market for Monsanto's pesticide products is an international one, Monsanto could decide to forgo registration in the United States and sell a pesticide only in foreign markets. Presumably, it will do so in those situations where it deems the data to be protected from disclosure more valuable than the right to sell in the United States.

conditions are rationally related to a legitimate government interest, a voluntary submission of data by an applicant in exchange for the economic advantages of a registration can hardly be called a taking. See *Corn Products Refining Co. v. Eddy*, 249 U.S. 427, 431-432 (1919) ("The right of a manufacturer to maintain secrecy as to his compounds and processes must be held subject to the right of the State, in the exercise of its police power and in promotion of fair dealing, to require that the nature of the product be fairly set forth"); see also *Westinghouse Electric Corp. v. United States Nuclear Regulatory Commission*, 555 F.2d 82.95 (CA3 1977).

B

Prior to the 1972 amendments, FIFRA was silent with respect to EPA's authorized use and disclosure of data submitted to it in connection with an application for registration. Another statute, the Trade Secrets Act, 18 U.S.C. §1905, however, arguably is relevant. That Act is a general criminal statute that provides a penalty for any employee of the United States Government who discloses, in a manner not authorized by law, any trade secret information that is revealed to him during the course of his official duties. This Court has determined that §1905 is more than an "antileak" statute aimed at deterring government employees from profiting by information they receive in their official capacities. See *Chrysler Corp. v. Brown*, 441 U.S. 281, 298-301 (1979). Rather, §1905 also applies to formal agency action, *i.e.*, action approved by the agency or department head. *Ibid.*

It is true that, prior to the 1972 amendments, neither FIFRA nor any other provision of law gave EPA authority to disclose data obtained from Monsanto. But the Trade Secrets Act is not a guarantee of confidentiality to submitters of data, and, absent an express promise, Monsanto had no reasonable, investment-backed expectation that its information would remain inviolate in the hands of EPA. In an industry that long has been the focus of great public concern and significant government regulation, the possibility was substantial that the Federal Government, which had thus far taken no position on disclosure of health, safety, and environmental data concerning pesticides, upon focusing on the issue, would find disclosure to be in the public interest. Thus, with respect to data submitted to EPA in connection with an application for registration prior to October 22, 1972 the Trade Secrets Act provided no basis for a reasonable investment-backed expectation that data submitted to EPA would remain confidential.

The situation may be different, however, with respect to data submitted by Monsanto to EPA during the period from October 22, 1972, through September 30, 1978. Under the statutory scheme then in effect, a submitter was given an opportunity to protect its trade secrets from disclosure by designating them as trade secrets at the time of submission. When Monsanto provided data to EPA during this period, it was with the understanding, embodied in FIFRA, that EPA was free to use any of the submitted data that were not trade secrets in considering the application of another, provided that EPA required the subsequent applicant to pay "reasonable compensation" to the original submitter. §3(c)(1)(D), 86 Stat. 979. But the statute also gave Monsanto explicit assurance that EPA was prohibited from disclosing publicly, or considering in connection with the application of another, any data submitted by an applicant if both the applicant and EPA determined the data to constitute trade secrets. §10. 86 Stat. 989.

Thus, with respect to trade secrets submitted under the statutory regime in force between the time of the adoption of the 1972 amendments and the adoption of the 1978 amendments, the Federal Government had explicitly guaranteed to Monsanto and other registration applicants an extensive measure of confidentiality and exclusive use. This explicit governmental guarantee formed the basis of a reasonable investment-backed expectation. If EPA, consistent with the authority granted it by the 1978 FIFRA amendments, were now to disclose trade-secret data or consider that data in evaluating the application of a subsequent applicant in a manner not authorized by the version of FIFRA in effect between 1972 and 1978, EPA's actions would frustrate Monsanto's reasonable investment-backed expectation with respect to its control over the use and dissemination of the data it had submitted.

* * *

V

We must next consider whether any taking of private property that may occur by operation of the data-disclosure and data-consideration provisions of FIFRA is a taking for a "public use." We have recently stated that the scope of the "public use" requirement of the Taking Clause is "coterminus with the scope of a sovereign's police powers." *Hawaii Housing Authority* v. *Midkiff*, —U.S. —,—(1984)(slip op. 10); see *Berman* v.

Parker, 348 U.S. 26, 33 (1954). The role of the courts in second-guessing the legislature's judgment of what constitutes a public use is extremely narrow. *Midkiff, supra; Berman, supra*, at 32.

The District Court found that EPA's action pursuant to the data-consideration provisions of FIFRA would effect a taking for a private use, rather than a public use, because such action benefits subsequent applicants by forcing original submitters to share their data with later applicants. 564 F. Supp., at 566. It is true that the most direct beneficiaries of EPA actions under the data-consideration provisions of FIFRA will be the later applicants who will support their applications by citation to data submitted by Monsanto or some other original submitter. Because of the data-consideration provisions, later applicants will not have to replicate the sometimes intensive and complex research necessary to produce the requisite data. This Court, however, has rejected the notion that a use is a public use only if the property taken is put to use for the general public. *Midkiff,*—U.S., at—(slip op. 13); *Rindge Co.* v. *Los Angeles*, 262 U.S. 700, 707 (1923); *Block* v. *Hirsh*, 256 U.S. 135, 155 (1921).

So long as the taking has a conceivable public character, "the means by which it will be attained is . . . for Congress to determine." *Berman*, 348 U.S., at 33. Here, the public purpose behind the data-consideration provision is clear from the legislative history. Congress believed that the provisions would eliminate costly duplication of research and streamline the registration process, making new end-use products available to consumers more quickly. Allowing applicants for registration, upon payment of compensation, to use data already accumulated by others, rather than forcing them to go through the time-consuming process repeating the research, would eliminate a significant barrier to entry into the pesticide market, thereby allowing greater competition among producers of end-use products. S. Rep. No. 95-334, at 30-31, 40-41; 124 Cong. Rec. 29756-29757 (1978) (remarks of Sen. Leahy). Such a procompetitive purpose is well within the police power of Congress. See *Midkiff,*—U.S., at—(slip op. 11-12).[18]

[18]Monsanto argues that EPA and, by implication, Congress misapprehended the true "barriers to entry" in the pesticide industry and that the challenged provisions of the law create, rather than reduce, barriers to entry. Such economic arguments are better directed to Congress. The proper inquiry before this Court is not whether the provisions in fact will accomplish their stated objectives.

Because the data-disclosure provisions of FIFRA provide for disclosure to the general public, the District Court did not find that those provisions constituted a taking for a private use. Instead, the court found that the data-disclosure provisions served no use. It reasoned that because EPA, before registration, must determine that a product is safe and effective, and because the label on a pesticide, by statute, must set forth the nature, contents, and purpose of the pesticide, the label provided the public with all the assurance it needed that the product is safe and effective. 564 F. Supp., at 567 and n.4. It is enough for us to state that the optimum amount of disclosure to the public is for Congress, not the courts, to decide, and that the statute embodies

Our review is limited to determining that the purpose is legitimate and that Congress rationally could have believed that the provisions would promote that objective. *Midkiff*—U.S., at—(slip op. 12); *Western & Southern Life Ins. Co. v. State Bd. of Equalization*, 451 U.S. 648, 671-672 (1981).

Congress' judgment on that question. See 123 Cong. Rec., at 25756 (remarks of Sen. Leahy). We further observe, however, that public disclosure can provide an effective check on the decisionmaking processes of EPA and allows members of the public to determine the likelihood of individualized risks peculiar to their use of the product. See H.R. Rep. No. 95-343, at 8 (remarks of Douglas M. Costle); S. Rep. No. 95-334, at 13.

We therefore hold that any taking of private property that may occur in connection with EPA's use or disclosure of data submitted to it by Monsanto between October 22, 1972, and September 30, 1978, is a taking for a public use.

* * *

[WHITE, J. took no part in the consideration or decision of this case.]

Dissenting and concurring opinion of O'CONNOR, J. omitted.

◇ NOTES

1. Where there is a "taking" of private property for a public purpose, the government must provide just compensation to the owner of the property.

2. Does the reasoning of this case mean that a manufacturer has no constitutional protection against a right to know law that requires disclosure of trade secrets unless there is a relevant state or federal law that *expressly* protects those secrets from disclosure? This has been the position of two courts that have examined the issue. In the *Steelworkers* case, discussed above, the Third Circuit Court of Appeals held that OSHA, which had attempted to include a very broad "trade secret" exception to the disclosure provisions in its Hazard Communication Standard, had no federal constitutional duty to protect a manufacturer's trade secrets.

United Steelworkers of America, AFL-CIO v. Auchter

763 F.2d 728 (3d Cir. 1985)

GIBBONS, Circuit Judge.

* * *

It appears that the unarticulated premise of the Secretary's concern about protecting trade secrets is that they may be constitutionally protected from the regulatory process. In *Westinghouse Electric Corp. v.*

United States Nuclear Regulatory Commission, 555 F.2d 82, 95 (3d Cir. 1977), we rejected a constitutional challenge to rules requiring disclosure of proprietary information as a condition to licensing. The *Westinghouse* holding is supported by the Supreme Court's holding in *Ruckelshaus v. Monsanto Co.*, 104 S.Ct. 2862, 52 U.S.L.W. 4886, 4891 (June 26, 1984), that a trade secret voluntarily disclosed to the E.P.A.

in exchange for a registration to sell pesticides is not protected if the owner had notice at the time of disclosure that the E.P.A. was authorized to use all data submitted. These cases suggest that a regulation requiring the disclosure even of formula or process information as a precondition for the sale of hazard-

ous products for use in the workplace would be valid. *See* McGarity and Shapiro *The Trade Secret Status of Health and Safety Testing Information: Reforming Agency Disclosure Policies*, 93 Harv. L. Rev. 837, 864-67 (1980).

* * *

An even more dramatic interpretation of *Monsanto* was offered by Federal District Court Judge Dickinson R. Debevoise of New Jersey in a case dealing with the constitutionality of that state's right to know law.

New Jersey State Chamber of Commerce v. Hughey

600 F. Supp. 606 (D.N.J. 1985)

DEBEVOISE, D. J.

* * *

D. *Trade Secrets:* Plaintiffs contend that the requirement of the Right to Know Act that employers disclose special health hazard substances without trade secret protection will deprive them of property without due process of law.

It will be recalled that the Act contains a procedure by which employers may claim that the presence of designated substances constitutes a trade secret and that if the trade secret claim can be substantiated the substance will not be revealed in the labels and lists to which the public has access. It will also be recalled that, unlike the federal standard, the Right to Know Act provides for a category of particularly dangerous chemicals designated special health hazard substances, as to which employers are not allowed to obtain trade secret protection. Disclosure of these substances, plaintiffs assert, will result in the loss of trade secrets which may have been the product of substantial and costly research endeavors. The forced disclosure of these trade secrets, it is said, will impair or destroy the employer's investment and endanger his ability to compete.

* * *

Following the Supreme Court's lead in *Monsanto*, it must be determined whether the mandated disclosures of trade secrets under the Right to Know Act are "takings" which will trigger a right to compensation. I conclude that they are not.

The factors to be considered in determining whether particular governmental regulation effects a taking include the character of the governmental

action, its economic impact and its interference with reasonable investment-backed expectations. Here the state is acting in an area of great public concern —worker health, environmental effects of hazardous substances in the workplace, community health and safety as it is effected by workplace chemical substances.

No evidence whatsoever has been submitted to show what effect, if any, the Right to Know Act trade secret provisions will have on employers in non-manufacturing sectors. The affidavits in this regard were submitted by employers in the manufacturing sector, and they were couched in vague and conclusory terms. Similarly the affidavits submitted by defendants and intervenors on the trade secret issue were highly generalized, utterly lacking in specificity.

As in *Monsanto*, reasonable investment backed expectations of employers would seem to be determinative. On this question plaintiffs focus on and would have this court rely on the Supreme Court's ruling in *Monsanto* with respect to data submitted during the period from October 22, 1972 through September 30, 1978. However, as described above, the situation during that period was totally dissimilar from the situation in the present case. There during the 1972-1978 period the statute itself gave a registrant the opportunity to protect its trade secrets from disclosure. Registrants submitted trade secret data relying on that statutory guarantee. The 1978 amendment of FIFRA stripped away the protection of the guarantee. That is what the Supreme Court characterized as a taking.

Nothing like that has happened in New Jersey. There has been no antecedent period of disclosure during which the state committed itself to protecting trade secrets. The state has simply adopted a statute and regulations in economic and social areas in which it unquestionably has the power to act. As part of the regulatory scheme disclosure is required which may result in a loss of trade secrets.

This is just the situation which prevailed in the pre-1972 period which was addressed in *Monsanto*. Prior to the 1972 amendment FIFRA was silent with respect to EPA's authorized use and disclosure of registrants' data submitted to it. As in the case of New Jersey's Right to Know Act, there was no preexisting legislation protecting trade secrets submitted by registrants. In such a situation the entity submitting data cannot have a "reasonable investment-backed expectation" that the agency receiving the data will maintain it in confidence. Consequently disclosure of the data ia not a taking for which the state must pay compensation under the Fifth Amendment or the Fourteenth Amendment.

Employers may face the unpleasant choice of disclosing trade secrets or limiting or shutting down operations in New Jersey. This may be a more onerous dilemma than Monsanto faced, but the reasoning in the *Monsanto* case is nevertheless applicable here: as long as the employer is aware of the conditions under which the data are submitted and as long as the conditions are rationally related to a legitimate government interest, a submission under the Right to Know Act does not constitute a taking. 81 L.Ed.2d at 835.

*   *   *

◇ NOTES

1. If "reasonable investment-backed expectation" is to be the criterion, at what point in a manufacturer's innovation process must that expectation be in existence? At the point the product is first formulated? At the point the product is first marketed? At any point at which the manufacturer makes a significant investment decision related to the product?

2. Look again at Section 14(b) of TSCA. Would the release of trade secret information under this section be a taking?

3. Restrictions on Disclosure Once Access Has Been Granted.

One of the ways in which right to know laws may protect interests such as employee confidentiality or employer trade secrets is simply to deny access to the information in question. Quite often, however, some disclosure of the information will be necessary if the underlying purposes of the right to know law are to be fulfilled. In these situations, it may be preferable to condition access to the information on an agreement by the person receiving it to refrain from disclosing the information to others or to refrain from using the information in certain ways. This is a common method of protecting information. The OSHA Hazard Communication Standard, for example, permits employers to require that workers (and their treating physicians) who receive trade secret information under the standard sign confidentiality agreements, in which they promise not to disclose the information to others and promise to pay specific monetary damages if they do. 29 CFR §1910.1200 (i)(3)(v). If restrictions of this nature are narrowly drawn to serve a substantial government interest—such as the protection of an employer's legitimate trade secrets or an employee's confidential medical information—they likely will be upheld by the courts. As the following case illustrates, however, restrictions that attempt a broader sweep may well run afoul of the constitutional right to freedom of speech.

Lawlor v. Shannon

18 Env. Law Rptr. 21496 (D. Mass 1988)

McNAUGHT, District Judge.

The plaintiffs here are six individuals and the Massachusetts Public Interest Research Group ("MASSPIRG"), a non-profit corporation organized under the laws of Massachusetts. They seek declaratory and injunctive relief against the Attorney General of the Commonwealth of Massachusetts, the Commissioner of the Massachusetts Department of Public Health ("DPH"), the Commissioner of the Massachusetts Department of Labor and Industries ("DLI") and the Commissioner of the Massachusetts Department of Environmental Quality Engineering ("DEQE") in their official and individual capacities.

The underlying action centers on Massachusetts General Laws c. 111F, the so-called "right to know" law, which went into effect September, 1984. In general, c. 111F requires all Massachusetts employers, whether public or private, who manufacture, process, use or store any "toxic or hazardous" substances to inform their "employees" and, in certain circumstances "community residents" (as those terms are defined in sec. 1 of c. 111F) of the nature and effect of such substances present in their workplace or community.

* * *

Specifically at issue here is the validity of section 21(b) of c. 111F, which places certain restrictions on the disclosure of information obtained under the law. Section 21(b) reads:

> (b) Any person who has obtained possession of or has access to information pursuant to the provisions of this chapter shall not disclose said information to any person not specifically authorized to receive it hereunder. Violation of this provision shall be punishable by a fine of not more than five thousand dollars or imprisonment for not more than one year, or both. Provided, however, that disclosure by an employee to his or her spouse, or to a fellow employee exposed to the same toxic or hazardous substance as is involved in the disclosure, shall not constitute a violation of this chapter.

Plaintiffs contend that unlike any of the other jurisdictions that have enacted "right to know" laws, such as New Jersey, Philadelphia and Cincinnati,

Massachusetts forbids those who use the law from communicating health and safety information to others. For example, they argue that under the terms of the statute's broad disclosure restriction, those who obtain information under c. 111F may not disclose such to their doctor, lawyer, neighbor or selectman. Claiming that section 21(b) was inserted to prevent open public discussion of issues related to the industrial use of hazardous chemicals, plaintiffs assert that the provision's prohibition and corresponding criminal penalties violate the First Amendment to the United States Constitution. They also contend that the section violates the Equal Protection clause of the Fourteenth Amendment and that it conflicts with regulations promulgated by the federal Occupational Health and Safety Administration (OSHA) and with a community right to know statute recently passed by Congress. Plaintiffs seek a declaratory ruling that the disclosure restriction is unconstitutional and an injunction preventing defendants from enforcing it against plaintiffs or others.

Terming this "essentially a First Amendment case," plaintiffs argue that the Constitution's strong presumption in favor of free and open communication of information and ideas renders section 21(b) invalid. They note that there is no interest in using right to know information for commercial purposes. Rather, plaintiffs claim to seek free communication of the information so that they and others may use it for public education regarding the potential dangers from hazardous chemical substances, for public discussion and debate of issues related to the use of these substances in Massachusetts and for organizing community campaigns directed toward these issues. The underlying focus of this communication, plaintiffs contend, is therefore of undeniable public importance.

That being the case, plaintiffs assert that the free speech interests in this case strike close to the heart of the rights clearly intended to be protected by the First Amendment. They argue, accordingly, that the restriction of these interests through section 21(b) of c. 111F must be soundly and persuasively justified. According to plaintiffs, defendants cannot meet the burden of justification. Plaintiffs contend that defendants have offered no concrete evidence to indicate that this disclosure restriction actually serves any legitimate governmental interest. Rather, the evidence strongly suggests that the primary purpose of section 21(b) is to discourage and suppress precisely the kind of public discussion and debate that plaintiffs seek to promote.

In support of this position, plaintiffs point to what they term is defendant's principal evidence regarding the nature of the governmental interests allegedly served by section 21(b), the affidavit of Richard E. Mastrangelo, General Counsel of the Associated Industries of Massachusetts (AIM). Plaintiffs argue that the affidavit tells nothing of the legislative intent behind section 21(b) and that it in fact confirms the suspicion that the chief purpose of the restriction is the curtailment of activities protected by the First Amendment. For example, at page four, Mr. Mastrangelo warns that:

> uncontrolled dissemination of MSDS's could exacerbate the already overly emotional issue of the use of chemicals in a community and prevent a rational focus on future regulation of such chemicals.

Such a statement, argue the plaintiffs, compels the conclusion that the business community sought to have section 21(b) added to the right to know law because it wanted to control the "focus" of political discussion in the Commonwealth on an issue of clear public concern. Later, Mr. Mastrangelo cites concerns that the dissemination of MSDS information "could promote frivolous lawsuits which would be economically damaging to employers in the Commonwealth." (p. 5) Plaintiffs call such an assertion "paternalistic disdain for the common sense of the average citizen of Massachusetts." Sixth Amendment and due process considerations aside, they argue that this is but another attempt to suppress protected speech.

Defendants take issue with a number of plaintiffs' contentions. They refer to c. 111F as a "carefully crafted statute that delicately balances, on the one hand, the interests of employees and community residents in having access to certain information which would enable them to protect their own health and safety . . . and, on the other hand, the legitimate proprietary, privacy, and public interests of business and government by protecting against the unauthorized disclosure of information to the general public." Defendants argue that section 21(b) is valid on its face and as applied because it furthers an important government interest unrelated to the suppression of expression and that the restrictions in section 21(b) are no greater than are necessary to further that government interest.

According to defendants, the line of First Amendment cases which is most analogous to the present action is that dealing with protective orders entered regarding discovery under F.R. Cv. P. 26. In support of this position, defendants rely most heavily on *Seattle Times Co. v. Rhinehart*, 467 U.S. 20 (1984).

Seattle Times involved the entry of a protective order during discovery to protect one of the parties' financial and privacy interests. The order prohibited the defendant newspaper from publishing or using the proscribed information in any way except as necessary to prepare and try the case. The newspaper argued that the protective order violated its First Amendment rights to free speech. The Court disagreed. It noted that the newspaper had obtained the information not because of any First Amendment right of access, but rather through the discovery processes, which are "a matter of legislative grace." 467 U.S. at 32 Concluding that the protective order rule furthers a substantial government interest and rejecting a heightened First Amendment scrutiny, the Court upheld the protective order.

Applying this "bitter with the sweet" principle to the present action, defendants argue that even if section 21(b) is considered a restraint on free expression, it nonetheless is true, as in *Seattle Times*, that an employee or community resident who obtains information under c. 111F does so not because of any First Amendment right of access, but by legislative grace. Defendants note that prior to the enactment of the right to know law, citizens had no right under Massachusetts law to such information because it was considered to be "not traditionally public", to borrow a phrase from the *Seattle Times* decision. From this point of view, defendants argue that the present action "is not the classic prior restraint that requires exacting First Amendment scrutiny." *Seattle Times*, 467 U.S. at 32.

Under this less exacting First Amendment scrutiny, defendants claim that in its dual role as a government and an employer, the Commonwealth has legitimate concerns regarding section 21(b). Among them are:

1. the disruption of essential services if information fell into the wrong hands;

2. that the alteration of MSDS's would misinform and/or endanger the public and perhaps injure a business' reputation;

3. that public safety could be jeopardized by crime and industrial sabotage; and

4. that fear of unwarranted disclosure might discourage employers from speaking early and honestly with DPH, DLI and/or DEQE in a

potentially dangerous situation affecting public health or safety.

Arguing that all of these concerns show a substantial governmental interest, and noting all the while that c. 111F is not a freedom of information act, defendants argue that the court should defer to the legislature's weighing of the competing needs and interests of the parties affected by that statute and rule that c. 111F is valid.

Upon review and consideration, I agree with the plaintiffs that the disclosure restriction in the Massachusetts right to know law must be evaluated on its own terms, and not by analogy to the civil discovery process addressed in *Seattle Times.*

The contrast between the two situations is striking. In the case of civil discovery, the potential for abuse is significant. As the Court noted, "[t]his abuse is not limited to matters of delay and expense; discovery may also seriously implicate privacy interests of litigants and third persons." *Seattle Times,* 467 U.S. at 34-5. In upholding the protective order in *Seattle Times,* the Court was careful to note that "judicial limitations on a party's ability to disseminate information discovered in advance of trial implicates the First Amendment rights of the restricted party to a lesser extent than would restraints on dissemination of information *in a different context.*" 467 U.S. at 34. (emphasis supplied)

We have here a "different context," for the nature of the information sought to be disseminated under c. 111F is contained in the statute itself, and involves no substantial privacy interest of the employer. In fact, under sec. 5 of c. 111F, manufacturers may request trade secret status from the DPH for certain information they wish to have omitted from the MSDS. According to the agreed statement of facts submitted by the parties, DPH received 1,134 such requests between November 26, 1984 and May 1, 1987. It has denied but two.

Another factor mitigating against the application of *Seattle Times* to this set of facts is the mechanism for restricting the distribution of information. Referring to "repeated expressions of concern about undue and uncontrolled discovery," the Court therein noted that "reliance must be had on what in fact and in law are ample powers of the district judge to prevent abuse." 467 U.S. at 35 n. 20, *quoting Herbert v. Lando,* 441 U.S. 153 (1979). Rule 26(c) protective orders, then, are to be issued for good cause on a case-by-case basis in the sound discretion of the trial judge. Such is not the case with the right to know law which, in section 21(b), contains what is essentially a blanket prohibition against disclosure, with no opportunity

for the sound discretion of the trial judge to enter the formula on a case-by-case basis. The attendant potential for the subjugation of First Amendment interests is, indeed, far greater here than in the *Seattle Times* scenario.

Having determined that the present action does not fall within the ambit of a protective order under Rule 26(c), a more traditional First Amendment analysis requires a finding that section 21(b) of c. 111F is an unconstitutional abridgment of protected speech.

The restriction on dissemination of information set out in section 21(b) is undeniably a regulation of the content of speech; it therefore "must be scrutinized more carefully" than mere time, place or manner restrictions. *Consolidated Edison Company v. Public Service Commission of New York,* 447 U.S. 529 (1980). In determining whether section 21(b) survives such scrutiny, the court applies the test set forth in *United States v. O'Brien,* 391 U.S. 367 (1967). The Court therein wrote:

To characterize the quality of the government interest which must appear, the Court has employed a variety of descriptive terms: compelling; substantial; subordinating; paramount; cogent; strong. Whatever imprecision inheres in these terms, we think it clear that a government regulation is sufficiently justified if it is within the Constitutional power of the Government; if it furthers an important or substantial government interest; if the government interest is unrelated to the suppression of free expression; and if the incidental restriction on alleged First Amendment freedoms is not greater than is essential to the furtherance of that interest.

391 U.S. at 376-77.

The first element of this four-pronged test is met here. Plaintiffs concede that promulgation of section 21(b) is within the police power of the Commonwealth. Based on the record and counsel's argument, however, defendants utterly fail to demonstrate that section 21(b) "furthers an important or substantial government interest." They repeatedly refer to the risk of sabotage or robbery without offering evidence that such concerns are legitimate. Plaintiffs point out that if risks of this nature actually existed, one could assume that other states with right to know laws would by now have experienced them. *See e.g., New Jersey Worker and Community Right to Know Act,* N.J. Stat. Ann., sec. 34:5A-1 to -31 (West 1984). Yet defendants offer no such evidence. In addition, several of the reasons articulated by defendants in support of the restriction are not "unrelated to the suppression of expression." For example, defendants argue that section 21(b) helps insure compliance with the whole of c. 111F. However, they offer no evidence to suggest that the traditional methods of insuring compliance, such as enforcement, education, and penalties for non-compliance, would be unworkable within the framework of the statute. The affidavit of Mr. Mastrangelo, supra, is likewise devoid of any factual bases for the largely conclusory statements contained therein.

Based on the foregoing, defendants' motion to dismiss is hereby denied; there being no genuine issues of material fact in controversy, plaintiffs' motion for summary judgment is granted. Furthermore, I hold section 21(b) of c. 111f to be unconstitutional on its face and order defendants to cease its enforcement.

I am left to determine the effect of this order on the remaining enforceable sections of the statute. The legislature included the following clause in the overall statutory scheme of c. 1117:

> If any section of this act shall be held unconstitutional either on its face or as applied, the unconstitutionality of that section shall not affect the remaining sections of this act.

Section 3 of chapter 470 of the Acts of 1983.

[Given] such legislative intent, coupled with the fact that the remaining sections of c. 111F "form a coherent and comprehensible regulatory system" when standing alone (*see Planned Parenthood League of Massachusetts v. Bellotti* 641 F.2d 1006, 1023 (1st Cir. 1981), I hold that section 21(b) is severable from c. 111F and that the remainder of the statute shall remain unaffected by this order.

CHAPTER 8

Retaliatory Employment

Practices

Improving workplace health and safety depends to a large extent on the willingness of employees to speak out, in various ways, in favor of making the necessary changes in working conditions. Similarly, attaining product safety and environmental protection depends to no small degree on the willingness of employees—especially engineers, scientists, and those in managerial positions—to insist that safety and environmental considerations play a significant role in product and process design and development. Experience and common sense tell us, however, that employees will be far less willing to take such action if their employer is free to fire or otherwise discriminate against them in return. This chapter explores the legal protections available to employees in these situations.

In addition, this chapter examines the protections available to workers who are discriminated against because they are deemed to be particularly susceptible to occupational disease or injury. This situation raises an issue that is central to the debate over occupational safety and health: Must the employer reduce workplace hazards to a level that is safe for all workers, or may the employer simply exclude from the workforce all those workers deemed especially sensitive to certain hazards?

A. A STARTING POINT: FOUR POTENTIAL WORKPLACE FREEDOMS

In addressing the general topic of retaliatory employment practices, we begin by considering the kind of freedom from retaliation that the "rational" employee might desire. This employee would probably suggest at least four basic freedoms.

The first would be *the freedom to avoid or to abate workplace hazards to oneself and one's fellow workers.* This would include both the right to remove oneself from the job site in the event that a serious health or safety hazard was present and the right to take lawful action in an attempt to eliminate the hazard, without fear of employer retaliation.

The second would be *the freedom to speak out about company policies with which one disagrees.* Conceivably, this could include a wide array of issues touching on the various internal and external activities of the employer. For the purposes of this text, however, we are concerned mainly with the freedom to speak out on issues pertaining to workplace technology and its effects on safety and health, to the company's environmental practices, or to the safety of the company's products. At its fullest expression, this freedom would include both the ability to speak out *within the company* and the ability to *take one's message to the public* (often characterized as a "whistleblower" situation).

The third freedom would be *the freedom to refuse to work on projects that one believed to be unethical or socially irresponsible.* Many electrical engineers, for example, have insisted that their employment contracts with engineering consulting firms specify that they are not required to work on projects funded by the United States Department of Defense. At *its* fullest expression, presumably, this freedom would encompass the right to refuse to work on any project that offended one's own sense of morality or social propriety.

These three freedoms all involve some affirmative decision on the part of the employee to take, or to refrain from taking, a particular course of action. Such decisions are more or less "issue-oriented," as they all involve a choice by the employee to do something designed to better his or her own lot or the lot of society generally. But there is a fourth freedom that, while it does not fit well into this general paradigm, might nonetheless be important to our hypothetical employee. This would be *the freedom from arbitrary or unjustified management decisions regarding one's job security.*

With these four freedoms in mind, then, let us now turn to the law. To what extent do legal protections serve to secure these freedoms for the American worker?

B. THE DECLINE OF THE EMPLOYMENT AT WILL DOCTRINE

To answer this question, we must begin with the venerable employment at will doctrine. Under this doctrine, which was the indisputable common law rule in this country until rather recently, *none* of our four freedoms would be present in an employment contract of an unspecified term. Simply put, the employment at will rule presumes that employment for an unspecified period may be terminated by either party, at any time, for any reason whatsoever. Although employment at will was not fully articulated as a judicial "rule" until the waning years of the nineteenth century, it was the prevailing common law doctrine throughout the United States by the early twentieth century.

For a time, it even enjoyed *constitutional* support. In *Adair v. United States,* 208 U.S. 161 (1908), the Supreme Court invalidated a federal law prohibiting employment discrimination based on union membership. Such an infringement of the employer's right of contract, said the court, violated the guarantees of "personal liberty [and] the right of property" found in the due process clause of the Fifth Amendment. Similarly, in *Coopage v. Kansas,* 236 U.S. 1 (1915), the court struck down a state law prohibiting employers from requiring their workers to agree (in what is commonly called a "yellow dog" contract) not to join a union. This wholesale constitutional protection of employer rights was short-lived, however. In 1930, the court upheld provisions in the federal

Railway Labor Act prohibiting the discharge of employees for union activity. Shortly thereafter, it upheld similar provisions in the National Labor Relations Act. See: *Texas & New Orleans Railroad v. Brotherhood of Railway Clerks*, 281 U.S. 548 (1930); and *NLRB v. Jones & Laughlin Steel Corp.*, 301 U.S. 1 (1937). Although these decisions did not do away with the employment at will rule, they made it clear that the rule could be modified by appropriate state or federal action.

Interestingly, the development of the employment at will doctrine in the United States was due in large part to one anomalous law review article. In fact, the concept ran counter to the law of employment contracts in Great Britain.

The Development of the Employment at Will Rule

J. M. Feinman

American law originally adopted the rules of English law on duration of service contracts. Toward the end of the nineteenth century, however, English and American law diverged. While the English used presumptions of long-term hiring and required reasonable notice of termination, American lawyers and courts developed the rule of termination at will.

* * *

English courts at an early point developed a relatively sophisticated approach to the termination of master-servant relationships. They identified two questions: What is the duration of the relation presumed to be when none is specifically stated? What length of notice must be given before the relation can be terminated? The English, unlike the Americans, saw that the questions were not the same and eventually developed a response to the second that mitigated somewhat the strictness of the early response to the first.

The duration of service relationships was a concern in early stages of English law, but the law was best formulated and made prominent only with the statement of a rule and policy by Blackstone:

> If the hiring be general, without any particular time limited, the law construes it to be a hiring for a year; upon a principle of natural equity, that the servant shall serve, and the master maintain him, throughout all the revolutions of the respective seasons, as well when there is work to be done as when there is not.

The rule thus stated expressed a sound principle: injustice would result if, for example, masters could

Source: From 20 *Am. J. Legal Hist.* 118 (1976). Temple University School of Law, Philadelphia, PA.

have the benefit of servants' labor during planting and harvest seasons but discharge them to avoid supporting them during the unproductive winter, or if servants who were supported during the hard season could leave their masters' service when their labor was most needed. But the source of the yearly hiring rule was not solely, as might be supposed from Blackstone's statement, in the judges' concern for fairness between master and servant. The rule was also shaped by the requirements of the Statutes of Labourers, which prescribed a duty to work and prohibited leaving an employment or discharging a servant before the end of a term, and by the Poor Laws, which used a test of residence and employment to determine which community was responsible for the support of a person. Thus, despite a concern with the "revolution of the seasons," the rule articulated by Blackstone was not restricted to agricultural and domestic workers. The presumption that an indefinite hiring was a hiring for a year extended to all classes of servants. Because the rule was designed for domestic servants broadly construed, however, those who were clearly not in that group were sometimes excluded. The types of employment now considered usual—where the hours or days of work were limited, or the employment only for a certain job—would sometimes be held not to import a yearly hiring, indicating some sophistication by the law in not extending a concept designed for one purpose beyond its reasonable reach.

* * *

As the law was faced with an increasing variety of employment situations, mostly far removed from the domestic relations which had shaped the earlier law, the importance of the duration of contract question diminished and the second issue, the notice required to terminate the contract, moved to the fore. Even when they recognized hirings as yearly ones, the courts refused to consider the contracts as entire and instead developed the rule that, unless specified otherwise, service contracts could be terminated on reasonable notice.

What constituted reasonable notice was a question of fact to be decided anew in each case, but certain conventions grew up. Domestic servants, who presumably no longer needed the benefit of the seasons, could be given a month's notice. Other types of employees could also be given a month's notice; three months was another common term, although some special cases required six or even twelve months' notice. Although notice was a separate question in each case, the custom of the trade was often determinative.

*   *   *

While English law followed a relatively clear path, American law at the same time exhibited a confusion of principles and rules. Through the middle of the nineteenth century, American courts and lawyers relied heavily on English precedents but often came to different results.

In colonial times some hirings, such as of day laborers, were conventionally terminable at will. Agricultural and domestic service relations often followed the English rule of yearly hirings. In the nineteenth century, however, whatever consensus existed about the state of the law dissolved. For example, Tapping Reeve's pioneering domestic relations treatise at mid-century stated the English presumption of yearly hiring but noted that no such rule existed in Connecticut. But, in the same year that Reeve's second edition was published, the New York Court of Appeals held that the English rule of yearly hiring was still in effect in New York, even giving Blackstone's "benefit of the seasons" rationale. Shortly thereafter Charles Manly Smith's treatise on master and servant, the first devoted solely to the subject in the United States, was published in Philadelphia. Covering English law with reference to American cases, Smith's treatise was noted for its exhaustive discussion of the intricacies of the law and it had significant impact for that reason. Smith stated a presumption that a general hiring was a yearly hiring for all servants, the presumption was rebuttable by custom or other evidence, and, in spite

of a yearly hiring, the relation was terminable on notice where that was customary.

The confusion over the nature of the field of master and servant law contributed to the confusion over the duration of service contracts issue. Master and servant law was originally classed as a domestic relation. The master-servant relation was personal, often familial; servants were described as "menial" not derogatorily but because they resided *intra moenia*, within the walls of the master's house. As the nineteenth century progressed, however, the true master-servant relation became overshadowed by the number of employees whose relationship to their employers was essentially commercial and therefore did not fit the pattern. The resulting tension influenced the direction of the law, with the earlier perception acting as a force delaying accommodations to new economic conditions.

*   *   *

By the 1870's the dissolution of the earlier law was apparent. Although the presumption of yearly hiring was recognized as anachronistic, the concept of reasonable notice had not caught on. Attempts were made to provide new, more fitting rules. . . .

*   *   *

Thus the time was ripe for a sure resolution of the problem; it was achieved by an Albany lawyer and prolific treatise writer named Horace Gray Wood. Wood sliced through the confusion and stated the employment at will doctrine in absolutely certain terms:

> With us the rule is inflexible, that a general or indefinite hiring is prima facie a hiring at will, and if the servant seeks to make it out a yearly hiring, the burden is upon him to establish it by proof. . . . [I]t is an indefinite hiring and is determinable at the will of either party, and in this respect there is no distinction between domestic and other servants.

The puzzling question is what impelled Wood to state the rule that has since become identified with his name. Wood's master and servant treatise, like his other works, won him acclaim for his painstaking scholarship, but that comprehensiveness and concern for detail were absent in his treatment of the duration of service contracts. First, the four American cases he cited in direct support of the rule were in fact far off the mark. Second, his scholarly disingenuity was extraordinary; he stated incorrectly that no American courts in recent years had approved the English rule, that the employment at will rule was inflexibly applied in the United States, and that the English rule was only for a yearly hiring, making no mention of notice. Third, in the absence of valid legal support, Wood offered no policy grounds for the rule he proclaimed.

Whatever its origin and the inadequacies of its explanation, Wood's rule spread across the nation until it was generally adopted.

Even at the height of the influence of the employment at will doctrine, employers and employees have been free to agree to include in the employment contract terms providing greater protection for the employee. Most important among these contractual exceptions to the employment at will rule, at least historically, have been the provisions secured by organized labor and memorialized in collective bargaining agreements. More recently, however, there has been a flurry of activity on the legislative and common law fronts. Indeed, it has now become fashionable to question whether the employment at will rule actually remains the prevailing common law doctrine. Much of the remainder of this chapter explores a number of the important legislative and common law exceptions to the rule.

The following article from *Business Week* provides an excellent—if somewhat topical—introduction to this analysis. Perhaps not surprisingly for a business-oriented journal, the authors tie the "revolution in employee rights" to a corresponding decline in the influence of unions.

Beyond Unions: A Revolution in Employee Rights Is in the Making

J. Hoerr, W.G. Glaberson, D.B. Moskowitz, V. Cahan, M.A. Pollock, and J. Tasini

The growth of unions . . . means that the United States is gradually shifting from a capitalistic community to a laboristic one—that is, to a community in which employees rather than businessmen are the strongest single influence.

—Sumner H. Slichter, Harvard University, 1946

Once hailed for its foresight, this provocative statement by the most prominent labor economist of his time now seems merely another example of clouded vision in academia. Although unions had organized 30% of the nonfarm work force by 1946—up from only 13.5% in 1935—they declined just as fast in the 1970s and 1980s. Obviously, Slichter didn't anticipate the huge shift of employment from manufacturing to services, among other things. But his assertion that an "employee class"—union and non-

Source: Reprinted from July 8, 1985, issue of *Business Week* by special permission, copyright © 1985 by McGraw-Hill, Inc.

union—would be influential in shaping public policy on employment practices is truer today than ever. The weakening of unions is stimulating the growth of an embryonic employee-rights movement that will forge revolutionary changes in the workplace and in the way companies manage people.

Workers are not yet clamoring for these rights through collective action. The movement has no picket lines and no Walter Reuthers or Jimmy Hoffas. But employee dissatisfaction with the boss-worker relationship in many companies is showing up clearly in polls, surveys, and a growing volume of court suits. In today's nonunion climate, the courts and state legislatures are becoming the most effective champions of employee rights.

UPPER HAND

Organized labor's share of the work force is down to 19% and dropping fast. If this trend persists—and nothing on the horizon appears likely to change it—by the year 2000 unions will represent only 18%

of all nonfarm workers. Labor's political and bargaining muscle will continue to atrophy.

Employers have gained the upper hand in part by using sophisticated tactics, sometimes illegal, to defeat organizing attempts. But a rude shock is in store for managers who think that fading union strength will enable them to deal with workers in a free and unfettered way. Says Paul C. Weiler, a labor law specialist at Harvard: "Some business leaders think they will get a union-free environment, but what they may get is a legalized environment."

*   *   *

WHISTLEBLOWERS

Many states are passing laws protecting whistleblowers—workers who report legal violations by their employer or who refuse to carry out orders that violate a public policy, such as an environmental law. Right-to-know laws requiring employers to identify hazardous substances used on the job are spreading. Nineteen states have already decreed that companies may not force workers to retire at any age. Dozens of states have statutes protecting employee privacy, including restrictions on using polygraph tests to screen job applicants.

These state laws add an important new layer to the major antidiscrimination laws and other labor legislation of the 1960s and 1970s. Employers may no longer hire, fire, and promote people on the basis of skin color, sex, age, religion, or—in some states and cities—marital status or sexual orientation. Other laws require that companies meet government safety and health standards, vest and fund benefit plans under federal and state rules, and adhere to complex rules for scheduling hours of work and bargaining with unions.

At the same time, state courts are opening up a new area of company decision-making to review by judges and juries: Management's ability to fire employees for reasons unrelated to discrimination. The time is coming when nonunion employees will no longer serve entirely at the employer's will—the so-called employment-at-will doctrine that has prevailed in the U.S. since the late 1800s. California is likely to pass the first state law prohibiting unjust dismissal by private employers, and similar legislation is pending in four other states. Slowly but inexorably, judicial and legislative law is recognizing that even nonunion employees have an implicit employment contract that is enforceable in the courts.

REPUGNANT NOTION

Judges, in particular, appear to be changing the law to reflect the values of "a nation of employees," as Slichter put it in the 1940s. Indeed, the New Jersey Supreme Court two months ago used this precise phrase to help explain why it departed from the employment-at-will doctrine in ruling that an employee was wrongly discharged. Concludes John T. Dunlop, the economist and former Labor Secretary, who was a colleague of Slichter: "I think the notion that an employer can get out of bed and fire anybody for any old reason is repugnant to a society of employees, whether they are organized into unions or not."

This trend is moving the U.S. closer to the European practice of legislating many of the terms of employment. In the process, the U.S. is drifting away from the collective bargaining system that has dominated employee relations and wage patterns since the 1930s. It was set in place by the National Labor Relations Act of 1935, which encouraged peaceful bargaining as a national policy. Government acted only as an external regulator of the system, while unions and companies negotiated the terms of their employment relationship in private. And organized labor became so firmly entrenched in leading industries that its wage bargains tended to set pay levels throughout the economy.

The new state statutes and court rulings deepen government involvement in the workplace. Except for setting the minimums for hourly wages and overtime rates, the government still doesn't determine pay levels. But all the other rules combined do more than simply nibble around the edges of employer decision-making: They add up to important limits on management. "This represents a very dramatic shift in philosophy," says Theodore J. St. Antoine, a law professor at the University of Michigan. "Now government is imposing substantive terms on the conditions of employment." Furthermore, warns John T. Joyce, president of the International Union of Bricklayers, employers face "a choice between collective bargaining or public laws regulating the workplace."

Employers both gain and lose flexibility because of this shift. On the one hand, they need no longer base their wage and salary decisions primarily on remaining competitive with union patterns. But increased government regulation in other areas will be costly. Unjust-dismissal laws alone could have an enormous impact on management methods, forcing changes in hiring, evaluation, and termination procedures. "The cost to employers would be gargantuan," says Philadelphia attorney David F. Girard-diCarlo of Blank, Rome, Comisky & McCauley, which represents about 100 large corporations.

LITTLE HEADWAY

Despite the steady erosion of union power, organized labor will rmain deeply embedded in industries such as autos, mining, steel, construction, retail food, railroads, airlines, and trucking. But while these old bastions of strength are crumbling, unions are making little headway in organizing such growth industries as financial services and high technology. Union membership in private industry has been dropping even faster than in the economy as a whole, plunging from 85.7% of the nongovernmental work force in 1958 to less than 17% today.

Labor's membership problem can be simply put: Its share of the nonfarm, private work force is shrinking annually by 8% per year as members retire or lose their jobs in the smokestack industries, according to labor economist Richard B. Freeman of Harvard. But unions are winning only 0.3% of the work force every year in representation elections. Unless something unanticipated—such as another depression or dramatic breakthroughs in organizing white-collar and clerical employees—broadens labor's appeal, the long-term decline will continue.

Several factors explain the unions' dwindling ability to organize new members, according to Freeman and Henry S. Farber of the Massachusetts Institute of Technology. Unions currently are winning only 45% to 50% of representation elections, down from 75% in 1950. The shift of employment from manufacturing to services, and from the Northeast to the South, along with the increase in white-collar and female employment, accounts for about 40% of this declining success rate, the economists say. Some 20% is attributable to the unions' reduced organizing efforts. But 40% of the problem, Freeman says, comes from increased management opposition. This includes the illegal firing of union activists to defeat organizing campaigns.

The perception that unions are declining as a force in society may be one reason state courts are increasingly willing to put the employee above the private property interests of employers. Rarely before, for example, had company officials been held criminally liable for job-related health problems. But in an unprecedented verdict on June 14, a county judge in Chicago found three company executives guilty of murder in the death of a worker who had inhaled cyanide fumes.

Growing numbers of well-educated, white-collar employees are adding to the pressures for protective labor laws, notes Denver attorney Warren L. Tomlinson in the May issue of the *Labor Law Journal*. He predicts an increase in state laws on health and safety and employee privacy. In mid-June, Oregon became the first state to pass a law setting health guidelines for the use of video display terminals by state and municipal employees. The Service Employees International Union and 9 to 5, National Association of Working Women campaigned for the law and are pushing similar legislation in 19 other states.

DEEP DISCONTENT

However, nonunion employees are also making themselves heard on work issues, through company surveys and public opinion polls. A recent Business Week/Harris Poll revealed a startling undercurrent of dissatisfaction with management, even among professional employees and executives. Only 36% of the professionals and 35% of executives who approached management with work-related problems were "very satisfied" with the solution. Indeed, 38% of the professional people polled said they had protested work conditions as members of a group, but only 24% had received real satisfaction. "The lack of satisfaction is ample evidence that the seeds of discontent are very deep," Harris says.

Company surveys show that employees want more than pay, benefits, and job security. Today's workers, says Walton E. Burdick, personnel vice-president at International Business Machines Corp., are interested in "participation in the destiny of the company, the public responsibility of the firm, the quality of management, and the content of the job." These demands are converging with competitive pressures to produce a trend that is related to the employee-rights push. It involves reorganizing work and giving workers more control over their jobs. The movement is taking place in thousands of factories, stores, and offices, and it is likely to become a permanent part of the U.S. industrial relations system.

Among nonunion companies, meanwhile, more and more managements are patterning their labor policies on models provided by such leaders as IBM and Hewlett-Packard Co., which have been committed to good employee relations from their founding. This means paying competitive salaries and benefits, providing amenities such as recreation facilities, involving workers in decision-making, and treating them fairly. Nonunion companies, in particular, are instituting formal systems, similar to union grievance procedures, that enable workers to resolve job-related problems. And no-layoff policies are spreading.

Materials Research Corp., of Orangeburg, N.Y., a supplier to such computer manufacturers as IBM, adopted a no-layoff policy four years ago. During the recession of 1981–82, MRC kept 100 excess employees on the staff at a cost of $4 million, and the company vows to retain the policy during the current computer slump. The increased employee loyalty to the company has reduced turnover among the 1,000 employees to less than 1% per month, thereby cutting the cost of hiring new people. "I tell every new employee that we have a contractual relationship," declares Sheldon Weinig, chairman and founder of MRC. "If they do a first-class job, pursue further education, and are willing to work heavy overtime, they have a job for life."

The U.S. has by no means seen the end to management exploitation of workers and labor conflict. However, Alan F. Westin of Columbia University, who has studied employee-rights issues for 30 years, says companies won't be able to return to the practices of the 1940s and 1950s as the threat of union organizing diminishes. Even then, the employment-at-will concept was accepted by workers and managers alike. But, comments Westin, "there has been such a deepening of the concept of employee rights that even most managers believe in it." Westin adds: "Today the ordinary American feels that fair treatment means that he can be fired for lack of business but not for lack of cause by abusive managers."

Some 60 million nonunion workers are employed "at-will," according to Jack Stieber, director of Michigan State University's School of Labor & Industrial Relations. About 100,000 to 150,000 of these workers are unfairly fired each year, Stieber calculates, but few have enough money or access to legal advice to undertake a court fight. Most court suits are brought by executives and other white-collar employees.

'HOTTEST TOPIC'

Nevertheless, wrongful-discharge suits are proliferating in the courts. Richard G. Moon, a Portland (Me.) management lawyer, estimates that 5,000 to 10,000 such cases are initiated each year. The majority are dismissed, but not without some expense for the employer. "This is the hottest topic in the law field," Moon adds. Meanwhile, the federal Equal

Employment Opportunity Commission and state and local human relations agencies in fiscal 1983 received 120,000 complaints of job discrimination, a 30% increase over 1982.

The increasing litigiousness on the part of workers is sending a message to the courts. "We get our social status; our pension, and our ability to live a decent life from work," says Columbia's Westin. "So the courts are saying there's a stigma to being discharged, and today they're more and more willing to protect individual employees from the capricious actions of management."

The employment-at-will doctrine is likely to remain stubbornly alive in many states. But the increasing court action is putting pressure on legislatures to codify the judicial law in statues. Bills prohibiting firings except for "just cause"—criminal acts, drunkeness, incompetence, excessive absenteeism, and the like—are pending in California, New York, Pennsylvania, Tennessee, and Massachusetts. One will be introduced soon in Michigan. The bills are much broader than the technical decisions so far handed down by the state courts, and most employers argue that this legislation would force them to keep incompetents on the job.

WARDING OFF DAMAGES

Kenneth J. McCulloch, of the New York law firm of Townley & Updike, believes that "a large number" of states will pass unjust-dismissal laws during the next decade. By the mid-1990s, he predicts, there will be a federal unjust-dismissal law. Other experts doubt that the trend will go so far so fast. Yet all agree that increasingly large punitive damages awarded to wrongly fired employees add to the pressures for legislation. An unjust-dismissal law would probably let workers appeal to a mediation or arbitration agency but would also bar them from suing the employer. "The smart employers are going to start agitating for some kind of legislation to manage these cases before they find themselves socked with a lot of damages," says Andria S. Knapp, a University of Pittsburgh law professor. . . .

Indeed, notes Samuel Estreicher of New York University, experience under the wrongful-discharge laws in European countries indicates that employers generally make only modest payments to illegally fired people. In Britain, for example, employees are seldom reinstated to their jobs; instead, they receive a monetary award, which in 1978 amounted to only five weeks' pay for the average wage earner. The widespread acceptance of these laws by employers, Estreicher says, "may well be attributed to their ability to insure themselves without excessive costs." In the U.S., employers decades ago similarly escaped huge damage awards for job-related injuries by supporting passage of state workers' compensation laws.

The campaign for unjust-dismissal laws is furthest along in California, where juries have granted generous damage awards to many discharged employees. During a five-year period ended in 1983, employees won 48 or 74 jury verdicts, according to the State Bar of California. Six of the plaintiffs were awarded damages in excess of $1 million each. Townley & Updike's McCulloch, who represents company defendants in California employment-at-will cases, describes the employment climate there this way: "Our clients have more problems with employees in California than any place else, because people there are more oriented toward individual rights. You could call them free spirits." In addition, he says, the liberal court attitude on at-will cases draws lawyers to this increasingly lucrative practice, and the lawyers promote even more cases.

To avoid the possibility of such awards, employer groups such as the California Chamber of Commerce are supporting one of two unjust-dismissal bills pending in the legislature. This bill reaffirms the principle of employment-at-will, prohibits only the whistleblower type of discharge, puts the burden of proof on the employee, and prevents employees from suing for damages in most cases. The second and more liberal bill pending in California is sponsored by the American Civil Liberties Union and allows an employer to fire people only for just cause. Moreover, the worker is presumed innocent and stays on the job until the employer proves wrongdoing before a mediator or an arbitrator.

. . . But the very introduction of the bills raised a crucial and highly sensitive question for organized labor. Could it afford to refuse support for legislation that would help 7 million nonunion workers, even though an unjust dismissal law might hurt union organizing efforts? In the end, the California

AFL-CIO decided to support the ACLU bill, although it is not the top legislative priority. Some union activists believe the bill would help organizing campaigns because it would prevent dismissal for union activity. In New York, Massachusetts, and Tennessee, state AFL-CIO officials say they are interested in supporting unjust-dismissal provisions, but the bills are nowhere near passage. Pennsylvania AFL-CIO leaders are still studying the bill in their state.

PART-TIME LEADERS

It seems unlikely, however, that organized labor will actually lead the fight anywhere for such legislation. Athough opinion polls make it clear that Americans favor a broadening of employee rights, there is as yet no organization behind it. But such groups may eventually spring up.

In New York State, an organization called the National Congress of Employees was incorporated less than a year ago. Its two part-time officers drafted the unjust dismissal bill now pending in the New York legislature, and the NCE has recruited a few dozen members. It could well be the first nonunion association founded to represent all employees, although not as a bargaining agency. Members' dues would go primarily for lobbying activities.

It is still questionable whether organizations such as NCE can persuade diverse groups of employees that they have mutual interests. In February the AFL-CIO's Committee on the Evolution of Work suggested that unions might help form such organizations, but some union leaders strongly oppose the idea. As for unjust-dismissal laws, the national AFL-CIO hasn't developed a position on support or opposition. Top federation officials refuse to accept the premise that legislation is an acceptable substitute for bargaining, or that unions are in a state of permanent decline.

But with or without unions, an employee-rights movement will continue to gather force. One way or another, employees will gain more influence over their work lives. Companies that fail to see that will be in trouble.

◇ NOTES

1. For an interesting discussion of the employment at will rule and its implications, see: C. W. Summers, "The Rights of Individual Workers: The Contract of Employment and the Rights of Individual Employees: Fair Representation and Employment at Will," 52 *Fordham LR* 1082 (1984). Professor Summers argues that "the employment at will doctrine had no legal precedent ... (and) was contrary to basic contract principles in creating a legal presumption that overrode the parties' unexpressed intent." *Id*, at 1084.

2. For a defense of the doctrine, see: S. Catler, "The Case Against Proposals to Eliminate the Employment at Will Rule," 5 *Ind. Rel. LJ* 471 (1983).

C. THE RIGHT TO REFUSE HAZARDOUS WORK UNDER FEDERAL LABOR STATUTES

There is a limited right to refuse hazardous working conditions under federal labor law. For workers in private industry, the most important of these are the rights available under the National Labor Relations Act and the Occupational Safety and Health Act.

1.   The National Labor Relations Act

(a) Generally

Unsafe Working Conditions: Employee Rights Under the Labor Management Relations Act and the Occupational Safety and Health Act

N.A. Ashford and J.I. Katz

I. INTRODUCTION

Occupational health and safety have long been an issue in the law governing labor-management relations. One of the clearest examples is the employee's right to refuse hazardous work, which is embodied in §7 and §502 of the Labor Management Relations Act (LMR Act).

* * *

II. HEALTH AND SAFETY PROTECTION UNDER THE LMR ACT

A. Refusal of Hazardous Work as a §7 Protected Concerted Activity

The LMR Act [§7] guarantees that: "Employees shall have the right ... to engage in ... concerted activities for the purpose of collective bargaining or other

Source: 52 *Notre Dame Lawyer*, 802–808, 815–819 (1977). Reprinted with permission, © by the *Notre Dame Law Review*, University of Notre Dame.

mutual aid or protection." [29 U.S.C. §157 (1970) states:

Employees shall have the right to self-organization, to form, join, or assist labor organizations, to bargain collectively through representatives of their own choosing, and to engage in other concerted activities for the purpose of collective bargaining or other mutual aid or protection, and shall also have the right to refrain from any or all of such activities except to the extent that such right may be affected by an agreement requiring membership in a labor organization as a condition of employment as authorized in section [8(a)(3)].]

Concerted activities are protected by §8(a)(1) of the Act, which forbids employers "to interfere with, restrain, or coerce employees in the exercise of the rights guaranteed in section [7]." [29 U.S.C. §158(a)(1)(1970)] Hence, discharge of employees or other adverse action by a company against employees constitutes an unfair labor practice if it is in response to protected §7 rights.

In *NLRB v. Washington Aluminum Co.* [370 U.S. 9

(1962)], the Supreme Court recognized that a walkoff by employees for unsafe working conditions should be afforded full §7 protection. Seven of eight non-union machine shop employees left the job in protest when the furnace broke in 11° weather and the company failed to supply adequate heat. The Court upheld the Board ruling that the company's discharge of the employees was a violation of §8(a)(1) despite the company's argument that the machinists violated a company rule forbidding employees to leave their work without the foreman's permission. It is noteworthy that the Court placed special emphasis on the unorganized status of these employees in affirming their work stoppage: "They had no bargaining representative, and in fact, no representative of any kind to present their grievances to their employer. Under these circumstances, they had to speak for themselves as best they could." It is precisely this philosophy through which unorganized workers have traditionally been encouraged to obtain improved working conditions. [This was true until the passage of the OSH Act in 1970, which covers both organized and unorganized employees.]

In order for a refusal to accept hazardous work to qualify as concerted activity protected by §7, it must meet several essential requirements which include a requirement that the refusal constitute a labor dispute under §2(9) of the LMR Act. As the Supreme Court noted in *Washington Aluminum,* it has long been settled that "the reasonableness of workers' decisions to engage in concerted activity is irrelevant to the determination of whether a labor dispute exists or not." Although the leaving of a job may be objectively "unnecessary and unwise," the objective determination of that fact cannot be used to establish whether a safety controversy exists. Thus it appears that neither the Board nor the courts need give scrutiny to the actual substance of the §7 safety walkoff nor to whether the conditions which the employees protest are, in fact, unsafe. The Board's decision concerning §7, affirmed by the Fourth Circuit, speaks directly to this point:

> [T]he issue here is not the objective measure of safety conditions, it is whether these employees left their jobs because they thought conditions were unsafe. They could not have been penalized for so doing just because, other employees tolerated such conditions or because by some external standard, they were too safety conscious.

This is in sharp contrast to specific standards of safety and types of evidence required under §502 of the LMR Act and §11(c) of the OSH Act.

While non-union employees have successfully used their §7 rights over unsafe working conditions, employees represented by unions have also availed themselves of §7 protection regarding health and safety, often with the support or sanction of the union. Where a collective bargaining agreement may contain a no-strike clause, union-sanctioned walkoffs over health and safety are generally restricted under §7. Walkoffs by a minority of employees in derogation of the union and in conflict with the union's exclusive representation would not be protected by §7. Where health and safety have been excluded from a no-strike provision, §7 is fully available and individual employees should be free to walk off the job when concerned over unsafe conditions so long as doing so is for the mutual benefit of other employees as well.

B. Section 502 and the Right to Strike over "Abnormally Dangerous Conditions"

In addition to the concerted activities protected under §7, §502 of the Taft-Hartley Act explicitly protects work stoppages by employees who refuse work because of abnormally dangerous conditions from being considered a "strike." There are two major tests pertinent to coverage under §502: what constitutes "good faith" and what hazards come within the scope of "abnormally dangerous conditions."

The courts have let the "good faith" test die a slow and subtle death. The case most cited as authority for the subjective test of good faith is *NLRB v. Knight Morley Corp.* [251 F.2d 753 (6th Cir. 1957)] In that case, the improper repair of an exhaust fan in a buffing department created an increased accumulation of dust and heat, causing buffers to walk off the job. The testimony from the buffers as to the physical conditions in the buffing room was held to be "competent" by the appeals court. Laymen may testify as to the physical conditions which they have observed.

The central focus of the court's opinion in *Knight Morley* lay not in the existence of good faith but the competency of the evidence to prove "abnormally dangerous conditions." Although the court alluded to the good faith of the employees in walking out, its decision is premised upon the existence of the evidence presented. It would seem that where such abnormally dangerous conditions exist, good faith must implicitly be present.

The subjective good faith test which appeared to emerge in *Knight Morley,* however, received its death-blow in the NLRB ruling in *Redwing Carriers* [130 N.L.R.B. 1208 (1961)], in which the Board stated:

> It is necessary first to clarify the meaning of the term "abnormally dangerous conditions" as used in Section 502. We are of the opinion that the term contemplates, and is intended to insure, an objective, as opposed to a subjective, test. What controls is not the state of mind of the employee or employees concerned, but whether the actual working conditions shown to exist by competent evidence might in the circumstances reasonably be considered "abnormally dangerous."

This test has been utilized in all cases since *Redwing Carriers* where §502 has been raised to defend activity which otherwise would be unprotected by a collective bargaining no-strike clause. It finally received Supreme Court approval in *Gateway Coal v. UMW.* [414 U.S. 368 (1974)]

Using the objective test, courts have refined the definition of abnormally dangerous conditions. They have found to be abnormally dangerous, situations in which the risk of danger has increased over that considered to be normal for the specific job assignment. Where temporary exhaust system breakdowns occurred creating "highly unpleasant" working conditions, the Board refused to provide §502 protection to employees who left the job, because the conditions were caused by periodic shutdowns, and hence a "normal" part of the job. In *Anaconda Aluminum* the Board applied a similar approach to an inherently dangerous job:

> Absent the emergence of new *factors or circumstances which change the character of the danger,* work which is recognized and accepted by employees as inherently dangerous does not become "abnormally dangerous" merely because employee patience with prevailing conditions wears thin or their forbearance ceases.

So it can be concluded that the task of proving the existence of abnormally dangerous conditions in order to come under §502 protection is by no means easy; and it is particularly difficult in inherently dangerous jobs. The burden of the objective test means that employees risk penalties if, in walking off the job, their judgment as to the extent of the hazard and the increase in the risk prove inconsistent with an objectively determined standard.

Finally it would seem that because §502 authorizes the quitting of work by "an employee or employees" it is arguably an individual right; moreover, the section does not mention a labor union. Hence the right would extend only to those employees actually faced with the abnormally dangerous conditions, and a safety issue could not be used as justification for a union walkout unless everyone were threatened by the hazard. However, this question has never been answered.

The relationship between §7 and §502 remains unclear in many respects. They could be read as establishing separate rights, without coterminous application. On the other hand, §502 defines the scope of the concerted activities protected by §7 in the context of a no-strike clause in the collective bargaining contract. Looking at the types of activities protected, it is clear that §7 is broader than §502 since the latter applies only to abnormally dangerous conditions while §7 does not depend on a determination concerning the nature or extent of the hazard. Thus, when employees wish to quit work because of unsafe (but perhaps less than abnormally dangerous) conditions, the preference of §7 protection over that of §502 is obvious. Section 7 protection would probably be chosen over §502 protection where a collective bargaining agreement had a clause reserving the right to strike, since the more limited §502 protection is not necessary because strikes are not forbidden. After *Boys Markets Inc. v. Retail Clerks Local 770* [398 U.S. 235 (1970)] the policy favoring arbitration where an express no-strike clause exists was extended such that an agreement to arbitrate gave rise to an *implied* promise not to strike. This severely restricted the scope of §7 concerted activity in general. What remained unresolved was the extent to which §502 kept the §7 protection open for abnormally dangerous conditions, thus making employer reprisal against the protected walkoff an unfair labor practice. This issue was addressed in *Gateway Coal* where a requirement that health and safety disputes be arbitrated does not necessarily extinguish a §502 right to strike, even where an express or implied no-strike clause exists. However, broad §7 concerted activities in the area of health and safety do not appear to be available where no-strike clauses, express or implied, are present.

* * *

D. The Right to Refuse Hazardous Work after *Gateway Coal*

In *Gateway Coal* the "presumption of arbitrability" was extended to safety disputes so that, absent explicit contractual exclusion of safety matters from arbitration, a union can be compelled to arbitrate the issue. The availability of a limited right to strike over safety afforded employees by §502 does not by itself remove the safety dispute from arbitrability. In *Gateway Coal,* the Supreme Court was careful to say that:

> [A]n arbitration agreement is usually linked with a concurrent no-strike obligation, but the two issues remain *analytically distinct.* Ultimately, each depends on the intent of the contracting parties. It would be unusual, but certainly permissible, for the parties to agree to a broad mandatory arbitration provision yet expressly negate any implied no-strike obligation. Such a contract would reinstate the situation commonly existing before our decision in *Boys Markets.*

Thus, contractually reserving the right to strike over safety may afford more protection against an employer's attempt to force employees back to work than reliance on the statutory protection afforded by §502 in "abnormally dangerous conditions." This may be the case both because the contractual right may explicitly extend the right to quit to other than "abnormally dangerous conditions" and because it renders §502 irrelevant to the issue of whether or not a strike has occurred since it may be contracted that a work stoppage for safety is not a strike. If a work stoppage for safety is not a strike, it does not violate the collective bargaining agreement. Therefore, employer retaliation in the form of an injunction may be prevented, or employer discriminatory practices may be the basis for an unfair labor practice charge by the NLRB, both without going first to arbitration. The irony of the latter situation, however, is that if the quitting of work occurs over a safety matter which itself qualified as a labor dispute for arbitration purposes, the NLRB may not wish to decide a highly technical safety question, but instead delay consideration of the alleged unfair labor practice until the arbitrator has made factual determinations necessary for the settlement of the arbitrable dispute and germane to the reasonableness of the work stoppage. While arguably foreclosed by *Gateway Coal,* this delay could conceivably be accompanied by a temporary restraining order enjoining the work stoppage until the factual determinations are made by the arbitrator. Alternatively, the work stoppage could be permitted, but in the event the arbitrator decides against the employees, a suit against the union for damages under §301 of the LMR Act could succeed unless either the collective

bargaining agreement explicitly limited the unions' liability in this situation or through their equitable jurisdiction, courts determine §301 damages against the union to be unavailable so long as the employees' claim was not unreasonable. It is clear that while an agreement to arbitrate and an agreement permitting or prohibiting a right to strike are "analytically distinct," the resulting judicial consequences are not.

If, as in *Gateway Coal,* a court orders both arbitration of a safety dispute *and* an immediate correction by the employer of the situation giving rise to the alleged unsafe condition, then it may be argued that the work stoppage is no longer necessary. *Gateway Coal* raises the question of whether an employer seeking an injunction of a work stoppage over health and safety might be required to arbitrate *and* to eliminate the alleged hazard pending arbitration as the *quid pro quo* for the union agreeing not to cease work, since the Supreme Court upheld the district court's injunction to do that. However, if the parties have already taken steps to arbitrate, can it be further argued that an employer's promise to correct an unsafe condition is the *quid pro quo* for the agreement not to cease work? Given the jurisdiction over health and safety by the Federal Coal Mine Health and Safety Act and the OSH Act, can the courts act in the context of an injunction proceeding to require employers to abate alleged hazards in return for employees returning to their jobs? If the courts can fashion a remedy to abate a hazard in this way, should a union not also be able to obtain an injunction under §301 of the LMR Act to force an employer to abate the hazard? Unions, of course, can contract for an employer's abatement of the alleged hazard as the *quid pro quo* for an agreement not to stop work pending the outcome of arbitration.

Irony lies in the possibility that unions may find it more in their interest to both exclude safety disputes from arbitration (as some unions have done) and to either (1) reserve the right to strike over health and safety issues, thus acquiring more protection against employer retaliation than §502 provides, or (2) not agree to an express no-strike clause, thus obtaining the greater protection afforded safety walkouts under §7. Thus, the "public policy favoring arbitration" of safety disputes may be defeated by the very decision which articulated its desirability—*Gateway Coal.*

The ramifications of the principles laid out in the foregoing article, and the extent of the interrelationship between Section 7 and Section 502, are explored in the following case involving acute exposure to ammonia fumes.

National Labor Relations Board v. Tamara Foods, Inc.

692 F.2d 1171 (8th Cir. 1982)

BECKER, S. D. J.

The National Labor Relations Board (NLRB) has filed in this Court an application for enforcement of its order issued on September 30, 1981, against Tamara Foods, Inc. (Tamara) pursuant to Section 10(a)-(c), 29 U.S.C. §160 (a)-(c) of the National Labor Relations Act (Act) as amended, 29 U.S.C. §§151, *et seq.* That Decision and Order of the NLRB is reported at 258 NLRB 180, 108 LRRM 1218. The Decision and Order of the NLRB disagreed with the Decision of an Administrative Law Judge (ALJ) recommending an order of dismissal of the complaint against Tamara by the Regional Director of the NLRB, charging unfair labor practices in violation of Section 8(a)(1) and Section 7 of the Act.

*   *   *

DECISION AND ORDER OF NLRB

Counsel for the General Counsel of the NLRB filed before the NLRB, timely exceptions to the Decision of the ALJ. After a hearing on the exceptions, briefs and record before the ALJ, a three member panel of the NLRB entered a unanimous Decision and Order concluding that Tamara violated Section 8(a)(1) of the Act, and ordered Tamara to cease and desist, and to take affirmative corrective action. It is enforcement of this Decision and Order that the NLRB seeks in this Court. The pertinent parts of the Decision and Order are as follows:

The Board has considered the record and the attached Decision in light of the exceptions and briefs and has decided to affirm only so much of that Decision as is consistent with this Decision and Order.

The complaint in this proceeding alleges that Respondent violated Section 8(a)(1) of the Act by unlawfully threatening to discharge and discharging 11 employees for engaging in a strike over unhealthy working conditions. The Administrative Law Judge dismissed the complaint on the ground that the activity of the 11 employees,

although concerted, was not protected by Section 7 of the Act. As further explained below, we find that the Act protects the rights of employees to strike over what they honestly believe to be unsafe and unhealthy working conditions, and we find, consequently, that Respondent's threat of discharge and its discharge of these employees for exercising those rights violated Section 8(a)(1). . . .

Respondent, which is engaged in the business of preparing and selling frozen foods, uses an ammonia refrigeration system to freeze its products. On several occasions ammonia gas from the refrigeration system leaked into the production area where Respondent's employees work. Employees testified without contradiction that the ammonia fumes had caused them to experience nausea, burning sensations in their noses and throats, headaches, tightness in their chests, and difficulty in breathing.

Respondent recognized the problem of occasional ammonia leaks and had unilaterally promulgated work rules to be followed in such situations.[2] [2 Respondent's employees are not represented by a labor organization and Respondent's plant rules, therefore, are not part of any collective-bargaining agreement.] Under these rules, sick leave was to granted automatically to any employee upon request; any employee who believed that an unsafe or unhealthy condition existed at the plant would be permitted to leave his or her work station and remain in the lunchroom or directly outside the plant, with full pay, until the condition was corrected; and any employee who clocked out prior to the end of the shift would be discharged.

On the morning of September 11, 1980, 15 or 20 minutes into the start of the 7 a.m. shift, the employees in the production area smelled ammonia leaking from the freezer system and began to feel its noxious effects. The employees, on their own initiative, left the production area and congregated in the lunchroom. Supervisor Helen Bury was then informed of the leakage and the employees were told to wait in the lunchroom until the problem was corrected.

Approximately 45 minutes later, Supervisor Bury informed the employees that the ammonia leakage had been stopped and directed them back into the production area. When the employees returned to the production area they found that it still contained ammonia fumes and was, in fact, worse then (sic) before. After working for 10 or 15 minutes, the employees retreated to the lunchroom for a second time and again reported the problem to Supervisory Bury. A number of employees went outside the plant for fresh air. Shortly thereafter, the employees were again assured that the ammonia problem had been corrected and were directed to return to the production area.

Upon their return to work for a third time, the employees continued to feel the harmful effects of the ammonia fumes. At this point, the entire complement of employees left their work stations. Many of the 50 employees began to clock out.

Supervisor Bury tried to convince the employees to stay at the plant and told them they could remain in the lunchroom and be paid for waiting until the ammonia problem was corrected. Several employees objected, however, and said that they were not going to wait at the plant. When Plant Manager A. J. Schopp noticed that employees were waiting in line to clock out, and was told that several employee had already done so, he told them that they had "better go clock back in" and that those who clocked out should not "bother coming in tomorrow." Schopp further stated that any employees who left work should consider themselves fired.

After Schopp's statement that clocking out would mean discharge, most of the employees decided to remain at the plant. Some of those who had already clocked out, clocked back in. However, 11 employees remained clocked out and left the plant, for which they were discharged.

The Administrative Law Judge found, and we agree, that the action of these 11 discharged employees was clearly "concerted activity" within the meaning of the Act. However, unlike the Administrative Law Judge, we find that the conduct was also clearly protected.

*   *   *

In *NLRB v. Washington Aluminum supra,* a case which closely parallels the instant proceeding,

the U.S. Supreme Court held that employees have the right under Section 7 of the Act to walk off their jobs, without prior notice to their employer and without following established plant rules forbidding employees from leaving their work stations without permission, if their action is a means of protesting what they perceive to be intolerable working conditions. the general rule is that the protections of Section 7 do "not depend on the *manner* in which the employees choose to press the dispute, but rather on the *matter* that they are protesting," *Plastilite Corporation,* 153 NLRB 180, 184 (1965), enfd. in pertinent part 375 F.2d 343 (8th Cir. 1967). Inquiry into the objective reasonableness of employees' concerted activity is neither necessary nor proper in determining whether that activity is protected. As we stated in *Plastilite Corporation, supra,* "we must respectfully disagree with any rule which would base the determination of whether a strike is protected upon its reasonableness in relation to the subject matter of the labor dispute. When subject matter of the labor dispute. When a labor dispute exists, the Act allows employees to engage in concerted activity which they decide is appropriate for their mutual aid and protection, including a strike, unless . . . that activity is specifically banned by another part of the statute, or unless it falls within certain other well-established proscriptions." Whether the protested working condition was actually as objectionable as the employees believed it to be, or whether their objection could have been pressed in a more efficacious or reasonable manner, is irrelevant to whether their concerted activity is protected by the Act.

Nor does the fact that employees fail to make a specific demand to the employer automatically render their conduct unprotected. Particularly where the employees are not represented by a labor organization which may speak to the employer on their behalf, "if from surrounding circumstances the employer should reasonably see that improvement of working conditions is behind the walkoff, it may not penalize the employees involved without running afoul of Section 8(a)(1)," *South Central Timber Development, Inc.,* [1977–78 CCH NLRB ¶18,310] 230 NLRB 468, 472 (1977); *NLRB v. Washington Aluminum, supra.*

Applying these principles to the instant case, we find that the 11 employees who clocked out to protest the presence of ammonia fumes in their work environment were clearly engaged in protected concerted activity. The presence of ammonia fumes in the work environment is, obviously, a working condition. The uncontested testimony of these employees demonstrates that their walkout was caused by concern over these fumes and their desire to see that "something might be done about it." The record in this case permits no doubt that their concerns were made known to Respondent. The Administrative Law Judge was seemingly of the view that the employees were not entitled to leave the plant because such action was in derogation of an existing plant rule and because Respondent had provided a procedure which adequately dealt with the problem. We disagree.

Of particular significance here is the fact that Respondent's employees are not represented by a labor organization or covered by a collective-bargaining agreement containing a "no-strike" clause. The fact that Respondent has unilaterally established and promulgated a rule restricting this activity is insufficient to deprive employees of a statutory right. Nor can it be said that because Respondent has provided an alternative solution to the problem, the employees are required to accept it. No matter how reasonable the alternative might seem, if the employees choose to exercise their statutory rights, they can not be penalized for doing so.

We must also reject the Administrative Law Judge's conclusion that the employees' walkout was unprotected because the Occupational Safety and Health Administration found Respondent's plant not to violate its regulations, or because another statutory scheme might have afforded the employees some form of relief. The rights guaranteed to employees under the National Labor Relations Act are distinct from and are not subordinate to the provisions of the Occupational Safety and Health Act, *Du-Tri Displays, supra*. To hold otherwise might seriously diminish the rights of employees to engage in concerted activity for their mutual aid and protection and would constitute an abdication of the role that Congress has assigned to the National Labor Relations Board in protecting those rights.

Accordingly, for the reasons stated above, we find that Respondent violated Section 8(a)(1) of the Act by discharging the 11 employees because they engaged in a strike over working conditions. We also find that Plant Manager A. J. Schoop's statement to employees that "If you clock out and

go home today, don't come back tomorrow . . . consider yourself fired," was a threat of discharge for engaging in protected activity and therefore in violation of Section 8(a)(1) of the Act.[3] [3 See *Vic Tanney International*, [1977–78 CCH NLRB ¶18,657] 232 NLRB 353 (1977), enfd. 622 F.2d 237 (6th Cir. 1980).]

Lastly, we agree with the General Counsel that the 11 discharged employees are entitled to full and immediate reinstatement and backpay. Employees who are unlawfully discharged while engaged in a lawful strike are entitled to backpay from the date of the discharge until the date that they are offered reinstatement. Abilities and Goodwill, Inc., [1978-79 CCH NLRB ¶15,634] 241 NLRB 27 (1979), enforcement denied on other grounds, 612 F.2d 6 (1st Cir. 1979). The Administrative Law Judge noted that the 11 discharged employees did not request reinstatement. However, under our current decisions, there is no requirement that employees who are unlawfully discharged during a protected strike must request reinstatement; "since it is the employer who has acted unlawfully in discharging the employee, the burden is on that employer to undo its unfair labor practice by offering immediate reinstatement to the employee, and by reimbursing the employee for all losses suffered from the date of its discriminatory action," *Abilities and Goodwill, Inc., supra.*

The Remedy

Having found that Respondent has engaged in unfair labor practices within the meaning of Section 8(a)(1) of the Act, we shall order that it cease and desist therefrom and that it take certain affirmative action designed to effectuate the policies of the Act.

Having found that on September 11, 1980, Respondent unlawfully threatened its employees with discharge for engaging in a lawful strike, we shall order that it cease and desist from such unlawful conduct.

Having found that, on September 11, 1980, Respondent unlawfully discharged 11 of its employees . . . for engaging in a lawful strike over working conditions and has thereafter refused and failed to offer these employees full and immediate reinstatement, we shall order that it cease and desist from such unlawful conduct and offer the 11 unlawfully discharged employees full and immediate reinstatement to their jobs or, if such positions no longer exist, to substantially equivalent positions, without prejudice to their seniority or other rights and privileges. We shall also order that Respondent make whole these 11 employees for any loss of pay they may have suffered because of Respondent's unlawful discharge, by payment to them of a sum equal to that which they would have earned from the date of their discharge until they are reinstated or receive valid offers of reinstatement, less any net interim earnings. Backpay shall be computed in accordance with the formula set forth in *F. W. Woolworth Company*, 90 NLRB 289 (1950), with interest thereon to be computed in the manner prescribed in *Florida Steel Corporation*, [1978 CCH NLRB ¶18,841] 231 NLRB 651 (1977)[5] [5 See, generally, *Isis Plumbing & Heating Co.*, [1962 CCH NLRB ¶11,607] 138 NLRB 717 (1962). In accordance with his dissent in *Olympic Medical Corporation*, [1980 CCH NLRB ¶17,116] 250 NLRB 146 (1980), Member Jenkins would award interest on the backpay due based on the formula set forth therein.]

Here followed the conclusions of law and order of the NLRB in detail.

* * *

DECISION GRANTING APPLICATION FOR ENFORCEMENT OF ORDER OF NLRB

After considering the briefs, arguments and the record, including the transcript of the hearing before the ALJ, we conclude that the application of the NLRB for an order enforcing its Decision and Order, quoted above, shall be granted.

* * *

Discussion of Contentions of Respondent

A. Tamara first contends that the NLRB erred in concluding from the uncontroverted facts that the walkout of the eleven employees was concerted activity (or strike) protected by Section 7 of the Act, 29 U.S. §157. Tamara contends in detail that this conclusion of the NLRB is based on an erroneous "subjective standard" to determine whether an "employee safety" protest is protected, rather than whether the "protest" activity was directed to "actual or objectively perceived danger" in the workplace. Further, Tamara argues that Section 7 protection of "safety protest" matters is limited to circumstances, said to be absent in this case, in which the protesting employ-

ees had no other alternative but to perform work perceived to be dangerous, or under conditions perceived to be dangerous.

Next Tamara contends that the NLRB erred in failing to balance and accommodate the employees' rights of protest against the claimed superior legitimate interests of Tamara to conduct business with loyalty of the employees, and without harassment or other disruptive conduct.

These contentions and arguments are lacking in merit under the controlling decisions of the Supreme Court of the United States in *National Labor Relations Board v. Washington Aluminum Co.*, [45 LC ¶17,637] 370 U.S. 9, 82 S. Ct. 1009, 8 L. Ed. 2d 298 (1962), and of this Court in *First National Bank of Omaha v. National Labor Relations Board* [60 LC ¶10,245] (C.A. 8 1969) 413 F.2d 921. The following quotations from the opinion of the Supreme Court of the United States in *National Labor Relations Board v. Washington Aluminum Co., supra*, demonstrate that the NLRB did not err in concluding that the walkout of the eleven employees of Tamara was concerted activity protected under Section 7 of the Act unless forbidden by other lawful prohibitions of the Occupational Safety and Health Act or of the Labor Management Relations Act:

> The Court of Appeals . . . refused to enforce an order of the [NLRB] directing respondent . . . to reinstate and make whole seven employees whom the company had discharged for leaving their work in the machine shop without permission on claims that the shop was too cold to work in.

* * *

> We cannot agree that employees necessarily lose their right to engage in concerted activities under §7 merely because they do not present a specific demand upon their employer to remedy a condition they find objectionable. The language of §7 is broad enough to protect concerted activities whether they take place before, after, or at the same time such a demand is made. To compel the Board to interpret and apply that language in the restricted fashion suggested by the respondent here would only tend to frustrate the policy of the Act to protect the right of workers to act together to better their working conditions. Indeed, as indicated by this very case, such an interpretation of §7 might place burdens on employees so great that it would effectively nullify the right to engage in concerted activities which that section protects. The seven employees here were part of a small group of employees

who were wholly unorganized. They had no bargaining representative and, in fact no representative of any kind to present their grievances to their employer. Under these circumstances, they had to speak for themselves as best they could. As pointed out above, prior to the day they left the shop, several of them had repeatedly complained to company officials about the cold working conditions in the shop.

*   *   *

Although the company contends to the contrary, we think that the walkout involved here did grow out of a "labor dispute" within the plain meaning of the definition of that term in §2(9) of the Act which declares that it includes "any controversy concerning terms, tenure or *conditions* of *employment.* . . . " The findings of the Board . . . show a running dispute between the machine shop employees and the company over the heating of the shop on cold days—a dispute which culminated in the decision of the employees to act concertedly in an effort to force the company to improve that condition of their employment. The fact that the company was already making every effort to repair the furnace and bring heat into the shop that morning does not change the nature of the controversy that caused the walkout. At the very most, that fact might tend to indicate that the conduct of the men in leaving was unnecessary and unwise, and it has been long settled that the reasonableness of the workers' decision to engage in concerted activity is irrelevant to the determination of whether a labor dispute exists or not. . . .

*   *   *

Nor can we accept the company's contention that because it admittedly had an established plant rule which forbade employees to leave their work without permission of the foreman, there was justifiable "cause" for discharging these employees, wholly separate and apart from any concerted activities in which they engaged in protest against the poorly heated plant.

*   *   *

It is of course true that §7 does not protect all concerted activities, but that aspect of the section is not involved in this case. The activities

engaged in here do not fall within the normal categories of unprotected concerted activities such as those that are unlawful, violent or in breach of contract. Nor can they be brought under this Court's more recent pronouncement which denied the protection of §7 to activities characterized as "indefensible" because they were found to show a disloyalty to the workers' employer which this Court deemed unnecessary to carry on the workers' legitimate concerted activities. The activities of these seven employees cannot be classified as "indefensible" by any recognized standard of conduct. Indeed, concerted activities by employees for the purpose of trying to protect themselves from working conditions as uncomfortable as the testimony and Board findings showed them to be in this case are unquestionably activities to correct conditions which modern labor-management legislation treats as too bad to have to be tolerated in a humane and civilized society like ours.

The striking similarities of the facts in the *Washington Aluminum Co.* case, *supra,* and in this case make it clear that the NLRB did not err in concluding that the walkout in this case was concerted activity protected by Section 7 of the Act.

*   *   *

We examine next the contentions of Tamara concerning the Occupational Safety and Health Act (OSH Act) and the Labor Management Relations Act (LMRA).

B. Tamara further contends that the NLRB erred in ignoring (1) the provisions and standards of the OSH Act, 29 U.S.C. §§651 *et seq.,* and (2) Section 502 of the LMRA, 29 U.S.C. §143. These contentions will be discussed separately.

1. The OSH Act and Standards. Tamara complains of the failure of the NLRB to cite and apply the OSH Act and standards promulgated thereunder, 29 C.F.R. §§1910.1000(a)(2) and 1977.12, as prohibiting the action of the eleven employees in walking off the job, particularly in the absence of a reasonable apprehension of death or serious injury by objective evidence, and in the absence of no choice other than a refusal to work while exposed to death or serious injury. This complaint of Tamara is without merit under the controlling decision of *Whirlpool*

Corporation v. Marshall, 445 U.S. 1, 100 S. Ct. 883, 63 L. Ed. 2d 154 (1980).

* * *

We conclude that the NLRB was not required to deny relief because of the existence of the OSH Act. The employees had sought relief in 1979 and in July, August and September 1980 under the OSH Act without relief from the conditions that continued repeatedly until the walkout. Thus the NLRB properly affirmed the findings of these facts of the ALJ. The NLRB also found that the ammonia fumes caused the employees to experience nausea, burning sensations in their noses, headaches, tightness in their chests and difficulty in breathing. These symptoms and the apprehension of injury were reasonable and objectively proven in light of the commonly known effects of exposure to ammonia. See part of the description of ammonia in *The New Columbia Encyclopedia* (Columbia University Press 1975) at page 92 as follows:

ammonia, chemical compound, NH_3, colorless gas that is about one half as dense as air at ordinary temperatures and pressures. It has a characteristic pungent, penetrating odor. It is extremely soluble in water; one volume of water dissolves about 1,200 volumes of the gas at 0°C (90 grams of ammonia in 100 cc of water), but only about 700 volumes at room temperature and still less at higher temperatures. The solution is alkaline because much of the dissolved ammonia reacts with water, H_2O, to form ammonium hydroxide, NH_4OH, a weak base. The ammonia sold for household use is a dilute water solution of ammonia in which ammonium hydroxide is the active cleansing agent. It should be used with caution since it can attack the skin and eyes. The vapors are especially irritating—prolonged exposure and inhalation cause serious injury and may be fatal.

The OSH Act is not designed to provide relief in these urgent circumstances. See the sometimes protracted deliberate proceedings possible under the unusual procedures of the OSH Act, described in *Donovan, Secretary of Labor v. Anheuser-Busch, Inc.* (C.A.8 1981) 666 F.2d 315 at 323-324.

The NLRB did not err in concluding that the rights guaranteed to these employees by the Act are superior to the provisions of the OSH Act.

Nor does the failure of the employees to expressly

demand specific corrective action affect the rights of these unorganized employees. As the NLRB found, the repeated complaints of the employees under the OSH Act and the experiences on the day of the walkout made evident the desires of the employees for action by Tamara.

2. *Section 502 of the LMRA.* Tamara argues that under Section 502 of the LMRA, 29 U.S.C. §143, "objective evidence" and "good faith belief" of "abnormally dangerous conditions of work" are required to support a finding of activity protected by Section 7 of the Act; and that this was not considered by the NLRB in this action. To the extent required by Section 502, the facts found by the NLRB and the ALJ were supported by objective evidence, and good faith belief was proven.

Section 502 (29 U.S.C. §143) is as follows:

Saving provision
    Nothing in this Act shall be construed to require an individual employee to render labor or service without his consent, nor shall anything in this Act be construed to make the quitting of his labor by an individual employee an illegal act; nor shall any court issue any process to compel the performance by an individual employee of such labor or service, without his consent; nor shall the quitting of labor by an employee or employees in good faith because of abnormally dangerous conditions for work at the place of employment of such employee or employees be deemed a strike under this Act [LMRA].

This provision does not modify Section 7 of the Act and is not relevant to the action of the NLRB in this case. The purpose of Section 502 was described, by the Supreme Court of the United States, in footnote 29 of *Whirlpool Corporation v. Marshall, supra,* as follows:

Similarly, Section 502 of the Labor Management Relations Act, 29 U.S.C. §143 provides that "the quitting of labor by an employee or employees in good faith because of abnormally dangerous conditions for work at the place of employment of such employee or employees [shall not] be deemed a strike." The effect of this section is to create an exception to a no-strike obligation in a collective-bargaining agreement. *Gateway Coal Co. v. Mine Workers,* [72 LC ¶14,192] 414 U.S. 368, 385, 38 L. Ed. 2d 583, 94 S. Ct. 629.

    The existence of these statutory rights also

makes clear that the Secretary's regulation [29 C.F.R. §1977.12(b)(1)] does not conflict with the general pattern of federal labor legislation in the area of occupational safety and health. See also 29 C.F.R. §1977.18 (1979).

There was no collective bargaining agreement in this case, to make Section 502 applicable.

C. Remaining Contentions of Tamara. Except for the complaint about the remedy, the remaining contentions of Tamara that the NLRB erred in its Decision and Order are determined to be lacking in merit by disposition of its primary contentions A and B above.

The remedy of the NLRB requiring reinstatement of the employees and awarding back pay was authorized, although the employees did not request reinstatement or offer to return to work. *Abilities and Goodwill, Inc.,* 241 NLRB, No. 5, 27 (1979), reversed on other grounds (C.A.1 1979) 612 F.2d 6; *National Labor Relations Board v. Lyon & Ryan Ford, Inc.,* [91 LC ¶12,755] (C.A.7 1981) 647 F.2d 745 at 755-757, *cert. denied,*—U.S.,—102 S. Ct. 391, 70 L. Ed. 2d 209; *National Labor Relations Board v. Mars Sales & Equipment Co.* [89 LC ¶12,184] (C.A.7 1980) 626 F.2d 567 at 573-575; *National Labor Relations Board v. Trident Seafoods Corp.* [90 LC ¶12,663] (C.A.9 1981) 642 F.2d 1148 at 1149-1150.

CONCLUSION

For the foregoing reasons the application of the NLRB for enforcement of its Decision and Order is hereby,

Granted.

◇ **NOTES**

1. Note that the employer had instituted a policy with regard to workplace hazards that many of us might find quite reasonable. If employees felt the workplace to be unsafe, they were allowed either to take sick leave or to vacate the immediate work area and remain on the premises (with pay) while the hazard was being abated. What rationale was there, then, for these employees to clock out and leave the job site completely?

2. Would the result in this case have been different if the workers had been unionized and were operating under a collective bargaining agreement?

3. Would the result have been different if only *one* of the employees had walked off the job and refused to come back? (See the discussion of the *Prill* case, below.)

4. Note that the court concludes that the existence of the OSHAct, and of a right to refuse hazardous work under that Act, do not affect the employee's right to refuse hazardous work under the provisions of Section 7. As with the right to obtain information from the employer regarding workplace health and safety (see Chapter 7), the rights afforded by the OSHAct and the NLRA are not mutually exclusive, they are complementary.

◇

(b) "Concerted activity" under Section 7

Among other things, Section 7 of the NLRA gives employees the right "to engage in . . . concerted activities for the purpose of collective bargaining and other mutual aid and protection." Section 8(a)(1) forbids employers "to interfere with, restrain, or coerce employees in the exercise of rights guaranteed in Section 7."

Activities regarding workplace health and safety can easily be construed as affording mutual aid and protection to workers. The problem has been trying to interpret how one defines "concerted" activities pursuant to Section 7. If the actions of an individual worker or group of workers regarding health and safety are considered concerted, they are protected under the Act and the employer cannot interfere with these activities (through retaliation or discrimination) under the law.

It has been relatively easy to extend protection to two or more workers when they engage in health and safety activities for the purpose of mutual aid and protection. When a group of workers refuse hazardous work and walk off a job, section 7 has been used to protect them for retaliatory discharge. *NLRB v. Washington Aluminum*, 370 U.S. 9 (1962). A more difficult situation emerges when an *individual* worker refuses hazardous work or takes other actions in the name of workplace health and safety. Can apparently individual actions ever be considered concerted activity and thus receive protection under Section 7 of the NLRA? While the NLRB was once inclined to interpret the notion of concerted activity broadly enough to encompass a range of activities undertaken by individuals, it has recently narrowed its scope and applied a more stringent interpretation.

In *Alleluia Cushion Co.*, 221 N.L.R.B. 999 (1975), an employee was fired for filing a safety complaint with California OSHA. The administrative law judge hearing the case dismissed the employee's complaint, finding no group action involved. The NLRB disagreed:

> Safe working conditions are matters of great and continuing concern for all within the workforce [and filing an OSHA complaint] was an action taken in furtherance of guaranteeing Respondent's employees their rights under the California Occupational Safety and Health Act. . . . Accordingly, where an employee speaks up and seeks to enforce statutory provisions relating to occupational safety designed for the benefit of all employees, in the absence of any evidence that fellow employees disavow such representation, we will find an implied consent thereto and deem such activity to be concerted.

Thus, the concept of concerted activity was extended to individuals who acted alone on matters that would benefit coworkers. Indeed, if the individual was simply asserting a right guaranteed by statute, concerted activity could be presumed unless coworkers specifically spoke against it. The rationale applied in *Alleluia* came to be known as the *per se* standard of concerted activity. That is, individual actions on issues of group concern and mutual benefit were deemed to be concerted, *per se,* absent proof to the contrary.

The case that turned the tide against this interpretation began in 1981, when an administrative law judge relied on *Alleluia* to conclude that Meyers Industries violated Section 8(a)(1) of the NLRA when it fired employee Kenneth Prill. Prill drove a truck for Meyers, hauling boats from Meyers's manufacturing facility in Michigan to dealers throughout the country. Prill had complained many times to his supervisor and to the president of the company about problems with his truck, and he had received a citation for driving a defective vehicle. During one two-week period, another employee drove Prill's truck and, upon completion of a trip and only incidentally in Prill's presence, complained about its steering and brakes. This second employee said he would not drive the truck again until it was repaired.

Several weeks later, the malfunctioning brakes caused an accident while Prill was driving the truck through Tennessee. Prill called the company president to advise him of the incident and of the extensive damage; he was told to drive the rig back to Michigan for repairs. Prill stated that it would be unsafe to do so and repeated this opinion in a subsequent phone call the following morning. Relying on a manual published by the Federal Highway Administration, Prill then contacted the Tennessee Public Service Commission to arrange an inspection of the truck. A citation was issued, and the truck was put out of service because of bad brakes and other damage. When Prill reported for work two days later, he was discharged because "we can't have you calling the cops like this all the time."

Although Prill won his case before the administrative law judge, the NLRB reversed the decision. *Meyers Industries Inc. v. Prill*, 268 N.L.R.B. No. 73 (1984). Overruling *Alleluia* and all its progeny, the NLRB dismissed the complaint:

> We are persuaded that the *per se* standard of concerted activity, by which the Board determines what *ought to be* of group concern and then artificially presumes it *is* of group concern, is at odds with the Act. The Board and courts always considered first, whether the activity is concerted, and only then, whether it is protected. This approach is mandated by the statute itself, which requires that a particular form of individual activity be both "concerted" and "protected." A Board finding that a particular form of individual activity warrants group support is not a sufficient basis for labeling the activity "concerted" within the meaning of Section 7.

The NLRB proceeded to adopt what is now called the *Meyers* standard:

> To find an employee's activity to be "concerted," we shall require that it be engaged in with or on the authority of other employees, and not solely by and on behalf of the employee himself. . . . [U]nder the standard we now adopt, the question of whether an employee engaged in concerted activity is, at its heart, a factual one, the fate of a particular case rising or falling on the record evidence.

Applying this new standard, the NLRB found that Prill was acting alone.

> Prill alone refused to drive the truck and trailer; he alone contacted the Tennessee Public Service Commission. . . . Prill acted solely on his own behalf. . . . [*I*]*ndividual* employee concern, even if openly manifested by several employees on an *individual* basis, is not sufficient to prove concerted for action.
>
> Although it might be argued that a solitary over-the-road truckdriver would be hard pressed to enlist the support of coworkers while away from the home terminal, the Board . . . is neither God nor the Department of Transportation. . . . We are not empowered to correct all immorality or even illegality arising under the total fabric of Federal and state laws.

The NLRB decision was reviewed by the D.C. Court of Appeals in *Prill v. NLRB*, 755 F.2d 941 (D.C. Cir. 1985). The Court concluded that the NLRB erred when it decided that its new definition of concerted activity was *mandated* by the NRLA. While noting that the NLRB has broad authority to construe the NLRA, and that "in appropriate circumstances, the Board even may elect to abandon or modify established precedent," the Court found that the Board misconstrued the bounds of the law, based its opinion on a faulty legal premise, and failed to provide an adequate rationale for its new standard. Citing a recent Supreme Court case, *NLRB v. City Disposal Systems*, 465 U.S. 822 (1984), the Circuit Court held that

Contrary to the Board's view in *Meyers*, neither the language nor the history of section 7 requires that the term "concerted activities" be interpreted to protect only the most narrowly defined forms of common action by employees. . . . [In addition] the Board failed even to consider whether the discharge of an employee because of safety complaints would discourage other employees from engaging in collective activity to improve working conditions.

On remand, the NLRB reconsidered the case in light of the D.C. Circuit Court's opinion, but declined to change its mind. The Meyers definition of concerted activity, the Board concluded, represents a reasonable construction of Section 7 of the Act. "Consistent with *City Disposal*, . . . we have exercised our discretion and have chosen [this] definition over other possibly permissible standards." 281 N.L.R.B. No. 118 (1986).

◇ NOTES

1. Since the *Meyers* decision, it has become more difficult for an individual to find protection from retaliatory discharge under Section 7 of the NLRA. When time permits it, however, it is still possible for the employee to contact his or her fellow workers, obtain their support, and transform an individual action into a collective one.

2. In addition, other avenues are available. A protection more recently put into place will help truckers like Mr. Prill and others who drive commercial vehicles. In 1982, Congress enacted the Surface Transportation Assistance Act (STAA), 49 U.S.C. §2305. The Act protects employees in the commercial motor transportation industry from being discriminated against in retaliation for refusing to operate a motor vehicle that does not comply with applicable state and federal safety regulation or for filing complaints alleging noncompliance. The statute empowers the Secretary of Labor to conduct an initial investigation of the employee's discharge and, upon a finding of reasonable cause to believe that the employee was discharged in violation of the STAA, requires the Secretary to issue an order directing the employer to reinstate the employee. The employer may request an evidentiary hearing and a final decision from the Secretary, but this request does not operate as a stay of the preliminary order to reinstate. The Supreme Court has upheld the authority of the Secretary to order reinstatement *prior* to an evidentiary hearing. *Brock v. Roadway Express, Inc.*, 481 U.S. 252 (1987).

2. The Occupational Safety and Health Act

Workers covered by OSHA also have a right to refuse hazardous work under the OSHAct. Rather than being explicitly set forth in the statute itself, however, the right is the creation of administrative rulemaking. The Whirlpool Corporation challenged OSHA's authority to issue such a regulation, but the agency's action was upheld by a unanimous decision of the Supreme Court in 1980.

Whirlpool Corporation v. Marshall

445 U.S. 1 (1980)

Mr. Justice STEWART delivered the opinion of the Court.

The Occupational Safety and Health Act of 1970 (Act) prohibits an employer from discharging or discriminating against any employee who exercises "any right afforded by" the Act.[2] The Secretary of Labor (Secretary) has promulgated a regulation providing that, among the rights that the Act so protects, is the right of an employee to choose not to perform his assigned task because of a reasonable apprehension of death or serious injury coupled with a reasonable belief that no less drastic alternative is available.[3] The question presented in the case before us is whether this regulation is consistent with the Act.

[2]Section 11(c)(1) of the Act, 29 U.S.C. §660(c)(1), provides in full:

No person shall discharge or in any manner discriminate against any employee because such employee has filed any complaint or instituted or caused to be instituted any proceeding under or related to this Act or has testified or is about to testify in any such proceeding or because of the exercise by such employee on behalf of himself or other of any right afforded by this chapter.

[3]The regulation, 29 CFR §1977.12, 38 Fed. Reg. 2681, 2683 (1973), as corrected 38 Fed. Reg. 4577 (1973), provides in full:

(a) In addition to protecting employees who file complaints, institute proceedings, or testify in proceedings under or related to the Act, section 11(c) also protects employees from discrimination occurring because of the exercise 'of any right afforded by this Act.' Certain rights are explicitly provided in the Act; for example, there is a right to participate as a party in enforcement proceedings (sec. 10). Certain other rights exist by necessary implication. For example, employees may request information from the Occupational Safety and Health Administration; such requests would constitute the exercise of a right afforded by the Act. Likewise, employees interviewed by agents of the Secretary in the course of inspections or investigations could not subsequently be discriminated against because of their cooperation.

(b)(1) On the other hand, review of the Act and examination of the legislative history discloses that, as a general matter, there is no right afforded by the Act which would entitle employees to walk off the job

I

The petitioner company maintains a manufacturing plant in Marion, Ohio, for the production of household appliances. Overhead conveyors transport appliance components throughout the plant. To protect employees from objects that occasionally fall from these conveyors, the petitioner has installed a horizontal wire mesh guard screen approximately 20 feet above the plant floor. This mesh screen is welded to angle-iron frames suspended from the building's structural steel skeleton.

Maintenance employees of the petitioner spend several hours each week removing objects from the screen, replacing paper spread on the screen to catch grease drippings from the material on the conveyors, and performing occasional maintenance

because of potential unsafe conditions at the workplace. Hazardous conditions which may be violative of the Act will ordinarily be corrected by the employer, once brought to his attention. If corrections are not accomplished, or if there is dispute about the existence of a hazard, the employee will normally have opportunity to request inspection of the workplace pursuant to section 8(f) of the Act, or to seek the assistance of other public agencies which have responsibility in the field of safety and health. Under such circumstances, therefore, an employer would not ordinarily be in violation of section 11(c) by taking action to discipline an employee for refusing to perform normal job activities because of alleged safety or health hazards.

(b)(2) However, occasions might arise when an employee is confronted with a choice between not performing assigned tasks or subjecting himself to serious injury or death arising from a hazardous condition at the workplace. If the employee, with no reasonable alternative, refuses in good faith to expose himself to the dangerous condition, he would be protected against subsequent discrimination. The condition causing the employee's apprehension of death or injury must be of such a nature that a reasonable person, under the circumstances then confronting the employee, would conclude that there is a real danger of death or serious injury and that there is insufficient time, due to the urgency of the situation, to eliminate the danger through resort to regular statutory enforcement channels. In addition, in such circumstances, the employee, where possible, must also have sought from his employer, and been unable to obtain, a correction of the dangerous condition.

work on the conveyors themselves. To perform these duties, maintenance employees usually are able to stand on the iron frames, but sometimes find it necessary to step onto the steel mesh screen itself.

In 1973 the company began to install heavier wire in the screen because its safety had been drawn into question. Several employees had fallen partly through the old screen, and on one occasion an employee had fallen completely through to the plant floor below but had survived. A number of maintenance employees had reacted to these incidents by bringing the unsafe screen conditions to the attention of their foremen. The petitioner company's contemporaneous safety instructions admonished employees to step only on the angle-iron frames.

On June 28, 1974, a maintenance employee fell to his death through the guard screen in an area where the newer, stronger mesh had not yet been installed. Following this incident, the petitioner effectuated some repairs and issued an order strictly forbidding maintenance employees from stepping on either the screens or the angle-iron supporting structure. An alternative but somewhat more cumbersome and less satisfactory method was developed for removing objects from the screen. This procedure required employees to stand on power-raised mobile platforms and use hooks to recover the material.

On July 7, 1974, two of the petitioner's maintenance employees, Virgil Deemer and Thomas Cornwell, met with the plant maintenance superintendent to voice their concern about the safety of the screen. The superintendant disagreed with their view, but permitted the two men to inspect the screen with their foreman and to point out dangerous areas needing repair. Unsatisfied with the petitioner's response to the results of this inspection, Deemer and Cornwell met on July 9 with the plant safety director. At that meeting, they requested the name, address, and telephone number of a representative of the local office of the Occupational Safety and Health Administration (OSHA). Although the safety director told the men that they "had better stop and think about what [they] were doing," he furnished the men with the information they requested. Later that same day, Deemer contacted an official of the regional OSHA office and discussed the guard screen.

The next day, Deemer and Cornwell reported for the night shift at 10:45 p.m. Their foreman, after himself walking on some of the angle-iron frames, directed the two men to perform their usual maintenance duties on a section of the old screen. Claim-

ing that the screen was unsafe, they refused to carry out this directive. The foreman then sent them to the personnel office, where they were ordered to punch out without working or being paid for the remaining six hours of the shift. The two men subsequently received written reprimands, which were placed in their employment files.

A little over a month later, the Secretary filed suit in the United States District Court for the Northern District of Ohio, alleging that the petitioner's actions against Deemer and Cornwell constituted discrimination in violation of §11(c)(1) of the Act. As relief, the complaint prayed, *inter alia,* that the petitioner be ordered to expunge from its personnel files all references to the reprimands issued to the two employees, and for a permanent injunction requiring the petitioner to compensate the two employees for the six hours of pay they had lost by reason of their disciplinary suspensions.

Following a bench trial, the District Court found that the regulation in question justified Deemer's and Cornwell's refusals to obey their foreman's order on July 10, 1974. The court found that the two employees had "refused to perform the cleaning operation because of a genuine fear of death or serious bodily harm," that the danger presented had been "real and not something which [had] existed only in the minds of the employees," that the employees had acted in good faith, and that no reasonable alternative had realistically been open to them other than to refuse to work. The District Court nevertheless denied relief, holding that the Secretary's regulation was inconsistent with the Act and therefore invalid. *Usery v. Whirlpool Corp.,* 416 F. Supp. 30, 32-34.

The Court of Appeals for the Sixth Circuit reversed the District Court's judgment. *Marshall* v. *Whirlpool Corp.,* 593 F.2d 715. Finding ample support in the record for the District Court's factual determination that the actions of Deemer and Cornwell had been justified under the Secretary's regulation, *id.,* at 719, n.5, the appellate court disagreed with the District Court's conclusion that the regulation is invalid. *Id.,* at 721-736. It accordingly remanded the case to the District Court for further proceedings. *Id.,* at 736. We granted certiorari, 444 U.S. 823, because the decision of the Court of Appeals in this case conflicts with those of two other Courts of Appeals on the important question in issue. See *Marshall* v. *Daniel Construction Co.,* 563 F.2d 707 [6 OSHC 1031] (CA5 1977); *Marshall* v. *Certified Welding Corp.,*—F.2d—[7 OSHC 1069] (CA10 1978). That question, as stated at the outset of this opinion, is whether the Secretary's regulation authorizing employee "self-help" in some circumstances, 29 CFR §1977.12(b)(2), is permissible under the Act.

II

The Act itself creates an express mechanism for protecting workers from employment conditions believed to pose an emergent threat of death or serious injury. Upon receipt of an employee inspection request stating reasonable grounds to believe that at imminent danger is present in a workplace, OSHA must conduct an inspection. 29 U.S.C. §657(f)(1). In the event this inspection reveals workplace conditions or practices that "could reasonably be expected to cause death or serious physical harm immediately or before the imminence of such danger can be eliminated through the enforcement procedures otherwise provided by" the Act, 29 U.S.C. §662(a), the OSHA inspector must inform the affected employees and the employer of the danger and notify them that he is recommending to the Secretary that injunctive relief be sought. 29 U.S.C. §662(c). At this juncture, the Secretary can petition a federal court to restrain the conditions or practices giving rise to the imminent danger. By means of a temporary restraining order or preliminary injunction, the court may then require the employer to avoid, correct, or remove the danger or to prohibit employees from working in the area. 29 U.S.C. §662(a).

To ensure that this process functions effectively, the Act expressly accords to every employee several rights, the exercise of which may not subject him to discharge or discrimination. An employee is given the right to inform OSHA of an imminently dangerous workplace condition or practice and request that OSHA inspect that condition or practice. 29 U.S.C. §657(f)(1). He is given a limited right to assist the OSHA inspector in inspecting the workplace, 29 U.S.C. §§657(a)(2), (e) and (f)(2), and the right to aid a court in determining whether or not a risk of imminent danger in fact exists. See 29 U.S.C. §660(c)(1). Finally, an affected employee is given the right to bring an action to compel the Secretary to seek injunctive relief if he believes the Secretary has wrongfully declined to do so. 29 U.S.C. §662(d).

In the light of this detailed statutory scheme, the Secretary is obviously correct when he acknowledges in his regulation that, "as a general matter, there is no right afforded by the Act which would entitle employees to walk off the job because of

potential unsafe conditions at the workplace." By providing for prompt notice to the employer of an inspector's intention to seek an injunction against an imminently dangerous condition, the legislation obviously contemplates that the employer will normally respond by voluntarily and speedily eliminating the danger. And in the few instances where this does not occur, the legislative provisions authorizing prompt judicial action are designed to give employees full protection in most situations from the risk of injury or death resulting from an imminently dangerous condition at the worksite.

As this case illustrates, however, circumstances may sometimes exist in which the employee justifiably believes that the express statutory arrangement does not sufficiently protect them from death or serious injury. Such circumstances will probably not often occur, but such a situation may arise when (1) the employee is ordered by his employer to work under conditions that the employee reasonably believes pose an imminent risk of death or serious bodily injury, and (2) the employee has reason to believe that there is not sufficient time or opportunity either to seek effective redress from his employer or to apprise OSHA of the danger.

Nothing in the Act suggests that those few employees who have to face this dilemma must rely exclusively on the remedies expressly set forth in the Act at the risk of their own safety. But nothing in the Act explicitly provides otherwise. Against this background of legislative silence, the Secretary has exercised his rulemaking power under 29 U.S.C. §657(g) (2) and has determined that, when an employee in good faith finds himself in such a predicament, he may refuse to expose himself to the dangerous condition, without being subjected to "subsequent discrimination" by the employer.

The question before us is whether this interpretative regulation[15] constitutes a permissible gloss on the Act by the Secretary, in light of the Act's language, structure, and legislative history. Our inquiry is informed by an awareness that the regulation is entitled to deference unless it can be said not to be a reasoned and supportable interpretation of the Act. *Skidmore* v. *Swift & Co.*, 323 U.S. 134, 139–140. See

---

[15]The petitioner has raised no issue concerning whether or not this regulation was promulated in accordance with the procedural requirements of the Administrative Act (APA), 5 U.S.C. §553. Thus, we accept the Secretary's designation of the regulation as "interpretative," and do not consider whether it qualifies as an "interpretative rule" within the meaning of the APA, 5 U.S.C. §553(b)(A).

Ford Motor Credit Co. v. *Milhollin*, 444 U.S. 555; *Mourning* v. *Family Publications Service, Inc.*, 411 U.S. 356.

A

The regulation clearly conforms to the fundamental objective of the Act—to prevent occupational deaths and serious injuries.[16] The Act, in its preamble, declares that its purpose and policy is "to assure so far as possible every working man and woman in the Nation safe and healthful working conditions and to *preserve* our human resources. . . ." 29 U.S.C. §651(b). (Emphasis added.)

To accomplish this basic purpose, the legislation's remedial orientation is prophylactic in nature. See *Atlas Roofing Co.* v. *Occupational Safety Comm'n*, 430 U.S. 442, 444-445. The Act does not wait for an employee to die or become injured. It authorizes the promulgation of health and safety standards and the issuance of citations in the hope that these will act to prevent deaths or injuries from ever occurring. It would seem anomalous to construe an Act so directed and constructed as prohibiting an employee, with no other reasonable alternative, the freedom to withdraw from a workplace environment that he reasonably believes is highly dangerous.

Moreover, the Secretary's regulation can be viewed as an appropriate aid to the full effectuation of the Act's "general duty" clause. That clause provides that "[e]ach employer . . . shall furnish to each of his employees employment and a place of employment which are free from recognized hazards that are causing or are likely to cause death or serious physical harm to his employees." 29 U.S.C. §654(a)(1). As

the legislative history of this provision reflects, it was intended itself to deter the occurrence of occupational deaths and serious injuries by placing on employers a mandatory obligation independent of the specific health and safety standards to be promulgated by the Secretary. Since OSHA inspectors cannot be present around the clock in every workplace, the Secretary's regulation ensures that employees will in all circumstances enjoy the rights afforded them by the "general duty" clause.

The regulation thus on its face appears to further the overriding purpose of the Act, and rationally to complement its remedial scheme.[18] In the absence of some contrary indication in the legislative history, the Secretary's regulation must, therefore, be upheld, particularly when it is remembered that safety legislation is to be liberally construed to effectuate the congressional purpose. *United States* v. *Bacto-Unidisk*, 394 U.S. 784, 798; *Lilly* v. *Grand Trunk R. Co.*, 317 U.S. 481, 486.

B

In urging reversal of the judgment before us, the petitioner relies primarily on two aspects of the Act's legislative history.

1. Representative Daniels of New Jersey sponsored one of several House bills that led ultimately to the passage of the Act. As reported to the House by the Committee on Education and Labor, the Daniels bill contained a section that was soon dubbed the "strike with pay" provision. This section provided that employees could request an HEW examination of the toxicity of any materials in their workplace. If that examination revealed a workplace substance that had "potentially toxic or harmful effects in such concentration as used or found," the employer was given 60 days to correct the potentially dangerous condition. Following the expiration of that period, the employer could not require that an employee be

---

[16]The Act's legislative history contains numerous references to the Act's preventive purpose and to the tragedy of each individual death or accident. See, *e.g.*, S. Rep. No. 1282, 91st Cong., 2d Sess., 2 (1970), Leg. Hist. 142; 116 Cong. Rec. 37628 (1970), Leg. Hist. 516-517 (Sen. Nelson); 116 Cong. Rec. 37628, 37630 (1970), Leg. Hist. 518, 522 (Sen. Cranston); 116 Cong. Rec. 37630 (1970), Leg. Hist. 522-523 (Sen. Randolph); H. R. Rep. No. 1291, 91st Cong., 2d Sess., 14, 23 (1970), Leg. Hist. 844, 853; 116 Cong. Rec. 38366 (1970), Leg. Hist. 978 (Rep. Young); 116 Cong. Rec. 38367-38368 (1970), Leg. Hist. 981 (Rep. Anderson); 116 Cong. Rec. 38386 (1970), Leg. Hist. 1031, 1032 (Rep. Dent); 116 Cong. Rec. 42203 (1970), Leg. Hist. 1210 (Rep. Daniels). As stated by Senator Yarborough, a sponsor of the Senate bill:

> We are talking about people's lives, not the indifference of some cost accountants. We are talking about assuring the men and women who work in our plants and factories that they will go home after a day's work with their bodies intact.

[18]It is also worth noting that the Secretary's interpretation of 29 U.S.C. §660(c)(1) conforms to the interpretation that Congress clearly wished the courts to give to the parallel antidiscrimination provision of the Federal Mine Safety and Health Amendments Act of 1977, 30 U.S.C. §801 *et seq.* The legislative history of that provision, 30 U.S.C. §815(c)(1), establishes that Congress intended it to protect "the refusal to work in conditions which are believed to be unsafe or unhealthful." S. Rep. No. 181, 95th Cong., 1st Sess., 35 (1977). See *id.*, at 36; 123 Cong. Rec. S10287-S10288 (daily ed. June 21, 1977) (remarks of Sen. Church, Sen. Williams, Sen. Javits).

exposed to toxic concentrations of the substance unless the employee was informed of the hazards and symptoms associated with the substance, the employee was instructed in the proper precautions for dealing with the substance, and the employee was furnished with personal protective equipment. If these conditions were not met, an employee could "absent himself from such risk of harm for the period necessary to avoid such danger without loss of regular compensation for such period."

This provision encountered still opposition in the House. Representative Steiger of Wisconsin introduced a substitute bill containing no "strike with pay" provision. In response, Representative Daniels offered a floor amendment that, among other things, deleted his bill's "strike with pay" provision. He suggested that employees instead be afforded the right to request an immediate OSHA inspection of the premises, a right which the Steiger bill did not provide. The House ultimately adopted the Steiger Bill.

The bill that was reported to and, with a few amendments, passed by the Senate never contained a "strike with pay" provision. It did, however, give employees the means by which they could request immediate Labor Department inspections. These two characteristics of the bill were underscored on the floor of the Senate by Senator Williams, the bill's sponsor.

After passage of the Williams bill by the Senate, it and the Steiger bill were submitted to a conference committee. There, the House acceded to the Senate bill's inspection request provisions.

The petitioner reads into this legislative history a congressional intent incompatible with an administrative interpretation of the Act such as is embodied in the regulation at issue in this case. The petitioner argues that Congress' overriding concern in rejecting the "strike with pay" provision was to avoid giving employees a unilateral authority to walk off the job which they might abuse in order to intimidate or harass their employer. Congress deliberately chose instead, the petitioner maintains, to grant employees the power to request immediate administrative inspections of the workplace which could in appropriate cases lead to coercive judicial remedies. As the petitioner views the regulation, therefore, it gives to workers precisely what Congress determined to withhold from them.

We read the legislative history differently. Congress rejected a provision that did not concern itself at all with conditions posing real and immediate threats of death or severe injury. The remedy which

the rejected provision furnished employees could have been invoked only after 60 days had passed following HEW's inspection and notification that improperly high levels of toxic substances were present in the workplace. Had that inspection revealed employment conditions posing a threat of imminent and grave harm, the Secretary of Labor would presumably have requested, long before expiration of the 60-day period, a court injunction pursuant to other provisions of the Daniels bill. Consequently, in rejecting the Daniels bill's "strike with pay" provision, Congress was not rejecting a legislative provision dealing with the highly perilous and fast-moving situations covered by the regulation now before us.

It is also important to emphasize that what primarily troubled Congress about the Daniels bill's "strike with pay" provision was its requirement that employees be paid their regular salary after having properly invoked their right to refuse to work under the section.[29] It is instructive that virtually every time the issue of an employee's right to absent himself from hazardous work was discussed in the legislative debates, it was in the context of the employee's right to continue to receive his usual compensation.

When it rejected the "strike with pay" concept, therefore, Congress very clearly meant to reject a law unconditionally imposing upon employers an obligation to continue to pay their employees their regular pay checks when they absented themselves from work for reasons of safety. But the regulation at issue

[29]Congress' concern necessarily was with the provision's compensation requirement. The law then, as it does today, already afforded workers a right, under certain circumstances, to walk off their jobs when faced with hazardous conditions. See 116 Cong. Rec. 42208 (1970), Leg. Hist. 1223-1224 (Rep. Scherle) (reference to Taft-Hartley Act), Under Section 7 of the National Labor Relations Act (NLRA), 29 U.S.C. §157, employees have a protected right to strike over safety issues. See *NLRB* v. *Washington Aluminum Co.*, 370 U.S. 9. Similarly, Section 502 of the Labor Management Relations Act, 29 U.S.C. §143, provides that "the quitting of labor by an employee or employees in good faith because of abnormally dangerous conditions for work at the place of employment of such employee or employees [shall not] be deemed a strike." The effect of this section is to create an exception to a no-strike obligation in a collective-bargaining agreement. *Gateway Coal Co.* v. *United Mine Workers,* 414 U.S. 368, 385 [1 OSHC 1461, 1469].

The existence of these statutory rights also make clear that the Secretary's regulation does not conflict with the general pattern of federal labor legislation in the area of occupational safety and health. See also 29 CFR §1977.18.

here does not require employers to pay workers who refuse to perform their assigned tasks in the face of imminent danger. It simply provides that in such cases the employer may not "discriminate" against the employees involved. An employer "discriminates" against an employee only when he treats that employee less favorably than he treats other similarly situated.

2. The second aspect of the Act's legislative history upon which the petitioner relies is the rejection by Congress of provisions contained in both the Daniels and the Williams bills that would have given Labor Department officials, in imminent danger situations, the power temporarily to shut down all or part of an employer's plant. These provisions aroused considerable opposition in both Houses of Congress. The hostility engendered in the House of Representatives led Representative Daniels to delete his version of the provision in proposing amendments to his original bill. The Steiger bill that ultimately passed the House gave the Labor Department no such authority. The Williams Bill, as approved by the Senate, did contain an administrative shutdown provision, but the conference committee rejected this aspect of the Senate bill.

The petitioner infers from these events a congressional will hostile to the regulation in question here. The regulation, the petitioner argues, provides employees with the very authority to shut down an employer's plant that was expressly denied a more expert and objective United States Department of Labor.

As we read the pertinent legislative history, however, the petitioner misconceives the thrust of Congress' concern. Those in Congress who prevented passage of the administrative shutdown provisions in the Daniels and Williams bills were opposed to the unilateral authority those provisions gave to federal officials, without any judicial safeguards, drastically to impair the operation of an employer's business. Congressional opponents also feared that the provisions might jeopardize the Government's otherwise neutral role in labor-management relations.

Neither of these congressional concerns is implicated by the regulation before us. The regulation accords no authority to government officials. It simply permits private employees of a private employer to avoid workplace conditions that they believe pose grave dangers to their own safety. The employees have no power under the regulation to order their employer to correct the hazardous condition or to clear the dangerous workplace of others. Moreover, any employee who acts in reliance on the regulation runs the risk of discharge or reprimand in the event a court subsequently finds that he acted unreasonably or in bad faith. The regulation, therefore, does not remotely resemble the legislation that Congress rejected.

C

For these reasons we conclude that 29 CFR §1977.12(b)(2) was promulgated by the Secretary in the valid exercise of his authority under the Act. Accordingly, the judgment of the Court of Appeals is affirmed.

◇ NOTES

1. The OSHA regulation was promulgated without public notice and comment; why was this deemed permissible? (See footnote 15 of the Supreme Court's opinion.)

2. The regulation was promulgated under Section 11(c) of the Act (discussed below), which prohibits discrimination against workers for exercising a "right" granted them by the OSHAct. What is the "right" being protected here? Where—if at all—is it found in the Act?

3. For an analysis of *Whirlpool* and the distinctions between Medical Removal Protection (MRP), "strike with pay," and the right to refuse hazardous work, see note 69 in the D.C. Circuit lead case in the section on Medical Removal Protection in Chapter 3.

4. To what extent is the OSHA right to refuse hazardous work likely to be available as a means of avoiding occupational exposures that cause latent disease?

5. On remand in the *Whirlpool* case, the district court ordered the employer to 1) expunge from the employees' personnel files the written reprimands that had been issued to them for their refusal to climb out onto the screen and 2) pay them the wages they lost for the time they were absent from work. However, the court declined to issue an injunction preventing Whirlpool from violating Section 11(c) in the future. *Marshall v. Whirlpool Corp.,* 9 OSHC 1038 (N.D. Ohio 1980).

D.   PROTECTIONS FROM OTHER RETALIATORY EMPLOYMENT PRACTICES

Beyond the limited legal protections available to workers who wish to avoid hazardous working conditions, other laws protect employees from specific types of employer retaliation. Several are discussed below. None of these laws are all-encompassing, however. Before assuming protection under one of these legal mechanisms, the employee must carefully determine if the "right" being protected actually covers his or her particular situation.

1.   Section 11(c) Of The OSHAct

For the OSHA-covered employee concerned about workplace safety and health, the logical starting point is Section 11(c) of the OSHAct. As discussed in the *Whirlpool* case, 11(c) specifies that no employer

> shall discharge or in any manner discriminate against any employee because such employee has [a] filed any complaint or caused to be instituted any proceeding under or related to this Act or [b] has testified or is about to testify in any such proceeding or [c] because of the exercise by such employee on behalf of himself or other of any right afforded by this chapter.

This provision was designed to prevent employers from retaliating against workers who take actions in support of safety and health under the Act. But how does it work, and how broad is its coverage?

(a)   The scope of the protection

By its terms, Section 11(c) only protects an "employee." What about the *prospective* employee, such as the workplace health and safety activist who finds that she cannot get a job in a new community because she has been "blacklisted" throughout the industry? OSHA has interpreted 11(c) as including prospective and former employees, as well as present employees. This would certainly appear to be consistent with the overall purpose of the OSHAct. At first blush, however, it might also appear inconsistent with the definition of "employee" found in Section 3(6) of the Act. There, an "employee" is described, rather curtly, as "an employee of an employer [who is covered by OSHA]." This might be read to suggest that an employee must be currently employed at the time of the alleged discriminatory treatment. Especially in light of the broad remedial purpose of Section 11(c), such a narrow reading of this language is not warranted. Rather, a close reading of the language and purpose of the Act indicates that Section

11(c) applies if the employee was employed either *while the discriminatory conduct occurred* or *while engaging in the action at which the discrimination is directed.*

There is no stated limitation on the type or source of the discrimination against which the employee is protected. Section 11(c) applies broadly to any "person" who discriminates "in any manner" against an employee. Interestingly, then, the prohibition against discrimination extends, beyond employers, to anyone who takes retaliatory action against the employee for having engaged in a protected activity under 11(c).

What are these protected activities? In essence, they relate to the initiation of or participation in proceedings "under or related to" the OSHAct, or the exercise "of any right" afforded by the Act. But what are proceedings "related to" the OSHAct? How close must the "relation" be? Clearly, actions directly mentioned in the Act or in an OSHA regulation are protected. These include filing an OSHA complaint, requesting or participating in an OSHA inspection, or refusing hazardous work in a manner consistent with the OSHA regulation. One might also argue more broadly that as long as the *subject matter* of the proceeding is the same as that of the OSHAct (i.e., workplace safety and health), the action is protected. See *Marshall v. Springville Poultry Farm,* 445 F. Supp. 2 (M.D. Pa. 1977); *Donovan v. R.D. Anderson Constr. Co.,* 552 F. Supp. 249 (D. Kan. 1982)

◇ **NOTE**

1. Consider the situation of a workplace health and safety activist who engages in her advocacy activity while she is *not* an "employee" under the OSHAct—when she is a governmental employee, for example. If later she is discriminated against when she seeks employment in an OSHA-covered workplace, we could say that she *would have been an "employee" under the Act but for the discrimination.* Under OSHA's interpretation of 11(c), she would be protected. Construct an argument for and against this interpretation of the Act.

◇

(b) Proof of discrimination

How does one prove that the discrimination was actually retaliatory—that it was directed at the employee's exercise of a protected activity under the statute and not motivated by employee performance or unrelated business concerns? This is a question for the trier of fact.

Marshall v. Commonwealth Aquarium

611 F.2d 1 (1st Cir. 1979)

CAMPBELL, Circuit Judge.

Commonwealth Aquarium (Aquarium), a pet store, appeals a decision of the United States District Court for the District of Massachusetts finding it in violation of Section 11(c)(1) of the Occupational Safety and Health Act of 1970, (OSHA), 29 U.S.C. §660(c)(1).

*   *   *

This case arises from the discharge of Jeffrey Boxer as manager of Aquarium on May 15, 1976. The complaint filed by the Secretary of Labor alleged that Aquarium, acting through one of its owners, Richard Lerner, discharged Boxer because he had engaged in activity protected by OSHA by reporting a potential health hazard at the store to health and labor authorities.[1] The hazard was created by the likelihood that several birds in Aquarium's inventory had contracted psittacosis or "parrot fever," a respiratory ailment which may be fatal to humans.

In determining that Aquarium had violated Section 11(c), the district court followed the format laid out in *Mt. Healthy City Board of Education v. Doyle,* 429 U.S. 274, 97 S. Ct. 568, 50 L. Ed. 2d 471 (1977). The court found Boxer's notification of health and labor authorities to be "a significant and substantial proximate causative factor for [Aquarium's] decision to discharge him," and next determined that Aquarium had "failed to prove by a preponderance of the evidence that [it] would have discharged Boxer even in the absence of the protected activity." Aquarium now complains that "[w]hile Boxer's complaints to health and labor authorities may have been the 'last straw' leading to the termination of his employment, which is not conceded," the evidence did not sufficiently demonstrate that Boxer's engaging in protected activity was a substantial factor precipitating his discharge. According to Aquarium, Boxer's discharge was the culmination of five months of employer dissatisfaction with his job performance.

Even, however, if we were so inclined, we would not be free to redetermine these matters. The district court's findings must stand unless clearly erroneous. *Cf. Sweeney v. Board of Trustees of Keene State College,* 604 F.2d 106 (1st Cir. 1979). Examination of the record indicates that the court's findings as to Aquarium's motivation in discharging Boxer are amply founded and not clearly erroneous.

Aquarium's next contention is that the district court altogether ignored certain evidence of purported legitimate business reasons for Boxer's discharge. In support of this, Aquarium isolates the following passage from the court's opinion:

> Apart from Lerner's testimony as to the deterioration of his relationship with Boxer prior to May 1976 which has been discussed and disbelieved above there is nothing in the record upon which the Court could base a finding that Lerner would have reached the same decision even in the absence of Boxer's complaints regarding the safety of his working conditions.

This passage is said to demonstrate a failure to consider, 1) the testimony of Martha Parks, an employee

[1]The underlying incident is explored in detail by the district court. *See* Marshall v. Commonwealth Aquarium, 469 F. Supp. 690 (D. Mass. 1979). Aquarium was sold by its owners approximately one year prior to trial.

of Boston Pet, one of Aquarium's "sister" stores, which revealed that she had encountered some problems with Boxer concerning the joint ordering of birds and that she had related those problems to Lerner; and 2) a so-called list of grievances which, according to Lerner, he drew up some four months prior to Boxer's discharge, itemizing employer complaints with Boxer's performance.

The court's comment does not, however, compel the conclusion that it totally overlooked this evidence. Nothing else in the record or opinion suggests such an extraordinary oversight. All the court seems to have meant to say is that the other evidence of employer dissatisfaction was too weak to constitute the basis for a finding of legitimate discharge. This was a reasonable conclusion, given the court's express discrediting of Lerner. Martha Parks' testimony by itself scarcely indicated conduct serious enough to lead to discharge; and the undated grievance list was not only of questionable weight and significance, but depended largely on Lerner's rejected credibility. We do not, therefore, regard the court's statement as indicating more than a warranted, if imperfectly phrased, discrediting of defendant's evidence.

Aquarium's final points concern the district court's exclusion of evidence allegedly relevant on the issue of Lerner's true motivation in firing Boxer. On direct examination Boxer had testified that after learning of the potential health hazard in the store he informed Lerner and suggested the taking of various precautions. Lerner, according to Boxer, responded by stating "Either conduct, just conduct business as usual or I will find somebody else who will conduct business as usual." Boxer further testified that Lerner similarly warned him on two subsequent occasions. Lerner, on taking the stand, denied threatening Boxer with discharge if he failed to conduct business as usual.[2] Lerner, however, did admit to having a discussion with Boxer concerning the feeding of birds. Lerner testified that "In the beginning of the psitta-

[2]Aquarium suggests in its brief that:
> "Even assuming he credibility of Boxer's own testimony that Lerner had admonished him that he would be fired if he did not conduct business as usual, those statements were, in each instance, directed at Boxer's reluctance to sell birds and not his complaints to health and labor authorities. Refusal to fulfill obligations as an employee, even in light of a potential or real health hazard, has been held not to constitute a protected activity."

To support this last proposition, Aquarium cites Usery v. Whirlpool Corp., 416 F. Supp. 30 (N.D. Ohio 1976). In *Whirlpool* the district court held invalid a regulation promulgated by the Secretary of Labor which interpreted

cosis situation [Boxer] refused to water and feed the birds and I told him that I would get someone from Boston Pet to water and feed them, if he didn't do it." Lerner was questioned by Aquarium's counsel as to his intent in making that remark. The district court

OSHA's retaliatory discharge provision as protecting an employee who reasonably withdraws from danger on the job where that employee has sought correction of the dangerous condition and where normal OSHA enforcement procedures would be inadequate. Without becoming involved in this issue, which is not essential to the resolution of the present case, we point out that Aquarium's reliance on *Whirlpool* is entirely misplaced since that decision was reversed by the Sixth Circuit. Marshall v. Whirlpool Corp., 593 F.2d 715 (1979). *But see* Marshall v. Daniel Construction Co., 563 F.2d 707 (5th Cir. 1977), cert. denied, 439 U.S. 880, 99 S. Ct. 216, 58 L. Ed. 2d 192 (1978).

sustained an objection to this question despite counsel's insistence that intent was "one of the primary issues in this case." We would agree that in general, where a defendant's intent is in issue, he should be given latitude to testify concerning his intent, however much the court may think such testimony "'rationalization.'" *Whiting v. United States*, 296 F.2d, 512, 519 (1st Cir. 1961); *United States v. Hayes*, 477 F.2d 868, 873 (10th Cir. 1973); Wigmore, Evidence §581 (3d ed. 1940). In the present case, however, Lerner testified at other times on the subject of his intent and, especially in the absence of an offer of proof, *see* Fed. R. Evid. 103(a)(2), we have no reason to believe that the excluded testimony would have been more than cumulative on the subject. *See* Fed. R. Evid. 403. We can discern no prejudice in any event.

*   *   *

◇ NOTES

1. Two important points emerge from this opinion. One concerns the nature of the factual burdens borne by each party. The approach taken here by the district court, and approved by First Circuit Court of Appeals, was adapted from the approach that had been followed in discrimination cases brought under Title VII of the Civil Rights Act. The burden is initially on the Labor Department (on behalf of the employee) to prove that the employee's exercise of the protected activity was, in the words of the district court, "a significant and substantial proximate causative factor" in the employee's discharge. If this burden is carried, the burden then *shifts* to the employer to prove that it would have discharged the employee even in the absence of the protected activity. However, the Supreme Court has recently held that the plaintiff in so-called "disparate impact" cases brought under Title VII always retains the ultimate burden of persuasion (see page 432 for a discussion of this case). Although the relevance of this decision to OSHA 11(c) cases is not clear, it may mean that, while the burden is still on the employer to *come forward with evidence* to indicate that the employee was fired for reasons other than the exercise of protected activity, the employee has the burden of convincing the judge (or jury) that this evidence is not persuasive.

2. The second point worth noting is that the trial court has considerable discretion in weighing factual matters such as these. As noted in this case, the appellate court is not to disturb the *factual* findings of the trial court unless they are "clearly erroneous."

3. In the *Commonwealth Aquarium* case there is testimony (albeit disputed) about a "smoking gun"—that is, a statement by the employer that the discrimination was motivated by the employee's notification of health and labor authorities. Quite often, however, there is no direct evidence of this nature. Instead, the plaintiff's case will rest on circumstantial evidence of discriminatory intent. In many cases, the strongest bit of circumstantial evidence is the fact that the discrimination occurred shortly after the employee took some action that is protected under the statute.

(c) Procedural requirements

Donovan v. Square D Company

709 F.2d 335 (5th Cir. 1983)

WILLIAMS, Circuit Judge.

On August 7, 1978, Dorothy Fugitt was terminated by her employer, Square D Company. Fugitt filed a complaint with the Secretary of Labor, alleging that her discharge had been in retaliation for safety related activities[1] and in violation of Section 11(c) of the Occupational Safety and Health Act of 1970 (OSHA), 29 U.S.C. §660(c). After investigation, the Department of Labor determined that the company acted unlawfully. Thereafter, OSHA investigatory personnel met with the company and requested that Fugitt be reinstated to her former position. The company refused. No further action was taken until more than two years later when the Secretary of Labor filed this action in the district court pursuant to Section 11(c)(2) of OSHA, 29 U.S.C. §660(c)(2), seeking reinstatement for Fugitt with back pay, as well as injunctive relief against future violations of Section 11(c).

On motion for summary judgment, the district court dismissed the action as barred by the two-year Texas statute of limitations for actions in tort. Tex. Rev. Stat. Ann. art. 5526 (Vernon Supp. 1982-93).

The Secretary appealed, arguing that state statutes of limitations were inapplicable to suits under OSHA brought by the Secretary to vindicate public rights and to implement national policy. The issue is one of first impression in this Court.[4] We find that the state limitations statute is not applicable, and we remand the case to the district court for consideration on the merits.

OSHA does not state a limitations period for actions brought under Section 11(c). In the absence of federally-prescribed limitations periods, the courts have frequently inferred that Congress intended to "borrow" the most analogous state statutes of limitations.

* * *

Thus it has been held that suits brought by the EEOC under Title VII of the Civil Rights Act of 1964 and by the NLRB in enforcing the National Labor Relations Act are not subject to state limitations periods. We conclude that OSHA suits brought by the Secretary of Labor under Section 11(c) similarly may not be barred by state statutes of limitations.

OSHA was enacted "to assure so far as possible every working man and woman in the Nation safe and healthful working conditions and to preserve our human resources," 29 U.S.C. §651(b). Towards this public goal, the statute envisions mandatory safety standards and establishes reporting, investigating and enforcement procedures to guarantee such standards are met. Section 11(c) operates in tandem with these provisions. It encourages employee reporting of OSHA violations and cooperation in agency investigative efforts without fear of employer retaliation.

---

[1]Ms. Fugitt began working for Square D in December 1975. In May 1977, she lost the ends of her fingers in an accident allegedly resulting from the malfunctioning of a company machine. After returning to work in February 1978, she allegedly was concerned with various hazards to which company employees were exposed. In June 1978, OSHA inspected the Square D premises; Fugitt contends that her supervisor directed employees to clean things up in preparation for the OSHA inspection.

On August 7, Ms. Fugitt took photographs of various hazards she believed to be presented by the plant machinery with the intention of forwarding these photographs to OSHA. She was informed by the company that there was a rule against photographing equipment, purportedly to protect trade secrets. When Fugitt refused to turn over her film she was discharged.

[4]The Tenth Circuit, in Marshall v. Intermountain Electric Co., 614 F.2d 260 [7 OSHC 2149] (10th Cir. 1980), was presented with this issue and held that state limitations periods did not apply to governmental suit under OSHA §11(c).

As stated by the Supreme Court in the contest of the Fair Labor Standards Act's (FLSA) anti-discrimination provision, "effective enforcement could . . . only be expected if employees [feel] free to approach officials with their grievances." *Mitchell v. Robert De Mario Jewelry, Inc.,* 361 U.S. 288, 292, 80 S. Ct. 332, 335, 4 L. Ed. 2d 323 (1960). Similarly, as the Court noted with respect to the National Labor Relations Act (NLRA), freedom from retaliation is necessary " 'to prevent the Board's channels of information from being dried up by employee intimidation of prospective complainants and witnesses.' " *NLRB v. Scrivener,* 405 U.S. 117, 122, 92 S. Ct. 798, 801, 31 L. Ed. 2d 79 (1972) (quoting *John Hancock Mutual Life Insurance Co. v. NLRB,* 191 F.2d 483, 485 (D.C. Cir. 1951)). Thus, in the case of OSHA, like that of the FLSA and NLRA, the long-term effect and primary purpose of anti-retaliation suits is to promote effective enforcement of the statute by protecting employee communication with federal authorities.[7]

OSHA Section 11(c) provides for individual relief such as reinstatement with back pay. While remedial, this provision operates primarily toward furthering the public statutory goals. "The fact that these proceedings operate to confer an incidental benefit on private persons does not detract from this public purpose." *Nabors v. NLRB, supra,* 323 F.2d at 688-89 (addressing the NLRB's enforcement proceedings, which provide back pay awards).

The public nature of these individual remedies is emphasized by the fact that the government alone possesses the right to bring suit under Section 11(c). A private cause of action does not exist. This governmental vindication of the public interest in suit under Section 11(c) of OSHA militates against an inference that congressional failure to specify a limitations period signals an adoption of state statutes of limitations.

This conclusion is further supported by the fact that Congress did specify time limitations for other procedural steps covered by Section 11(c). Section 11(c) explicitly provides a thirty-day limitations period in which an aggrieved individual may file a complaint with the Secretary, followed by a ninety-day period in which the Secretary must render his determination. 29 U.S.C. §660(c)(2), (3). This expressed concern for prompt action at the initial complaint and investigation stages stands in marked contrast to the lack of any time limitation upon a suit brought after a determination that a violation has occurred. By adopting these contrasting provisions within the same statutory section we must conclude that Congress intended that there be prompt initial action followed by a flexible period in which to bring suit. The most reasonable inference to draw is that Congress desired expeditious action at the outset of proceedings to preserve the parties' interests, yet in recognition of both the need for agency flexibility and the reality of administrative backlog, Congress elected to forego placing an inflexible timetable upon the Secretary for bringing suit. Neither appellee nor our own research discloses any evidence of contrary intent.

Appellee argues that this open-ended period for suit could not have been intended, as OSHA defendants would thereby be subject to the unreasonable surprise and prejudice which can result from the prosecution of stale claims. We disagree with appellee's contention in part because of safeguards contained in the statute. Section 11(c)(2) grants the district courts jurisdiction to order "all appropriate relief" in each anti-retaliation case, including reinstatement with back pay. This discretionary power requires the courts " 'to locate "a just result" in light of the circumstances peculiar to the case.' " *Occidental Life Insurance Co. v. EEOC,* 432 US. 355, 373, 97 S. Ct. 2447, 2458, 53 L. Ed. 2d 402 (1977) (quoting *Albemarle Paper Co. v. Moody,* 422 U.S. 405, 424-25, 95 S. Ct. 2362, 2374-75, 45 L. Ed. 2d 280 (1975)). If cases should arise in which an inordinate and inexcusable delay results in prejudice to a defendant's ability to present his defense, the district courts may restrict or even deny back pay relief.[11] *Ibid. Accord*

---

[7]The Act's legislative history indicates that employee reporting of violations was meant to play an essential role in OSHA's scheme because government inspectors alone would never adequately police the millions of workplaces covered by the statute. *See e.g.,* Committee Print, Legislative History of the Occupational Safety and Health Act of 1970, 92d Cong., 1st Sess. 152, 1032-33 (June 1971).

OSHA covers over 2.5 million establishments. In 1979 and 1980, the federal government employed only 1581 compliance and safety officers, creating a situation where approximately two percent of the covered firms could be inspected in any one year. *See* The President's Report on Occupational Safety and Health for 1980, at p. 42 (Aug. 1981).

[11]In determining the effects of delayed prosecution upon a defendant, the district courts may consider whether the company was on notice that a violation had been determined and that prosecution was intended. Depending upon the length of the delay, and the peculiar circumstances of the individual case, the availability of such notice might properly afford a defendant sufficient opportunity to gather and preserve evidence in anticipation of court action.

Marshall v. Intermountain Electric Co., 614 F.2d 260, 263 n.8 [7 OSHC 2149, 2151] (10th Cir. 1980).

Finally, the most persuasive consideration of all. We find the application of the diverse state statutes of limitations to OSHA anti-retaliation suits would 'frustrate" or "interfere with" national policy. This conclusion strongly negates any inferences that their application was intended by Congress. *Occidental Life Insurance Co. v. EEOC,* 432 U.S. 355, 97 S. Ct. 2447, 53 L. Ed. 2d 402 (1977). As the Supreme Court cautioned in *Occidental,* "[s]tate legislatures do not devise their limitations periods with national interests in mind, and it is the duty of the federal courts to assure that importation of state law will not frustrate or interfere with the implementation of national policies." 432 U.S. at 367, 97 S. Ct. at 2455.

Were state limitations periods applicable, suits by the Secretary of Labor would be governed by different and, in some cases, uncertain time limitations. Relatively short limitations periods of a year or two, as in the immediate case, might operate to bar suit in some states. In other states lengthier limitations periods, e.g., six years, might apply. If these diverse periods were applicable, the Secretary might be compelled to shift enforcement activities disproportionately among the states. This could lead to concentrating all immediate activity to these states with short limitations periods, occasioning lengthy periods of delay in those states with longer limitations periods.

*   *   *

We thus conclude that application of state limitations periods to OSHA Section 11(c) suits would frustrate the implementation of national safety policy. This specific conclusion, considered along with the well-reasoned general principle that governmental suits vindicating public interests are not barred by state limitations periods, leads us to find that Congress did not intend in its silence to subject OSHA Section 11(c) anti-retaliation suits to the vagaries of state limitations law.

CONCLUSION

The district court erred in dismissing the suit as time-barred by the two-year Texas tort limitations period. Accordingly, we reverse and remand for consideration on the merits.

◇ NOTES

1. The 30-day notice requirement has been held to be jurisdictional. That is, the employee who fails without good cause to file his or her 11(c) claim within 30 days will find that the protections of this section are no longer available. However, some courts have held that the 30-day notification requirement will be "tolled" (held in abeyance) when the fact of the discrimination was hidden from the employee. Others have held that the filing of the 11(c) claim with some other appropriate official (such as a union representative), rather than with OSHA, is permissible.

2. As noted by the court, there is no private cause of action under 11(c). See *Taylor v. Brighton Corp.,* 616 F.2d 256 (6th Cir. 1980). The action must be filed by the Labor Department on behalf of the employee. What happens if the Labor Department refuses to file a complaint? Section 11(c)(2) specifies that if "the Secretary determines that the provisions of [11(c)] have been violated, he shall bring an action." In the appropriate situation, then, an employee may be able to secure a writ of mandamus ordering the Labor Department to proceed with a case.

3. What happens if the Labor Department takes the case, but does a poor job of prosecuting it? In at least one case, a U.S. District Court has held that the employee has a remedy against the United States.

Chadsey v. United States of America

11 OSHC 1198 (D. Ore. 1983)

FRYE, District Judge.

This matter is before the court upon the complaint of William W. Chadsey, plaintiff, against the United States of America, defendant, under the Federal Tort Claims Act, 28 U.S.C. §1346(b). The case was tried to the court. The court has considered the evidence, arguments, trial memoranda, and post-trial memoranda of plaintiff and defendant, and now being fully advised makes the following Findings of Fact and Conclusions of Law:

FINDINGS OF FACT

William W. Chadsey (Chadsey) was employed by Lumco, Inc. (Lumco) from June 8, 1971, until he was involuntarily terminated on June 7, 1972. He was a chipper operator, first on the night shift and later on the day shift.

Chadsey was a thorn in the side of Lumco's management. While working on the night shift he complained to the State of Oregon Safety Division that an extension should be placed on the chipper to keep the chips from flying out. As a result of this complaint, an extension was added to the chipper, and management ordered chipper operators to wear a hard hat with a face mask *at all times*, whether they were in the chipper room or not. Chadsey disliked wearing the face mask and felt this requirement went beyond the boundaries of safety and was imposed as a result of his complaint to the State of Oregon. Management later modified the orders so that the mask portion was to be worn only while the chipper operators were in the chipper room.

In the early spring of 1972 Chadsey was moved to the day shift. Shortly thereafter he complained to his foreman about the unsafe condition of the walkway around the top of the hopper on the chipper machine. The walkway had no inside guardrail and Chadsey was afraid he would fall into the chipper. Management placed the lack of a guardrail on its "safety list" and the matter was discussed in safety meetings, but no inside railing was installed at that time.

Chadsey did not perform as well on the day shift as his predecessor. He was unable to handle both the job of chipper operator and cleanup man as his predecessor had done. In May, 1972, Lumco added a backup man to help Chadsey with his work. Lumco also decided that since there were two men on the

day shift chipper operator's job, the day shift chipper operators would have to babbit the knives rather than the night shift chipper operators. Chadsey wrote a letter on June 4, 1972, to management documenting reasons why he believed the knives should be babbitted on the night shift. This letter was a reasonable statement of Chadsey's position. Although it was not welcomed by management, it was not a hostile or offensive letter, and it was not the precipitating cause of Chadsey's discharge.

Sometime during May, 1972, Chadsey made a late night telephone call to the home of one of Lumco's management, complaining about his job. This behavior further irritated his supervisors, but again, did not result in his termination.

On June 6, 1972, while he was working on top of the chipper hopper, Chadsey nearly fell into the hopper. He made no complaint on that day, but on June 7, 1972, upon arriving at work, he went to his foreman's office to report the incident. He could not find his foreman, and so he wrote a note, dating it 8:20 a.m. and left the note on the foreman's desk. The note read as follows:

> Pete:
> I almost fell in the Chip hopper yesterday while trying to clear plug up. Before some one is killed I am turning this in to the Dept. of Labor as a death hazard.
>
> /s/Bill Chadsey

Chadsey was terminated at 9:30 a.m. and given his final paycheck. Thereafter, management ordered that an inside railing be installed on the chipper. The note written to Chadsey's foreman, with its threat to report Lumco to the Department of Labor, was the precipitating cause of and a substantial factor in Chadsey's termination.

On that same day, June 7, 1972, Chadsey filed a complaint with the Department of Labor, alleging that he was discharged in violation of 29 U.S.C. §660(c), in that his termination was in retaliation for reporting or threatening to report workplace hazards.

In July, 1973, an investigation was made, and in December, 1975, the Department of Labor denied Chadsey's claim. He protested the denial, which was treated as an appeal, and the initial determination was reversed. As a result, a lawsuit *(Marshall v. Lumco)* was filed by the United States Attorney on behalf of the Secretary of Labor on March 11, 1977, nearly five years after Chadsey's discharge. The suit sought an injunction, reinstatement of Chadsey, and back pay plus interest. Because Lumco went out of business on December 31, 1975, the remedies of injunction and reinstatement were dismissed from this suit. Back pay plus interest from June 7, 1972 through December 31, 1975, was ultimately sought.

In this lawsuit the Department of Labor was represented by William Kates, an attorney with the Office of Solicitor, United States Department of Labor, assigned to the Seattle, Washington, office. Kates was at all times the lawyer responsible for the handling of the case. At no time was Chadsey represented by his nephew, attorney Phillip Chadsey.

During the course of the lawsuit filed by the Department of Labor against Lumco, Lumco requested production of Chadsey's unemployment records, records of compensation Chadsey had received from the insurance company as a result of an automobile accident, and medical and claim records relating to the Social Security disability claim. These records were necessary because on May 26, 1975, Chadsey had been involved in an automobile accident. As a result of the accident, he had filed a claim with the Social Security Administration alleging total disability. Chadsey had also received unemployment compensation for several months following his termination with Lumco. He also had been co-owner with his wife of a small grocery store after his employment with Lumco ended. All of this information was crucial on the issue of Chadsey's damages.

Pursuant to this request, Kates asked Chadsey to provide certain information. Chadsey complied as far as he was able and authorized the Oregon Employment Division and the Social Security Administration to supply information directly to Kates. Kates failed to satisfactorily respond to Lumco's request for production.

Thereafter, a motion for production of documents was filed by Lumco's attorneys. An order for production was issued by the court. In response, Kates wrote to Lumco's lawyers giving certain information, but supplying no documents. The lawyers for Lumco, attempting to enforce the order of production, filed a motion for sanctions, a motion for costs for failure to provide discovery, and additional motions to compel production. Between April 2, 1979, and the date of the hearing on the motions to compel on October 22, 1979, Kates did not take appropriate action to obtain the documents and/or records the court had ordered produced, although he had obtained an affidavit from Chadsey which set forth insurance information on Chadsey's claim. Kates filed a response to the Order Requiring Production and filed a motion to amend the pretrial order and limit the relief claimed on behalf of Chadsey.

After the court hearing on October 22, 1979, Kates wrote to the Oregon Employment Division and obtained a response that the records sought did not exist. Kates supplied that letter to Lumco's lawyers. At no time, however, did Kates subpoena the Social Security records or the medical records supporting the Social Security claims, nor did he attempt to obtain any records on Chadsey's automobile insurance claim.

Ultimately Lumco's attorneys sought dismissal of the claim for failure to produce documents. Kates did not respond to the motion for dismissal, and the magistrate recommended dismissal. Thereafter, Kates filed objections to the magistrate's recommendation, but the case was dismissed. In addition to failing to issue subpoenas to obtain the information which the Court had ordered turned over to Lumco, Kates failed to communicate with Lumco's attorneys or with the court in order to prevent dismissal of the case.

Chadsey complied with the requests of Kates to the best of his lay person's ability. Chadsey's actions were not uncooperative or negligent, nor were his actions the cause of Kate's failure to produce documents. Furthermore, had the case of *Marshall v. Lumco* been tried, Chadsey's interests would have prevailed.

* * *

2. Section 8 of the NLRA

Section 8(a)(1) of the NLRA makes it unlawful for an employer "to interfere with, restrain, or coerce employees in the exercise of their rights guaranteed under Section 7." As discussed in Chapter 6, such interference, restraint, or coercion constitutes an unfair labor practice. In addition to the right to refuse hazardous work, the language of Section 8 clearly envisions broad protection of all activities—especially those involving labor organizing—that are authorized under the Act. As with Section 11(c) of the OSHAct, the remedy for the aggrieved employee is to file a claim with the appropriate governmental representative, here the NLRB. In contrast to the 30-day requirement under 11(c), however, the employee has six months within which to file a claim under Section 8(a).

But although the language of 8(a) is rather strong, its actual protective effect—in practice—appears far weaker. The following article paints a bleak picture of the usefulness of the NLRA protections in discouraging retaliatory actions by employers.

Promises to Keep: Securing Workers' Rights to Self-Organization Under the NLRA

P. Weiler

III. THE REGULATORY APPROACH

The figures in Table [8-1] reveal a remarkable growth in unfair labor practices, particularly in discriminatory discharges of union supporters, by American employers. The [conclusions of other authors] notwithstanding, it is reasonable to suppose that this

Source: From 96 *Harv. L.R.* 1769 (1983). Copyright © 1983 by the Harvard Law Review Association.

spiraling increase in coercion by employers has had something to do with the declining success rate of unions in NLRB elections. The current state of affairs is entirely at odds with the promise of protection made to American workers through the passage of the Wagner Act nearly fifty years ago. Over the last two decades, there has been a sustained search for a legal cure for this industrial relations pathology, a search that culminated in the near-enactment of the Labor Reform Act in 1978. I shall argue that we need far more radical surgery than any considered so far.

Table 8-1. Unfair Labor Practices by Employers, 1950–1980

Year	Certification Elections	Charges Against Employers	§8(a)3 Charges	Fraction Found Meritorious	Backpay Awards (Average Amount)	Reinstatees
1950	5619	4472	3213	NA	2259 ($477)	2111
1955	4215	4362	3089	NA	1836 ($428)	1275
1957	4729	3655	2789	NA	1457 ($354)	922
1960	6380	7723	6044	29.1% (overall)	3110 ($335)	1885
1965	7576	10,931	7367	35.5% (overall)	4644 ($599)	5875
1970	7773	13,601	9290	34.2% (overall)	6828 ($403)	3779
1975	8061	20,311	13,426	32.3% (employer) 30.2% (overall)	7405 ($1524)	3816
1980	7296	31,281	18,315	39.0% (employer) 35.7% (overall)	15,642 ($2054)	10,033

Before I do so, however, it is necessary to review the deficiencies in the current regime and in our restricted vision of how it might be refurbished.

A. Current Remedies Under the NLRA

1. The Remedial Philosophy of the NLRA. A tacit assumption of American labor law is that regulation is the appropriate means of containing coercive employer actions. Certain forms of harmful and illegitimate behavior are prohibited on pain of legal measures imposed through the administrative procedures of the NLRB. Although both practicing and academic lawyers tend to be preoccupied with the substantive definitions of legal and illegal behavior at the margins of labor jurisprudence, no one would suggest that the current crisis lies there. Rather, the failure of the system to prevent unfair practices is generally attributed to the weakness of the sanctions for even the crudest forms of retaliation against union supporters, and to delays in the administration of the law. Remedying these problems, it is supposed, will cure the ills of the entire system.

*   *   *

Consider, as an illustration of the difference between protecting a group right to self-organization and repairing the harms done to individuals, the case of an employee discharged in response to his union organizing activities. The law might repair the harm to this individual by restoring him to his job and making up the income that he has lost. If the employer's purpose had simply been to punish the worker for supporting the union, the fact that the law would effectively undo this damage at the employer's expense might discourage the use of the tactic in the future.

But the real purpose of such discharges is to break the momentum of the union's organizing campaign. By the time the discharged employee has been reinstated, much of the union's support may have melted away, and the election may thus have been lost. Unlike the discharged employees' loss of wages, this setback to the employees' quest for a collective voice in their workplace cannot easily be repaired. To protect the employees' group rights, the NLRA must rely on the preventive force of its sanctions. But the traditional remedies for discriminatory discharge —backpay and reinstatement—simply are not effective deterrents to employers who are tempted to trample on their employees' rights.

2. The Backpay Award. At first blush, the backpay award might seem to serve both remedial and deterrent functions. Although from the employees' point of view the award is merely compensation for what has been lost, from the employer's point of view it is a financial penalty: the employer is required to pay for services it has not received. The problem is that this "fine"—paid to the worker rather than the state—is far too small to be a significant deterrent. Looking again at Table [7-1], one finds that the average backpay award in 1980 was only $2000. The small size of the average award is partly attributable to cases in which employers settle quickly and reinstate the employees to limit the monetary loss. But even in troublesome cases in which reinstatement is not forthcoming for months or years, the employer's potential liability is inherently limited. Early in the life of the Wagner Act, the principle was settled that the proper measure of the backpay award is not the wages the guilty employer failed to pay, but rather the net loss suffered by the employee after the deduction of any wages earned in the interim in another job. Indeed, the law requires the employee to take all reasonable steps necessary to mitigate his loss by finding another job. If he fails to take such steps, his potential earnings may be deducted from the backpay award even if they are not actually earned.

The combination of the net loss and mitigation doctrines is the most telling illustration of the supremacy of the reparative policy over the deterrent policy in American labor law. If the backpay remedy were designed to deter the employer's unlawful conduct, there would be no reason to deduct any wages that the employee earned, or could have earned, in another job. By minimizing the employer's potential liability, such a deduction removes most of the deterrent effect of the backpay award.

*   *   *

3. Reinstatement. What about reinstatement, the other standard form of relief? In principle, such an in-kind remedy seems nicely designed to play both a reparative and a preventive role. The dismissed employee gets his job back, and his fellow workers see the power of collective action and labor law protection. Not only is the employer deprived of the fruits of its illegal behavior, but it also suffers a serious erosion of its hitherto absolute sway within its own plant. The prospect of such a result would seem to be a major disincentive to flouting the law in the first place.

The reinstatement remedy, however, has proved to

be far less effective in practice than in theory. It is one thing for the NLRB to calculate the amount of wages lost and see that this sum is paid to the discharged employee. It is quite another thing for an outside agency to try to reconstruct an enduring employment relationship. An employer that is sufficiently antiunion to break the law by firing a union supporter is also likely to feel quite vindictive when forced to take the employee back, and may well start looking for an excuse to get rid of him again. The employee, fully aware of this attitude, will often be reluctant to return, especially if returning means giving up another job that he has obtained in the interim.

There have been two systematic studies of the efficacy of the reinstatement remedy, one from New England in the early 1960's[80] and the other from Texas in the early 1970's.[81] Both studies reached remarkably similar conclusions. First, only about 40% of the employees for whom reinstatement was ordered actually took their jobs back. The major reason given by employees for declining reinstatement was fear of employer retaliation, a factor that may well also have affected some of those who said that they had simply found better jobs. The employees' fears were apparently well founded; of employees who did go back, nearly 80% were gone within a year or two, and most blamed their departure on vindictive treatment by the employer. If the reinstatement remedy is judged by its ability to reestablish an enduring employment relationship, the verdict is clear: this goal is achieved in only about 10% of the meritorious cases. Small wonder that the prospect of a reinstatement order does not loom large for the antiunion employer making a coldblooded assessment of the legal sanctions it faces.

Time is the crucial variable in reinstatement cases. If the Board can secure reinstatement quickly, the

wrongfully discharged employee is likely to accept the offer. In time, however, the employee generally obtains another job, which he will be reluctant to leave for the bleak prospects at his old position. Thus, he becomes progressively less likely to assert his reinstatement rights. Only about 5% of reinstatement offers obtained after six months are in fact accepted.

The timing of reinstatement is even more significant insofar as this remedy is designed to undo the chilling effect of a discriminatory discharge on the group impulse toward collective bargaining. If the dismissed employee were returned to his job immediately, the employer's attempt at intimidation would likely backfire. The message the employees would receive instead would be a demonstration of how effective a trade union could be in asserting and defending employees' rights against oppressive management. But if that lesson is to be at all influential, reinstatement must come before the representation election is actually conducted—normally about two months after the union has filed its petition. Although delay is unfortunate for the individual worker, he usually finds another job and will eventually receive all his lost pay. But delay is fatal to the viability of a union organizing drive.

* * *

B. Delay in Unfair Labor Practice Cases

I. Delay Under the NLRA.
We have seen that time is of the essence in implementing NLRB remedies. If either a reinstatement order or a *Gissel* bargaining order [see Chapter 6] is to be effective, it must come quickly. The question that arises, then, is how long an employer can forestall an enforceable order in an unfair labor practice proceeding. The answer is distressing: nearly 1000 days as of 1980.

Table [8-2] sets out the various stages in the Board's process and gives the median time taken at each stage, as these figures fluctuated between 1960 and 1980. In the first stage, someone files an unfair labor practice charge with one of the Board's regional offices, where charges are investigated and spurious complaints are screened out. If the charge is found to be meritorious, the Board staff will try to settle the case; the regional office will issue a formal complaint if settlement efforts are unsuccessful. In 1980, this first stage typically took a month and a half.

In the second stage, formal complaints are tried before an administrative law judge (ALJ). The independent, tenured ALJ conducts a full-scale eviden-

---

[80]The New England study concerned 194 employees for whom the Boston regional office of the NLRB obtained offers of reinstatement from 71 employers between July 1, 1962, and July 1, 1964. The results of the study are summarized in *Hearings on H.R. 11725 before the Special Subcomm. on Labor of the Comm. on Education and Labor,* 90th Cong., 1st Sess. 3–12 (1967) [hereinafter cited as *Hearings*].

[81]The Texas study concerned 217 employees for whom the Ft. Worth regional office of the NLRB obtained offers of reinstatement from 86 employers in 1971 and 1972. It is published in Stephens & Chaney, *A Study of the Reinstatement Remedy under the National Labor Relations Act,* 25 Lab. L. J. 31 (1974), and also in Chaney, *The Reinstatement Remedy Revisited,* 32 Lab. L. J. 357 (1981).

Table 8-2. Delay in Processing Unfair Labor Practice Charges, 1960–1980 (all times in days)

Year	Filing to Complaint	Complaint To Close of Hearing	Close of Hearing to ALJ Decision	ALJ to Board Decision	Filing to Board Decision	Bd. Decis. to C.A. Opinion
1960	52	66	88	149	426	
1965	59	67	123	122	390	
1970	57	58	84	124	348	
1975	54	55	72	134	332	359
1980	46	155	158	133	484	485

tiary hearing near the location of the complaint, analyzes the transcript and posthearing briefs, and ultimately produces a lengthy written opinion. All of this takes another ten months or more. The ALJ, however, does not have the authority to make a legally binding ruling. If any party takes exceptions to the ALJ's preliminary ruling, the final decision must be made by the Board itself.

The third stage, in which the ALJ's ruling is reviewed by the Board in Washington, D.C., consumes another four to five months. But even the painstaking process within the NLRB structure does not end the arduous journey of an unfair labor practice charge. Without an enforcement order from a federal court of appeals, a Board order is entirely without teeth. The enforcement proceeding in the court of appeals provides an employer who is prepared to spend the requisite legal fees an opportunity to obtain extensive review of the Board's conclusions and to put off for another sixteen months the point at which the employer is finally faced with a legal directive that it must obey under penalty of sanctions for contempt of court.

One should not assume that this lengthy route must be traveled in all cases. Most employers do not resist to the bitter end. In fact, the regional offices dispose of over 80% of the charges filed each year without issuing formal complaints. The vast majority of cases handled in this manner, however, are unfounded charges that are withdrawn or administratively dismissed. Nearly half of the "meritorious" charges—those that eventually secure some relief for the complainant by way of settlement or order—do enter the formal process through the issuance of a complaint. Even among these, however, the majority are settled before the ALJ hearing actually takes place. Only a tiny fraction of the caseload, 500 decisions each year, reaches the circuit courts. Such a small number of cases, it might be argued, does not warrant major surgery on the system.

That upbeat conclusion would, however, be misleading. Any informal settlement of the case requires the employer's consent. The fact that an enforceable order for relief will not be available for almost three years lurks in the background of the settlement negotiations and reduces the employee's bargaining power. An employer will be reluctant to settle a discriminatory discharge case if the result will be reinstatement of a key union supporter during the representation campaign. The prospect of long delay in prosecuting a formal charge increases the likelihood that the employer will be able to persuade the aggrieved employee to accept a cash payment and waive his right to reinstatement. When the effects of procedural delay on the dynamics of settlement in unfair labor practice cases are taken into account, the magnitude of the problem is clear. Delay is the Achilles heel of the regulatory approach to the representation process under the NLRA.

* * *

3. Reducing Delay: Interim Relief. No one suggests that a trial-type procedure could produce a final decision in fewer than two or three months in even a fraction of the tens of thousands of unfair labor practice cases processed annually under the NLRA. Recognition of the delay inherent in such a procedure has led to a different proposal: providing interim relief from unfair labor practices committed during the organizational campaign.

The present NLRA authorizes, but does not require, the use of an expedited remedy for unfair labor practices committed by employers. Section 10(j) empowers the Board to petition a federal district court for an interim order restraining an unfair

Table 8-3. Injunction Petitions for Unfair Labor Practices, 1960–1980

	Against Unions Under Section 10(*l*)			Against Employers Under Section 10(j)		
Years Averaged	(1) Charges Subject to Section 10(*l*)	(2) §10(*l*) Petitions	(3) (Col. 2)/ (Col. 1)	(4) Charges Subject to Section 10(j)	(5) §10(j) Petitions	(6) (Col. 5)/ (Col. 4)
1950–1959	350	66	18.9%	4760	1	0.021%
1960–1969	1617	221	13.7%	10,234	12	0.117%
1970–1974	2393	247	10.3%	16,428	11	0.067%
1975–1980	2647	229	8.7%	26,212	41	0.156%

labor practice pending the Board's own leisurely proceedings toward a final verdict. In contrast with this permissive provision is section 10(*l*), which was enacted in 1947 and applies only to certain unfair labor practices committed by unions. Section 10(*l*) requires the Board to petition for interim relief when it has reasonable cause to believe that a union has violated the provisions of the Act limiting organizational picketing and prohibiting secondary boycotts —two traditional union weapons—and assigns first priority to the investigation of such charges. The theory behind section 10(*l*) is that a union must be immediately restrained from using self-help to force recognition by an employer, because such "top-down organizing" obviously threatens employee self-determination. But if the law acts quickly to protect employees from illegal coercion by the union—as, in my view, it should—should it not respond with the same alacrity when the coercion is by the employer?

The actual practice under section 10(j) has been startlingly different from that under section 10(*l*). Table [8-3] summarizes the historical trends. Even in the most recent period, when the General Counsel of the NLRB was firmly committed to the use of immediate relief under section 10(j) as an antidote to the acceleration of unfair labor practices and when the pace at which section 10(j) petitions were filed more than doubled to an average of about fifty a year, the employer's risk of facing a section 10(j) petition in connection with a section 8(a)(3) complaint was still just one-fiftieth of the union's exposure under section 10(*l*). Only one section 10(j) proceeding was instituted for every 1700 section 8(a)(3) charges of antiunion discrimination by employers. The Board thus failed to use what was and still is, for preventive as well as for reparative purposes, the most effective weapon in its arsenal.

*   *   *

◇ NOTES

1. The preface to this article notes that, "Professor Weiler argues that the traditional response of labor law reformers—calling for improvement of the regulatory framework through the provision of speedier and stiffer sanctions—is simply incapable of stemming the rise in antiunion activities by employers. He contends that the initial promise of the NLRA can be redeemed only by elimination of the protracted representation campaign through a system of 'instant elections' modeled on the Canadian labor law regime." 96 Harv. L.R. at 1769. What "reforms," if any, would you propose to address this situation? Would making Section 8 a more effective tool for aggrieved employees be consistent with the view of the purpose of the NLRA as enunciated by the Supreme Court in the *First National Maintenance* case, discussed in Chapter 6?

2. To what extent are the problems described in the Weiler article likely to be encountered by employees seeking protection under the other statutory and common law mechanisms described in this chapter?

3.   Federal Environmental and Safety Statues

Most federal regulatory statues dealing with health, safety, or the environment contain some protection for employees who are discriminated against for seeking to enforce the statute. Although these provisions are similar in concept to Section 11 of the OSHAct and Section 8 of the NLRA, there are significant differences. There are also important differences among the various statutes themselves. One such provision is set forth below. This is a section of the Energy Reorganization Act, which applies to employees at nuclear power plants and nuclear power plant construction projects. Note particularly that, in contrast to Section 11(c) of the OSHAct, the worker has a clear judicial remedy if the Secretary of Labor denies the claim as well as a right to recover attorney's fees and costs.

Section 210 of the Energy Reorganization Act

42 U.S.C. §5851

EMPLOYEE PROTECTION

(a) Discrimination against employee

No employer, including a Commission licensee, an applicant for a Commission license, or a contractor or a subcontractor of a Commission licensee or applicant, may discharge any employee or otherwise discriminate against any employee with respect to his compensation, terms, conditions, or privileges of employment because the employee (or any person acting pursuant to a request of the employee)—

(1) commenced, caused to be commenced, or is about to commence or cause to be commenced a proceeding under this chapter or the Atomic Energy Act of 1954, as amended [42 U.S.C. 2011 et seq.], or a proceeding for the administration or enforcement of any requirement imposed under this chapter or the Atomic Energy Act of 1954, as amended;

(2) testified or is about to testify in any such proceeding or;

(3) assisted or participated or is about to assist or participate in any manner in such a proceeding or in any other manner in such a proceeding or in any other action to carry out the purposes of this chapter or the Atomic Energy Act of 1954, as amended [42 U.S.C. 2011 et seq.].

(b) Complaint, filing and notification

(1) Any employee who believes that he has been discharged or otherwise discriminated against by any person in violation of subsection (a) of this section may, within thirty days after such violation occurs, file (or have any person file on his behalf) a complaint with the Secretary of Labor (hereinafter in this subsection referred to as the "Secretary") alleging such discharge or discrimination. Upon receipt of such a complaint, the Secretary shall notify the person named in the complaint of the filing of the complaint and the Commission.

(2)(A) Upon receipt of a complaint filed under paragraph (1), the Secretary shall conduct an investigation of the violation alleged in the complaint. Within thirty days of the receipt of such complaint, the Secretary shall complete such investigation and shall notify in writing the complainant (and any person acting in his behalf) and the person alleged to have committed such violation of the results of the investigation conducted pursuant to this subparagraph. Within ninety days of the receipt of such complaint the Secretary shall, unless the proceeding on the complaint is terminated by the Secretary on the basis of a settlement entered into by the Secretary and the person alleged to have committed such violation, issue an order either providing the relief prescribed by subparagraph (B) or denying the complaint. An order of the Secretary shall be made on the record after notice and opportunity for public hearing. The Secretary may not enter into a settlement terminating a proceeding on a complaint without the participation and consent of the complainant.

(B) If, in response to a complaint filed under paragraph (1), the Secretary determines that a violation of subsection (a) of this section has occurred, the Secretary shall order the person who committed such violation to (i) take affirmative action to abate the violation, and (ii) reinstate the complainant to his former position together with the compensation (including back pay), terms, conditions, and privileges of his employment, and the Secretary may order such person to provide compensatory damages to the complainant. If an order is issued under this paragraph, the Secretary, at the request of the complainant shall assess against the person against whom the order is issued a sum equal to the aggregate amount of all costs and expenses (including attorneys' and expert witness fees) reasonably incurred, as determined by the Secretary, by the complainant for, or in connection with, the bringing of the complaint upon which the order was issued.

◇ NOTES

1. For comparable provisions in other federal health, safety, and environmental statutes, see e.g., Section 23 of the Toxic Substances Control Act, 42 U.S.C. §2622; Section 7001 of the Solid Waste Disposal Act (also known as the Resource Conservation and Recovery Act), 42 U.S.C. §6971; Section 507 of the Clean Water Act, 33 U.S.C. §1367; Section 322 of the Clean Air Act, 42 U.S.C. §7622; and Section 110 of the Comprehensive Environmental Response, Compensation, and Liability Act, 42 U.S.C. §9610.

2. The Supreme Court has held that this statute does *not* preempt a worker's state law claim for intentional infliction of emotional distress allegedly stemming from disciplinary action taken against the worker after she made safety complaints at a nuclear facility. *English v. General Electric,* —U.S.—, 110 S. Ct. 2270 (1990).

◇

4. General Whistleblower Statutes

"Whistleblowers" are individuals who, often at risk of their jobs, report their employers for violating the law or for taking actions that endanger public health or the environment. The term came into common usage in the late 1960s when a Department of Defense employee "blew the whistle" on his employer by reporting billion dollar cost overruns on certain transportation and weapons programs. This individual lost his job. (He fought back and was reinstated in 1982, more than 12 years later.) More recently, Morton Thiokol engineers attempted to stop the fateful launch of the Challenger shuttle by alerting NASA to the rocket's defective O-rings. These public acts of courage overshadow the uncelebrated but equally heroic acts of many more public and private sector employees who also report evidence of wrongdoing or hazardous conditions.

California, Connecticut, Maine, and Michigan grant broad statutory protection to individuals who report violations of laws, rules, and regulations. The Michigan Whistleblowers' Protection Act, M.C.L. §15.361 et seq., provides as follows:

> An employer shall not discharge, threaten, or otherwise discriminate against an employee regarding the employee's compensation, terms, conditions, location, or privileges of employment because the employee, or a person acting on behalf of the employee, reports or is about to report, verbally, or in writing, a violation or a suspected violation of a law or regulation or rule promulgated pursuant to law of this state, a political subdivision of this state, or the United States to a public body, unless the employee knows that the report is false, or because an employee is requested by a public body to participate in an investigation, hearing, or inquiry held by that public, or a court action.

No similarly comprehensive protection exists at the federal level. Congress is considering bills that would consolidate more than a dozen laws affecting private sector whistleblowers, establish a uniform statute of limitations, and consolidate the administrative process of hearing complaints. Congress is also considering a bill to expand protection for public sector employees who expose environmental problems that violate federal statutes. The Government Accountability Project (GAP), a Washington based clearinghouse and resource center for whistleblowers seeking redress, has proposed model legislation that would protect *all* employees.

5. Common Law Protections

By far the most variable, and the least predictable, protections for retaliatory employment practices are those developed under the common law. As they are creatures of the state courts, their availability and scope tend to vary widely from state to state. Nonetheless, they are an important adjunct to statutory protections, both because they can provide protection for retaliatory employment practices where no statutory protection is applicable and because they can provide some assurances of job security *beyond* protection from retaliation. Common law actions for retaliatory, unreasonable, or unfair employment practices can be divided into four basic categories: actions based on alleged violations of workplace personnel policies; actions based on alleged breach of implied promise of job security; actions based on alleged breach of implied covenant of good faith and fair dealing; and actions based on alleged violations of public policy. Depending on the nature of the allegations, the claim will be brought either as a tort claim, as a breach of contract claim, or as both. Although the fourth category—violation of public policy—is the most relevant to the concerns of this text, a word or two about each of them is appropriate.

(a) Violation of personnel policies

This category—perhaps the most straightforward of them all—is not really a deviation from the old employment at will rule, as it relies on an explicit contract between employer and employee. Rather, it holds that personnel policies established by the employer and made known to the employee form a part of the employment contract and are enforceable as such. These may be oral statements made by the employer to the employees, or, more commonly, written policies incorporated in employee handbooks, manuals, and the like. When for example, an employee manual states the general policy that employees will be fired only for just cause, courts have treated this as a binding part of the employment contract.

One argument against this approach is that, unless they form part of a collective bargaining agreement, personnel policies are not *bargained for* by the employee. Rather, they typically are issued unilaterally by the employer and arguably may be *changed* unilaterally by the employer. In contract law terms, then, it is sometimes said that the employee has offered no *consideration* (i.e., has offered nothing of value) in return for the adoption of the policy by the employer. Thus, the argument runs, there is no contractual relationship as to the existence of the stated policy and thus no action for breach of contract.

Once the policy is issued, however, the employee presumably makes his or her decision to stay with this employer—and not to move to another job—in part on the basis of the statements set forth in the personnel policy. Logically, this decision to remain with the employer constitutes sufficient consideration for the formation of a contract as to the terms of the policy. When the employer changes the policy, the employee again faces the decision to stay or to leave. (There may, however, be an implied promise not to change the policy.)

(b) Breach of implied promise of job security

Where there is no express employer policy on the point, some courts have held that certain circumstances give rise to an *implied* promise of job security. Such implied promises—when they are found to exist—are said to flow from the entire course of dealing between the employer and the employee. Their existence, then, is dependent on the facts of the particular case. Normally, an implied promise of job security will be associated with situations in which the employee has spent several years with the same employer, has an excellent employment record, and has received considerable encouragement and support from the employer.

In *Pugh v. See's Candies, Inc.*, 116 Cal. App. 311, 171 Cal. Rptr. 917 (1981), for example, an employee was terminated, for no stated reason, after 29 years of employment. There had been no criticism of his work during the entire period of his employment, and he had been told by the president of the company that "if you are loyal to [the company] and do a good job, your future is secure." Under these circumstances, the California Court of Appeals held that "there were facts in evidence from which the jury could determine the existence of . . . an implied promise [of job security]."

(c) Breach of implied covenant of good faith and fair dealing

A related concept is the implied covenant of good faith and fair dealing. Where it is recognized, the existence of such a covenant is not said to depend on the particular facts of the case at hand. Rather, it is said to be an implied part of *every* commercial contract, including employment contracts. It does not promise job security, but it does promise that one's job security will not be jeopardized in ways that offend notions of fundamental fairness. (What is "fair" in any particular case, obviously, will depend on the facts of that case.)

(d) Violation of public policy

Finally, some courts have developed common law protections for employees who are discharged—or otherwise discriminated against—because they sought to uphold some important public policy. Although this body of law is still developing, it appears roughly analogous to the statutory protections discussed earlier in this chapter. That is, in order to be protected, one must be able to show that the discrimination was in retaliation for one's furtherance of a clearly articulated public policy, such as a statute or regulation. In general, these cases can be divided into two categories: those where the employee has taken some action in order to further a public policy and those where the employee has *refused* to carry out an action that would *violate* a public policy.

(i) Discharge for carrying out some articulated public policy

Lally v. Copygraphics

Supreme Court of New Jersey
85 N.J. 668 (1981)

Per Curiam.

We affirm the Appellate Division's determination, 173 N.J. Super. 162, that a plaintiff has a common law right of action for wrongful discharge based upon an alleged retaliatory firing attributable to the filing of a workers' compensation claim.

*   *   *

In particular we endorse the conclusion of the Appellate Division that there exists a common law cause of action for civil redress for a retaliatory firing that is specifically declared unlawful under N.J.S.A. 34:15-39.1 and 39.2. The statutory declaration of the illegality of such a discharge underscores its wrongful and tortious character for which redress should be available. Such a cause of action is strongly founded in public policy which, in this case, is reflected in the statutory prohibitions themselves. See *Pierce v. Ortho Pharmaceutical Corp.*, 84 N.J. 58, 66-73 (1980). Moreover, the penal and administrative remedies that are provided by N.J.S.A. 34:15-39.1 and 39.2 to rectify this form of illegal employment practice will clearly be augmented by recognition of an alternative or supplemental judicial right to secure civil redress. A common law action for wrongful discharge in this context will effectuate statutory objectives and complement the legislative and administrative policies which undergird the workers' compensation laws. The determination of the Appellate Division that the statutory treatment of this kind of retaliatory firing is not preemptive of a civil right of redress is sound. 173 N.J. Super. at 170-172, 179.

*   *   *

The court below apparently felt impelled to find that the administrative relief provided by N.J.S.A. 34:15-39.1 and 39.2 is strictly limited. It did so,

seemingly, to strengthen its conclusion that there exists a viable common law cause of action for civil redress which has not been superseded by the legislative treatment. Such a civil cause of action, however, is firmly anchored as a matter of public policy upon the unlawful, wrongful, and tortious character of the proscribed conduct. *Cf. Pierce v. Ortho Pharmaceutical Corp., supra*, 84 N.J. at 66-73. If the Legislature had wanted to foreclose a judicial cause of action, it would have done so expressly. *Cf. Kaczmarek v. N.J. Turnpike Authority*, 77 N.J. 329 (1978) (under N.J.S.A. 34:13A-5.5(c) PERC has "exclusive power" to deal with unfair labor practices). Thus, reliance by the Appellate Division upon the alleged "inadequacy" of the administrative remedy, as proof of legislative intent not to abridge or preempt a common law remedy, was unnecessary.

We need not pass upon the soundness of the limited view of the administrative remedies expressed below. For the reasons stated below, 173 N.J. Super. at 177-178, we conclude that the forms of relief specifically enumerated in the statute, namely, restoration to employment and compensation for lost wages, are clearly available to the Commissioner. This enumeration, however, is not necessarily one of strict limitation as suggested by the Appellate Division. 173 N.J. Super. at 180. The pertinent part of the statute reads,

> As an alternative to any other sanctions herein or *otherwise provided by law*, the Commissioner of Labor and Industry may impose a penalty not exceeding $1,000.00 for any violation of this act (emphasis added) [N.J.S.A. 34:15-39.2].

*   *   *

With these observations and for all of the reasons expressed, we affirm the judgment below.

Schreiber, J., dissenting.

This case has its roots in an accident on March 18, 1975 in which plaintiff Jo Ann Lally was injured while working for defendant Copygraphics at its plant. She returned to work March 31 and was discharged April 11, after having been told her medical bills would not be paid.

Plaintiff filed a workers' compensation petition on the day of her discharge. Subsequently, she received a compensation award of $1870 for permanent disability and $114.29 for temporary disability.

* * *

Until *Pierce v. Ortho Pharmaceutical Corp.*, 84 N.J. 58 (1980), we adhered to the common law proposition that "an employer had the unbridled authority to discharge, with or without cause, an employee in the absence of contractual . . . restrictions," *English v. College of Medicine and Dentistry of N.J.*, 73 N.J. 20, 23 (1977), statutory limitations, *Nicoletta v. No. Jersey District Water Supply Comm'n*, 77 N.J. 145, 150 (1978), or constitutional safeguards, *Perry v. Sindermann*, 408 U.S. 593, 92 S. Ct. 2694, 33 L. Ed. 2d 570 (1972). In *Pierce*, we adopted the principle that "an employee has a cause of action for wrongful discharge when the discharge is contrary to a clear mandate of public policy." 84 N.J. at 72. Declarations of such public policy could be found in legislation, adminsitrative rules and regulations, and judicial decisions. *Id.* at 72. *Pierce* also held that a wrongfully discharged employee could maintain an action in contract or in tort in which compensatory and punitive damages might be recovered. *Id.* at 72.

Plaintiff sought damages, compensatory and punitive, because she was discharged in violation of N.J.S.A. 34:15-39.1. This statute enunciates a policy condemning reprisals, either by discharge or discrimination, against employees who have claimed or attempted to claim workers' compensation. The policy having been fixed and announced by the Legislature, the *Pierce* principle supports the proposition that an employee has a cause of action when discharged in retaliation for filing or asserting a right to file a claim petition under the Workers' Compensation Act.[1]

However, there must still be considered the nature and extent of that cause of action. In addition to promulgating the policy against reprisals, N.J.S.A. 34:15-39.1 also provides a remedy, namely, restoration to the job and compensation for lost wages. The

Legislature, having established the standard of conduct which employers must follow, determined the civil relief available to employees when that standard is violated. The Legislature has fixed that relief to promote the purpose of the statute. There is nothing in the legislative history indicating a legislative intent that recovery for a judicially cognizable tort be expanded beyond recovery of loss of wages and reinstatement. The public policy enunciated in section 39.1 did not exist before enactment of the statute. In *English v. College of Medicine and Dentistry of N.J.*, 73 N.J. 20, 23 (1977), we held that an employer had "unbridled authority to discharge, with or without cause," in the absence of contractual, statutory or constitutional restraints. Our research discloses no decision in New Jersey or elsewhere before enactment of section 39.1 in 1966 which recognized a common law cause of action arising out of a discharge because of reprisal for filing a workers' compensation claim. The first opinion acknowledging such a cause of action was *Frampton v. Central Indiana Gas Company*, 260 Ind. 249, 297 N.E.2d 425 (1973), a decision which has been rejected by many jurisdictions. See *Loucks v. Star City Glass Co.*, 551 F.2d 745 (7 Cir. 1977); *Martin v. Tapley*, 360 So. 2d 708 (Ala.1978); *Segal v. Arrow Industries Corp.*, 364 So. 2d 89 (Fla. App.1978); *Kelsay v. Motorola, Inc.*, 51 Ill. App. 3d 1016, 9 Ill. Dec. 630, 366 N.E. 2d 1141 (1977), rev'd 74 Ill. 2d 172, 23 Ill. Dec. 559, 384 N.E. 2d 353 (1978); *Stephens v. Justiss-Mears Oil Co.*, 300 So. 2d 510 (La.App.1974); *Dockery v. Lampart Table Co.*, 36 N.C. App. 293, 244 S.E. 2d 272 (1978), *cert.* den. 295 N.C. 465, 246 S.E. 2d 215 (1978). The cause of action here was created by the statute and therefore the remedy provided therein would appear to be exclusive. *Transamerica Mortgage Advisors v. Lewis*, 444 U.S. 11, 100 S. Ct. 242, 62 L. Ed. 2d 146 (1979) (where statutory remedy for violation of the Investments Advisers Act is rescission, an additional private cause of action for damages will not be implied). This position is harmonious with the general proposition that where a statute expressly provides for a particular remedy, a court should not read others into it in the absence of clear legislative intent. See *id.* at 19-20, 100 S. Ct. at 247, 62 L. Ed. 2d at 154-155; *Botany Mills v. United States*, 278 U.S. 282, 289, 49 S. Ct. 129, 132, 73 L. Ed. 2d 379, 385 (1928) ("[w]hen a statute limits a thing to be done in a particular mode, it includes the negative of any other mode"); *Martin v. Althoff*, 27 Ariz. App. 588, 557 P.2d 187, 190 (1976); 2A *Sutherland*, Statutory Construction (4 ed. Supp. 1980) §55.03 at 63.

---

[1] In *Pierce* we held that, even in the absence of a specific remedy in the statute, an employee would have a valid cause of action. 84 N.J. at 68.

◇ **NOTES**

1. As is usually the situation with successful "wrongful discharge" cases brought under this theory, the actions taken by the discharged employee in *Lally* were directly related to a particular statute and were consistent with the policies underlying that statute. It is somewhat more unusual, however, that the court here was willing to find a common law cause of action even where certain statutory protections already existed. As a practical matter, do you think the New Jersey court would have been as willing to find a common law remedy here if the existing statutory remedy had been more adequate?

2. In general, where there is an existing statutory remedy, the employer may well be able to argue that the availability of statutory relief obviates the need for a common law cause of action or that the existence of the statutory remedy indicates a legislative intent to preempt common law remedies. This latter argument may be especially effective in the case of federal protective statutes, such as those discussed above. (But see *English v. General Electric*, discussed in note 2 on page 407.)

◇

The action taken by the employee in *Lally*—filing a worker's compensation claim—likely would not be seen by most people as a direct challenge to the integrity of the employer. Often, however, the employee action leading to the discharge is more directly confrontational. The clearest example of this is the whistleblower, the employee who calls attention to the employer's alleged wrongdoing. In general, courts have been less willing to extend common law protection to employees who air their employer's "dirty linen" in public. The majority and dissenting opinions in the following case point out some of the arguments for, and against, providing common law protection to whistleblowers.

Palmateer v. International Harvester Co.

Supreme Court of Illinois
85 Ill. 2d 124, 421 N.E.2d 876 (1981)

SIMON, Justice.

The plaintiff, Ray Palmateer, complains of his discharge by International Harvester Company (IH). He had worked for IH for 16 years, rising from a unionized job at an hourly rate to a managerial position on a fixed salary. Following his discharge, Palmateer filed a four-count complaint against IH, alleging in count II that he had suffered a retaliatory discharge. According to the complaint, Palmateer was fired both for supplying information to local law-enforcement authorities that an IH employee might be involved in a violation of the Criminal Code of 1961 (Ill. Rev. Stat. 1979, ch. 38, par. 1-1 et seq.) and for agreeing to

assist in the investigation and trial of the employee if requested. The circuit court of Rock Island County ruled the complaint failed to state a cause of action and dismissed it; the appellate court affirmed in a divided opinion. We granted Palmateer leave to appeal to determine the contours of the tort of retaliatory discharge approved in Kelsay v. Motorola, Inc. (1978), 74 Ill. 2d 172, 384 N.E.2d 353.

*   *   *

By recognizing the tort of retaliatory discharge, *Kelsay* acknowledged the common law principle that parties to a contract may not incorporate in it rights and obligations which are clearly injurious to the public. This principle is expressed forcefully in cases which insist that an employer is in contempt for discharging an employee who exercises the civic right and duty of serving on a jury. But the Achilles heel of the principle lies in the definition of public policy. When a discharge contravenes public policy in any way the employer has committed a legal wrong. However, the employer retains the right to fire workers at will in cases "where no clear mandate of public policy is involved." But what constitutes clearly mandated public policy?

There is no precise definition of the term. In general, it can be said that public policy concerns what is right and just and what affects the citizens of the State collectively. It is to be found in the State's constitution and statutes and, when they are silent, in its judicial decisions. Although there is no precise line of demarcation dividing matters that are the subject of public policies from matters purely personal, a survey of cases in other States involving retaliatory discharges shows that a matter must strike at the heart of a citizen's social rights, duties, and responsibilities before the tort will be allowed. . . .

The cause of action is allowed where the public policy is clear, but is denied where it is equally clear that only private interests are at stake. It is clear that Palmateer has here alleged that he was fired in violation of an established public policy. The claim is that he was discharged for supplying information to a local law-enforcement agency that an IH employee might be violating the Criminal Code, for agreeing to gather further evidence implicating the employee, and for intending to testify at the employee's trial, if it came to that. . . . There is no public policy more basic, nothing more implicit in the concept of ordered liberty, than the enforcement of a State's criminal code. There is no public policy more important or more fundamental than the one favoring the effective protection of the lives and property of citizens.

No specific constitutional or statutory provision requires a citizen to take an active part in the ferreting out and prosecution of crime, but public policy nevertheless favors citizen crime-fighters. "Public policy favors the exposure of crime, and the cooperation of citizens possessing knowledge thereof is essential to effective implementation of that policy. Persons acting in good faith who have probable cause to believe crimes have been committed should not be deterred from reporting them by the fear of unfounded suits by those accused." (Joiner v. Benton Community Bank (1980), 82 Ill. 2d 40, 44, 411 N.E.2d 229.) Although *Joiner* involved actions for malicious prosecution, the same can be said for the citizen

employee who fears discharge. Public policy favors Palmateer's conduct in volunteering information to the law-enforcement agency. Once the possibility of crime was reported, Palmateer was under a statutory duty to further assist officials when requested to do so. Public policy thus also favors Palmateer's agreement to assist in the investigation and prosecution of the suspected crime.

The foundation of the tort of retaliatory discharge lies in the protection of public policy, and there is a clear public policy favoring investigation and prosecution of criminal offenses. Palmateer has stated a cause of action for retaliatory discharge.

IH contends that even if there is a public policy discouraging violations of the Criminal Code, that public policy has too wide a sweep. IH points out that the crime here might be nothing more than the theft of a $2 screwdriver. It feels that in the exercise of its sound business judgment it ought to be able to properly fire a managerial employee who recklessly and precipitously resorts to the criminal justice system to handle such a personnel problem. But this response misses the point. The magnitude of the crime is not the issue here. It was the General Assembly, the People's representatives, who decided that the theft of a $2 screwdriver was a problem that should be resolved by resort to the criminal justice system. IH's business judgment, no matter how sound, cannot override that decision. "[T]he employer is not so absolute a sovereign of the job that there are not limits to his prerogative." (Tameny v. Atlantic Richfield Co. (1980), 27 Cal. 3d 167, 178, 164 Ca. Rptr. 839, 845, 610 P.2d 1330, 1336.) The law is feeble indeed if it permits IH to take matters into its own hands by retaliating against its employees who cooperate in enforcing the law.

*   *   *

Appellate Court affirmed in part and reversed in part.

RYAN, Justice, dissenting.

Although I authored the opinion in Kelsay v. Motorola, Inc., I cannot agree to extend the cause of action for retaliatory discharge approved in that case into the nebulous area of judicially created public policy, as has been done by the opinion in this case. I fear that the result of this opinion will indeed fulfill the prophesy of Mr. Justice Underwood's dissent in *Kelsay.* "Henceforth, no matter how indolent, insubordinate or obnoxious an employee may be, . . . [the]

employer may thereafter discharge him only at the risk of being compelled to defend a suit for retaliatory discharge and unlimited punitive damages. . . . "

Kelsay relied on the fact that the legislature had clearly established the public policy that injured workers had a right to file claims for compensation with the Industrial Commission. We there held that discharging the employee for filing such a claim violated that public policy. Here the public policy supporting the cause of action cannot be found in any expression of the legislature, but only in the vague belief that public policy requires that we all become "citizen crime-fighters."

Many of our cases state that public policy is to be found in the constitution and the statutes of this State and, when these are silent, in the decisions of the courts. There are other opinions of this court which simply say that the public policy of this State is to be found on its constitution and statutes and make no mention of the role of judicial decisions. Whatever the accepted role of the judiciary may be in declaring public policy, it is generally acknowledged that the question of public policy is first and foremost a matter of legislative concern. . . .

. . . In attempting to define public policy, this court has stated that it is that principle of law which declares that no one may lawfully do that *which has a tendency* to be injurious to the public welfare or to be against the public good. Stating the converse of this definition, it can be said that public policy favors that *which has a tendency* to be beneficial to the public welfare or to be for the public good. In view of such a general definition, the correctness of the statement that public policy is a vague expression and is not subject to precise definition cannot be questioned. Certainly, no employer should be subject to suit and unlimited punitive damages based on a nebulous charge that he discharged an employee for doing that *which has a tendency* to be beneficial to the public welfare or for the public good. To sustain a cause of action for retaliatory discharge, the limitations on the employer's right to discharge employees must be more precisely defined.

* * *

It is indeed praiseworthy that the plaintiff in our case is interested in ferreting out crime. His complaint, however, does not allege conduct on his part that will bring it within the area of any public policy that has been articulated by the legislature. The plaintiff was not discharged for failing to violate or for complying with the requirements of our obstruction-of-justice statute (Ill. Rev. Stat. 1979, ch. 38, par. 31-4), or of the

section of our statute concerning refusing to aid an officer (Ill. Rev. Stat. 1979, ch. 38, par. 31-8). These sections were referred to in the majority opinion. If the plaintiff would have been discharged for such a reason, *strong, clear, fundamental* articulated public policy would have been contravened and an action in tort would then be appropriate. The complaint, however, does not even allege that a crime had been committed or that the plaintiff reported to the law-enforcement agency that a crime had been committed. It only alleges that plaintiff was discharged because he reported to a law-enforcement agency that an employee of the defendant *might* be involved in a violation of the criminal code and that he had agreed to assist the law-enforcement agency in gathering further information. It should be remembered that the plaintiff was not a unionized employee, but held a position in management. By assuming the role of a "citizen crime-fighter" undertaking to ferret out crime for the police the plaintiff, through his spying, could seriously affect labor relations of his employer. Also, his conduct, without consulting with the proper management personnel, could impair the company's internal security program. In other words, the plaintiff here had taken it upon himself to become involved in crime fighting when it was neither required by law, nor by his employment, and obviously was against the wishes of his employer.

By departing from the general rule that an at-will employment is terminable at the discretion of the employer, the courts are attempting to give recognition to the desire and expectation of an employee in continued employment. In doing so, however, the courts should not concentrate solely on promoting the employee's expectations. The courts must recognize that the allowance of a tort action for retaliatory discharge is a departure from, and an exception to, the general rule. The legitimate interest of the employer in guiding the policies and destiny of his operation cannot be ignored. The new tort for retaliatory discharge is in its infancy. In nurturing and shaping this remedy, courts must balance the interests of employee and employer with the hope of fashioning a remedy that will accommodate the legitimate expectations of both. In the process of emerging from the harshness of the former rule, we must guard against swinging the pendulum to the opposite extreme. In Percival v. General Motors Corp. (8th Cir. 1976), 539 F.2d 1126, 1130, the court stated:

> It should be kept in mind that as far as an employment relationship is concerned, an employer as well as an employee has rights; . . .

*   *   *

The deteriorating business climate in this State is a topic of substantial interest. A general discussion of that subject is not appropriate to this dissent. It must be acknowledged, however, that Illinois is not attracting a great amount of new industry and business and that industries are leaving the State at a troublesome rate. I do not believe that this court should further contribute to the declining business environment by creating a vague concept of public policy which will permit an employer to discharge an unwanted employee, one who could be completely disruptive of labor-management relations through his police spying and citizen crime-fighter activities, only at the risk of being sued in tort not only for compensatory damages, but also for punitive damages.

*   *   *

In order to establish the necessary balance between employer and employee interests, I would hold that the employee may maintain an action for retaliatory discharge only when the discharge has been violative of some *strong* public policy that has been *clearly* articulated. Usually, that clear articulation would be found in legislative enactment. I do not think that an employer should be compelled to defend a tort action and possibly, be forced to pay a disgruntled discharged employee compensatory, and possibly substantial, punitive damages because of a violation of some vague concept of public policy that has never been articulated by anyone except four members of this court.

I therefore respectfully dissent.

(ii) Discharge for refusing to carry out an action that would have contravened an articulated public policy

In general, courts that have recognized the "public policy" exception to the employment-at-will rule have been most receptive to cases in which the employee was discharged for failing to take some action that would have violated the law. Perhaps easiest for a court to decide are those cases that involve employees who suffered retaliation because they would not follow their employer's instructions to take actions that would have violated *criminal* laws. Courts have also been sympathetic to employees who have been discharged for refusing to violate health, safety, or environmental laws. As the following case points out, however, a refusal to carry out an employer's directive will be protected under this theory only if it can be tied to an accepted and articulated public policy.

Pierce v. Ortho Pharmaceutical Corp.

Supreme Court of New Jersey
84 N.J. 61 (1980)

POLLOCK, J.

This case presents the question whether an employee at will has a cause of action against her employer to recover damages for the termination of her employment following her refusal to continue a project she viewed as medically unethical. Resolution of this question involves an examination of the common law doctrine of at will employment to determine whether we should adopt an exception to the rule allowing an employer to discharge an at will employee without cause.

Plaintiff, Dr. Grace Pierce, sued for damages after termination of her employment with defendant, Ortho Pharmaceutical Corporation. The trial judge granted defendant's motion for summary judgment. The Appellate Division reversed and remanded for a full trial. 166 N.J. Super. 335 (1979). We granted defendant's petition for certification. 81 N.J. 266 (1979). We now reverse the Appellate Division and reinstate the summary judgment granted by the Law Division.

Since the matter involves a motion for a summary judgment, we glean the facts from the pleadings, affidavits, and depositions before the court on the motion, giving plaintiff the benefit of all reasonable inferences that may be drawn in her favor. R. 4:46-2.

Ortho specializes in the development and manufacture of therapeutic and reproductive drugs. Dr. Pierce is a medical doctor who was first employed by Ortho in 1971 as an Associate Director of Medical Research. She signed no contract except a secrecy agreement, and her employment was not for a fixed term. She was an employee at will. In 1973, she became the Director of Medical Research/Therapeutics, one of three major sections of the Medical Research Department. Her primary responsibilities were to oversee development of therapeutic drugs and to establish procedures for testing those drugs for safety, effectiveness, and marketability. Her immediate supervisor was Dr. Samuel Pasquale, Executive Medical Director.

In the spring of 1975, Dr. Pierce was the only medical doctor on a project team developing loperamide, a liquid drug for treatment of diarrhea in infants, children, and elderly persons. The proposed formulation contained saccharin. Although the concentration was consistent with the formula for loperamide marketed in Europe, the project team agreed that the formula was unsuitable for use in the United States. An alternative formulation containing less saccharin might have been developed within approximately three months.

By March 28, however, the project team, except for Dr. Pierce, decided to continue with the development of loperamide. That decision was made apparently in response to a directive from the Marketing Division of Ortho. This decision meant that Ortho would file an investigational new drug application (IND) with the Federal Food and Drug Administration (FDA), continuing laboratory studies on loperamide and begin work on a formulation. FDA approval is required before any new drug is tested clinically on humans. 21 *U.S.C.* §355; 21 *C.F.R.* §§310.3 *et seq.* Therefore, loperamide would be tested on patients only if the FDA approved the saccharin formulation.

Dr. Pierce knew that the IND would have to be filed with and approved by the FDA before clinical testing could begin. Nonetheless, she continued to oppose the work being done on loperamide at Ortho. On April 21, 1975, she sent a memorandum to the project team expressing her disagreement with its decision to proceed with the development of the drug. In her opinion, there was no justification for seeking FDA permission to use the drug in light of medical controversy over the safety of saccharin.

Dr. Pierce met with Dr. Pasquale on May 9 and informed him that she disagreed with the decision to file an IND with the FDA. She felt that by continuing to work on loperamide she would violate her interpretation of the Hippocratic oath. She concluded that the risk that saccharin might be harmful should preclude testing the formula on children or elderly persons, especially when an alternative formulation might soon be available.

Dr. Pierce recognized that she was joined in a difference of "viewpoints" or "opinion" with Dr. Pasquale and others at Ortho concerning the use of a formula containing saccharin. In her opinion, the safety of saccharin in loperamide pediatric drops was medically debatable. She acknowledged that Dr. Pasquale was entitled to his opinion to proceed with the IND. On depositions she testified concerning the reason for her difference of opinion about the safety of using saccharin in loperamide pediatric drops:

Q   That was because in your medical opinion that was an unsafe thing to do. Is that so?
A   No. I didn't know. The question of saccharin was one of potential harm. It was controversial. Even though the rulings presently look even less favorable for saccharin it is still a controversial issue.

After their meeting on May 9, Dr. Pasquale informed Dr. Pierce that she would no longer be assigned to the loperamide project. On May 14, Dr. Pasquale asked Dr. Pierce to choose other projects. After Dr. Pierce returned from vacation in Finland, she met on June 16 with Dr. Pasquale to discuss other projects, but she did not choose a project at that meeting. She felt she was being demoted, even though her salary would not be decreased. Dr. Pierce summarized her impression of that meeting in her letter of resignation submitted to Dr. Pasquale the following day. In that letter, she stated:

Upon learning in our meeting June 16, 1975, that you believe I have not 'acted as a Director', have displayed inadequacies as to my competence, responsibility, productivity, inability to relate to the Marketing Personnel, that you, and reportedly Dr. George Braun and Mr. Verne Willaman consider me to be non-promotable and that I am now or soon will be demoted, I find it impossible to continue my employment at Ortho.

The letter made no specific mention of her difference of opinion with Dr. Pasquale over continuing the work on loperamide. Nonetheless, viewing the matter most favorably to Dr. Pierce, we assume the sole reason for the termination of her employment was the dispute over the loperamide project. Dr. Pasquale accepted her resignation.

In her complaint, which was based on principles of tort and contract law, Dr. Pierce claimed damages for the termination of her employment. Her complaint alleged:

> The Defendant, its agents, servants and employees requested and demanded Plaintiff follow a course of action and behavior which was impossible for Plaintiff to follow because of the Hippocratic oath she had taken, because of the ethical standards by which she was governed as a physician, and because of the regulatory schemes, both federal and state, statutory and case law, for the protection of the public in the field of health and human well-being, which schemes Plaintiff believed she should honor.

However, she did not specify that testing would violate any state or federal statutory regulation. Similarly, she did not state that continuing the research would violate the principles of ethics of the American Medical Association. She never contended her participation in the research would expose her to a claim for malpractice.

*   *   *

In the last century, the common law developed in a laissez-faire climate that encouraged industrial growth and approved the right of an employer to control his own business, including the right to fire without cause an employee at will. *See* Comment, 26 *Hastings L.J.* 1434, 1441 (1975). The twentieth century has witnessed significant changes in socioeconomic values that have led to reassessment of the common law rule. Businesses have evolved from small and medium size firms to gigantic corporations in which ownership is separate from management. Formerly there was a clear delineation between employers, who frequently were owners of their own businesses, and employees. The employer in the old sense has been replaced by a superior in the corporate hierarchy who is himself an employee. We are a nation of employees. Growth in the number of employees has been accompanied by increasing recognition of the need for stability in labor relations.

Commentators have questioned the compatibility of the traditional at will doctrine with the realities of modern economics and employment practices. *See, e.g.,* Blades, *Employment at Will vs. Individual Freedom: On Limiting the Abusive Exercise of Employer Power,* 67 *Colum. L. Rev.* 1404 (1967) [hereinafter cited as Blades].

*   *   *

Consistent with this policy, many states have recognized the need to protect employees who are not parties to a collective bargaining agreement or other contract from abusive practices by the employers.

Recently those states have recognized a common law cause of action for employees at will who were discharged for reasons that were in some way "wrongful." The courts in those jurisdictions have taken varied approaches, some recognizing the action in tort, some in contract. *See* Comment, 93 *Harv. L. Rev.* 1816, 1818-1824 (1980). Nearly all jurisdictions link the success of the wrongful discharged employee's action to proof that the discharge violated public policy.

*   *   *

This Court has long recognized the capacity of the common law to develop and adapt to current needs. *Jersey Shore Medical Center—Fitkin Hospital v. Baum,* 84 *N.J.* 137, 149 (1980); *Collopy v. Newark Eye and Ear Infirmary,* 27 *N.J.* 29, 43-44 (1958). The interests of employees, employers, and the public lead to the conclusion that the common law of New Jersey should limit the right of an employer to fire an employee at will.

IV

In recognizing a cause of action to provide a remedy for employees who are wrongfully discharged, we must balance the interests of the employee, the employer, and the public. Employees have an interest in knowing they will not be discharged for exercising their legal rights. Employers have an interest in knowing they can run their businesses as they see fit as long as their conduct is consistent with public policy. The public has an interest in employment stability and in discouraging frivolous lawsuits by dissatisfied employees.

Although the contours of an exception are important to all employees at will, this case focuses on the special considerations arising out of the right to fire an employee at will who is a member of a recognized profession. One writer has described the predicament that may confront a professional employed by a large corporation:

> Consider, for example, the plight of an engineer who is told that he will lose his job unless he falsifies his data or conclusions, or unless he approves a product which does not conform to

specifications or meet minimum standards. Consider also the dilemma of a corporate attorney who is told, say in the context of an impending tax audit or antitrust investigation, to draft backdated corporate records concerning events which never took place or to falsify other documents so that adverse legal consequences may be avoided by the corporation; and the predicament of an accountant who is told to falsify his employer's profit and loss statement in order to enable the employer to obtain credit. [Blades, *supra* at 1408-1409 (footnotes omitted).]

Employees who are professionals owe a special duty to abide not only by federal and state law, but also by the recognized codes of ethics of their professions. That duty may oblige them to decline to perform acts required by their employers. However, an employee should not have the right to prevent his or her employer from pursuing its business because the employee perceives that a particular business decision violates the employee's personal morals, as distinguished from the recognized code of ethics of the employee's profession. *See* Comment, 28 *Vand. L. Rev.* 805, 832 (1975).

We hold that an employee has a cause of action for wrongful discharge when the discharge is contrary to a clear mandate of public policy. The sources of public policy include legislation; administrative rules, regulations or decisions; and judicial decisions. In certain instances, a professional code of ethics may contain an expression of public policy. However, not all such sources express a clear mandate of public policy. For example, a code of ethics designed to serve only the interests of a profession or an administrative regulation concerned with technical matters probably would not be sufficient. Absent legislation, the judiciary must define the cause of action in case-by-case determinations. An employer's right to discharge an employee at will carries a correlative duty not to discharge an employee who declines to perform an act that would require a violation of a clear mandate of public policy. However, unless an employee at will identifies a specific expression of public policy, he may be discharged with or without cause.

An employee who is wrongfully discharged may maintain a cause of action in contract or tort or both. An action in contract may be predicated on the breach of an implied provision that an employer will not discharge an employee for refusing to perform an act that violates a clear mandate of public

policy. *Cf. Vasquez v. Glassboro Services, Inc.*, 83 *N.J.* 86 (1980).

An action in tort may be based on the duty of an employer not to discharge an employee who refused to perform an act that is a violation of a clear mandate of public policy. In a tort action, a court can award punitive damages to deter improper conduct in an appropriate case. *DiGiovanni v. Pessel.* 55 *N.J.* 188, 190-191 (1970); Prosser, *Torts* §2 at 9 (1971); 28 *Vand. L. Rev, supra* at 836. That remedy is not available under the law of contracts. *See, e.g.,* Corbin, *Contracts* §1077 at 367 (1951). Our holding should not be construed to preclude employees from alleging a breach of the express terms of an employment agreement. Despite the dissent's unaccountable suggestion to the contrary, Dr. Pierce did not assert the breach of any specific contractual provision as a basis for relief.

Employees will be secure in knowing that their jobs are safe if they exercise their rights in accordance with a clear mandate of public policy. On the other hand, employers will know that unless they act contrary to public policy, they may discharge employees at will for any reason. Finally, our holding protects the interest of the public in stability of employment and in the elimination of frivolous lawsuits. Courts allowing at will employees to sue for wrongful discharge have expressed concern that employees will file groundless suits. *See, e.g., Geary v. United States Steel Co.,* 319 A.2d at 179. Commentators have also noted that disgruntled employees may be encouraged to bring vexatious suits. *See, e.g.,* Blades, *supra* at 1428. However, the standard enunciated above provides a workable means to screen cases on motions to dismiss for failure to state a cause of action or for summary judgment. If an employee does not point to a clear expression of public policy, the court can grant a motion to dismiss or for summary judgment.

V

We now turn to the question whether Dr. Pierce was discharged for reasons contrary to a clear mandate of public policy. As previously stated, granting Ortho's motion for summary judgment is appropriate at this juncture only if there is no genuine issue as to any material fact.

The material facts are uncontroverted. In opposing the motion for summary judgment, Dr. Pierce did not contend that saccharin was harmful, but that it was controversial. Because of the controversy, she said she could not continue her work on loperamide. Her supervisor, Dr. Pasquale, disagreed and thought that research should continue.

As stated above, before loperamide could be tested on humans, an IND had to be submitted to the FDA to obtain approval for such testing. 21 *U.S.C.* §355. The IND must contain complete manufacturing specifications, details of pre-clinical studies (testing on animals) which demonstrate the safe use of the drug, and a description of proposed clinical studies. The FDA then has 30 days to withhold approval of testing. 21 *C.F.R.* §312.1. Since no IND had been filed here, and even giving Dr. Pierce the benefit of all doubt regarding her allegations, it is clear that clinical testing of loperamide on humans was not imminent.

Dr. Pierce argues that by continuing to perform research on loperamide she would have been forced to violate professional medical ethics expressed in the Hippocratic oath. She cites the part of the oath that reads: "I will prescribe regimen for the good of my patients according to my ability and my judgment and never do harm to anyone." Clearly, the general language of the oath does not prohibit specifically research that does not involve tests on humans and that cannot lead to such tests without governmental approval.

We note that Dr. Pierce did not rely on or allege violation of any other standards, including the "codes of professional ethics" advanced by the dissent. Similarly, she did not allege that continuing her research would constitute an act of medical malpractice or violate any statute, including *N.J.S.A.* 45:9-16(h).

In this case, Dr. Pierce has never contended that saccharin would necessarily cause harm to anyone. She alleged that the current controversy made continued investigation an unnecessary risk. However when she stopped work on loperamide, there was no risk. Our point here is not that participation in unethical conduct must be imminent before an employee may refuse to work. *See post* at 84. The more relevant consideration is that Dr. Pierce does not allege that preparation and filing of the IND was unethical. Further Dr. Pierce does not suggest that Ortho would have proceeded with human testing without FDA approval. The case would be far different if Ortho had filed the IND, the FDA had disapproved it, and Ortho insisted on testing the drug on humans. The actual facts are that Dr. Pierce could not have harmed anyone by continuing to work on loperamide.

Viewing the matter most favorably to Dr. Pierce, the controversy at Ortho involved a difference in medical opinions. Dr. Pierce acknowledged that Dr. Pasquale was entitled to his opinion that the oath did not forbid work on loperamide. Nonetheless, implicit in Dr. Pierce's position is the contention that Dr. Pasquale and Ortho were obliged to accept her opinion. Dr. Pierce contends, in effect, that Ortho should have stopped research on loperamide because of her opinion about the controversial nature of the drug.

Dr. Pierce espouses a doctrine that would lead to disorder in drug research. Under her theory, a professional employee could redetermine the propriety of a research project even if the research did not involve a violation of a clear mandate of public policy. Chaos would result if a single doctor engaged in research were allowed to determine, according to his or her individual conscience, whether a project should continue. *Cf. Report of the Ad Hoc Committee on the Principles of Medical Ethics,* American Medical Association 3 (1979). An employee does not have a right to continued employment when he or she refuses to conduct research simply because it would contravene his or her personal morals. An employee at will who refuses to work for an employer in answer to a call of conscience should recognize that other employees and their employer might heed a different call. However, nothing in this opinion should be construed to restrict the right of an employee at will to refuse to work on a project that he or she believes is unethical. In sum, an employer may discharge an employee who refuses to work unless the refusal is based on a clear mandate of public policy.

As stated above, the thrust of Dr. Pierce's complaint is not that saccharin is dangerous, but that it is controversial. At oral argument, Dr. Pierce's attorney conceded that she did not intend to submit the question of the safety of saccharin to the jury. That is, plaintiff did not intend to adduce expert testimony demonstrating the dangers of the formulation of loperamide containing the proposed level of saccharin. *Cf. Jackson v. Muhlenberg Hospital, supra,* 53 *N.J.* at 142-143. As a matter of law, there is no public policy against conducting research on drugs that may be controversial, but potentially beneficial to mankind, particularly where continuation of the research is subject to approval by the FDA. Consequently, although we recognize an employee may maintain an action for wrongful discharge, we hold there are no issues of material fact to be resolved at trial.

Under these circumstances, we conclude that the Hippocratic oath does not contain a clear mandate of public policy that prevented Dr. Pierce from continuing her research on loperamide. To hold otherwise would seriously impair the ability of drug manufacturers to develop new drugs according to their best judgment. *See Percival v. General Motors Corp., supra,* 539 *F.2d* at 1130; *Geary v. United States Steel Corp., supra,* 319 *A.2d* at 179-180.

The legislative and regulatory framework pertaining to drug development reflects a public policy that research involving testing on humans may proceed with FDA approval. The public has an interest in the development of drugs, subject to the approval of a responsible management and the FDA, to protect and promote the health of mankind. Research on new drugs may involve questions of safety, but courts should not preempt determination of debatable questions unless the research involves a violation of a clear mandate of public policy. Where pharmaceutical research does not contravene a clear mandate of public policy, the extent of research is controlled by regulation through the FDA, liability in tort, and corporate responsibility.

Accordingly, we reverse the judgment of the Appellate Division and remand the cause to the trial court for entry of judgment for defendant.

◇   NOTES

1. Even though it ultimately disagreed with her position, the court was more receptive to Dr. Pierce's claim than many courts would have been. Indeed, this decision suggests that, at least in New Jersey, the guidelines issued by a professional society might be a sufficiently clear and articulated reflection of "public policy" to support a suit for wrongful discharge. Construct policy arguments for and against the use of professional society standards as evidence of public policy in this context.

2. This case preceded *Lally v. Copygraphics* by one year. Does the latter decision define this cause of action differently than the *Pierce* case? Pay particular attention to the language used by the dissenting judge in *Lally.*

3. Go back to the four freedoms postulated at the outset of this chapter. Which would you say is the most broadly protected under the law? Which is the most narrowly protected? What advice would you give to a prospective employee who wanted the broadest possible legal "right" to act consistently with his or her own personal moral code while on the job?

E.   EXCLUSION FROM THE WORKPLACE BECAUSE OF PRESUMED SUSCEPTIBILITY TO WORKPLACE HAZARDS

Up to this point, we have focused on the legal protections available to employees who encounter retaliation because they took, or refused to take, a particular course of action. We now turn to an examination of the protections available to employees who face exclusion from the workplace, not for anything they have done or failed to do, but because of their presumed susceptibility to health hazards known to be present in the workplace. This issue leads to a deeper, more fundamental inquiry: Must the employer make the workplace safe for every worker, or need the employer only ensure that the workplace is safe for those who are allowed to work in it?

This is obviously a value-laden question. Whether one looks at it from an ethical, political, or economic perspective, the answer is not always an easy one to formulate.

Perhaps predictably, then, the law has had some difficulty with this question as well. Even with the intricate network of federal and state laws pertaining to the general issue of workplace safety and health, answers to this fundamental question have appeared, if at all, in piecemeal fashion. Most of the legal commentary has addressed employer policies that either exclude female employees to protect their offspring from reproductive hazards or exclude some employees on the basis of monitoring data suggesting that they are at particular risk from workplace chemical exposures.

1.   Exclusion of Pregnant or Fertile Female Employees in Order to Protect the Fetus

As of the early 1980s, a number of employers had implemented policies—commonly known as "fetal protection" policies—excluding fertile females from jobs involving exposure to certain reproductive hazards. (A few have excluded nonsterilized males as well.) These policies appear to have grown out of a combination of moral and economic concerns. While they provide certain protections for future offspring, the economic benefit of such policies inures largely to the employer, who avoids the cost of removing the reproductive hazard from the workplace while at the same time ensuring against future tort liability for damages to the offspring of the exposed workers. These policies have been the subject of employee challenge under both the OSHAct and Title VII of the Federal Civil Rights Act.

(a) The OSHAct

American Cyanamid Company imposed a fetus protection policy at its Willow Springs, West Virginia plant. Under the policy, female workers of child-bearing age could not work in the lead pigment department unless they had been surgically sterilized. Five women underwent sterilization to retain their jobs. The remaining two women in the lead pigment department chose not to be sterilized and were transferred to other departments in the plant with lower rates of pay.

Acting on a worker complaint, OSHA inspected the plant and issued a citation under the general duty clause, having determined that the policy itself was a hazard to the reproductive capacity of the affected women. The OSHA Review Commission disagreed and vacated the citation. When OSHA failed to appeal this decision, the Oil, Chemical and Atomic Workers Union sought review in the D.C. Circuit.

Oil, Chemical and Atomic Workers International Union v. American Cyanamid Company

741 F.2d 444 (D.C. Cir. 1984)

BORK, Circuit Judge.

Petitioners Oil, Chemical and Atomic Workers International Union and Local 3-499, Oil Chemical and Atomic Workers (together, "OCAW") seek reversal of an order by the Occupational Safety and Health Review Commission. The Commission held that respondent American Cyanamid Company's fetus protection policy "is not a hazard cognizable under the Occupational Safety and Health Act."

*   *   *

For reasons discussed below, we think the language of the Act cannot be stretched so far as to hold that the sterilization option of the fetus protection policy is a "hazard" of "employment" under the general duty clause. Consequently, we affirm.

I

It is important to understand the context in which this case arose and the task that is set for this court. American Cyanamid found, and the administrative law judge agreed, that it could not reduce ambient lead levels in one of its departments sufficiently to eliminate the risk of serious harm to fetuses carried by women employees. The company was thus confronted with unattractive alternatives: It could remove all women of childbearing age from that department, a decision that would have entailed discharging some of them and giving others reduced pay at other jobs, or the company could attempt to mitigate the severity of this outcome by offering continued employment in the department to women who were surgically sterilized. The company chose the latter alternative, and the women involved were thus faced with a distressing choice. Some chose sterilization, some did not.

As we understand the law, we are not free to make a legislative judgment. We may not, on the one hand, decide that the company is innocent because it chose to let the women decide for themselves which course was less harmful to them. Nor may we decide that the company is guilty because it offered an option of sterilization that the women might ultimately regret choosing. These are moral issues of no small complexity, but they are not for us. Congress has enacted a statute and our only task is the mundane one of interpreting its language and applying its policy.

In January and February of 1978, Glen Mercer, the plant Director of Industrial Relations, conducted a series of meetings for small groups of the Willow Island plant's female employees. At these meetings Mercer informed the women that hundreds of chemicals used at the plant were harmful to fetuses and that, consequently, the company had decided to exclude women of "childbearing capacity" from all departments of the plant where such chemicals were used. Mercer further declared that the company would deem any women between the ages of 16 and 50 to be of childbearing capacity unless she presented proof that she had been surgically sterilized.

A company doctor and nurse accompanied Mercer to these meetings and addressed the women. They explained to the women that such "buttonhole surgery" was simple and that it could be obtained locally in several places. The women were also told that the company's medical insurance would pay for the procedure, and that sick leave would be provided to those undergoing the surgery.

Mercer told the women that once the fetus protection policy was fully implemented the plant would have only about seven jobs for fertile women in the entire facility. Approximately thirty women were then employed at the Willow Island plant. Apart from the women who obtained those seven positions, Mercer said that female employees who failed to undergo surgical sterilization by May 1, 1978, would be terminated. The company extended the May 1, 1978, deadline several times. In September, 1978, the company informed the women of changes in its policy. The deadline had been extended to October 2, 1978, the Inorganic Pigments Department was the only department affected, and the only material covered by the policy was lead. It is undisputed that lead poses a severe danger to fetuses. OSHA's lead standard states the agency's belief that "the fetus is at risk from exposure to lead throughout the gestation period . . ." and is "susceptible to neurological damage." 43 Fed. Reg. 54,422 (1978). OSHA concluded that "blood lead levels should be kept below 30 μg/100 g." *Id.* That level is 30 micrograms of lead per 100 grams of whole blood. American Cyanamid was unable to reduce the lead hazard in its Inorganic Pigments Department to the level required to protect fetuses. The Administrative Law Judge ("ALJ") determined that it was economically infeasible to reduce ambient air lead levels to 200 micrograms (μg) of lead per cubic meter (m^3) of air at the Inorganic Pigments Department of the Willow Island plant. *See American Cyanamid Co.,* 1980 O.S.H. Dec. (CCH) ¶24,828 (Sept. 30, 1980). The preamble to the lead standard predicts that air lead levels of 200 μg/m^3 would lead to 83.8% of blood lead levels above 40 μg/100 grams (g) at any given time. An ambient air lead level of even 50 μg/m^3 would result in a predicted 20.3% of blood lead levels above 40 μg/100g at any one time. *See* 43 Fed. Reg. 52,963 (1978). In *United Steelworkers of America v. Marshall,* 647 F.2d 1189 [8 OSHC 1810] (D.C. Cir. 1980), *cert. denied,* 453 U.S. 913 [11 OSHC 1264] (1981), without reaching the question of economic feasibility, this court found that OSHA had failed to establish the technological feasibility of reducing ambient air lead levels to 50 μg/m^3 in the lead pigment manufacturing industry. 647 F.2d at 1295 [8 OSHC at 1902].

American Cyanamid apparently concluded, after considering these facts, that the only realistic and clearly lawful possibility left open to it was to remove women capable of bearing children from the Inorganic Pigments Department.

Between February and July, 1978, five women employed in the Inorganic Pigments Department underwent surgical sterilization at a hospital not connected with the company. Two female employees in that department did not choose sterilization. The company transferred them into other departments and, after ninety days, lowered their rate of pay to correspond to the rates characteristic of their new jobs.

II

The Occupational Safety and Health Act of 1970 provides, in relevant part:

> The Congress declares it to be its purpose and policy . . . to assure so far as possible every working man and woman in the Nation safe and healthful working conditions.

Section 2(b), 29 U.S.C. §651(b) (1982). This statement of purpose and policy must be considered in construing other provisions of the Act. The general duty clause of the Occupational Safety and Health Act of 1970 provides:

> (a) Each employer—
>
> (1) shall furnish to each of his employees employment and a place of employment which are free from recognized hazards that are causing or are likely to cause death or serious physical harm to his employees.

29 U.S.C. §654(a)(1).

There is no doubt that the words of the general duty clause can be read, albeit with some semantic distortion, to cover the sterilization exception contained in American Cyanamid's fetus protection policy. As OCAW points out, the rule bears upon "employment," and may be described as a "condition of employment." The policy may be characterized as a "hazard" to female employees who opted for sterilization in order to remain in the Inorganic Pigments Department, though it requires some stretching to call the offering of a choice a "hazard" to the person who is given the choice. Sterilization can, of course, be a "serious physical harm." For these reasons, OCAW contends that the sterilization exception falls within the plain meaning of the statutory language. That conclusion is necessary, however, only if the words of the statute inescapably have the meaning petitioners find in them and are unaffected by precedent, usage, and congressional intent. The words of the statute—in particular, the terms "working conditions" and "hazards"—are not so plain that they foreclose all interpretation.

Indeed, these words have been interpreted by courts, and in a way that strongly suggests affirmance of the Commission. *Corning Glass Workers v. Brennan,* 417 U.S. 188 (1974), though it involved the Equal Pay Act of 1963, spoke to the meaning of the phrase, "working conditions" and decided it should be given content by "the language of industrial relations." *Id.* at 202. The language of industrial relations, of course, is as relevant to the OSH Act as to the Equal Pay Act. In *Corning Glass Workers,* the Supreme Court stated:

> [T]he element of working conditions encompasses two subfactors: "surroundings" and "hazards." "Surroundings" measures the elements, such as toxic chemicals or fumes, regularly encountered by a worker, their intensity, and their frequency. *"Hazard" takes into account the physical hazards regularly encountered, their frequency, and the severity of injury they can cause.* This definition of "working conditions" is . . . well accepted across a wide range of American industry.

Id. (emphasis added) (footnotes omitted). The narrowness of this definition is emphasized by the fact that, using it, the Court determined that the difference between a night shift and a day shift was not a "working condition."

This definition was applied by the Fourth Circuit to the OSH Act in *Southern Ry. v. OSHRC,* 539 F.2d 335 [3 OSHC 1940], *cert. denied,* 429 U.S. 999 [4 OSHC 1936] (1976):

> We think this aggregate of "surroundings" and "hazards" contemplates an area broader in its contours than . . . "particular, discrete hazards" . . . but something less than the employment relationship in its entirety . . . [W]e are of the opinion that the term "working conditions" as used in Section 4(b)(1) means the environmental area in which an employee customarily goes about his daily tasks.

539 F.2d at 339 (emphasis added). Although a different section of the Act was involved in *Southern Ry.,* we can think of no reason why the definition given of "working conditions" should not apply to section 2(b) and hence influence the concept of "hazards" in the general duty clause. Congress may be presumed to have legislated about industrial relations with the "language of industrial relations" in mind. It follows, therefore, that the general duty clause does not apply to a policy as contrasted with a physical condition of the workplace.

We might rest the outcome of this case upon this case law, but petitioners advance an argument by analogy that should be discussed: "In this case, employees had in effect two choices: to undergo sterilization, or to quit. That such a 'choice' offends the Act can be seen from the fact that if, instead of a sterilization *policy,* the hazard at issue in this case had consisted of exposure to a sterilizing *chemical,* employees would have had the same choices: to undergo sterilization, or to quit. . . . [T]he Commission would not suggest that the presence of such factors would remove the sterilizing *chemical* from the ambit of the Act, and there is no logical basis for suggesting that the presence of such factors should remove the sterilization *policy* from the ambit of the Act." Brief for Petitioners at 33 (footnote omitted) (emphasis in original). To make the analogy complete, we should also suppose that in the hypothetical case, the employer could not possibly remove enough of the chemical to eliminate its sterilizing effect. We have no desire to decide this hypothetical case, so we will assume for the sake of the argument that an employer could not, under the OSH Act, permit workers to choose continued employment and sterilization by the chemical. It remains true, nonetheless, that the "hazards" in the two cases are not identical unless it can be said that anything, no matter what its nature or how it operates, is a "hazard" within the meaning of the general duty clause if it has a harmful effect. A chemical is not the same thing as a policy and a congressional decision to deal with one does not necessarily constitute a decision to deal with the other.

This, essentially, was the reasoning of the Review Commission. The Commission pointed out that the Act does not define the term "hazard" and turned to the legislative history for guidance. "Congressional floor debates, committee reports, and individual and minority views . . . are replete with discussions of air pollutants, industrial poisons, combustibles and explosives, noise, unsafe work practices and inadequate training and the like." J.A. at 507 (footnote omitted). From this, and other evidence of a similar nature, the Commission concluded that "Congress conceived of occupational hazards in terms of processes and materials which cause injury or disease by operating directly upon employees as they engage in work or work-related activities." *Id.* at 508 (footnote omitted). The fetus protection policy, by contrast, does not affect employees while they are engaged in work or work-related activities. The decision to be sterilized "grows out of economic and social factors which operate primarily outside the workplace," and hence the fetus protection policy "is not a hazard within the meaning of the general duty clause." *Id.* We agree with this conclusion. Were we to decide otherwise, we would have to adopt a broad principle of unforseeable scope: any employer pol-

icy which, because of employee economic incentives, left open an option exercised outside the workplace that might be harmful would constitute a "hazard" that made the employer liable under the general duty clause. It might be possible to legislate limitations upon such a principle but that is a task for Congress rather than courts. As it now stands, the Act should not be read to make an employer liable for every employee reaction to the employer's policies. There must be some limit to the statute's reach and we think that limit surpassed by petitioner's contentions. The kind of "hazard" complained of here is not, as the Commission said, sufficiently comparable to the hazards Congress had in mind in passing this law.

We are not prepared to speculate that, although Congress was thinking only about tangible hazards such as chemicals, it would, had it considered the subject, have decided that any employer-offered choice which leads to injury rather than discharge is a violation of the Act. That conclusion would have required a great deal of thought about unforeseen liabilities for employers and how far to let employees decide what is in their own best interest. It is not possible to say that, in all circumstances imaginable, Congress would have made employers liable for giving employees an option where the only feasible alternative was discharge. It seems to us safer, therefore, to confine the term "hazards" under the general duty clause to the types of hazards we know Congress had in mind.

Petitioner's argument may reveal a degree of uneasiness about the implications of their position. It is clear that American Cyanamid had to prevent exposure to lead of women of childbearing age, and, furthermore, that the company could not have been charged under the Act if it had accomplished that by discharging the women or by simply closing the Department, thus putting all employees who worked there, including women of childbearing age, out of work. The company was charged only because it offered the women a choice. Perhaps uncomfortable with the position that it was the offering of a choice that made the company liable, counsel for OCAW stated at oral argument that there would have been no violation if the company had simply stated that "only sterile women" would be employed in the Department because there would then have been no "requirement" of sterilization. We agree that such an announcement would not have involved a violation of the general duty clause, but we fail to see how that policy differs under the statute from the policy American Cyanamid adopted. An "only sterile women" announcement would also have given women of childbearing age the option of surgical sterilization. The only difference between this case and the hypothetical is that here the company pointed out the option and provided information about it. As petitioners frame the issue, therefore, violation of the general duty clause depends on the explicitness with which an employer phrases an option made available by its policy. It cannot be that the employer is better shielded from liability the less information it provides. It would, in any event, be difficult to find that distinction in the words of the general duty clause.

The case might be different if American Cyanamid had offered the choice of sterilization in an attempt to pass on to its employees the cost of maintaining a circumambient lead concentration higher than that permitted by law. But that is not this case. The company could not reduce lead concentrations to a level that posed an acceptable risk to fetuses. The sterilization exception to the requirement of removal from the Iorganic Pigments Department was an attempt not to pass on cost of unlawful conduct but to permit the employees to mitigate costs to them imposed by unavoidable physiological facts.

The women involved in this mater were put to a most unhappy choice. But no statute redresses all grievances, and we must decide cases according to law. Reasoning from precedent, congressional intent, and the unforeseeable consequences of a contrary holding, we conclude that American Cyanamid's fetus protection policy did not constitute a "hazard" within the meaning of the OSH Act.

Affirmed.

◇ NOTES

1. Would the result have been the same here had there been no OSHA standard for lead?

2. The court places considerable emphasis on the finding that it would not have been "feasible" for the company to have reduced workplace lead levels any lower. What is the statutory justification for the court's reliance on feasibility here?

◇

(b) The civil rights act

Title VII of the Civil Rights Act of 1964 prohibits employment discrimination on the basis of sex, unless the employer can demonstrate that sex is a "bona fide occupational qualification [BFOQ] reasonably necessary to the normal operation of that particular business." The Pregnancy Discrimination Act, added to Title VII in 1978, requires that "women affected by pregnancy, childbirth, or related medical conditions shall be treated the same for all employment-related purposes . . . as other persons not so affected, but similar in their ability or inability to work." As women-only fetal protection policies discriminate against women on the basis of their capacity to become pregnant, and as Title VII's BFOQ defense has been construed narrowly by the Supreme Court, many commentators (including the authors of this text) predicted that the courts would rule that such policies constitute unlawful sex discrimination. One commentator, Wendy W. Williams of Georgetown University, argued that Title VII would permit fetal protection policies only if they were applied equally to fertile employees of *both* sexes. This, she reasoned, would lead to healthier work environments.

> The option of excluding workers at risk may well seem less attractive in light of such evidence than it did when the employer assumed that only women workers

transmitted fetal hazards. A workplace composed exclusively of sterile men and women and post-menopausal women will be most unappealing to most employers. Under these circumstances, the employer may be inspired to develop solutions short of exclusions, thus not only protecting itself from liability and advancing the health of offspring but promoting the employment interests of workers as well. [W. Williams, Firing the Woman to Protect the Fetus: The Reconciliation of Fetal Protection with Employment Opportunity Goals Under Title VII, 69 Geo. L.J. 641, 703-704 (1981).]

As of late 1990, the results in the federal courts have not borne out this prediction. Four circuit courts have addressed the applicability of Title VII to fetal protection policies, and all have held that such policies do not violate Title VIII if the employer can establish that exclusion of the women in question is a *business necessity*.

The business necessity defense was developed by the Supreme Court in conjunction with the court's recognition that an employment practice that is neutral *on its face*, and that was developed without any discriminatory intent, could still violate Title VII if it was discriminatory in its effect. In situations of unintentional (so-called disparate impact) discrimination of this nature, reasoned the Supreme Court, an employer should be able to avoid Title VII liability upon a lesser showing than is required under the BFOQ defense. Thus, the court created the business necessity defense, which requires that the disputed employment practice have a "manifest relation to the employment in question." See *Griggs v. Duke Power Co.*, 401 U.S. 424, 431 (1971). In applying the "business necessity" analysis to fetal protection policy cases, however, the circuit courts have gone considerably beyond the Supreme Court's stated rationale for creating the defense.

This has engendered considerable controversy, as is amply reflected in the following decision of a hotly divided Seventh Circuit Court of Appeals in a case involving a fetus protection policy imposed by Johnson Controls of Wisconsin. It has also brought Supreme Court review. The *Johnson Controls* case has been argued to the Supreme Court and, as this book goes to press, a decision is expected in early 1991.

United Auto Workers v. Johnson Controls, Inc.

886 F.2d 871 (7th Cir. 1989)

COFFEY, Circuit Judge.

Since 1982 Johnson Controls, Inc. (hereinafter "Johnson Controls" or "Johnson") has maintained a fetal protection policy designed to prevent unborn children and their mothers from suffering the adverse effects of lead exposure. International Union, United Automobile, Aerospace and Agricultural Implement Workers of America, UAW (hereinafter "UAW"), several UAW local unions and a group of individual employees brought suit alleging that this policy violated Title VII, 42 U.S.C. §2000e, et seq. The district court granted summary judgment in favor of John-

son Controls and the plaintiffs appealed. This case was originally argued before a panel of this court and the panel's opinion was circulated among all the members of the court pursuant to Circuit Rule 40(f). Prior to publication of the panel opinion, a majority of the members voted to hear the case before an *en banc* court and, following rehearing *en banc*, a majority of the court voted to affirm the decision of the district court.

I

The Battery Division of Johnson Controls, Inc., was created upon Johnson Controls' 1978 purchase of Globe Union, Inc. (hereinafter "Globe" or "Globe

Union"). Globe Union was formed through the consolidation of two battery companies and had been in the battery business for almost fifty years before Johnson's purchase. Globe Union and Johnson Controls have maintained ongoing efforts to improve industrial safety through measures designed to minimize the risk lead poses to those directly involved in the manufacturing of batteries.

The steps that Globe Union and Johnson Controls have taken to regulate lead exposure have not been focused merely on complying with governmental safety regulations, but originate from their longstanding corporate concern for the danger lead poses to the health and welfare of their employees, their employees' families and the general public. During the period of the 1970's when OSHA's regulation of employee exposure to lead was virtually non-existent, Johnson Controls' predecessor, Globe Union, initiated a large number of innovative programs in an attempt to control and regulate industrial lead exposure. For example, in 1969, Dr. Charles Fishburn, M.D., who later became one of the primary proponents of Johnson Controls' fetal protection policy, instituted programs for monitoring employee blood lead levels. In an attempt to manage lead exposure, other safety programs were initiated at Globe and Johnson including a lead hygiene program, respirator program, biological monitoring program, medical surveillance program and a program regulating the type, use and disposal of employee work clothing and footwear to minimize lead exposure. Globe Union also transferred employees out of high lead environments whenever a physician's medical evaluation report established that the individual had a high blood lead level. In the case of such transfers, medical removal benefits were provided to the employee before OSHA required such compensation. Globe Union and Johnson Controls have continued to address their serious concern for industrial safety through efforts to design and regulate lead manufacturing areas to reduce employee lead exposure. For example, laminar flow pumps constantly supply a down draft of low velocity clean air to improve the environment of workstations where employees deal with lead. Central vacuum systems and powered floor scrubbers and sweepers are used to keep the manufacturing area as clear of lead dust as possible. Since Johnson Controls' purchase of Globe Union in 1978, it has spent approximately $15 million on environmental engineering controls at its battery division plants.

* * *

Johnson adopted its current fetal protection program in 1982 following its determination, based upon scientific research, that it was medically necessary to bar women from working in high lead exposure positions in the battery manufacturing division. The fetal protection policy applies to work environments in which any current employee has recorded a blood lead level exceeding 30 μg/dl during the preceding year or in which the work site has yielded an air sample during the past year containing a lead level in excess of 30 μg per cubic meter. The policy recites that women with child-bearing capacity will neither be hired for nor allowed to transfer into those jobs in which lead levels are defined as excessive. A grandfather clause in Johnson's fetal protection policy permits fertile women who were assigned to high lead exposure positions at the time of the adoption of the policy to remain in those job assignments if they are able to maintain blood lead levels below 30 μg/dl.[9] Those employees who are removed from positions because of excessive lead levels are transferred to another job in Johnson's employ without suffering either a loss of pay or benefits.

The major reason Johnson adopted its current fetal protection policy was the inability of the previous voluntary policy to achieve the desired purpose: protecting pregnant women and their unborn children from dangerous blood lead levels. Between 1979 and 1983, at least six Johnson Controls employees in high lead exposure positions became pregnant while maintaining blood lead levels in excess of 30 micrograms.

* * *

Prior to adopting its updated fetal protection policy, Johnson seriously considered alternatives to the exclusion of women with childbearing capacity from high lead exposure positions, but after research and consultation with medical and scientific experts found

[9]Under the fetal protection policy an incumbent female employee with a blood lead level reading above 30 μg/dl is permitted a period of time to reduce her blood level to 30 μg/dl. If the blood level of a fertile female employee is in excess of 40 μg/dl, she is transferred at the earliest possible date. The record does not disclose the number, if any, of female employees who remain in high lead exposure positions or who were transferred as a result of the fetal protection policy.

itself unable to structure and implement any alternatives which would adequately protect the unborn child from the risks associated with excessive lead exposure. Johnson's experience demonstrated that the voluntary exclusion program was ineffective. To date neither Johnson nor any other battery manufacturer has been able to produce a lead free battery, or to utilize engineering research and technology to implement a system or procedure capable of reducing the lead exposure of its employees to acceptable levels for fertile women. Limitation of the fetal protection policy to women actually pregnant was found ineffective because there is the very definite possibility that lead exposure will occur between conception and the time the woman discovers her pregnancy.[11] Such a limitation is further inadequate because reduction of blood lead levels following removal from a lead exposure area requires a significant length of time that frequently extends well into the pregnancy term. Limitations of the policy to women planning pregnancy also was not found to be a suitable alternative because of one of the exigencies of life, the frequency of unplanned or undetected pregnancies. Permitting fertile female employees to attempt to maintain a blood lead level below 30 μg/dl or utilizing the mean or median blood lead levels of current workers as a measure of whether a woman should be

---

[11]There will normally be some delay in diagnosis of pregnancy:

> The first sign of pregnancy and the first reason most pregnant women see a physician is absence of an expected menstrual period. If a patient's periods are usually regular, absence of menses for 1 wk or more is presumptive evidence of pregnancy. Pregnancies are usually dated in weeks, starting from the first day of the last menstrual period. Thus if the patient's menses were regular and if ovulation did occur on day 14 of the cycle, obstetric dates are about 2 wk longer than embryologic dates. If the patient's periods are irregular, the difference will be greater or less than 2 wk. Usually, 2 wk after missing a period the patient is considered to be six wk pregnant and the uterus is correspondingly enlarged.

R. Berkow, *The Merck Manual of Diagnosis and Therapy,* 1744-45 (14th ed. 1987). Thus, even in ideal cases, there is normally some time lag between pregnancy's onset and diagnosis. In other cases, a mother's failure to perceive a pregnancy or a delay in receiving prompt medical care can result in a pregnancy diagnosis later in pregnancy. Under a policy requiring removal only on discovery of pregnancy, the unborn child would be subject to lead exposure throughout the period prior to diagnosis of pregnancy.

permitted in a position would also not effectively protect the unborn child. The reason these actions would be inadequate is that an employee's risk of high lead levels is usually greatest immediately after commencement of work in a high lead environment.

* * *

II

Proper analysis of the Title VII issues this case presents requires a thorough understanding of the following fundamental question: Does lead pose a health risk to the offspring of Johnson's female employees? In considering the evidence in the record on this subject it is important to note that *both* the UAW and Johnson Controls agree on appeal that a *substantial* health hazard to the unborn child in the womb has been established. The UAW admits in its brief that "it is clear that . . . substantial risk of harm to the fetus . . . has been established." UAW Brief at 33. Similarly, Johnson Controls states that "[t]he evidence in the record on [substantial risk of harm to the fetus] is overwhelming."

* * *

The chief reason why an unborn child's lead exposure is of such great concern is that it has been medically established that lead attacks the fetus' central nervous system and retards cognitive development.

* * *

Unlike physical birth defects, such as those associated with thalidomide, lead's sometimes subtle damaging effects may not fully manifest themselves until the child is diagnosed as having learning problems in a school setting some five to six years after birth:

> What we are worried about are very subtle things, the ability to really affect learning ability. And so far as impairing the child's progress, they really aren't evident until he gets into school. He discovers that he can't remember, that his brain cannot pay attention, what our psychologists here called deficits in auditory processing, which is a fancy way of saying they can't understand what they hear, can't process it, and use it effectively. And those things will impair a child perhaps toward the end of the first grade, particularly in the second grade.

Deposition of Dr. J. Julian Chisholm, Jr., M.D., Director, Lead Program, John F. Kennedy Institute and Associate Professor of Pediatrics, Johns Hopkins School of Medicine (hereinafter Chisholm Dep.), at 27.

Probably the worst aspect of lead's influence upon an unborn child's future intellectual development is that its effects have frequently been found to be irreversible. Further, the most recent research suggests that the unborn child may be affected at lead levels previously believed safe. *See* J.M. Davis & D. Svendsgaard, Lead and Child Development, 329 Nature 297 (1987) (Collecting results of recent studies in this area).

Lead exposure can also pose other physical threats to the unborn child such as reduction of the infant's birth weight, premature delivery, and stillbirth. *See* Dr. Chisholm Aff. ¶6. Lead may also affect the other vital fetal organs including, but not limited to, the liver and kidneys.

The danger resulting from lead exposure cannot simply be avoided through removing a pregnant woman from lead exposure promptly after the discovery of pregnancy. Dr. Chisholm, a recognized expert in the research field of treatment and prevention of lead poisoning in young children, observed that "excluding only women who are actually pregnant from work areas where there are elevated blood lead levels would not sufficiently protect the health and safety of the unborn child" Dr. Chisholm Aff. at ¶10. This is true because *lead continued to exert an effect upon the mother and her unborn child for a significant period of time after she has been removed from lead exposure.*

* * *

The overwhelming evidence in this record establishes that an unborn child's exposure to lead creates a substantial health risk involving a danger of permanent harm. This evidence clearly approaches a "general consensus within the scientific community," and certainly "suffices to show that within that community there is [a] considerable body of opinion that significant risk exists to the unborn child from exposure to lead." *Wright v. Olin Corp.*, 697 F.2d 1172, 1191 (4th Cir. 1982). Next we consider the proper legal standards to be applied when employees bring a Title VII sex discrimination action challenging an employer's response to this serious health risk.

III

Having considered both the nature of the risk of harm that lead exposure presents to the unborn child and the mother and the policies Johnson implemented in response to this problem, we now turn to the question of the proper legal analysis to be applied to Johnson's fetal protection program under Title VII. The question presented is should we follow the lead of the Fourth Circuit, the Eleventh Circuit and the EEOC in determining that these policies can be justified with a "business necessity" defense or must we conclude that these policies may only be justified with a bona fide occupational qualification defense.

In approaching this issue we are cognizant of the mandates the United States Supreme Court has recited on two occasions concerning the necessity of avoiding rigid application of proof patterns to particular factual situations. The Court's concern was first set forth in *Furnco Construction Corp. v. Waters*, 438 U.S. 567, 576 (1978), where the Court noted that the formula it had devised for demonstrating a prima facie case of disparate treatment in *McDonnell Douglas Corp. v. Green*, 411 U.S. 792, 802 (1973) "was not intended to be an inflexible rule." The Court expanded on this same subject in *International Brotherhood of Teamsters v. United States*, 431 U.S. 324, 358 (1977), when it observed:

> Our decision in [*McDonnell Douglas*] did not purport to create an inflexible formulation. We expressly noted that "[t]he facts necessarily will vary in Title VII cases, and the specification . . . of the prima facie proof required from [a plaintiff] is not necessarily applicable in every respect to differing factual situations." [*McDonnell Douglas*, 411 U.S. at 802 n.13]."

(Footnote omitted). The thrust of these repeated Supreme Court pronouncements is that courts are required to avoid inflexible application of judicially devised proof patterns in cases that present factual circumstances different from those encountered previously. *See also Wright v. Olin Corp.*, 697 F.2d 1172, 1184 (4th Cir. 1982). Any proof scheme a federal court applies is useful only if it assists the court in properly identifying the employment practices Congress intended to prohibit under Title VII. These concerns are particularly important in a case of this nature where the interest in financial reward is balanced against a medically established risk of the birth of a medically or physically deprived baby and where the challenged distinction is based upon the reality that only the female of the human species is capable of childbearing.

Two other federal courts of appeals and the Equal Employment Opportunity Commission have addressed the question of the defenses available to an employer under Title VII in a case challenging a fetal protection program. The first court of appeals to address this question was the Fourth Circuit in *Wright v. Olin Corp.*, 697 F.2d 1172 (4th Cir. 1982). That case involved a fetal protection program very similar to the one Johnson instituted, in that it forbade any fertile woman from working in a job which " 'may require contact with and exposure to known or suspected abortifacient or teratogenic agents.' " *Olin*, 697 F.2d at 1182. In considering which of several possible theories of claim and defense should apply in a Title VII analysis of fetal protection policy, the Fourth Circuit observed:

> We must start by conceding that the fact situation [the fetal protection policy] presents does not fit with absolute precision into any of the developed theories. It differs in some respects—either in its claim or defense elements—from each of the paradigmatic fact situations with which the different theories have been centrally concerned. This of course accounts for the conflict on the point between the parties. That there would be such fact situations in Title VII litigation has always been recognized by the Supreme Court as it has developed and applied the different theories. *The Court has continually admonished, and indeed demonstrated in its own decisions, that these theories were not expected nor intended to operate with rigid precision with respect to the infinite variety of factual patterns that would emerge in Title VII litigation.* So has this court.

697 F.2d at 1184 (emphasis added, footnotes omitted).

The court applied the disparate impact/business necessity theory of claim and defense that normally is applied only in cases in which an employer's policy is "facially neutral." Even though the court recognized that the facial neutrality of a fetal protection policy "might be subject to logical dispute, the dispute would involve mere semantic quibbling having no relevance to the underlying principle that gave rise to this theory." 597 F.2d at 1186. Because a fetal protection policy involves motivations and consequences most closely resembling a disparate impact case, the Fourth Circuit felt it should be analyzed under the disparate impact/business necessity theory. *See id.* The Fourth Circuit defined the business necessity defense in the context of a fetal protection policy as requiring a demonstration that "significant risks of harm to the unborn children of women workers from their exposure during pregnancy to toxic hazards in the workplace make necessary, for the safety of the unborn children, that fertile women workers, though not men workers, be appropriately

restricted from exposure to those hazards. . . ." 697 F.2d at 1190 (footnote omitted). However, the Fourth Circuit permitted this evidentiary demonstration to be rebutted with proof that "there are 'acceptable alternative policies or practices which would better accomplish the business purpose . . . [or protect against the risk of harm], or accomplish it equally well with a lesser differential . . . impact [between women and men workers].' "

The Eleventh Circuit utilized a similar analysis in *Hayes v. Shelby Memorial Hospital*, 726 F.2d 1543 (11th Cir. 1984) (Tuttle, J.). In *Hayes* a hospital terminated a pregnant woman's employment upon discovering her pregnancy. In *Hayes* the Court utilized the elements of the business necessity defense found in *Olin* to establish that the involved policy was not "facially discriminatory." The Eleventh Circuit stated: "In other words, the employer must show (1) that there is a substantial risk of harm to the fetus or potential offspring of women employees from the women's exposure, either during pregnancy or while fertile, to toxic hazards in the workplace and (2) that the hazard applies to fertile or pregnant women, but not to men." 726 F.2d at 1548 (footnote omitted). The theory underlying the facial neutrality analysis utilized in *Hayes* is that a policy meeting the above criteria "is neutral in the sense that it effectively and equally protects the offspring of all employees." 726 F.2d at 1548. If facial neutrality is established, the court proceeds to a disparate impact/business necessity analysis. 726 F.2d at 1552. Under the Eleventh Circuit's analysis, if facial neutrality is not established, the employer must present a bona fide occupational qualification defense to justify its fetal protection policy.

The Eleventh Circuit went on to set out the disparate impact/business necessity analysis it would apply in cases where facial neutrality was established. The court recognized that a fetal protection policy, even if "facially neutral," "clearly has a disproportionate impact on women since only they are affected by it." *Id.* However "the employer's business necessity defense applies automatically, just as the employee's prima facie case of disparate impact applies automatically. That is because to reach the disparate impact stage of analysis in a fetal protection case, the employer has *already* proved—to overcome the presumption of facial discrimination—that its policy is justified on a scientific basis and addresses a harm that does not affect men." *Id.* at 1553. As in *Olin*, "the employer's business necessity defense may be rebutted by proof that there are acceptable alternative policies that would better accomplish the purposes of promoting fetal health, or that would accomplish the purpose with less adverse impact on one sex." *Id.*

Although *Olin* and *Hayes* present somewhat different analyses, both cases, in essence, determine that a business necessity defense in a fetal protection policy case requires (1) a demonstration of the existence of a substantial health risk to the unborn child, and (2) establishment that transmission of the hazard to the unborn child occurs only through women. Both cases also allow the employee to present evidence of less discriminatory alternatives equally capable of preventing the health hazard to the unborn.

On October 3, 1988, the Equal Employment Opportunity Commission, the agency responsible for the administration of Title VII, issued a *Policy Statement on Reproductive and Fetal Hazards Under Title VII* that, in substance, endorsed the approaches that the Fourth and Eleventh Circuits have taken to fetal protection cases. Equal Employment Opportunity Commission, *Policy Statement on Reproductive and Fetal Hazards Under Title VII* (October 3, 1988) (found in Fair Empl. Prac. Manual (BNA) 401:6013). As the Supreme Court has recognized, while such EEOC pronouncements "do not have the force of law, . . . still they ' "constitute a body of experienced and informed judgment to which courts and litigants may resort for guidance." ' " *Local No. 93, International Association of Fire Fighters v. City of Cleveland*, 478 U.S. 501, 518 (1988) (quoting *General Electric Co. v. Gilbert*, 429 U.S. 125, 142 (1976) which quoted, in turn, *Skidmore v. Swift & Co.*, 323 U.S. 134, 140 (1944)). A fair reading of the EEOC's Policy Statement reflects that the EEOC thoroughly considered the various interests under Title VII and followed earlier judicial decisions only after concluding that these decisions properly implemented Title VII policies. The EEOC noted that fetal protection "*cases do not fit neatly into the traditional Title VII analytical framework and, therefore, must be regarded as a class unto themselves.*" *Policy Statement* (found in Fair Empl. Prac. Manual (BNA) 401:6013, 6015 n.11) (emphasis added). The EEOC then candidly recognized that fetal protection policies that "exclude only women constitute *per se* violations of the Act." *Id.* at 401:6014 (footnote omitted). However, the EEOC went on to observe that

[a]lthough the BFOQ defense is normally the only one available in cases of overt discrimination, *the Commission follows the lead of every court of appeals to have addressed the question [in determining] that the business necessity defense applies to these cases. While business necessity has traditionally been limited*

to disparate impact cases, there is an argument that in this narrow class of cases the defense should be flexibly applied.

Id. at 401:6014-15 (emphasis added, footnote omitted). The EEOC concluded that:

The issues [in a fetal protection policy case to which the business necessity defense is applicable] are (1) whether there exists a substantial risk of harm to employees' offspring through the exposure of employees to a reproductive hazard in the workplace; (2) whether the harm to employees' offspring takes place through the exposure of employees of one sex but not employees of the opposite sex; and (3) whether the employer's policy effectively eliminates the risk of fetal or reproductive harm. Even if these elements are proved, the policy will not withstand scrutiny [if] it is shown that there exists a reasonable alternative policy that will protect employees' offspring from fetal or reproductive harm and that has a less discriminatory impact on employees of the restricted sex. Thus, an employer's reproductive or fetal protection policy must be neutrally designed to protect all employees' offspring from hazards existing in the workplace. Where substantial evidence exists that the risk of harm to employees' offspring takes place only through the exposure of one sex to a hazard existing in the workplace, an employer may exclude from the workplace employees of that sex, but only to the extent necessary to protect employees' offspring from reproductive or fetal hazards."

Id. at 401:6015-16 (footnotes omitted).

* * *

We are convinced that the components of the business necessity defense the courts of appeals and the EEOC have utilized in fetal protection cases balance the interests of the employer, the employee and the unborn child in a manner consistent with Title VII.

* * *

We now proceed to determine whether this defense can be utilized to sustain Johnson Controls' fetal protection policy.

IV

In *Wards Cove Packing Co. v. Atonio*, 109 S.Ct. 2115, 2125-26 (1989), the Supreme Court recently described the general policies underlying the business necessity defense that we utilize in considering Johnson Controls' fetal protection policy:

> Though we have phrased the query differently in different cases, it is generally well-established that at the justification stage of . . . a disparate impact case, the dispositive issue is whether a challenged practice serves, in a significant way, the legitimate employment goals of the employer. The touchstone of this inquiry is a reasoned review of the employer's justification for his use of the challenged practice. A mere insubstantial justification in this regard will not suffice, because such a low standard of review would permit discrimination to be practiced through the use of spurious, seemingly neutral employment practices. At the same time, though, there is no requirement that the challenged practice be "essential" or "indispensable" to the employer's business for it to pass muster: this degree of scrutiny would be almost impossible for most employers to meet, and would result in a host of evils.

(Citations omitted).

In *Wards Cove* the Court also clarified the proof burdens to be applied in addressing an employer's business necessity defense:

> [T]he employer carries the burden of producing evidence of a business justification for his employment practice. *The burden of persuasion, however, remains with the disparate-impact plaintiff.* . . . We acknowledge that some of our earlier decisions can be read as suggesting otherwise. But to the extent that those cases speak of an employers' (sic) "burden of proof" with respect to a legitimate business justification defense, they should have been understood to mean an employer's production—but not persuasion—burden. The persuasion burden here must remain with the plaintiff, for it is he who must prove that it was "because of such individual's race, color," etc., that he was denied a desired employment opportunity. See 42 U.S.C. §2000e-2(a).

Wards Cove, 109 S. Ct. at 2126 (citations omitted, emphasis added).

The allocation of the burden of proof under substantive Title VII law outlined in *Wards Cove* plays a significant role in summary judgment proceedings of this nature. We have previously recognized that: "Summary judgment is properly entered in favor of a party when the opposing party is unable to make a showing sufficient to prove an essential element of a case on which the opposing party bears the burden of proof."

* * *

"Where the record taken as a whole could not lead a rational trier of fact to find for the non-moving party, there is no 'genuine issue for trial.' " *Matsushita Electric Industrial Co. v. Zenith Radio Corp.,* 475 U.S. 574, 587 (1986) (quoting *First National Bank of Arizona v. Cities Service Co.,* 391 U.S. 253, 288-89 (1968)).

* * *

Our inquiry must be based on the underlying premise that the creation of a record adequate to meet legal challenges is the responsibility of the parties litigating the case. We may neither add nor subtract from the record; we must accept it as it is.

* * *

We now turn to whether the UAW has established a genuine issue of material fact concerning any of the elements of the business necessity defense upon which it bears the burden of persuasion.

A. Substantial Risk of Harm to the Unborn Child

Both the UAW and Johnson Controls agree on appeal that the significant evidence of risks to the health of the fetus contained in the record establishes a *substantial* health risk to the unborn child.

* * *

Accordingly, we are convinced that there is no genuine issue of material fact with respect to this component of Johnson Controls' business necessity defense.[28]

B. Exposure Through a Single Sex

The UAW's efforts in this case have primarily been devoted toward negating the second element of Johnson's business necessity defense, that the risk of

[28]There might be a suggestion that the unborn child would be harmed if his or her mother were deprived of insurance benefits or wages that could be utilized for prenatal care as a result of the application of Johnson Controls' fetal protection policy. This issue bears no relevance to Johnson Controls' employment practices for any female employees deprived of jobs in high lead environments under Johnson's fetal protection policy, instituted in 1982, are transferred to other positions in Johnson Controls' employ without *any* loss of either wages or benefits.

transmission of potentially harmful lead exposure to unborn children is substantially confined to fertile female employees. On this issue, as with the question of substantial risk of harm to the unborn child, "it is not necessary to prove the existence of a general consensus on the [issue] within the qualified scientific community." *Olin*, 697 F.2d at 1191.

In this case Johnson Controls' experts, without exception, testified that a male worker's exposure to lead at levels within the 50 μg/dl maximum set forth in OSHA's current (1978) lead exposure guidelines did not pose a substantial risk of genetically transmitted harm from the male to the unborn child. Moreover, Johnson's experts took the position that because this data dealt exclusively with animals, the results of these studies were not scientifically established as being applicable to humans. In contrast, the UAW witnesses posited that animal studies had demonstrated that there was a possible risk of genetic damage to human offspring as a result of male lead exposure. The UAW witnesses attempt to bridge the wide chasm between the results of animal studies and a conclusion of genetic harm allegedly transmitted through the male human being with human studies merely establishing a correlation between male lead exposure and changes in sperm shape. It is interesting to note that the UAW has not presented any medical evidence in the record of any human study scientifically documenting genetic defects in human beings resulting from male lead exposure. It is this lack of convincing scientific data that the plaintiffs attempt to gloss over and cast aside in ignoring the differences between the effect of lead on the human and animal reproductive systems.

* * *

Because scientific data available as of this date reflects that the risk of transmission of harm to unborn children is confined to fertile female employees, the sex-based distinction present in Johnson Controls' fetal protection policy is based upon real physical differences between men and women relating to child bearing capacity and is consistent with Title VII.

C. Adequate But Less Discriminatory Alternatives

We are cognizant of the fact that Johnson's fetal protection policy might very well not have been sustainable had the UAW presented facts and reasoning sufficient for the trier of fact to conclude that "there are 'acceptable alternative policies or practices which would better accomplish the business purpose . . . [of protecting against the risk of harm], or accomplish equally well with a lesser differential . . . impact [between women and men workers].' " *Olin* 697 F.2d at 1191 (quoting *Robinson v. Lorillard Corp.* 444 F.2d 791, 798 (4th Cir.), *cert. dismissed,* 404 U.S. 1006 (1971)).

* * *

Even were we to conclude that the UAW had preserved this issue for appeal, we would be constrained to hold that the UAW failed to present facts sufficient for a trier of fact to conclude that less discriminatory alternatives would equally effectively achieve an employer's legitimate purpose of protecting unborn children from the substantial risk of harm lead exposure creates. In *Wards Cove Packing Co. v. Atonio,* 109 S. Ct. 2115, 2126-27 (1989), the Supreme Court recently explained the burden a Title VII plaintiff must carry in order to establish that an employer's policy is invalid on the basis of the availability of less discriminatory alternatives:

[I]f . . . [plaintiffs] cannot persuade the trier of fact on the question of [the employer's] business necessity defense, [plaintiffs] may still be able to prevail. To do so, [plaintiffs] will have to persuade the factfinder that "other tests or selection devices, without a similarly undesirable racial effect, would also serve the employer's legitimate [hiring] interest[s];" by so demonstrating, [plaintiffs] would prove that "[the employers were] using [their] test merely as a "pretext" for discrimination." . . . Of course, any alternative practices which [plaintiffs] offer up in this respect must be equally effective as [the employers'] chosen hiring procedures in achieving [the employers'] legitimate employment goals. Moreover, "[f]actors such as the cost or other burdens of proposed alternative selection devices are relevant in determining whether they would be equally as effective as the challenged practices in serving the employer's legitimate business goals." *Watson* [*v. Fort Worth Bank & Trust*, 108 S. Ct. 2777, 2790 (1988)] (O'Connor, J.). "Courts are generally less competent than employers to restructure business practices," *Furnco Construction Corp. v. Waters*, 438 U.S. 567, 578 (1978); consequently, the judiciary should proceed with care before mandating that an employer must adopt a plaintiff's alternate selection or hiring process in response to a Title VII suit.

* * *

The UAW, in its briefs and argument, has failed to present even one specific alternative to Johnson's fetal protection policy, much less a demonstration of how any particular economically and technologically feasible alternative would effectively achieve Johnson's purpose of preventing the risk of fetal harm associated with the exposure to lead of fertile female employees.

The record also demonstrates that viable alternatives to the fetal protection program were not presented to the court that would equally effectively further Johnson's legitimate interests. As detailed in Section I, *supra,* Johnson Controls itself considered various possible less discriminatory alternatives prior to its adoption of the current fetal protection policy in 1982. In considering these alternatives, Johnson realized that lead could not be eliminated as a battery component. Furthermore, technically and economically feasible alternatives in the manufacturing process are incapable of reducing lead exposure to acceptable levels for pregnant women.

*   *   *

V

Having just held that the business necessity defense shields an employer from liability for sex discrimination under Title VII in a fetal protection policy involving the type of facts present herein, we are also convinced that Johnson Controls' fetal protection policy could be upheld under the bona fide occupational qualification defense.

In addressing the bona fide occupational qualification question, we have observed that: "It is universally recognized that this exception to Title VII was 'meant to be an extremely narrow exception to the general prohibition of discrimination. . . .' " *Torres v. Wisconsin Dept. of Health & Social Services,* 859 F.2d 1523, 1527 (7th Cir. 1988) (quoting *Dothard v. Rawlinson,* 433 U.S. 321, 334 (1977)). Nonetheless, this formulation should not be treated as inviting a black letter conclusion that the employer automatically loses in a case in which it is required to demonstrate a bona fide occupational qualification. The bona fide occupational qualification defense, like other Title VII defenses, must be construed in a manner which gives meaningful and thoughtful consideration to the *interests of all those affected by a company's policy, in this case the employer, the employee and the unborn child.* Indeed, the fact that Johnson's fetal protection policy applies exclusively to the high lead exposure areas of its battery division demon-

strates why the policy is drafted with sufficiently definite terminology as to constitute a "narrow exception to the general prohibition of discrimination. . . ."[35]

*   *   *

Torres bears particular relevance to our discussion of the description of Johnson Controls' business. At a broad level, Johnson's business, insofar as relevant to this case, is the manufacture of batteries. Johnson's business is "unique" because it requires the use of lead, an extremely toxic substance that has been scientifically established to pose very serious dangers to young children and, in particular, to the offspring of female employees. In order to respond to the problems accompanying its unique battery manufacturing operation, Johnson Controls has properly made it part of its business to attempt to manufacture batteries in as safe a manner as possible. This safety interest is every bit as critical to the mission of Johnson's battery manufacturing business as rehabilitation of prisoners is to the mission of the prison facility at issue in *Torres.* Furthermore, like the prison in *Torres,* Johnson has found it necessary to "innovate" to achieve its essential goal of manufacturing batteries safely through the adoption of a fetal protection policy that would address the health/safety problems related to its female employees significantly more effectively than the alternative policies it had considered. *See* Sections II and IV-C, *supra.*

Having established that industrial safety (preventing hazards to health) is legitimately part of the "essence" of the "business" of a battery manufacturer, as it is of any manufacturing enterprise, the next inquiry under *Torres* is whether Johnson Controls' fetal protection policy is "directly related" to industrial safety. See *Torres,* 859 F.2d at 1530. Certainly a policy is directly related to industrial safety when it protects unborn children from a substantial risk of devastating and permanent impairment or loss of intellectual ability or injury to vital organs resulting from exposure to a toxic industrial chemical.

As in *Torres,* "[t]he more difficult question is whether the proposed BFOQ [is] 'reasonably necessary' to furthering the objective of [industrial safety]." *Torres,* 859 F.2d at 1530. In "unique" businesses, like the

---

[35] *Torres,* 859 F.2d at 1527 (quoting *Dothard,* 433 U.S. at 334). The battery division is only a small segment of Johnson Controls' entire business operation. Johnson Controls employs 25,700 employees, only 425 of whom work in its battery division. *See* Dan's Marketing Services, Inc., 1 *America's Corporate Families* 694 (1988).

living areas of the women's prison in *Torres* or Johnson Control's battery manufacturing operation, where an employer adopts an employment policy designed to address a difficult societal problem, *Torres* requires that courts reviewing such a determination under Title VII give some deference to the employer's decisions. As we noted in *Torres:*

> We believe . . . that the defendants were required to meet an unrealistic, and therefore unfair burden when they were required to produce "objective evidence, either from empirical studies or otherwise, displaying the validity of their theory." *Torres* [*v. Wisconsin Dept. of Health and Social Services,* 639 F. Supp. 271, 280 (E.D. Wis. 1986)]. Given the nature of their "business"—administering a prison for female felons—the defendants, of necessity, had to innovate. Therefore, their efforts ought to be evaluated on the basis of the totality of the circumstances as contained in the entire record. In the Title VII context, the decision of penal administrators need not be given as much deference as accorded their decisions in constitutional cases. However, their judgments still are entitled to substantial weight when they are the product of a reasoned decision-making process, based on available information and experience.

859 F.2d at 1532 (citations omitted). *Cf. Wards Cove,* 109 S. Ct. at 2127 (" 'Courts are generally less competent than employers to restructure business practices,' *Furnco Construction Corp. v. Waters,* 438 U.S. 567, 578 (1978); consequently, the judiciary should proceed with care before mandating that an employer must adopt a plaintiff's alternate [employment practice] in response to a Title VII suit.").

In resolving the question of whether Johnson Controls' BFOQ is reasonably necessary to industrial safety, we recognize that Title VII establishes the general propositions that determination of whether a proposed BFOQ is "reasonably necessary" to furthering the objective of industrial safety requires that Johnson Controls " 'had reasonable cause to believe, that is, a factual basis for believing that all or believing that all or substantially all [women capable of pregnancy] would be unable to perform safely and efficiently the duties of the job involved,' " *Dothard v. Rawlinson,* 433 U.S. 321, 333 (1977) (quoting *Weeks v. Southern Bell Telephone & Telegraph Co.,* 408 F.2d 228, 235 (5th Cir. 1969)), and that "[i]n the *usual case,* the argument that a particular job is too dangerous for *women* may appropriately be met by the rejoinder

that it is the purpose of Title VII to allow the *individual woman* to make that choice for herself." *Dothard,* 433 U.S. at 335 (footnote omitted, emphasis added). *It is important to remember, however, that while Dothard established these propositions as general rules, the Supreme Court determined that Dothard was an unusual case justifying a departure from this general maxim.* The Court stated: "More is at stake in this case, however, than an individual woman's decision to weigh and accept the risks of employment in a 'contact' position in a maximum security male prison." *Id.* The Court concluded that a bona fide occupational qualification excluding women from such positions was justified because a woman's sex could create a risk of sexual assaults which would undermine prison security. *See Id.* at 335-37.

Similarly, "[m]ore is at stake in this case . . . than an individual woman's decision to weigh and accept the risks of employment." *Id.* at 335. A female's decision to work in a high lead exposure job risks the intellectual and physical development of the baby she may carry. The status of women in America has changed both in the family and in the economic system. Since they have become a force in the workplace as well as in the home because of their desire to better the family's station in life, it would not be improbable that a female employee might somehow rationally discount this clear risk in her hope and belief that her infant would not be adversely affected from lead exposure. The unborn child has no opportunity to avoid this grave danger, but bears the definite risk of suffering permanent consequences. This situation is much like that involved in blood transfusion cases. There courts have held that individuals may choose for themselves whether to refuse to personally acquiesce in a blood transfusion that had been established as medically necessary, but that parents may not always rely upon parental rights or religious liberty rights to similarly refuse to consent to such a medically necessary transfusion for their minor children.[38] The risks to the unborn child from lead are also shared by society in the form of government financed programs to train or maintain a handicapped child in non-institutional or institutional environments and to provide the

---

[38] In these cases a court will commonly appoint a guardian *ad litem* with the authority to consent for the child to the required transfusion and will hold a hearing to determine whether the child has been medically neglected as a result of the denial of the transfusion. *See generally In re E.G.,* 161 Ill. App. 3d 765, 515 N.E.2d 286, 287 (1987), *appeal allowed,* 520 N.E.2d 385 (1988).

child with the training necessary to overcome the mental and physical harm attributable to lead exposure. Thus, since "more is at stake" than the individual woman's decision to risk her own safety, *Dothard* supports, rather than bars, a conclusion that an employer's fetal protection policy constitutes a bona fide occupational qualification. In such circumstances, "given the reasonable objectives of the employer, the very womanhood . . . of the employee undermines . . . her capacity to perform a job satisfactorily." *Torres,* 859 F.2d at 1528 (citing *Dothard,* 433 U.S. at 336).

Against this substantive background, we hold that Johnson has carried its burden of demonstrating that its fetal protection plan is reasonably necessary to further industrial safety, a matter we have determined to be part of the essence of Johnson Controls' business. Initially, there can be no doubt that the exclusion of women who are actually pregnant from positions involving high levels of lead exposure sets forth a bona fide occupational qualification. As established in section II, *supra,* there is clear and unrefuted evidence in the record of a substantial and irreversible risk to the unborn child's mental development from lead exposure in the womb. This danger is "hardly a '[m]yth or purely habitually assumption.' " *Torres,* 859 F.2d at 1531 (quoting *Los Angeles Dept. of Water & Power v. Manhart,* 435 U.S. 702, 707 (1978)). The convincing scientific evidence of this risk and the very serious consequences of this danger combine to make this health risk quite different from the concerns in *Muller v. Oregon,* 208 U.S. 412, 421-22 (1908), which we would currently characterize as stereotypical rather than real. *Compare Dothard,* 433 U.S. at 333 (noting "that it is impermissible under Title VII to refuse to hire an individual woman or man on the basis of stereotyped characterizations of the sexes"); *Torres,* 859 F.2d at 1527-28 (distinguishing between "stereotyped characterizations of the sexes" and "real . . . differences between men and women").

We are also of the opinion that Johnson Controls' well reasoned and scientifically documented decision to apply this policy to all fertile women employed in high lead exposure positions constitutes a bona fide occupational qualification. The evidence presented concerning the lingering effects of lead in a woman's body, combined with the magnitude of medical difficulties in detecting and diagnosing early pregnancy, lead us to agree with Johnson Controls that there exists a reasonable basis in fact to conclude that an extension of this policy to all fertile women is proper and reasonably necessary to fur-

ther the industrial safety concern of preventing the unborn child's exposure to lead.

Based upon the current status of research into lead's hazardous effects, we also agree that Johnson Controls has demonstrated to our satisfaction that exclusion of fertile women from positions in any area of its battery plant in which an employee has reported a blood lead level in excess of 30 μg/dl or where an air lead measurement has been in excess of 30 is reasonably necessary to the industrial safety-based concern of protecting the unborn child from lead exposure. At the time Johnson Controls adopted its policy, the 30 μg/dl lead exposure level coincided with the Centers for Disease Control's determination of acceptable blood lead levels for children. However, it is becoming increasingly clear that the 30 μg/dl lead exposure level once believed to be safe for unborn children is no longer medically accepted as risk free. As mentioned previously, the Centers for Disease Control, in 1985, based upon "current knowledge concerning screening, diagnosis, treatment, follow-up, and environmental intervention for children with elevated blood levels," revised the level of elevated lead exposure from 30 to 25 μg/dl and suggested that an unborn child's blood lead level remain below 25 μg/dl. As also noted previously, recent lead studies suggest that harm may be present at levels even lower than those earlier believed to be safe. Thus, lead absorption levels such as those mandated by OSHA, which were thought to have been sufficiently protective of the unborn child when they were enacted over ten years ago, are now considered insufficient.

* * *

There is a reasonable basis in fact, grounded in medical and scientific research data, for concluding that Johnson Controls' has met its burden of establishing that the fetal protection policy is reasonably necessary to industrial safety.[43] Thus, the fetal pro-

---

[43]Judge Easterbrook suggests that "by one estimate 20 million industrial jobs could be closed to women," if "the majority is right," "for many substances in addition to lead pose fetal risks." Easterbrook Dissent at 96. This assertion is based upon the following language in Bureau of National Affairs' Special Report, *Pregnancy and Employment* p. 57 (1987): "One government source estimates that 15 million to 20 million jobs in the United States expose workers to chemicals that *may* cause reproductive injury." (Emphasis supplied). This speculative statement, taken at its face value, merely suggests a possibility of reproductive

tection policy should be recognized as establishing a bona fide occupational qualification protecting the policy against claims of sex discrimination.

* * *

CUDAHY, *Circuit Judge,* dissenting.

I respectfully dissent from the majority opinion. I would be pleased to join almost all of Judge Easterbrook's eloquent dissent except for its disposition of the case. Here I join Judge Posner's equally cogent statement, which adopts the BFOQ standard but advocates remand for a full trial on that basis. It may (and should) be difficult to establish a BFOQ here but I would afford the defendant an opportunity to try.

* * *

It is a matter of some interest that, of the twelve federal judges to have considered this case to date, none has been female. This may be quite significant because this case, like other controversies of great potential consequence, demands, in addition to command of the disembodied rules, some insight into social reality. What is the situation of the pregnant woman, unemployed or working for the minimum wage and unprotected by health insurance, in relation to her pregnant sister, exposed to an indeterminate lead risk but well-fed, housed and doctored? Whose fetus is at greater risk? Whose decision is this to make? We, who are unfortunately all male, must address these and other equally complex questions through the clumsy vehicle of litigation. At least let it be complete litigation focusing on the right standard.

POSNER, *Circuit Judge,* dissenting.

Johnson Controls refuses to employ any woman to make batteries unless she presents medical evidence of sterility. Today this court holds the refusal lawful under Title VII. A reader of the majority opinion

injury from unidentified and undefined toxic substances. Before our decision could be applied to any of these unidentified substances, obviously they would have to be subjected to the myriad tests and research that have conclusively established the grave risk from lead substances. Thus, an employer presenting a business necessity or bona fide occupational qualification defense would have to establish that the substance had undergone the same rigid testing and research. In addition, if ever a lead-free battery were developed, the problems in this case would fall by the wayside. We hope that this is achieved tomorrow.

might be excused for thinking that the case had been fully tried—and before this court—rather than decided by a district judge on a motion of summary judgment. I think it a mistake to suppose that we can decide this case once and for all on so meager a record. It is a mistake whether we affirm, on the ground that the evidence of danger to the fetus of a woman working in an environment dense with airborne lead, combined with evidence of the difficulty of reducing the amount of lead any further, conclusively establishes the lawfulness of Johnson Controls' policy; or reverse, with directions to enter judgment for the plaintiffs, on the ground that Title VII outlaws all fetal protection policies because all bear more heavily on female than on male workers.

Title VII forbids an employer deliberately to exclude a worker from a particular job because of the worker's sex unless sex is a "bona fide occupational qualification reasonably necessary to the normal operation of that particular business or enterprise." 42 U.S.C. §2000e-2(e)(1). This defense is central to the appeal and we should attend carefully to its scope and meaning. It is written narrowly and has been read narrowly. See, e.g. *Dothard v. Rawlinson,* 433 U.S. 321, 33237 (1977); *Torres v. Wisconsin Dept. of Health & Social Services,* 859 F.2d 1523, 1527-28 (7th Cir. 1988). There is no useful legislative history concerning the defense, and—no doubt because the prohibition of sex discrimination was added to Title VII at the last minute—no reference at all to the application of the defense to sex discrimination was added to Title VII at the last minute—no reference at all to the application of the defense to sex discrimination. A narrow reading is, nevertheless, inevitable. A broad reading would gut the statute. For it is unlikely that most employment discrimination in the private sector is irrational. Few private employers discriminate without having some reason for doing so; competition tends to drive from the market firms that behave irrationally. See Becker, The Economics of Discrimination (2d ed. 1971). If the defense of bona fide occupational qualification were broadly construed—for example, to excuse all sex discrimination that the employer could show was cost-justified—very little sex discrimination in employment, as well as very little employment discrimination based on religion or national origin (forms of discrimination that, like sex discrimination but unlike discrimination based on race or color, are also excused if a bona fide occupational qualification is established), would be forbidden. Title VII's reach would be shortened drastically.

Two courts of appeals faced with challenges under Title VII to fetal protection policies have concluded that such policies can never satisfy the stringent requirements of the occupational qualification defense. See *Wright v. Olin Corp.*, 697 F.2d 1172, 1185 (4th Cir. 1982); *Hayes v. Shelby Memorial Hospital*, 726 F.2d 1543, 1549 (11th Cir. 1984). But this conclusion, rather than resulting in instant victory for the plaintiffs, led those courts to stitch a new defense expressly for fetal protection cases. See 697 F.2d at 1183-92; 726 F.2d at 1548-54. I am not myself deeply shocked that courts sometimes rewrite statutes to address problems that the legislators did not foresee.

* * *

I do not think judges must or should ratify absurd results by sticking doggedly to the plain meaning of statutory language.

But we do not need to bite this bullet here, because the wording of the occupational qualification provision is not so cramped that it has to be stretched to bring (some) fetal protection policies within its scope. *Cf. Pime v. Loyola University*, 803 F.2d 351, 356-57 (7th Cir. 1986) (concurring opinion). Nor is a defensible way of stretching it to recast what is plainly a disparate treatment case—that is, a case of intentional discrimination against a protected group—as a disparate impact case, and then invoke the recent decision in which the Supreme Court expanded the "business necessity" defense. See *Wards Cove Packing Co. v. Atonio*, 109 S.Ct. 2115 (1989); also *Allen v. Seidman*, No. 88-1811 (7th Cir. July 27, 1989). This legerdemain is as unnecessary as it is questionable. "[R]easonably necessary," one of the key terms of the occupational qualification defense, means more than just reasonable but less than absolutely necessary. On the way to concluding that the defense is unavailable in fetal protection cases the court in *Wright* misquoted the provision by leaving out the word "reasonably," see 697 F.2d at 1185 n.21, and the misquotation is faithfully repeated in *Hayes*, see 726 F.2d at 1549. The other key words of the defense, "*normal* operation" (emphasis added), should dispel concern that consideration of all interests other than the employer's interest in selling a quality product at the lowest possible price is precluded. It is possible to make batteries without considering the possible consequences for people who might be injured in the manufacturing process, just as it would be possible to make batteries with slave laborers, but neither mode of operation would be normal. To confine the occupational qualification defense to concerns with price and product quality would deny a defense

to Johnson Controls even if the company excluded only pregnant women, as distinct from all women who might become pregnant, from making batteries. I do not understand the plaintiffs to be arguing that Title VII requires Johnson Controls to permit women known to be pregnant to continue working in an atmosphere dense with lead. If on the other hand a fetal protection policy that excludes women from a given job classification cannot be said to be reasonably necessary to the employer's normal operation, I do not see why we should *want* to save it from condemnation under Title VII.

* * *

The defense is applicable to this case and although it is of limited scope it is not the proverbial eye of a needle. In particular, the "normal operation" of a business encompasses ethical, legal, and business concerns about the effects of an employer's activities on third parties. An employer might be validly concerned on a variety of grounds both practical and ethical with the hazards of his workplace to the children of his employees. A pregnant employee exposed to heavy concentrations of lead in the air may absorb the lead into her bloodstream and from there transmit it to her fetus through the placenta, causing, years later, mental retardation or other injury to the child. The parties agree that there is a solid medical basis for concern with fetal injury from airborne lead in the concentration found in battery plants, and this concern could in turn cause the employer to worry about being sued by injured children of his employees. Such a suit would not be preempted by workers' compensation law, because the plaintiff would not be the worker. The employer would therefore be exposed to full common law damages, punitive as well as compensatory. The mother's own negligence—for if she had been clearly warned of the hazard, but voluntarily became pregnant anyway and continued to work making batteries, she would be acting negligently with regard to the fetus—would not be imputed to the child and therefore would not reduce the employer's liability.

* * *

We should not dismiss the concern over tort liability as a narrow, selfish "bottom line" concern irrelevant to the purposes of Title VII. The potential cost of tort liability to Johnson Controls is an approximation of the potential cost to the children who have suffered prenatal injury from the airborne lead absorbed into their mothers' bloodstreams. That is a social cost that Title VII does not require a company to ignore. At some point it may become large enough to affect the company's normal method of operation and supply the ground for a bona fide occupational qualification of infertility.

A related point is that an employer might have moral qualms about endangering children or might fear the effect on his public relations. The ethical concern cannot be wholly dismissed, as could an ethical conviction that a woman's place is in the home. We know from the controversy over abortion that many people are passionately protective of fetal welfare, and they cannot all be expected—perhaps they cannot be required—to park their passions at the company gate. That "strong [state] interest in protecting the potential life of the fetus" of which the Supreme Court spoke in *Maher v. Roe*, 432 U.S. 464, 478 (1977), and other cases is not a judicial invention; it is the product of a groundswell of powerful emotion by a significant part of the community, and is only indirectly, although possibly substantially, in conflict with women's workplace aspirations.

*   *   *

I conclude that Title VII even as amended by the Pregnancy Discrimination Act does not outlaw all fetal protection policies. Whether a particular policy is lawful is a question of fact, and since the burden of proof is on the defendant it will be the rare case where the lawfulness of such a policy can be decided on the defendant's motion for summary judgment. This is not that rare case. Even if we accept that the amount of airborne lead in Johnson Controls' battery-making operation is dangerous to the fetuses of female employees and that the company cannot reduce the danger further without shutting down the operation, a host of unanswered questions remains. The first concerns the feasibility of warnings as a substitute for a blanket exclusion of all fertile women.

*   *   *

We do not know what other manufacturers of batteries do about the hazards of airborne lead to the fetus—whether they are content to rely on warnings, for example, and if so of what kind and with what effect.

*   *   *

We also do not know how profitable the business of manufacturing batteries is, and therefore how vulnerable it is to fears, as yet speculative, of litiga-

tion arising from fetal damage. (The case at this stage is a tissue of speculation.)

*   *   *

Even on the limited record before us, however, it is clear that the defendant's fetal protection policy is excessively cautious in two regards; first in presuming that any women under the age of 70 is fertile, and second in excluding a presumptively fertile woman from any job from which she might ultimately be promoted into battery making, even if her present job does not expose her to lead. Since these aspects of the policy are severable from the rest of it I do not think their deficiencies need condemn the entire policy, especially since the first is harmless because a woman too old to bear children has only to submit a letter to that effect from her doctor to be permitted to work in the battery plant.

*   *   *

But although the defendant did not present enough evidence to warrant the grant of summary judgment in its favor, there is no ground for barring it from presenting additional evidence at trial. Therefore it would be equally precipitate for us to direct the entry of judgment in the plaintiffs' favor.

*   *   *

We should be as hesitant to endanger the health of children by condemning all fetal protection policies as we should be hesitant to endanger the jobs of women by placing our imprimatur on such policies. We should vacate the district court's judgment and remand for further proceedings to enable the compilation of an adequate evidentiary record.

EASTERBROOK, Circuit Judge, with whom FLAUM, Circuit Judge, joins, dissenting.

Whether employers should restrain adults from engaging in acts hazardous to their children is an ethical, medical, economic, and political problem of great complexity. But this is a statutory case, and we must implement the law rather than give our own answer. Johnson's policy is sex discrimination, forbidden unless sex is a "bona fide occupational qualification"—which it is not.

*   *   *

General Electric Co. v. Gilbert, 429 U.S. 125 (1976), held that a rule distinguishing on account of pregnancy is not sex discrimination, because women are in both the "pregnant" and "non-pregnant" groups. See also *Geduldig v. Aiello*, 417 U.S. 484 (1974). The Court saw the line as one between pregnant employees and all others, a line based on something other than sex (or at least something in addition to sex). Johnson's line based on *ability* to become pregnant, however, is assuredly based on sex. That would be ground for distinguishing *Gilbert*, but Congress interred *Gilbert* in 1978 by enacting the Pregnancy Discrimination Act, 42 U.S.C. §2000e(k) (the PDA), which provides:

> The terms "because of sex" or "on the basis of sex" [in Title VII] include, but are not limited to, because of or on the basis of pregnancy, childbirth, or related medical conditions; and women affected by pregnancy, childbirth, or related medical conditions shall be treated the same for all employment-related purposes . . . as other persons not so affected but similar in their ability or inability to work.

This amendment to Title VII makes distinctions based on women's ability to bear children sex discrimination. It also has a built-in BFOQ standard: unless pregnant employees differ from others "in their ability or inability to work," they must be treated "the same" as other employees "for all employment-related purposes." Although located in a definitional provision, the language after the semicolon is substantive and governs Johnson's plan.

Wright v. Olin Corp., 697 F.2d 1172 (4th Cir. 1982), the only other appellate decision that has dealt with a fetal protection policy similar to Johnson's, took a different view. *Wright* observed that a policy using sex as a ground of decision may cause women no more injury than a policy neutral with regard to sex, yet having a disparate impact. A policy designed to promote the health of offspring of both sexes is neutral in objective. A sex-neutral policy is judged under an approach more lenient than the BFOQ standard. Believing that a fetal protection policy rests on strong justifications, *Wright* treated the policy as sex-neutral so that it could sustain a rule functionally identical to Johnson's. 697 F.2d at 1184-92.

This makes things turn not on whether the employer uses sex as a ground of decision but on whether the employer uses sex to serve a "good" policy. If the policy is beneficent and the injury to women "tolerable" in light of the interests served, the court changes the standard of inquiry. Yet whether a policy is "good" is a statutory question, governed by the BFOQ test stated in §2000e-2(e)(1) and the supplemental rule of the PDA that women and men who are "similar in their ability or inability to work" must be treated the same. A court's belief that a good end is in view does not justify departure from the statutory framework; it is an occasion for applying the statutory framework. *Wright* ignored the PDA and inverted ordinary rules of statutory interpretation when stating (with echoes in the majority's opinion today): "The inappropriateness of applying the overt discrimination/b.f.o.q. theory of claim and defense . . . is that, properly applied, it would prevent the employer from asserting a justification defense which under developed Title VII [disparate impact] doctrine it is entitled to present." 697 F.2d at 1182 n.21. In other words, this *must* be a disparate impact case because an employer couldn't win it as a disparate treatment case. If the rigors of the BFOQ suggest the need for a fresh approach, that is a job for another branch.

* * *

Johnson defends its fetal policy on the basis of concern for the welfare of the next generation, an objective unrelated to its ability to make batteries (§2000e-2(e)(1) speaks of the "operation of the business") or to any woman's "ability or inability to work" (the standard of the PDA). Johnson allowed women to work until 1982, without ill effects on its business; for all we know (the record is silent), other firms in the same business employ women in the kinds of jobs from which Johnson excludes them. The majority does not mention the PDA, which, added to the BFOQ rules, puts out of bounds the justification Johnson offers.

At oral argument before the panel counsel offered a new defense of Johnson's policy: that it is morally required to protect children from their parents' mistakes. This justification is redolent of *Muller v. Oregon*, 208 U.S. 412 (1908), which sustained a statute curtailing women's hours of work on the ground that maternal functions unsuited women for long hours. The Court wrote:

> [B]y abundant testimony of the medical fraternity continuance for a long time on her feet at work, repeating this from day to day, tends to injurious effects upon the body, and as healthy mothers are essential to vigorous offspring, the physical well-being of woman becomes an object

of public interest and care in order to preserve the strength and vigor of the race. . . . [H]er physical structure and a proper discharge of her maternal functions—having in view not merely her own health, but the well-being of the race—justify legislation to protect her from the greed as well as the passion of man. The limitations which this statute places upon her contractual powers . . . are not imposed solely for her benefit, but also largely for the benefit of all.

208 U.S. at 421-22. The "abundant testimony of the medical fraternity" turned out to the triumph of imagination over data. Dangers decried in *Muller* are today perceived as chimerical, excuses for blockading women as effective competitors of men in the labor force. Legislation of the sort allowed by *Muller* "protected" women out of their jobs by making women less attractive as employees. An employer that needed flexibility in assigning hours of work had to hire men; women were consigned to jobs with regular hours but lower wages. See Elisabeth M. Landes, *The Effect of State Maximum Hours Laws on the Employment of Women in 1920*, 88 J. Pol. Econ. 476 (1980) (finding that "protective" legislation reduced women's hours, hourly wages, and annual income). Such laws also treat women in a stereotypical way. State laws requiring or allowing employers to treat women differently, on the assumption that women are less able than men to take the precautions essential for healthy children, are preempted by Title VII—not because of an express preemption clause, but because no state law may require or excuse a violation of federal law. *Rosenfeld v. Southern Pacific Co.*, 444 F.2d 1219, 1225-27 (9th Cir. 1971). Statutes of the sort sustained in *Muller*, are museum pieces, reminders of wrong turns in the law. It is not enough to say that Johnson is a private employer while *Muller* dealt with state laws. Title VII is addressed to private employers. The question is whether a justification of a particular kind is an acceptable defense of sex discrimination. This justification is not. No legal or ethical principle compels or allows Johnson to assume that women are less able than men to make intelligent decisions about the welfare of the next generation, that the interests of the next generation always trump the interests of living woman, and that the only acceptable level of risk is zero. "[T]he purpose of Title VII is to allow the individual woman to make that choice for herself." *Dothard*, 433 U.S. at 355.

Although some women may become pregnant, and a subset of their children might suffer, Johnson

cannot exclude all fertile women from its labor force on their account. Most women in an industrial labor force do not become pregnant;[3] most of these will have blood lead levels under 30 μg/dl (only about ⅓ of the employees exposed to lead at Johnson's plants have higher levels); most of those who become pregnant with levels exceeding 30 μg/dl will bear normal children (Johnson reports no birth defects or other abnormalities in the eight pregnancies among its employees).[4] Concerns about a tiny minority of women cannot set the standard by which all are judged. An employer establishes a BFOQ only if there is "a factual basis for believing that all or substantially all women would be unable to perform safely and efficiently the duties of the job involved." *Weeks v. Southern Bell Telephone & Telegraph Co.*, 408 F.2d 228, 235 (5th Cir. 1969), quoted with approval in *Dothard*, 433 U.S. at 333, and adopted as the law of this circuit in *Torres*, 859 F.2d at 1527, 1530. Fear of prenatal injury (which has not happened in the history of the employer) is a far cry from something that prevents "all or substantially all" women from doing their jobs.

* * *

II

Having adopted the *Wright-Hayes* approach, we still should not affirm the district court's judgment. *Hayes* opined that a fetal protection policy applicable only to women violates Title VII

unless the employer shows (1) that a substantial risk of harm exists and (2) that the risk is borne only by members of one sex; and (3) the employee fails to show that there are acceptable

---

[3]Although some 9% of all fertile women become pregnant each year, the birth rate for blue collar women over 30 is about 2%, and of working women 45-49 only 1 in 5,000 becomes pregnant in a given year. The data are collected in Mary E. Becker, *From* Muller v. Oregon *to Fetal Vulnerability Policies*, 53 U. Chi. L. Rev. 1219, 1233 (1986). The record does not reveal the birth rate for Johnson's female laborers, but it must be lower given Johnson's strenuous efforts to discourage pregnancy among those exposed to lead.

[4]One of the children has an elevated level of lead in the blood, but this has not produced an identifiable problem. It is hard, however, to link outcomes such as learning disabilities to lead, since learning disabilities could have many other causes, and it is therefore hard to show in an individual case (as opposed to statistically) that lead injured the child.

alternative policies that would have a lesser impact on the affected sex.

726 F.2d at 1554, at the time *Wright* and *Hayes* were decided, and when the EEOC issued its policy statement, courts believed that "business necessity" in a disparate impact case is a *defense*. "Business necessity" and "BFOQ" were not so distinct. We know from *Wards Cove Packing Co. v. Atonio* 109 S. Ct. 2115 (1989), however, that the plaintiff bears the burden of persuasion on all questions in every disparate impact case, as the majority today emphasizes. So the *Wright-Hayes* standard has been watered down. The court's "adoption" of *Wright, Hayes,* and the EEOC's policy statement is thus in practice more favorable to employers than the Fourth and Eleventh Circuits (and the EEOC) anticipated their approach would be. The plaintiff won in *Hayes* she would lose under the majority's approach. [sic]

Even on the majority's un-demanding standard, however, there are material disputes. The EEOC's policy statement properly criticized the district court for granting summary judgment despite evidentiary disputes material to the application of the *Wright-Hayes* standard.

Substantial Risk. Is there a substantial risk of harm to the offspring of female employees? That lead in the blood is dangerous no one doubts, although experts dispute how much is too much and whether lead is more risky to the fetus and developing infants than to adults. The extent to which lead in a mother's blood, *at levels to which Johnson expose* [sic] *its employees,* endangers the fetus is a subject on which there is additional dispute. Showing great injury at 100 μg/dl is one thing; showing risks when no one has levels over 50 μg/dl (and, per OSHA's rules, anyone wanting to get below 30 μg/dl is entitled to a respirator) is another. The record does not quantify the risks at the levels OSHA permits. It also does not reveal the extent to which lead crosses the placenta. Johnson's chief medical consultant, Dr. Fishburn, believes that risk is greatest in the first weeks of pregnancy, before women can withdraw to lead-free environments. Experts take the contrary view that lead does not cross the placenta until late in pregnancy, and that because lead levels in the blood fall substantially within 12 weeks of the last exposure, the danger is reduced if women are removed from contact with lead promptly on becoming pregnant.

It is painful to see conflicts of this kind settled by litigation; judges cannot unravel medical mysteries by observing scientists' demeanor on the stand. This is not how scientists resolve disputes among themselves. Demeanor tells the judge only whether the scientist *believes* what he says, something almost irrelevant to a scientific dispute. Scientists formulate hypotheses, collect data, and apply statistical methods to assess them; judges and jurors find this process alien. Yet so long as the substantive rule of law requires a court to resolve scientific disagreements—which the *Wright-Hayes* standard does, though the BFOQ standard avoids the problem—the judge must follow the rules, which means that material disputes must be resolved at trial.

A small risk, even if compellingly documented, is not enough to exclude women from employment. How great is too great? Most women do not become pregnant in any given year, and most female employees do not have blood lead levels exceeding 30 μg/dl, so average risk to the employee population may be small. If a woman becomes pregnant with a blood lead level of 40 μg/dl, is the risk one learning disability in two pregnancies? One in two thousand? One in two million? These figures imply different policies, yet we do not know which is correct. How risky is a blood lead level exceeding 30 μg/dl compared with other hazards? Most comparisons show that smoking and drinking are quite dangerous to fetuses, more so than many contaminants found in the workplace. Some 2,500 children under the age of one die every year because of their parents' smoking; others never make it to birth. The hazards created by occupational chemicals span many orders of magnitude: some are safer than the sweeteners we wolf down, some are dangerous indeed. Where does lead fit on that spectrum? I cannot believe that Johnson would be entitled to fire female employees who smoke or drink during pregnancy—let alone to fire all female employees because some might smoke or drink—which makes it hard to exclude women to curtail risk from other substances.

How does the risk attributable to lead compare, say, to the risk to the next generation created by driving a taxi? A female bus or taxi driver is exposed to noxious fumes and the risk of accidents, all hazardous to a child she carries. Would it follow that taxi and bus companies can decline to hire women? That an employer could forbid pregnant employees to drive cars, because of the risk accidents pose to fetuses? For all we can tell, accepting Johnson's argument compels us to answer "yes" to these questions, which simply points up the need to quantify the hazard. It also points up the political nature of the dispute. How much risk is too much is a moral or economic or political question, one ill suited to the processes of litigation and not the sort of question Title VII puts to a judge.[15]

The most concerted effort to estimate the risks lead poses to offspring was conducted by OSHA in the course of promulgating its lead rules. OSHA considered and rejected a proposal to exclude women capable of bearing children from jobs in which blood lead levels may exceed 30 μg/100 g. It stated:

> Industry testimony further suggests that women of childbearing potential could be "protected" by excluding them from employment in many parts of the lead industry. . . . The record in this rulemaking is clear that male workers may be adversely effected [sic] by lead as well as women. Male workers may be rendered infertile or impotent, and both men and women are subject to genetic damage which may affect both the course and outcome of pregnancy. Given the data in this record, OSHA believes there is no basis whatsoever for the claim that women of child-bearing age should be excluded from the workplace in order to protect the fetus or the course of pregnancy. Effective compliance with all aspects of these standards will minimize risk to all persons and should therefore insure equal employment for both men and women.

43 Fed. Reg. 52953, 52966 (1978). OSHA's regulations require employers to make respirators available, so that both men and women desiring children can reduce their blood lead levels. 29 C.F.R. §1910.1025 (f)(1)(iii). They also require employers to conduct annual educational programs "with particular attention to the adverse reproductive effects on both males and females," §1910.1025(l)(1)(v)(D). In order to say that the risk to offspring at firms complying with OSHA's rules (as Johnson says it does) is so "substantial" that a woman should be allowed to work at any job where there is the slightest chance of a blood lead level exceeding 30 μg/dl, a court must disagree with the judgment of OSHA that the 50 μg/100 g limit, plus the availability of respirators to employees seeking to attain a level of 30 μg/100 g, is enough. My colleagues essentially take judicial notice that OSHA is wrong, slip op. 52-53, an extraordinary step. The record does not contain evidence sufficient to contradict OSHA's conclusion, let alone evidence so lopsided that summary judgment is appropriate.

One more observation. "Substantial risk" must mean substantial *net* risk. Excluding women from industrial jobs at Johnson may reduce risk attributable to lead at the cost of increasing other hazards. There is a strong correlation between the health of the infant and prenatal medical care; there is also a powerful link between the parents' income and infants' health, for higher income means better nutrition, among other things. See Aaron Wildavsky, *Searching for Safety* 59-72 (1988); Victor R. Fuchs, *How We Live* 31-40 (1983). Removing women from well-paying jobs (and the attendant health insurance), or denying women access to these jobs, may reduce the risk from lead while also reducing levels of medical care and the quality of nutrition. The net effect of lower income and less medical care could be a reduction in infants' prospects.[16] Mary E. Becker, *From* Muller v. Oregon *to Fetal Vulnerability Policies*, 53 U. Chi. L. Rev. 1219, 1229-31 (1986).

Nothing in the record shows the net risks. When asked at oral argument whether Johnson attempted to determine net effects before adopting its policy, counsel said no. He continued, in effect: "If there is a greater risk, that's not our concern." A "not on *our* watch" position is the bureaucrat's shelter and not becoming if, as Johnson so earnestly maintains, it is concerned about the welfare of unborn children. These helpless beings do not know or care about the

---

[15]There is a tradition in public health of resolving doubts by assuming that risks exist until they can be disproved. As the uncertainties are substantial, this process often produces measures of risk that appear to be substantial. Yet whether to assume that the maximum likely hazard will come to pass—a process known as "conservative risk assessment"—is itself a political question. A court would be obliged to try to produce the most accurate, rather than the most conservative, assessment, for resolving doubts in one direction only produces inaccurate comparisons of poorly-understood risks (which will be overstated) against well-understood risks (which would be accurately stated). See Albert L. Nichols & Richard J. Zeckhauser, *The Perils of Prudence: How Conservative Risk Assessments Distort Regulation*, Regulation 13 (Nov/Dec 1986). Yet in an effort to measure risk accurately a court would get little aid from existing studies, often tailored to fulfilling regulatory demands for a "conservative" bias.

[16]The majority says that "[t]his issue bears no relevance to Johnson Controls' employment practices", slip op. 32 n.28, because Johnson protected the salary and benefits of women transferred out of jobs in 1982. But Johnson does not offer women *excluded* from these jobs in 1983 or later the salary and benefits they could have earned in them, and it also does not protect the income and benefits of employees who because of the policy cannot exercise their seniority (bumping) rights to avoid layoffs.

source of risk; they would care only about its aggregate level. Surely Title VII does not allow an employer to adopt a policy that simultaneously makes both women and their children worse off.

Mediation through a Single Sex. Is the risk to the child transmitted by one sex only? If "the risk" is defined as risk caused by lead entering the fetus's blood via the placenta, it is by definition confined to one sex. But if we ask instead "Does the presence of lead in a parent's blood pose a risk to the fetus?," then the evidence conflicts. The broader perspective is the correct one when aggregate levels of risk are the proper concern.

Three affidavits in the record, and papers in medical journals, maintain that lead in the blood creates risks for offspring of both male and female employees. The American Public Health Association and other medical groups have filed a brief as *amici curiae* marshaling an impressive array of studies linking lead with injury to the male reproductive system, and thence to offspring. E.g., Christopher Winder, Reproductive Effects of Occupational Exposures to Lead: Policy Considerations, 8 NeuroToxology 411 (1987); Herbert L. Needleman & David Bellinger, Commentary, Environmental Research (1988). Most of the data come from animal studies, but some human studies suggest that the effects occur in our species too. OSHA concluded that lead in men as well as women is hazardous to the unborn. The medical surveillance guidelines published as an appendix to the agency's lead control regulations states:

> Exposure to lead can have serious effects on reproductive function in both males and females. In male workers exposed to lead there can be a decrease in sexual drive, impotence, decreased ability to produce healthy sperm, and sterility. Malformed sperm (teratospermia), decreased number of sperm (hypospermia), and sperm with decreased motility (asthenospermia) can all occur. Teratospermia has been noted at mean blood levels of 53 μg/100 g and hypospermia and asthenospermia at 41 μg/100 g. Furthermore, there appears to be a dose-response relationship for teratospermia in lead exposed workers. . . . [B]ecause of the demonstrated adverse effects of lead on reproductive function in both the male and female as well as the risk of genetic damage of lead on both the ovum and sperm, OSHA recommends a 30 μg/100 g maximum permissible blood lead level in both males and females who wish to bear children.

29 C.F.R. Part 1910, pp. 833-34 (1987). Perhaps OSHA is wrong; its findings do not bind Johnson in this litigation. But its view has been sustained once after rigorous attention. *United Steelworkers of America v. Marshall*, 647 F.2d 1189, 1256-58 [8 OSHC 1810, 1857-58] (D.C. Cir. 1980) (describing as "abundant" the evidence supporting the belief that lead injures the reproductive systems of both sexes and sustaining as rational OSHA's decision that the evidence does not justify excluding women from the workplace). Affidavits and professional articles describing the latest evidence also support OSHA's judgment. The district court could not properly reject it without holding a trial.

Least of all could a court reject it, as my colleagues do, on the ground that as a matter of law animal studies are not "solid scientific data," slip op. 33. The District of Columbia Circuit concluded that "virtually undisputed evidence [shows] that at 30 μg/100 g and above, men suffer . . . interference with their ability to produce normal sperm", 647 F.2d at 1249 [8 OSHC at 1851]. The medical profession, like the Food and Drug Administration, will be stunned to discover that animal studies are too "speculative," slip op. 33, to be the basis of conclusions about risks. Often animal studies are the best foundation for decision. The Supreme Court has concluded that they may be used, e.g., *Industrial Union Department v. American Petroleum Institute*, 448 U.S. 607, 657 n.64 [8 OSHC 1586, 1604] (1980) (plurality opinion). OSHA relied on these very animal studies when establishing its lead rules, as the D.C. Circuit held it may. 647 F.2d at 1257 & n.97 [8 OSHC at 1857]. See also, e.g., *Simpson v. Young*, 854 F.2d 1429 (D.C. Cir. 1988); *Public Citizen Health Research Group v. Tyson*, 796 F.2d 1479, 1489-90, 1492 [12 OSHC 1905, 1922] (D.C. Cir. 1986), sustaining risk assessments based on animal studies. Indeed, on occasion animal studies *compel* responsible agencies to act. *Public Citizen v. Young*, 831 F.2d 1108 (D.C. Cir. 1987).

Less Restrictive Alternatives. An employer cannot close employment opportunities to women in order to protect the next generation if some more modest alternative would do (nearly) as well at protecting the unborn. Johnson's policy has a striking sweep: *no* fertile woman can be hired for a job in which *any* employee has had a blood lead level exceeding 30 μg/dl *anytime* during the last year, or in *any* job that might lead to a promotion to such a job. As a practical matter, this means every industrial job at Johnson's battery plants. The firm advised its hiring offices to tell women that "we have no openings for women capable of bearing children". To state the policy is to reveal many less stringent options that might be almost as good at protecting the interests of children.

Women over 40 rarely have children. Why are they forbidden to work? (One of the plaintiffs, 50 years old and divorced when the suit was filed, had nonetheless been excluded from jobs covered by the policy.) Covering them reduces the lead risk to zero, but "zero" is not the only acceptable level.

Many workers in jobs in which some employee has a level exceeding 30 μg/dl have levels less than that. Only ⅓ of Johnson's industrial employees exceed the 30 μg/dl figure. The levels of lead in the blood depend not only on lead in the air but also on personal hygiene. Some workers allow particles to remain on their clothes or person, staging points from which they can be swallowed. Why not allow a woman to enter jobs in which significant fractions of workers sustain levels less than 30 μg/dl, to see whether she can too? Why treat all women as unable to follow good industrial hygiene, or as unwilling to use a respirator, just because ⅓ of men have elevated levels of lead in the blood? Johnson replies that lead levels usually are greatest shortly after entering a new job, before the employee learns how to reduce the level; the women might become pregnant during these initial weeks or months. True enough, but again this answer assumes that the only acceptable level of risk is zero.

Some workers who start at entry-level jobs with low exposure to lead will be promoted to higher-lead jobs; others will not (or will leave before then). Johnson excludes women even from these safer jobs, although they pose no appreciable risk to off-spring. Doubtless Johnson believes that unimpeded lines of progression make its operation more efficient, since it can invest more in training women in skills that are transferrable to new jobs within the firm. Yet this form of savings does not count under Title VII. Let us suppose that Johnson had light-lifting and heavy-lifting jobs, joined in a line of progression. Would the inability of many or even all women to perform the heavy-lifting job permit Johnson to exclude all women from the light-lifting job? The exclusion might be desirable from the perspective of business efficiency, but Title VII would not allow it. So, too, Title VII forbids a firm to exclude all women from a line of progression on the ground that some will drop out to have children, rendering their training worthless to the firm. Women have less attachment to the work force than men, making it privately rational for firms to invest less in training them for high-skill jobs; Title VII just as clearly

forbids firms to do this. What Johnson has done is little different, and at some cost to itself it can change the fetal protection plan to enlarge women's opportunities at no cost whatever in risk to the unborn.

III

The *Wright-Hayes* standard is the wrong one. Johnson needed to, and did not, establish that sex is a BFOQ for employment at its battery plants. Yet even given the majority's decision to adopt the *Wright-Hayes* standard, the plaintiffs are entitled to a trial. Seven judges of this court have analyzed the conflicting medical evidence and reached their own conclusions about its significance, conclusions at variance with those drawn by the American Public Health Association and the Occupational Safety and Health Administration from the same kind of evidence.[17] Judges may be astute students of medical findings, but the presence of thoughtful persons on the other side suggests caution—and at all events appellate judges should not be resolving scientific disputes.

This is the most important sex-discrimination case this circuit has ever decided. It is likely the most important sex-discrimination case in any court since 1964, when Congress enacted Title VII. If the majority is right, then by one estimate 20 million industrial jobs could be closed to women, for many substances in addition to lead pose fetal risks. See note 7 above. Whether that would happen is of course a separate question; legal entitlements need not translate to action. But the law would allow employers to consign more women to "women's work" while reserving better-paying but more hazardous jobs for men. Title VII was designed to eliminate rather than perpetuate such matching of sexes to jobs.

Title VII requires employers to evaluate applicants and employees as individuals rather than as

---

[17]The *amici curiae* whose view of the evidence differs from that of the majority include the American Public Health Association, the American Society of Law and Medicine, the Planned Parenthood Federation of America, several toxicologists and physicians who have published scholarly papers on the subject (some of which Johnson relied on, to the consternation of the authors who appear before us to denounce its interpretation of their research), and Eula Bingham, a professor of environmental health who was also the Assistant Secretary of Labor for OSHA at the time that agency published its lead regulations. Whether we think these persons and groups right or wrong, we should not suspect them of insufficient devotion to the health and welfare of the next generation.

members of a group defined by sex. The statute has its cost; prenatal injuries are among these. Appeals to the "flexibility" with which the Supreme Court has allocated burdens of proof and persuasion get us nowhere. No amount of "flexibility" justifies sex discrimination without a BFOQ, unless by "flexibility" we mean a prerogative to disregard the statute when it requires decisions antithetical to our beliefs. Although my colleagues refer to many constitutional cases, such as *Rostker v. Goldberg,* 453 U.S. 57 (1981), for the proposition that sex discrimination sometimes is permissible, cases showing that Congress *may* authorize sex-based decisions hardly shows that in this instance it *did.* Title VII forbids rather than requires resort to sex as a basis of decision.

Risk to the next generation is incident to all activity, starting with getting out of bed. (Staying in bed all day has its own hazards.) To insist on *zero* risk, which the court says Johnson may do, is to exclude women from the industrial jobs that have been a male preserve. By all means let society bend its energies to improving the prospects of those who come after us. Demanding zero risk produces not progress but paralysis. Defining tolerable risk, and seeking to reduce that limit, is more useful—but it is a job for Congress or OSHA in conjunction with medical and other sciences. Laudable though its objective be, Johnson may not reach its goal at the expense of women.

◇ **NOTES**

1. In addition to the *Wright* and *Hayes* cases discussed in the above opinion, the other circuit court case to address the issue is *Zuniga v. Kleberg County Hospital,* 692 F2 and 986 (5th Cir 1982). Like *Hayes, Zuniga* involved a woman who was fired from her job as an x-ray technician when it was learned she was pregnant.

2. Among the commentators who predicted that the courts would find women-only fetal protection policies discriminatory are Ashford and Caldart, The Control of Reproductive Hazards in the Workplace: A Prescription for Prevention, 5 Indus. Rel. L. J. 523, 543-547 (1983); Mattson, The Pregnancy Amendment: .Fetal rights and the Workplace, 86 Case & Cont., No. 6, at 33 (1981); Nothstein and Ayres, Sex-Based Considerations of Differentiation in the Workplace: Exploring the Biomedical Interface Between OSHA and Title VII, 26 Vill. L. Rev. 239 (1981); Williams, "Firing the Woman to Protect the Fetus: The Reconciliation of Fetal Protection with Employment Opportunity Goals Under Title VII," 69 Geo. L.J. 641 (1981); J. Bertin, Discrimination Against Women of Childbearing Capacity, presented at the Hastings Center (January 8, 1982): Stillman, The Law in Conflict: Accommodating Equal Employment and Occupational Health Obligations, presented at the American Occupational Health Conference, Anaheim, California (May 2, 1979).

3. Ronald Bayer of the Hastings Center has argued that underlying the Title VII furor over female exclusionary policies is "a recognition that the American economy so limits the possibilities of its women workers that they would demand, as a sign of liberation, the right to share with men access to reproductive risks." Bayer, Women, Work and Reproductive Hazards, Hastings Center Report 14 (1982).

4. In implementing its fetal protection policy, Johnson Controls did not fire any workers. Rather, it moved all of its affected female workers to other jobs and protected their pay and seniority status. As Judge Easterbrook notes in his dissent, however, the main long-term impact of the policy is to exclude *prospective* workers from these workplaces. In the *Hayes* case, the Eleventh Circuit stated that if the hospital had proven that Ms. Hayes's duties posed an unreasonable risk to her fetus, it would still "be under a heavy burden to examine all alternative possibilities for keeping Hayes employed in some capacity."

2.   Exclusion of Employees on the Basis of Human Monitoring Data

Many of these same legal principles are relevant to a consideration of the extent to which employees may be excluded because of information suggesting that chemicals present in the workplace pose a particular danger to them.

Monitoring the Worker for Exposure and Disease: Scientific, Legal and Ethical Considerations in the Use of Biomarkers

N. A. Ashford, C. Spadafor, D. Hattis, and C. Caldart

To what extent may employers use health and expo-sure information to limit or terminate the employ-ment status of individual employees or to deny employment to a prospective employee?

UNDER COMMON LAW

At early common law, an employer had the right to take an employee's health into account in determin-ing whether to continue to employ that person. If the employment contract was "open," with no defi-nite term, the employee could be discharged for any reason, including health status, at the will of the employer. If the contract of employment was for a definite term, the employee could be discharged for "just cause." Typically, significant illness or disability constituted "just cause." Although federal labor law, workers' compensation and recent common law lim-itations on the doctrine of "employment at will" have profoundly affected the nature of employee-employer relations in this century, courts continue to recog-nize an employer's interest in discharging employ-ees who cannot perform their work safely. Thus, if the worker has no statutory or contractual protection, an employer likely retains a presumptive common law right to discharge the worker whose health status makes continued employment dangerous or whose health status prevents him or her from performing his or her job.

Source: Excerpts from *Monitoring the Worker for Exposure and Disease: Scientific, Legal and Ethical Considerations in the Use of Biomarkers.* Johns Hopkins University Press, Baltimore, MD., 1990 pp. 145–154.

Medical and exposure monitoring, however, place the issue in a somewhat different light. In a broad sense, monitoring is designed to reveal whether an employee has been, or in the future may be, harmed by the workplace itself. When the employer seeks to discharge an employee on the basis of such data, it will be because the employee was, or may be, harmed by a situation *created by the employer.* The right of the employer to discharge the employee is not as clear here as in the general case.

Suppose that an employer is complying with an existing OSHA standard for a particular toxic expo-sure and monitoring reveals that one of the firm's employees is likely to suffer serious and irreparable health damage unless he or she is removed from the workplace. In this situation, the employer is comply-ing with public policy as enunciated by OSHA and, absent a mandatory MRP provision, arguably is free to discharge the employee. If an employer fails to comply with the applicable OSHA standards, however, or if no standard exists and the employer permits workplace exposure levels that violate state and fed-eral requirements to maintain a safe place of employ-ment, the employer is violating public policy. Arguably, to permit the employer to take advantage of that violation by discharging the employee is to permit a further violation of public policy. The courts would be loath to allow the employer who negligently breaks the arm of an employee subsequently to fire that employee because of a resultant inability to do heavy lifting. Although the analogy is not perfect, one who subjects employees to toxic substances commits sub-stantially the same act. An employer's use of human monitoring data to discharge employees in such a circumstance may well be impermissible as a matter of public policy. The employer may be obliged *at common law* to find safe assignments for the workers at comparable pay or bear the cost of their removal as part of doing business.

UNDER THE OSHACT GENERAL DUTY CLAUSE

The use of monitoring data to limit or deny employment opportunities also raises issues under the general duty clause of the OSHAct. When monitoring information reveals that an employee risks serious health damage from continued exposure to a workplace toxicant, it may also indicate that the employer is in violation of the general duty clause. When a workplace exposure constitutes a "recognized hazard" likely to cause death or serious physical harm, an employer violates the general duty clause if he or she does not take appropriate steps to eliminate the hazard. When an employer asserts that an employee cannot work without injury to health, the employer admits that the workplace is unsafe. That admission triggers the remedial provisions of the OSHAct. In the case of toxic substances, this would appear to require reduction of the exposure itself, not mere removal of presumptively sensitive employees from the site of exposure.

Section 11(c)(1) of the OSHAct prohibits employers from discharging or otherwise discriminating against any employee "because of the exercise by such employee on behalf of himself or others of any right afforded by this chapter." If an employee insists on retaining his or her job in the face of medical data indicating that continued exposure to a workplace chemical will likely pose a danger to health, the employee may well be asserting a "right" afforded by the general duty clause. That is, in insisting on retaining employment, the employee is asserting a right to a workplace that comports with the requirements of the general duty clause. Accordingly, an employer who discharges or otherwise discriminates against a worker because of perceived susceptibility to a toxic exposure arguably violates the section 11(c) prohibition.

Support for this position is found in an OSHA regulation, issued under section 11(c) and upheld in a unanimous 1980 Supreme Court decision. *Whirlpool Corp. v. Marshall*, 445 U.S. 1 (1980). The regulation gives individual workers a limited right to leave the workplace when they face a situation likely to cause "serious injury or death." Where an employee exercises this right to refuse hazardous work, the employer may not take discriminatory action against the employee by discharging the employee or by

issuing a reprimand to be included in the employment file. According to the district court to which the issue was remanded for consideration, withholding the employee's pay during the period in which the employee exercises the right is also prohibited. As a worker may absent him or herself from a hazardous work assignment under certain conditions without loss of pay or job security, it would be anomalous to allow an employer to discharge or remove the employee without pay because of the same hazardous condition. This would make the result depend on whether the employee asserted a right to refuse hazardous work before the employer took action to discharge him or her from employment.

The issue has not yet been faced directly by a court. In a 1984 decision, however, the District of Columbia Circuit Court of Appeals did suggest that a general duty clause violation might exist if an employer removed susceptible employees "in an attempt to pass on to its employees the cost of maintaining a circum-ambient [toxic substance] concentration higher than that permitted by law," or in an attempt to avoid the cost of reducing concentrations "to a level that posed an acceptable risk." *O.C.A.W. v. American Cyanamid* 741 F.2d 444 (D.C. Cir. 1984). The best approach may be a regulatory solution. The implementation of mandatory MRP for toxic substances exposure in general, as OSHA has done for lead, might be accomplished by a generic MRP standard. An employer's compliance with a mandatory MRP provision for a particular exposure would remove the threat of a general duty clause citation.

UNDER ANTI-DISCRIMINATION LAWS

In addition to potential liability under the common law and the OSHAct general duty clause, an employer who uses monitoring information to limit employment opportunities may also face liability under anti-discrimination laws. Although not all workplace discrimination is prohibited, state and federal law forbid certain *bases* for discrimination. Many of these may apply to an employer's use of human monitoring information. A detailed discussion of the relevant discrimination laws is beyond the scope of this book, but an outline of their potential impact on human monitoring is set forth below.

HANDICAP DISCRIMINATION

Congress and most states have passed laws barring discrimination against handicapped individuals in certain employment situations. The laws, which vary widely among the jurisdictions, all place potential limitations on the use of human monitoring data. Although the courts have adopted a case-by-case approach, the worker who is denied employment opportunities on the basis of monitoring results often falls within the literal terms of many handicap discrimination statutes. In general, two issues will be determinative: whether the workplace in question is covered by a state or federal handicap act and, if so, whether the worker in question is "handicapped" under that act.

The Federal Rehabilitation Act 29 U.S.C. Section 701-795(q) provides handicapped persons with two potential avenues of protection against job discrimination. Section 503 prohibits private employers with federal contracts of $2500 or more from discriminating against a present or prospective employee on the basis of handicap. Courts have generally held, however, that section 503 does not create a private right of action on the part of the aggrieved individual. A private right of action *is* available under a companion provision, section 504. Nonetheless, the scope of the federal act is far from all-inclusive. The various state acts offer a potential for more extensive coverage. Most extend beyond public contractors and apply to most of the major employers within the state.

A worker excluded from a workplace or job assignment because of the results of human monitoring has been removed because he or she is ostensibly at higher risk of injury or illness than the majority of workers. The worker is perceived as having a physical condition that sets him or her apart from others. Although this clearly is discrimination on the basis of physical status, an applicable handicap discrimination statute will not prohibit the action unless the relevant definitional criteria are met. The stated criteria do not differ widely among most jurisdictions, but judicial interpretations of these criteria have varied substantially. Some state courts have interpreted handicap discrimination laws broadly, taking positions that appear to limit significantly the use of monitoring data for employee exclusion. Others have taken much more restrictive positions. Some federal courts have adopted a middle-ground approach.

At present, the general applicability of handicap discrimination statutes to the use of human monitoring information is unclear. Examining the definitional criteria in the federal act, on which many of the state statutes are based, will illustrate the issues facing courts—and the potential range of logical interpretations. The Rehabilitation Act of 1973 defines a "handicapped" individual as "any person who (i) has a physical or mental impairment which substan-

tially limits one or more of such person's major life activities, (ii) has a record of such an impairment, or (iii) is regarded as having such an impairment."

In the great majority of cases, the persons facing reduced employment opportunity as a result of human monitoring data do not *presently* have a substantially debilitating medical condition and thus do not satisfy either the first or second clauses of the federal definition. Rather, they are *perceived* as having an increased risk of developing such a condition in the future. Are they, then, "regarded" as having a substantial impairment under the third clause? A narrow reading of the statute might lead to a negative conclusion. In a literal sense, such persons are regarded as having the impairment itself. Arguably, however, they are being treated *as if* they had a substantial impairment by being denied employment opportunities normally extended to those without such a disability. In this sense, they are regarded as substantially impaired. This latter interpretation finds support in the Senate Committee Report presented before the insertion of this language into the Act. The Senate Report explained that the third clause of the definition applies both to "persons who do not in fact have the condition which they are perceived as having" *and* to "persons whose mental or physical condition does not substantially limit their life activities." This second provision appears broad enough to cover persons excluded on the basis of monitoring information.

The one federal district court that directly examined the issue has affirmed the applicability of the 1973 Rehabilitation Act to pre-employment screening of perceived high-risk individuals. In *E.E. Black, Ltd. v. Marshall* [497 F. Supp. 1088 (D. Hawaii 1980)], the federal district court for Hawaii held that a twenty-nine-year-old who had been denied employment as a carpenter's apprentice as a result of positive findings in lower back x-rays was protected by section 503. The court rejected the suggestion that employers may avoid the Act's proscriptions merely by establishing that they have discriminated against a worker on the basis of an *insubstantial* physical disability. In this regard, the opinion noted that the purpose of the Act is not to permit an employer to "be rewarded if his reason for rejecting the applicant were ridiculous enough."

Nonetheless, the court in *E. E. Black* also emphasized that not all high risk individuals would be treated as "handicapped" under the Act. Addressing the requirement that the actual or perceived disability must "substantially limit" a major life activity, the court read into the Act a requirement that the actual or perceived impairment be "a substantial handicap *to employment.*" In determining whether a particular condition meets this criterion, the court indicated that one must first assume that all similar employers within the relevant geographic area use the disputed pre-employment screen (or other discriminatory practice) and then weigh that against the physical and mental capabilities of the particular applicant. If the resultant employment limitations appear "substantial," the person will be deemed "handicapped."

Although perhaps not wholly consistent with the literal terms of the Act, this construction of the statute appears to be an attempt to fashion a viable framework for evaluating the treatment of perceived high-risk individuals within the context of "handicap" discrimination. The Act seems designed primarily to protect the seriously handicapped, but its language is broad enough to cover discriminatory practices based on data obtained through human monitoring. The middle-ground adopted in *E. E. Black* imposes a reasonable limitation on an employer's use of monitoring data.

Even in cases in which handicap discrimination is established, an employer may escape liability if the discriminatory practice is reasonably necessary for efficient operation of the business. The Rehabilitation Act provides employers with no affirmative defense, but does require the handicapped individual to prove that he or she is "qualified" for the job. Thus, if a handicap prevents a worker from safely or effectively performing the job, an exclusionary practice may be permissible under the Act. Most state handicap statutes include some form of affirmative defense. Although these vary among jurisdictions, many appear analogous to the familiar defenses that have developed under Title VII of the Civil Rights Act.

CIVIL RIGHTS AND AGE DISCRIMINATION

Employers who exclude workers on the basis of monitoring information may also run afoul of the more general laws against discrimination. Title VII of the federal Civil Rights Act [42 U.S.C. Section 2000e] prohibits employment discrimination on the basis of race, color, religion, sex or national origin. The scope of the act is substantially broader than that of the federal handicap discrimination act, and it affords protection for the great majority of the nation's employees. In addition, many states extend similar protection to employees not covered by the federal act. The Age Discrimination in Employment Act [29 U.S.C. Sections 621–678] and some state acts provide protection of comparable breadth against discrimination on the basis of age.

As with handicap discrimination, the applicability of these laws to the use of human monitoring information is not yet clear. In the ordinary case, exclusionary practices based on monitoring data will not be *per se* discriminatory on the basis of race, sex, national origin or age. Nor are they likely to involve *disparate treatment* of one of these protected classes. That is, they will not be part of a policy that, while neutral on its face, masks a specific employer intent to discriminate on one or more of these impermissible bases. The practical impact of an exclusionary practice, however, may fall disproportionately on a particular race, sex, ethnic or age group.

The Supreme Court has long held that a claim of *disparate impact* states a viable cause of action under the Civil Rights Act. A similar rationale has been applied in the area of age discrimination. In a 1975 decision, the Court held that job applicants denied employment on the basis of a pre-employment screen establish a *prima facie* case of racial discrimination when they demonstrate that "the tests in question select applicants for hire or promotion in a racial pattern significantly different from that of the pool of applicants." *Albermarle Paper Co. v. Moody,* 422 U.S. 405, 425 (1975). Proof of disparate impact thus requires statistical analysis demonstrating a "significantly" disproportionate effect on a protected class. The cases provide no clear guidance, however, as to the level of disproportion that is required before an effect is deemed "significant."

The potential for disparate impact inheres in many uses of human monitoring data. A genetic screen for sickle-cell anemia, for example, will disproportionally exclude blacks and certain ethnic groups because they have much higher incidence of this trait than does the general population. Similarly, tests that consistently yield a higher percentage of positive results in one gender than the other may give rise to exclusionary practices that discriminate on the basis of sex. Finally, a wide variety of exclusionary practices based on monitoring data may have a disparate impact on older workers. Older workers have been in the workforce longer and usually have been exposed to hazardous work environments much more often than their younger colleagues. Their prior exposure may have impaired their health or left them more vulnerable to current workplace hazards. They may, for example, have a pre-existing illness as a result of previous workplace exposures. Their age alone may account for a certain degree of body deterioration.

When the employee establishes a *prima facie* case of disparate impact, the employer will have an opportunity to produce evidence indicating that the exclusionary practice constitutes a "business necessity." If such proof is offered, the employee must carry the burden of proving that the practice is *not* a business necessity. *Wards Cove Packing Co. v. Atonio*, 109 S.Ct. 2115 (1989). The Supreme Court has characterized "business necessity" as requiring "a manifest relation to the employment in question." *Griggs v. Duke Power*, 401 U.S. 424, 431 (1971). In the words of an often cited Fourth Circuit Court of Appeals opinion, this means that the practice must be "necessary to the safe and efficient operation of the business." *Robinson v. Lorillard Corp.*, 444 F.2d 791, 798 (4th Cir.) *cert. denied*, 404 U.S. 1006 (1971). Further, if the employee can establish that another, less discriminatory practice "would be equally as effective as the challenged practice in serving the [employer's] legitimate business goals," the business necessity defense will not stand.

There are two principal reasons why "business necessity" may be difficult to establish for exclusionary practices based on human monitoring data. The first is that the great majority of these practices are not designed to protect the health and safety of the public or of other workers. Instead, their "business purpose" is the protection of the excluded worker and, not incidentally, the protection of the employer from the anticipated costs associated with the potential illness of that worker. The Supreme Court has indicated that employer cost is a factor to be considered in the evaluation of disparate impact cases. Nonetheless, as noted in one analysis, "the courts are usually skeptical of an employer's argument that it refuses to hire qualified applicants for their own good, and they often require a higher level of justifi-

cation in these cases than in cases in which public safety is at stake." *McGarrity and Schroeder, Risk-Oriented Employment Screening*, 59 Tex. L. Rev. 999, 1049 (1981).

Another, and probably more serious, obstacle to the successful assertion of a business necessity defense will be the unreliability of the screening procedures themselves. If the exclusion of susceptible (i.e., high-risk) individuals truly *is* a "business necessity," its rationale disappears if the test used as the basis for such exclusion cannot provide reasonable assurance that those excluded actually are susceptible (i.e., at high risk). Indeed, without such assurance, the test becomes little more than an instrument for arbitrariness and only adds to the discriminatory nature of the exclusionary practice. To the extent that the tests in question are not reliable, the availability of the business necessity defense is questionable.

The foregoing discussion of discrimination has presupposed that the "screened" worker will be excluded from the workplace. As suggested throughout this book, however, employers may have another option. In many cases, the employer will be in a position to provide these workers with other jobs in workplaces that do not involve exposures to the substances from which they may suffer adverse health effects. If such alternative positions were supplied, at benefit levels comparable to those of the positions from which exclusion was sought, employers might avoid the proscriptions of the various discrimination laws. Providing an alternative position would certainly remove much of the incentive for filing a discrimination claim. Further, even if such a claim were filed, courts might find that an adequate MRP program obviated the charge of discrimination. This could be one area where good law and good social policy coincide.

◇ NOTES

1. The authors (who include the authors of this text) suggest that the common law may afford some protection to the worker who is excluded from the unsafe workplace. If such a state common law right were to be developed as a "public policy" exception to the employment at will rule, what would be the clear, articulated public policy on which this exception would be based? If the basis is the OSHAct, would the common law cause of action be preempted by Section 11(c) of the Act?

2. The authors rely for support on the D.C. Circuit's opinion in the *American Cyanamid* fetal protection policy case. Does the reasoning applied in that case appear to support the arguments offered here regarding the OSHAct general duty clause or does it make those arguments appear less plausible?

3. As suggested in this discussion of the right to refuse hazardous work, an employee facing discrimination of this nature may be able to characterize the discrimination as "retaliatory" by asserting one of the "rights" discussed in the opening parts of this chapter. Why might the employee want to do this?

4. In *Hayes v. Shelby Memorial Hospital,* discussed above in the section on fetal protection policies, the Eleventh Circuit noted that "potential litigation costs [cannot] form the basis for the business necessity [defense, as] potential liability is too contingent and too broad a factor to amount to a business necessity." 726 F.2d at 1553, n.15. Judge Posner's dissent in *Auto Workers v. Johnson Controls Inc.,* however, suggests that concern over tort liability to third parties (here the yet unborn children of exposed female workers) might satisfy Title VII's BFOQ defense. It is not clear that he would make the same suggestion if the concern were compensation claims from the workers themselves.

5. As this book goes to press, there is legislation pending before Congress that would undo the doctrine of *Wards Cove Packing.* Under *Wards Cove,* the employee has the burden in a disparate impact case of proving that the employer's rationale for a discriminatory policy is *not* a business necessity. Under the proposed legislation, the burden would be on the employer to prove that it is. A version of this legislation was passed by Congress in 1990, but was vetoed by President Bush.

6. Another federal statute that may provide protection to "screened" workers is the Americans with Disabilities Act (ADA), which was signed into law on July 26, 1990. See 42 U.S.C. §1201, *et. seq.* The ADA requires employers to provide reasonable accommodations in order to allow disabled workers to perform their jobs. Its scope is much broader than that of the Rehabilitation Act. After two years, it will cover all employers with 25 or more employees. After two more years, all employers with 15 or more employees will be covered.

CHAPTER 9

Compensation for

Occupational Injury

and Disease

In prior chapters, we have focused primarily on the use of legal mechanisms to prevent occupational injury and disease. Obviously, however, some workers *are* injured on the job, and others *do* contract workplace-related diseases. How, and to what extent, do they receive compensation when this happens? The potential sources of compensation for occupational injury and disease, and for resultant loss of income, vary widely from situation from situation. Principal among them are private disability insurance, private health insurance, the Social Security system, the welfare system, workers' compensation, and lawsuits in tort. In this chapter we focus on the last two of these. We do so because, of all the social mechanisms designed to aid injured workers, these two alone have as one of their stated goals—in addition to providing compensation—the prevention of workplace injury and disease.

A. STATE WORKERS' COMPENSATION

As the result of a legislative campaign begun in the early 1900s, all states today have broad-based workers' compensation statutes. Interestingly, initial support for workers' compensation legislation came from business groups, not from labor.

Preventing Illness and Injury in the Workplace

Office of Technology Assessment, U.S. Congress

PROGRESSIVE ERA AIMS

In the early 1900s a number of Progressive Era humanitarian efforts underlined the plight of the injured worker and paved the way for workers' compensation programs. Crystal Eastman conducted the now-famous "Pittsburgh Survey" of 1907-08. She examined the economic conditions of the families of workers who had been killed or injured. In over half the cases, she found that "the employers assumed absolutely no share of the inevitable income loss. The costs of work accidents fell directly, almost wholly, and in likelihood finally, upon the injured workmen and their dependents." She concluded that a system of compensation was necessary to achieve equity, social expediency, and prevention.

At about the same time, a State commission in Illinois reported that most court awards for industrial accidents were small, and that the families of the injured were often forced to live on charity. Moreover, for employers who had liability insurance, only 42 percent of payments went to medical care. The remaining 58 percent went for administration, claims investigation, and legal expenses.

EMPLOYERS' ATTITUDES

The apparently small awards made to most workers was not the only reason for dissatisfaction with legal remedies. Employers, who as a group supported workers' compensation legislation before labor unions did, also found advantages in compensation programs. There is some evidence that just prior to the creation of workers' compensation laws, injured workers, at least in some circumstances, won a substantial portion of lawsuits against their employers.

Moreover, workers' compensation substituted a regular, fixed, and predictable compensation payment for uncertain, potentially ruinous liability judgments. Employers also feared that without a workers' compensation system, the courts would start making more awards to injured employees, especially if a worker could show that his/her employer had violated one of the increasing number of State safety regulations.

Finally, employers advocated workers' compensation in order to remove one source of hostility from labor-management relations and possibly to prevent more fundamental changes in the worker-employer relationship. They specifically opposed the passage of liability law reforms that would have eliminated the common law defenses of employers. Some large companies had already established company benefit plans that provided payments for work injuries. Smaller manufacturers favored creation of such plans, but lacked the resources to do so privately. Larger manufacturers feared that if such plans were not created, legislators might act to change employer and employee rights. In the absence of changes, it was feared that the nascent unions would be given a boost.

For these reasons, some of the initial advocates of workers' compensation included groups like the National Association of Manufacturers, the National Civic Federation, the American Association for Labor Legislation, and a number of the leading industrialists of the day.

LABOR UNION REACTIONS

Unions, on the other hand, initially opposed workers' compensation. They generally wanted workers to retain the right to sue employers and advocated abolition of the three common law defenses. They held this position in part because they thought injured workers would receive larger payments under such a plan and because, at the time, they generally mistrusted the government and feared that governmental intervention would weaken unions.

Union opposition was also based on their perception that workers' compensation was "palliative and not preventive." The belief that workers' compensation could provide an economic incentive for prevention was, according to MacLaury, important in changing labor's position; it "seemed to tip the scales."*

Source: Office of Technology Assessment, U.S. Congress, *Preventing Illness and Injury in the Workplace,* 1985, pp. 207–209.

*MacLaury, J., The Job Safety Law of 1970: Its Passage Was Perilous, *Monthly Labor Review* 104(3):18–24, March, 1981.

Workers' Compensation

L. Boden

DESCRIPTION OF WORKER'S COMPENSATION

Workers' compensation provides income benefits, medical payments, and rehabilitation payments to workers injured on the job as well as benefits to survivors of fatally injured workers. There are 50 state and three federal workers' compensation jurisdictions, each with its own statute and regulations.

While state and federal systems are different in numerous ways, they have several characteristics in common. Benefit formulas are prescribed by law. Generally, medical care and rehabilitation expenses are fully covered, but lost wages are only partially reimbursed. Employers are legally responsible for paying benefits to injured workers. Some large employers pay these benefits themselves, but most pay yearly premiums to insurance companies, which process all claims and pay compensation to injured workers. Workers' compensation is a no-fault system. Injured workers do not need to prove that their injuries were caused by employer negligence. In fact, employers are generally required to pay benefits even if the injury is entirely the worker's fault.

The change to a no-fault system was established to minimize litigation. For a worker to qualify for workers' compensation benefits only three conditions must be met: There must be an injury or illness; the injury or illness must "arise out of and in the course of employment"; and there must be medical costs, rehabilitation costs, lost wages, or disfigurement. Clearly, these conditions are much easier for the injured worker to demonstrate than is employer negligence. For example, if a worker falls at work and breaks a leg, all three conditions are easily demonstrated. Unusual cases sometimes arise in which the question of the relationship of an injury to employment is difficult to resolve, and there may be questions about when a worker is ready to return to work. Such issues may result in litigation, but they are the exception not the rule. In most cases a worker files a claim for compensation with the employer, and the claim is accepted and paid either

Source: From Levy and Wegman, Eds., *Occupational Health: Recognizing and Preventing Work-Related Disease,* 2nd ed. Boston: Little, Brown and Co., 1988, pp. 149–162.

directly by the employer or by the workers' compensation insurance carrier of the employer.

The following case is typical of the events that follow many minor claims for workers' compensation:

> Mr. Fisher had a painful muscle strain while lifting a heavy object at work on Monday afternoon. He went to the plant nurse and described the injury. He was sent home and was unable to return to work until the following Friday morning. On Tuesday the nurse sent an industrial accident report to the workers' compensation carrier and a copy to the state workers' compensation agency. Two weeks after he returned to work Mr. Fisher received a check from the insurance company covering part of his lost wages—as mandated by state statute—and all of his out-of-pocket medical expenses related to the muscle strain.

Workers' compensation has wider coverage than did the common law system. Under workers' compensation workplace injuries and illnesses are compensable even if they are only in part work-related. Suppose, for example, a worker with preexisting chronic low back pain becomes permanently disabled as a result of lifting a heavy object at work. In this case the worker's preexisting condition may just as easily have been aggravated by carrying out the garbage at home, but the fact that the disabling event occurred at work is generally sufficient for compensation to be awarded.

Cases in which an occupational injury or illness becomes disabling as a result of non-work exposures are more complicated. A worker with non-disabling silicosis may leave a granite quarry job for warehouse work. Without further exposure the silicosis will probably never become disabling. However, the worker begins to smoke cigarettes and loses lung function until partial disability results. In most states this worker should receive compensation from the owner of the granite quarry if the work relationship can be demonstrated.

Generally, diseases are considered eligible for compensation if occupational exposure is the *sole cause* of the disease, is *one of several causes* of the disease, was *aggravated* by or aggravates a nonoccupational exposure, or *hastens* the onset of disability (Table [9-1]). Several states, including California and Florida, allow disability to be apportioned between occupational and nonoccupational causes. While at first

Table 9-1. Likelihood of Compensation, by Source of Preexisting Condition and Source of Ultimate Disability

Source of Ultimate Disability	Source of Preexisting Condition	
	Work-related	Non-work-related
Work-related	Compensable	Generally compensable
Non-work-related	Generally compensable	Not compensable

Source: Adapted from P.S. Barth, H.A. Hunt. *Workers' compensation and work-related illnesses.* Cambridge: MIT Press, 1980.

this may seem like a sensible approach, apportionment creates some difficult decisions for workers' compensation administrators. Many disabilities are not additively caused by two separable exposures. With silica exposure or cigarette exposure alone, the worker in the above example would not have become disabled. Often, as in the case of lung cancer caused by asbestos exposure and smoking, the contribution to disability or death of two factors is many times greater than that of one alone. Such issues make the apportionment of disability very difficult if not impossible.

When workers' compensation was introduced, workers gained a swifter, more certain, and less litigious system than existed before. In return, however, covered workers waived their right to sue employers through common law. They also accepted lower awards than those given by juries in negligence suits: Worker's compensation provides no payments for "pain and suffering" as there might be in a common law settlement. In addition, disability payments under workers' compensation are often much less than lost income, especially for more severe injuries.

The United States does not have a unified workers' compensation law. Each state has its own system with its own standards and idiosyncracies. In addition, federal systems cover federal employees, longshoremen and harbor workers, and workers employed in the District of Columbia. Almost all states require employers either to purchase insurance or to demonstrate that they are able to pay any claims that might be made by their employees. In most states private insurers underwrite workers' compensation insurance paid for by premiums from individual employers. In some states a non-profit state workers' compensation fund has been establishes; the state

government therefore acts as an insurance carrier, collecting premiums and disbursing benefits. State funds seem to be very effective in delivering benefits: They disburse a higher percentage of premiums in the form of benefits than do private insurance carriers.

*   *   *

THE ROLE OF THE PHYSICIAN IN WORKERS' COMPENSATION

Workers' compensation is basically a legal system, not a medical system. The decision points for claims in this complex system are shown in [Figure 9-1]. If a claim is rejected by the workers' compensation carrier or self-insured employer, it will generally be necessary for the injured worker to hire a lawyer. The worker's lawyer may then bargain with the lawyers for the insurance carrier in an attempt to settle the dispute informally. If this bargaining does not result in agreement, the claim must either be dropped or taken before an administrative board—a quasi-judicial body established by state statute—for a hearing. To the worker or to a physician who may be called to testify in such a hearing, such a proceeding may be indistinguishable from a formal trial: Witnesses are sworn, rules of evidence are followed, and testimony is recorded.

As part of this legal proceeding medical questions are often raised. There may be disagreement about the degree of disability of a worker, when an injured worker is ready to return to work, or whether a particular injury or illness is work-related. To settle these disputes physicians may be called on to give their medical opinions about employees' disabilities. Many physicians do not like to testify in such hearings,

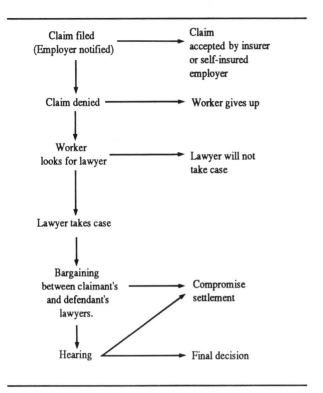

Figure 9-1. Decision points for workers' compensation claims.

and most are not prepared by their training or experience to assume this role. Their expertise may be challenged; moreover, they may be confused by the different meanings of legal and medical terminology.

In workers' compensation, decisions are based on legal definitions, and the legal distinction between disability and impairment is often unclear to physicians. . . . A physician called in to testify about whether or not a worker is permanently and totally disabled may understand total disability as a state of physical helplessness and may therefore testify that the injured worker is not totally helpless. However, this standard is not what a workers' compensation board would apply; the term *disability* as used in workers' compensation proceedings means that wages have been lost, while *total disability* means that the injured worker loses wages as a result of not being able to perform gainful employment. This definition is in contrast to total *impairment,* which might imply that the injured worker could, in addition, not feed him- or herself or get out of bed. For example, a worker who has been exposed to silica at work may

have substantially reduced pulmonary function and therefore impairment. However, if the worker continues to work at the same job, no wages have been lost and therefore no disability payment is made. Many states, however, offer specified payments for disfigurement or losses of sight, hearing, or limbs, with compensation based on impairment and not on disability (Table [9-2]).

ESTABLISHING WORK-RELATEDNESS FOR COMPENSATION

The burden of proving that disease is occupational in origin lies with workers. They must find physicians who are convinced that their illnesses are occupational in origin or that their illnesses were aggravated or hastened by occupational exposures. Physicians must then be able to convince referees who hear the cases that the diseases are indeed work-related.

The burden of proof might at first seem to be impossible for those diseases that are not uniquely occupational in origin. For example, lung cancer

Table 9-2. Income Benefits for Scheduled Injuries in Selected Jurisdictions (as of January 1, 1986)*

Jurisdiction	Arm at Shoulder	Hand	Thumb	First Finger	Foot
Alabama	$ 48,840	$ 37,400	$13,640	$ 9,460	$ 30,580
Alaska	59,000	45,400	14,000	8,700	39,700
California	58,975	53,540	7,595	3,360	33,740
Connecticut	123,864	100,044	37,715	21,438	74,636
Delaware	58,923	51,852	17,677	11,785	37,710
Georgia	34,875	24,800	9,300	6,200	20,925
Illinois	120,275	97,244	35,827	20,472	79,331
Mississippi	26,600	19,950	7,980	4,655	16,625
New York	46,800	36,600	11,250	6,900	30,750
Washington	36,000	32,400	12,960	8,100	25,200
Wisconsin	50,000	44,800	17,920	6,270	28,000
Federal employees	305,729	239,096	73,493	45,075	200,880
U.S. longshoremen	185,715	145,239	44,643	27,381	122,024

*Amounts in table reflect maximum potential entitlement.
Source: Reprinted with the permission of the Chamber of Commerce of the United States of America from *Analysis of Workers' Compensation 1986* © 1986 Chamber of Commerce of the United States of America.

may be caused by smoking, air pollution (although not definitely established), occupational or nonoccupational radiation exposure, or all of these factors.

Suppose that a worker with lung cancer has smoked cigarettes, has had diagnostic x-rays, and has also been occupationally exposed to ionizing radiation in a uranium mine. Since occupational lung cancer does not have distinctive clinical features, an expert medical witness, using clinical judgment, cannot say that the disease is without question occupational in origin. The expert witness cannot even say with certainty that occupational exposure to ionizing radiation was one of several causes or if it hastened the onset of the cancer. At best all that can be determined is that the worker has an increased risk of lung cancer as a result of the job. In this case the legal standard is that there must be a "preponderence of evidence" that the disease is occupational in origin, or the case is unlikely to be settled in favor of the disabled worker. A "preponderence of evidence" means that it is more likely than not (probability greater than 50 percent) that the illness in question was caused by, aggravated by, or hastened by workplace exposure.

In some cases workers' compensation laws have been written so that payment of a claim may be denied even though convincing evidence is presented that the illness was caused by or aggravated by the worker's employment. Some states require that a disease not be "an ordinary disease of life." In other words, diseases such as emphysema and hearing loss may not be compensable because they often occur among people with no occupational exposure. More than 20 states have a related requirement that diseases are only compensable if they are "peculiar to" or "characteristic of" a worker's occupation.

All jurisdictions have a statute of limitations (often one or two years) on claims for workers' compensation. A two-year statute of limitation means that the claim must be filed by the worker within two years of a given event. A time limit of two years after the worker has learned that a disease is work-related imposes no particular hardship on occupational disease victims. In some states, however, the time period begins when the disease becomes symptomatic, even if this takes place before the disease is diagnosed or determined to be work-related. The latter policy for starting the statute of limitation may be a special problem if the

Table 9-3. Time Limits on Filing Occupational Disease Claims in Selected Jurisdictions (as of January 1, 1986)

Jurisdiction	Time Limit on Claim Filing
Alabama	Within 2 years after injury, death, or last payment; radiation—within 2 years after disability or death and claimant knows or should know relation to employment.
California	Disability—within 1 year from injury or last payment; death—within 1 year after death and in no case more than 240 weeks after injury, except for asbestos-related disease claims. Date of injury is defined as when claimant is disabled and knows or should know of relation to employment.
Colorado	Within 3 years after commencement of disability or death; within 5 years in case of ionizing radiation, asbestosis, silicosis, or anthracosis.
Massachusetts	Within 1 year after injury or death; delay excusable.
Michigan	Within 2 years after claimant knows or should know of relation to employment.
New York	Within 2 years after disability or death, or 2 years after claimant knows or should know of relation to employment.
North Carolina	Within 2 years after disablement, death, or last payment; radiation—within 2 years after incapacity and claimant knows or should know relation to employment.
Oregon	Within 5 years after last exposure and within 180 days after disablement or physician informs claimant of disablement. 10 years after last exposure for radiation disease.
Utah	Within 1 year after claimant knows or should know relation to employment, but no later than 3 years after death; permanent partial disability—within 2 years.
Virginia	Within 2 years after diagnosis is first communicated to worker or within 5 years after exposure, whichever is first, within 3 years after death, occurring within limit for disability.
Federal employees	Within three years after injury, death, or disability and claimant knows or should know relation to employment; delay excusable.
U.S. longshoremen	Within 1 year after injury, death, last payment, or knowledge of relation to employment.

Source: Reprinted with permission of the Chamber of Commerce of the United States of America from *Analysis of Workers' Compensation 1986* © 1986 Chamber of Commerce of the United States of America.

worker's physician is not familiar with the occupational disease. The most burdensome statutes require that a claim be filed one or two years after exposure. Since chronic occupational diseases commonly do not manifest themselves until five, ten, twenty, or more years after exposure, such rules effectively eliminate the possibility of compensation for workers with these illnesses. In a recent study, occupational disease compensation among a group of asbestos insulators was only half as great in states with these restrictive statutes of limitation as in other states. Time limits for filing workers' compensation claims are described in Table [9-3].

THE PROBLEM OF COMPROMISE SETTLEMENTS

A workers' compensation claim that is denied by the employer or insurer does not automatically go to a hearing. The injured worker must first find a lawyer who will take the case (see Fig. [9-1]). The lawyer's fee is often based on the portion of the award attributed to lost wages, which means that the lawyer's fee will be small in a small award and that the lawyer will receive nothing if the claim is denied. Thus, it is hard for injured or ill workers to find lawyers to represent them when claims are small or success is unlikely.

Lawyers generally prefer to bargain informally with the defendant's lawyers rather than go to trial. If a compromise settlement can be reached prior to trial, no preparation for a hearing is necessary, and the lawyer will therefore have more time to work on other cases and earn additional income. A settlement reached outside the courtroom is called a compromise settlement because the amount paid to the claimant generally is a compromise between the maximum amount that the claimant could receive in a court decision and the amount (if any) of the settlement if the claimant lost.

In the face of protracted litigation with uncertain results, a compromise settlement may seem very attractive to an injured worker who may have no wage income for a considerable period and may be facing large medical bills. The injured worker may therefore prefer a small settlement paid immediately to a much larger but uncertain settlement that would not be available for one or two years. Especially where the worker does not foresee a quick return to work, a settlement may be accepted that might seem quite small to an outside observer. Insurers may use their knowledge of the financial pressures on the claimant to obtain a small settlement; they will thus contest, delaying the time when the case is closed in the hope of obtaining a small compromise settlement.

The compromise settlement will usually be paid in a lump sum to the injured worker and the attorney. This lump sum settlement will take the place of future payments for lost earnings and medical and rehabilitation costs. Many compromise settlements also release the insurer from future liability: If the claimant's condition should change at a later date or if future medical needs were inadequately estimated, the insurer would not incur the costs of any increased disability or medical or rehabilitation expenses. The injured worker who has accepted a "compromise and release" settlement may later need additional medical care but not have the resources to pay for that care.

For example, a worker with back injury was denied compensation by his employer, who claimed that

the injury was not work-related. He then took action that led to his being offered a lump sum settlement:

> I went to my union representative and filled out the forms for the industrial accident board, and about three weeks later they sent me an award which was about $600 . . . and I wouldn't take it.
>
> But then I applied for an attorney and talked to my attorney, and then filed suit. They turned around and told my attorney that they would consider [the injury] an industrial accident. So, I never did go to court. All they did was talk to may lawyer. They settled out of court. My lawyer told me while I was in the hospital that they wanted to settle it for $7,500. The fee for him [would be] $2,500.*

If the settlement of $7,500 was the result of a compromise and release agreement, the insurer or employer will not be liable for any future disability or medical costs resulting from this injury.

RECOMMENDATIONS OF THE NATIONAL COMMISSION

As part of the Occupational Safety and Health Act of 1970 Congress established the National Commission on State Workmen's Compensation Laws to "undertake a comprehensive study and evaluation of State workmen's compensation laws in order to determine if such laws provide an adequate, prompt, and equitable system of compensation." In 1972 the Commission released its report, which described many problems of workers' compensation and made many recommendations for improving state workers' compensation systems including these seven "essential" recommendations:

1. Compulsory coverage. Employees could not lose coverage by agreeing to waive their rights to benefits.

2. No occupational or numerical exemptions to coverage. All workers, including agricultural and domestic workers, should be covered. All employers, even if they have only one employee, should be covered.

3. Full coverage of work-related diseases—the elimination of arbitrary barriers to coverage

*Adapted from Subcommittee on Labor, Committee on Labor, and Public Welfare, U.S. Senate. Hearings on the National Workers' Compensation Standards Act, 1974. Statement of Lawrence Barefield.

such as highly restrictive time limits, occupational disease schedules, and exclusion of "ordinary diseases of life."

4. Full medical and physical rehabilitation services without arbitrary limits.

5. Employees' choice of jurisdiction for filing interstate claims.

6. Adequate weekly cash benefits for temporary total disability, permanent total disability, and fatal cases.

7. No arbitrary limits on duration or sum of benefits.

Since this report was issued, many states have changed their statutes to follow the recommendations of the National Commission. By increasing coverage and raising benefits, they have substantially improved the value of workers' compensation to injured employees. However, general changes in coverage have done little to discourage the litigation of occupational disease claims and costly occupational injury claims.

B.   THE LEVEL OF COMPENSATION FOR OCCUPATIONAL INJURY AND DISEASE

One important measure of the effectiveness of workers' compensation—both as a remedial and as a preventive mechanism—is the extent of the remuneration provided to those who experience occupational injury or disease. As the foregoing discussion suggests, the workers' compensation system appears to be at its most efficient when dealing with occupational injury. The compensation rate is substantially higher for cases of occupational injury than for cases of occupational disease. Further, victims of occupational injury receive a substantially larger percentage of their actual losses than do victims of occupational disease. In neither case, however, do the workers' compensation benefits constitute even half of the actual wage loss incurred. Nor do they comprise the majority of the overall compensation received.

In 1980, in response to a requirement in the federal Black Lung Benefits Reform Act of 1977, the U.S. Department of Labor presented a report to Congress on occupational disease. Included in that report was a comprehensive estimate of compensation levels for occupational injury and disease.

An Interim Report to Congress on Occupational Diseases

U.S. Department of Labor (Washington D.C.: U.S. Government Printing Office, 1980 1-4)

MAGNITUDE AND SEVERITY OF OCCUPATIONAL DISEASES

The long latency period associated with most *chronic* diseases and cancer and the interaction of multiple causal factors make it difficult to estimate the magnitude of the occupational disease problem. However, epidemiologic statistics show that workers exposed to certain substances such as asbestos, cotton dust, and silica experience excess disability and/or pre-mature death. Traditional sources of data include: employer records required by OSHA, state workers' compensation records, and physician reports. Employer records understate the occupational disease problem because affected workers may no longer be employed by the firm where the exposure occurred. In addition, some employers are reluctant to report occupational diseases when they are diagnosed because of their potential financial liability. Workers' compensation data understate the problem for similar reasons. The lack of a universal data system and limitations imposed by state compensation laws also reduce the usefulness of workers' compensation data. Physician-based records also understate the problem because few physicians are trained to recognize chronic occupational diseases and cancer.

Because the traditional approaches understate the magnitude of the occupational disease problem, an analysis was undertaken of individual self-reporting of work-related illness. This analysis assumes that an individual worker with a diagnosed medical condition is able to determine the work-relatedness of that condition. Information was obtained from an interview survey which has no direct link to compensation benefits. There was no financial incentive therefore for individuals to overstate the occupational origin of their illness. These data are limited to disabled individuals between the ages of 20 and 65. The data do not include individuals over the age of 65 or those who may have died from occupational diseases. Occupational related cancer deaths are therefore omitted from these data.

Almost two million workers reported that they were severely or partially disabled from an occupationally-related disease. Approximately 700,000 of these occupational disease victims suffer long-term total disability. The 1.2 million workers partially disabled from an occupational disease were either temporarily out of the labor force because of their disability or limited in the kind and/or amount of work they could perform. These data include chronic cases of totally disabling byssinosis and asbestosis, as well as partially disabling conditions, such as varicose veins, arthritis and ulcers. The major disabling conditions reported by workers as being related to their jobs include: back and spinal conditions (34 percent); heart and circulatory illnesses (26 percent); respiratory conditions (13 percent); mental illnesses (9 percent); and digestive conditions (8 percent). In 1978, the lost income for disabling occupational diseases amounted to $11.4 billion.

There is a long lag between exposure and compensation claims due to the long-latency period between exposure to hazardous substances and the onset of disability and/or death. Analysis of workers exposed to 10 substances known to cause chronic respiratory diseases and cancer of the respiratory tract in large groups of workers indicates that future deaths and disabling conditions due to past exposures will present a major problem for our health care systems and income maintenance programs. This analysis also indicates the need to collect more and better data on occupational diseases.

ADEQUACY OF EXISTING COMPENSATION SYSTEMS

A general measure of the adequacy of compensation for occupational disease victims can be ascertained by examining the extent to which lost earnings are replaced. Such data show that public and private income support programs replace about 40 percent of the wages lost by individuals who are severely disabled from an occupational disease, compared with a 60 percent replacement rate for occupational injury victims. The major source of extra income support received by occupational injury victims is workers' compensation.

The major sources of income support for those severely disabled from an occupational disease are: social security (53 percent), pensions (21 percent), veterans benefits (17 percent), welfare (16 percent), workers' compensation (5 percent), and private insurance (1 percent).

Not all of those severely disabled from an occupational disease receive income support, while some receive support from more than one program. Among those severely disabled from an occupational disease, one out of every four receive no income support payments. One in every three receive multiple benefits.

WORKERS' COMPENSATION

State workers' compensation benefits are supposed to replace two-thirds of the employee's lost income for a work-related injury or illness. Employers agreed to provide "no-fault" workers' compensation benefits as part of an historic compromise in which workers gave up their rights to sue employers. A worker, therefore is not required to prove employer negligence, but is required to prove that an injury or illness was caused by the job.

Only five percent of those severely disabled from an occupational disease receive workers' compensation benefits. This lack of coverage is due, in part, to the difficulties involved in establishing the work-relatedness of disabling illnesses because of: (a) the length of time which has elapsed between the hazardous exposure and the onset of disability and/or death, (b) the multiple causes of diseases, and (c) associating specific firms. Therefore, claims are often not fully compensated and litigation is more common.

*   *   *

SOCIAL SECURITY AND WELFARE

Two of the major sources of income for those severely disabled from an occupational disease are social security and welfare programs—53 percent receive social security benefits while 16 percent are welfare recipients. Those eligible for social security disability benefits in 1978 received an average annual payment of $3,900; recipients of supplemental security income received an average annual payment of about $1,900. Occupational diseases are costing the social security and welfare systems about $2.2 billion annually.

The social security disability insurance program spends about 5 cents out of every dollar on administrative and legal expenses. This compares very favorably with State workers' compensation programs.

While the social security system is the major source of income support for those severely disabled from an occupational disease, not all severely disabled workers are eligible for social security benefits. The major reason for this is that they cannot meet the "recency of employment" requirement. Even for those disabled individuals eligible for SSDI benefits, there is a five-month waiting period for cash payments and an additional waiting period of two years for Medicare benefits.

Due in large part to the difficulty of proving the causal connection between exposure and disease—a subject that is addressed later in this chapter—the general problems in obtaining compensation for occupational disease are exacerbated when the disease arises from exposure to toxic substances.

Are Workers Adequately Compensated for Injury Resulting from Exposure to Toxic Substances? An Overview of Worker Compensation and Suits in Tort

C. Caldart

A central purpose of the Occupational Safety and Health Act of 1970 (OSH Act) is to place the primary burden for insuring workplace health and safety on the employer—and thus indirectly on all of us through higher prices—rather than on the individual worker.[1] One measure of the Act's effectiveness, then, is how that burden is allocated when the workplace is *not* safe. Not surprisingly, we find that it is still the worker who shoulders the major portion of this cost. What is surprising, however, is the *extent* to which the burden is currently being borne by the worker.

Source: From Homburger and Marquis, Eds., *Chemical Safety Regulation and Compliance*. Basel: S. Karger AG, 1985, pp. 92–98.
[1]The philosophy underlying the OSH Act is well-stated in the Senate Committee report describing the rationale for the Act's general duty clause (Sec. 5): 'Employers have primary control of the work environment and should insure that it is safe and healthful.' S. Rep. No. 1282, 91st Cong., 2nd sess. p.9 (1970).

OVERALL COMPENSATION FOR OCCUPATIONAL DISABILITY

A worker who is injured as a result of an occupational accident or occupational disease is, depending on the nature of the injury, susceptible to three potential categories of "damage": medical expenses; wage loss; and physical pain and emotional suffering ("pain and suffering"). The various state and federal worker compensation systems are designed to provide compensation for the first and second of these items, but compensation for the third item—pain and suffering—is specifically precluded. Compensation for pain and suffering *is* available under the tort system, but as worker suits for employer negligence are expressly barred by all state worker compensation laws, compensation for pain and suffering is largely non-existent. Wage loss, then, is the major item for which our current system is designed to provide compensation. And it is here that the system appears to be failing.

Based on figures from 1975, a research team from

Syracuse University concluded in a comprehensive study of injured workers that *all* compensation for occupational disability from *all* major sources (worker compensation, social insurance, and private insurance) covered only 28% of all wage loss caused by occupational accident and disease.[2] That is, 72% of this wage loss went wholly uncompensated. Of the portion for which compensation was provided, only 30% of that compensation (representing 9% of the overall wage loss) came from worker compensation. Indeed, the figures lead one to conclude that, in general, workers might well be in a better position if the worker compensation system were abandoned entirely, and they were allowed to return to their traditional tort remedies.

COMPENSATION FOR EXPOSURE TO TOXIC SUBSTANCES

Where does exposure to workplace toxins fit into this picture? *Acute* toxic exposure—generally arising out of industrial accidents—usually involve situations of rather immediate injury, and can generally be expected to fall into the overall pattern set forth above. But the more insidious aspect of toxic substance exposure, of course, is its propensity to cause latent, and serious, *disease*. If our system does a poor job of compensating workers for all wage loss growing out of occupational disability, it does an abysmal job of compensating them for that portion of the wage loss that is caused by occupational disease.

In a study completed in 1980 [and excerpted above], the United States Department of Labor concluded that only 5% of all cases of serious occupational disease receive benefits through the worker compensation system. Perhaps more directly relevant to toxic substance exposure is a report completed in 1983 for the Minister of Labour for the Province of Ontario, Canada.[4] Based on the Doll and Peto estimate that approximately 4% of all cancer deaths in the United States are due to workplace exposures,[5] the report estimates that only 1 out of

17 occupational cancer fatalities receives any award under the Ontario worker compensation system. The United States experience would appear to be much worse. Based on United States Labor Department calculations of the ratio of the number of industrial disease awards relative to the size of the working population for both the United States and Ontario,[6] one can estimate that roughly 1 out of 79 occupational cancer fatalities in the United States receives worker compensation. This is an estimated compensation rate of 1.3%.

*   *   *

THE IMPACT OF THE TORT SYSTEM

And even when worker compensation *can* be obtained, the benefits in toxic substance cases tend to be disproportionally low. Worker compensation is commonly designed to provide approximately $2/3$ of actual wage loss, up to a scheduled limit.[7] In cases of serious disability—such as occupational cancer—the amount paid is usually insufficient. Indeed, the United States Labor Department reported in 1980 that, among those cases of occupational disease in the United States that do receive worker compensation, the average award is $9,700, or an estimated 13% of the average wage loss caused by occupational disease.

This has led workers to seek compensation through the tort system, where there is no prescribed limit on wage loss recovery, and where recovery for pain and suffering is allowed. Though, as noted, the *quid pro quo* for the implementation of worker compensation systems was a ban on the worker's common law right to sue his/her employer in negligence, courts in some states have allowed workers to bring suit against their employers for intentionally exposing them to toxic substances. Further, workers have in some instances been successful in bringing products liability claims against the manufacturers of toxic substances. But even here, the results are not encouraging.

A general comparison of the tort and worker compensation systems was provided in a recent analysis of the Syracuse University data.[11] Comparing wage

[2]Syracuse University, The Health Studies Program: Wage Losses and Workers' Compensation Benefits—The Survey of Workers' Compensation Recipients (1976).

[4]Paul C. Weiler: Protecting Workers from Disability: Challenges for the Eighties. p.21-23 (1983). (Available from Ontario Government Bookstore, 880 Bay Street, Toronto, Ontario M7A IN8.)

[5]Doll and Peto: The Causes of Cancer: Quantitative Estimates of Avoidable Risk of Cancer in the United States, Natn. Cancer Inst, 66:6 (1981).

[6]United States Department of Labor, [see pps. 461–463 above] at p.72. The comparison ratio is 4.67 to 1.

[7]For a survey of the various state laws, see United States Chamber of Commerce, Analysis of Workers' Compensation Laws. (This is an annual compilation.)

[11]Ashford and Johnson: Negligence vs. No-Fault Liability: An Analysis of the Workers Compensation Example. Seton Hall L. Rev. 12:725, 764 (1982).

loss recoveries under the present worker compensation system with expected recoveries of workers were instead allowed to sue their employers in negligence, the authors concluded that "the negligence approach to occupational injury provides a higher expected wage loss recovery than worker compensation for most types of injuries." But for "high loss" injuries—which must certainly include occupational cancers and other serious occupational diseases—they found the opposite to be true. Further, they concluded that neither the worker compensation nor the negligence approach compensates the "high loss" victims for more than 10% of their actual wage loss.

A more telling comment may be the recent Rand corporation study of the compensation received by asbestos victims in products liability claims.[12] To date, the average recovery for all such claims of asbestos-related disease and death has been $35,000. While this may appear, at first glance, to be a reasonable figure, it is a figure worth examining. The recoveries received in the asbestos cases arguably represent the best that our system has to offer to the victims of toxic substance exposure. These were tort cases, with no limitation on potential recovery. Further, proof of causation in asbestos cases—especially those involving asbestosis and mesothelioma—is much easier to establish than in most situations involving latent occupational disease. Finally, the liability of the asbestos manufacturers appears—in general—to be fairly clear. Yet how many among us would settle for $35,000 in exchange for asbestosis or cancer? (And the Johns-Manville bankruptcy, of course, looms as a spectre on the horizon.)

[12]Rand Corporation Institute of Civil Justice: Costs of Asbestos Litigation (1983), as reported in Disease Victims Get One-Third of Claims Sought in Lawsuits. According to Study, 13 Occupational Safety and Health Reporter (BNA) 213 (1983).

A FINAL LOOK AT ADEQUACY

Let us return to our original question: is there an adequate level of compensation for the victims of toxic substance exposure? Certainly, the numerical estimates taken from the various cited studies are open to question. These must be seen as no more than a rough approximation of the overall picture, as the margin for error is clearly substantial. Some will argue that these figures understate the actual problem, while others will find overstatement. Even allowing for a healthy margin of error, however, the data do indicate that worker compensation provides only a small percentage of the wage loss caused by workplace exposure to toxic substances. Assessments of "adequacy," of course, tend to vary with individual notions of fairness. Nonetheless, the data lead almost inexorably to the conclusion that our present compensation system does not deal adequately with the effects of workplace toxins.

If any one policy implication springs from this data, it is the continuing need for preventive measures. When we speak of the cost of implementing the technological process changes necessary to meaningfully reduce worker exposure to toxic substances, we must be mindful as well of the very real cost—in terms of wage loss and human suffering—of *not* implementing them. At present, our compensation system appears to be doing little to assist the OSHAct in effecting prevention. With recoveries so low, there would appear to be inadequate economic incentive for employers to implement preventive measures outside of what is required by OSHA.

In sum, whether by design or by neglect, we have made a rather low level of social commitment to compensating workers for occupational disability, and have instead kept the burden for toxic substance exposure directly on the worker. Before the promise of the OSHAct is to be fulfilled, this is a *status quo* that must be changed.

◇ NOTES

1. Most certainly, one could draw a different set of policy conclusions from these data. For example, if one believes that occupational disease is currently at or near its socially optimum level, then the extent of compensation might well be deemed adequate. Or one might believe that it is more efficient to put the burden of workplace illness on the employee, rather than the employer. Indeed, this was the position taken by the Supreme Judicial Court of Massachusetts in *Farwell v. Boston and Worcester Rail Road Co.*, 45 Mass. (4 Metc) 49 (1842). In this light, the *status quo* may appear considerably more appropriate.

2. The Doll and Peto estimate (that 4 percent of all cancers are caused by occupational exposures), on which some of the foregoing article is based, is thought by many to be an underestimate. See Epstein, Losing the War Against Cancer: Who's to Blame and What to Do About It, 20 *International Journal of Health Services* 53-71 (1990); Schneiderman, Davis, and Wagener, Lung cancer that is not attributable to smoking [letter]. 261 *J. Amer. Med. Assoc.* 2635-2636 (1989). (Published correction appears in 262 *J. Amer. Med. Assoc.* 904 (1989)); O. Axelson and L. Forrestiere. Estimated increases in lung cancer in non-smokers in Italy. Workshop on Increasing Cancers in Industrial Countries, Collelgium Ramazzini, Carpi, Italy. October 20-22 (1989).

3. The Manville Corporation, the world's largest manufacturer of asbestos, has been operating under the protection of Chapter 11 of the federal bankruptcy law since August 26, 1982. Manville cited as the reason for its bankruptcy filing its involvement in some 15,000 asbestos-exposure lawsuits and its estimate that some 32,000 more would be filed by the end of the century. Since that time, the Manville Property Damage Settlement Trust has been established to pay out funds to asbestos claimants. As of August 1989, the trust was reported to be "winding up" settlements with the original 15,000 prebankruptcy claimants, at an average of $40,000 per claim. The White Lung Association, an asbestos victim's group, has charged that the trust will not have enough money on hand to pay the tens of thousands of claimants who filed their claims after the bankruptcy filing in 1982. Reportedly, these postbankruptcy claims are being filed at the rate of "11,000 to 12,000 per month." 4 *Toxics Law Reporter* (BNA) 449 (1989).

C.   THE USE OF TORT LAW TO AUGMENT WORKERS' COMPENSATION

As discussed above, there has been an increasing use of tort suits in an attempt to recover compensation for occupational injury and disease. Given the substantial commitment of time and expense necessary to bring suit, workplace tort suits have tended to involve cases of serious injury and disease, where workers' compensation benefits are likely to be relatively low. These suits have encountered a mixed reception in the various state courts, and results have varied widely from state to state. Broadly speaking, they have failed more often than they have succeeded. Nonetheless, especially because they carry with them the potential for much higher damage awards than are available under workers' compensation, workplace tort suits have succeeded often enough to capture the attention of corporate America.

The common law tort system uses a deceptively simple framework to address the exceptionally complex issue of assigning responsibility for human injury and determining the level at which such injury should be compensated. In general, the injured party (the plaintiff) must prove three basic things in order to prevail: 1) that the defendant is *liable* for the injury in question (generally, either because the defendant has committed negligence, or because strict liability applies), 2) that the actions of the defendant *proximately caused* the injury, and 3) that the plaintiff suffered *actual damages* (economic, physical, or psychological) as a result of the injury. The burden is on the plaintiff to prove all three of these, and the plaintiff must carry this burden by "a preponderance of the evidence." If the plaintiff is successful, it is then up to the "trier of fact" (the jury, or if it is not a jury trial, the judge) to place a monetary value on the plaintiff's damages.

Beyond these basic touchstones, the nature of tort remedies tends to vary from state to state. Although close attention to the specific laws of the particular state is essential, certain general principles can be identified. The worker seeking to augment workers' compensation through a common law tort action has two potential arguments, both of which seek to circumvent the "exclusivity" clauses found in all state workers' compensation statutes. The first argument is that, even though compensation imposes a general prohibition on employee suits against the employer, that prohibition is not applicable to the particular case at hand. Thus, the argument runs, *this* worker may bring a tort suit directly against the employer, under some articulated exception to the general rule. The second argument depends on the worker's ability to identify some *third party* on whom to impose liability. According to this argument, the third party is not protected by the workers' compensation exclusivity clause and may be sued in tort. Thus far, the majority of the suits have been against employers and third-party manufacturers.

1.   Suits against Employers

One theory of employer tort liability relies on the argument that exclusivity clauses in workers' compensation statutes were not designed to shield the employer from responsibility for "intentional" acts. This theory is explored in the following two decisions.

Johns-Manville Products Corporation v. Contra Costa Superior Court (Reba Rudkin, Real Party in Interest)

California Supreme Court
27 Cal. 3d.465, 165 Cal.
Rptr. 858, 612 P.2d 948 (1980)

MOSK, Justice.

Section 3600 of the Labor Code provides that an employer is liable for injuries to its employees arising out of and in the course of employment, and section 3601 declares that where the conditions of workers' compensation exist, the right to recover such compensation is the exclusive remedy against an employer for injury or death of an employee. The issue to be decided in this proceeding is whether an employee is barred by these provisions from prosecuting an action at law against his employer for the intentional torts of fraud and conspiracy in knowingly ordering the employee to work in an unsafe

environment, concealing the risk from him, and, after the employee had contracted an industrial disease, deliberately failing to notify the state, the employee, or doctors retained to treat him, of the disease and its connection with the employment, thereby aggravating the consequences of the disease.

We conclude that while the workers' compensation law bars the employee's action at law for his initial injury, a cause of action may exist for aggravation of the disease because of the employer's fraudulent concealment of the condition and its cause.

Reba Rudkin, real party in interest (hereinafter plaintiff), brought an action against Johns-Manville Products corporation, his employer for 29 years (defendant) and others, alleging as follows:

Defendant is engaged in mining, milling, manufacturing, and packaging asbestos. Plaintiff worked in its Pittsburg, California, plant for 29 years beginning in February 1946, and he was continuously exposed to asbestos during that period. As a result of the exposure, he developed pneumonoconiosis, lung cancer, or other asbestos-related illnesses.

The defendant corporation has known since 1924 that long exposure to asbestos or the ingestion of that substance is dangerous to health, yet it concealed this knowledge from plaintiff, and advised him that it was safe to work in close proximity to asbestos. It failed to provide him with adequate protective devices and did not operate the plant in accordance with state and federal regulations governing dust levels.

In addition, the doctors retained by defendant to examine plaintiff were unqualified, and defendant did not provide them with adequate information regarding the risk of asbestos exposure. It failed to advise these doctors of the development of pulmonary disease in plaintiff or of the fact that the disease was the result of the working conditions at the plant, although it knew that his illness was caused by exposure to asbestos. Finally, defendant willfully failed to file a First Report of Occupational Injury or Illness with the State of California regarding plaintiff's injury, as required by law. Had this been done, and if the danger from asbestos had been revealed, plaintiff would have been protected. Each of these acts and omissions was done falsely and fraudulently by defendant, with intent to induce plaintiff to continue to work in a dangerous environment. Plaintiff was ignorant of the risks involved, and would not have continued to work in such an environment if he had known the facts.

* * *

The primary focus of the dispute between the parties centers upon the question whether section 4553 is intended to cover the intentional acts of employers which cause employee injuries. The section provides that "compensation otherwise recoverable shall be increased one-half where the employee is injured by reason of the serious and willful misconduct" of the employer, not to exceed $10,000. Defendant urges that the penalty imposed upon employers by this section is a substitute for a common law right of action against an employer whose intentional misconduct results in injury, while plaintiff argues that such misconduct is distinguishable from the "serious and willful misconduct" described in section 4553, and therefore his complaint alleg-

ing intentional acts by defendant is cognizable in an action at law.

Defendant relies upon both the legislative history of the Workers' Compensation Act and cases interpreting the words "serious and willful misconduct" in support of its position.

Prior to 1917, the law allowed an employee a choice of remedies if an injury was caused by an employer's gross negligence or willful misconduct. he could either claim workers' compensation benefits or maintain an action at law for damages. In that year, however, this provision was deleted and a new section added specifying a one-half increase in compensation in the event of serious and willful misconduct by the employer. This history, contends defendant, demonstrates that the right to seek additional compensation for injuries caused by the serious and willful misconduct of the employer was intended by the Legislature as a substitute for the right to seek damages in an action at law for such conduct.

Plaintiff claims that the reason for the amendment was a desire by the Legislature to equalize the treatment of employer and employee with regard to the commission of the acts of serious and willful misconduct. However, the argument is not convincing because it does not account for the repeal of the provision allowing an employee to bring an action at law for the employer's willful misconduct.

We find the historical background cited by defendant to be persuasive. The clear implication is that the addition in 1917 of the "exclusive remedy" limitation and the provision for a penalty for the willful misconduct of the employer was a substitute for the previous right of an employee to bring an action at law.

* * *

However, while the case law cannot be described as consistent, it reveals that in some exceptional circumstances the employer is not free from liability at law for his intentional acts even if the resulting injuries to his employees are compensable under workers' compensation. Indeed, in one unusual situation, despite the "exclusive remedy" provision of section 3601, an action at law was allowed for injuries incurred in the employment where the employer's conduct was negligent rather than intentional.

First we consider cases in which the intentional acts of the employer have been held not to justify an action at law. Compensation was determined to be the exclusive remedy for injuries suffered in a case in which the employer concealed the dangers inherent in the material the employees were required to handle *(Wright v. FMC Corporation* (1978) 81 Cal. App. 3d 777, 779, 146 Cal. Rptr. 740) or made false representations in that regard *(Buttner v. American Bell Tel. Co.* (1940) 41 Cal. App. 2d 581, 584, 107 P.2d 439.) The same conclusion was reached on the basis of allegations that the employer was guilty of malicious misconduct in allowing an employee to use a machine without proper instruction. *(Law v. Dartt* (1952) 109 Cal. App. 2d 508, 509, 240 P.2d 1013).

The reason for the foregoing rule seems obvious. It is not uncommon for an employer to "put his mind" to the existence of a danger to an employee and nevertheless fail to take corrective actions. (See, e.g., *Rogers Materials Co. V. Ind. Acc. Com., supra,* 63 Cal. 2d 717, 723, 48 Cal. Rptr. 129, 408 P.2d 737.) In many of these cases, the employer does not warn the employee of the risk. Such conduct may be characterized as intentional or even deceitful. Yet if an action at law were allowed as a remedy, many cases cognizable under workers' compensation would also be prosecuted outside that system. The focus of the inquiry in a case involving work-related injury would often be not whether the injury arose out of and in the course of employment, but the state of knowledge of the employer and the employee regarding the dangerous condition which caused the injury. Such a result would undermine the underlying premise upon which the workers' compensation system is based. That system balances the advantage to the employer of immunity from liability at law against the detriment of relatively swift and certain compensation payments. Conversely, while the employee receives expeditious compensation, he surrenders his right to a potentially larger recovery in a common law action for the negligence or willful misconduct of his employer. This balance would be significantly disturbed if we were to hold, as plaintiff urges, that any misconduct of an employer which may be characterized as intentional warrants an action at law for damages. It seems clear that section 4553 is the sole remedy for additional compensation against an employer whose employee is injured in the first instance as the result of a deliberate failure to assure that the physical environment of the work place is safe.

Thus, if the complaint alleged only that plaintiff contracted the disease because defendant knew and concealed from him that his health was endangered by asbestos in the work environment, failed to supply adequate protective devices to avoid disease, and violated governmental regulations relating to dust levels at the plant, plaintiff's only remedy would be to prosecute his claim under the workers' compensation law.

* * *

In the present case, plaintiff alleges that defendant fraudulently concealed from him, and from doctors retained to treat him, as well as from the state, that he was suffering from a disease caused by ingestion of asbestos, thereby preventing him from receiving treatment for the disease and inducing him to continue to work under hazardous conditions. These allegations are sufficient to state a cause of action for aggravation of the disease, as distinct from the hazards of the employment which caused him to contract the disease.

* * *

We conclude the policy of exclusivity of workers' compensation as a remedy for injuries in the employment would not be seriously undermined by holding defendant liable for the aggravation of this plaintiff's injuries, since we cannot believe that many employers will aggravate the effects of an industrial injury by not only deliberately concealing its existence but also its connection with the employment. Nor can we believe that the Legislature in enacting the workers' compensation law intended to insulate such flagrant conduct from tort liability. Finally, although plaintiff filed an application for workers' compensation and may receive an award in that proceeding, double recovery may be avoided by allowing the employer a set-off in the event plaintiff is awarded compensation for the aggravation of his injury in that proceeding and in the present case as well.

BIRD, C.J., TOBRINER, Manuel and NEWMAN, JJ., concur.

CLARK, Justice, dissenting.

I dissent.

The net effect of the majority opinion is to discourage employers from engaging in medical programs designed to minimize the risk and effects of occupational disease. By imposing on the employer tort liability for compensatory and punitive damages in addition to liability for workers' compensation benefits, today's decision can only deter employer initiation and maintenance of such programs. Moreover, sound public policy expressed in constitutional and statutory provisions make[s] clear that imposition of such tort liability in favor of a worker covered by compensation benefits is not permitted unless there is specific constitutional or statutory exemption.

The majority do not even attempt to find language permitting such exception.

The portions of the complaint held by the majority to state a cause of action may be summarized as follows: Aware since 1924 of the dangers of working with asbestos, Johns-Manville retained physicians and surgeons to examine plaintiff and other of its employees. Johns-Manville failed to advise the retained physicians "of the development of chest pathology and/or pulmonary disease in plaintiff and other employees or that such condition was the result of working conditions at said defendant's plant." The foregoing were fraudulent acts in that Johns-Manville intended to induce plaintiff to continue working in a dangerous environment. Had plaintiff been aware of his condition, he would have terminated his employment, avoiding further injury.

PUBLIC POLICY

There are a number of valuable and necessary industries in our society that—unfortunately—involve substantial risk of occupational disease. Aware of these risks, many employers provide special medical programs designed to minimize the risk. Our workers' compensation law provides benefits for those who, despite such programs, become disabled by occupational disease. Ordinarily, those benefits are in lieu of employer liability in tort for damages.

Today's decision holds that when an employer adopts a special medical program to minimize the risk of an occupational disease, his potential exposure for liability for the disease will include liability in tort for compensatory and punitive damages in addition to the compensation benefits ordinarily available to injured workers. I am afraid that today's lesson will not be lost on employers: Should they try to remain ignorant of employees' health problems and thereby disengage themselves from special medical programs for fear of triggering today's tort liability?

The employers' potential loss and thus the effect of today's decision cannot be minimized. Losses from punitive awards may not be insured against, and tort liability insurers traditionally exempt claims for willful misconduct and for claims like that asserted by plaintiff where workers' compensation is available. [Citations omitted.]

The majority have drawn a classification between employers who do not engage in medical programs to discover occupational illnesses and thus will have their liability limited to workers' compensation, and employers who maintain such programs and thereby run the risk of tort liability in addition to compensation liability. If such a classification is warranted, does not public policy require the exposure to liability be reversed? From the employee point of view there is little reason to discriminate between victims of industrial injury based on concealment of the illness. Extent of disability should determine recovery in all cases.

*   *   *

Blankenship v. Cincinnati Milacron Chemicals

Ohio Supreme Court
69 Ohio St. 2d 608, 433 N.E. 2d, 572 (1982)

William B. BROWN, J.

The sole issue raised in this appeal is whether the trial court properly granted [defendant's] motion to dismiss [plaintiffs'] complaint on the grounds that an employee is barred by Section 35, Article II of the Ohio Constitution, and R.C. 4123.74 and 4123.741 from prosecuting an action at law for an intentional tort.

*   *   *

The primary focus of the dispute between the parties centers upon the question of whether the Workers' Compensation Act (R.C. 4123.35 *et seq.*) is intended to cover an intentional tort committed by employers against their employees. Section 35, Article II of the Ohio Constitution, serves as a basis for legislative enactments in the area of workers' compensation by providing, in pertinent part:

> For the purpose of providing compensation to workmen and their dependents, for death, injuries or occupational disease, occasioned in the course of such workmen's employment, laws may be passed establishing a state fund to be created by compulsory contribution thereto by employers, and administered by the state, determining the terms and conditions upon which payment shall be made therefrom. Such com-

pensation shall be in lieu of all other rights to compensation, or damages, for such death, injuries, or occupational disease, and any employer who pays the premium or compensation provided by law, passed in accordance herewith, shall not be liable to repond in damages at common law or by statute for such death, injuries or occupational disease.

The constitutional mandate has been implemented by R.C. 4123.74 which provides:

Employers who comply with section 4123.35 of the Revised Code shall not be liable to respond in damages at common law or by statute for any injury, or occupational disease, or bodily condition, *received or contracted by any employee in the course of or arising out of his employment* . . . whether or not such injury, occupational disease [or] bodily condition . . . is compensable under section 4123.01 to 4123.94, inclusive, of the Revised Code. (Emphasis added.)

Clearly, neither the relevant constitutional language nor the pertinent statutory language expressly extend the grant of immunity to actions alleging intentional tortious conduct by employers against their employees. The General Assembly, however, in enacting R.C. 4123.95, established a rule of construction which is clearly of assistance in determining the scope of employer immunity. This section provides that:

Sections 4123.01 to 4123.94, inclusive, of the Revised Code, shall be liberally construed in favor of employees and the dependents of deceased employees.

It is with this requirement in mind that we address the language in R.C. 4123.74. The emphasized language in R.C. 4123.74 quoted above, as was noted in *Delamotte v. Midland Ross* (1978), 64 Ohio App. 2d 159, 161, " . . . clearly limits the categories of injuries for which the employer is exempt from civil liability." By designating as compensable only those injuries " . . . received or contracted . . . in the course of or arising out of . . . employment . . . ," The General Assembly has expressly limited the scope of compensability. In so doing, the General Assembly surely did not intend to remove all remedies from the employee whose injury is not compensable under the Act. And, by its use of this phrase, the General Assembly has seemingly allowed the judiciary the freedom to determine what risks are incidental to employment in light of the humanitarian purposes which underlie the Act.

In this regard, this court further agrees with the *Delamotte* court that where an employee asserts in his complaint a claim for damages based on an intentional tort, " . . . the substance of the claim is not an 'injury . . . received or contracted by any employee in the course of or arising out of his employment' within the meaning of R.C. 4123.74. . . ." *Id.* No reasonable individual would equate intentional and unintentional conduct in terms of the degree of risk which faces an employee nor would such individual contemplate the risk of an intentional tort as a natural risk of employment.[8] Since an employer's intentional conduct does not arise out of employment, R.C. 4123.74 does not bestow upon employers immunity from civil liability for their intentional torts and an employee may resort to a civil suit for damages. Accord *Barley v. Harrison Manufacturing* (No. E-80-75, May 22, 1981), Sixth District Court of Appeals, unreported; *Pariseau v. Wedge Products, Inc.* (No. 43195, May 7, 1981), Eighth District Court of Appeals, unreported.[9]

This holding not only comports with constitutional and statutory requirements, but it is also consistent with the legislative goals which underlie the Workers' Compensation Act.

The workers' compensation system is based on the premise that an employer is protected from a suit for negligence in exchange for compliance with the Workers' Compensation Act. The Act operates as a balance of mutual compromise between the interests of the employer and the employee whereby employees relinquish their common law remedy and accept lower benefit levels coupled with the

[8]In *Toth v. Standard Oil Co.* (1953), 160 Ohio St. 1, 5, this court stated that "injury" as used in the Ohio Workers' Compensation Act, comprehends a physical or traumatic damage or harm, accidental in its character in the sense of being unforeseen, unexpected and unusual.

An intentional tort, then, is clearly not an "injury" arising out of the course of employment. This very point was recognized in *Boek. v. Wong Hing* (1930), 180 Minn. 470, 231 N.W. 233, wherein it was stated, at page 471, that it would be a "perversion of" the Workmen's Compensation Act's purpose to allow employers immunity from intentional torts. Indeed, it would be travesty on the use of the English language to allow someone who intentionally inflicts an injury on another to call the injury a work incident.

[9]See, also, *Bowk v. Wong Hing, supra; Antonio v. Hirsch* (1957), 163 N.Y. Supp. 2d 489, 3 A.D. 2d 939; *Cohen v. Lion Products Co.* (D.C. Mass. 1959), 177 F. Supp. 486; *Skelton v. W.T. Grant Co.* (C.A. 5, 1964), 331 F.2d 593; 2A Larson, Workmen's Compensation Law 13-1, Section 68.

greater assurance of recovery and employers give up their common law defenses and are protected from unlimited liability. But the protection afforded by the Act has always been for negligent acts and not for intentional tortious conduct.[11] Indeed, workers' compensation Acts were designed to improve the plight of the injured worker, and to hold that intentional torts are covered under the Act would be tantamount to encouraging such conduct, and this clearly cannot be reconciled with the motivating spirit and purpose of the Act.

It must also be remembered that the compensation scheme was specifically designed to provide less than full compensation for injured employees. Damages such as pain and suffering and loss of services on the part of a spouse are unavailable remedies to the injured employee. Punitive damages cannot be obtained. Yet, these damages are available to individuals who have been injured by intentional tortious conduct of third parties, and there is no legitimate reason why an employer should be able to escape from such damages simply because he committed an intentional tort against his employee.

In addition, one of the avowed purposes of the Act is to promote a safe and injury-free work environment. (R.C. 1101.11 and 4101.12.) Affording an employer immunity for his intentional behavior certainly would not promote such an environment, for an employer could commit intentional acts with impunity with the knowledge that, at the very most, his workers' compensation premiums may rise slightly.

Moreover, as this court noted in *State, ex rel. Crawford, v. Indus. Comm.* (1924), 110 Ohio St. 271, 274, workers' compensation "... is founded upon the principle of insurance...." An insurance policy does not protect the policy holder from the consequences of his intentional tortious act. Indeed, it would be against public policy to permit insurance against the intentional tort. See, generally, *Northwestern National Cas. Co. v. McNulty* (C.A. 5, 1962), 307 F.2d 432.

The use of the element of intent in workers' compensation is in no way novel. Under R.C. 4123.54, an employee is denied benefits when he inflicts an injury upon himself intentionally. This section thus illustrates that intent plays an important role in the determination of whether an injury is compensable.

* * *

CELEBREZZE, C.J., concurring.

I enthusiastically agree with the syllabus, content and conclusion of the majority opinion that workers injured by an employer's intentional or malicious action may sue the employer for common law damages and that the Workers' Compensation remedy is not the exclusive avenue open to employees seeking redress for injuries arising out of workplace hazards.

I am troubled by the language in the dissenting opinions that workers who are intentionally chemically poisoned on-the-job should not be able to recover damages from their employers because the elimination of health risks would cost too much money, thus decreasing the profits of corporations. I submit that anyone who believes that injuries or death from gases, fumes, impure air or dust should not be eliminated because a manufacturer will suffer a competitive disadvantage is an enemy of all workers. The dissenters position is one that I would expect to be championed by a 19th century "robber baron," not a justice of this court who is duty-bound to serve all the people of Ohio.

The minority opinions are remarkably insensitive to the particularly egregious behavior on the part of employers—fraudulently misrepresenting or concealing workplace hazards—which the state has an interest in preventing. Indeed, under the theory articulated by the dissenters, an employer could intentionally cause an employee to suffer disability or death and, yet, remain immune from tort liability.

* * *

CLIFFORD F. BROWN, J., concurring.

The decision in this case, reflected in the thorough analysis by Justice William B. Brown and the perspicacity displayed by Chief Justice Celebrezze in his concurring opinion, establishes that this court has not yet reached a state of institutionalized impotence. Rather, it retains vitality and an appreciation of the need for the growth of the law in the field of workers' compensation.

A decision other than that we reach today would not only reject the legislative mandate of liberal

---

[11] As was stated in *Mandolidis v. Elkins Industries, Inc.* (W. Va. 1978), 246 S.E. 2d 907, 913:

> The workmen's compensation system completely supplanted the common law tort system only with respect to *negligently* caused industrial accidents, and employers and employees gained certain advantages and lost certain rights they had heretofore enjoyed. Entrepreneurs were not given the right to carry on their enterprises without any regard to the life and limb of the participants in the endeavor and free from all common law liability. (Emphasis *sic*.)

construction in favor of claimants, it would be a display of judicial anemia and necrosis. All law is justice, and justice is law. It is the adoption and promotion of what is good, and the avoidance of evil. Our construction of the law complies with this definition of justice, while an opposite decision would constitute injustice.

Although the legal issue in this case is one of first impression for this court, the just result we reach merely adopts and amplifies the rule already recognized in Ohio in the well-reasoned opinion for a unanimous Sixth District Court of Appeals in *Delamotte v. Midland Ross* (1978), 64 Ohio App. 2d 159, and followed by the Sixth and Eighth District Courts of Appeal in *Barley v. Harrison Manufacturing* and *Pariseau v. Wedge Products, Inc.*, cited by the majority. The product of such experienced and well intentioned judicial minds of our appellate courts should not be given a short shrift.

Furthermore, our decision in this case should be considered in light of the fact that for occupational diseases incurred in the work place, such as appellants' chemical poisoning, there is no workers' compensation paid unless there is total disability. R.C. 4123.68 (BB) and (Y). *State, ex rel. Miller, v. Mead Corp.* (1979), 58 Ohio St. 2d 405.

This court has never yet ruled that an employer may intentionally harm an employee and remain immune to civil suit. Nor have we yet ruled that a fellow employee may intentionally harm another employee with such impunity. The legislature does not abrogate or change the common law, unless its statutory language is clear and concise and not subject to any other reasonable construction. *Triff v. National Bronze & Aluminum Foundry* (1939), 135 Ohio St. 191, 202. Accordingly, the common law right of action against an employer or fellow employee for intentional torts remains.

I note that the decision we reach today is attacked on the basis that "goods manufactured in this state will thereby suffer a competitive disadvantage, and a less hospitable climate is created to attract and maintain industry in this state." This is a scare tactic to create an illusion that industry will leave Ohio and establish itself in other states because Ohio does not grant immunity to employers who intentionally harm their employees. This tactic and illusion assumes without foundation that other states grant such immunity to employers who intentionally maim and kill their employees. Minnesota, New York, Massachusetts and West Virginia, as we have done here, do not grant such immunity. See footnotes 8, 9 and 11. The critics of our decision have not called attention to any other state that has granted such immunity to employers. Employers bent upon chemically poisoning their employees have no place to go in the United States to perform their perfidy with impunity.

* * *

KRUPANSKY, J., dissenting.

The majority opinion, while appearing on the surface to be a humanitarian gesture, in effect undermines the beneficent purposes for which the Ohio Workers' Compensation Act was created. I must, therefore, respectfully dissent.

In my opinion, the Ohio Constitution and the Revised Code mean precisely what they say: workers' compensation shall be *in lieu of all other rights to compensation, or damages* for . . . injuries, or occupational disease, and any employer who pays the premium . . . *shall not be liable to respond in damages at common law* or by statute for such . . . injuries or occupational disease." Section 35, Article II, Ohio Constitution. (Emphasis added.) Appellants have convinced the majority that "all" does not mean "all," but instead means "all, except for rights to compensation for intentional torts."

Nowhere in the language of the Constitution or the statute is there support for the conclusion that the workers' compensation system was designed to compensate employees solely for employer negligence. The majority reason neither the code nor the Constitution "expressly extend the grant of immunity to actions alleging intentional tortious conduct." However, neither is immunity expressly granted to actions alleging simple negligence. Does it follow, therefore, that a liberal construction of the Act calls also for an exception for simple negligence, and that we should allow double recovery for employer negligence since it was not expressly excluded under the Act? The majority claims, if intentional torts are covered under the Act then intentional torts are encouraged. If this is true, then are not negligence and industrial accidents similarly encouraged, since they are covered under the Act?

Such reasoning leads ultimately to releasing the floodgates to a whole vista of lawsuits, each claiming exceptions to the all-inclusive language of the Act. The majority opinion represents a foot in the door policy to encourage workers to sue their employers for damages in addition to compensation provided under the Act.

The intent of the General Assembly, in enacting R.C. 4123.35 was to eliminate *all damage suits* outside the Act for injury or disease arising out of employment, including suits based on intentional tort. If the General Assembly desired to create an exception for intentional misconduct it surely could have done so. R.C. 4123.54, which disallows recovery for employees whose injuries were "purposely self-inflicted," illustrates the legislature's awareness of the possibility of excepting intentional wrongdoing.

As the majority notes, workers' compensation does not provide full compensation for employees suffering from occupational diseases or injuries. The Act, however, does not differentiate between injury or disease caused negligently and injury or disease caused intentionally. If, as the majority concludes, "there is no legitimate reason why an employer should be able to escape from" providing full compensation for intentional torts, then there is likewise no reason to allow the employer to "escape" from paying full damages for simple negligence.

The majority's myopic approach disrupts the delicate balance struck by the Act between the interests of labor, management and the public and signals the erosion of a valuable system which has served its purpose of providing a common fund for the benefit of all workers.

While R.C. 4123.95 provides for a liberal construction of the Act, it must not be used as a panacea to justify reasoning which suffers from logical malnutrition. One of the long-range effects of permitting recovery in these types of cases is the additional costs that will ultimately have to be borne by the consumer through increased product prices. Goods manufactured in this state will thereby suffer a competitive disadvantage, and a less hospitable climate is created to attract and maintain industry in this state. Since industry provides jobs, the labor force has an interest in encouraging industry. Thus, while some workers may benefit from recovery against the employer for intentional torts in addition to collecting workers' compensation benefits, we would be ill-advised to engage in such irresponsibility for the benefit of a few at the detriment of so many.

To avoid such catastrophic results, I would therefore affirm the judgments of the lower courts.

HOLMES, J., concurs in the foregoing dissenting opinion.

◇ NOTES

1. Although many states have not yet addressed the issue, *Blankenship* currently is the exception rather than the rule. The courts have not been quick to expand the "intentional tort" exception to the workers' compensation bar on worker lawsuits. Some states have been unwilling to apply the exception unless the employer *actually intended* harm to employee. In these states, it is not enough that the employee's injury can be said to be a natural and foreseeable consequence of the employer's actions (as in the *Blankenship* "failure to warn" situation).

2. In general, the courts have had less trouble accepting the more limited rationale of *Rudkin,* with its focus on the exacerbation of occupational disease through the fraudulent concealment of the effects of occupational exposure.

3. An example of employer reaction to *Blankenship* and related cases can be found in the following excerpt from a 1982 *New York Times* article:

> "It used to be that anyone who was injured would get their workmens' compensation and that was that," said Sheila Birnbaum [a New York lawyer who represents corporate defendants in products liability suits]. "Now they get their workmens' comp and then they sue the employer on some other theory of law. Different state courts have used different legal theories, but taken all together the recent decisions are making some real inroads against the traditional employers' immunity shield. It is a situation that has the manufacturers very, very upset." Robert Gillespy, a Cleveland lawyer who filed a friend-of-the-court brief in the Blankenship case on behalf of the Ohio Chamber of Commerce and several large companies that think workmens' compensation should be

the exclusive remedy for injured workers, said he was expecting "a huge flood" of civil suits against employers.

[Tamar Lewin, Workers Are Filing Suit in Job Disability Cases, *The New York Times*, Thursday, April 29, 1982, p. D1.] Subsequent to *Blankenship* the Ohio legislature amended that state's workers' compensation law to narrow significantly the availability of this type of lawsuit. The courts have held that the retrospective application of this provision violates the Ohio constitution, see *Pratt* v. *National Distillers and Chemical Corp.*, 853 F.2d 1329 (6th Cir. 1987), and they are divided on the permissibility of its prospective application, see *Brady* v. *Safety Kleen Corp.*, 710 F. Supp. 684 (S.D. Ohio 1989), and cases cited therein.

4. Under common law, the supreme court of a state is free to change or expand tort law in order to serve social policy goals, and differences in social policy are very much at the heart of the dispute between the majority and dissent in *Blankenship*. In both *Rudkin* and *Blankenship*, however, the courts were first being asked to interpret the will of the legislature (by interpreting the state's workers' compensation statute). Do the judges appear to be chaffing a bit under this restriction? To what extent do you think that social policy concerns affected their interpretation of the statute? (Obviously, if the legislature disagrees, it can amend the statute, as was done in Ohio.)

5. What will be the effect on this line of cases of the various state and federal right to know laws?

6. In general, the exclusivity clause will not bar suits by the *children* of workers, who have *themselves* been injured as a result of the occupational exposures of a parent. Nor will it bar suits brought on behalf of a fetus that did not reach full term as a result of an occupational exposure to a parent. These suits may be brought on a negligence theory. See generally, Ashford and Caldart, The Control of Reproductive Hazards in the Workplace: A Prescription for Prevention, 5 *Indus. Rel. L.J.* 523, 552-560 (1983).

A related theory of employer liability is known as the "dual capacity" doctrine. In some states, where the employer acts in a secondary capacity, apart from the role of employer, the employee may have a cause of action in tort for workplace injuries that flow from that secondary capacity. According to one oft-cited statement of the doctrine, the employer must possess "a second persona so completely independent from and unrelated to his status as employer that by established standards the law recognizes it as a separate legal person." L.W. Larson, 2A *Larson's Workmen's Compensation Law* §72.81, 14-229 (1982). In this sense, then, the employer is treated as a third party to the employer/employee relationship and is thus considered to be outside the protection of the exclusivity clause. Some courts have interpreted this doctrine as imposing *manufacturer's* liability on employers who manufacture products that are used by their workers in the workplace. In general, such liability has been imposed only where the product is one that is also sold by the employer to parties outside the workplace. For example, in *Mercer v. Uniroyal Inc.*, 49 Ohio Ap. 2d 279, 361 N.E. 2d 492 (1976), the Ohio Court of Appeals allowed a Uniroyal employee who had been injured as a result of a defective Uniroyal tire during the course of employment to bring a tort suit directly against his employer. Similarly, in *Douglas v. E.J. Gallo Winery*, 169 Cal. App. 3d 103, 137 Cal. Rpt.

797 (1977), a Gallo employee was allowed to sue the company for injuries caused by scaffolding similar to that manufactured by the employer for the general public. Other courts have rejected this approach and have held that the mere fact that the employer also acts as a manufacturer is not enough to warrant an application of the dual capacity doctrine. See, e.g., *Longever v. Revere Copper & Brass, Inc.*, 108 N.E.2d 857 (Mass. 1980), and *Kottis v. United States Steel Corp.*, 543 F.2d 22 (7th Cir. 1976) (applying Indiana law). In general, the dual capacity doctrine has been rejected in more states than it has been adopted.

2. Suits against Manufacturers

As noted, workers' compensation provides no immunity to *third-party* manufacturers. In general, the worker may sue the manufacturer under a negligence theory, or under a theory of strict products liability. Proof of negligence may be difficult if the manufacturer's methods of production comport with industry standards. To succeed in a negligence suit the worker may well find it necessary to advocate a common law standard of care that is more stringent than industry practice would dictate. If the elements of strict liability can be established, however, negligence need not be proven. Although the principle of strict products liability has been adopted in some form by the courts of all 50 states, the specific criteria for its application tend to vary somewhat from jurisdiction to jurisdiction. Perhaps the best general statement of the doctrine is found in Section 402A of the Restatement (Second) of Torts (1965):

> (1) One who sells any product in a defective condition unreasonably dangerous to the user or consumer . . . is subject to liability for the physical harm thereby caused to the ultimate user or consumer . . . if
> (a) the seller engaged in the business of selling such a product, and
> (b) it is expected to and does reach the user or consumer without substantial change in the condition in which it is sold.
> (2) The rule stated in Subsection (1) applies although
> (a) the seller has exercised all possible care in the preparation and sale of [the] product.

The key phrase here is "defective condition unreasonably dangerous." The meaning of these terms is explored in the following two decisions.

Barker v. Lull Engineering Co.

California Supreme Court
20 Cal. 3d 413, 143 Cal.
Rptr. 225, 573 P.2d 443 (1978)
JOBRINER, Acting C. J.

As this court has recognized on numerous occasions, the term defect as utilized in the strict liability context is neither self-defining nor susceptible to a single definition applicable in all contexts.

* * *

Resort to the numerous product liability precedents in California demonstrates that the defect or defectiveness concept has embraced a great variety of injury-producing deficiencies, ranging from products that cause injury because they deviate from the manufacturer's intended result (e.g., the one soda bottle in ten thousand that explodes without explanation (*Escola v. Coca Cola Bottling Co.* (1944) 24 Cal. 2d 453)), to products which, though "perfectly" manufactured, are unsafe because of the absence of a safety device (e.g., a paydozer without rear view mirrors (*Pike v. Frank G. Hough Co., supra*, 2 Cal. 3d 465)), and including products that are dangerous because they lack adequate warnings or instructions (e.g., a telescope that contains inadequate instructions for assembling a "sun filter" attachment (*Midgley v. S.S. Kresge Co.* (1976) Cal. App. 3d 67).

* * *

In general, a manufacturing or production defect is readily identifiable because a defective product is one that differs from the manufacturer's intended result or from other ostensibly identical units of the same product line. For example, when a product comes off the assembly line in a substandard condition it has incurred a manufacturing defect. (E.g., *Lewis v. American Hoist & Derrick Co.* (1971) 20 Cal. App. 3d 570, 580). A design defect, by contrast, cannot be identified simply by comparing the injury-producing product with the manufacturer's plans or with other units of the same product line, since by definition the plans and all such units will reflect the same design. Rather than applying any sort of deviation-from-the-norm test in determining whether a product is defective in design for strict liability purposes, our cases have employed two alternative criteria in ascertaining, in Justice Traynor's words, whether there is something "wrong, if not in the manufacturer's manner of production, at least in his product." (Traynor, The Ways and Meanings of Defective Products and Strict Liability, 32 *Tenn. L. Rev.* 363, 366.)

First, our cases establish that a product may be found defective in design if the plaintiff demonstrates that the product failed to perform as safely as an ordinary consumer would expect when used in an intended or reasonably foreseeable manner. This initial standard, somewhat analogous to the Uniform Commercial Code's warranty of fitness and merchantability (Cal. U. Com. Code, §2314), reflects the warranty heritage upon which California product liability doctrine in part rests. As we noted in *Greenman*, "implicit in [a product's] presence on the market . . . [is] a representation that it [will] safely do

the jobs for which it was built." (59 Cal. 2d at p.64.) When a product fails to satisfy such ordinary consumer expectations as to safety in its intended or reasonably foreseeable operation, a manufacturer is strictly liable for resulting injuries. [Citations omitted.] Under this standard, an injured plaintiff will frequently be able to demonstrate the defectiveness of a product by resort to circumstantial evidence, even when the accident itself precludes identification of the specific defect at fault. [Citations omitted.]

As Professor Wade has pointed out, however, the expectation of the ordinary consumer cannot be viewed as the exclusive yardstick for evaluating design defectiveness because "[i]n many situations . . . the consumer would not know what to expect, because he would have no idea how safe the product could be made." (Wade, On the Nature of Strict Tort Liability for Products, supra 44 *Miss. L.J.* 825, 829.) Numerous California decisions have implicitly recognized this fact and have made clear, through varying linguistic formulations, that a product may be found defective in design, even if it satisfies ordinary consumer expectations, if through hindsight the jury determines that the product's design embodies "excessive preventable danger," or, in other words, if the jury finds that the risk of danger inherent in the challenged design outweighs the benefits of such design. [Citations omitted.]

A review of past cases indicates that in evaluating the adequacy of a product's design pursuant to this latter standard, a jury may consider, among other relevant factors, the gravity of the danger posed by the challenged design, the likelihood that such danger would occur, the mechanical feasibility of a safer alternative design, the financial cost of an improved design, and the adverse consequences to the product and to the consumer that would result from an alternative design.

* * *

Beshada v. Johns-Manville Products Corporation

Supreme Court of New Jersey
90 N.J. 191, 447 A.2d
539 (1982)

PASHMAN, J.

The sole question is whether defendants in a product liability case based on strict liability for failure to warn may raise a "state of the art" defense. Defen-

dants assert that the danger of which they failed to warn was undiscovered at the time the product was marketed and that it was undiscoverable given the state of scientific knowledge at that time. The case comes to us on appeal from the trial court's denial of plaintiffs' motion to strike the state-of-the-art defense. For the reasons stated below, we reverse the trial court judgment and strike the defense.

I

These six consolidated cases are personal injury and wrongful death actions brought against manufacturers and distributors of asbestos products. Plaintiffs are workers, or survivors of deceased workers, who claim to have been exposed to asbestos for varying periods of time. They allege that as a result of that exposure they contracted asbestosis (a non-malignant scarring of the lungs), mesothelioma (a rare cancer of the lining of the chest, the pleura, or the lining of the abdomen, the peritoneum) and other asbestos-related illnesses.

These cases involve asbestos exposure dating back perhaps as far as the 1930's. The suits are first arising now because of the long latent period between exposure and the discernible symptoms of asbestosis and mesothelioma. See *Borel v. Fibreboard Paper Products Corporation*, 493 F.2d 1076, 1083 (5th Cir. 1973). Plaintiffs have raised a variety of legal theories to support their materials in the course of their work. They allege that they were given no warning, handling instructions or safety equipment to protect them from the dangers of asbestos.

*   *   *

III

Our inquiry starts with the principles laid down in *Freund v. Cellofilm Properties, Inc.*, 87 N.J. 229 (1981), *Suter v. San Angelo Foundry & Machine Company*, 81 N.J. 150 (1979) and *Cepeda v. Cumberland Engineering Company, Inc.*, 76 N.J. 152 (1978). In *Suter*, we summarized the principle of strict liability as follows:

> If at the time the seller distributes a product, it is not reasonably fit, suitable and safe for its intended or reasonably foreseeable purposes so that users or others who may be expected to come in contact with the product are injured as a result thereof, then the seller shall be responsible for the ensuing damages. [*Id.* at 169 (footnote omitted).]

The determination of whether a product is "reasonably fit, suitable and safe" depends on a comparison of its risks and its utility (risk-utility equation).

> Central to this theory is the risk-utility equation for determining liability. The theory is that only safe products should be marketed—a safe product being one whose utility outweighs its inherent risk, provided that the risk has been reduced to the greatest extent possible consistent with the product's continued utility. [*Freund*, 87 N.J. at 238, n.1.]

In *Cepeda*, we explained that in the context of design defect liability, strict liability is identical to liability for negligence, with one important caveat: "The only qualification is as to the requisite of foreseeability by the manufacturer of the dangerous propensity of the chattel manifested at the trial—this being imputed to the manufacturer." *Cepeda*, 76 N.J. at 172. *See Freund v. Cellofilm Properties, Inc.*, 87 N.J. at 239. In so holding, we adopted the explication of strict liability offered by Dean Wade:

> The time has now come to be forthright in using a tort way of thinking and tort terminology [in cases of strict liability in tort]. There are several ways of doing it, and it is not difficult. The simplest and easiest way, it would seem, is to assume that the defendant knew of the dangerous condition of the product and ask whether he was then negligent in putting it on the market or supplying it to someone else. In other words, the scienter is supplied as a matter of law, and there is no need for the plaintiff to prove its existence as a matter of fact. Once given this notice of the dangerous condition of the chattel, the question then becomes whether the defendant was negligent to people who might be harmed by that condition if they came into contact with it or were in the vicinity of it. Another way of saying this is to ask whether the magnitude of the risk created by the dangerous condition of the product was outweighed by the social utility attained by putting it out in this fashion. [Wade, On the Nature of Strict Tort Liability for Products, 44 *Minn. L.J.* 825, 834-35 (1973), quoted in *Cepeda*, 76 N.J. at 172.]

Stated differently, negligence is conduct-oriented, asking whether defendant's actions were reasonable; strict liability is product-oriented, asking whether the product was reasonably safe for its foreseeable purposes. *Freund*, 87 N.J. at 238.

"Warning" cases constitute one category of strict liability cases. Their relation to the strict liability principles set forth above can best be analyzed by focusing on the definition of safe products found in footnote 1 of *Freund*. *See supra* at 200. For purposes of analysis, we can distinguish two tests for determining whether a product is safe: (1) does its utility outweigh its risk? and (2) if so, has that risk been reduced to the greatest extent possible consistent with the product's utility? *Id.* at 238, n.1. The first question looks to the product as it was in fact marketed. If that product caused more harm than good, it was not reasonably fit for its intended purposes. We can therefore impose strict liability for the injuries it caused without having to determine whether it could have been rendered safer. The second aspect of strict liability, however, requires that the risk from the product be reduced to the greatest extent possible without hindering its utility. Whether or not the product passes the initial risk-utility test, it is not reasonably safe if the same product could have been made or marketed more safely.

Warning cases are of this second type. When plaintiffs urge that a product is hazardous because it lacks a warning, they typically look to the second test, saying in effect that regardless of the overall cost-benefit calculation the product is unsafe because a warning could make it safer at virtually no added cost and without limiting its utility. *Freund* recognized this, noting that in cases alleging "an inadequate warning as to safe use, the utility of the product, as counter-balanced against the risks of its use, is rarely at issue." *Id.* at 242.

Freund is our leading case on strict liability for failure to warn. In *Freund*, Justice Handler applied the principles set forth above, initially laid down in *Suter* and *Cepeda,* to warning cases. The issue there was whether there is any difference between negligence and strict liability in warning cases. We stated unequivocally that there is. That difference is the same difference that we noted in *Suter* and *Cepeda* concerning other design defect cases:

> when a plaintiff sues under strict liability, there is no need to prove that the manufacturer knew or should have known of any dangerous propensities of its product—such knowledge is imputed to the manufacturer. [*Freund v. Cellofilm Properties, Inc.,* 87 N.J. at 239.]

Thus, we held in *Freund* that it was reversible error for the trial judge to instruct the jury only with a negligence charge.

With these basic principles of design defect strict liability in New Jersey as our framework for analysis, we turn now to a discussion of the state-of-the-art defense.

IV

As it relates to warning cases, the state-of-the-art defense asserts that distributors of products can be held liable only for injuries resulting from dangers that were scientifically discoverable at the time the product was distributed. Defendants argue that the question of whether the product can be made safer must be limited to consideration of the available technology at the time the product was distributed. Liability would be absolute, defendants argue, if it

could be imposed on the basis of a subsequently discovered means to make the product safer since technology will always be developing new ways to make products safer. Such a rule, they assert, would make manufacturers liable whenever their products cause harm, whether or not they are reasonably fit for their foreseeable purposes.

Defendants conceptualize the scientific unknowability of the dangerous propensities of a product as a technological barrier to making the product safer by providing warnings. Thus, a warning was not "possible" within the meaning of the *Freund* requirement that risk be reduced "to the greatest extent possible."

In urging this position, defendants must somehow distinguish the *Freund* holding that knowledge of the dangers of the product is imputed to defendants as a matter of law. A state-of-the-art defense would contravene that by requiring plaintiffs to prove at least that knowledge of the dangers was scientifically available at the time of manufacture.

Defendants argue that *Freund* did not specify precisely what knowledge is imputed to defendants. They construe *Freund* to impute only that degree of knowledge of the product's dangerousness that existed at the time of manufacture or distribution.[6]

---

[6]Defendants seek support in this regard from the following passage in *Suter:*

> Depending upon the proofs, some factors which may be considered by the jury in deciding the reasonableness of the manufacturer's conduct include the technological feasibility of manufacturing a product whose design would have prevented or avoided the accident, given the known state of the art; and the likelihood that the product will cause injury and the probable seriousness of the injury. We observe in passing that the state of the art refers not only to the common practice and standards in the industry but also to other design alternatives within practical and technological limitations at the time of distribution. [81 N.J. 171 72 (citation omitted).]

Defendants argue on the basis of this passage that *Suter* has explicitly accepted the state-of-the-art defense, and that *Freund* should be construed to be consistent with that reading of *Suter.*

We note, first, that this passage in *Suter* is dictum. *Suter* did not involve a state-of-the-art defense. The situation addressed in the case is not one in which the state of the art has improved between the time the product was distributed and the time of trial. Second, *Suter,* unlike *Freund* and this case, was not a warning case. The passage from *Suter* suggests only that state-of-the-art may be relevant to a safety device case. . . .

We hold in this case that a state-of-the-art defense should not be allowed in failure to warn cases. There are

While we agree that *Freund* did not explicitly address this question, the principles laid down in *Freund* and our prior cases contradict defendants' position. Essentially, state-of-the-art is a negligence defense. It seeks to explain why defendants are not culpable for failing to provide a warning. They assert, in effect, that because they could not have known the product was dangerous, they acted reasonably in marketing it without a warning. But in strict liability cases, culpability is irrelevant. The product was unsafe. That it was unsafe because of the state of technology does not change the fact that it was unsafe. Strict liability focuses on the product, not the fault of the manufacturer. "If the conduct is unreasonably dangerous, then there should be strict liability without reference to what excuse defendant might give for being unaware of the danger." Keeton, 48 *Tex. L. Rev.* at 408.

When the defendants argue that it is unreasonable to impose a duty on them to warn of the unknowable, they misconstrue both the purpose and effect of strict liability. By imposing strict liability, we are not requiring defendants to have done something that is impossible. In this sense, the phrase "duty to warn" is misleading. It implies negligence concepts with their attendant focus on the reasonableness of defendant's behavior. However, a major concern of strict liability—ignored by defendants—is the conclusion that if a product was in fact defective, the distributor of the product should compensate its victims for the misfortune that it inflicted on them.

If we accepted defendant's argument, we would create a distinction among fact situations that defies common sense. Under the defendants' reading of *Freund*, defendant would be liable for failure to warn if the danger was knowable even if defendants were not negligent in failing to discover it. Defendants would suffer no liability, however, if the danger was undiscoverable. But, as Dean Keeton explains,

> if a defendant is to be held liable for a risk that is discoverable by some genius but beyond the defendant's capacity to do so, why should he not also be liable for a risk that was just as great but was not discoverable by anyone? [Keeton, 48 *Tex. L. Rev.* at 409.]

We are buttressed in our conclusion that the state-of-the-art defense is inconsistent with *Freund* by the

strong conceptual similarities between warning and safety device cases. . . . Thus, the reasoning applied here may be equally applicable to cases involving failure to include appropriate safety devices in products. However, there may be distinctions between the two situations. We specifically decline to address whether a state-of-the-art defense is appropriate to safety device cases.

recent decision of Judge Ackerman in *Marcucci v. Johns-Manville Sales Corp.*, Nos. 76-414, 76-604 and 76-1510 (D.N.J. Feb. 19, 1982), in which he applied New Jersey law to strike defendant's state-of-the-art defense.

The most important inquiry, however, is whether imposition of liability for failure to warn of dangers which were undiscoverable at the time of manufacture will advance the goals and policies sought to be achieved by our strict liability rules. We believe that it will.

Risk Spreading. One of the most important arguments generally advanced for imposing strict liability is that the manufacturers and distributors of defective products can best allocate the costs of the injuries resulting from those products. The premise is that the price of a product should reflect all of its costs, including the cost of injuries caused by the product. This can best be accomplished by imposing liability on the manufacturer and distributors. Those persons can insure against liability and incorporate the cost of the insurance in the price of the product. In this way, the costs of the product will be borne by those who profit from it: the manufacturers and distributors who profit from its sale and the buyers who profit from its use. "It should be a cost of doing business that in the course of doing that business an unreasonable risk was created." Keeton, 48 *Tex. L. Rev.* at 408. *See* Prosser, *The Law of Torts*, §75, p.495 (4th ed. 1971).

Defendants argue that this policy is not forwarded by imposition of liability for unknowable hazards. Since such hazards by definition are not predicted, the price of the hazardous product will not be adjusted to reflect the costs of the injuries it will produce. Rather, defendants state, the cost "will be borne by the public at large and reflected in a general, across the board increase in premiums to compensate for unanticipated risks." There is some truth in this assertion, but it is not a bad result.

First, the same argument can be made as to hazards which are deemed scientifically knowable but of which the manufacturers are unaware. Yet is is well established under our tort law that strict liability is imposed even for defects which were unknown to the manufacturer. It is precisely the imputation of knowledge to the defendant that distinguishes strict liability from negligence. *Freund*, 87 N.J. at 240. Defendants advance no argument as to why risk spreading works better for unknown risks than for unknowable risks.

Second, spreading the costs of injuries among all those who produce, distribute and purchase manufactured products is far preferable to imposing it on the innocent victims who suffer illnesses and disability from defective products. This basic normative premise is at the center of our strict liability rules. It is unchanged by the state of scientific knowledge at the time of manufacture.

Finally, contrary to defendants' assertion, this rule will not cause the price and production level of manufactured products to diverge from the so-called economically efficient level. Rather, the rule will force the price of any particular product to reflect the cost of insuring against the possibility that the product will turn out to be defective.

Accident Avoidance. In *Suter*, we stated:

> Strict liability in a sense is but an attempt to minimize the costs of accidents and to consider who should bear those costs. *See* the discussion in Calabresi & Hirschoff, Toward a Test for Strict Liability in Torts, 81 *Yale L.J.* 1055 (1972), in which the authors suggest that the strict liability issue is to decide which party is the "cheapest cost avoider" or who is in the best position to make the cost-benefit analysis between accident costs and accident avoidance costs and to act on that decision once it is made. *Id.* at 1060. Using this approach, it is obvious that the manufacturer rather than the factory employee is "in the better position both to judge whether avoidance costs would exceed foreseeable accident costs and to act on that judgment." *Id.* [*Suter v. San Angelo Foundry*, 81 N.J at 173-74.]

Defendants urge that this argument has no force as to hazards which by definition were undiscoverable. Defendants have treated the level of technological knowledge at a given time as an independent variable not affected by defendants' conduct. But this view ignores the important role of industry in product safety research. The "state-of-the-art" at a given time is partly determined by how much industry invests in safety research. By imposing on manufacturers the costs of failure to discover hazards, we create an incentive for them to invest more actively in safety research.

Fact finding process. The analysis thus far has assumed that it is possible to define what constitutes "undiscoverable" knowledge and that it will be reasonably possible to determine what knowledge was technologically discoverable at a given time. In fact, both assumptions are highly questionable. The vast confusion that is virtually certain to arise from any attempt to deal in a trial setting with the concept of scientific knowability constitutes a strong reason for avoiding the concept altogether by striking the state-of-the-art defense.

Scientific knowability, as we understand it, refers not to what in fact was known at the time, but to what *could have been* known at the time. In other words, even if no scientist had actually formed the belief that asbestos was dangerous, the hazards would be deemed "knowable" if a scientist could have formed that belief by applying research or performing tests that were available at the time. Proof of what could have been known will inevitably be complicated, costly, confusing and time-consuming. Each side will have to produce experts in the history of science and technology to speculate as to what knowledge was feasible in a given year. We doubt that juries will be capable of even understanding the concept of scientific knowability, much less be able to resolve such a complex issue. Moreover, we should resist legal rules that will so greatly add to the costs both sides incur in trying a case.

The concept of knowability is complicated further by the fact, noted above, that the level of investment in safety research by manufacturers is one determinant of the state-of-the-art at any given time. Fairness suggests that manufacturers not be excused from liability because their prior inadequate investment in safety rendered the hazards of their product unknowable. Thus, a judgment will have to be made as to whether defendants' investment in safety research in the years preceding distribution of the product was adequate. If not, the experts in the history of technology will have to testify as to what would have been knowable at the time of distribution if manufacturers had spent the proper amount on safety in prior years. To state the issue is to fully understand the great difficulties it would engender in a courtroom.

In addition, discussion of state-of-the-art could easily confuse juries into believing that blameworthiness is at issue. Juries might mistakenly translate the confused concept of state-of-the-art into the simple question of whether it was defendants' fault that they did not know of the hazards of asbestos. But that would be negligence, not strict liability.

For precisely this reason, Professor Keeton has urged that negligence concepts be carefully avoided in strict liability cases.

> My principal thesis is and has been that theories of negligence should be avoided altogether in the products liability area in order to simplify the law, and that if the sale of a product is made under circumstances that would subject someone to an unreasonable risk in fact, liability for

harm resulting from those risks should follow. [Keeton, 48 *Tex. L. Rev.* at 409 (footnote omitted).]

This Court has expressed the same concern in *Freund,* reversing the trial court's jury charge because the "terminology employed by the trial judge was riddled with references to negligence, knowledge and reasonable care on the part of a manufacturer." 87 N.J. at 243. "[W]e must be concerned with the effect of the trial judge's articulation upon the jury's deliberative processes." *Id.* at 244.

V

For the reasons expressed above, we conclude that plaintiffs' position is consistent with our holding in *Freund* and prior cases and will achieve the various policies underlying strict liability. The burden of illness from dangerous products such as asbestos should be placed upon those who profit from its production and, more generally, upon society at large, which reaps the benefits of the various products our economy manufactures. That burden should not be imposed exclusively on the innocent victim.

Although victims must in any case suffer the pain involved, they should be spared the burdensome financial consequences of unfit products. At the same time, we believe this position will serve the salutary goals of increasing product safety research and simplifying tort trials.

Defendants have argued that it is unreasonable to impose a duty on them to warn of the unknowable. Failure to warn of a risk which one could not have known existed is not unreasonable conduct. But this argument is based on negligence principles. We are not saying that defendants' products were not reasonably safe because they did not have a warning. Without a warning, users of the product were unaware of its hazards and could not protect themselves from injury. We impose strict liability because it is unfair for the distributors of a defective product not to compensate its victims. As between those innocent victims and the distributors, it is the distributors—and the public which consumes their products—which should bear the unforeseen costs of the product.

The judgment of the trial court is reversed; the plaintiff's motion to strike the state-of-the-art defense is granted.

◇ NOTES

1. The *Beshada* case provides an excellent discussion of the rationale behind, and the precepts of, strict products liability. Note that one of the goals of strict liability is to encourage the development of safer technology. A court-imposed negligence standard may also incorporate a "technology-forcing" goal of this nature. As noted by Judge Learned Hand in his influential *T.J. Hooper* decision, one of the functions of negligence law is to measure an industry's technological performance against its technological potential.

> [I]n most cases reasonable prudence is in fact common prudence; but strictly it is never the measure; a whole calling may have unduly lagged in the adoption of new and available devices. . . . *Courts* must in the end say what is required; there are precautions so imperative that even their universal disregard will not excuse their omission.

[*The T.J. Hooper,* 60 F.2d 737, 740 (2d Cir. 1932) (emphasis supplied).] To the extent that tort law imposes design requirements that are stricter than those currently in use—or to the extent that industry *perceives* that it will do so—it has the potential to encourage technological innovation or adaptation. Of course, if the state legislature —or Congress—believes that courts have gone to far in setting standards for industry performance in a particular field, it can always pass a law modifying (or nullifying) the common law.

2. Although some state courts have followed the lead of the *Beshada* decision on the availability of the state-of-the-art defense, most do not treat "failure to warn" cases under strict liability. More often, failure to warn cases are treated under negligence principles, and the state-of-the-art defense is available. Dean Keeton, on whom the *Beshada* court relies for conceptual support, subsequently changed his mind on the appropriateness of the state-of-the-art defense. *See*: W.P. Keeton, D. Dobbs, R. Keeton, and D. Owen, *Prosser and Keeton on the Law of Torts*, 697-698, n.21 (5th ed. 1984). As a matter of social policy, do you think the *Beshada* court was wise to make this a strict liability situation? What kind of incentives does it create? Would we want to have it applied to, say, the development of a cure for AIDS? In a later opinion, the New Jersey Supreme Court held that state of the art *is* a valid defense in failure to warn cases involving the sale of pharmaceuticals, although it placed the burden on the manufacturer to prove what the "state of the art" *was* at the time of sale. *Feldman v. Lederle Laboratories*, 97 N.J. 429, 479 A.2d 374 (1984).

3. In *Barker v. Lull* (in a part of the opinion not reprinted above) the California Supreme Court announced it was rejecting the strict liability formulation of Section 402(A) of the *Restatement* to the extent that it requires plaintiffs in a strict liability action to prove both that a product was "defective" and "unreasonably dangerous." It is enough, held the Court, that they prove that the product was defective. Compare the definition of "reasonably safe" (also stated as "*not* unreasonably dangerous") in *Bashada* with the second definition of "design defect" enunciated in *Barker*. Is there a meaningful difference? A more significant difference is that *Barker* (again in a part of the opinion not reprinted here) placed the burden of proof on the manufacturer to prove that the benefits of the product outweighed the risks. This position has not been widely followed. See *Knitz v. Minister Machine Co.*, 69 Ohio St. 2d 460, 432 N.E. 2d 814 (1982).

4. Whether the suit is based on negligence or strict liability, the worker must establish that he or she was a "user" of the product in question and that such use of the product was reasonably forseeable at the time it was distributed. In the average case, this will not be a difficult burden. It was reasonably forseeable, for example, that construction workers would "use" asbestos when they were putting insulation into buildings. However, a federal district court in Texas has held that lead smelter workers poisoned by lead taken from automobile batteries could not sue the battery manufacturers, because the workers were not "users" of the batteries. Instead, the court held, they were users of the lead *once it was removed from the batteries*, and the lead itself was not the "product" sold by the manufacturers. See *Johnson v. Murph Metals, Inc.*, 562 F. Supp. 246 (N.D. Tex. 1983).

5. Where the defendant is a chemical manufacturer, breach of an applicable TSCA, OSHA, or Consumer Product Safety Act regulation may well be persuasive evidence of negligence, and could establish negligence as a matter of law. See Heaton, Allar and Maier, The Uses of Regulatory Evidence in Tort Actions: Automobiles, Consumer Products, and Occupational Safety and Heallth, 4*J. Prod. Liab.* 231 (1981).

6. There are defenses to strict products liability. As strict liability is not a cause of action for negligence, many courts have held that the *contributory negligence* of the plaintiff is not a defense. See *McCown v. International Harvester Co.*, 663 Pa. 13, 342 A.2d 381 (1975). However, there appears to be a trend toward reducing the plaintiff's award in proportion to the plaintiff's negligence, under the concept of comparative negligence. See *Daly v. General Motors*, 20 Cal. 3d 725, 575 P.2d 1162

(1978). Assumption of the risk—which requires proof that the plaintiff knew of, appreciated, and voluntarily exposed him- or herself to the danger—is generally held to be a defense to strict products liability. Restatement (Second) of Torts, §402A, Comment n (1965). As noted by the Seventh Circuit Court of Appeals, "one who stands under the guillotine blade with knowledge of the fact that there will be a descent of the operational part upon activation of the controlling lever would scarcely seem to be in a position to claim that the blade fell faster than he anticipated." *Moran v. Raymond Corp.*, 484 F.2d 1008 (7th Cir. 1973), cert. denied, 415 U.S. 932 (1974). Some state legislatures have passed laws codifying, modifying, or adding to these common law defenses.

7. Assumption of risk is not always a clear-cut doctrine, and its use as a defense may not always be allowed. As noted by the Court of Appeals of the State of Washington, the policy implications of the defense may be at odds with the policy goals of strict products liability:

> It seems to us that a rule which excludes the manufacturer from liability if the defect of the design of his product is patent but applies the duty if such a defect is latent is somewhat anomalous. The manufacturer of the obviously defective product ought not to escape because the product was obviously a bad one. The law, we think, ought to discourage misdesign rather than encouraging it in its obvious form."

[*Palmer v. Massey-Ferguson, Inc.*, 3 Wash. App. 508, 517, 476 P.2d 713, 718-719 (1970).] See also *Micallef v. Miehle Co.*, 39 N.Y.S.2d 376, 384 N.Y.S.2d 115 (1976).

8. Where the workplace itself is especially dangerous, the degree of caution expected of the worker may be lessened. Further, the *employer's* knowledge of the danger will not be imputed to the worker. See W.P. Keeton, D. Owen, J. Montgomery, and M. Green, *Products Liability and Safety* 533 (2d ed. 1989)

3.   Suits against Other Third Parties

Depending on the particular facts of the case and the law of the particular state, a wide variety of other parties may find themselves subject to a lawsuit for workplace injury or disease.

(a) The government

Although a worker may well have a good argument that a particular government regulation—such as an OSHA standard—was set too leniently, the government generally is not liable in tort for the policy choices it makes in promulgating regulation. Under the doctrine of *sovereign immunity* incorporated into the Eleventh Amendment to the United States Constitution, the federal government is not liable unless it *consents* to be. Such consent has been given in the Federal Tort Claims Act, which provides that the United States "shall be liable . . . in the same manner and to the same extent as a private individual under like circumstances," 28 U.S.C. §2674, but *not* where the governmental act or omission constitutes "a discretionary function or duty on the part of a federal agency or an employee of the government, whether or not the discretion involved be abused." 28 U.S.C. §2680.

Cunningham v. United States

786 F.2d 1445 (9th Cir. 1986)

SKOPIL, Circuit Judge.

Georgia Cunningham appeals from a district court's dismissal of her cause of action brought under the Federal Torts Claims Act, 28 U.S.C. §2674 ("FTCA"). We affirm.

FACTS AND PROCEEDINGS BELOW

Mitchell Cunningham was an employee at the Stauffer Chemical Company Phosphate Plant ("Stauffer Plant") in Silverbow County, Montana. On August 14, 1982 Cunningham was sprayed with raw phosphorous. He died the following day.

The Stauffer Plant is subject to inspection and regulation by the Occupational Safety and Health Administration ("OSHA"), Department of Labor. OSHA conducted two safety and ten health inspections of the Stauffer Plant between 1973 and 1982. The Stauffer Plant received no OSHA citations prior to 1982. Thereafter, OSHA conducted a post-accident inspection and cited the Stauffer Plant for several deficiencies.

Georgia Cunningham brought a cause of action against the United States under the FTCA, contending OSHA safety inspectors failed to exercise reasonable care in performing their inspections of the Stauffer Plant. She argued the deficiencies found in the post-accident inspection would have been identified and remedied by the plant prior to the fatal accident if OSHA had conducted its prior inspections properly.

The district court granted the government's motion to dismiss on the grounds that the action was barred by the discretionary function exception. Cunningham appeals.

DISCUSSION

A party may bring a cause of action against the United States only to the extent it has waived its sovereign immunity. *United States v. Orleans*, 425 U.S. 807, 814 (1976). A party bringing a cause of action against the federal government bears the burden of demonstrating an unequivocal waiver of immunity. *Holloman v. Watt*, 708 F.2d 1399, 1401 (9th Cir. 1983), *cert. denied*, 466 U.S. 958 (1984).

A court lacks jurisdiction over a cause of action if the federal government's alleged negligence is "based on the exercise or performance or the failure to exercise or perform a discretionary function or duty on the part of a federal agency or an employee of the Government, whether or not the discretion involved be abused." 28 U.S.C. §2680(a). *See Dalehite v. United States*, 346 U.S. 15, 33 (1953).

The nature of the conduct involved governs whether the so-called discretionary function exception applies. *United States v. S.A. Empresa de Viacao Aerea Rio Grandense ("Varig Airlines")*, 467 U.S. 797, 104 S. Ct. 2755, 2764-65 (1984). "[T]he basic inquiry . . . is whether the challenged acts of a government employee—whatever his or her rank—are of the nature and quality that Congress intended to shield from tort liability." *Id.* The purpose of the exception is to prevent judicial second-guessing of administrative decisionmaking based on social, economic, and political policy. *Id.* at 2765. " '[I]f judicial review would encroach upon this type of balancing done by an agency, then the exception would apply.' " *Chamberlin v. Isen*, 779 F.2d 522, 523 (9th Cir. 1985) (quoting *Begay v. United States*, 768 F.2d 1059, 1064 (9th Cir. 1985)).

Georgia Cunningham claims OSHA was negligent in conducting its inspections. OSHA safety inspections are similar to those discussed in *Varig Airlines*. In *Varig Airlines*, the Civil Aeronautics Agency allegedly negligently inspected and certified an aircraft that did not meet minimum fire safety standards. 104 S. Ct. at 2758. The Supreme Court found that a negligent failure to inspect falls within the discretionary function exception. *Varig Airlines*, 104 S. Ct. at 2768. The court emphasized that the manufacturer has "the duty to ensure that an aircraft conforms to FAA safety regulations . . . while the FAA retains the responsibility for policing compliance." *Id.* at 2768. The same is true for companies operating under the directives of OSHA. The employer has the statutory responsibility for maintaining a safe workplace. 29 U.S.C. §654(a).

This court, in *Natural Gas Pipeline Co. v. United States*, 742 F.2d 502, 504-05 (9th Cir. 1984), found that FAA's alleged failure to discover aircraft safety was protected by the discretionary function exception. The same is true in this case for OSHA's alleged failure to adequately monitor the Stauffer Plant. Both OSHA's decision to review the employer's compliance with safety standards and its actual inspections of the Stauffer Plant are discretionary functions. Congress has left to OSHA's discretion the establishment of safety standards and the enforcement of those standards. "When an agency determines the extent to which it will supervise the safety procedures of private individuals, it is exercising discretionary regulatory authority of the most basic kind." *Varig Airlines,* 104 S. Ct. at 2768. *See also Begay,* 768 F.2d at 1064 (decision whether to implement safety regulations in uranium mines is within *Varig's* coverage).

The acts of OSHA inspectors in executing agency directives are protected by the discretionary function exception. *See Varig Airlines,* 104 S. Ct. at 2768; *Dalehite,* 346 U.S. at 36 ("acts of subordinates in carrying out the operations of government in accordance with official directions cannot be actionable"); *see also Begay,* 768 F.2d at 1064.

◇ **NOTES**

1. Does this interpretation of "discretionary" appear unduly broad? Can one draw a meaningful distinction between the discretion involved in *promulgating* a standard and the discretion involved in *enforcing* a standard? What would be the policy implications of holding the government liable for negligence in the performance of its statutory duties?

2. The United States has settled lawsuits brought against it in its role as the supplier of asbestos to shipyards. See Settlement of Asbestos Suit is Final; Government Agrees to Pay $5.75 Million, Occupational Safety and Health Rep., Vol. 8, Feb. 16, 1978. As a general rule of thumb, the more the government does something (such as acting as the supplier of a product) that appears "nongovernmental" in nature, the more likely it is to be subject to suit under the Federal Tort Claims Act.

3. The United States has also been found liable for the Navy's failure to warn bystanders of the dangers of airborne asbestos at the Portsmouth Naval Shipyard. See *Dube v. Pittsburgh Corning,* 870 F.2d 790 (1st Cir. 1989), where the daughter of a worker who brought the substance home on his clothes was allowed to sue the Navy for asbestos-related disease.

4. In *Caldwell v. United States,* 6 OSHC 1411 (D.D.C. 1978), the plaintiff worker argued that a right to sue OSHA for negligence was implied in the OSHAct itself. In rejecting this argument, the court noted that an earlier draft of the Act had included a provision allowing an injured worker to sue when OSHA had "unreasonably" failed to act against an imminent hazard, but that this language was deleted from the final version.

◇

(b) Unions

Where a union assumes a role in policing workplace health and safety, it becomes a potential target for worker suit. To prevail against the union, however, the worker must contend with the preemptive effect of the National Labor Relations Act.

United Steelworkers of America v. Rawson

U.S. Supreme Court
_____ U.S. _____, 110 S.Ct. 1904 (1990)

WHITE, Justice

We granted certiorari in this case because the decisions of the Supreme Court of Idaho, holding that petitioner may be liable under state law for the negligent inspection of a mine where respondents' decedents worked, raised important questions about the operation of federal and state law in defining the duties of a labor union acting as a collective bargaining agent.

I

This dispute arises out of an underground fire that occurred on May 2, 1972, at the Sunshine Mine in Kellogg, Idaho, and caused the deaths of 91 miners. Respondents, the survivors of four of the deceased miners, filed this state-law wrongful death action in Idaho state court. Their complaint alleged that the miners' deaths were proximately caused by fraudulent and negligent acts of petitioner United Steelworkers of America (Union), the exclusive bargaining representative of the miners working at the Sunshine Mine. As to the negligence claim, the complaint specifically alleged that the Union "undertook to act as accident prevention representative and enforcer of an agreement negotiated between [*sic*] [the Union] on behalf of the decreased miners," App. 53-54, and "undertook to provide representatives who inspected [the Sunshine Mine] and pretended to enforce the contractual accident prevention clauses." *id.*, at 54. Respondents' answers to interrogatories subsequently made clear that their suit was based on contentions that the Union had, through a collective-bargaining agreement negotiated with the operator of the Sunshine Mine, caused to be established a joint management/labor safety committee intended to exert influence on management on mine safety measures; that members of the safety committee designated by the Union had been inadequately trained on mine safety issues; and that the Union, through its representatives on the safety committee, had negligently performed inspections of the mine that it had promised to conduct, failing to uncover obvious and discoverable deficiencies, *Id.*, at 82-83.

The trial court granted summary judgment for the Union, accepting the Union's argument that "federal law has preempted the field of union representation and its obligation to its membership," App. to Pet. for Cert. 164a, and that "[n]egligent performance of [a union's] contractual duties does not state a claim under federal law for breach of fair representation," *id.*, at 163a. The Supreme Court of Idaho reversed. *Dunbar v. United Steelworkers of America*, 100 Idaho 523, 602 P.2d 21 (1979). In the view of the Supreme Court of Idaho, although federal law unquestionably imposed on the Union a duty of fair representation of the miners, respondents' claims were "not necessarily based on the violation of the duty of fair representation and such is not the only duty owed by a union to its members." *Id.*, at 526, 602 P. 2d, at 24. Three of the five justices concurred specially to emphasize that "the precise nature of the legal issues raised by [respondents'] wrongful death action is not entirely clear at the present procedural posture of the case," and that "a final decision whether the wrongful death action . . . is preempted . . . must therefore await a full factual development." *Id.*, at 547, 602 P.2d, at 25 (Bakes, J., specially concurring). We denied the Union's petition for certiorari. *Steelworkers v. Dunbar*, 446 U.S. 983 (1980).

After extensive discovery, the trial court again granted summary judgment for the Union. App. to Pet. for Cert. 89a-106a. As to respondents' fraud claim, the court concluded that the record was devoid of evidence supporting the contentions that the Union had made misrepresentations of fact, that the Union had intended to defraud the miners, or that the miners had relied on Union representations. *Id.*, at 96a. On the negligence count, the trial court first noted that, in its view, respondents' claims centered on the collective-bargaining contract between the Union and the Sunshine Mine, especially Article IX of the agreement, which established the joint labor/management safety committee. *Id.*, at 90a-91a. The trial court urged the State Supreme Court to reconsider its conclusion that respondents' state-law negligence claim was not pre-empted by federal labor law, reasoning that "[repondents] are complaining about the manner in which the Union carried out the collective bargaining agreement, essentially saying the Union advisory committee should have done more," and that respondents "are attempting to hold the [Union] liable on the basis of its representational duties." *Id.*, at 103a-104a.

The Supreme Court of Idaho originally affirmed the grant of summary judgment on appeal. *Id.,* at 49a-88a. On rehearing, however, the Idaho Supreme Court withdrew its prior opinion and concluded that respondents had stated a valid claim under Idaho law that was not pre-empted by federal labor law. *Rawson v. United Steelworkers of America,* 111 Idaho 630, 726 P.2d 742 (1986). Distinguishing this Court's decision in *Allis-Chalmers Corp. v. Lueck,* 471 U.S. 202 (1985), which held that resolution of a state-law tort claim must be treated as a claim arising under federal labor law when it is substantially dependent on construction of the terms of a collective-bargaining agreement, the Supreme Court of Idaho stated that "in the instant case, the provisions of the collective bargaining agreement do not require interpretation, . . . but rather the provisions determine only the nature and scope of the Union's duty." 111 Idaho, at 640, 726 P.2d, at 752. The court continued: "Our narrow holding today is that the Union, having inspected, assumed a duty to use due care in inspecting and, from the duty to use due care in inspecting arose the further duty to advise the committee of any safety problems the inspection revealed." *Ibid.* The court also affirmed the trial court's conclusion that summary judgment for the Union was proper on respondents' fraud claim. *Id.,* at 633, 726 P.2d, at 745.

The Union again petitioned for certiorari. While that petition was pending, we decided *Electrical Workers v. Hechler,* 481 U.S. 851 (1987), in which it was held that an individual employee's state-law tort suit against her union for breach of the union's duty of care to provide the employee with a safe workplace must be treated as a claim under federal labor law, when the duty of care allegedly arose from the collective-bargaining agreement between the union and the employer. Six days later, we granted the Union's petition, vacated the judgment of the Supreme Court of Idaho, and remanded this case for further consideration in light of *Hechler. Steelworkers v. Rawson,* 482 U.S. 901 (1987).

On remand, the Supreme Court of Idaho "adhere[d] to [its] opinion as written. 115 Idaho 785, 788, 770 P.2d 794, 797 (1988). The court also distinguished *Hechler,* stressing that there we had considered a situation where the alleged duty of care arose from the collective-bargaining agreement, whereas in this case "the activity was concededly undertaken and the standard of care is imposed by state law without reference to the collective bargaining agreement." 115 Idaho, at 786, 770 P.2d, at 795. The court further

stated that it was "not faced with looking at the Collective Bargaining Agreement to determine whether it imposes some new duty upon the union— rather it is conceded that the union undertook to inspect and, thus, the issue is solely whether that inspection was negligently performed under traditional Idaho tort law." *Id.,* at 787, 770 P.2d, at 796.

We granted certiorari, 493 U.S.— (1990), and we now reverse.

II

Section 301 of the Labor Management Relations Act, 1947 (LMRA), 61 Stat. 156, 29 U.S.C. §185(a), states:

> Suits for violation of contracts between an employer and a labor organization representing employees in an industry affecting commerce as defined in this Act, or between any such labor organizations, may be brought in any district court of the United States having jurisdiction of the parties, without respect to the amount in controversy or without regard to the citizenship of the parties.

Over 30 years ago, this Court held that §301 not only provides the federal courts with jurisdiction over controversies involving collective-bargaining agreements but also authorizes the courts to fashion "a body of federal law for the enforcement of these collective bargaining agreements." *Textile Workers v. Lincoln Mills of Alabama,* 353 U.S. 448, 451 (1957). Since then, the Court has made clear that §301 is a potent source of federal labor law, for though state courts have concurrent jurisdiction over controversies involving collective-bargaining agreements, *Charles Dowd Box Co. v. Courtney,* 368 U.S. 502 (1962), state courts must apply federal law in deciding those claims, *Teamsters v. Lucas Flour Co.,* 369 U.S. 95 (1962), and indeed any state-law cause of action for violation of collective-bargaining agreements is entirely displaced by federal law under §301, see *Avco Corp. v. Machinists,* 390 U.S. 557 (1968). State law is thus "preempted" by §301 in that only the federal law fashioned by the courts under §301 governs the interpretation and application of collective-bargaining agreements.

In recent cases, we have recognized that the pre-emptive force of §301 extends beyond state-law contract actions. In *Allis-Chalmers Corp. v. Lueck,* 471 U.S. 202 (1985), we held that a state-law tort action against an employer may be preempted by §301 if the duty to the employee of which the tort is a violation is created by a collective bargaining agreement and without existence independent of the agreement. Any other result, we reasoned, would "allow parties to evade the requirements of §301 by relabeling their contract claims as claims for tortious breach of contract." *Id.*, at 211. We extended this rule of preemption to a tort suit by an employee against her union in *Electrical Workers v. Hechler, supra.* There Hechler alleged that her union had by virtue of its collective-bargaining agreement with the employer and its relationship with her assumed the duty to ensure that she was provided with a safe workplace, and that the union had violated this duty. As in *Allis-Chalmers,* the duty relied on by Hechler was one without existence independent of the collective-bargaining agreement (unions not, under the common law of Florida, being charged with a duty to exercise reasonable care in providing a safe workplace, see 481 U.S., at 859-860) but was allegedly created by the collective-bargaining agreement, of which Hechler claimed to be a third-party beneficiary, *see id.*, at 861. Because resolution of the tort claim would require a court to "ascertain, first, whether the collective-bargaining agreement in fact placed an implied duty of care on the Union . . . , and second, the nature and scope of that duty," *id.*, at 862, we held that the tort claim was not sufficiently independent of the collective-bargaining agreement to withstand the preemptive force of §301.

At first glance it would not appear difficult to apply these principles to the instant case. Respondents alleged in their complaint that the Union was negligent in its role as "enforcer of an agreement negotiated between [sic][the Union] on behalf of the deceased miners," App. 53-54, a plain reference to the collective-bargaining agreement with the operator of the Sunshine Mine. Respondents' answers to interrogatories gave substance to this allegation by stating that "by the contract language" of the collective-bargaining agreement, the Union had caused the establishment of the joint safety committee with purported influence on mine safety issues, and that members of the safety committee had failed reasonably to perform inspections of the mine or to uncover obvious and discoverable deficiencies in the mine safety program. App. 82-83. The only possible interpretation of these pleadings, we believe, is that the duty on which respondents relied as the basis of their tort suit was one allegedly assumed by the Union in the collective-bargaining agreement. Prior to our remand, the Supreme Court of Idaho evidently was of this view as well. The court noted then

that the Union could be liable under state tort law because it allegedly had contracted to inspect, and had in fact inspected the mine "pursuant to the provisions of the collective bargaining agreement." 111 Idaho, at 638, 726 P.2d, at 750. Although the Idaho Supreme Court believed that resolution of the tort claim would not require interpretation of the terms of the collective-bargaining agreement, it acknowledged that the provisions of that agreement determined "the nature and scope of the Union's duty," *id.,* at 640, 726 P.2d, at 752.

The situation is complicated, however, by the Idaho Supreme Court's opinion after our remand. Although the court stated that it adhered to its prior opinion as written, 115 Idaho, at 788, 770 P.2d, at 797, it also rejected the suggestion that there was any need to look to the collective-bargaining agreement to discern whether it placed any implied duty on the union. Rather, Idaho law placed a duty of care on the Union because the Union did, in fact, actively inspect the mine, and the Union could be held liable for the negligent performance of that inspection. *Id.,* at 787, 770 P.2d, at 796. According to the Supreme Court of Idaho, the Union may be liable under state tort law because its duty to perform that inspection reasonably arose from the fact of the inspection itself rather than the fact that the provision for the Union's participation in mine inspection was contained in the labor contract.

As we see it, however, respondents' tort claim cannot be described as independent of the collective-bargaining agreement. This is not a situation where the Union's delegates are accused of acting in a way that might violate the duty of reasonable care owed to every person in society. There is no allegation, for example, that members of the safety committee negligently caused damage to the structure of the mine, an act that could be unreasonable irrespective of who committed it and could foreseeably cause injury to any person who might possibly be in the vicinity.

Nor do we understand the Supreme Court of Idaho to have held that any casual visitor in the mine would be liable for violating some duty to the miners if the visitor failed to report obvious defects to the appropriate authorities." Indeed, the court did not disavow its previous opinion, where it acknowledged that the Union's representatives were participating in the inspection process pursuant to the provisions of the collective-bargaining agreement and that the agreement determined the nature and scope of the Union's duty. If the Union failed to perform a duty in connection with inspection, it was a duty arising out of the collective-bargaining agreement signed by the Union as the bargaining agent for the miners. Clearly, the enforcement of that agreement and the remedies for its breach are matters governed by federal law. "[Q]uestions relating to what the parties to a labor agreement agreed, and what legal consequences were intended to flow from breaches of that agreement, must be resolved by reference to uniform federal law, whether such questions arise in the context of a suit for breach of contract or in a suit alleging liability in tort." *Allis-Chalmers Corp. v. Lueck*, 471 U.S., at 211. Preemption by federal law cannot be avoided by characterizing the Union's negligent performance of what it does on behalf of the members of the bargaining unit pursuant to the terms of the collective-bargaining contract as a state-law tort. Accordingly, this suit, if it is to go forward at all, must proceed as a case controlled by federal rather than state law.

III

The Union insists that the case against it may not go forward even under federal law. It argues first that only the duty of fair representation governs the exercise of its representational functions under the collective-bargaining contract, and that a member may not sue it under §301 for breach of contract. Second, the Union submits that even if it may be sued under §301, the labor agreement contains no enforceable promise made by it to the members of the unit in connection with inspecting the mine. Third, the Union asserts that as the case now stands, it is charged with only negligence which is insufficient to prove a breach of its duty of fair representation.

"It is now well established that, as the exclusive bargaining representative of the employees, . . . the Union had a statutory duty fairly to represent all of those employees, both in its collective bargaining . . . and in its enforcement of the resulting collective bargaining agreement." *Vaca v. Sipes*, 386 U.S. 171, 177 (1967). "Under this doctrine, the exclusive agent's statutory authority to represent all members of a designated unit includes a statutory obligation to serve the interests of all members without hostility or discrimination toward any, and to exercise its discretion with complete good faith and honesty, and to avoid arbitrary conduct." *Ibid.* This duty of fair representation is of major importance, but a breach occurs "only when a union's conduct toward a member of the collective bargaining unit is arbitrary, discriminatory, or in bad faith. *Id.*, at 190. The courts

have in general assumed that mere negligence, even in the enforcement of a collective-bargaining agreement, would not state a claim for breach of the duty of fair representation, and we endorse that view today.

The Union's duty of fair representation arises from the National Labor Relations Act itself. *See Breininger v. Sheet Metal Workers*, 493 U.S.——, —— (1989); *Del-Costello v. Teamsters*, 462 U.S. 151, 164 (1983); *United Parcel Service, Inc. v. Mitchell*, 451 U.S. 56, 66 (1981) (Stewart, J., concurring in judgment). The duty of fair representation is thus a matter of status rather than contract. We have never held, however, that as a matter of federal law, a labor union is *prohibited* from voluntarily assuming additional duties to the employees by contract. Although at one time it may have appeared most unlikely that unions would be called upon to assume such duties, see *Humphrey v. Moore*, 375 U.S. 335, 346-357 (1964) (Goldberg, J., concurring in result), nonetheless "it is of the utmost importance that the law reflect the realities of industrial life and the nature of the collective bargaining process." *id.*, at 358, and it may well be that if unions begin to assume duties traditionally viewed as the prerogatives of management, cf. *Breininger, supra*, at ——; *Electrical Workers v. Hechler*, 481 U.S., at 859-860, employees will begin to demand that unions be held more strictly to account in their carrying out those duties. Nor do we know what the source of law would be for such a prohibition, for "when neither the collective-bargaining process nor its end product violates any command of Congress, a federal court has no authority to modify the substantive terms of a collective-bargaining contract." *United Mine Workers Health & Retirement Funds v. Robinson*, 455 U.S. 562, 576 (1982); cf. *H.K. Porter Co. v. NLRB*, 397 U.S. 99, 106-108 (1970).

Our decision in *Electrical Workers v. Hechler, supra*, is relevant here. There we were presented with a claim by an employee that the union had breached its duty to provide her with a safe workplace. The alleged duty was plainly based on the collective-bargaining agreement that the union had negotiated with the employer; Hechler argued that she was a third-party beneficiary of that agreement. *Id.*, at 861, 864-865. Hechler carefully distinguished her §301 claim from a fair representation claim, *id.*, at 864, and so did we, for the distinction had a significant effect; the statutes of limitations for the two claims are different. *Id.*, at 863-865. We therefore accepted, and again accept, that "a labor union . . . may assume a responsibility towards employees by accepting a duty of care through a contractual agreement," *id.*, at 860, even if that contractual agreement is a collective-bargaining contract to which only the union and the employer are signatories.

But having said as much, we also think it necessary to emphasize caution, lest the courts be precipitate in their efforts to find unions contractually bound to employees by collective-bargaining agreements. The doctrine of fair representation is an important check on the arbitrary exercise of union power, but it is a purposefully limited check, for a "wide range of reasonableness must be allowed a statutory bargaining representative in serving the unit it represents." *Ford Motor Co. v. Huffman*, 345 U.S. 330, 338 (1953). If an employee claims that a union owes him a more far-reaching duty, he must be able to point to language in the collective-bargaining agreement specifically indicating an intent to create obligations enforceable against he union by the individual employees. Cf. *Republic Steel Corp. v. Maddox*, 379 U.S. 650, 653 (1965).

Applying this principle to the case at hand, we are quite sure that respondents may not maintain a §301 suit against the Union. Nothing in the collective-bargaining agreement suggests that it creates rights directly enforceable by the individual employees against the Union. The pertinent part of the collective-bargaining agreement, Article IX, consists entirely of agreements between the Union and the employer and enforceable only by them. App. 20-22. Section 2 of the Article provides that "a committee consisting of two (2) supervisory personnel and two (2) reliable employees, approved by the Union, shall inspect" the mine if an employee complains to the shift boss that he is being forced to work in unusually unsafe conditions but receives no redress, *id.*, at 20, but even if this section might be interpreted as obliging the Union to inspect the mine in such circumstances, the promise is not one specifically made to, or enforceable by, individual employees. Nor have respondents placed anything in the record indicating that any such complaints were made or that the Union failed to act on them. Section 4 of the Article states that a Union member may accompany the state mine safety inspection team on its inspections of the mine, and Section 5 states that a Union designate and the Safety Engineer "shall make a tour of a section of the mine" once each month, *id.*, at 22, but again the agreement gives no indication that these obligations, if such is what they are, may be enforced by an individual employee.

Moreover, under traditional principles of contract interpretation, respondents have no claim, for with exceptions under federal labor law not relevant here,

see Lewis v. Benedict Coal Corp., 361 U.S. 459, 468-471 (1960), third-party beneficiaries generally have no greater rights in a contract than does the promisee. For the respondents to have an enforceable right as third-party beneficiaries against the Union, at the very least the employer must have an enforceable right as promisee. But the provisions in the collective-bargaining agreement relied on by respondents are not promises by the Union to the employer. Cf. *Teamsters v. Lucas Flour Co.*, 369 U.S., at 104-106. They are, rather, concessions made by the employer to the Union, a limited surrender of the employer's exclusive authority over mine safety. A violation by the employer of the provisions allowing inspection of the mine by Union delegates might form the basis of a §301 suit against the employer, but we are not presented with such a case.

IV

In performing its functions under the collective-bargaining agreement, the Union did, as it concedes, owe the miners a duty of fair representation, but we have already noted that respondents' allegation of mere negligence will not state a claim for violation of that dury. *Supra*, at _____. Indeed, respondents have never specifically relied on the federal duty of fair representation, nor have they alleged that the Union improperly discriminated among its members or acted in arbitrary and capricious fashion in failing to exercise its duties under the collective-bargaining agreement. Cf. *Vaca v. Sipes*, 386 U.S., at 177. Respondents did, of course, allege that the Union had committed fraud on the membership in violation of state law, a claim that might implicate the duty of fair representation. The Supreme Court of Idaho held, however, that summary judgment was properly entered on this claim because respondents had failed to demonstrate specific facts showing the existence of a genuine issue for trial. 111 Idaho, at 633, 726 P.2d, at 745. Respondents did not cross-petition to challenge this aspect of the Idaho Supreme Court's judgment, and we are in no position to question it.

It follows that the judgment of the Supreme Court of Idaho must be reversed.

DISSENTING OPINION

Justice KENNEDY, with whom The Chief Justice and Justice SCALIA join, dissenting.

The Idaho Supreme Court held that summary judgment was improper and that Tharon Rawson and the other respondents could proceed to trial against the United Steelworkers of America on a state law tort theory. Although the respondents have not yet established liability under Idaho law, the Union argues that federal law must govern and bar their suit. To support this position, the Union relies on both §301 of the Labor Management Relations Act, 29 U.S.C. §185(a), and the duty of fair representation implicit in §9(a) of the National Labor Relations Act (NLRA), 49 Stat. 453, as amended, 29 U.S.C. §159(a). The Court accepts the Union's contentions with respect to §301 and does not reach the issue of pre-emption by the duty of fair representation. With all respect, I dissent. Neither of the Union's arguments for displacing Idaho law without any trial on the merits has validity.

I

The Union bases its §301 argument on our decisions in *Lingle v. Norge Division of Magic Chef, Inc.,* 486 U.S. 399, 405-406 (1988); *Electrical Workers v. Hechler,* 481 U.S. 851, 854 (1987); and *Allis-Chalmers Corp. v. Lueck,* 471 U.S. 202, 211 (1985). These cases hold that §301 pre-empts state law causes of action that require interpretation of a collective-bargaining agreement. In my view, they have no application here. The Idaho Supreme Court, whose determination of state law supersedes that of the trial court, has declared that the respondents' case rests on allegations of the Union's active negligence in a voluntary undertaking, not its contractual obligations.

Adopting verbatim a standard from the Restatement (Second) of Torts §323 (1965), the Idaho Court expressed the law governing the respondents' claims as follows:

> One who undertakes, gratuitously or for consideration, to render services to another which he should recognize as necessary for the protection of the other's person or things, is subject to liability to the other for physical harm resulting from his failure to exercise reasonable care to perform his undertaking, if
>
> '(a) his failure to exercise such care increases the risk of harm, [or]
>
> '(b) the harm is suffered because of the other's reliance upon the undertaking.' [*Rawson v. United Steelworkers of America,* 111 Idaho 630, 637, 726 P.2d 742, 749 (1986).]

According to the Idaho Supreme Court's second opinion, the respondents can prove the elements of the tort described in §323 without relying on the

Union's collective-bargaining agreement. The Court states:

> In the instant case, we are not faced with looking at the Collective Bargaining Agreement to determine whether it imposes some new duty upon the union—rather it is conceded the union undertook to inspect and, thus, the issue is solely whether that inspection was negligently performed under traditional Idaho tort law. [115 Idaho 785, 787, 770 P.2d 794, 796 (1989).]

Placing this analysis of state law in the context of our precedents, the Idaho court explains:

> [T]he instant case is clearly distinguishable from *Hechler* in that here the state tort basis of the action was not abandoned, but has been pursued consistently both at the trial and appellate levels and the tort exists without reference to the collective bargaining agreement. [*Id.,* at 787-788, 770 P.2d, at 796-797.]

The court states further:

> [As in *Lingle v. Norge Division of Magic Chef, Inc.*], no interpretation of the collective-bargaining agreement is required to determine whether the union member of the inspection team committed a tort when he committed various acts and omissions such as failure to note the self-rescuers were stored in boxes with padlocks or that the activating valves of the oxygen-breathing-apparatuses were corroded shut. Rather, such alleged acts of negligence are measured by state tort law. [*Id.,* at 788, 770 P.2d, at 797.]

These statements reveal that the Idaho Supreme Court understood the federal preemption standards and interpreted state law not to implicate them. Because we have no basis for disputing the construction of state law by a state supreme court, see *Clemons v. Mississippi* 494 U.S. ____, ____ (1990), I submit that, at this stage of the proceedings, we must conclude that §301 does not govern the respondents' claims.

The Court reaches a different conclusion because it doubts that the Idaho Supreme Court means what it seems to have said. The Court bases its view, to a large extent, on the Idaho court's expressed intention to "adhere to [its first] opinion as written." 115 Idaho, at 788, 770 P.2d, at 797. The first opinion says: "Because the union, pursuant to the provisions of the collective bargaining agreement, had contracted to inspect and *in fact, inspected* the mine, it owed the (minimal) duty to its members to exercise due care in inspecting and in reporting the findings of its inspection." 111 Idaho, at 638, 726 P.2d, at 750. The

Court construes the remark to negate the unequivocal statements quoted above. I cannot accept this labored interpretation.

The Idaho Supreme Court's adherence to the first opinion does not implicate §301 because it does not require interpretation of a collective-bargaining agreement. The first opinion suggests that the respondents may refer to the collective-bargaining agreement. It does not eliminate the possibility, identified three times in the second opinion, that the respondents may prove the elements of §323 without relying on the collective-bargaining agreement. Even the Union concedes:

> After *Hechler*, as we understand matters, both plaintiffs and the Idaho court would locate the source of the union's duty to inspect [in a non-negligent manner] in the union's action of accompanying company and state inspectors on inspections of the mine, and not in any contractual agreement by the union to inspect. [Brief for Petitioner 27-28.]

The Court, thus, reads too much into the last sentence of the Idaho Supreme Court's second opinion.

I see no reason not to allow this case to go forward with a simple mandate: the respondents may press their state claims so long as they do not rest upon the collective-bargaining agreement. To the extent that any misunderstanding might exist, this approach would preserve all federal interests. If the Idaho Supreme Court, after a trial on the merits, were to uphold a verdict resting on the Union's obligations under the collective-bargaining agreement, we could reverse its decision. But for now we must take the case as the Idaho Supreme Court has given it to us. According to the second opinion, the respondents may prove the elements of §323 without relying on the Union's contractual duties.

The Court also rules against the respondents because it surmises that §323 has no general applicability. The Court assumes that only union members could recover from the Union for its negligence in inspecting the mine and that union members could not recover from anyone else for comparable negligence. See *ante* at 7-8. I agree that a State cannot circumvent our decisions in *Lingle*, *Hechler*, and *Allis-Chalmers*, by the mere "relabeling" as a tort claim an action that in law is based upon the collective-bargaining process. *Allis-Chalmers*, 471 U.S., at 211. We must have the ultimate responsibility for deciding whether a state law depends on a collective-bargaining agreement for the purposes of §301. In this case, however, I see no indication that the tort theory pressed by the respondents has the limited application presumed by the Court.

The Idaho Supreme Court did not invent, for the purposes of this case, the theory underlying the respondents' claims. As Cardozo put it: "It is ancient learning that one who assumes to act, even though gratuitously, may thereby become subject to the duty of acting carefully, if he acts at all." *Glanzer v. Shepard*, 233 N.Y. 236, 239, 135 N.E. 275, 276 (1922). Restatement §323, upon which the Idaho Court relies, embodies this principle and long has guided the interpretation of Idaho tort law. See, *e.g.*, *Steiner Corp. v. American District Telegraph*, 106 Idaho 787, 791, 683 P.2d 435, 439 (1984) (fire alarm failure); *S.H. Kress & Co. v. Godman*, 95 Idaho 614, 616, 515 P. 2d 561, 563 (1973) (boiler explosion); *Fagundes v. State*, 116 Idaho 173, 176, 774 P.2d 343, 346 (App. 1989) (helicopter crash); *Carroll v. United Steelworkers of America*, 107 Idaho 717, 723, 692 P.2d 361, 367 (1984) (Bistline, J., dissenting) (machinery accident). The Court has identified no basis for its assumption that §323 has a narrower scope than its plain language and these cases indicate. I thus would not find pre-emption on the mere supposition that the Union's duty runs only to the union members.

II

The Union also argues that the duty of fair representation immunizes it from liability under §323. Allowing the States to impose tort liability on labor organizations, it contends, would upset the balance of rights and duties that federal law has struck between unions and their members. I disagree because nothing in the NLRA supports the Union's position.

Section 9(a) of the NLRA, 29 U.S.C. §159(a), grants a duly elected union the exclusive authority to represent all employees in a collective-bargaining unit. We have reasoned:

> The fair interpretation of the statutory language is that the organization chosen to represent a craft is chosen to represent all its members, the majority as well as the minority, and it is to act for and not against those whom it represents. It is a principle of general application that the exercise of a granted power to act in behalf of others involves the assumption toward them, of a duty to exercise the power in their interest and behalf, and that such a grant of power will not be deemed to dispense with all duty toward those from whom it exercised unless so expressed. [*Steele v. Louisville & Nashville R. Co.*, 323 U.S. 192, 202 (1944) (footnote omitted) (interpretation

of §2(a) of the Railway Labor Act, 45 U.S.C. §152 (1982 ed.), adopted for §9(a) of the NLRA in *Ford Motor Co. v. Huffman,* 345 U.S. 330, 337 (1953)).]

As a result, we have read §9(a) to establish a duty of fair representation requiring a union "to serve the interests of all members without hostility or discrimination toward any, to exercise its discretion with complete good faith and honesty, and to avoid arbitrary conduct." *Vaca v. Sipes,* 386 U.S. 171, 177 (1967).

Although we have inferred that Congress intended to impose a duty of fair representation in §9(a), I see no justification for the further conclusion that Congress desired to grant unions as immunity from all state tort law. Nothing about a union's status as the exclusive representative of a bargaining unit creates a need to exempt it from general duties to exercise due care to avoid injuring others. At least to some extent, therefore, I would conclude that Congress "by silence indicate[d] a purpose to let state regulation be imposed." *Retail Clerks v. Schermerhorn,* 375 U.S. 96, 104 (1963).

Our decision in *Farmer v. Carpenters,* 430 U.S. 290 (1977), confirms this view. *Farmer* held that the NLRA did not preempt a union member's action against his union for intentional infliction of emotional distress. *See id.,* at 305. The union member complained that his union ridiculed him in public and refused to refer jobs to him in accordance with hiring hall rules. *See id.,* at 293. In analyzing this claim, we ruled that the NLRA's pre-emption of state tort law depends on two factors: "the state interests in regulating the conduct in question and the potential for interference with the federal regulatory scheme." *Id.,* at 297. Both of these factors militated against pre-emption in *Farmer.* Noting that "our cases consistently have recognized the historic state interest in 'such traditionally local matters as public safety and order,'" id., at 299 (quoting *Allen-Bradley Local v. Wisconsin Employment Relations Bd.,* 315 U.S. 740, 749 (1942)), we ruled that the tort law addressed proper matters of state concern. We further observed that, although the tort liability for intentional infliction of emotional distress might interfere with the federal prohibition against discrimination by a union, that "potential for interference is insufficient to counterbalance the legitimate and substantial interest of the State in protecting its citizens." 430 U.S., at 304.

The *Farmer* analysis reveals that Idaho may hold the union liable for negligence in inspecting the mine. The strength and legitimacy of the State's interests in mine safety stand beyond question; the union's failure to exercise due care, according to the allegations, caused or contributed to the deaths of 91 Idaho miners. Allowing this case to proceed to trial, moreover, would pose little threat to the federal regulatory scheme. State courts long have held unions liable for personal injuries under state law. See, *e.g., DiLuzio v. United Electrical, Radio, and Machine Workers of America,* 386 Mass. 314, 318, 435 N.E. 2d 1027, 1030 (1982) (assault at workplace); *Brawner v. Sanders,* 244 Ore. 302, 307, 417 P.2d 1009, 1012 (1966) (in banc) (personal injuries); *Marshall v. International Longshoremen's and Warehousemen's Union,* 57 Cal. 2d 781, 787, 371 P.2d 987, 991 (1962) (stumble in union hall parking lot); *Inglis v. Operating Engineers Local Union No. 12,* 58 Cal.2d 269, 270, 373 P.2d 467, 468 (1962) (assault at union meeting); *Hulahan v. Sheehan,* 522 S.W. 2d 134, 139-141 (Mo. App. 1975) (slip and fall on union hall stairs). The Union presents no argument that this long-standing practice has interfered with federal labor regulation. Indeed, as the Court itself holds, nothing in the federal statutory scheme addresses the Union's conduct or provides redress for the injuries that it may have produced. See *ante,* at 8-12.

The Union's position also deviates from the well-established position of the courts of appeals. These courts have found pre-emption by the duty of fair representation in two situations. First, the courts have said that the duty of fair representation preempts state duties that depend on a collective bargaining agreement or on the union's status as the exclusive collective bargaining agent. See, *e.g., Richardson v. United Steelworkers of America,* 864 F.2d 1162, 1165-1167 (CA5 1989); *Condon v. Local 2944, United Steelworkers of America,* 683 F.2d 590, 595 (CA1 1982). As noted above, however, the Union's duties in this case do not stem from a contract or from its status as a union. Second, other courts have found the federal duty of fair representation to supplant equivalent state law duties. See, *e.g., Jones v. Truck Drivers Local Union No. 299,* 838 F.2d 856, 861 (CA6 1988) (sex discrimination); *Maynard v. Revere Copper Products,* 773 F.2d 733, 735 (CA6 1985) (handicapped discrimination); *Peterson v. Air Line Pilots Assn., International,* 795 F.2d 1161, 1170 (CA4 1985) (blacklisting). In this case, state law differs from federal law in that the duty of fair representation does not address the conduct in question. The Union, as a result, has shown no support for its contention that the duty of fair representation pre-empts the Idaho tort law. For these reasons, I dissent.

◇ NOTES

1. The Sunshine Mine disaster was a compelling example of workplace safety gone awry, and it received considerable national attention.

2. After *Rawson,* what is left of the worker negligence suit against the union? If there is no language creating a duty of care in the collective bargaining agreement, how would the worker prove that the union had assumed such a duty *independent of the collective agreement?* Is the dissent's argument that the Court should have deferred to the Idaho Supreme Court's finding on this point a persuasive one?

◇

(c) Health and safety contractors

There is a growing cadre of industrial health and safety professionals who specialize in workplace inspections and evaluations. Typically, they contract with a variety of employers to conduct on-site health and safety inventories, and to recommend steps to reduce occupational injury and disease. In assuming these contractual duties to the employer, they may also be assuming a duty of care to the worker.

Canipe v. National Loss Control Service Corp.

736 F.2d 1055 (5th Cir. 1984).

WISDOM, Circuit Judge.

This diversity action involves the applicability of the principles of section 324A of the American Law Institute's Restatement (Second) of Torts (1965) to a tort case subject to Tennessee law. Section 324A deals with the liability of an actor, rendering services to one person, for failure to exercise reasonable care to protect a third person.

Billy Canipe sues to recover for a severe personal injury that he sustained at work. The defendant, National Loss Control Service Corporation (National Loss), has contracted with the plaintiff's employer (Kraft, Inc.) to provide safety inspections and related accident-prevention services at the plant in which the plaintiff worked. Canipe alleges that the defendant performed its contractual duties negligently, and that this negligence was a proximate cause of the plaintiff's injury. After a long period of discovery, the district court ruled that the plaintiff had presented no genuine issue of material fact, and therefore granted the defendant's motion for summary judgment. *Canipe v. National Loss Control Service Corp.,* N.D. Miss. 1983, 566 F. Supp. 521. Because of the thoroughness and thoughtfulness of the district court's opinion, we are hesitant to overturn the court's

decision. Nevertheless, we have concluded that the decision rests upon an erroneous legal premise. Accordingly, we affirm in part, reverse in part, and remand.

I

Summary judgment is appropriate, of course, only when the movant has demonstrated the lack of any genuine issue of material fact. In reviewing a grant of summary judgment for the defendant, we must examine the record in the light most favorable to the plaintiff and draw all reasonable inferences in the plaintiff's favor. *See, e.g., Gulf Mississippi Marine Corp. v. George Engine Co.,* 5 Cir. 1983, 687 F.2d 668, 670-71. Viewed from that perspective, the facts of this case are as follows.

The defendant is a national corporation engaged in the business of helping other companies improve workplace safety. National Loss entered into its contractual relationship with Kraft in January 1978.[1] The contract between these two parties was national in scope and was intended to establish only the

[1]The contracts between Kraft and National Loss were each for a one-year term. The parties renewed their original agreement, without substantial modification, every year between 1978 and 1982.

general framework for National Loss's provision of services to the various Kraft plants around the country. The primary purpose of this contract, therefore, was to set the hourly rates for the various services that National Loss could provide to Kraft plants. The service to be performed at a particular plant was to be determined on a "by request" basis. Under the arrangement between National Loss and Kraft's Humko[2] plant in Memphis, Tennessee, National Loss agreed to conduct quarterly safety inspections of the plant and to provide the plant management with loss experience analyses.[3] Humko plant management, in addition to contracting for these services, conducted its own periodic safety inspections of the plant.

On July 5, 1979, Kraft's Corporate Safety Manager, Jack Hansen, sent the following memorandum to all of Kraft's safety managers:

> Attached for your information is a copy of an OSHA-Gram from the *National Safety News*. It explains the change in OSHA's position on defining a serious violation. This new definition will probably result in our locations receiving more serious violations and higher penalties. Obviously, the answer is to survey all of our respective locations to make sure that the locations do not have any OSHA violations.

On July 10, Allen Jamison, who was the Kraft official in charge of coordinating safety inspections at the Humko plant, wrote to the National Loss manager of the Kraft account. That letter states in pertinent part:

> In reference to Jack Hansen's memo of July 5, 1979[,] explaining the change in OSHA's position on defining a serious violation, I would like to request that your Loss Control people during their third quarter visit concentrate more on an OSHA type inspection, and if any serious violations are observed, to include them in the recommendation section. The third quarter is the only quarter that we would like the format changed.

During their next visit to the Humko plant, which occurred in September 1979, the National Loss inspectors conducted a thorough four-day search of the plant, concentrating on detecting violations of OSHA regulations. The two inspectors later sent

plant management a detailed, forty-four page report on the violations they had discovered.

Canipe worked at the Humko plant as Assistant Operator of two machines known as "flake roll machines," which are capable of producing different types of flaky chemical products. As a part of his duties, he was required to clean each machine whenever production on that machine shifted from one product to another. To clean the machine, he had to lift a transparent plastic dust cover and then use an air hose to force out of the machine's auger trough any matter remaining in the trough. Canipe contends that he was taught a method of cleaning the machine while the auger was in motion. On January 31, 1980, while cleaning one of the machines in this manner, Canipe caught his shirt sleeve on a large pin at the end of the auger. He was pulled into the machine and his right arm was amputated.

Canipe sued National Loss on November 20, 1981. He based his primary theory of recovery upon section 324A of the Restatement (Second) of Torts (1965) which states:

> One who undertakes, gratuitously or for consideration, to render services to another which he should recognize as necessary for the protection of a third person or his things, is subject to liability to the third person for physical ham resulting from his failure to exercise reasonable care to protect his undertaking, if
>
> (a) his failure to exercise reasonable care increases the risk of such harm, or
>
> (b) he has undertaken to perform a duty owed by the other to the third person, or
>
> (c) the harm is suffered because of reliance by the other or the third person upon the undertaking.

*   *   *

On April 28, 1983, the district court granted the defendant's motion for summary judgment on the ground that Tennessee law did not recognize the applicability of the principles underlying section 324A. The plaintiff moved for reconsideration. The court then reversed its position on the applicability of section 324A under Tennessee law, but granted summary judgment for the defendant nonetheless, holding that the plaintiff had not presented any facts that would justify applying section 324A. The plaintiff appeals that decision.

*   *   *

---

[2]Humko Products is a division of Kraft.
[3]Loss experience analyses are retrospective analyses of previous accidents. The objective of these analyses is to identify safety problems by noting statistical trends in, for example, the locations or types of injuries experienced at the plant.

[O]ur independent review of Tennessee case law convinces us that Tennessee has adopted the principle of section 324A.

* * *

We hold that the district court was correct to conclude that section 324A, in principle, provides a valid basis for recovery under Tennessee law. We affirm that part of the district court's decision.

1

To determine whether the record contains facts indicating that National Loss performed its undertaking negligently, it is first necessary to ascertain the scope of the undertaking, for the scope of the defendant's undertaking determines the scope of its duty. *See Blessing v. United States*, E.D. Pa.1978, 447 F. Supp. 1160, 1189-90; *Fireman's Fund American Insurance Co. v. Coleman*, Ala. 1980, 394 So. 2d 334, 349 (Jones, J., concurring in the result); *Evans v. Otis Elevator Co.*, 1961, 403 Pa. 13, 19, 168 A.2d 573, 576. One of the parties' principal disputes on appeal concerns the scope of National Loss's undertaking. Canipe contends that National Loss contracted to inspect and evaluate work practices and procedures. He argues, therefore, that National Loss was negligent for not recognizing the unsafe nature of the cleaning procedure taught to the plaintiff. The district court held, and the defendant asserts on appeal, that the contract made the defendant responsible only for inspecting for physical hazards.

We find it unnecessary to resolve this dispute. Accepting the defendant's characterization of its contractual obligations, we find that the record contains facts that at least establish as a genuine issue whether the defendant performed these obligations negligently. The parties agree that National Loss was obligated to inspect for physical hazards, and that in September 1979 its agents were looking specifically for violations of OSHA regulations. Canipe has pointed to evidence in the record implying that negligence by National Loss in its inspection for physical hazards and OSHA violations proximately caused his injury. First, the depositions of the two inspectors who conducted the September 1979 survey indicate clearly that they examined the two flake roll machines the plaintiff operated. Second, the plaintiff submitted an affidavit by Dr. Louis B. Trucks, an expert in safety engineering, listing a number of physical hazards in the flake roll machines. Dr. Trucks found that the machines were inadequately guarded and did not

contain adequate warnings, that the on-off switch was located on a different level and 30 feet away from the machines, and that the machines did not contain an emergency switch-off for workers who find themselves in peril. In Dr. Trucks' opinion, some of these hazards constituted violations of OSHA regulations. All of these facts support the plaintiff's allegation of negligence on the part of the defendant.

Similarly, the record contains evidence that the alleged negligence was the proximate cause of the plaintiff's injury. Allen Jamison, the Kraft official in charge of coordinating safety policy at the Humko plant, stated in his deposition that Kraft would remedy any OSHA violations uncovered by the National Loss inspectors regardless whether the remedies were cost justified. It cannot be disputed that Canipe might not have been injured had National Loss detected, and Kraft remedied, the physical hazards that Dr. Trucks identified in the flake roll machines. There was, therefore, at least a genuine issue as to whether the defendant negligently performed its contractual undertaking and whether this negligence proximately caused the plaintiff's injury.

2

A plaintiff seeking to recover under section 324A must demonstrate that the facts of his case come within one of the three subsections of section 324A. Therefore, to paraphrase section 324A, we cannot affirm the grant of summary judgment for the defendant in the present case unless the record shows that there is no genuine issue whether

> (a) the defendant's failure to exercise reasonable care increased the risk of the injury suffered by the plaintiff, or
> (b) the defendant undertook to perform a duty owed by Kraft to the plaintiff, or
> (c) the plaintiff's injury resulted from reliance by Kraft or the plaintiff upon the defendant's undertaking.

We find the district court's holding on this point erroneous.

The district court correctly determined that subsection (a) of section 324A does not apply to this case. This subsection requires some change in conditions that increases the risk of harm to the plaintiff over the level of risk that existed before the defendant became involved. *See Stacy v. Aetna Casualty & Surety Co.*, 5 Cir. 1973, 484 F.2d 289, 293 n. 4; *Blessing v. United States*, E.D. Pa. 1978, 447 F. Supp. 1160, 1197 n.53. A failure to detect a hazardous condition does not by itself implication subsection (a).

The district court erred, however, by taking an unduly narrow approach to subsections (b) and (c). Implicit in the district court's treatment of subsection (b) is the assumption that this subsection applies only if Kraft had delegated *entirely* "the responsibility to provide its employees, including plaintiff, with a safe working environment". 566 F. Supp. at 527. Similarly, the district court's discussion of subsection (c) apparently assumes that Kraft's reliance on National Loss must have been *wholesale*—that is, preclusive of Kraft's undertaking any similar safety services itself—for subsection (c) to apply. *See id.* at 528.

Such a restrictive approach to subsections (b) and (c) is not supported by the relevant cases. Subsection (b) comes into play as long as the party who owes the plaintiff a duty of care has delegated to the defendant any particular part of that duty.

*   *   *

Similarly, an employer's partial reliance on the defendant's undertaking will suffice to trigger subsection (c). An employer need not forsake completely a particular aspect of its safety program. The reliance element of subsection (c) is satisfied if, in relying on the defendant's undertaking, the employer "neglect[s] or reduce[s]" its own safety program. *Bussey v. Travelers Insurance Co.,* 5 Cir. 1981 (per curiam) (summary calendar), 643 F.2d 1075, 1078.

The record in the present case contains evidence from which a jury could legitimately infer that Kraft delegated to National Loss its duty to identify OSHA violations at the Humko plant, and that Kraft relied on National Loss to identify OSHA violations at that plant. On July 5, 1979, Kraft's Corporate Safety Director, Jack Hansen, sent a memorandum to all of Kraft's safety managers instructing them to ensure that there were no OSHA violations at Kraft plants. Five days later, Allen Jamison wrote to National Loss

to request the simulated OSHA inspection.[13] These facts themselves imply delegation and reliance. In addition, Jamison stated in his deposition that Kraft had not itself inspected the Humko plant for OSHA violations, and there is nothing in the record to show that the Humko plant kept any safety experts on its payroll.

The nature of the defendant's business and of its contractual relationship with Kraft is also significant. Many of the cases dealing with section 324A involve the employer's workmen's compensation or liability carrier, who performed inspections to reduce the loss experience of its insured and accordingly to benefit itself by reducing the claims it had to pay. In the case before us, the defendant is not an insurer at all. It is a company that holds itself out as an expert in safety services and that provides those services for a fee. Kraft hired the defendant specifically to find work hazards. These facts, in combination with those noted in the preceding paragraph, establish as a genuine dispute the applicability of section 324A(b) and (c).

*   *   *

The record contains evidence that would support a jury finding that the defendant performed its undertaking negligently, that such negligence proximately caused the plaintiff's injury, and that Kraft either (1) delegated to the defendant part of Kraft's duty to maintain a safe workplace, or (2) relied at least partly on the defendant to discover unsafe working conditions. The grant of summary judgments in favor of the defendant was erroneous.

[13]The report generated from this inspection states that "[t]he purpose of the survey was to identify problem areas requiring corrective action necessary to bring the operations into compliance with Federal Standards." Supp.-Record, exh. P-4, at 1.

◇ NOTES

1. About half the states also allow worker suits against worker compensation insurance companies that perform health and safety inspections (for the purpose of setting insurance rates). See M. Rothstein, *Occupational Safety and Health Law* §508, (2d ed. 1983).

2. A third-party architect or engineer whose negligence contributes to a workplace injury or disease would appear to be liable under the same general theory.

(d) Supervisory personnel

Some commentators (including the authors of this text) have argued that those in management who are directly responsible for occupational health and safety, perhaps even including corporate officers, may be susceptible to worker tort suits in some situations. See Ashford and Caldart, The Control of Reproductive Hazards in the Workplace: A Prescription for Prevention, 5 Indus. Rel. L. J., 523, 557-558 (1983); Martin, Sue Who? Chem Tech, 159 (1977). However, there is a substantial barrier to such liability. The exclusivity provisions of workers' compensation statutes will provide considerable protection to supervisory personnel. If they are employees of the employer and are acting on the employer's behalf, their on-the-job actions ordinarily will be treated as the actions of the employer. Presumptively, then, workers' compensation is the exclusive remedy. However, where the supervisor is an independent contractor or is acting outside the scope of his or her employment, this logic no longer applies. See *Miller v. Reed*, 27 O. App. 3d 70, 499 NE 2d 919 (1986). In some states, the extent to which supervisors and other agents of the employer can be held personally liable for workplace injuries will be spelled out in state statutes. As the following case illustrates, the state legislature is free to narrow the scope of its workers' compensation exclusivity provisions.

Streeter v. Sullivan

Supreme Court of Florida
509 So. 2d 268 (Fla. 1987)

KOGAN, Justice.

*   *   *

STREETER

Suzanne Sullivan was employed as a branch manager of the Davie branch of Atlantic Federal Savings and Loan (Atlantic). In 1981 Atlantic, through Streeter and Melcher, made the economic decision to remove the armed guard from the Davie branch, despite persistent requests from the employees of that branch to maintain the guard. The Davie branch was robbed once in the fall of 1981 and again in June of 1982. Throughout this period the Davie branch employees stepped up their requests to Streeter and Melcher to reassign the armed guard to the branch.

During the June, 1982 robbery, the perpetrator threatened Suzanne Sullivan's life. In July of 1982 the same man returned to the Davie branch and killed Suzanne Sullivan.

Mark Sullivan, Suzanne's husband, brought suit against Atlantic, as well as Streeter and Melcher, alleging that the defendants had acted with gross negligence in failing to provide adequate security, and that this failure proximately caused Suzanne's death.

*   *   *

STANLICK

Stanlick, a truck driver for Kaplan Industries, was injured when he fell asleep while driving Kaplan's truck. Stanlick brought an action against Kaplan Industries, and Donald and John Kaplan, individually. Stanlick alleged that the Kaplans required him to work excessively long hours in violation of federal law and that the Kaplans required Stanlick to falsify his driving records in order to evade detection by federal authorities. He thus alleged that the Kaplans were guilty of willful and wanton misconduct resulting in foreseeable injury to Stanlick.

The liability or immunity of all defendants rests upon our interpretation of section 440.11(1), Florida Statutes (1981). The statute grants immunity to employers and employees for simple negligence but imposes liability on employees who act with gross negligence with respect to their fellow employees. The statute reads:

> The same immunities from liability enjoyed by an employer shall extend as well to each employee of the employer when such employee is acting in furtherance of the employer's business and the injured employee is entitled to receive benefits under this chapter. *Such fellow employee immunities shall not be applicable to an*

employee who acts, with respect to a fellow employee, with willful and wanton disregard or unprovoked physical aggression or with gross negligence when such acts result in injury or death or such acts proximately cause such injury or death, nor shall such immunities be applicable to employees of the same employer when each is operating in the furtherance of the employer's business but they are assigned primarily to unrelated works within private or public employment (emphasis added).

We believe the emphasized portion of the statute to be an unambiguous statement of the legislature's desire to impose liability on all employees who act with gross negligence with respect to their fellow employees, regardless of the grossly negligent employee's corporate status.

The defendants request this Court to define the term "employee," for the purpose of this statute, to exclude corporate officers who are performing the employer's nondelegable duty to maintain a safe workplace. In defining the term "employee," as used in section 440.11(1), we turn to the definitional section of the Worker's Compensation Act, section 440.02. That statute reads, in pertinent part:

> When used in this chapter, unless the context clearly requires otherwise, the following terms shall have the following meanings: . . .
>
> (2)(b) The term "employee" includes any person who is an officer of a corporation and who performs services for renumeration for such corporation within this state, whether or not such services are continuous.

We believe that the plain language of this statute definitively brings corporate officers within the scope of the term "employee."

The defendants argue that a distinction be drawn between classes of employees. In support of their arguments, defendants rely primarily on their interpretation of the legislative intent behind the statute, as well as case law having roots that extend to a period before the statute was significantly amended.

The first argument espoused by the defendants is that in order to be liable to a fellow employee, a corporate officer must have committed some affirmative act going beyond the scope of the employer's nondelegable duty to provide a safe workplace. *Kaplan v. Tenth Judicial Circuit,* 495 So. 2d 231 (Fla.2d DCA 1986), *West v. Jessop,* 339 So. 2d 1136 (Fla.2d DCA 1976). Defendants argue that if any affirmative acts were committed in either of these cases they did not go beyond the employer's nondelegable duty to

provide a safe place to work. On this basis, the defendants contend that for purposes of providing a safe workplace, the corporate officer is not an employee, but rather an "alter-ego" of the employer, deserving the benefits of the employer's immunity. See *Zurich Insurance Co. v. Scofi,* 366 So. 2d 1193 (Fla.2d DCA), *cert. denied,* 378 So. 2d 348 (Fla. 1979).

*   *   *

The affirmative act doctrine has its roots in cases interpreting section 440.11(1) before it was amended in 1978.[4] Those cases did not have the benefit of the legislature's statement expressly imposing liability on grossly negligent employees who injure other employees. The basis of those opinions was legislatively abrogated by section 440.11(1). Thus, to the extent that those cases conflict with this opinion (as well as section 440.11(1), as amended), we disapprove of them.

*   *   *

The 1978 amendment to section 440.11(1) authorizes actions against all fellow employees for acts of gross negligence resulting in injury to other employees. To separate corporate officers from this rule requires a highly convoluted and logistically suspect construction of the statute. The defendants, at oral argument, contended that the term "fellow employee" is a term of art, applying solely to employees with whom the injured person work on an every day basis. Therefore, the defendants contend, the term "fellow employee" could only refer to nonsupervisory employees. We are equally disinclined to follow this line of reasoning. Again, we must stress that the plain language of section 440.11(1) fully precludes any such interpretation. By "fellow employees," the statute clearly is intended to include *all* employees, not just nonsupervisory employees.

*   *   *

OVERTON, Justice, dissenting.

I dissent. The majority's interpretation of section 440.11(1), Florida Statutes (1983), was clearly not the intent of the legislature when it brought corporate officers, executives, and supervisors within the ben-

[4]It has no roots in the common law, where a corporate officer was without doubt liable for gross negligence, and perhaps even simple negligence. See Frantz v. McBee Co., 77 So. 2d 796 (Fla.1955).

efits of the worker's compensation act. The majority's decision will have a substantial economic impact, since it will provide employees a means to litigate a decision by corporate officers to reduce the work force by filing a civil action alleging that the decision is a grossly negligent act because it results in an unsafe work place.

* * *

I would apply worker's compensation immunity to corporate officers. . . .

The majority decision appears to require immediate legislative review.

McDONALD, C.J., concurs.

(e) Medical professionals

A final party worthy of note is the medical professional who treats the worker for occupational injury or disease.

The Control of Reproductive Hazards in the Workplace: A Prescription for Prevention

N. Ashford and C. Caldart

Typically, a physician or nurse practitioner will see an employee periodically during the course of his or her employment. Where the employment involves exposure to a reproductive hazard, and where damage ultimately results as a consequence of such exposure, a cause of action for professional negligence may arise. A case for malpractice would require proof of three factors: that the medical professional knew, or reasonably should have known, that the worker was exposed to the hazard; that he or she had a duty to render proper medical advice or take action with regard to the exposure; and that he or she breached that duty.

Source: From 5 *Indus. Rel. L. J.* 523-563, 558 (1983).

In many cases, the existence of these requisite elements may depend on the relationship between the medical professional and the workplace. A physician who treats a number of workers from a particular workplace, such as a company doctor or an employee of a health-maintenance organization that has contracted to provide medial services as part of a workplace health-care plan, is likely to be in a better position to know about the various chemicals present in the workplace. Additionally, a physician or nurse who specialized in occupational health may be held to a higher standard of care than the average practitioner. The professional duty will vary with the circumstances of the case, but will probably encompass an obligation to adequately warn workers of the potential consequences of further exposure.

The nature of the relationship between the medical professional and the workplace will be important for other reasons as well. If the physician or nurse is clearly identified as serving the interests of the *employer* only and as not providing medical treatment to the worker, the requisite professional/patient relationship does not really exist. If this is the case, the medical professional likely will have no legal duty either to warn the worker of medical risk or to recommend medical treatment. In addition, when the medical professional is an employee of the patient's employer, the worker seeking to bring an action for medical malpractice must contend with the exclusivity provisions of the workers' compensation statute. In theory, the medical professional is no different from any other co-employee, and workers' compensation will be the exclusive remedy.

As with the liability of supervisory personnel, however, the courts have begun to chip away at the personal immunity of the employee-physician. A few courts have applied the *dual capacity* doctrine and have held that the company doctor's role as the worker's medical provider is sufficiently separate from the doctor's role as the worker's co-employee to permit the imposition of malpractice liability outside the bounds of workers' compensation. In *Tatrai v. Presbyterian University Hospital,* 497 Pa. 247, 439 A.2d 1162 (1982), for example, a woman who worked as an operating room technician at a hospital was permitted to sue her employer for medical malpractice for injury she incurred when she went to the hospital's emergency room for treatment. Although the injury occurred when she was on duty, the court reasoned that her presence in the emergency room was related solely to the hospital's role as a medical provider and not to its role as her employer. In this regard, said the court, she was no different from any patient who had come to the hospital for treatment. Two courts have taken the dual capacity doctrine a bit farther. Both Indiana and Louisiana have held that a company physician will always be treated as an *independent contractor*—rather than as an employee —for purposes of medical malpractice liability, on the theory that a corporation cannot practice medicine. *Ross v. Schubert,* 388 N.E.2d 623 (Ind. App. 1979); *Ducote v. Albert,* 503 So. 2d 85 (La. App. 1987). If the corporation cannot provide the treatment, the courts have reasoned, it cannot be liable for any resultant malpractice. Liability, then, must rest with the individual physician.

In all states, a medical professional who treats a worker will not be able to invoke the protection of workers' compensation exclusivity if the medical professional cannot establish that he or she truly is an employee of the worker's employer. The fact that a physician or nurse has a continuing arrangement with the employer to treat the workers likely will not, by itself, be enough to establish the necessary employment relationship. If the medical professional maintains a separate office or sees other patients, his or her status as an "employee" may be questionable. The team physician for the National Football League's Chicago Bears, for example, has been held to be an independent contractor for purposes of a malpractice suit by a player. See *Bryant v. Fox,* 162 Ill. App. 3d 46, 515 N.E.2d 775 (1987).

◇ **NOTES**

1. As noted earlier, the dual capacity doctrine remains the minority rule among the states. However, courts appear more willing to apply the doctrine here than in the products liability context, and one commentator has observed that its use in the medical malpractice context "appears to be somewhat of a trend." Postol, Suing the Doctor: Lawsuits by Injured Workers Against the Occupational Physician, 31 *J. of Occupational Med.* 891, 892 (1989). The doctrine has been applied to the company doctor in California, *Duprey v. Shane,* 39 Cal. 2d 781, 249 P.2d 8 (1952); Georgia, *Davis v. Stover,* 366 S.E. 2d 670 (Ga. 1988); and Ohio, *Guy v. Arthur H. Thomas Co.,* 55 Ohio St. 2d 183, 378 N.E.2d 488 (1978). It has been specifically rejected in this context in other states, including New Jersey, *Boyle v. Breme,* 187 N.J. Super. 129, 453 A.2d 1335 (App.Div. 1982), and Tennessee, *McAlister v. Methodist Hospital,* 550 S.W.2d 240 (Tenn. 1977).

2. An employee who would ordinarily be protected by the exclusivity provisions of the workers' compensation law (such as a company doctor or supervisor) may be held to be acting outside the scope of employment (and thus beyond the protection of the workers' compensation law) if he or she *intentionally* harms the worker. (Even without its fellow servant statute, Florida might well apply this rule in the case of gross negligence as well; see footnote 4 to the *Streeter* case, *supra.*) Further, if such intentional harm was authorized (explicitly or implicitly) by the employer, it may subject the *employer* to tort liability in states that recognize the "intentional act" exception to workers' compensation exclusivity.

3. This is not an exhaustive compendium of all the potential third-party defendants in a workplace liability suit. For example, it is generally held that a property owner owes a duty to maintain a safe premises and that this duty extends to the employees of an independent contractor doing work at the site. For a discussion of this and other types of third-party suits, see M. Rothstein, *Occupational Safety and Health Law* §506 (2d ed. 1983).

◇

D. PROOF OF CAUSATION IN OCCUPATIONAL DISEASE CASES GROWING OUT OF EXPOSURE TO TOXIC SUBSTANCES

Whether the worker seeks recovery under workers' compensation or tort law, he or she must bear the burden of proving that the injury or illness in question was caused by a workplace condition. In the case of occupational accidents, this commonly involves the kinds of traditional fact-based inquiries with which the law has long been comfortable. In the case of occupational disease, however, the evidentiary record commonly is much more uncertain and the resultant factual inquiry much more complicated. Indeed, when the disease in question is one with several potential etiologies (as is the case with many diseases or conditions caused by toxic substance exposure), proof of causation in the traditional sense may not be possible.

Monitoring Workers for Exposure and Disease: Scientific, Legal and Ethical Considerations in the Use of Biomarkers

N. A. Ashford, C. Spadafor, D. Hattis, and C. Caldart

Regardless of whether the case is brought in tort or under workers' compensation, the worker must satisfy a causation requirement in order to prevail. In a common law tort action, the plaintiff must prove—by a preponderance of the evidence—that the defendant's actions were the *proximate cause* of the injury in question. In its simplest terms, this means that a worker must be able to demonstrate that it is *more likely than not* that his or her disease was caused by a

Source: From *Monitoring Workers for Exposure and Disease: Scientific, Legal and Ethical Considerations in the Use of Biomarkers.* Baltimore: Johns Hopkins University Press, 1990, pp. 154–158.

workplace condition, and not by some other factor. The burden in a workers' compensation case is essentially the same. As noted, the worker must prove that the injury was job-related. Here again, this requires proof that it is more likely than not that a workplace condition led to the disease in question.

Although this does not require scientific certainty, it does require particularity. In situations of toxic substance exposure, it will not be sufficient to establish that it is more probable than not that the chemical in question causes a disease. Rather, the worker must prove by a greater than fifty percent probability that the chemical caused his or her particular case of that disease. Even in the best-prepared cases, this is a difficult and often insurmountable burden. The chief reasons for this difficulty are by now all too familiar.

In the first place, the probability that exposure to a toxic substance, especially at the relatively low levels commonly encountered, will produce disease or other long-term cellular damage in any given exposed person is usually quite low. It thus is likely that many actual correlations between medical effects and chemical exposure simply go undetected. Further, there commonly are a number of other (so-called "confounding") factors, beyond the suspect toxic exposure itself, which also can cause the injury in question. The statistical probability that any given exposed person who contracts the condition did so as a result of the exposure, then, usually is less than fifty percent. Finally, many of the results of toxic substance exposure (including most cancers) have long latency periods, and do not become manifest until many years after exposure. Accordingly, an already difficult proof problem is exacerbated by the vagaries of time, distance, and memory.

* * *

THE PROMISE OF BIOMARKERS

Obviously, to the extent that increased use of human monitoring adds to the existing data base on the observed correlations between particular diseases and particular chemicals, it will provide increased evidence for use in compensation proceedings generally. More than this, though, human monitoring has the potential to bring about a change in the *nature* of the evidence used in these cases.

Typically, the evidence offered to prove causation in chemical exposure cases is premised on a *statis-tical* correlation between disease and exposure. Whether the underlying data are from epidemiologic studies, from toxicological experiments, or from the results of a complicated risk assessment model, they usually are *population*-based. This places the plaintiff at the mercy of the attributable risk (expressed as the percentage of cases of the disease attributable to the exposure) for the study population. Unless the attributable risk is greater than 50%—that is, unless the incidence rate among those exposed to the chemical is more than double the background rate—the plaintiff cannot prove *on the basis of the available statistical evidence,* that it is more likely than not that his or her particular case of the disease was caused by the chemical exposure.

[Figure 9-2]—representing a hypothetical population of 200 workers—illustrates the point. Here, 40 of the workers who were exposed to the chemical contracted the disease. Of those, there is a statistical probability that 10 (25%) contracted the disease as a result of the exposure. On the basis of this evidence alone, however, there is no way of identifying *which* 10 of the 40 are in that category. Without additional evidence, none of the 40 diseased workers who was exposed to the chemical can prove that it is more likely than not that his or her case of the disease was caused by the exposure. Accordingly, even though it is likely that 10 of the 40 are entitled to compensation, the causation requirement will prevent them from receiving it under either the tort or workers' compensation system.[176]

---

[176]Of course, if the attributable risk were *higher* than 50%, *all* of the exposed workers with the disease theoretically would recover.

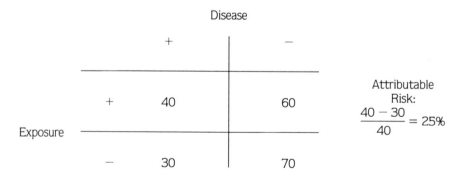

$$\text{Attributable Risk:} \quad \frac{40 - 30}{40} = 25\%$$

Figure 9-2. Disease–exposure risk matrix.

The developing science of human monitoring may offer a way to distinguish individual claimants from the pack. Conceivably, the data generated by various human monitoring procedures will:

- increase our knowledge of the "sub-clinical" effects of toxic substances, thus permitting us to track the effect of a chemical exposure over time, and also expanding the universe of "medical conditions" for which compensation may be provided;

- eventually enable us to establish that a particular person has been exposed to a particular chemical (or class of chemicals); and

- eventually enable us to establish that a particular person's medical condition (or sub-clinical effect) was caused by exposure to a particular chemical (or class of chemicals).

Already, human monitoring data are being used in some situations to show sub-clinical changes thought to be associated with particular chemical exposures.[177] This can have many applications in toxic substance compensation cases. In the long term, evidence of sub-clinical changes occurring between the time of exposure and the time of disease may be a way of distinguishing those whose disease was caused by the exposure from those who contracted the disease because of other factors. Such evidence may also give rise to more immediate legal relief. There is a growing trend toward allowing those who can establish that they have been exposed to a toxic substance—and that they thus have been placed at increased risk of future harm—to recover the costs of medical surveillance from the responsible party.[178] Proof of certain sub-clinical effects, such as DNA damage, would tend to support an allegation of increased risk, and would make the claim for medical surveillance all the more compelling. Further, such evidence may support a separate claim for damages *for*

having been put at increased risk of future harm, although it is not clear that such a claim would be viable in most states.[179]

Finally, some evidence of "sub-clinical" effects may give rise to a right to recover compensation *for those effects themselves*. For example, human monitoring can detect certain changes in the immune system. There is a body of literature suggesting that chemical exposures can harm the immune system,[180] and evidence of immune system damage has been offered in recent cases involving toxic substance exposure.[181] Thus far, allegations of immune system damage have met with mixed success in the courts, both because the relationship between chemical exposure and immune system damage is not yet clear, and because the evidence of immune system damage was not always considered persuasive. Although human monitoring may not be able to tie particular immune system deficiencies to particular exposures, it should be able to establish with greater certainty whether immune system damage has, in fact, occurred.

[179]The emerging rule seems to be that one cannot recover for increased risk unless one can establish that the risk of contracting the medical condition in question is greater than 50%. *See e.g.*, Sterling v. Velsicol Chemical Corp., 855 F.2d 1188, 1205 (6th Cir. 1988); Hagerty v. L & L Marine Services, Inc., 788 F.2d 315, modified *en banc*, 797 F.2d 256 (5th Cir. 1986); Herber v. Johns-Manville Corp., 785 F.2d 79 (3d Cir. 1986). Some courts have allowed recovery where the risk was less than 50%; *see* Brafford v. Susquehanna Corp., 586 F.Supp. 14 (D. Colo. 1984); Herskovits v. Group Health Cooperative of Puget Sound, 664 P.2d 474 (Wash. 1983) (decreased chance of survival for a cancer patient). *See* J. King, Causation, Valuation, and Chance in Personal Injury Torts Involving Preexisting Conditions and Future Consequences, Yale Law Journal (1981) 90:1353, for a theoretical justification for this approach. It does involve something of a conceptual leap, because a court must be able to recognize that being placed at a moderately higher risk—say 20%—of contracting a fatal disease, *is* a harm in and of itself. Recovery for *emotional damages for having been put at risk* is a separate potential source of compensation; *see* Sterling v. Velsicol Chemical Corp., 855 F.2d at 1207, where the court *did* allow recovery on this basis, up to a fixed maximum of $9000 per year per person.

[180]Descotes, J., *Immunotoxicity of Drugs and Chemicals* (New York: Elsevier, 1986).

[181]*See* R. Rothman and A. Maskin, Defending Immunotoxicity Cases, Toxics Law Reporter (BNA) (1989) 4:1219, and the cases cited therein. (This is very much written from the corporate defendant's point of view.)

[177]The plaintiffs reportedly were prepared to offer such evidence in the suit brought by the families of eight leukemia victims against W.R. Grace & Co. and Beatrice Companies, Inc. over contaminated drinking water in Woburn, Massachusetts. *See* M. Pacelle, Contaminated Verdict, *American Lawyer*, (December 1986) 1.

[178]*See* L. Gara, "Medical Surveillance Damages: Using Common Sense and the Common Law to Mitigate the Dangers Posed by Environmental Hazards," *Harvard Environmental Law Review* (1988) 12:265.

Looking farther to the future, it is quite possible that further developments in the science of biomarkers will permit the identification of "chemical footprints"—a distinctive change in the DNA that can be tied to a particular chemical or class of chemicals. At the very least, this should make it much easier to distinguish those who have been exposed to a particular chemical in the workplace from those who have not, and to identify which of the many potential defendants was responsible. More importantly, it should eventually permit the correlation of particular cases of diseases such as cancer with exposure to particular chemicals (or classes of chemicals). To the extent that this happens, it will narrow the scope of the evidence from the population to the individual, and will place the deliberations in these cases on firmer scientific footing.

◇ **NOTES**

1. As one of the authors of this text has noted in a somewhat different context, the underlying issue is that the tort and workers' compensation systems have not yet developed the "tools" that will allow them to assimilate the kind of evidence typically presented in toxic substance exposure cases.

> Adjustment of the tort system to meet the challenges posed by these problems is entirely appropriate. Clearly, the underlying purpose of the causation element of tort law is not to preserve any particular mathematical concept of probability or certainty, but rather to preserve fundamental notions of fairness and justice. The principles of proximate causation built into our current tort law appear to be based on the paradigms of classic physics: the traditional formulations of tort law largely treat causation as a simple linear pathway between two phenomena, and largely treat damage as an acute, readily identifiable event. Neither concept is adequate to describe the manifestation of latent diseases such as cancer or inter-generational mutation. It is thus necessary that we move toward a concept of tort liability that is relevant to modern biochemical science. [Caldart and Hattis, Testimony before the Massachusetts Special Legislative Commission on Liability for Releases of Oil and Hazardous Materials (Dec. 5, 1984).]

2. One approach to the causation problem in cases of this nature—until the available evidence permits differentiation among plaintiffs—would be to require the defendant to pay damages *in proportion to the attributable risk* to each exposed plaintiff who contracts the disease. Thus, in the example given in the preceding excerpt, each of the 40 workers in the "exposed and diseased" category would receive 25 percent of his or her damages. See, e.g., Trauberman, Statutory Reform of "Toxic Torts": Relieving Legal, Scientific, and Economic Burdens on the Chemical Victim, 7 *Harv. Env. L.R.* 177, 226 (1983).

3. A second approach would be to allow each exposed person to recover damages for *having been put at increased risk of harm.* The extent of these damages would be measured by the jury, according to the statistical probability of increased harm. (See footnote 179 in the preceding excerpt.)

4. A related approach would be to allow the exposed worker to recover for the *emotional damages* of having been put at increased risk of harm. Is there a potential for fraud or overstatement with this approach? Is it offset by equity and social policy considerations? Compare *Payton v. Abbot Laboratories,* 386 Mass. 540 (1982), with *Sterling v. Velsicol Chemical Corp.,* 855 F.2d 1188, 1207 (6th Cir. 1988).

5. A fourth approach would be to require all users of hazardous substances to contribute to a fund from which all workers suffering from particular diseases would be compensated, regardless of whether they could prove etiology. See Barth, A Proposal for Dealing with Occupational Diseases, and Priest, Law and Economics and Law Reform: Comment on Barth's Cancer Compensation Proposal, 13 *J. Legal Stud.* 569 (1984).

6. All of these approaches would involve something of a shift in social policy. Is this warranted? Who is responsible for the conditions giving rise to scientific uncertainty, and who bears the burden?

APPENDIX A

The Occupational

Safety and Health Act

Public Law 91-596
91st Congress, S. 2193
December 29, 1970
29 U.S.C. §651 et seq.

AN ACT

To assure safe and healthful working conditions for working men and women; by authorizing enforcement of the standards developed under the Act; by assisting and encouraging the States in their efforts to assure safe and healthful working conditions; by providing for research, information, education, and training in the field of occupational safety and health; and for other purposes.

Be it enacted by the Senate and House of Representatives of the United States of America in Congress assembled, That this Act may be cited as the "Occupational Safety and Health Act of 1970."

Occupational Safety and Health Act of 1970.

Congressional Findings and Purpose

SEC. (2) The Congress finds that personal injuries and illnesses arising out of work situations impose a substantial burden upon, and are a hindrance to, interstate commerce in terms of lost production, wage loss, medical expenses, and disability compensation payments.

29 U.S.C. 651.

(b) The Congress declares it to be its purpose and policy, through the exercise of its powers to regulate commerce among the several States and with foreign nations and to provide for the general welfare, to assure so far as possible every working man and woman in the Nation safe and healthful working conditions and to preserve our human resources—

(1) by encouraging employers and employees in their efforts to reduce the number of occupational safety and health hazards at their places of employment,

and to stimulate employers and employees to institute new and to perfect existing programs for providing safe and healthful working conditions;

(2) by providing that employers and employees have separate but dependent responsibilities and rights with respect to achieving safe and healthful working conditions;

(3) by authorizing the Secretary of Labor to set mandatory occupational safety and health standards applicable to businesses affecting interstate commerce, and by creating an Occupational Safety and Health Review Commission for carrying out adjudicatory functions under the Act;

(4) by building upon advances already made through employer and employee initiative for providing safe and healthful working conditions;

(5) by providing for research in the field of occupational safety and health, including the psychological factors involved, and by developing innovative methods, techniques, and approaches for dealing with occupational safety and health problems;

(6) by exploring ways to discover latent diseases, establishing causal connections between diseases and work in environmental conditions, and conducting other research relating to health problems, in recognition of the fact that occupational health standards present problems often different from those involved in occupational safety;

(7) by providing medical criteria which will assure insofar as practicable that no employee will suffer diminished health, functional capacity, or life expectancy as a result of his work experience;

(8) by providing for training programs to increase the number and competence of personnel engaged in the field of occupational safety and health;

(9) by providing for the development and promulgation of occupational safety and health standards;

(10) by providing an effective enforcement program which shall include a prohibition against giving advance notice of any inspection and sanctions for any individual violating this prohibition;

(11) by encouraging the States to assume the fullest responsibility for the administration and enforcement of their occupational safety and health laws by providing grants to the States to assist in identifying their needs and responsibilities in the area of occupational safety and health, to develop plans in accordance with the provisions of this Act, to improve the administration and enforcement of State occupational safety and health laws, and to conduct experimental and demonstration projects in connection therewith;

(12) by providing for appropriate reporting procedures with respect to occupational safety and health which procedures will help achieve the objectives of this Act and accurately describe the nature of the occupational safety and health problem;

(13) by encouraging joint labor-management efforts to reduce injuries and disease arising out of employment.

Definitions

29 U.S.C. 652.   SEC. 3. For the purposes of this Act—

(1) The term "Secretary" means the Secretary of Labor.

(2) The term "Commission" means the Occupational Safety and Health Review Commission established under this Act.

(3) The term "commerce" means trade, traffic, commerce, transportation, or communication among the several States, or between a State and any place outside thereof, or within the District of Columbia, or a possession of the United States (other than the Trust Territory of the Pacific Islands), or between points in the same State but through a point outside thereof.

(4) The term "person" means one or more individuals, partnerships, associations, corporations, business trusts, legal representatives, or any organized group of persons.

(5) The term "employer" means a person engaged in a business affecting commerce who has employees, but does not include the United States or any State or political subdivision of a State.

(6) The term "employee" means an employee of an employer who is employed in a business of his employer which affects commerce.

(7) The term "State" includes a State of the United States, the District of Columbia, Puerto Rico, the Virgin Islands, American Samoa, Guam, and the Trust Territory of the Pacific Islands.

(8) The term "occupational safety and health standard" means a standard which requires conditions, or the adoption or use of one or more practices, means, methods, operations, or processes, reasonably necessary or appropriate to provide safe or healthful employment and places of employment.

(9) The term "national consensus standard" means any occupational safety and health standard or modification thereof which (1) has been adopted and promulgated by a nationally recognized standards-producing organization under procedures whereby it can be determined by the Secretary that persons interested and affected by the scope or provisions of the standard have reached substantial agreement on its adoption, (2) was formulated in a manner which afforded an opportunity for diverse views to be considered and (3) has been designated as such a standard by the Secretary, after consultation with other appropriate Federal agencies.

(10) The term "established Federal standard" means any operative occupational safety and health standard established by any agency of the United States and presently in effect, or contained in any Act of Congress in force on the date of enactment of this Act.

(11) The term "Committee" means the National Advisory Committee on Occupational Safety and Health established under this Act.

(12) The term "Director" means the Director of the National Institute for Occupational Safety and Health.

(13) The term "Institute" means the National Institute for Occupational Safety and Health established under this Act.

(14) The term "Workmen's Compensation Commission" means the National Commission on State Workmen's Compensation Laws established under this Act.

Applicability of This Act

SEC. 4 (a) This Act shall apply with respect to employment performed in a workplace in a State, the District of Columbia, the Commonwealth of Puerto Rico, the Virgin Islands, American Samoa, Guam, the Trust Territory of the Pacific Islands, Wake Island, Outer Continental Shelf lands defined in the Outer Continental Shelf Lands Act, Johnston Island, and the Canal Zone. The Secretary of the Interior shall, by regulation, provide for judicial enforcement of this Act by the courts established for areas in which there are no United States district courts having jurisdiction.

29 U.S.C. 653.

67 Stat. 462.
43 USC 1331
note.

(b) (1) Nothing in this Act shall apply to working conditions of employees with respect to which other Federal agencies, and State agencies acting under section 274 of the Atomic Energy Act of 1954, as amended (42 U.S.C. 2021), exercise statutory authority to prescribe or enforce standards or regulations affecting occupational safety or health.

73 Stat. 688.

(2) The safety and health standards promulgated under the Act of June 30, 1936, commonly known as the Walsh-Healey Act (41 U.S.C. 35 et seq.), the Service Contract Act of 1965 (41 U.S.C. 351 et seq.), Public Law 91-54, Act of August 9, 1969 (40 U.S.C. 333), Public Law 85-742, Act of August 23, 1958 (33 U.S.C. 941), and the National Foundation on Arts and Humanities Act (20 U.S.C. 951 et seq.) are superseded on the effective date of corresponding standards, promulgated under this Act, which are determined by the Secretary to be more effective. Standards issued under the laws listed in this paragraph and in effect on or after the effective date of this Act shall be deemed to be occupational safety and health standards issued under this Act, as well as under such other Acts.

49 Stat. 2036.
79 Stat. 1034.
83 Stat. 96.
72 Stat. 835.
79 Stat. 845.

(3) The Secretary shall, within three years after the effective date of this Act, report to the Congress his recommendations for legislation to avoid unnecessary duplication and to achieve coordination between this Act and other Federal laws.

Report to
Congress.

(4) Nothing in this Act shall be construed to supersede or in any manner affect any workmen's compensation law or to enlarge or diminish or affect in any other manner the common law or statutory rights, duties, or liabilities of employers and employees under any law with respect to injuries, diseases, or death of employees arising out of, or in the course of, employment.

Duties

29 U.S.C. 654.

SEC. 5. (a) Each employer—

(1) shall furnish to each of his employees employment and a place of employment which are free from recognized hazards that are causing or are likely to cause death or serious physical harm to his employees;

(2) shall comply with occupational safety and health standards promulgated under this Act.

(b) Each employee shall comply with occupational safety and health standards and all rules, regulations, and orders issued pursuant to this Act which are applicable to his own actions and conduct.

Occupational Safety and Health Standards

29 U.S.C. 655.

80 Stat. 381;
81 Stat. 195.
5 USC 500.

SEC. 6. (a) Without regard to chapter 5 of title 5, United States Code, or to the other subsections of this section, the Secretary shall, as soon as practicable during the period beginning with the effective date of this Act and ending two years after such date, by rule promulgate as an occupational safety or health standard any national consensus standard, and any established Federal standard, unless he determines that the promulgation of such a standard would not result in improved safety or health for specifically designated employees. In the event of conflict among any such standards, the Secretary shall promulgate the standard which assures the greatest protection of the safety or health of the affected employees.

(b) The Secretary may by rule promulgate, modify, or revoke any occupational safety or health standard in the following manner:

(1) Whenever the Secretary, upon the basis of information submitted to him in writing by an interested person, a representative of any organization of employers or employees, a nationally recognized standards-producing organization, the Secretary of Health, Education, and Welfare, the National Institute for Occupational Safety and Health, or a State or political subdivision, or on the basis of information developed by the Secretary or otherwise available to him, determines that a rule should be promulgated in order to serve the objectives of this Act, the Secretary may request the recommendations of an advisory committee appointed under section 7 of this Act. The Secretary shall provide such an advisory committee with any proposals of his own or of the Secretary of Health, Education, and Welfare, together with all pertinent factual information developed by the Secretary or the Secretary of Health, Education, and Welfare, or otherwise available, including the results of research, demonstrations, and experiments. An advisory committee shall submit to the Secretary its recommendations regarding the rule to be promulgated within ninety days from the date of its appointment or within such longer or shorter period as may be prescribed by the Secretary, but in no event for a period which is longer than two hundred and seventy days. *Advisory committee, recommendations.*

(2) The Secretary shall publish a proposed rule promulgating, modifying, or revoking an occupational safety or health standard in the Federal Register and shall afford interested persons a period of thirty days after publication to submit written data or comments. Where an advisory committee is appointed and the Secretary determines that a rule should be issued, he shall publish the proposed rule within sixty days after the submission of the advisory committee's recommendations or the expiration of the period prescribed by the Secretary for such submission. *Publication in Federal Register.*

(3) On or before the last day of the period provided for the submission of written data or comments under paragraph (2), any interested person may file with the Secretary written objections to the proposed rule, stating the grounds therefor and requesting a public hearing on such objections. Within thirty days after the last day for filing such objections, the Secretary shall publish in the Federal Register a notice specifying the occupational safety or health standard to which objections have been filed and a hearing requested, and specifying a time and place for such hearing. *Hearing, notice.* *Publication in Federal Register.*

(4) Within sixty days after the expiration of the period provided for the submission of written data or comments under paragraph (2), or within sixty days after the completion of any hearing held under paragraph (3), the Secretary shall issue a rule promulgating, modifying, or revoking an occupational safety or health standard or make a determination that a rule should not be issued. Such a rule may contain a provision delaying its effective date for such period (not in excess of ninety days) as the Secretary determines may be necessary to insure that affected employers and employees will be informed of the existence of the standard and of its terms and that employers affected are given an opportunity to familiarize themselves and their employees with the existence of the requirements of the standard.

(5) The Secretary, in promulgating standards dealing with toxic materials or harmful physical agents under this subsection, shall set the standard which most adequately assures, to the extent feasible, on the basis of the best available evidence, that no employee will suffer material impairment of health or functional capacity even if such employee has regular exposure to the hazard dealt with by such standard for the *Toxic materials.*

period of his working life. Development of standards under this subsection shall be based upon research, demonstrations, experiments, and such other information as may be appropriate. In addition to the attainment of the highest degree of health and safety protection for the employee, other considerations shall be the latest available scientific data in the field, the feasibility of the standards, and experience gained under this and other health and safety laws. Whenever practicable, the standard promulgated shall be expressed in terms of objective criteria and of the performance desired.

Temporary variance order.

(6) (A) Any employer may apply to the Secretary for a temporary order granting a variance from a standard or any provision thereof promulgated under this section. Such temporary order shall be granted only if the employer files an application which meets the requirements of clause (B) and establishes that (i) he is unable to comply with a standard by its effective date because of unavailability of professional or technical personnel or of materials and equipment needed to come into compliance with the standard or because necessary construction or alteration of facilities cannot be completed by the effective date, (ii) he is taking all available steps to safeguard his employees against the hazards covered by the standard, and (iii) he has an effective program for coming into compliance with the standard as quickly as practicable. Any temporary order issued under this paragraph shall prescribe the practices, means, methods, operations, and processes which the employer must adopt and use while the order is in effect and state in detail his program for coming into compliance with the standard. Such a temporary order may be granted only after notice to employees and an opportunity for a hearing: *Provided,* That the Secretary may issue one interim order to be effective until a decision is made on the basis of the hearing. No temporary order may be in effect for longer than the period needed by the employer to achieve compliance with the standard or one year, whichever is shorter, except that such an order may be renewed not more than twice (I) so long as the requirements of this paragraph are met and (II) if an application for renewal is filed at least 90 days prior to the expiration date of the order. No interim renewal of an order may remain in effect for longer than 180 days.

Notice, hearing.

Renewal.

Time limitation.

(B) An application for a temporary order under this paragraph (6) shall contain:

(i) a specification of the standard or portion thereof from which the employer seeks a variance,

(ii) a representation by the employer, supported by representations from qualified persons having firsthand knowledge of the facts represented, that he is unable to comply with the standard or portion thereof and a detailed statement of the reasons therefor,

(iii) a statement of the steps he has taken and will take (with specific dates) to protect employees against the hazard covered by the standard,

(iv) a statement of when he expects to be able to comply with the standard and what steps he has taken and what steps he will take (with dates specified) to come into compliance with the standard, and

(v) a certification that he has informed his employees of the application by giving a copy thereof to their authorized representative, posting a statement giving a summary of the application and specifying where a copy may be examined at the place or places where notices to employees are normally posted, and by other appropriate means.

A description of how employees have been informed shall be contained in the certification. The information to employees shall also inform them of their right to petition the Secretary for a hearing.

(C) The Secretary is authorized to grant a variance from any standard or portion thereof whenever he determines, or the Secretary of Health, Education, and Welfare certifies, that such variance is necessary to permit an employer to participate in an experiment approved by him or the Secretary of Health, Education, and Welfare designed to demonstrate or validate new and improved techniques to safeguard the health or safety of workers.

(7) Any standard promulgated under this subsection shall prescribe the use of labels or other appropriate forms of warning as are necessary to insure that employees are apprised of all hazards to which they are exposed, relevant symptoms and appropriate emergency treatment, and proper conditions and precautions of safe use or exposure. Where appropriate, such standard shall also prescribe suitable protective equipment and control or technological procedures to be used in connection with such hazards and shall provide for monitoring or measuring employee exposure at such locations and intervals, and in such manner as may be necessary for the protection of employees. In addition, where appropriate, any such standard shall prescribe the type and frequency of medical examinations or other tests which shall be made available, by the employer or at his cost, to employees exposed to such hazards in order to most effectively determine whether the health of such employees is adversely affected by such exposure. In the event such medical examinations are in the nature of research, as determined by the Secretary of Health, Education, and Welfare, such examinations may be furnished at the expense of the Secretary of Health, Education, and Welfare. The results of such examinations or tests shall be furnished only to the Secretary or the Secretary of Health, Education, and Welfare, and, at the request of the employee, to his physician. The Secretary, in consultation with the Secretary of Health, Education, and Welfare, may by rule promulgated pursuant to section 553 of title 5, United States Code, make appropriate modifications in the foregoing requirements relating to the use of labels or other forms of warning, monitoring or measuring, and medical examinations, as may be warranted by experience, information, or medical or technological developments acquired subsequent to the promulgation of the relevant standard. *(Labels, etc. Protective equipment, etc. Medical examinations.)*

(8) Whenever a rule promulgated by the Secretary differs substantially from an existing national consensus standard, the Secretary shall, at the same time, publish in the Federal Register a statement of the reasons why the rule as adopted will better effectuate the purposes of this Act than the national consensus standard. *(Publication in Federal Register.)*

(c) (1) The Secretary shall provide, without regard to the requirements of chapter 5, title 5, United States Code, for an emergency temporary standard to take immediate effect upon publication in the Federal Register if he determines (A) that employees are exposed to grave danger from exposure to substances or agents determined to be toxic or physically harmful or from new hazards, and (B) that such emergency standard is necessary to protect employees from such danger. *(Temporary standard. Publication in Federal Register. 80 Stat. 381; 81 Stat. 195. 5 USC 500.)*

(2) Such standard shall be effective until superseded by a standard promulgated in accordance with the procedures prescribed in paragraph (3) of this subsection. *(Time limitation.)*

(3) Upon publication of such standard in the Federal Register the Secretary shall commence a proceeding in accordance with section 6(b) of this Act, and the standard

as published shall also serve as a proposed rule for the proceeding. The Secretary shall promulgate a standard under this paragraph no later than six months after publication of the emergency standard as provided in paragraph (2) of this subsection.

Variance rule. (d) Any affected employer may apply to the Secretary for a rule or order for a variance from a standard promulgated under this section. Affected employees shall be given notice of each such application and an opportunity to participate in a hearing. The Secretary shall issue such rule or order if he determines on the record, after opportunity for an inspection where appropriate and a hearing, that the proponent of the variance has demonstrated by a preponderance of the evidence that the conditions, practices, means, methods, operations, or processes used or proposed to be used by an employer will provide employment and places of employment to his employees which are as safe and healthful as those which would prevail if he complied with the standard. The rule or order so issued shall prescribe the conditions the employer must maintain, and the practices, means, methods, operations, and processes which he must adopt and utilize to the extent they differ from the standard in question. Such a rule or order may be modified or revoked upon application by an employer, employees, or by the Secretary on his own motion, in the manner prescribed for its issuance under this subsection at any time after six months from its issuance.

Publication in Federal Register. (e) Whenever the Secretary promulgates any standard, makes any rule, order, or decision, grants any exemption or extension of time, or compromises, mitigates, or settles any penalty assessed under this Act, he shall include a statement of the reasons for such action, which shall be published in the Federal Register.

Petition for judicial review. (f) Any person who may be adversely affected by a standard issued under this section may at any time prior to the sixtieth day after such standard is promulgated file a petition challenging the validity of such standard with the United States court of appeals for the circuit wherein such person resides or has his principal place of business, for a judicial review of such standard. A copy of the the petition shall be forthwith transmitted by the clerk of the court to the Secretary. The filing of such petition shall not, unless otherwise ordered by the court, operate as a stay of the standard. The determinations of the Secretary shall be conclusive if supported by substantial evidence in the record considered as a whole.

(g) In determining the priority for establishing standards under this section, the Secretary shall give due regard to the urgency of the need for mandatory safety and health standards for particular industries, trades, crafts, occupations, businesses, workplaces or work environments. The Secretary shall also give due regard to the recommendations of the Secretary of Health, Education, and Welfare regarding the need for mandatory standards in determining the priority for establishing such standards.

Advisory Committees; Administration

29 U.S.C. 656. Establishment; membership. SEC. 7. (a) (1) There is hereby established a National Advisory Committee on Occupational Safety and Health consisting of twelve members appointed by the Secretary, four of whom are to be designated by the Secretary of Health, Education, and Welfare, without regard to the provisions of title 5, United States Code, governing appointments in the competitive service, and composed of representatives of management, labor, occupational safety and occupational health professions, and of the public. The Secretary shall designate one of the public members as Chairman. The members shall

be selected upon the basis of their experience and competence in the field of occupational safety and health.

(2) The Committee shall advise, consult with, and make recommendations to the Secretary and the Secretary of Health, Education, and Welfare on matters relating to the administration of the Act. The Committee shall hold no fewer than two meetings during each calendar year. All meetings of the Committee shall be open to the public and a transcript shall be kept and made available for public inspection. *Public transcript.*

(3) The members of the Committee shall be compensated in accordance with the provisions of section 3109 of title 5, United States Code. *80 Stat. 416.*

(4) The Secretary shall furnish to the Committee an executive secretary and such secretarial, clerical, and other services as are deemed necessary to the conduct of its business.

(b) An advisory committee may be appointed by the Secretary to assist him in his standard-setting functions under section 6 of this Act. Each such committee shall consist of not more than fifteen members and shall include as a member one or more designees of the Secretary of Health, Education, and Welfare, and shall include among its members an equal number of persons qualified by experience and affiliation to present the viewpoint of the employers involved, and of persons similarly qualified to present the viewpoint of the workers involved, as well as one or more representatives of health and safety agencies of the States. An advisory committee may also include such other persons as the Secretary may appoint who are qualified by knowledge and experience to make a useful contribution to the work of such committee, including one or more representatives of professional organizations of technicians or professionals specializing in occupational safety or health, and one or more representatives of nationally recognized standards-producing organizations, but the number of persons so appointed to any such advisory committee shall not exceed the number appointed to such committee as representatives of Federal and State agencies. Persons appointed to advisory committees from private life shall be compensated in the same manner as consultants or experts under section 3109 of title 5. United State Code. The Secretary shall pay to any State which is the employer of a member of such a committee who is a representative of the health or safety agency of that State, reimbursement sufficient to cover the actual cost to the State resulting from such representative's membership on such committee. Any meeting of such committee shall be open to the public and an accurate record shall be kept and made available to the public. No member of such committee (other than representatives of employers and employees) shall have an economic interest in any proposed rule. *80 Stat. 416.* *Recordkeeping.*

(c) In carrying out his responsibilities under this Act, the Secretary is authorized to—

(1) use, with the consent of any Federal agency, the services, facilities, and personnel of such agency, with or without reimbursement, and with the consent of any State or political subdivision thereof, accept and use the services, facilities, and personnel of any agency of such State or subdivision with reimbursement; and

(2) employ experts and consultants or organizations thereof as authorized by section 3109 of title 5, United States Code, except that contracts for such employment may be renewed annually; compensate individuals so employed at rates not in excess of the rate specified at the time of service for grade GS-18 under section 5332 of title 5, United States Code, including traveltime, and allow them while away from

80 Stat. 499;
83 Stat. 190.
their homes or regular places of business, travel expenses (including per diem in lieu of subsistence) as authorized by section 5703 of title 5, United States Code, for persons in the Government service employed intermittently, while so employed.

Inspections, Investigations, and Recordkeeping

29 U.S.C. 657.
Sec. 8. (a) In order to carry out the purposes of this Act, the Secretary, upon presenting appropriate credentials to the owner, operator, or agent in charge, is authorized—

(1) to enter without delay and at reasonable times any factory, plant, establishment, construction site, or other area, workplace or environment where work is performed by an employee of an employer; and

(2) to inspect and investigate during regular working hours and at other reasonable times, and within reasonable limits and in a reasonable manner, any such place of employment and all pertinent conditions, structures, machines, apparatus, devices, equipment, and materials therein, and to question privately any such employer, owner, operator, agent or employee.

Subpoena power.
(b) In making his inspections and investigations under this Act the Secretary may require the attendance and testimony of witnesses and the production of evidence under oath. Witnesses shall be paid the same fees and mileage that are paid witnesses in the courts of the United States. In case of a contumacy, failure, or refusal of any person to obey such an order, any district court of the United States or the United States courts of any territory or possession, within the jurisdiction of which such person is found, or resides or transacts business, upon the application by the Secretary, shall have jurisdiction to issue to such person an order requiring such person to appear to produce evidence if, as, and when so ordered, and to give testimony relating to the matter under investigation or in question, and any failure to obey such order of the court may be punished by said court as a contempt thereof.

Recordkeeping.
(c) (1) Each employer shall make, keep and preserve, and make available to the Secretary or the Secretary of Health, Education, and Welfare, such records regarding his activities relating to this Act as the Secretary, in cooperation with the Secretary of Health, Education, and Welfare, may prescribe by regulation as necessary or appropriate for the enforcement of this Act or for developing information regarding the causes and prevention of occupational accidents and illnesses. In order to carry out the provisions of this paragraph such regulations may include provisions requiring employers to conduct periodic inspections. The Secretary shall also issue regulations requiring that employers, through posting of notices or other appropriate means, keep their employees informed of their protections and obligations under this Act, including the provisions of applicable standards.

Work-related deaths, etc.; reports.
(2) The Secretary, in cooperation with the Secretary of Health, Education, and Welfare, shall prescribe regulations requiring employers to maintain accurate records of, and to make periodic reports on, work-related deaths, injuries and illnesses other than minor injuries requiring only first aid treatment and which do not involve medical treatment, loss of consciousness, restriction of work or motion, or transfer to another job.

(3) The Secretary, in cooperation with the Secretary of Health, Education, and Welfare, shall issue regulations requiring employers to maintain accurate records of employee exposures to potentially toxic materials or harmful physical agents which are

required to be monitored or measured under section 6. Such regulations shall provide employees or their representatives with an opportunity to observe such monitoring or measuring, and to have access to the records thereof. Such regulations shall also make appropriate provision for each employee or former employee to have access to such records as will indicate his own exposure to toxic materials or harmful physical agents. Each employer shall promptly notify any employee who has been or is being exposed to toxic materials or harmful physical agents in concentrations or at levels which exceed those prescribed by an applicable occupational safety and health standard promulgated under section 6, and shall inform any employee who is being thus exposed of the corrective action being taken.

(d) Any information obtained by the Secretary, the Secretary of Health, Education, and Welfare, or a State agency under this Act shall be obtained with a minimum burden upon employers, especially those operating small businesses. Unnecessary duplication of efforts in obtaining information shall be reduced to the maximum extent feasible.

(e) Subject to regulations issued by the Secretary, a representative of the employer and a representative authorized by his employees shall be given an opportunity to accompany the Secretary or his authorized representative during inspection of any workplace under subsection (a) for the purpose of aiding such inspection. Where there is no authorized employee representative, the Secretary or his authorized representative shall consult with a reasonable number of employees concerning matters of health and safety in the workplace.

(f) (1) Any employees or representative of employees who believe that a violation of a safety or health standard exists that threatens physical harm, or that an imminent danger exists, may request an inspection by giving notice to the Secretary or his authorized representative of such violation or danger. Any such notice shall be reduced to writing, shall set forth with reasonable particularity the grounds for the notice, and shall be signed by the employees or representative of employees, and a copy shall be provided the employer or his agent no later than at the time of inspection, except that, upon the request of the person giving such notice, his name and the names of individual employees referred to therein shall not appear in such copy or on any record published, released, or made available pursuant to subsection (g) of this section. If upon receipt of such notification the Secretary determines there are reasonable grounds to believe that such violation or danger exists, he shall make a special inspection in accordance with the provisions of this section as soon as practicable, to determine if such violation or danger exists. If the Secretary determines there are no reasonable grounds to believe that a violation or danger exists he shall notify the employees or representative of the employees in writing of such determination.

(2) Prior to or during any inspection of a workplace, any employees or representative of employees employed in such workplace may notify the Secretary or any representative of the Secretary responsible for conducting the inspection, in writing, of any violation of this Act which they have reason to believe exists in such workplace. The Secretary shall, by regulation, establish procedures for informal review of any refusal by a representative of the Secretary to issue a citation with respect to any such alleged violation and shall furnish the employees or representative of employees requesting such review a written statement of the reasons for the Secretary's final disposition of the case.

Reports,
publication.

Rules and
regulations.

(g) (1) The Secretary and Secretary of Health, Education, and Welfare are authorized to compile, analyze, and publish, either in summary or detailed form, all reports or information obtained under this section.

(2) The Secretary and the Secretary of Health, Education, and Welfare shall each prescribe such rules and regulations as he may deem necessary to carry out their responsibilities under this Act, including rules and regulations dealing with the inspection of an employer's establishment.

Citations

29 U.S.C. 658.

SEC. 9. (a) If, upon inspection or investigation, the Secretary or his authorized representative believes that an employer has violated a requirement of section 5 of this Act, of any standard, rule or order promulgated pursuant to section 6 of this Act, or of any regulations prescribed pursuant to this Act, he shall with reasonable promptness issue a citation to the employer. Each citation shall be in writing and shall describe with particularity the nature of the violation, including a reference to the provision of the Act, standard, rule, regulation, or order alleged to have been violated. In addition, the citation shall fix a reasonable time for the abatement of the violation. The Secretary may prescribe procedures for the issuance of a notice in lieu of a citation with respect to de minimis violations which have no direct or immediate relationship to safety or health.

(b) Each citation issued under this section, or a copy or copies thereof, shall be prominently posted, as prescribed in regulations issued by the Secretary, at or near each place a violation referred to in the citation occurred.

Limitation.

(c) No citation may be issued under this section after the expiration of six months following the occurrence of any violation.

Procedure for Enforcement

29 U.S.C. 659.

SEC. 10. (a) If, after an inspection or investigation, the Secretary issues a citation under section 9(a), he shall, within a reasonable time after the termination of such inspection or investigation, notify the employer by certified mail of the penalty, if any, proposed to be assessed under section 17 and that the employer has fifteen working days within which to notify the Secretary that he wishes to contest the citation or proposed assessment of penalty. If, within fifteen working days from the receipt of the notice issued by the Secretary the employer fails to notify the Secretary that he intends to contest the citation or proposed assessment of penalty, and no notice is filed by any employee or representative of employees under subsection (c) within such time, the citation and the assessment, as proposed, shall be deemed a final order of the Commission and not subject to review by any court or agency.

(b) If the Secretary has reason to believe that an employer has failed to correct a violation for which a citation has been issued within the period permitted for its correction (which period shall not begin to run until the entry of a final order by the Commission in the case of any review proceedings under this section initiated by the employer in good faith and not solely for delay or avoidance of penalties), the Secretary shall notify the employer by certified mail of such failure and of the penalty proposed to be assessed under section 17 by reason of such failure, and that the employer has fifteen working days within which to notify the Secretary that he wishes to contest the

Secretary's notification or the proposed assessment of penalty. If, within fifteen working days from the receipt of notification issued by the Secretary, the employer fails to notify the Secretary that he intends to contest the notification or proposed assessment of penalty, the notification and assessment, as proposed, shall be deemed a final order of the Commission and not subject to review by any court or agency.

(c) If an employer notifies the Secretary that he intends to contest a citation issued under section 9 (a) or notification issued under subsection (a) or (b) of this section, or if, within fifteen working days of the issuance of a citation under section 9 (a), any employee or representative of employees files a notice with the Secretary alleging that the period of time fixed in the citation for the abatement of the violation is unreasonable, the Secretary shall immediately advise the Commission of such notification, and the Commission shall afford an opportunity for a hearing (in accordance with section 554 of title 5, United States Code, but without regard to subsection (a) (3) of such section). 80 Stat. 384. The Commission shall thereafter issue an order, based on findings of fact, affirming, modifying, or vacating the Secretary's citation or proposed penalty, or directing other appropriate relief, and such order shall become final thirty days after its issuance. Upon a showing by an employer of a good faith effort to comply with the abatement requirements of a citation, and that abatement has not been completed because of factors beyond his reasonable control, the Secretary, after an opportunity for a hearing as provided in this subsection, shall issue an order affirming or modifying the abatement requirements in such citation. The rules of procedure prescribed by the Commission shall provide affected employees or representatives of affected employees an opportunity to participate as parties to hearings under this subsection.

Judicial Review

SEC. 11. (a) Any person adversely affected or aggrieved by an order of the Commission 29 U.S.C. 660. issued under subsection (c) of section 10 may obtain a review of such order in any United States court of appeals for the circuit in which the violation is alleged to have occurred or where the employer has its principal office, or in the Court of Appeals for the District of Columbia Circuit, by filing in such court within sixty days following the issuance of such order a written petition praying that the order be modified or set aside. A copy of such petition shall be forthwith transmitted by the clerk of the court to the Commission and to the other parties, and thereupon the Commission shall file in the court the record in the proceeding as provided in section 2112 of title 28, United States Code. Upon such filing, the court shall have jurisdiction of the proceeding and 72 Stat. 941; of the question determined therein, and shall have power to grant such temporary 80 Stat. 1323. relief or restraining order as it deems just and proper, and to make and enter upon the pleadings, testimony, and proceedings set forth in such record a decree affirming, modifying, or setting aside in whole or in part, the order of the Commission and enforcing the same to the extent that such order is affirmed or modified. The commencement of proceedings under this subsection shall not, unless ordered by the court, operate as a stay of the order of the Commission. No objection that has not been urged before the Commission shall be considered by the court, unless the failure or neglect to urge such objection shall be excused because of extraordinary circumstances. The findings of the Commission with respect to questions of fact, if supported by substantial evidence on the record considered as a whole, shall be conclusive. If any party shall apply to the court for leave to adduce additional evidence and shall show to the satisfaction of the court that such additional evidence is material and that there were reasonable grounds for the failure to adduce such evidence in the hearing before the Commission, the court may order such additional evidence to be taken before the Commission and to be made a part of the record. The Commission may modify its findings as to the facts, or make new findings, by reason of additional evidence so taken and filed, and it shall file such modified or new findings, which findings with respect to questions of fact, if supported by substantial evidence on the record considered as a whole, shall be conclusive, and its recommendations, if any, for the modification or setting aside of its original order. Upon the filing of the record with it, the jurisdiction of the court shall be exclusive and its judgment and decree shall be final, except that the same shall be subject to review by the Supreme Court of the United States, as provided in section 1254 of title 28, United States Code. Petitions filed under this subsection shall be heard expeditiously.

(b) The Secretary may also obtain review or enforcement of any final order of the Commission by filing a petition for such relief in the United States court of appeals for the circuit in which the alleged violation occurred or in which the employer has its principal office, and the provisions of subsection (a) shall govern such proceedings to the extent applicable. If no petition for review, as provided in subsection (a), is filed within sixty days after service of the Commission's order, the Commission's findings of fact and order shall be conclusive in connection with any petition for enforcement which is filed by the Secretary after the expiration of such sixty-day period. In any such case, as well as in the case of a noncontested citation or notification by the Secretary which has become a final order of the Commission under subsection (a) or (b) of section 10, the clerk of the court, unless otherwise ordered by the court, shall forthwith enter a decree enforcing the order and shall transmit a copy of such decree to the Secretary and the employer named in the petition. In any contempt proceeding brought to enforce a decree of a court of appeals entered pursuant to this subsection or subsection (a), the court of appeals may assess the penalties provided in section 17, in addition to invoking any other available remedies.

(c) (1) No person shall discharge or in any manner discriminate against any employee because such employee has filed any complaint or instituted or caused to be instituted any proceeding under or related to this Act or has testified or is about to testify in any such proceeding or because of the exercise by such employee on behalf of himself or others of any right afforded by this Act.

(2) Any employee who believes that he has been discharged or otherwise discrimi-nated against by any person in violation of this subsection may, within thirty days after such violation occurs, file a complaint with the Secretary alleging such discrimination. Upon receipt of such complaint, the Secretary shall cause such investigation to be made as he deems appropriate. If upon such investigation, the Secretary determines that the provisions of this subsection have been violated, he shall bring an action in any appropriate United States district court against such person. In any such action the United States district courts shall have jurisdiction, for cause shown to restrain viola-tions of paragraph (1) of this subsection and order all appropriate relief including rehiring or reinstatement of the employee to his former position with back pay.

(3) Within 90 days of the receipt of a complaint filed under the subsection the Secretary shall notify the complainant of his determination under paragraph 2 of this subsection.

The Occupational Safety and Health Act   523

The Occupational Safety and Health Review Commission

SEC. 12. (a) The Occupational Safety and Health Review Commission is hereby established. The Commission shall be composed of three members who shall be appointed by the President, by and with the advice and consent of the Senate, from among persons who by reason of training, education, or experience are qualified to carry out the functions of the Commission under this Act. The President shall designate one of the members of the Commission to serve as Chairman. *29 U.S.C. 661. Establishment; membership.*

(b) The terms of members of the Commission shall be six years except that (1) the members of the Commission first taking office shall serve, as designated by the President at the time of appointment, one for a term of two years, one for a term of four years, and one for a term of six years, and (2) a vacancy caused by the death, resignation, or removal of a member prior to the expiration of the term for which he was appointed shall be filled only for the remainder of such unexpired term. A member of the Commission may be removed by the President for inefficiency, neglect of duty, or malfeasance in office. *Terms.*

(c)(1) Section 5314 of title 5, United States Code, is amended by adding at the end thereof the following new paragraph: *80 Stat. 460.*

"(57) Chairman, Occupational Safety and Health Review Commission."

(2) Section 5315 of title 5, United States Code, is amended by adding at the end thereof the following new paragraph:

"(94) Members, Occupational Safety and Health Review Commission."

(d) The principal office of the Commission shall be in the District of Columbia. Whenever the Commission deems that the convenience of the public or of the parties may be promoted, or delay or expense may be minimized, it may hold hearings or conduct other proceedings at any other place. *Location.*

(e) The Chairman shall be responsible on behalf of the Commission for the administrative operations of the Commission and shall appoint such hearing examiners and other employees as he deems necessary to assist in the performance of the Commission's functions and to fix their compensation in accordance with the provisions of chapter 51 and subchapter III of chapter 53 of title 5, United States Code, relating to classification and General Schedule pay rates: *Provided,* That assignment, removal and compensation of hearing examiners shall be in accordance with sections 3105, 3344, 5362, and 7521 of title 5, United States Code.

(f) For the purpose of carrying out its functions under this Act, two members of the Commission shall constitute a quorum and official action can be taken only on the affirmative vote of at least two members. *Quorum.*

(g) Every official act of the Commission shall be entered of record, and its hearings and records shall be open to the public. The Commission is authorized to make such rules as are necessary for the orderly transaction of its proceedings. Unless the Commission has adopted a different rule, its proceedings shall be in accordance with the Federal Rules of Civil Procedure. *Public records.* *28 USC app.*

(h) The Commission may order testimony to be taken by deposition in any proceedings pending before it at any state of such proceeding. Any person may be compelled to appear and depose, and to produce books, papers, or documents, in the same manner as witnesses may be compelled to appear and testify and produce like documentary evidence before the Commission. Witnesses whose depositions are taken under this subsection, and the persons taking such depositions, shall be entitled to the same fees as are paid for like services in the courts of the United States.

61 Stat. 150;

(i) For the purpose of any proceeding before the Commission, the provisions of section 11 of the National Labor Relations Act (29 U.S.C. 161) are hereby made applicable to the jurisdiction and powers of the Commission.

Report.

(j) A hearing examiner appointed by the Commission shall hear, and make a determination upon, any proceeding instituted before the Commission and any motion in connection therewith, assigned to such hearing examiner by the Chairman of the Commission, and shall make a report of any such determination which constitutes his final disposition of the proceedings. The report of the hearing examiner shall become the final order of the Commission within thirty days after such report by the hearing examiner, unless within such period any Commission member has directed that such report shall be reviewed by the Commission.

(k) Except as otherwise provided in this Act, the hearing examiners shall be subject to the laws governing employees in the classified civil service, except that

80 Stat. 453.

appointments shall be made without regard to section 5108 of title 5, United States Code. Each hearing examiner shall receive compensation at a rate not less than that prescribed for GS-16 under section 5332 of title 5, United States Code.

Procedures to Counteract Imminent Dangers

29 U.S.C. 662.

SEC. 13. (a) The United States district courts shall have jurisdiction, upon petition of the Secretary, to restrain any conditions or practices in any place of employment which are such that a danger exists which could reasonably be expected to cause death or serious physical harm immediately or before the imminence of such danger can be eliminated through the enforcement procedures otherwise provided by this Act. Any order issued under this section may require such steps to be taken as may be necessary to avoid, correct, or remove such imminent danger and prohibit the employment or presence of any individual in locations or under conditions where such imminent danger exists, except individuals whose presence is necessary to avoid, correct, or remove such imminent danger or to maintain the capacity of a continuous process operation to resume normal operations without a complete cessation of operations, or where a cessation of operations is necessary, to permit such to be accomplished in a safe and orderly manner.

(b) Upon the filing of any such petition the district court shall have jurisdiction to grant such injunctive relief or temporary restraining order pending the outcome of an enforcement proceeding pursuant to this Act. The proceeding shall be as provided by

28 USC app.

Rule 65 of the Federal Rules, Civil Procedure, except that no temporary restraining order issued without notice shall be effective for a period longer than five days.

(c) Whenever and as soon as an inspector concludes that conditions or practices described in subsection (a) exist in any place of employment, he shall inform the affected employees and employers of the danger and that he is recommending to the Secretary that relief be sought.

(d) If the Secretary arbitrarily or capriciously fails to seek relief under this section, any employee who may be injured by reason of such failure, or the representative of such employees, might bring an action against the Secretary in the United States district court for the district in which the imminent danger is alleged to exist or the employer has its principal office, or for the District of Columbia, for a writ of mandamus to compel the Secretary to seek such an order and for such further relief as may be appropriate.

Representation in Civil Litigation

Sec. 14. Except as provided in section 518(a) of title 28, United States Code, relating to litigation before the Supreme Court, the Solicitor of Labor may appear for and represent the Secretary in any civil litigation brought under this Act but all such litigation shall be subject to the direction and control of the Attorney General.

29 U.S.C. 663.
80 Stat. 613.

Confidentiality of Trade Secrets

Sec. 15. All information reported to or otherwise obtained by the Secretary or his representative in connection with any inspection or proceeding under this Act which contains or which might reveal a trade secret referred to in section 1905 of title 18 of the United States Code shall be considered confidential for the purpose of that section, except that such information may be disclosed to other officers or employees concerned with carrying out this Act or when relevant in any proceeding under this Act. In any such proceeding the Secretary, the Commission, or the court shall issue such orders as may be appropriate to protect the confidentiality of trade secrets.

29 U.S.C. 664.

62 Stat. 791.

Variations, Tolerances, and Exemptions

Sec. 16. The Secretary, on the record, after notice and opportunity for a hearing may provide such reasonable limitations and may make such rules and regulations allowing reasonable variations, tolerances, and exemptions to and from any or all provisions of this Act as he may find necessary and proper to avoid serious impairment of the national defense. Such action shall not be in effect for more than six months without notification to affected employees and an opportunity being afforded for a hearing.

29 U.S.C. 665.

Penalties

Sec. 17. (a) Any employer who willfully or repeatedly violates the requirements of section 5 of this Act, any standard, rule, or order promulgated pursuant to section 6 of this Act, or regulations prescribed pursuant to this Act, may be assessed a civil penalty of not more than $10,000 for each violation.

(b) Any employer who has received a citation for a serious violation of the requirements of section 5 of this Act, of any standard, rule, or order promulgated pursuant to section 6 of this Act, or of any regulations prescribed pursuant to this Act, shall be assessed a civil penalty of up to $1,000 for each such violation.

(c) Any employer who has received a citation for a violation of the requirements of section 5 of this Act, of any standard, rule, or order promulgated pursuant to section 6 of this Act, or of regulations prescribed pursuant to this Act, and such violation is specifically determined not to be of a serious nature, may be assessed a civil penalty of up to $1,000 for each such violation.

(d) Any employer who fails to correct a violation for which a citation has been issued under section 9(a) within the period permitted for its correction (which period shall not begin to run until the date of the final order of the Commission in the case of

29 U.S.C. 666.

any review proceeding under section 10 initiated by the employer in good faith and not solely for delay or avoidance of penalties), may be assessed a civil penalty of not more than $1,000 for each day during which such failure or violation continues.

(e) Any employer who willfully violates any standard, rule, or order promulgated pursuant to section 6 of this Act, or of any regulations prescribed pursuant to this Act, and that violation caused death to any employee, shall, upon conviction, be punished by a fine of not more than $10,000 or by imprisonment for not more than six months, or by both; except that if the conviction is for a violation committed after a first conviction of such person, punishment shall be by a fine of not more than $20,000 or by imprisonment for not more than one year, or by both.

(f) Any person who gives advance notice of any inspection to be conducted under this Act, without authority from the Secretary or his designees, shall, upon conviction, be punished by a fine of not more than $1,000 or by imprisonment for not more than six months, or by both.

(g) Whoever knowingly makes any false statement, representation, or certification in any application, record, report, plan, or other document filed or required to be maintained pursuant to this Act shall, upon conviction, be punished by a fine of not more than $10,000, or by imprisonment for not more than six months, or by both.

65 Stat. 721;
79 Stat. 234.

(h) (1) Section 1114 of title 18, United States Code, is hereby amended by striking out "designated by the Secretary of Health, Education, and Welfare to conduct investigations, or inspections under the Federal Food, Drug, and Cosmetic Act" and inserting in lieu thereof "or of the Department of Labor assigned to perform investigative, inspection, or law enforcement functions."

(2) Notwithstanding the provisions of sections 1111 and 1114 of title 18, United States Code, whoever, in violation of the provisions of section 1114 of such title, kills a person while engaged in or on account of the performance of investigative, inspection, or law enforcement functions added to such section 1114 by paragraph (1) of this subsection, and who would otherwise be subject to the penalty provisions of such section 1111, shall be punished by imprisonment for any term of years or for life.

62 Stat. 756.

(i) Any employer who violates any of the posting requirements, as prescribed under the provisions of this Act, shall be assessed a civil penalty of up to $1,000 for each violation.

(j) The Commission shall have authority to assess all civil penalties provided in this section, giving due consideration to the appropriateness of the penalty with respect to the size of the business of the employer being charged, the gravity of the violation, the good faith of the employer, and the history of previous violations.

(k) For purposes of this section, a serious violation shall be deemed to exist in a place of employment if there is a substantial probability that death or serious physical harm could result from a condition which exists, or from one or more practices, means, methods, operations, or processes which have been adopted or are in use, in such place of employment unless the employer did not, and could not with the exercise of reasonable diligence, know of the presence of the violation.

(l) Civil penalties owed under this Act shall be paid to the Secretary for deposit into the Treasury of the United States and shall accrue to the United States and may be recovered in a civil action in the name of the United States brought in the United States district court for the district where the violation is alleged to have occurred or where the employer has its principal office.

State Jurisdiction and State Plans

SEC. 18. (a) Nothing in this Act shall prevent any State agency or court from asserting   29 U.S.C. 667.
jurisdiction under State law over any occupational safety or health issue with respect to
which no standard is in effect under section 6.

(b) Any State which, at any time, desires to assume responsibility for development
and enforcement therein of occupational safety and health standards relating to any
occupational safety or health issue with respect to which a Federal standard has been
promulgated under section 6 shall submit a State plan for the development of such
standards and their enforcement.

(c) The Secretary shall approve the plan submitted by a State under subsection (b),
or any modification thereof, if such plan in his judgment—

(1) designates a State agency or agencies as the agency or agencies responsible
for administering the plan throughout the State,

(2) provides for the development and enforcement of safety and health standards relating to one or more safety or health issues, which standards (and the
enforcement of which standards) are or will be at least as effective in providing safe
and healthful employment and places of employment as the standards promulgated
under section 6 which relate to the same issues, and which standards, when applicable
to products which are distributed or used in interstate commerce, are required by
compelling local conditions and do not unduly burden interstate commerce,

(3) provides for a right of entry and inspection of all workplaces subject to the
Act which is at least as effective as that provided in section 8, and includes a
prohibition on advance notice of inspections,

(4) contains satisfactory assurances that such agency or agencies have or will
have the legal authority and qualified personnel necessary for the enforcement of
such standards,

(5) gives satisfactory assurances that such State will devote adequate funds to the
administration and enforcement of such standards,

(6) contains satisfactory assurances that such State will, to the extent permitted
by its law, establish and maintain an effective and comprehensive occupational safety
and health program applicable to all employees of public agencies of the State and
its political subdivisions, which program is as effective as the standards contained in
an approved plan,

(7) require employers in the State to make reports to the Secretary in the same
manner and to the same extent as if the plan were not in effect, and

(8) provides that the State agency will make such reports to the Secretary in such
form and containing such information, as the Secretary shall from time to time
require.

(d) If the Secretary rejects a plan submitted under subsection (b), he shall afford   Notice of
the State submitting the plan due notice and opportunity for a hearing before so doing.   hearing.

(e) After the Secretary approves a State plan submitted under subsection (b), he
may, but shall not be required to, exercise his authority under sections 8, 9, 10, 13, and
17 with respect to comparable standards promulgated under section 6, for the period
specified in the next sentence. The Secretary may exercise the authority referred to
above until he determines, on the basis of actual operations under the State plan, that
the criteria set forth in subsection (c) are being applied, but he shall not make such

determination for at least three years after the plan's approval under subsection (c). Upon making the determination referred to in the preceding sentence, the provisions of sections 5 (a) (2), 8 (except for the purpose of carrying out subsection (f) of this section), 9, 10, 13, and 17, and standards promulgated under section 6 of this Act, shall not apply with respect to any occupational safety or health issues covered under the plan, but the Secretary may retain jurisdiction under the above provisions in any proceeding commenced under section 9 or 10 before the date of determination.

Continuing evaluation.

(f) The Secretary shall, on the basis of reports submitted by the State agency and his own inspections make a continuing evaluation of the manner in which each State having a plan approved under this section is carrying out such plan. Whenever the Secretary finds, after affording due notice and opportunity for a hearing, that in the administration of the State plan there is a failure to comply substantially with any provision of the State plan (or any assurance contained therein), he shall notify the State agency of his withdrawal of approval of such plan and upon receipt of such notice such plan shall cease to be in effect, but the State may retain jurisdiction in any case commenced before the withdrawal of the plan in order to enforce standards under the plan whenever the issues involved do not relate to the reasons for the withdrawal of the plan.

Plan rejection, review.

(g) The State may obtain a review of a decision of the Secretary withdrawing approval of or rejecting its plan by the United States court of appeals for the circuit in which the State is located by filing in such court within thirty days following receipt of notice of such decision a petition to modify or set aside in whole or in part the action of the Secretary. A copy of such petition shall forthwith be served upon the Secretary, and thereupon the Secretary shall certify and file in the court the record upon which the decision complained of was issued as provided in section 2112 of title 28, United States Code. Unless the court finds that the Secretary's decision in rejecting a proposed State plan or withdrawing his approval of such a plan is not supported by substantial evidence the court shall affirm the Secretary's decision. The judgment of the court shall be subject to review by the Supreme Court of the United States upon certiorari or certification as provided in section 1254 of title 28, United States Code.

72 Stat. 941; 80 Stat. 1323.

62 Stat. 928.

(h) The Secretary may enter into an agreement with a State under which the State will be permitted to continue to enforce one or more occupational health and safety standards in effect in such State until final action is taken by the Secretary with respect to a plan submitted by a State under subsection (b) of this section, or two years from the date of enactment of thus Act, whichever is earlier.

Federal Agency Safety Programs and Responsibilities

29 U.S.C. 668.

SEC. 19. (a) It shall be the responsibility of the head of each Federal agency to establish and maintain an effective and comprehensive occupational safety and health program which is consistent with the standards promulgated under section 6. The head of each agency shall (after consultation with representatives of the employees thereof)—

(1) provide safe and healthful places and conditions of employment, consistent with the standards set under section 6;

(2) acquire, maintain, and require the use of safety equipment, personal protective equipment, and devices reasonably necessary to protect employees;

Recordkeeping.

(3) keep adequate records of all occupational accidents and illnesses for proper evaluation and necessary corrective action;

(4) consult with the Secretary with regard to the adequacy as to form and content of records kept pursuant to subsection (a)(3) of this section; and

(5) make an annual report to the Secretary with respect to occupational acci- Annual report. dents and injuries and the agency's program under this section. Such report shall include any report submitted under section 7902(e)(2) of title 5, United States Code. 80 Stat. 530.

(b) The Secretary shall report to the President a summary or digest of reports Report to President. submitted to him under subsection (a)(5) of this section, together with his evaluations of and recommendations derived from such reports. The President shall transmit annually to the Senate and the House of Representatives a report of the activities of Report to Congress. Federal agencies under this section.

(c) Section 7902(c)(1) of title 5, United States Code, is amended by inserting after "agencies" the following: "and of labor organizations representing employees".

(d) The Secretary shall have access to records and reports kept and filed by Federal Records, etc.; availability. agencies pursuant to subsections (a)(3) and (5) of this section unless those records and reports are specifically required by Executive order to be kept secret in the interest of the national defense or foreign policy, in which case the Secretary shall have access to such information as will not jeopardize national defense or foreign policy.

Research and Related Activities

SEC. 20. (a)(1) The Secretary of Health, Education, and Welfare, after consultation with 29 U.S.C. 669. the Secretary and with other appropriate Federal departments or agencies, shall conduct (directly or by grants or contracts) research, experiments, and demonstrations relating to occupational safety and health, including studies of psychological factors involved, and relating to innovative methods, techniques, and approaches for dealing with occupational safety and health problems.

(2) The Secretary of Health, Education, and Welfare shall from time to time consult with the Secretary in order to develop specific plans for such research, demonstrations, and experiments as are necessary to produce criteria, including criteria identifying toxic substances, enabling the Secretary to meet his responsibility for the formulation of safety and health standards under this Act; and the Secretary of Health, Education, and Welfare, on the basis of such research, demonstrations, and experiments and any other information available to him, shall develop and publish at least annually such criteria as will effectuate the purposes of this Act.

(3) The Secretary of Health, Education, and Welfare, on the basis of such research, demonstrations, and experiments, and any other information available to him, shall develop criteria dealing with toxic materials and harmful physical agents and substances which will describe exposure levels that are safe for various periods of employment, including but not limited to the exposure levels at which no employee will suffer impaired health or functional capacities or diminished life expectancy as a result of his work experience.

(4) The Secretary of Health, Education, and Welfare shall also conduct special research, experiments, and demonstrations relating to occupational safety and health as are necessary to explore new problems, including those created by new technology in occupational safety and health, which may require ameliorative action beyond that which is otherwise provided for in the operating provisions of this Act. The Secretary of Health, Education, and Welfare shall also conduct research into the motivational and behavioral factors relating to the field of occupational safety and health.

Toxic sub-
stances,
records.

(5) The Secretary of Health, Education, and Welfare, in order to comply with his responsibilities under paragraph (2), and in order to develop needed information regarding potentially toxic substances or harmful physical agents, may prescribe regulations requiring employers to measure, record, and make reports on the exposure of employees to substances or physical agents which the Secretary of Health, Education, and Welfare reasonably believes may endanger the health or safety of employees. The

Medical
examinations.

Secretary of Health, Education, and Welfare also is authorized to establish such programs of medical examinations and tests as may be necessary for determining the incidence of occupational illnesses and the susceptibility of employees to such illnesses. Nothing in this or any other provision of this Act shall be deemed to authorize or require medical examination, immunization, or treatment for those who object thereto on religious grounds, except where such is necessary for the protection of the health or safety of others. Upon the request of any employer who is required to measure and record exposure of employees to substances or physical agents as provided under this subsection, the Secretary of Health, Education, and Welfare shall furnish full financial or other assistance to such employer for the purpose of defraying any additional expense incurred by him in carrying out the measuring and recording as provided in this subsection.

Toxic sub-
stances,
publication.

(6) The Secretary of Health, Education, and Welfare shall publish within six months of enactment of this Act and thereafter as needed but at least annually a list of all known toxic substances by generic family or other useful grouping, and the concentrations at which such toxicity is known to occur. He shall determine following a written request by any employer or authorized representative of employees, specifying with reasonable particularity the grounds on which the request is made, whether any substance normally found in the place of employment has potentially toxic effects in such concentrations as used or found; and shall submit such determination both to employers and affected employees as soon as possible. If the Secretary of Health, Education, and Welfare determines that any substance is potentially toxic at the concentrations in which it is used or found in a place of employment, and such substance is not covered by an occupational safety or health standard promulgated under section 6, the Secretary of Health, Education, and Welfare shall immediately submit such determination to the Secretary, together with all pertinent criteria.

Annual
studies.

(7) Within two years of enactment of this Act, and annually thereafter the Secretary of Health, Education, and Welfare shall conduct and publish industrywide studies of the effect of chronic or low-level exposure to industrial materials, processes, and stresses on the potential for illness, disease, or loss of functional capacity in aging adults.

Inspections.

(b) The Secretary of Health, Education, and Welfare is authorized to make inspections and question employers and employees as provided in section 8 of this Act in order to carry out his functions and responsibilities under this section.

Contract
authority.

(c) The Secretary is authorized to enter into contracts, agreements, or other arrangements with appropriate public agencies or private organizations for the purpose of conducting studies relating to his responsibilities under this Act. In carrying out his responsibilities under this subsection, the Secretary shall cooperate with the Secretary of Health, Education, and Welfare in order to avoid any duplication of efforts under this section.

(d) Information obtained by the Secretary and the Secretary of Health, Education, and Welfare under this section shall be disseminated by the Secretary to employers and employees and organizations thereof.

(e) The functions of the Secretary of Health, Education, and Welfare under this Act shall, to the extent feasible, be delegated to the Director of the National Institute for Occupational Safety and Health established by section 22 of this Act. Delegation of functions.

Training and Employee Education

SEC. 21. (a) The Secretary of Health, Education, and Welfare, after consultation with the Secretary and with other appropriate Federal departments and agencies, shall conduct, directly or by grants or contracts (1) education programs to provide an adequate supply of qualified personnel to carry out the purposes of this Act, and (2) informational programs on the importance of and proper use of adequate safety and health equipment. 29 U.S.C. 670.

(b) The Secretary is also authorized to conduct, directly or by grants or contracts, short-term training of personnel engaged in work related to his responsibilities under this Act.

(c) The Secretary, in consultation with the Secretary of Health, Education, and Welfare, shall (1) provide for the establishment and supervision of programs for the education and training of employers and employees in the recognition, avoidance, and prevention of unsafe or unhealthful working conditions in employments covered by this Act, and (2) consult with and advise employers and employees, and organizations representing employers and employees as to effective means of preventing occupational injuries and illnesses.

National Institute for Occupational Safety and Health

SEC. 22. (a) It is the purpose of this section to establish a National Institute for Occupational Safety and Health in the Department of Health, Education, and Welfare in order to carry out the policy set forth in section 2 of this Act and to perform the functions of the Secretary of Health, Education, and Welfare under sections 20 and 21 of this Act. 29 U.S.C. 671. Establishment.

(b) There is hereby established in the Department of Health, Education, and Welfare a National Institute for Occupational Safety and Health. The Institute shall be headed by a Director who shall be appointed by the Secretary of Health, Education, and Welfare, and who shall serve for a term of six years unless previously removed by the Secretary of Health, Education, and Welfare. Director, appointment, term.

(c) The Institute is authorized to—

(1) develop and establish recommended occupational safety and health standards; and

(2) perform all functions of the Secretary of Health, Education, and Welfare under sections 20 and 21 of this Act.

(d) Upon his own initiative, or upon the request of the Secretary or the Secretary of Health, Education, and Welfare, the Director is authorized (1) to conduct such research and experimental programs as he determines are necessary for the development of criteria for new and improved occupational safety and health standards, and (2) after consideration of the results of such research and experimental programs make recommendations concerning new or improved occupational safety and health standards. Any occupational safety and health standard recommended pursuant to this section shall immediately be forwarded to the Secretary of Labor, and to the Secretary of Health, Education, and Welfare.

(e) In addition to any authority vested in the Institute by other provisions of this section, the Director, in carrying out the functions of the Institute, is authorized to—

(1) prescribe such regulations as he deems necessary governing the manner in which its functions shall be carried out;

(2) receive money and other property donated, bequeathed, or devised, without condition or restriction other than that it be used for the purposes of the Institute and to use, sell, or otherwise dispose of such property for the purpose of carrying out its functions;

(3) receive (and use, sell, or otherwise dispose of, in accordance with paragraph (2), money and other property donated, bequeathed, or devised to the Institute with a condition or restriction, including a condition that the Institute use other funds of the Institute for the purposes of the gift;

(4) in accordance with the civil service laws, appoint and fix the compensation of such personnel as may be necessary to carry out the provisions of this section;

80 Stat. 416.
(5) obtain the services of experts and consultants in accordance with the provisions of section 3109 of title 5, United States Code;

(6) accept and utilize the services of voluntary and noncompensated personnel and reimburse them for travel expenses, including per diem, as authorized by
83 Stat. 190.
section 5703 of title 5, United States Code;

(7) enter into contracts, grants or other arrangements, or modifications thereof to carry out the provisions of this section, and such contracts or modifications thereof may be entered into without performance or other bonds, and without regard to section 3709 of the Revised Statutes, as amended (41 U.S.C. 5), or any other provision of law relating to competitive bidding;

(8) make advance, progress, and other payments which the Director deems necessary under this title without regard to the provisions of section 3648 of the Revised Statutes, as amended (31 U.S.C. 529); and

(9) make other necessary expenditures.

Annual report to HEW, President, and Congress.
(f) The Director shall submit to the Secretary of Health, Education, and Welfare, to the President, and to the Congress an annual report of the operations of the Institute under this Act, which shall include a detailed statement of all private and public funds received and expended by it, and such recommendations as he deems appropriate.

Grants to the States

29 U.S.C. 672.
SEC. 23. (a) The Secretary is authorized, during the fiscal year ending June 30, 1971, and the two succeeding fiscal years, to make grants to the States which have designated a State agency under section 18 to assist them—

(1) in identifying their needs and responsibilities in the area of occupational safety and health,

(2) in developing State plans under section 18, or

(3) in developing plans for—

(A) establishing systems for the collection of information concerning the nature and frequency of occupational injuries and diseases;

(B) increasing the expertise and enforcement capabilities of their personnel engaged in occupational safety and health programs; or

(C) otherwise improving the administration and enforcement of State occupational safety and health laws, including standards thereunder, consistent with the objectives of this Act.

(b) The Secretary is authorized, during the fiscal year ending June 30, 1971, and the two succeeding fiscal years, to make grants to the States for experimental and demonstration projects consistent with the objectives set forth in subsection (a) of this section.

(c) The Governor of the State shall designate the appropriate State agency for receipt of any grant made by the Secretary under this section.

(d) Any State agency designated by the Governor of the State desiring a grant under this section shall submit an application therefor to the Secretary.

(e) The Secretary shall review the application, and shall, after consultation with the Secretary of Health, Education, and Welfare, approve or reject such application.

(f) The Federal share for each State grant under subsection (a) or (b) of this section may not exceed 90 per centum of the total cost of the application. In the event the Federal share for all States under either such subsection is not the same, the differences among the States shall be established on the basis of objective criteria.

(g) The Secretary is authorized to make grants to the States to assist them in administering and enforcing programs for occupational safety and health contained in State plans approved by the Secretary pursuant to section 18 of this Act. The Federal share for each State grant under this subsection may not exceed 50 per centum of the total cost to the State of such a program. The last sentence of subsection (f) shall be applicable in determining the Federal share under this subsection.

(h) Prior to June 30, 1973, the Secretary shall, after consultation with the Secretary of Health, Education, and Welfare, transmit a report to the President and to the Congress, describing the experience under the grant programs authorized by this section and making any recommendations he may deem appropriate. *Report to President and Congress.*

Statistics

Sec. 24. (a) In order to further the purposes of this Act, the Secretary, in consultation *29 U.S.C. 673.* with the Secretary of Health, Education, and Welfare, shall develop and maintain an effective program of collection, compilation, and analysis of occupational safety and health statistics. Such program may cover all employments whether or not subject to any other provisions of this Act but shall not cover employments excluded by section 4 of the Act. The Secretary shall compile accurate statistics on work injuries and illnesses which shall include all disabling, serious, or significant injuries and illnesses, whether or not involving loss of time from work, other than minor injuries requiring only first aid treatment and which do not involve medical treatment, loss of consciousness, restriction of work or motion, or transfer to another job.

(b) To carry out his duties under subsection (a) of this section, the Secretary may—

(1) promote, encourage, or directly engage in programs of studies, information and communication concerning occupational safety and health statistics;

(2) make grants to States or political subdivisions thereof in order to assist them in developing and administering programs dealing with occupational safety and health statistics; and

(3) arrange, through grants or contracts, for the conduct of such research and investigations as give promise of furthering the objectives of this section.

(c) The Federal share for each grant under subsection (b) of this section may be up to 50 per centum of the State's total cost.

(d) The Secretary may, with the consent of any State or political subdivision thereof, accept and use the services, facilities, and employees of the agencies of such

State or political subdivision, with or without reimbursement, in order to assist him in carrying out his functions under this section.

Reports.

(e) On the basis of the records made and kept pursuant to section 8(c) of this Act, employers shall file such reports with the Secretary as he shall prescribe by regulation, as necessary to carry out his functions under this Act.

(f) Agreements between the Department of Labor and States pertaining to the collection of occupational safety and health statistics already in effect on the effective date of this Act shall remain in effect until superseded by grants or contracts made under this Act.

Audits

29 U.S.C. 674.

SEC. 25. (a) Each recipient of a grant under this Act shall keep such records as the Secretary or the Secretary of Health, Education, and Welfare shall prescribe, including records which fully disclose the amount and disposition by such recipient of the proceeds of such grant, the total cost of the project or undertaking in connection with which such grant is made or used, and the amount of that portion of the cost of the project or undertaking supplied by other sources, and such other records as will facilitate an effective audit.

(b) The Secretary or the Secretary of Health, Education, and Welfare, and the Comptroller General of the United States, or any of their duly authorized representatives, shall have access for the purpose of audit and examination to any books, documents, papers, and records of the recipients of any grant under this Act that are pertinent to any such grant.

Annual Report

29 U.S.C. 675.

SEC. 26. Within one hundred and twenty days following the convening of each regular session of each Congress, the Secretary and the Secretary of Health, Education, and Welfare shall each prepare and submit to the President for transmittal to the Congress a report upon the subject matter of this Act, the progress toward achievement of the purpose of this Act, the needs and requirements in the field of occupational safety and health, and any other relevant information. Such reports shall include information regarding occupational safety and health standards, and criteria for such standards, developed during the preceding year; evaluation of standards and criteria previously developed under this Act, defining areas of emphasis for new criteria and standards; an evaluation of the degree of observance of applicable occupational safety and health standards, and a summary of inspection and enforcement activity undertaken; analysis and evaluation of research activities for which results have been obtained under governmental and nongovernmental sponsorship; an analysis of major occupational diseases; evaluation of available control and measurement technology for hazards for which standards or criteria have been developed during the preceding year; description of cooperative efforts undertaken between Government agencies and other interested parties in the implementation of this Act during the preceding year; a progress report on the development of an adequate supply of trained manpower in the field of occupational safety and health, including estimates of future needs and the efforts being made by Government and others to meet those needs; listing of all toxic substances in industrial usage for which labeling requirements, criteria, or standards have not yet been established; and such recommendations for additional legislation as are deemed necessary to protect the safety and health of the worker and improve the administration of this Act.

National Commission on State Workmen's Compensation Laws

SEC. 27. (a) (1) The Congress hereby finds and declares that—

> 29 U.S.C. 676.

(A) the vast majority of American workers, and their families, are dependent on workmen's compensation for their basic economic security in the event such workers suffer disabling injury or death in the course of their employment; and that the full protection of American workers from job-related injury or death requires an adequate, prompt, and equitable system of workmen's compensation as well as an effective program of occupational health and safety regulation; and

(B) in recent years serious questions have been raised concerning the fairness and adequacy of present workmen's compensation laws in the light of the growth of the economy, the changing nature of the labor force, increases in medical knowledge, changes in the hazards associated with various types of employment, new technology creating new risks to health and safety, and increases in the general level of wages and the cost of living.

(2) The purpose of this section is to authorize an effective study and objective evaluation of State workmen's compensation laws in order to determine if such laws provide an adequate, prompt, and equitable system of compensation for injury or death arising out of or in the course of employment.

(b) There is hereby established a National Commission on State Workmen's Compensation Laws.

> Establishment.

(c) (1) The Workmen's Compensation Commission shall be composed of fifteen members to be appointed by the President from among members of State workmen's compensation boards, representatives of insurance carriers, business, labor, members of the medical profession having experience in industrial medicine or in workmen's compensation cases, educators having special expertise in the field of workmen's compensation, and representatives of the general public. The Secretary, the Secretary of Commerce, and the Secretary of Health, Education, and Welfare shall be ex officio members of the Workmen's Compensation Commission:

> Membership.

(2) Any vacancy in the Workmen's Compensation Commission shall not affect its powers.

(3) The President shall designate one of the members to serve as Chairman and one to serve as Vice Chairman of the Workmen's Compensation Commission.

(4) Eight members of the Workmen's Compensation Commission shall constitute a quorum.

> Quorum.

(d)(1) The Workmen's Compensation Commission shall undertake a comprehensive study and evaluation of State workmen's compensation laws in order to determine if such laws provide an adequate, prompt, and equitable system of compensation. Such study and evaluation shall include, without being limited to, the following subjects: (A) the amount and duration of permanent and temporary disability benefits and the criteria for determining the maximum limitations thereon, (B) the amount and duration of medical benefits and provisions insuring adequate medical care and free choice of physician, (C) the extent of coverage of workers, including exemptions based on numbers or type of employment, (D) standards for determining which injuries or

> Study.

diseases should be deemed compensable, (E) rehabilitation, (F) coverage under second or subsequent injury funds, (G) time limits on filing claims, (H) waiting periods, (I) compulsory or elective coverage, (J) administration, (K) legal expenses, (L) the feasibility and desirability of a uniform system of reporting information concerning job-related injuries and diseases and the operation of workmen's compensation laws, (M) the resolution of conflict of laws, extraterritoriality and similar problems arising from claims with multistate aspects, (N) the extent to which private insurance carriers are excluded from supplying workmen's compensation coverage and the desirability of such exclusionary practices, to the extent they are found to exist, (O) the relationship between workmen's compensation on the one hand, and old-age, disability, and survivors insurance and other types of insurance, public or private, on the other hand, (P) methods of implementing the recommendations of the Commission.

Report to President and Congress.

(2) The Workmen's Compensation Commission shall transmit to the President and to the Congress not later than July 31, 1972, a final report containing a detailed statement of the findings and conclusions of the Commission, together with such recommendations as it deems advisable.

Hearings.

(e)(1) The Workmen's Compensation Commission or, on the authorization of the Workmen's Compensation Commission, any subcommittee or members thereof, may, for the purpose of carrying out the provisions of this title, hold such hearings, take such testimony, and sit and act at such times and places as the Workmen's Compensation Commission deems advisable. Any member authorized by the Workmen's Compensation Commission may administer oaths or affirmations to witnesses appearing before the Workmen's Compensation Commission or any subcommittee or members thereof.

(2) Each department, agency, and instrumentality of the executive branch of the Government, including independent agencies, is authorized and directed to furnish to the Workmen's Compensation Commission, upon request made by the Chairman or Vice Chairman, such information as the Workmen's Compensation Commission deems necessary to carry out its functions under this section.

(f) Subject to such rules and regulations as may be adopted by the Workmen's Compensation Commission, the Chairman shall have the power to—

(1) appoint and fix the compensation of an executive director, and such additional staff personnel as he deems necessary, without regard to the provisions of title 5, United States Code, governing appointments in the competitive service, and without regard to the provisions of chapter 51 and subchapter III of chapter 53 of such title relating to classification and General Schedule pay rates, but at rates not in excess of the maximum rate for GS-18 of the General Schedule under section 5332 of such title, and

80 Stat. 416. Contract authorization.

(2) procure temporary and intermittent services to the same extent as is authorized by section 3109 of title 5, United States Code.

(g) The Workmen's Compensation Commission is authorized to enter into contracts with Federal or State agencies, private firms, institutions, and individuals for the conduct of research or surveys, the preparation of reports, and other activities necessary to the discharge of its duties.

Compensation; travel expenses.

(h) Members of the Workmen's Compensation Commission shall receive compensation for each day they are engaged in the performance of their duties as members of the Workmen's Compensation Commission at the daily rate prescribed for GS-18 under section 5332 of title 5, United States Code, and shall be entitled to reimbursement for travel, subsistence, and other necessary expenses incurred by them in the performance of their duties as members of the Workmen's Compensation Commission.

(i) There are hereby authorized to be appropriated such sums as may be necessary to carry out the provisions of this section. Appropriation.

(j) On the ninetieth day after the date of submission of its final report to the President, the Workmen's Compensation Commission shall cease to exist. Termination.

Economic Assistance to Small Businesses

SEC. 28. (a) Section 7 (b) of the Small Business Act, as amended, is amended— 29 U.S.C. 677.

(1) by striking out the period at the end of "paragraph (5)" and inserting in lieu thereof "; and"; and 72 Stat. 387;
83 Stat. 802.
15 USC 636.

(2) by adding after paragraph (5) a new paragraph as follows:

"(6) to make such loans (either directly or in cooperation with banks or other lending institutions through agreements to participate on an immediate or deferred basis) as the Administration may determine to be necessary or appropriate to assist any small business concern in effecting additions to or alterations in the equipment, facilities, or methods of operation of such business in order to comply with the applicable standards promulgated pursuant to section 6 of the Occupational Safety and Health Act of 1970 or standards adopted by a State pursuant to a plan approved under section 18 of the Occupational Safety and Health Act of 1970, if the Administration determines that such concern is likely to suffer substantial economic injury without assistance under this paragraph."

(b) The third sentence of section 7 (b) of the Small Business Act, as amended, is amended by striking out "or (5)" after "paragraph (3)" and inserting a comma followed by "(5) or (6)".

(c) Section 4 (c)(1) of the Small Business Act, as amended, is amended by inserting "7 (b) (6) ," after "7 (b) (5),". 80 Stat. 132.
15 USC 633.

(d) Loans may also be made or guaranteed for the purposes set forth in section 7 (b)(6) of the Small Business Act, as amended, pursuant to the provisions of section 202 of the Public Works and Economic Development Act of 1965, as amended. 79 Stat. 556.
42 USC 3142.

Additional Assistant Secretary of Labor

SEC. 29. (a) Section 2 of the Act of April 17, 1946 (60 Stat. 91) as amended (29 U.S.C. 553) is amended by— 29 U.S.C. 678.
75 Stat. 338.

(1) striking out "four" in the first sentence of such section and inserting in lieu thereof "five"; and

(2) adding at the end thereof the following new sentence, "One of such Assistant Secretaries shall be an Assistant Secretary of Labor for Occupational Safety and Health.".

(b) Paragraph (20) of section 5315 of title 5, United States Code, is amended by striking out "(4)" and inserting in lieu thereof "(5)". 80 Stat. 462.

Additional Positions

SEC. 30. Section 5108(c) of title 5, United States Code, is amended by— 29 U.S.C. 679.

(1) striking out the word "and" at the end of paragraph (8);

(2) striking out the period at the end of paragraph (9) and inserting in lieu thereof a semicolon and the word "and"; and

(3) by adding immediately after paragraph (9) the following new paragraph:

"(10) (A) the Secretary of Labor, subject to the standards and procedures prescribed by this chapter, may place an additional twenty-five positions in the Department of Labor in GS-16, 17, and 18 for the purposes of carrying out his responsibilities under the Occupational Safety and Health Act of 1970;

"(B) the Occupational Safety and Health Review Commission, subject to the standards and procedures prescribed by this chapter, may place ten positions in GS-16, 17, and 18 in carrying out its functions under the Occupational Safety and Health Act of 1970."

Emergency Locator Beacons

29 U.S.C. 680.
72 Stat. 775.
49 USC 1421.

Sec. 31. Section 601 of the Federal Aviation Act of 1958 is amended by inserting at the end thereof a new subsection as follows:

Emergency Locator Beacons

"(d) (1) Except with respect to aircraft described in paragraph (2) of this subsection, minimum standards pursuant to this section shall include a requirement that emergency locator beacons shall be installed—

"(A) on any fixed-wing, powered aircraft for use in air commerce the manufacture of which is completed, or which is imported into the United States, after one year following the date of enactment of this subsection; and

"(B) on any fixed-wing, powered aircraft used in air commerce after three years following such date.

"(2) The provisions of this subsection shall not apply to jet-powered aircraft; aircraft used in air transportation (other than air taxis and charter aircraft); military aircraft; aircraft used solely for training purposes not involving flights more than twenty miles from its base; and aircraft used for the aerial application of chemicals."

Separability

29 U.S.C. 681.

Sec. 32. If any provision of this Act, or the application of such provision to any person or circumstance, shall be held invalid, the remainder of this Act, or the application of such provision to persons or circumstances other than those as to which it is held invalid, shall not be affected thereby.

Appropriations

29 U.S.C. 682.

Sec. 33. There are authorized to be appropriated to carry out this Act for each fiscal year such sums as the Congress shall deem necessary.

Effective Date

29 U.S.C. 683.

Sec. 34. This Act shall take effect one hundred and twenty days after the date of its enactment.

Approved December 29, 1970.

LEGISLATIVE HISTORY:

HOUSE REPORTS: No. 91-1291 accompanying H.R. 16785 (Comm. on Education and Labor) and No. 91-1765 (Comm. of Conference).

SENATE REPORT No. 91-1282 (Comm. on Labor and Public Welfare).

CONGRESSIONAL RECORD, Vol. 116 (1970):

Oct. 13, Nov. 16, 17, considered and passed Senate.

Nov. 23, 24, considered and passed House, amended, in lieu of H.R. 16785.

Dec. 16, Senate agreed to conference report.

Dec. 17, House agreed to conference report.

APPENDIX B

The Toxic Substances

Control Act

Public Law 94-469
94th Congress
Oct. 11, 1976
15 U.S.C. §2601, et seq.

AN ACT

To regulate commerce and protect human health and the environment by requiring   Oct. 11, 1976
   testing and necessary use restrictions on certain chemical substances, and for   [S. 3149]
   other purposes.
   Be it enacted by the Senate and House of Representatives of the United States of America in
Congress assembled,

Section 1. Short Title and Table of Contents

This Act may be cited as the "Toxic Substances Control Act".   Toxic
Table of Contents   Substances
     Control Act.
     15 USC 2601
Sec. 1.   Short title and table of contents.   note.
Sec. 2.   Findings, policy, and intent.
Sec. 3.   Definitions.
Sec. 4.   Testing of chemical substances and mixtures.
Sec. 5.   Manufacturing and processing notices.
Sec. 6.   Regulation of hazardous chemical substances and mixtures.
Sec. 7.   Imminent hazards.
Sec. 8.   Reporting and retention of information.
Sec. 9.   Relationship to other Federal laws.
Sec. 10. Research, development, collection, dissemination, and utilization of data.
Sec. 11. Inspections and subpoenas.

Sec. 12. Exports.

Sec. 13. Entry into customs territory of the United States.

Sec. 14. Disclosure of data.

Sec. 15. Prohibited acts.

Sec. 16. Penalties.

Sec. 17. Specific enforcement and seizure.

Sec. 18. Preemption.

Sec. 19. Judicial review.

Sec. 20. Citizens' civil actions.

Sec. 21. Citizens' petitions.

Sec. 22. National defense waiver.

Sec. 23. Employee protection.

Sec. 24. Employment effects.

Sec. 25. Studies.

Sec. 26. Administration of the Act.

Sec. 27. Development and evaluation of test methods.

Sec. 28. State programs.

Sec. 29. Authorization for appropriations.

Sec. 30. Annual report.

Sec. 31. Effective date.

Sec. 2. Findings, Policy, and Intent

15 USC 2601.

(a) **Findings.**—The Congress finds that—

(1) human beings and the environment are being exposed each year to a large number of chemical substances and mixtures;

(2) among the many chemical substances and mixtures which are constantly being developed and produced, there are some whose manufacture, processing, distribution in commerce, use, or disposal may present an unreasonable risk of injury to health or the environment; and

(3) the effective regulation of interstate commerce in such chemical substances and mixtures also necessitates the regulation of intrastate commerce in such chemical substances and mixtures.

(b) **Policy.**—It is the policy of the United States that—

(1) adequate data should be developed with respect to the effect of chemical substances and mixtures on health and the environment and that the development of such data should be the responsibility of those who manufacture and those who process such chemical substances and mixtures;

(2) adequate authority should exist to regulate chemical substances and mixtures which present an unreasonable risk of injury to health or the environment, and to take action with respect to chemical substances and mixtures which are imminent hazards; and

(3) authority over chemical substances and mixtures should be exercised in such a manner as not to impede unduly or create unnecessary economic barriers to technological innovation while fulfilling the primary purpose of this Act to assure that such

innovation and commerce in such chemical substances and mixtures do not present an unreasonable risk of injury to health or the environment.

(c) **Intent of Congress.**—It is the intent of Congress that the Administrator shall carry out this Act in a reasonable and prudent manner, and that the Administrator shall consider the environmental, economic, and social impact of any action the Administrator takes or proposes to take under this Act.

Sec. 3. Definitions

As used in this Act: 15 USC 2602.

(1) the term "Administrator" means the Administrator of the Environmental Protection Agency.

(2) (A) Except as provided in subparagraph (B), the term "chemical substance" means any organic or inorganic substance of a particular molecular identity, including—

(i) any combination of such substances occurring in whole or in part as a result of a chemical reaction or occurring in nature, and

(ii) any element or uncombined radical.

(B) Such term does not include—

(i) any mixture,

(ii) any pesticide (as defined in the Federal Insecticide, Fungicide, and Rodenti- 7 USC 136
cide Act) when manufactured, processed, or distributed in commerce for use as a   note.
pesticide,

(iii) tobacco or any tobacco product,

(iv) any source material, special nuclear material, or byproduct material (as such   42 USC 2011
terms are defined in the Atomic Energy Act of 1954 and regulations issued under   note.
such Act),   26 USC 4181.

(v) any article the sale of which is subject to the tax imposed by section 4181 of
the Internal Revenue Code of 1954 (determined without regard to any exemptions   26 USC 4182,
from such tax provided by section 4182 or 4221 or any other provision of such   4221.
Code), and

(vi) any food, food additive, drug, cosmetic, or device (as such terms are defined
in section 201 of the Federal Food, Drug, and Cosmetic Act) when manufactured,   21 USC 321.
processed, or distributed in commerce for use as a food, food additive, drug,
cosmetic, or device.

The term "food" as used in clause (vi) of this subparagraph includes poultry and poultry products (as defined in sections 4(e) and 4(f) of the Poultry Products Inspection Act), meat and meat food products (as defined in section 1(j) of the Federal Meat   21 USC 453.
Inspection Act), and eggs and egg products (as defined in section 4 of the Egg Products   21 USC 601.
Inspection Act).   21 USC 1033.

(3) The term "commerce" means trade, traffic, transportation, or other commerce (A) between a place in a State and any place outside of such State, or (B) which affects trade, traffic, transportation, or commerce described in clause (A).

(4) The terms "distribute in commerce" and "distribution in commerce" when used to describe an action taken with respect to a chemical substance or mixture or article containing a substance or mixture mean to sell, or the sale of, the substance, mixture, or article in commerce; to introduce or deliver for introduction into commerce,

or the introduction or delivery for introduction into commerce of, the substance, mixture, or article; or to hold, or the holding of, the substance, mixture, or article after its introduction into commerce.

(5) The term "environment" includes water, air, and land and the interrelationship which exists among and between water, air, and land and all living things.

(6) The term "health and safety study" means any study of any effect of a chemical substance or mixture on health or the environment or on both, including underlying data and epidemiological studies, studies of occupational exposure to a chemical substance or mixture, toxicological, clinical, and ecological studies of a chemical substance or mixture, and any test performed pursuant to this Act.

(7) The term "manufacture" means to import into the customs territory of the United States (as defined in general headnote 2 of the Tariff Schedules of the United States), produce, or manufacture.

19 USC 1202.

(8) The term "mixture" means any combination of two or more chemical substances if the combination does not occur in nature and is not, in whole or in part, the result of a chemical reaction; except that such term does include any combination which occurs, in whole or in part, as a result of a chemical reaction if none of the chemical substances comprising the combination is a new chemical substance and if the combination could have been manufactured for commercial purposes without a chemical reaction at the time the chemical substances comprising the combination were combined.

(9) The term "new chemical substance" means any chemical substance which is not included in the chemical substance list compiled and published under section 8(b).

(10) The term "process" means the preparation of a chemical substance or mixture, after its manufacture, for distribution in commerce—

(A) in the same form or physical state as, or in a different form or physical state from, that in which it was received by the person so preparing such substance or mixture, or

(B) as part of an article containing the chemical substance or mixture.

(11) The term "processor" means any person who processes a chemical substance or mixture.

(12) The term "standards for the development of test data" means a prescription of—

(A) the—

(i) health and environmental effects, and

(ii) information relating to toxicity, persistence, and other characteristics which affect health and the environment,

for which test data for a chemical substance or mixture are to be developed and any analysis that is to be performed on such data, and

(B) to the extent necessary to assure that data respecting such effects and characteristics are reliable and adequate—

(i) the manner in which such data are to be developed,

(ii) the specification of any test protocol or methodology to be employed in the development of such data, and

(iii) such other requirements as are necessary to provide such assurance.

(13) The term "State" means any State of the United States, the District of Columbia, the Commonwealth of Puerto Rico, the Virgin Islands, Guam, the Canal Zone,

American Samoa, the Northern Mariana Islands, or any other territory or possession of the United States.

(14) The term "United States", when used in the geographic sense, means all of the States.

Sec. 4. Testing of Chemical Substances and Mixtures

(a) **Testing Requirements.**—If the Administrator finds that— 15 USC 2603.

(1) (A) (i) the manufacture, distribution in commerce, processing, use, or disposal of a chemical substance or mixture, or that any combination of such activities, may present an unreasonable risk of injury to health or the environment,

(ii) there are insufficient data and experience upon which the effects of such manufacture, distribution in commerce, processing, use, or disposal of such substance or mixture or of any combination of such activities on health or the environment can reasonably be determined or predicted, and

(iii) testing of such substance or mixture with respect to such effects is necessary to develop such data; or

(B) (i) a chemical substance or mixture is or will be produced in substantial quantities, and (I) it enters or may reasonably be anticipated to enter the environment in substantial quantities or (II) there is or may be significant or substantial human exposure to such substance or mixture,

(ii) there are insufficient data and experience upon which the effects of the manufacture, distribution in commerce, processing, use, or disposal of such substance or mixture or of any combination of such activities on health or the environment can reasonably be determined or predicted, and

(iii) testing of such substance or mixture with respect to such effects is necessary to develop such data; and

(2) in the case of a mixture, the effects which the mixture's manufacture, distribution in commerce, processing, use, or disposal or any combination of such activities may have on health or the environment may not be reasonably and more efficiently determined or predicted by testing the chemical substances which comprise the mixture;

the Administrator shall by rule require that testing be conducted on such substance or Rules.
mixture to develop data with respect to the health and environmental effects for which there is an insufficiency of data and experience and which are relevant to a determination that the manufacture, distribution in commerce, processing, use, or disposal of such substance or mixture, or that any combination of such activities, does or does not present an unreasonable risk of injury to health or the environment.

(b) (1) **Testing Requirement Rule.**—A rule under subsection (a) shall include—

(A) identification of the chemical substance or mixture for which testing is required under the rule,

(B) standards for the development of test data for such substance or mixture, Standards for
and development of
test data. Data,

(C) with respect to chemical substances which are not new chemical substances submittal to
and to mixtures, a specification of the period (which period may not be of unreason- Administrator.
able duration) within which the persons required to conduct the testing shall submit

to the Administrator data developed in accordance with the standards referred to in subparagraph (B).

In determining the standards and period to be included, pursuant to subparagraphs (B) and (C), in a rule under subsection (a), the Administrator's considerations shall include the relative costs of the various test protocols and methodologies which may be required under the rule and the reasonably foreseeable availability of the facilities and personnel needed to perform the testing required under the rule. Any such rule may require the submission to the Administrator of preliminary data during the period prescribed under subparagraph (C).

(2) (A) The health and environmental effects for which standards for the development of test data may be prescribed include carcinogenesis, mutagenesis, teratogenesis, behavioral disorders, cumulative or synergistic effects, and any other effect which may present an unreasonable risk of injury to health or the environment. The characteristics of chemical substances and mixtures for which such standards may be prescribed include persistence, acute toxicity, subacute toxicity, chronic toxicity, and any other characteristic which may present such a risk. The methodologies that may be prescribed in such standards include epidemiologic studies, serial or hierarchical tests, in vitro tests, and whole animal tests, except that before prescribing epidemiologic studies of employees, the Administrator shall consult with the Director of the National Institute for Occupational Safety and Health.

Review of standards.

(B) From time to time, but not less than once each 12 months, the Administrator shall review the adequacy of the standards for development of data prescribed in rules under subsection (a) and shall, if necessary, institute proceedings to make appropriate revisions of such standards.

(3) (A) A rule under subsection (a) respecting a chemical substance or mixture shall require the persons described in subparagraph (B) to conduct tests and submit data to the Administrator on such substance or mixture, except that the Administrator may permit two or more of such persons to designate one such person or a qualified third party to conduct such tests and submit such data on behalf of the persons making the designation.

(B) The following persons shall be required to conduct tests and submit data on a chemical substance or mixture subject to a rule under subsection (a):

(i) Each person who manufactures or intends to manufacture such substance or mixture if the Administrator makes a finding described in subsection (a)(1)(A)(ii) or (a)(1)(B)(ii) with respect to the manufacture of such substance or mixture.

(ii) Each person who processes or intends to process such substance or mixture if the Administrator makes a finding described in subsection (a)(1)(A)(ii) or (a)(1)(B)(ii) with respect to the processing of such substance or mixture.

(iii) Each person who manufactures or processes or intends to manufacture or process such substance or mixture if the Administrator makes a finding described in subsection (a)(1)(A)(ii) or (a)(1)(B)(ii) with respect to the distribution in commerce, use, or disposal of such substance or mixture.

(4) Any rule under subsection (a) requiring the testing of and submission of data for a particular chemical substance or mixture shall expire at the end of the reimbursement period (as defined in subsection (c)(3)(B)) which is applicable to test data for such substance or mixture unless the Administrator repeals the rule before such date; and a rule under subsection (a) requiring the testing of and submission of data for a category of chemical substances or mixtures shall expire with respect to a chemical substance or

mixture included in the category at the end of the reimbursement period (as so defined) which is applicable to test data for such substance or mixture unless the Administrator before such date repeals the application of the rule to such substance or mixture or repeals the rule.

(5) Rules issued under subsection (a) (and any substantive amendment thereto or repeal thereof) shall be promulgated pursuant to section 553 of title 5, United States Code, except that (A) the Administrator shall give interested persons an opportunity for the oral presentation of data, views, or arguments, in addition to an opportunity to make written submissions; (B) a transcript shall be made of any oral presentation; and (C) the Administrator shall make and publish with the rule the findings described in paragraph (1) (A) or (1) (B) of subsection (a) and, in the case of a rule respecting a mixture, the finding described in paragraph (2) of such subsection.

Oral presentation and written submissions.

Transcript. Publication.

(c) Exemption.—(1) Any person required by a rule under subsection (a) to conduct tests and submit data on a chemical substance or mixture may apply to the Administrator (in such form and manner as the Administrator shall prescribe) for an exemption from such requirement.

Application.

(2) If, upon receipt of an application under paragraph (1), the Administrator determines that—

(A) the chemical substance or mixture with respect to which such application was submitted is equivalent to a chemical substance or mixture for which data has been submitted to the Administrator in accordance with a rule under subsection (a) or for which data is being developed pursuant to such a rule, and

(B) submission of data by the applicant on such substance or mixture would be duplicative of data which has been submitted to the Administrator in accordance with such rule or which is being developed pursuant to such rule,

the Administrator shall exempt, in accordance with paragraph (3) or (4), the applicant from conducting tests and submitting data on such substance or mixture under the rule with respect to which such application was submitted.

(3) (A) If the exemption under paragraph (2) of any person from the requirement to conduct tests and submit test data on a chemical substance or mixture is granted on the basis of the existence of previously submitted test data and if such exemption is granted during the reimbursement period for such test data (as prescribed by subparagraph (B)), then (unless such person and the persons referred to in clauses (i) and (ii) agree on the amount and method of reimbursement) the Administrator shall order the person granted the exemption to provide fair and equitable reimbursement (in an amount determined under rules of the Administrator)—

Fair and equitable reimbursement.

(i) to the person who previously submitted such test data, for a portion of the costs incurred by such person in complying with the requirement to submit such data, and

(ii) to any other person who has been required under this subparagraph to contribute with respect to such costs, for a portion of the amount such person was required to contribute.

In promulgating rules for the determination of fair and equitable reimbursement to the persons described in clauses (i) and (ii) for costs incurred with respect to a chemical substance or mixture, the Administrator shall, after consultation with the Attorney General and the Federal Trade Commission, consider all relevant factors, including the effect on the competitive position of the person required to provide reimbursement in

Rules.

relation to the person to be reimbursed and the share of the market for such substance or mixture of the person required to provide reimbursement in relation to the share of such market of the persons to be reimbursed. An order under this subparagraph shall, for purposes of judicial review, be considered final agency action.

Reimbursement period.

(B) For purposes of subparagraph (A), the reimbursement period for any test data for a chemical substance or mixture is a period—

(i) beginning on the date such data is submitted in accordance with a rule promulgated under subsection (a), and

(ii) ending—

(I) five years after the date referred to in clause (i), or

(II) at the expiration of a period which begins on the date referred to in clause (i) and which is equal to the period which the Administrator determines was necessary to develop such data,

whichever is later.

(4) (A) If the exemption under paragraph (2) of any person from the requirement to conduct tests and submit test data on a chemical substance or mixture is granted on the basis of the fact that test data is being developed by one or more persons pursuant to a rule promulgated under subsection (a), then (unless such person and the persons referred to in clauses (i) and (ii) agree on the amount and method of reimbursement) the Administrator shall order the person granted the exemption to provide fair and equitable reimbursement (in an amount determined under rules of the Administrator)—

(i) to each such person who is developing such test data, for a portion of the costs incurred by each such person in complying with such rule, and

(ii) to any other person who has been required under this subparagraph to contribute with respect to the costs of complying with such rule, for a portion of the amount such person was required to contribute.

In promulgating rules for the determination of fair and equitable reimbursement to the persons described in clauses (i) and (ii) for costs incurred with respect to a chemical substance or mixture, the Administrator shall, after consultation with the Attorney General and the Federal Trade Commission, consider the factors described in the second sentence of paragraph (3) (A). An order under this subparagraph shall, for purposes of judicial review, be considered final agency action.

(B) If any exemption is granted under paragraph (2) on the basis of the fact that one or more persons are developing test data pursuant to a rule promulgated under subsection (a) and if after such exemption is granted the Administrator determines that no such person has complied with such rule, the Administrator shall (i) after providing written notice to the person who holds such exemption and an opportunity for a hearing, by order terminate such exemption, and (ii) notify in writing such person of the requirements of the rule with respect to which such exemption was granted.

Publication in Federal Register.

(d) Notice.—Upon the receipt of any test data pursuant to a rule under subsection (a), the Administrator shall publish a notice of the receipt of such data in the Federal Register within 15 days of its receipt. Subject to section 14, each such notice shall (1) identify the chemical substance or mixture for which data have been received; (2) list the uses or intended uses of such substance or mixture and the information required by the applicable standards for the development of test data; and (3) describe the nature of the test data developed. Except as otherwise provided in section 14, such data shall be made available by the Administrator for examination by any person.

(e) Priority List.—(1) (A) There is established a committee to make recommendations to the Administrator respecting the chemical substances and mixtures to which the Administrator should give priority consideration for the promulgation of a rule under subsection (a). In making such a recommendation with respect to any chemical substance or mixture, the committee shall consider all relevant factors, including—

(i) the quantities in which the substance or mixture is or will be manufactured,

(ii) the quantities in which the substance or mixture enters or will enter the environment,

(iii) the number of individuals who are or will be exposed to the substance or mixture in their places of employment and the duration of such exposure,

(iv) the extent to which human beings are or will be exposed to the substance or mixture,

(v) the extent to which the substance or mixture is closely related to a chemical substance or mixture which is known to present an unreasonable risk of injury to health or the environment,

(vi) the existence of data concerning the effects of the substance or mixture on health or the environment,

(vii) the extent to which testing of the substance or mixture may result in the development of data upon which the effects of the substance or mixture on health or the environment can reasonably be determined or predicted, and

(viii) the reasonably foreseeable availability of facilities and personnel for performing testing on the substance or mixture.

The recommendations of the committee shall be in the form of a list of chemical substances and mixtures which shall be set forth, either by individual substance or mixture or by groups of substances or mixtures, in the order in which the committee determines the Administrator should take action under subsection (a) with respect to the substances and mixtures. In establishing such list, the committee shall give priority attention to those chemical substances and mixtures which are known to cause or contribute to or which are suspected of causing or contributing to cancer, gene mutations, or birth defects. The committee shall designate chemical substances and mixtures on the list with respect to which the committee determines the Administrator should, within 12 months of the date on which such substances and mixtures are first designated, initiate a proceeding under subsection (a). The total number of chemical substances and mixtures on the list which are designated under the preceding sentence may not, at any time, exceed 50.

(B) As soon as practicable but not later than nine months after the effective date of this Act, the committee shall publish in the Federal Register and transmit to the Administrator the list and designations required by subparagraph (A) together with the reasons for the committee's inclusion of each chemical substance or mixture on the list. At least every six months after the date of the transmission to the Administrator of the list pursuant to the preceding sentence, the committee shall make such revisions in the list as it determines to be necessary and shall transmit them to the Administrator together with the committee's reasons for the revisions. Upon receipt of any such revision, the Administrator shall publish in the Federal Register the list with such revision, the reasons for such revision, and the designations made under subparagraph (A). The Administrator shall provide reasonable opportunity to any interested person to file with the Administrator written comments on the committee's list, any revision of such list by the committee, and designations made by the committee, and shall make

Committee to make recommendations to Administrator.

Recommendations, list of chemical substances and mixtures.

Publication in Federal Register; transmittal to Administrator.

List revision, publication in Federal Register.

Comments.

Publication in
Federal Register.
such comments available to the public. Within the 12-month period beginning on the date of the first inclusion on the list of a chemical substance or mixture designated by the committee under subparagraph (A) the Administrator shall with respect to such chemical substance or mixture either initiate a rulemaking proceeding under subsection (a) or if such a proceeding is not initiated within such period, publish in the Federal Register the Administrator's reason for not initiating such a proceeding.

Membership.
(2) (A) The committee established by paragraph (1) (A) shall consist of eight members as follows:

(i) One member appointed by the Administrator from the Environmental Protection Agency.

(ii) One member appointed by the Secretary of Labor from officers or employees of the Department of Labor engaged in the Secretary's activities under the Occupational Safety and Health Act of 1970.

(iii) One member appointed by the Chairman of the Council on Environmental Quality from the Council or its officers or employees.

(iv) One member appointed by the Director of the National Institute for Occupational Safety and Health from officers or employees of the Institute.

(v) One member appointed by the Director of the National Institute of Environmental Health Sciences from officers or employees of the Institute.

(vi) One member appointed by the Director of the National Cancer Institute from officers or employees of the Institute.

(vii) One member appointed by the Director of the National Science Foundation from officers or employees of the Foundation.

(viii) One member appointed by the Secretary of Commerce from officers or employees of the Department of Commerce.

(B) (i) An appointed member may designate an individual to serve on the committee on the member's behalf. Such a designation may be made only with the approval of the applicable appointing authority and only if the individual is from the entity from which the member was appointed.

(ii) No individual may serve as a member of the committee for more than four years in the aggregate. If any member of the committee leaves the entity from which the member was appointed, such member may not continue as a member of the committee, and the member's position shall be considered to be vacant. A vacancy in the committee shall be filled in the same manner in which the original appointment was made.

(iii) Initial appointments to the committee shall be made not later than the 60th day after the effective date of this Act. Not later than the 90th day after such date the members of the committee shall hold a meeting for the selection of a chairperson from among their number.

(C) (i) No member of the committee, or designee of such member, shall accept employment or compensation from any person subject to any requirement of this Act or of any rule promulgated or order issued thereunder, for a period of at least 12 months after termination of service on the committee.

(ii) No person, while serving as a member of the committee, or designee of such member, may own any stocks or bonds, or have any pecuniary interest, of substantial value in any person engaged in the manufacture, processing, or distribution in commerce of any chemical substance or mixture subject to any requirement of this Act or of any rule promulgated or order issued thereunder.

(iii) The Administrator, acting through attorneys of the Environmental Protection Agency, or the Attorney General may bring an action in the appropriate district court of the United States to restrain any violation of this subparagraph.

(D) The Administrator shall provide the committee such administrative support services as may be necessary to enable the committee to carry out its function under this subsection.

(f) Required Actions.—Upon the receipt of—

(1) any test data required to be submitted under this Act, or

(2) any other information available to the Administrator,

which indicates to the Administrator that there may be a reasonable basis to conclude that a chemical substance or mixture presents or will present a significant risk of serious or widespread harm to human beings from cancer, gene mutations, or birth defects, the Administrator shall, within the 180-day period beginning on the date of the receipt of such data or information, initiate appropriate action under section 5, 6, or 7 to prevent or reduce to a sufficient extent such risk or publish in the Federal Register a finding that such risk is not unreasonable. For good cause shown the Administrator may extend such period for an additional period of not more than 90 days. The Administrator shall publish in the Federal Register notice of any such extension and the reasons therefor. A finding by the Administrator that a risk is not unreasonable shall be considered agency action for purposes of judicial review under chapter 7 of title 5, United States Code. This subsection shall not take effect until two years after the effective date of this Act.

Publication in Federal Register.

5 USC 701.

(g) Petition for Standards for the Development of Test Data.—A person intending to manufacture or process a chemical substance for which notice is required under section 5 (a) and who is not required under a rule under subsection (a) to conduct tests and submit data on such substance may petition the Administrator to prescribe standards for the development of test data for such substance. The Administrator shall by order either grant or deny any such petition within 60 days of its receipt. If the petition is granted, the Administrator shall prescribe such standards for such substance within 75 days of the date the petition is granted. If the petition is denied, the Administrator shall publish, subject to section 14, in the Federal Register the reasons for such denial.

Publication in Federal Register. Post, p. 2034.

Sec. 5. Manufacturing and Processing Notices

(a) In General.—(1) Except as provided in subsection (h), no person may—

15 USC 2604.

(A) manufacture a new chemical substance on or after the 30th day after the date on which the Administrator first publishes the list required by section 8 (b), or

(B) manufacture or process any chemical substance for a use which the Administrator has determined, in accordance with paragraph (2), is a significant new use, unless such person submits to the Administrator, at least 90 days before such manufacture or processing, a notice, in accordance with subsection (d), of such person's intention to manufacture or process such substance and such person complies with any applicable requirement of subsection (b).

(2) A determination by the Administrator that a use of a chemical substance is a

significant new use with respect to which notification is required under paragraph (1) shall be made by a rule promulgated after a consideration of all relevant factors, including—

(A) the projected volume of manufacturing and processing of a chemical substance,

(B) the extent to which a use changes the type or form of exposure of human beings or the environment to a chemical substance,

(C) the extent to which a use increases the magnitude and duration of exposure of human beings or the environment to a chemical substance, and

(D) the reasonably anticipated manner and methods of manufacturing, processing, distribution in commerce, and disposal of a chemical substance.

(b) Submission of Test Data.—(1) (A) If (i) a person is required by subsection (a) (1) to submit a notice to the Administrator before beginning the manufacture or processing of a chemical substance, and (ii) such person is required to submit test data for such substance pursuant to a rule promulgated under section 4 before the submission of such notice, such person shall submit to the Administrator such data in accordance with such rule at the time notice is submitted in accordance with subsection (a) (1).

(B) If—

(i) a person is required by subsection (a)(1) to submit a notice to the Administrator, and

(ii) such person has been granted an exemption under section 4 (c) from the requirements of a rule promulgated under section 4 before the submission of such notice,

such person may not, before the expiration of the 90 day period which begins on the date of the submission in accordance with such rule of the test data the submission or development of which was the basis for the exemption, manufacture such substance if such person is subject to subsection (a) (1) (A) or manufacture or process such substance for a significant new use if the person is subject to subsection (a) (1) (B).

(2) (A) If a person—

(i) is required by subsection (a)(1) to submit a notice to the Administrator before beginning the manufacture or processing of a chemical substance listed under paragraph (4), and

(ii) is not required by a rule promulgated under section 4 before the submission of such notice to submit test data for such substance,

such person shall submit to the Administrator data prescribed by subparagraph (B) at the time notice is submitted in accordance with subsection (a) (1).

(B) Data submitted pursuant to subparagraph (A) shall be data which the person submitting the data believes show that—

(i) in the case of a substance with respect to which notice is required under subsection (a) (1) (A), the manufacture, processing, distribution in commerce, use, and disposal of the chemical substance or any combination of such activities will not present an unreasonable risk of injury to health or the environment, or

(ii) in the case of a chemical substance with respect to which notice is required under subsection (a) (1) (B), the intended significant new use of the chemical substance will not present an unreasonable risk of injury to health or the environment.

(3) Data submitted under paragraph (1) or (2) shall be made available, subject to section 14, for examination by interested persons.

(4) (A) (i) The Administrator may, by rule, compile and keep current a list of chemical substances with respect to which the Administrator finds that the manufacture, processing, distribution in commerce, use, or disposal, or any combination of such activities, presents or may present an unreasonable risk of injury to health or the environment.

(ii) In making a finding under clause (i) that the manufacture, processing, distribution in commerce, use, or disposal of a chemical substance or any combination of such activities presents or may present an unreasonable risk of injury to health or the environment, the Administrator shall consider all relevant factors, including—

(I) the effects of the chemical substance on health and the magnitude of human exposure to such substance; and

(II) the effects of the chemical substance on the environment and the magnitude of environmental exposure to such substance.

(B) The Administrator shall, in prescribing a rule under subparagraph (A) which lists any chemical substance, identify those uses, if any, which the Administrator determines, by rule under subsection (a) (2), would constitute a significant new use of such substance.

(C) Any rule under subparagraph (A), and any substantive amendment or repeal of such a rule, shall be promulgated pursuant to the procedures specified in section 553 of title 5, United States Code, except that (i) the Administrator shall give interested persons an opportunity for the oral presentation of data, views, or arguments, in addition to an opportunity to make written submissions, (ii) a transcript shall be kept of any oral presentation, and (iii) the Administrator shall make and publish with the rule the finding described in subparagraph (A). *Oral presentation. Transcript. Publications.*

(c) Extension of Notice Period.—The Administrator may for good cause extend for additional periods (not to exceed in the aggregate 90 days) the period, prescribed by subsection (a) or (b) before which the manufacturing or processing of a chemical substance subject to such subsection may begin. Subject to section 14, such an extension and the reasons therefor shall be published in the Federal Register and shall constitute a final agency action subject to judicial review. *Publication in Federal Register.*

(d) Content of Notice; Publications in the Federal Register.—(1) The notice required by subsection (a) shall include—

(A) insofar as known to the person submitting the notice or insofar as reasonably ascertainable, the information described in subparagraphs (A), (B), (C), (D), (F), and (G) of section 8 (a) (2), and

(B) in such form and manner as the Administrator may prescribe, any test data in the possession or control of the person giving such notice which are related to the effect of any manufacture, processing, distribution in commerce, use, or disposal of such substance or any article containing such substance, or of any combination of such activities, on health or the environment, and

(C) a description of any other data concerning the environmental and health effects of such substance, insofar as known to the person making the notice or insofar as reasonably ascertainable.

Such a notice shall be made available, subject to section 14, for examination by interested persons.

(2) Subject to section 14, not later than five days (excluding Saturdays, Sundays and legal holidays) after the date of the receipt of a notice under subsection (a) or of data under subsection (b), the Administrator shall publish in the Federal Register a notice which—

(A) identifies the chemical substance for which notice or data has been received;

(B) lists the uses or intended uses of such substance; and

(C) in the case of the receipt of data under subsection (b), describes the nature of the tests performed on such substance and any data which was developed pursuant to subsection (b) or a rule under section 4.

A notice under this paragraph respecting a chemical substance shall identify the chemical substance by generic class unless the Administrator determines that more specific identification is required in the public interest.

(3) At the beginning of each month the Administrator shall publish a list in the Federal Register of (A) each chemical substance for which notice has been received under subsection (a) and for which the notification period prescribed by subsection (a), (b), or (c) has not expired, and (B) each chemical substance for which such notification period has expired since the last publication in the Federal Register of such list.

(e) Regulation Pending Development of Information.—(1) (A) If the Administrator determines that—

(i) the information available to the Administrator is insufficient to permit a reasoned evaluation of the health and environmental effects of a chemical substance with respect to which notice is required by subsection (a) ; and

(ii) (I) in the absence of sufficient information to permit the Administrator to make such an evaluation, the manufacture, processing, distribution in commerce, use, or disposal of such substance, or any combination of such activities, may present an unreasonable risk of injury to health or the environment, or

(II) such substance is or will be produced in substantial quantities, and such substance either enters or may reasonably be anticipated to enter the environment in substantial quantities or there is or may be significant or substantial human exposure to the substance,

Proposed order.

the Administrator may issue a proposed order, to take effect on the expiration of the notification period applicable to the manufacturing or processing of such substance under subsection (a), (b), or (c), to prohibit or limit the manufacture, processing, distribution in commerce, use, or disposal of such substance or to prohibit or limit any combination of such activities.

(B) A proposed order may not be issued under subparagraph (A) respecting a chemical substance (i) later than 45 days before the expiration of the notification period applicable to the manufacture or processing of such substance under subsection (a), (b), or (c), and (ii) unless the Administrator has, on or before the issuance of the proposed order, notified, in writing, each manufacturer or processor, as the case may be, of such substance of the determination which underlies such order.

(C) If a manufacturer or processor of a chemical substance to be subject to a proposed order issued under subparagraph (A) files with the Administrator (within the 30-day period beginning on the date such manufacturer or processor received the notice required by subparagraph (B) (ii) objections specifying with particularity the provisions of the order deemed objectionable and stating the grounds therefor, the proposed order shall not take effect.

(2) (A) (i) Except as provided in clause (ii), if with respect to a chemical substance with respect to which notice is required by subsection (a), the Administrator makes the determination described in paragraph (1) (A) and if— *Injunction, application.*

(I) the Administrator does not issue a proposed order under paragraph (1) respecting such substance, or

(II) the Administrator issues such an order respecting such substance but such order does not take effect because objections were filed under paragraph (1)(C) with respect to it,

the Administrator, through attorneys of the Environmental Protection Agency, shall apply to the United States District Court for the District of Columbia or the United States district court for the judicial district in which the manufacturer or processor, as the case may be, of such substance is found, resides, or transacts business for an injunction to prohibit or limit the manufacture, processing, distribution in commerce, use, or disposal of such substance (or to prohibit or limit any combination of such activities).

(ii) If the Administrator issues a proposed order under paragraph (1) (A) respecting a chemical substance but such order does not take effect because objections have been filed under paragraph (1)(C) with respect to it, the Administrator is not required to apply for an injunction under clause (i) respecting such substance if the Administrator determines, on the basis of such objections, that the determinations under paragraph (1) (A) may not be made.

(B) A district court of the United States which receives an application under subparagraph (A)(i) for an injunction respecting a chemical substance shall issue such injunction if the court finds that—

(i) the information available to the Administrator is insufficient to permit a reasoned evaluation of the health and environmental effects of a chemical substance with respect to which notice is required by subsection (a) ; and

(ii) (I) in the absence of sufficient information to permit the Administrator to make such an evaluation, the manufacture, processing, distribution in commerce, use, or disposal of such substance, or any combination of such activities, may present an unreasonable risk of injury to health or the environment, or

(II) such substance is or will be produced in substantial quantities, and such substance either enters or may reasonably be anticipated to enter the environment in substantial quantities or there is or may be significant or substantial human exposure to the substance.

(C) Pending the completion of a proceeding for the issuance of an injunction under subparagraph (B) respecting a chemical substance, the court may, upon application of the Administrator made through attorneys of the Environmental Protection Agency, issue a temporary restraining order or a preliminary injunction to prohibit the manufacture, processing, distribution in commerce, use, or disposal of such a substance (or any combination of such activities) if the court finds that the notification period applicable under subsection (a), (b), or (c) to the manufacturing or processing of such substance may expire before such proceeding can be completed.

(D) After the submission to the Administrator of test data sufficient to evaluate the health and environmental effects of a chemical substance subject to an injunction issued under subparagraph (B) and the evaluation of such data by the Administrator, the district court of the United States which issued such injunction shall, upon petition, dissolve the injunction unless the Administrator has initiated a proceeding for the issuance of a rule under section 6 (a) respecting the substance. If such a proceeding has been initiated, such court shall continue the injunction in effect until the effective date of the rule promulgated in such proceeding or, if such proceeding is terminated without the promulgation of a rule, upon the termination of the proceeding, whichever occurs first.

(f) **Protection Against Unreasonable Risks.**—(1) If the Administrator finds that there is a reasonable basis to conclude that the manufacture, processing, distribution in commerce, use, or disposal of a chemical substance with respect to which notice is required by subsection (a), or that any combination of such activities, presents or will present an unreasonable risk of injury to health or environment before a rule promulgated under section 6 can protect against such risk, the Administrator shall, before the expiration of the notification period applicable under subsection (a), (b), or (c) to the manufacturing or processing of such substance, take the action authorized by paragraph (2) or (3) to the extent necessary to protect against such risk.

Proposed rule.

(2) The Administrator may issue a proposed rule under section 6 (a) to apply to a chemical substance with respect to which a finding was made under paragraph (1)—

(A) a requirement limiting the amount of such substance which may be manufactured, processed, or distributed in commerce,

(B) a requirement described in paragraph (2), (3), (4), (5), (6), or (7) of section 6 (a), or

(C) any combination of the requirements referred to in subparagraph (B).

Publication in Federal Register.

Such a proposed rule shall be effective upon its publication in the Federal Register. Section 6 (d) (2) (B) shall apply with respect to such rule.

(3) (A) The Administrator may—

Proposed order.

(i) issue a proposed order to prohibit the manufacture, processing, or distribution in commerce of a substance with respect to which a finding was made under paragraph (1), or

Injunction application.

(ii) apply, through attorneys of the Environmental Protection Agency, to the United States District Court for the District of Columbia or the United States district court for the judicial district in which the manufacturer, or processor, as the case may be, of such substance, if found, resides, or transacts business for an injunction to prohibit the manufacture, processing, or distribution in commerce of such substance.

A proposed order issued under clause (i) respecting a chemical substance shall take effect on the expiration of the notification period applicable under subsection (a), (b), or (c) to the manufacture or processing of such substance.

(B) If the district court of the United States to which an application has been made under subparagraph (A) (ii) finds that there is a reasonable basis to conclude that the manufacture, processing, distribution in commerce, use, or disposal of the chemical substance with respect to which such application was made, or that any combination of such activities, presents or will present an unreasonable risk of injury to health or the environment before a rule promulgated under section 6 can protect against such risk, the court shall issue an injunction to prohibit the manufacture, processing, or distribution in commerce of such substance or to prohibit any combination of such activities.

(C) The provisions of subparagraphs (B) and (C) of subsection (e)(1) shall apply with respect to an order issued under clause (i) of subparagraph (A); and the provisions of subparagraph (C) of subsection (e)(2) shall apply with respect to an injunction issued under subparagraph (B).

(D) If the Administrator issues an order pursuant to subparagraph (A)(i) respecting a chemical substance and objections are filed in accordance with subsection (e)(1)(C), the Administrator shall seek an injunction under subparagraph (A)(ii) respecting such substance unless the Administrator determines, on the basis of such objections, that such substance does not or will not present an unreasonable risk of injury to health or the environment.

(g) Statement of Reasons for Not Taking Action. — If the Administrator has not initiated any action under this section or section 6 or 7 to prohibit or limit the manufacture, processing, distribution in commerce, use, or disposal of a chemical substance, with respect to which notification or data is required by subsection (a)(1)(B) or (b), before the expiration of the notification period applicable to the manufacturing or processing of such substance, the Administrator shall publish a statement of the Administrator's reasons for not initiating such action. Such a statement shall be published in the Federal Register before the expiration of such period. Publication of such statement in accordance with the preceding sentence is not a prerequisite to the manufacturing or processing of the substance with respect to which the statement is to be published.

Publication in Federal Register.

(h) Exemptions. —(1) The Administrator may, upon application, exempt any person from any requirement of subsection (a) or (b) to permit such person to manufacture or process a chemical substance for test marketing purposes—

(A) upon a showing by such person satisfactory to the Administrator that the manufacture, processing, distribution in commerce, use, and disposal of such substance, and that any combination of such activities, for such purposes will not present any unreasonable risk of injury to health or the environment, and

(B) under such restrictions as the Administrator considers appropriate.

(2)(A) The Administrator may, upon application, exempt any person from the requirement of subsection (b)(2) to submit data for a chemical substance. If, upon receipt of an application under the preceding sentence, the Administrator determines that—

(i) the chemical substance with respect to which such application was submitted is equivalent to a chemical substance for which data has been submitted to the Administrator as required by subsection (b)(2), and

(ii) submission of data by the applicant on such substance would be duplicative of data which has been submitted to the Administrator in accordance with such subsection.

the Administrator shall exempt the applicant from the requirement to submit such data on such substance. No exemption which is granted under this subparagraph with respect to the submission of data for a chemical substance may take effect before the beginning of the reimbursement period applicable to such data.

(B) If the Administrator exempts any person, under subparagraph (A), from submitting data required under subsection (b)(2) for a chemical substance because of

Fair and equitable reimbursement.

the existence of previously submitted data and if such exemption is granted during the reimbursement period for such data, then (unless such person and the persons referred to in clauses (i) and (ii) agree on the amount and method of reimbursement) the Administrator shall order the person granted the exemption to provide fair and equitable reimbursement (in an amount determined under rules of the Administrator)—

(i) to the person who previously submitted the data on which the exemption was based, for a portion of the costs incurred by such person in complying with the requirement under subsection (b) (2) to submit such data, and

(ii) to any other person who has been required under this subparagraph to contribute with respect to such costs, for a portion of the amount such person was required to contribute.

In promulgating rules for the determination of fair and equitable reimbursement to the persons described in clauses (i) and (ii) for costs incurred with respect to a chemical substance, the Administrator shall, after consultation with the Attorney General and the Federal Trade Commission, consider all relevant factors, including the effect on the competitive position of the person required to provide reimbursement in relation to the persons to be reimbursed and the share of the market for such substance of the person required to provide reimbursement in relation to the share of such market of the persons to be reimbursed. For purposes of judicial review, an order under this subparagraph shall be considered final agency action.

Reimbursement period. (C) For purposes of this paragraph, the reimbursement period for any previously submitted data for a chemical substance is a period—

(i) beginning on the date of the termination of the prohibition, imposed under this section, on the manufacture or processing of such substance by the person who submitted such data to the Administrator, and

(ii) ending—

(I) five years after the date referred to in clause (i), or

(II) at the expiration of a period which begins on the date referred to in clause (i) and is equal to the period which the Administrator determines was necessary to develop such data,

whichever is later.

(3) The requirements of subsections (a) and (b) do not apply with respect to the manufacturing or processing of any chemical substance which is manufactured or processed, or proposed to be manufactured or processed, only in small quantities (as defined by the Administrator by rule) solely for purposes of—

(A) scientific experimentation or analysis, or

(B) chemical research on, or analysis of such substance or another substance, including such research or analysis for the development of a product,

if all persons engaged in such experimentation, research, or analysis for a manufacturer or processor are notified (in such form and manner as the Administrator may prescribe) of any risk to health which the manufacturer, processor, or the Administrator has reason to believe may be associated with such chemical substance.

(4) The Administrator may, upon application and by rule, exempt the manufacturer of any new chemical substance from all or part of the requirements of this section if the Administrator determines that the manufacture, processing, distribution in commerce, use, or disposal of such chemical substance, or that any combination of such activities, will not present an unreasonable risk of injury to health or the

environment. A rule promulgated under this paragraph (and any substantive amendment to, or repeal of, such a rule) shall be promulgated in accordance with paragraphs (2) and (3) of section 6(c).

(5) The Administrator may, upon application, make the requirements of subsections (a) and (b) inapplicable with respect to the manufacturing or processing of any chemical substance (A) which exists temporarily as a result of a chemical reaction in the manufacturing or processing of a mixture or another chemical substance, and (B) to which there is no, and will not be, human or environmental exposure.

(6) Immediately upon receipt of an application under paragraph (1) or (5) the Administrator shall publish in the Federal Register notice of the receipt of such application. The Administrator shall give interested persons an opportunity to comment upon any such application and shall, within 45 days of its receipt, either approve or deny the application. The Administrator shall publish in the Federal Register notice of the approval or denial of such an application.

Publication in Federal Register. Comments.

Publication in Federal Register.

(i) Definition.—For purposes of this section, the terms "manufacture" and "process" mean manufacturing or processing for commercial purposes.

Sec. 6. Regulation of Hazardous Chemical Substances and Mixtures

(a) Scope of Regulation.—If the Administrator finds that there is a reasonable basis to conclude that the manufacture, processing, distribution in commerce, use, or disposal of a chemical substance or mixture, or that any combination of such activities, presents or will present an unreasonable risk of injury to health or the environment, the Administrator shall by rule apply one or more of the following requirements to such substance or mixture to the extent necessary to protect adequately against such risk using the least burdensome requirements:

15 USC 2605.

(1) A requirement (A) prohibiting the manufacturing, processing, or distribution in commerce of such substance or mixture, or (B) limiting the amount of such substance or mixture which may be manufactured, processed, or distributed in commerce.

(2) A requirement—

(A) prohibiting the manufacture, processing, or distribution in commerce of such substance or mixture for (i) a particular use or (ii) a particular use in a concentration in excess of a level specified by the Administrator in the rule imposing the requirement, or

(B) limiting the amount of such substance or mixture which may be manufactured, processed, or distributed in commerce for (i) a particular use or (ii) a particular use in a concentration in excess of a level specified by the Administrator in the rule imposing the requirement.

(3) A requirement that such substance or mixture or any article containing such substance or mixture be marked with or accompanied by clear and adequate warnings and instructions with respect to its use, distribution in commerce, or disposal or with respect to any combination of such activities. The form and content of such warnings and instructions shall be prescribed by the Administrator.

(4) A requirement that manufacturers and processors of such substance or mixture make and retain records of the processes used to manufacture or process

such substance or mixture and monitor or conduct tests which are reasonable and necessary to assure compliance with the requirements of any rule applicable under this subsection.

(5) A requirement prohibiting or otherwise regulating any manner or method of commercial use of such substance or mixture.

(6) (A) A requirement prohibiting or otherwise regulating any manner or method of disposal of such substance or mixture, or of any article containing such substance or mixture, by its manufacturer or processor or by any other person who uses, or disposes of, it for commercial purposes.

(B) A requirement under subparagraph (A) may not require any person to take any action which would be in violation of any law or requirement of, or in effect for, a State or political subdivision, and shall require each person subject to it to notify each State and political subdivision in which a required disposal may occur of such disposal.

(7) A requirement directing manufacturers or processors of such substance or mixture (A) to give notice of such unreasonable risk of injury to distributors in commerce of such substance or mixture and, to the extent reasonably ascertainable, to other persons in possession of such substance or mixture or exposed to such substance or mixture, (B) to give public notice of such risk of injury, and (C) to replace or repurchase such substance or mixture as elected by the person to which the requirement is directed.

Any requirement (or combination of requirements) imposed under this subsection may be limited in application to specified geographic areas.

(b) Quality Control.—If the Administrator has a reasonable basis to conclude that a particular manufacturer or processor is manufacturing or processing a chemical substance or mixture in a manner which unintentionally causes the chemical substance or mixture to present or which will cause it to present an unreasonable risk of injury to health or the environment—

(1) the Administrator may by order require such manufacturer or processor to submit a description of the relevant quality control procedures followed in the manufacturing or processing of such chemical substance or mixture; and

(2) if the Administrator determines—

(A) that such quality control procedures are inadequate to prevent the chemical substance or mixture from presenting such risk of injury, the Administrator may order the manufacturer or processor to revise such quality control procedures to the extent necessary to remedy such inadequacy; or

(B) that the use of such quality control procedures has resulted in the distribution in commerce of chemical substances or mixtures which represent an unreasonable risk of injury to health or the environment, the Administrator may order the manufacturer or processor to (i) give notice of such risk to processors or distributors in commerce of any such substance or mixture, or to both, and, to the extent reasonably ascertainable, to any other person in possession of or exposed to any such substance, (ii) to give public notice of such risk, and (iii) to provide such replacement or repurchase of any such substance or mixture as is necessary to adequately protect health or the environment.

Hearing.

A determination under subparagraph (A) or (B) of paragraph (2) shall be made on the record after opportunity for hearing in accordance with section 554 of title 5, United

States Code. Any manufacturer or processor subject to a requirement to replace or repurchase a chemical substance or mixture may elect either to replace or repurchase the substance or mixture and shall take either such action in the manner prescribed by the Administrator.

(c) Promulgation of Subsection (a) Rules.—(1) In promulgating any rule under subsection (a) with respect to a chemical substance or mixture, the Administrator shall consider and publish a statement with respect to—

Statement, publication.

(A) the effects of such substance or mixture on health and the magnitude of the exposure of human beings to such substance or mixture,

(B) the effects of such substance or mixture on the environment and the magnitude of the exposure of the environment to such substance or mixture,

(C) the benefits of such substance or mixture for various uses and the availability of substitutes for such uses, and

(D) the reasonably ascertainable economic consequences of the rule, after consideration of the effect on the national economy, small business, technological innovation, the environment, and public health.

If the Administrator determines that a risk of injury to health or the environment could be eliminated or reduced to a sufficient extent by actions taken under another Federal law (or laws) administered in whole or in part by the Administrator, the Administrator may not promulgate a rule under subsection (a) to protect against such risk of injury unless the Administrator finds, in the Administrator's discretion, that it is in the public interest to protect against such risk under this Act. In making such a finding the Administrator shall consider (i) all relevant aspects of the risk, as determined by the Administrator in the Administrator's discretion, (ii) a comparison of the estimated costs of complying with actions taken under this Act and under such law (or laws), and (iii) the relative efficiency of actions under this Act and under such law (or laws) to protect against such risk of injury.

(2) When prescribing a rule under subsection (a) the Administrator shall proceed in accordance with section 553 of title 5, United States Code (without regard to any reference in such section to sections 556 and 557 of such title), and shall also (A) publish a notice of proposed rulemaking stating with particularity the reason for the proposed rule; (B) allow interested persons to submit written data, views, and arguments, and make all such submissions publicly available; (C) provide an opportunity for an informal hearing in accordance with paragraph (3); (D) promulgate, if appropriate, a final rule based on the matter in the rulemaking record (as defined in section 19 (a)), and (E) make and publish with the rule the finding described in subsection (a).

5 USC 556, 557.

Notice, publication. Written data, views, arguments, submittal. Hearing. Final rule.

(3) Informal hearings required by paragraph (2) (C) shall be conducted by the Administrator in accordance with the following requirements:

Informal hearings.

(A) Subject to subparagraph (B), an interested person is entitled—

(i) to present such person's position orally or by documentary submissions (or both), and

(ii) if the Administrator determines that there are disputed issues of material fact it is necessary to resolve, to present such rebuttal submissions and to conduct (or have conducted under subparagraph (B) (ii)) such cross-examination of persons as the Administrator determines (I) to be appropriate, and (II) to be required for a full and true disclosure with respect to such issues.

(B) The Administrator may prescribe such rules and make such rulings con-

Rules.

cerning procedures in such hearings to avoid unnecessary costs or delay. Such rules or rulings may include (i) the imposition of reasonable time limits on each interested person's oral presentations, and (ii) requirements that any cross-examination to which a person may be entitled under subparagraph (A) be conducted by the Administrator on behalf of that person in such manner as the Administrator determines (I) to be appropriate, and (II) to be required for a full and true disclosure with respect to disputed issues of material fact.

(C) (i) Except as provided in clause (ii), if a group of persons each of whom under subparagraphs (A) and (B) would be entitled to conduct (or have conducted) cross-examination and who are determined by the Administrator to have the same or similar interests in the proceeding cannot agree upon a single representative of such interests for purposes of cross-examination, the Administrator may make rules and rulings (I) limiting the representation of such interest for such purposes, and (II) governing the manner in which such cross-examination shall be limited.

(ii) When any person who is a member of a group with respect to which the Administrator has made a determination under clause (i) is unable to agree upon group representation with the other members of the group, then such person shall not be denied under the authority of clause (i) the opportunity to conduct (or have conducted) cross-examination as to issues affecting the person's particular interests if (I) the person satisfies the Administrator that the person has made a reasonable and good faith effort to reach agreement upon group representation with the other members of the group and (II) the Administrator determines that there are substantial and relevant issues which are not adequately presented by the group representative.

Verbatim transcript. (D) A verbatim transcript shall be taken of any oral presentation made, and cross-examination conducted in any informal hearing under this subsection. Such transcript shall be available to the public.

Compensation. (4) (A) The Administrator may, pursuant to rules prescribed by the Administrator, provide compensation for reasonable attorneys' fees, expert witness fees, and other costs of participating in a rulemaking proceeding for the promulgation of a rule under subsection (a) to any person—

(i) who represents an interest which would substantially contribute to a fair determination of the issues to be resolved in the proceeding, and

(ii) if—

(I) the economic interest of such person is small in comparison to the costs of effective participation in the proceeding by such person, or

(II) such person demonstrates to the satisfaction of the Administrator that such person does not have sufficient resources adequately to participate in the proceeding without compensation under this subparagraph.

In determining for purposes of clause (i) if an interest will substantially contribute to a fair determination of the issues to be resolved in a proceeding, the Administrator shall take into account the number and complexity of such issues and the extent to which representation of such interest will contribute to widespread public participation in the proceeding and representation of a fair balance of interests for the resolution of such issues.

(B) In determining whether compensation should be provided to a person under subparagraph (A) and the amount of such compensation, the Administrator shall take into account the financial burden which will be incurred by such person in participating in the rulemaking proceeding. The Administrator shall take such action as may be necessary to ensure that the aggregate amount of compensation paid under this paragraph in any fiscal year to all persons who, in rulemaking proceedings in which they receive compensation, are persons who either—

(i) would be regulated by the proposed rule, or

(ii) represent persons who would be so regulated, may not exceed 25 per centum of the aggregate amount paid as compensation under this paragraph to all persons in such fiscal year.

(5) Paragraph (1), (2), (3), and (4) of this subsection apply to the promulgation of a rule repealing, or making a substantive amendment to, a rule promulgated under subsection (a).

(d) **Effective Date.**—(1) The Administrator shall specify in any rule under subsection (a) the date on which it shall take effect, which date shall be as soon as feasible.

(2) (A) The Administrator may declare a proposed rule under subsection (a) to be effective upon its publication in the Federal Register and until the effective date of final action taken, in accordance with subparagraph (B), respecting such rule if— *Publication in Federal Register.*

(i) the Administrator determines that—

(I) the manufacture, processing, distribution in commerce, use, or disposal of the chemical substance or mixture subject to such proposed rule or any combination of such activities is likely to result in an unreasonable risk of serious or widespread injury to health or the environment before such effective date; and

(II) making such proposed rule so effective is necessary to protect the public interest; and

(ii) in the case of a proposed rule to prohibit the manufacture, processing, or distribution of a chemical substance or mixture because of the risk determined under clause (i) (I), a court has in an action under section 7 granted relief with respect to such risk associated with such substance or mixture.

Such a proposed rule which is made so effective shall not, for purposes of judicial review, be considered final agency action.

(B) If the Administrator makes a proposed rule effective upon its publication in the Federal Register, the Administrator shall, as expeditiously as possible, give interested persons prompt notice of such action, provide reasonable opportunity, in accordance with paragraphs (2) and (3) of subsection (c), for a hearing on such rule, and either promulgate such rule (as proposed or with modifications) or revoke it; and if such a hearing is requested, the Administrator shall commence the hearing within five days from the date such request is made unless the Administrator and the person making the request agree upon a later date for the hearing to begin, and after the hearing is concluded the Administrator shall, within ten days of the conclusion of the hearing, either promulgate such rule (as proposed or with modifications) or revoke it. *Notice.*

(e) **Polychlorinated Biphenyls.**—(1) Within six months after the effective date of this Act the Administrator shall promulgate rules to— *Rules.*

(A) prescribe methods for the disposal of polychlorinated biphenyls, and

(B) require polychlorinated biphenyls to be marked with clear and adequate warnings, and instructions with respect to their processing, distribution in commerce, use, or disposal or with respect to any combination of such activities.

Requirements prescribed by rules under this paragraph shall be consistent with the requirements of paragraphs (2) and (3).

(2) (A) Except as provided under subparagraph (B), effective one year after the effective date of this Act no person may manufacture, process, or distribute in commerce or use any polychlorinated biphenyl in any manner other than in a totally enclosed manner.

(B) The Administrator may by rule authorize the manufacture, processing, distribution in commerce or use (or any combination of such activities) of any polychlorinated biphenyl in a manner other than in a totally enclosed manner if the Administrator finds that such manufacture, processing, distribution in commerce, or use (or combination of such activities) will not present an unreasonable risk of injury to health or the environment.

"Totally enclosed manner." (C) For the purposes of this paragraph, the term "totally enclosed manner" means any manner which will ensure that any exposure of human beings or the environment to a polychlorinated biphenyl will be insignificant as determined by the Administrator by rule.

(3) (A) Except as provided in subparagraphs (B) and (C)—

(i) no person may manufacture any polychlorinated biphenyl after two years after the effective date of this Act, and

(ii) no person may process or distribute in commerce any polychlorinated biphenyl after two and one-half years after such date.

Petition for exemption. (B) Any person may petition the Administrator for an exemption from the requirements of subparagraph (A), and the Administrator may grant by rule such an exemption if the Administrator finds that—

(i) an unreasonable risk of injury to health or environment would not result, and

(ii) good faith efforts have been made to develop a chemical substance which does not present an unreasonable risk of injury to health or the environment and which may be substituted for such polychlorinated biphenyl.

Terms and conditions. An exemption granted under this subparagraph shall be subject to such terms and conditions as the Administrator may prescribe and shall be in effect for such period (but not more than one year from the date it is granted) as the Administrator may prescribe.

(C) Subparagraph (A) shall not apply to the distribution in commerce of any polychlorinated biphenyl if such polychlorinated biphenyl was sold for purposes other than resale before two and one half years after the date of enactment of this Act.

(4) Any rule under paragraph (1), (2) (B), or (3) (B) shall be promulgated in accordance with paragraphs (2), (3), and (4) of subsection (c).

(5) This subsection does not limit the authority of the Administrator, under any other provision of this Act or any other Federal law, to take action respecting any polychlorinated biphenyl.

Sec. 7. Imminent Hazards

Civil action.
15 USC 2606.
(a) **Actions Authorized and Required.**—(1) The Administrator may commence a civil action in an appropriate district court of the United States—

(A) for seizure of an imminently hazardous chemical substance or mixture or any article containing such a substance or mixture,

(B) for relief (as authorized by subsection (b)) against any person who manufactures, processes, distributes in commerce, or uses, or disposes of, an imminently hazardous chemical substance or mixture or any article containing such a substance or mixture, or

(C) for both such seizure and relief.

A civil action may be commenced under this paragraph notwithstanding the existence of a rule under section 4, 5, or 6 or an order under section 5, and notwithstanding the pendency of any administrative or judicial proceeding under any provision of this Act.

(2) If the Administrator has not made a rule under section 6 (a) immediately effective (as authorized by subsection 6 (d) (2) (A) (i)) with respect to an imminently hazardous chemical substance or mixture, the Administrator shall commence in a district court of the United States with respect to such substance or mixture or article containing such substance or mixture a civil action described in subparagraph (A), (B), or (C) of paragraph (1).

(b) Relief Authorized.—(1) The district court of the United States in which an action under subsection (a) is brought shall have jurisdiction to grant such temporary or permanent relief as may be necessary to protect health or the environment from the unreasonable risk associated with the chemical substance, mixture, or article involved in such action. Jurisdiction.

(2) In the case of an action under subsection (a) brought against a person who manufactures, processes, or distributes in commerce a chemical substance or mixture or an article containing a chemical substance or mixture, the relief authorized by paragraph (1) may include the issuance of a mandatory order requiring (A) in the case of purchasers of such substance, mixture, or article known to the defendant, notification to such purchasers of the risk associated with it; (B) public notice of such risk; (C) recall; (D) the replacement or repurchase of such substance, mixture, or article; or (E) any combination of the actions described in the preceding clauses.

(3) In the case of an action under subsection (a) against a chemical substance, mixture, or article, such substance, mixture, or article may be proceeded against by process of libel for its seizure and condemnation. Proceedings in such an action shall conform as nearly as possible to proceedings in rem in admiralty.

(c) Venue and Consolidation.—(1) (A) An action under subsection (a) against a person who manufactures, processes, or distributes a chemical substance or mixture or an article containing a chemical substance or mixture may be brought in the United States District Court for the District of Columbia or for any judicial district in which any of the defendants is found, resides, or transacts business; and process in such an action may be served on a defendant in any other district in which such defendant resides or may be found. An action under subsection (a) against a chemical substance, mixture, or article may be brought in any United States district court within the jurisdiction of which the substance, mixture, or article is found.

(B) In determining the judicial district in which an action may be brought under subsection (a) in instances in which such action may be brought in more than one judicial district, the Administrator shall take into account the convenience of the parties.

(C) Subpeonas requiring attendance of witnesses in an action brought under subsection (a) may be served in any judicial district.

(2) Whenever proceedings under subsection (a) involving identical chemical substances, mixtures, or articles are pending in courts in two or more judicial districts,

they shall be consolidated for trial by order of any such court upon application reasonably made by any party in interest, upon notice to all parties in interest.

(d) Action Under Section 6.—Where appropriate, concurrently with the filing of an action under subsection (a) or as soon thereafter as may be practicable, the Administrator shall initiate a proceeding for the promulgation of a rule under section 6 (a).

(e) Representation.—Notwithstanding any other provision of law, in any action under subsection (a), the Administrator may direct attorneys of the Environmental Protection Agency to appear and represent the Administrator in such an action.

(f) Definition.—For the purposes of subsection (a), the term "imminently hazardous chemical substance or mixture" means a chemical substance or mixture which presents an imminent and unreasonable risk of serious or widespread injury to health or the environment. Such a risk to health or the environment shall be considered imminent if it is shown that the manufacture, processing, distribution in commerce, use, or disposal of the chemical substance or mixture, or that any combination of such activities, is likely to result in such injury to health or the environment before a final rule under section 6 can protect against such risk.

Sec. 8. Reporting and Retention of Information

Rules.
15 USC 2607.

(a) Reports.—(1) The Administrator shall promulgate rules under which—

(A) each person (other than a small manufacturer or processor) who manufactures or processes or proposes to manufacture or process a chemical substance (other than a chemical substance described in subparagraph (B) (ii)) shall maintain such records, and shall submit to the Administrator such reports, as the Administrator may reasonably require, and

(B) each person (other than a small manufacturer or processor) who manufactures or processes or proposes to manufacture or process—

(i) a mixture, or

(ii) a chemical substance in small quantities (as defined by the Administrator by rule) solely for purposes of scientific experimentation or analysis or chemical research on, or analysis of, such substance or another substance, including any such research or analysis for the development of a product,

shall maintain records and submit to the Administrator reports but only to the extent the Administrator determines the maintenance of records or submission of reports, or both, is necessary for the effective enforcement of this Act.

The Administrator may not require in a rule promulgated under this paragraph the maintenance of records or the submission of reports with respect to changes in the proportions of the components of a mixture unless the Administrator finds that the maintenance of such records or the submission of such reports, or both, is necessary for the effective enforcement of this Act. For purposes of the compilation of the list of chemical substances required under subsection (b), the Administrator shall promulgate rules pursuant to this subsection not later than 180 days after the effective date of this Act.

(2) The Administrator may require under paragraph (1) maintenance of records

and reporting with respect to the following insofar as known to the person making the report or insofar as reasonably ascertainable:

(A) The common or trade name, the chemical identity, and the molecular structure of each chemical substance or mixture for which such a report is required.

(B) The categories or proposed categories of use of each such substance or mixture.

(C) The total amount of each such substance and mixture manufactured or processed, reasonable estimates of the total amount to be manufactured or processed, the amount manufactured or processed for each of its categories of use, and reasonable estimates of the amount to be manufactured or processed for each of its categories of use or proposed categories of use.

(D) A description of the byproducts resulting from the manufacture, processing, use, or disposal or each such substance or mixture.

(E) All existing data concerning the environmental and health effects of such substance or mixture.

(F) The number of individuals exposed, and reasonable estimates of the number who will be exposed, to such substance or mixture in their places of employment and the duration of such exposure.

(G) In the initial report under paragraph (1) on such substance or mixture, the manner or method of its disposal, and in any subsequent report on such substance or mixture, any change in such manner or method.

To the extent feasible, the Administrator shall not require under paragraph (1), any reporting which is unnecessary or duplicative.

(3) (A) (i) The Administrator may by rule require a small manufacturer or processor of a chemical substance to submit to the Administrator such information respecting the chemical substance as the Administrator may require for publication of the first list of chemical substances required by subsection (b).

(ii) The Administrator may by rule require a small manufacturer or processor of a chemical substance or mixture—

(I) subject to a rule proposed or promulgated under section 4, 5 (b) (4), or 6, or an order in effect under section 5 (e), or

(II) with respect to which relief has been granted pursuant to a civil action brought under section 5 or 7,

to maintain such records on such substance or mixture, and to submit to the Administrator such reports on such substance or mixture, as the Administrator may reasonably require. A rule under this clause requiring reporting may require reporting with respect to the matters referred to in paragraph (2).

(B) The Administrator, after consultation with the Administrator of the Small Business Administration, shall by rule prescribe standards for determining the manufacturers and processors which qualify as small manufacturers and processors for purposes of this paragraph and paragraph (1). Standards.

(b) Inventory.—(1) The Administrator shall compile, keep current, and publish a list of each chemical substance which is manufactured or processed in the United States. Such list shall at least include each chemical substance which any person reports, under section 5 or subsection (a) of this section, is manufactured or processed in the United States. Such list may not include any chemical substance which was not manufactured or processed in the United States within three years before the effective date of the rules promulgated pursuant to the last sentence of subsection (a)(1). In the case of a chemical substance for which a notice is submitted in accordance with section 5, such chemical substance shall be included in such list as of the earliest date (as determined by the Administrator) on which such substance was manufactured or processed in the United States. The Administrator shall first publish such a list not later than 315 days after the effective date of this Act. The Administrator shall not include in such list any chemical substance which is manufactured or processed only in small quantities (as defined by the Administrator by rule) solely for purposes of scientific experimentation or analysis or chemical research on, or analysis of, such substance or another substance, including such research or analysis for the development of a product.

(2) To the extent consistent with the purposes of this Act, the Administrator may, in lieu of listing, pursuant to paragraph (1), a chemical substance individually, list a category of chemical substances in which such substance is included.

(c) **Records.**—Any person who manufactures, processes, or distributes in commerce any chemical substance or mixture shall maintain records of significant adverse reactions to health or the environment, as determined by the Administrator by rule, alleged to have been caused by the substance or mixture. Records of such adverse reactions to the health of employees shall be retained for a period of 30 years from the date such reactions were first reported to or known by the person maintaining such records. Any other record of such adverse reactions shall be retained for a period of five years from the date the information contained in the record was first reported to or known by the person maintaining the record. Records required to be maintained under this subsection shall include records of consumer allegations of personal injury or harm to health, reports of occupational disease or injury, and reports or complaints of injury to the environment submitted to the manufacturer, processor, or distributor in commerce from any source. Upon request of any duly designated representative of the Administrator, each person who is required to maintain records under this subsection shall permit the inspection of such records and shall submit copies of such records.

Rules.

(d) **Health and Safety Studies.**—The Administrator shall promulgate rules under which the Administrator shall require any person who manufactures, processes, or distributes in commerce or who proposes to manufacture, process, or distribute in commerce any chemical substance or mixture (or with respect to paragraph (2), any person who has possession of a study) to submit to the Administrator—

(1) lists of health and safety studies (A) conducted or initiated by or for such person with respect to such substance or mixture at any time, (B) known to such person, or (C) reasonably ascertainable by such person, except that the Administrator may exclude certain types or categories of studies from the requirements of this subsection if the Administrator finds that submission of lists of such studies are unnecessary to carry out the purposes of this Act; and

(2) copies of any study contained on a list submitted pursuant to paragraph (1) or otherwise known by such person.

(e) **Notice to Administrator of Substantial Risks.**—Any person who manufactures, processes, or distributes in commerce a chemical substance or mixture and who obtains information which reasonably supports the conclusion that such substance or mixture presents a substantial risk or injury to health or the environment shall immediately inform the Administrator of such information unless such person has actual knowledge that the Administrator has been adequately informed of such information.

(f) **Definitions.**—For purposes of this section, the terms "manufacture" and "process" mean manufacture or process for commercial purposes.

Sec. 9. Relationship to Other Federal Laws

(a) **Laws Not Administered by the Administrator.**—(1) If the Administrator has reasonable basis to conclude that the manufacture, processing, distribution in commerce, use, or disposal of a chemical substance or mixture, or that any combination of such activities, presents or will present an unreasonable risk of injury to health or the environment and determines, in the Administrator's discretion, that such risk may be prevented or reduced to a sufficient extent by action taken under a Federal law not administered by the Administrator, the Administrator shall submit to the agency which administers such law a report which describes such risk and includes in such description a specification of the activity or combination of activities which the Administrator has reason to believe so presents such risk. Such report shall also request such agency— 15 USC 2608. Report.

(A) (i) to determine if the risk described in such report may be prevented or reduced to a sufficient extent by action taken under such law, and

(ii) if the agency determines that such risk may be so prevented or reduced, to issue an order declaring whether or not the activity or combination of activities specified in the description of such risk presents such risk; and

(B) to respond to the Administrator with respect to the matters described in subparagraph (A).

Any report of the Administrator shall include a detailed statement of the information on which it is based and shall be published in the Federal Register. The Agency receiving a request under such a report shall make the requested determination, issue the requested order, and make the requested response within such time as the Administrator specifies in the request, but such time specified may not be less than 90 days from the date the request was made. The response of an agency shall be accompanied by a detailed statement of the findings and conclusions of the agency and shall be published in the Federal Register. Publication in Federal Register.

(2) If the Administrator makes a report under paragraph (1) with respect to a chemical substance or mixture and the agency to which such report was made either—

(A) issues an order declaring that the activity or combination of activities specified in the description of the risk described in the report does not present the risk described in the report, or

(B) initiates, within 90 days of the publication in the Federal Register of the response of the agency under paragraph (1), action under the law (or laws) administered by such agency to protect against such risk associated with such activity or combination of activities,

the Administrator may not take any action under section 6 or 7 with respect to such risk.

(3) If the Administrator has initiated action under section 6 or 7 with respect to a risk associated with a chemical substance or mixture which was the subject of a report made to an agency under paragraph (1), such agency shall before taking action under

the law (or laws) administered by it to protect against such risk consult with the Administrator for the purpose of avoiding duplication of Federal action against such risk.

(b) **Laws Administered by the Administrator.**—The Administrator shall coordinate actions taken under this Act with actions taken under other Federal laws administered in whole or in part by the Administrator. If the Administrator determines that a risk to health or the environment associated with a chemical substance or mixture could be eliminated or reduced to a sufficient extent by actions taken under the authorities contained in such other Federal laws, the Administrator shall use such authorities to protect against such risk unless the Administrator determines, in the Administrator's discretion, that it is in the public interest to protect against such risk by actions taken under this Act. This subsection shall not be construed to relieve the Administrator of any requirement imposed on the Administrator by such other Federal laws.

(c) **Occupational Safety and Health.**—In exercising any authority under this Act, the Administrator shall not, for purposes of section 4 (b)(1) of the Occupational Safety and Health Act of 1970, be deemed to be exercising statutory authority to prescribe or enforce standards or regulations affecting occupational safety and health.

29 USC 651 note.

(d) **Coordination.**—In administering this Act, the Administrator shall consult and coordinate with the Secretary of Health, Education, and Welfare and the heads of any other appropriate Federal executive department or agency, any relevant independent regulatory agency, and any other appropriate instrumentality of the Federal Government for the purpose of achieving the maximum enforcement of this Act while imposing the least burdens of duplicative requirements on those subject to the Act and for other purposes. The Administrator shall, in the report required by section 30, report annually to the Congress on actions taken to coordinate with such other Federal departments, agencies, or instrumentalities, and on actions taken to coordinate the authority under this Act with the authority granted under other Acts referred to in subsection (b).

Sec. 10. Research, Development, Collection, Dissemination, and Utilization of Data

15 USC 2609.

(a) **Authority.**—The Administrator shall, in consultation and cooperation with the Secretary of Health, Education, and Welfare and with other heads of appropriate departments and agencies, conduct such research, development, and monitoring as is necessary to carry out the purposes of this Act. The Administrator may enter into contracts and may make grants for research, development, and monitoring under this subsection. Contracts may be entered into under this subsection without regard to sections 3648 and 3709 of the Revised Statutes (31 U.S.C. 529, 14 U.S.C. 5).

(b) **Data Systems.**—(1) The Administrator shall establish, administer, and be responsible for the continuing activities of an interagency committee which shall design, establish, and coordinate an efficient and effective system, within the Environmental Protection Agency, for the collection, dissemination to other Federal departments and agencies, and use of data submitted to the Administrator under this Act.

(2)(A) The Administrator shall, in consultation and cooperation with the Secretary of Health, Education, and Welfare and other heads of appropriate departments and agencies design, establish, and coordinate an efficient and effective system for the retrieval of toxicological and other scientific data which could be useful to the Administrator in carrying out the purposes of this Act. Systematized retrieval shall be developed for use by all Federal and other departments and agencies with responsibilities in the area of regulation of study of chemical substances and mixtures and their effect on health or the environment.

(B) The Administrator, in consultation and cooperation with the Secretary of Health, Education, and Welfare, may make grants and enter into contracts for the development of a data retrieval system described in subparagraph (A). Contracts may be entered into under this subparagraph without regard to sections 3648 and 3709 of the Revised Statutes (31 U.S.C. 529, 41 U.S.C. 5).

(c) Screening Techniques.—The Administrator shall coordinate, with the Assistant Secretary for Health of the Department of Health, Education, and Welfare, research undertaken by the Administrator and directed toward the development of rapid, reliable, and economical screening techniques for carcinogenic, mutagenic, teratogenic, and ecological effects of chemical substances and mixtures.

(d) Monitoring.—The Administrator shall, in consultation and cooperation with the Secretary of Health, Education, and Welfare, establish and be responsible for research aimed at the development, in cooperation with local, State, and Federal agencies, of monitoring techniques and instruments which may be used in the detection of toxic chemical substances and mixtures and which are reliable, economical, and capable of being implemented under a wide variety of conditions.

(e) Basic Research.—The Administrator shall, in consultation and cooperation with the Secretary of Health, Education, and Welfare, establish research programs to develop the fundamental scientific basis of the screening and monitoring techniques described in subsections (c) and (d), the bounds of the reliability of such techniques, and the opportunities for their improvement.

(f) Training.—The Administrator shall establish and promote programs and workshops to train or facilitate the training of Federal laboratory and technical personnel in existing or newly developed screening and monitoring techniques.

(g) Exchange of Research and Development Results.—The Administrator shall, in consultation with the Secretary of Health, Education, and Welfare and other heads of appropriate departments and agencies, establish and coordinate a system for exchange among Federal, State, and local authorities of research and development results respecting toxic chemical substances and mixtures, including a system to facilitate and promote the development of standard data format and analysis and consistent testing procedures.

Sec. 11. Inspections and Subpoenas

(a) In General.—For purposes of administering this Act, the Administrator, and any duly designated representative of the Administrator, may inspect any establishment, facility, or other premises in which chemical substances or mixtures are manufactured, 15 USC 2610.

processed, stored, or held before or after their distribution in commerce and any conveyance being used to transport chemical substances, mixtures, or such articles in connection with distribution in commerce. Such an inspection may only be made upon the presentation of appropriate credentials and of a written notice to the owner, operator, or agent in charge of the premises or conveyance to be inspected. A separate notice shall be given for each such inspection, but a notice shall not be required for each entry made during the period covered by the inspection. Each such inspection shall be commenced and completed with reasonable promptness and shall be conducted at reasonable times, within reasonable limits, and in a reasonable manner.

(b) Scope.—(1) Except as provided in paragraph (2), an inspection conducted under subsection (a) shall extend to all things within the premises or conveyance inspected (including records, files, papers, processes, controls, and facilities) bearing on whether the requirements of this Act applicable to the chemical substances or mixtures within such premises or conveyance have been complied with.

(2) No inspection under subsection (a) shall extend to—
(A) financial data,
(B) sales data (other than shipment data),
(C) pricing data,
(D) personnel data, or
(E) research data (other than data required by this Act or under a rule promulgated thereunder),
unless the nature and extent of such data are described with reasonable specificity in the written notice required by subsection (a) for such inspection.

(c) Subpoenas.—In carrying out this Act, the Administrator may by subpoena require the attendance and testimony of witnesses and the production of reports, papers, documents, answers to questions, and other information that the Administrator deems necessary. Witnesses shall be paid the same fees and mileage that are paid witnesses in the courts of the United States. In the event of contumacy, failure, or refusal of any person to obey any such subpoena, any district court of the United States in which venue is proper shall have jurisdiction to order any such person to comply with such subpoena. Any failure to obey such an order of the court is punishable by the court as a contempt thereof.

Sec. 12. Exports

15 USC 2611.

(a) In General.—(1) Except as provided in paragraph (2) and subsection (b), this Act (other than section 8) shall not apply to any chemical substance, mixture, or to an article containing a chemical substance or mixture, if—
(A) it can be shown that such substance, mixture, or article is being manufactured, processed, or distributed in commerce for export from the United States, unless such substance, mixture, or article was, in fact, manufactured, processed, or distributed in commerce, for use in the United States, and
(B) such substance, mixture, or article (when distributed in commerce), or any container in which it is enclosed (when so distributed), bears a stamp or label stating that such substance, mixture, or article is intended for export.
(2) Paragraph (1) shall not apply to any chemical substance, mixture, or article if

the Administrator finds that the substance, mixture, or article will present an unreasonable risk of injury to health within the United States or to the environment of the United States. The Administrator may require, under section 4, testing of any chemical substance or mixture exempted from this Act by paragraph (1) for the purpose of determining whether or not such substance or mixture presents an unreasonable risk of injury to health within the United States or to the environment of the United States.

(b) Notice.—(1) If any person exports or intends to export to a foreign country a chemical substance or mixture for which the submission of data is required under section 4 or 5 (b), such person shall notify the Administrator of such exportation or intent to export and the Administrator shall furnish to the government of such country notice of the availability of the data submitted to the Administrator under such section for such substance or mixture.

(2) If any person exports or intends to export to a foreign country a chemical substance or mixture for which an order has been issued under section 5 or a rule has been proposed or promulgated under section 5 or 6, or with respect to which an action is pending, or relief has been granted under section 5 or 7, such person shall notify the Administrator of such exportation or intent to export and the Administrator shall furnish to the government of such country notice of such rule, order, action, or relief.

Sec. 13. Entry into Customs Territory of the United States

(a) In General.—(1) The Secretary of the Treasury shall refuse entry into the customs territory of the United States (as defined in general headnote 2 to the Tariff Schedules of the United States) of any chemical substance, mixture, or article containing a chemical substance or mixture offered for such entry if— *15 USC 2612. 19 USC 1202.*

    (A) it fails to comply with any rule in effect under this Act, or

    (B) it is offered for entry in violation of section 5 or 6, a rule or order under section 5 or 6, or an order issued in a civil action brought under section 5 or 7.

(2) If a chemical substance, mixture, or article is refused entry under paragraph (1), the Secretary of the Treasury shall notify the consignee of such entry refusal, shall not release it to the consignee, and shall cause its disposal or storage (under such rules as the Secretary of the Treasury may prescribe) if it has not been exported by the consignee within 90 days from the date of receipt of notice of such refusal, except that the Secretary of the Treasury may, pending a review by the Administrator of the entry refusal, release to the consignee such substance, mixture, or article on execution of bond for the amount of the full invoice of such substance, mixture, or article (as such value is set forth in the customs entry), together with the duty thereon. On failure to return such substance, mixture, or article for any cause to the custody of the Secretary of the Treasury when demanded, such consignee shall be liable to the United States for liquidated damages equal to the full amount of such bond. All charges for storage, cartage, and labor on and for disposal of substances, mixtures, or articles which are refused entry or release under this section shall be paid by the owner or consignee, and in default of such payment shall constitute a lien against any future entry made by such owner or consignee. *Notification.*

(b) Rules.—The Secretary of the Treasury, after consultation with the Administrator, shall issue rules for the administration of subsection (a) of this section.

Sec. 14. Disclosure of Data

15 USC 2613.

(a) In General.—Except as provided by subsection (b), any information reported to, or otherwise obtained by, the Administrator (or any representative of the Administrator) under this Act, which is exempt from disclosure pursuant to subsection (a) of section 552 of title 5, United States Code, by reason of subsection (b)(4) of such section, shall, notwithstanding the provisions of any other section of this Act, not be disclosed by the Administrator or by any officer or employee of the United States, except that such information—

(1) shall be disclosed to any officer or employee of the United States—

(A) in connection with the official duties of such officer or employee under any law for the protection of health or the environment, or

(B) for specific law enforcement purposes;

(2) shall be disclosed to contractors with the United States and employees of such contractors if in the opinion of the Administrator such disclosure is necessary for the satisfactory performance by the contractor of a contract with the United States entered into on or after the date of enactment of this Act for the performance of work in connection with this Act and under such conditions as the Administrator may specify;

(3) shall be disclosed if the Administrator determines it necessary to protect health or the environment against an unreasonable risk of injury to health or the environment; or

(4) may be disclosed when relevant in any proceeding under this Act, except that disclosure in such a proceeding shall be made in such manner as to preserve confidentiality to the extent practicable without impairing the proceeding.

In any proceeding under section 552 (a) of title 5, United States Code, to obtain information the disclosure of which has been denied because of the provisions of this subsection, the Administrator may not rely on section 552 (b)(3) of such title to sustain the Administrator's action.

(b) Data From Health and Safety Studies.—(1) Subsection (a) does not prohibit the disclosure of—

(A) any health and safety study which is submitted under this Act with respect to—

(i) any chemical substance or mixture which, on the date on which such study is to be disclosed has been offered for commercial distribution, or

(ii) any chemical substance or mixture for which testing is required under section 4 or for which notification is required under section 5, and

(B) any data reported to, or otherwise obtained by, the Administrator from a health and safety study which relates to a chemical substance or mixture described in clause (i) or (ii) of subparagraph (A).

This paragraph does not authorize the release of any data which discloses processes used in the manufacturing or processing of a chemical substance or mixture or, in the case of a mixture, the release of data disclosing the portion of the mixture comprised by any of the chemical substances in the mixture.

(2) If a request is made to the Administrator under subsection (a) of section 552 of title 5, United States Code, for information which is described in the first sentence of paragraph (1) and which is not information described in the second sentence of such

paragraph, the Administrator may not deny such request on the basis of subsection (b) (4) of such section.

(c) Designation and Release of Confidential Data.—(1) In submitting data under this Act, a manufacturer, processor, or distributor in commerce may (A) designate the data which such person believes is entitled to confidential treatment under subsection (a), and (B) submit such designated data separately from other data submitted under this Act. A designation under this paragraph shall be made in writing and in such manner as the Administrator may prescribe.

(2) (A) Except as provided by subparagraph (B), if the Administrator proposes to release for inspection data which has been designated under paragraph (1) (A), the Administrator shall notify, in writing and by certified mail, the manufacturer, processor, or distributor in commerce who submitted such data of the intent to release such data. If the release of such data is to be made pursuant to a request made under section 552(a) of title 5, United States Code, such notice shall be given immediately upon approval of such request by the Administrator. The Administrator may not release such data until the expiration of 30 days after the manufacturer, processor, or distributor in commerce submitting such data has received the notice required by this subparagraph.

(B) (i) Subparagraph (A) shall not apply to the release of information under paragraph (1), (2), (3), or (4) of subsection (a), except that the Administrator may not release data under paragraph (3) of subsection (a) unless the Administrator has notified each manufacturer, processor, and distributor in commerce who submitted such data of such release. Such notice shall be made in writing by certified mail at least 15 days before the release of such data, except that if the Administrator determines that the release of such data is necessary to protect against an imminent, unreasonable risk of injury to health or the environment, such notice may be made by such means as the Administrator determines will provide notice at least 24 hours before such release is made.

(ii) Subparagraph (A) shall not apply to the release of information described in subsection (b) (1) other than information described in the second sentence of such subsection.

(d) Criminal Penalty for Wrongful Disclosure.—(1) Any officer or employee of the United States or former officer or employee of the United States, who by virtue of such employment or official position has obtained possession of, or has access to, material the disclosure of which is prohibited by subsection (a), and who knowing that disclosure of such material is prohibited by such subsection, willfully discloses the material in any manner to any person not entitled to receive it, shall be guilty of a misdemeanor and fined not more than $5,000 or imprisoned for not more than one year, or both. Section 1905 of title 18, United States Code, does not apply with respect to the publishing, divulging, disclosure, or making known of, or making available, information reported or otherwise obtained under this Act.

(2) For the purposes of paragraph (1), any contractor with the United States who is furnished information as authorized by subsection (a) (2), and any employee of any such contractor, shall be considered to be an employee of the United States.

(e) Access by Congress.—Notwithstanding any limitation contained in this section or any other provision of law, all information reported to or otherwise obtained by the Administrator (or any representative of the Administrator) under this Act shall be made available, upon written request of any duly authorized committee of the Congress, to such committee.

Sec. 15. Prohibited Acts

15 USC 2614.   It shall be unlawful for any person to—

(1) fail or refuse to comply with (A) any rule promulgated or order issued under section 4, (B) any requirement prescribed by section 5 or 6, or (C) any rule promulgated or order issued under section 5 or 6;

(2) use for commercial purposes a chemical substance or mixture which such person knew or had reason to know was manufactured, processed, or distributed in commerce in violation of section 5 or 6, a rule or order under section 5 or 6, or an order issued in action brought under section 5 or 7;

(3) fail or refuse to (A) establish or maintain records, (B) submit reports, notices, or other information, or (C) permit access to or copying of records, as required by this Act or a rule thereunder; or

(4) fail or refuse to permit entry or inspection as required by section 11.

Sec. 16. Penalties.

15 USC 2615.   **(a) Civil.**—(1) Any person who violates a provision of section 15 shall be liable to the United States for a civil penalty in an amount not to exceed $25,000 for each such violation. Each day such a violation continues shall, for purposes of this subsection, constitute a separate violation of section 15.

Hearing.   (2) (A) A civil penalty for a violation of section 15 shall be assessed by the Administrator by an order made on the record after opportunity (provided in accordance with this subparagraph) for a hearing in accordance with section 554 of title 5, United States Code. Before issuing such an order, the Administrator shall give written notice to the person to be assessed a civil penalty under such order of the Administrator's proposal to issue such order and provide such person an opportunity to request, within 15 days of the date the notice is received by such person, such a hearing on the order.

(B) In determining the amount of a civil penalty, the Administrator shall take into account the nature, circumstances, extent, and gravity of the violation or violations and, with respect to the violator, ability to pay, effect on ability to continue to do business, any history of prior such violations, the degree of culpability, and such other matters as justice may require.

(C) The Administrator may compromise, modify, or remit, with or without conditions, any civil penalty which may be imposed under this subsection. The amount of such penalty, when finally determined, or the amount agreed upon in compromise, may be deducted from any sums owing by the United States to the person charged.

Petition for judicial review.   (3) Any person who requested in accordance with paragraph (2) (A) a hearing respecting the assessment of a civil penalty and who is aggrieved by an order assessing a civil penalty may file a petition for judicial review of such order with the United States Court of Appeals for the District of Columbia Circuit or for any other circuit in which such person resides or transacts business. Such a petition may only be filed within the 30-day period beginning on the date the order making such assessment was issued.

(4) If any person fails to pay an assessment of a civil penalty—

(A) after the order making the assessment has become a final order and if such person does not file a petition for judicial review of the order in accordance with paragraph (3), or

(B) after a court in an action brought under paragraph (3) has entered a final judgment in favor of the Administrator,

the Attorney General shall recover the amount assessed (plus interest at currently prevailing rates from the date of the expiration of the 30-day period referred to in paragraph (3) or the date of such final judgment, as the case may be) in an action brought in any appropriate district court of the United States. In such an action, the validity, amount, and appropriateness of such penalty shall not be subject to review.

(b) Criminal.—Any person who knowingly or willfully violates any provision of section 15 shall, in addition to or in lieu of any civil penalty which may be imposed under subsection (a) of this section for such violation, be subject, upon conviction, to a fine of not more than $25,000 for each day of violation, or to imprisonment for not more than one year, or both.

Sec. 17. Specific Enforcement and Seizure

(a) Specific Enforcement.—(1) The district courts of the United States shall have jurisdiction over civil actions to— 15 USC 2616.

(A) restrain any violation of section 15,

(B) restrain any person from taking any action prohibited by section 5 or 6 or by a rule or order under section 5 or 6,

(C) compel the taking of any action required by or under this Act, or

(D) direct any manufacturer or processor of a chemical substance or mixture manufactured or processed in violation of section 5 or 6 or a rule or order under section 5 or 6 and distributed in commerce, (i) to give notice of such fact to distributors in commerce of such substance or mixture and, to the extent reasonably ascertainable, to other persons in possession of such substance or mixture or exposed to such substance or mixture, (ii) to give public notice of such risk or injury, and (iii) to either replace or repurchase such substance or mixture, whichever the person to which the requirement is directed elects.

(2) A civil action described in paragraph (1) may be brought—

(A) in the case of a civil action described in subparagraph (A) of such paragraph, in the United States district court for the judicial district wherein any act, omission, or transaction constituting a violation of section 15 occurred or wherein the defendant is found or transacts business, or

(B) in the case of any other civil action described in such paragraph, in the United States district court for the judicial district wherein the defendant is found or transacts business.

In any such civil action process may be served on a defendant in any judicial district in which a defendant resides or may be found. Subpoenas requiring attendance of witnesses in any such action may be served in any judicial district.

(b) Seizure.—Any chemical substance or mixture which was manufactured, processed, or distributed in commerce in violation of this Act or any rule promulgated or order issued under this Act or any article containing such a substance or mixture shall be liable to be proceeded against, by process of libel for the seizure and condemnation of such substance, mixture, or article, in any district court of the United States within the jurisdiction of which such substance, mixture, or article is found. Such proceedings shall conform as nearly as possible to proceedings in rem in admiralty.

Sec. 18. Preemption

15 USC 2617. (a) **Effect on State Law.**—(1) Except as provided in paragraph (2), nothing in this Act shall affect the authority of any State or political subdivision of a State to establish or continue in effect regulation of any chemical substance, mixture, or article containing a chemical substance or mixture.

(2) Except as provided in subsection (b)—

(A) if the Administrator requires by a rule promulgated under section 4 the testing of a chemical substance or mixture, no State or political subdivision may, after the effective date of such rule, establish or continue in effect a requirement for the testing of such substance or mixture for purposes similar to those for which testing is required under such rule; and

(B) if the Administrator prescribes a rule or order under section 5 or 6 (other than a rule imposing a requirement described in subsection (a) (6) of section 6) which is applicable to a chemical substance or mixture, and which is designed to protect against a risk of injury to health or the environment associated with such substance or mixture, no State or political subdivision of a State may, after the effective date of such requirement, establish or continue in effect, any requirement which is applicable to such substance or mixture, or an article containing such substance or mixture, and which is designed to protect against such risk unless such requirement (i) is identical to the requirement prescribed by the Administrator, (ii) is adopted under the authority of the Clean Air Act or any other Federal law, or (iii) prohibits the use of such substance or mixture in such State or political subdivision (other than its use in the manufacture or processing of other substances or mixtures).

Application. (b) **Exemption.**—Upon application of a State or political subdivision of a State the Administrator may by rule exempt from subsection (a) (2), under such conditions as may be prescribed in such rule, a requirement of such State or political subdivision designed to protect against a risk of injury to health or the environment associated with a chemical substance, mixture, or article containing a chemical substance or mixture if—

(1) compliance with the requirement would not cause the manufacturing, processing, distribution in commerce, or use of the substance, mixture, or article to be in violation of the applicable requirement under this Act described in subsection (a) (2), and

(2) the State or political subdivision requirement (A) provides a significantly higher degree of protection from such risk than the requirement under this Act described in subsection (a) (2) and (B) does not, through difficulties in marketing, distribution, or other factors, unduly burden interstate commerce.

Sec. 19. Judicial Review

Petition.
15 USC 2618. (a) **In General.**—(1)(A) Not later than 60 days after the date of the promulgation of a rule under section 4 (a), 5 (a)(2), 5 (b)(4), 6 (a), 6 (e), or 8, any person may file a petition for judicial review of such rule with the United States Court of Appeals for the District

of Columbia Circuit or for the circuit in which such person resides or in which such person's principal place of business is located. Courts of appeals of the United States shall have exclusive jurisdiction of any action to obtain judicial review (other than in an enforcement proceeding) of such a rule if any district court of the United States would have had jurisdiction of such action but for this subparagraph.

(B) Courts of appeals of the United States shall have exclusive jurisdiction of any action to obtain judicial review (other than in an enforcement proceeding) of an order issued under subparagraph (A) or (B) of section 6 (b) (1) if any district court of the United States would have had jurisdiction of such action but for this subparagraph. Jurisdiction.

(2) Copies of any petition filed under paragraph (1) (A) shall be transmitted forthwith to the Administrator and to the Attorney General by the clerk of the court with which such petition was filed. The provisions of section 2112 of title 28, United States Code, shall apply to the filing of the rulemaking record of proceedings on which the Administrator based the rule being reviewed under this section and to the transfer of proceedings between United States courts of appeals. Petition copies, transmittal to Administrator and Attorney General.

(3) For purposes of this section, the term "rulemaking record" means— "Rulemaking record."

(A) the rule being reviewed under this section;

(B) in the case of a rule under section 4 (a), the finding required by such section, in the case of a rule under section 5 (b) (4), the finding required by such section, in the case of a rule under section 6 (a) the finding required by section 5 (f) or 6 (a), as the case may be, in the case of a rule under section 6 (a), the statement required by section 6 (c)(1), and in the case of a rule under section 6 (e), the findings required by paragraph (2) (B) or (3) (B) of such section, as the case may be;

(C) any transcript required to be made of oral presentations made in proceedings for the promulgation of such rule;

(D) any written submission of interested parties respecting the promulgation of such rule; and

(E) any other information which the Administrator considers to be relevant to such rule and which the Administrator identified, on or before the date of the promulgation of such rule, in a notice published in the Federal Register. Notice, publication in Federal Register.

(b) Additional Submissions and Presentations; Modifications.—If in an action

under this section to review a rule the petitioner or the Administrator applies to the court for leave to make additional oral submissions or written presentations respecting such rule and shows to the satisfaction of the court that such submissions and presentations would be material and that there were reasonable grounds for the submissions and failure to make such submissions and presentations in the proceeding before the Administrator, the court may order the Administrator to provide additional opportunity to make such submissions and presentations. The Administrator may modify or set aside the rule being reviewed or make a new rule by reason of the additional submissions and presentations and shall file such modified or new rule with the return of such submissions and presentations. The court shall thereafter review such new or modified rule. Review.

(c) Standard of Review.—(1) (A) Upon the filing of a petition under subsection (a)

(1) for judicial review of a rule, the court shall have jurisdiction (i) to grant appropriate relief, including interim relief, as provided in chapter 7 of title 5, United States Code,

and (ii) except as otherwise provided in subparagraph (B), to review such rule in accordance with chapter 7 of title 5, United States Code.

(B) Section 706 of title 5, United States Code, shall apply to review of a rule under this section, except that—

(i) in the case of review of a rule under section 4 (a), 5 (b) (4), 6 (a), or 6 (e), the standard for review prescribed by paragraph (2) (E) of such section 706 shall not apply and the court shall hold unlawful and set aside such rule if the court finds that the rule is not supported by substantial evidence in the rulemaking record (as defined in subsection (a) (3)) taken as a whole;

(ii) in the case of review of a rule under section 6 (a), the court shall hold unlawful and set aside such rule if it finds that—

(I) a determination by the Administrator under section 6 (c) (3) that the petitioner seeking review of such rule is not entitled to conduct (or have conducted) cross-examination or to present rebuttal submissions, or

(II) a rule of, or ruling by, the Administrator under section 6 (c) (3) limiting such petitioner's cross-examination or oral presentations,

has precluded disclosure of disputed material facts which was necessary to a fair determination by the Administrator of the rulemaking proceeding taken as a whole; and section 706 (2) (D) shall not apply with respect to a determination, rule, or ruling referred to in subclause (I) or (II); and

(iii) the court may not review the contents and adequacy of—

(I) any statement required to be made pursuant to section 6 (c) (1), or

(II) any statement of basis and purpose required by section 553 (c) of title 5, United States Code, to be incorporated in the rule

except as part of a review of the rulemaking record taken as a whole.

"Evidence."   The term "evidence" as used in clause (i) means any matter in the rulemaking record.

(C) A determination, rule, or ruling of the Administrator described in subparagraph (B) (ii) may be reviewed only in an action under this section and only in accordance with such subparagraph.

(2) The judgment of the court affirming or setting aside, in whole or in part, any rule reviewed in accordance with this section shall be final, subject to review by the Supreme Court of the United States upon certiorari or certification, as provided in section 1254 of title 28, United States Code.

(d) Fees and Costs.—The decision of the court in an action commenced under subsection (a), or of the Supreme Court of the United States on review of such a decision, may include an award of costs of suit and reasonable fees for attorneys and expert witnesses if the court determines that such an award is appropriate.

(e) Other remedies.—The remedies as provided in this section shall be in addition to and not in lieu of any other remedies provided by law.

Sec. 20. Citizens' Civil Actions

15 USC 2619.   **(a) In General.**—Except as provided in subsection (b), any person may commence a civil action—

(1) against any person (including (A) the United States, and (B) any other governmental instrumentality or agency to the extent permitted by the eleventh

amendment to the Constitution) who is alleged to be in violation of this Act or any rule promulgated under section 4, 5, or 6 or order issued under section 5 to restrain such violation, or

(2) against the Administrator to compel the Administrator to perform any act or duty under this Act which is not discretionary.

Any civil action under paragraph (1) shall be brought in the United States district court for the district in which the alleged violation occurred or in which the defendant resides or in which the defendant's principal place of business is located. Any action brought under paragraph (2) shall be brought in the United States District Court for the District of Columbia, or the United States district court for the judicial district in which the plaintiff is domiciled. The district courts of the United States shall have jurisdiction [Jurisdiction.] over suits brought under this section, without regard to the amount in controversy or the citizenship of the parties. In any civil action under this subsection process may be served on a defendant in any judicial district in which the defendant resides or may be found and subpoenas for witnesses may be served in any judicial district.

(b) **Limitation.**—No civil action may be commenced—

(1) under subsection (a) (1) to restrain a violation of this Act or rule or order under this Act—

(A) before the expiration of 60 days after the plaintiff has given notice of such [Notice.] violation (i) to the Administrator, and (ii) to the person who is alleged to have committed such violation, or

(B) if the Administrator has commenced and is diligently prosecuting a proceeding for the issuance of an order under section 16 (a) (2) to require compliance with this Act or with such rule or order or if the Attorney General has commenced and is diligently prosecuting a civil action in a court of the United States to require compliance with this Act or with such rule or order, but if such proceeding or civil action is commenced after the giving of notice, any person giving such notice may intervene as a matter of right in such proceeding or action; or

(2) under subsection (a) (2) before the expiration of 60 days after the plaintiff [Notice.] has given notice to the Administrator of the alleged failure of the Administrator to perform an act or duty which is the basis for such action or, in the case of an action under such subsection for the failure of the Administrator to file an action under section 7, before the expiration of ten days after such notification.

Notice under this subsection shall be given in such manner as the Administrator shall [Rule.] prescribe by rule.

(c) **General.**—(1) In any action under this section, the Administrator, if not a party, may intervene as a matter of right.

(2) The court, in issuing any final order in any action brought pursuant to subsection (a), may award costs of suit and reasonable fees for attorneys and expert witnesses if the court determines that such an award is appropriate. Any court, in issuing its decision in an action brought to review such an order, may award costs of suit and reasonable fees for attorneys if the court determines that such an award is appropriate.

(3) Nothing in this section shall restrict any right which any person (or class of persons) may have under any statute or common law to seek enforcement of this Act or any rule or order under this Act or to seek any other relief.

(d) Consolidation.—When two or more civil actions brought under subsection (a) involving the same defendant and the same issues or violations are pending in two or more judicial districts, such pending actions, upon application of such defendants to such actions which is made to a court in which any such action is brought, may, if such court in its discretion so decides, be consolidated for trial by order (issued after giving all parties reasonable notice and opportunity to be heard) of such court and tried in—

(1) any district which is selected by such defendant and in which one of such actions is pending.

(2) a district which is agreed upon by stipulation between all the parties to such actions and in which one of such actions is pending, or

(3) a district which is selected by the court and in which one of such actions is pending.

The court issuing such an order shall give prompt notification of the order to the other courts in which the civil actions consolidated under the order are pending.

Sec. 21. Citizens' Petitions

15 USC 2620.

(a) In General.—Any person may petition the Administrator to initiate a proceeding for the issuance, amendment, or repeal of a rule under section 4, 6, or 8 or an order under section 5 (e) or (6) (b) (2).

(b) Procedures.—(1) Such petition shall be filed in the principal office of the Administrator and shall set forth the facts which it is claimed establish that it is necessary to issue, amend, or repeal a rule under section 4, 6, or 8 or an order under section 5 (e), 6 (b) (1) (A), or 6 (b) (1) (B).

Public hearing.

(2) The Administrator may hold a public hearing or may conduct such investigation or proceeding as the Administrator deems appropriate in order to determine whether or not such petition should be granted.

(3) Within 90 days after filing of a petition described in paragraph (1), the Administrator shall either grant or deny the petition. If the Administrator grants such petition, the Administrator shall promptly commence an appropriate proceeding in accordance with section 4, 5, 6, or 8. If the Administrator denies such petition, the Administrator shall publish in the Federal Register the Administrator's reasons for such denial.

Publication in Federal Register.

Civil action.

(4) (A) If the Administrator denies a petition filed under this section (or if the Administrator fails to grant or deny such petition within the 90-day period) the petitioner may commence a civil action in a district court of the United States to compel the Administrator to initiate a rulemaking proceeding as requested in the petition. Any such action shall be filed within 60 days after the Administrator's denial of the petition or, if the Administrator fails to grant or deny the petition within 90 days after filing the petition, within 60 days after the expiration of the 90-day period.

(B) In an action under subparagraph (A) respecting a petition to initiate a proceeding to issue a rule under section 4, 6, or 8 or an order under section 5 (e) or 6 (b) (2), the petitioner shall be provided an opportunity to have such petition considered by the court in a de novo proceeding. If the petitioner demonstrates to the satisfaction of the court by a preponderance of the evidence that—

(i) in the case of a petition to initiate a proceeding for the issuance of a rule under section 4 or an order under section 5 (e)—

(I) information available to the Administrator is insufficient to permit a reasoned evaluation of the health and environmental effects of the chemical substance to be subject to such rule or order; and

(II) in the absence of such information, the substance may present an unreasonable risk to health or the environment, or the substance is or will be produced in substantial quantities and it enters or may reasonably be anticipated to enter the environment in substantial quantities or there is or may be significant or substantial human exposure to it; or

(ii) in the case of a petition to initiate a proceeding for the issuance of a rule under section 6 or 8 or an order under section 6 (b)(2), there is a reasonable basis to conclude that the issuance of such a rule or order is necessary to protect health or the environment against an unreasonable risk of injury to health or the environment. the court shall order the Administrator to initiate the action requested by the petitioner. If the court finds that the extent of the risk to health or the environment alleged by the petitioner is less than the extent of risks to health or the environment with respect to which the Administrator is taking action under this Act and there are insufficient resources available to the Administrator to take the action requested by the petitioner, the court may permit the Administrator to defer initiating the action requested by the petitioner until such time as the court prescribes.

(C) The court in issuing any final order in any action brought pursuant to subparagraph (A) may award costs of suit and reasonable fees for attorneys and expert witnesses if the court determines that such an award is appropriate. Any court, in issuing its decision in an action brought to review such an order, may award costs of suit and reasonable fees for attorneys if the court determines that such an award is appropriate.

(5) The remedies under this section shall be in addition to, and not in lieu of, other remedies provided by law.

Sec. 22. National Defense Waiver

The Administrator shall waive compliance with any provision of this Act upon a request and determination by the President that the requested waiver is necessary in the interest of national defense. The Administrator shall maintain a written record of the basis upon which such waiver was granted and make such record available for in camera examination when relevant in a judicial proceeding under this Act. Upon the issuance of such a waiver, the Administrator shall publish in the Federal Register a notice that the waiver was granted for national defense purposes, unless, upon the request of the President, the Administrator determines to omit such publication because the publication itself would be contrary to the interests of national defense, in which event the Administrator shall submit notice thereof to the Armed Services Committees of the Senate and the House of Representatives.

15 USC 2621.

Publication in Federal Register. Notice to congressional committee.

Sec. 23. Employee Protection

(a) **In General.**—No employer may discharge any employee or otherwise discriminate against any employee with respect to the employees' compensation, terms, conditions, or privileges of employment because the employee (or any person acting pursuant to a request of the employee) has—

15 USC 2622.

(1) commenced, caused to be commenced, or is about to commence or cause to be commenced a proceeding under this Act;

(2) testified or is about to testify in any such proceeding; or

(3) assisted or participated or is about to assist or participate in any manner in such a proceeding or in any other action to carry out the purposes of this Act.

(b) Remedy.—(1) Any employee who believes that the employee has been discharged or otherwise discriminated against by any person in violation of subsection (a) of this section may, within 30 days after such alleged violation occurs, file (or have any person file on the employee's behalf) a complaint with the Secretary of Labor (hereinafter in this section referred to as the "Secretary") alleging such discharge or discrimination. Upon receipt of such a complaint, the Secretary shall notify the person named in the complaint of the filing of the complaint.

Notification.

Investigation.

Notification.

(2) (A) Upon receipt of a complaint filed under paragraph (1), the Secretary shall conduct an investigation of the violation alleged in the complaint. Within 30 days of the receipt of such complaint, the Secretary shall complete such investigation and shall notify in writing the complainant (and any person acting on behalf of the complainant) and the person alleged to have committed such violation of the results of the investigation conducted pursuant to this paragraph. Within ninety days of the receipt of such complaint the Secretary shall, unless the proceeding on the complaint is terminated by the Secretary on the basis of a settlement entered into by the Secretary and the person alleged to have committed such violation, issue an order either providing the relief prescribed by subparagraph (B) or denying the complaint. An order of the Secretary shall be made on the record after notice and opportunity for agency hearing. The Secretary may not enter into a settlement terminating a proceeding on a complaint without the participation and consent of the complainant.

Notice, hearing.

(B) If in response to a complaint filed under paragraph (1) the Secretary determines that a violation of subsection (a) of this section has occurred, the Secretary shall order (i) the person who committed such violation to take affirmative action to abate the violation, (ii) such person to reinstate the complainant to the complainant's former position together with the compensation (including back pay), terms, conditions, and privileges of the complainant's employment, (iii) compensatory damages, and (iv) where appropriate, exemplary damages. If such an order issued, the Secretary, at the request of the complainant, shall assess against the person against whom the order is issued a sum equal to the aggregate amount of all costs and expenses (including attorney's fees) reasonably incurred, as determined by the Secretary, by the complainant for, or in connection with, the bringing of the complaint upon which the order was issued.

(c) Review.—(1) Any employee or employer adversely affected or aggrieved by an order issued under subsection (b) may obtain review of the order in the United States Court of Appeals for the circuit in which the violation, with respect to which the order was issued, allegedly occurred. The petition for review must be filed within sixty days from the issuance of the Secretary's order. Review shall conform to chapter 7 of title 5 of the United States Code.

(2) An order of the Secretary, with respect to which review could have been obtained under paragraph (1), shall not be subject to judicial review in any criminal or other civil proceeding.

(d) Enforcement.—Whenever a person has failed to comply with an order issued under subsection (b)(2), the Secretary shall file a civil action in the United States district court for the district in which the violation was found to occur to enforce such order. In actions brought under this subsection, the district courts shall have jurisdiction to grant all appropriate relief, including injunctive relief and compensatory and exemplary damages. Civil actions brought under this subsection shall be heard and decided expeditiously.

Civil action.

Jurisdiction.

(e) Exclusion.—Subsection (a) of this section shall not apply with respect to any employe who, acting without direction from the employee's employer (or any agent of the employer), deliberately causes a violation of any requirement of this Act.

Sec. 24. Employment Effects

(a) In General.—The Administrator shall evaluate on a continuing basis the potential effects on employment (including reductions in employment or loss of employment from threatened plant closures) of—

Evaluation.
15 USC 2623.

(1) the issuance of a rule or order under section 4, 5, or 6, or

(2) a requirement of section 5 or 6.

(b) (1) Investigations.—Any employee (or any representative of an employee) may request the Administrator to make an investigation of—

(A) a discharge or layoff or threatened discharge or layoff of the employee, or

(B) adverse or threatened adverse effects on the employee's employment,

allegedly resulting from a rule or order under section 4, 5, or 6 or a requirement of section 5 or 6. Any such request shall be made in writing, shall set forth with reasonable particularity the grounds for the request, and shall be signed by the employee, or representative of such employee, making the request.

(2) (A) Upon receipt of a request made in accordance with paragraph (1) the Administrator shall (i) conduct the investigation requested, and (ii) if requested by any interested person, hold public hearings on any matter involved in the investigation unless the Administrator, by order issued within 45 days of the date such hearings are requested, denies the request for the hearings because the Administrator determines there are no reasonable grounds for holding such hearings. If the Administrator makes such a determination, the Administrator shall notify in writing the person requesting the hearing of the determination and the reasons therefor and shall publish the determination and the reasons therefor in the Federal Register.

Public hearings.

Notification.

Publication in Federal Register.

(B) If public hearings are to be held on any matter involved in an investigation conducted under this subsection—

(i) at least five days' notice shall be provided the person making the request for the investigation and any person identified in such request,

(ii) such hearings shall be held in accordance with section 6 (c) (3), and

(iii) each employee who made or for whom was made a request for such hearings and the employer of such employee shall be required to present information respecting the applicable matter referred to in paragraph (1) (A) or (1) (B) together with the basis for such information.

(3) Upon completion of an investigation under paragraph (2), the Administrator shall make findings of fact, shall make such recommendations as the Administrator

Recommendations.

deems appropriate, and shall make available to the public such findings and recommendations.

(4) This section shall not be construed to require the Administrator to amend or repeal any rule or order in effect under this Act.

Sec. 25. Studies

15 USC 2624.

(a) **Indemnification Study.**—The Administrator shall conduct a study of all Federal laws administered by the Administrator for the purpose of determining whether and under what conditions, if any, indemnification should be accorded any person as a result of any action taken by the Administrator under any such law. The study shall—

(1) include an estimate of the probable cost of any indemnification programs which may be recommended;

(2) include an examination of all viable means of financing the cost of any recommended indemnification; and

Submittal to Congress. GAO review.

(3) be completed and submitted to Congress within two years from the effective date of enactment of this Act.

The General Accounting Office shall review the adequacy of the study submitted to Congress pursuant to paragraph (3) and shall report the results of its review to the Congress within six months of the date such study is submitted to Congress.

Consultation.

(b) **Classification, Storage, and Retrieval Study.**—The Council on Environmental Quality, in consultation with the Administrator, the Secretary of Health, Education, and Welfare, the Secretary of Commerce, and the heads of other appropriate Federal departments or agencies, shall coordinate a study of the feasibility of establishing (1) a standard classification system for chemical substances and related substances, and (2) a standard means for storing and for obtaining rapid access to information respecting such substances. A report on such study shall be completed and submitted to Congress not later than 18 months after the effective date of enactment of this Act.

Report to Congress.

Sec. 26. Administration of the Act

15 USC 2625.

(a) **Cooperation of Federal Agencies.**—Upon request by the Administrator, each Federal department and agency is authorized—

(1) to make its services, personnel, and facilities available (with or without reimbursement) to the Administrator to assist the Administrator in the administration of this Act; and

(2) to furnish to the Administrator such information, data, estimates, and statistics, and to allow the Administrator access to all information in its possession as the Administrator may reasonably determine to be necessary for the administration of this Act.

(b) **Fees.**—(1) The Administrator may, by rule, require the payment of a reasonable fee from any person required to submit data under section 4 or 5 to defray the cost of administering this Act. Such rules shall not provide for any fee in excess of $2,500 or, in the case of a small business concern, any fee in excess of $100. In setting a fee under this paragraph, the Administrator shall take into account the ability to pay of the person required to submit the data and the cost to the Administrator of reviewing such data.

Such rules may provide for sharing such a fee in any case in which the expenses of testing are shared under section 4 or 5.

(2) The Administrator, after consultation with the Administrator of the Small Business Administration, shall by rule prescribe standards for determining the persons which qualify as small business concerns for purposes of paragraph (1). Consultation. Rule.

(c) Action With Respect to Categories.—(1) Any action authorized or required to be taken by the Administrator under any provision of this Act with respect to a chemical substance or mixture may be taken by the Administrator in accordance with that provision with respect to a category of chemical substances or mixtures. Whenever the Administrator takes action under a provision of this Act with respect to a category of chemical substances or mixtures, any reference in this Act to a chemical substance or mixture (insofar as it relates to such action) shall be deemed to be a reference to each chemical substance or mixture in such category.

(2) For purposes of paragraph (1): Definitions.

(A) The term "category of chemical substances" means a group of chemical substances the members of which are similar in molecular structure, in physical, chemical, or biological properties, in use, or in mode of entrance into the human body or into the environment, or the members of which are in some other way suitable for classification as such for purposes of this Act, except that such term does not mean a group of chemical substances which are grouped together solely on the basis of their being new chemical substances.

(B) The term "category of mixtures" means a group of mixtures the members of which are similar in molecular structure, in physical, chemical, or biological properties, in use, or in the mode of entrance into the human body or into the environment, or the members of which are in some other way suitable for classification as such for purposes of this Act.

(d) Assistance Office.—The Administrator shall establish in the Environmental Protection Agency an identifiable office to provide technical and other nonfinancial assistance to manufacturers and processors of chemical substances and mixtures respecting the requirements of this Act applicable to such manufacturers and processors, the policy of the Agency respecting the application of such requirements to such manufacturers and processors, and the means and methods by which such manufacturers and processors may comply with such requirements. Establishment.

(e) Financial Disclosures.—(1) Except as provided under paragraph (3), each officer or employee of the Environmental Protection Agency and the Department of Health, Education, and Welfare who—

(A) performs any function or duty under this Act, and

(B) has any known financial interest (i) in any person subject to this Act or any rule or order in effect under this Act, or (ii) in any person who applies for or receives any grant or contract under this Act,

shall, on February 1, 1978, and on February 1 of each year thereafter, file with the Administrator or the Secretary of Health, Education, and Welfare (hereinafter in this subsection referred to as the "Secretary"), as appropriate, a written statement concerning all such interests held by such officer or employee during the preceding calendar year. Such statement shall be made available to the public.

(2) The Administrator and the Secretary shall—

(A) act within 90 days of the effective date of this Act—

(i) to define the term "known financial interests" for purposes of paragraph (1), and

(ii) to establish the methods by which the requirement to file written statements specified in paragraph (1) will be monitored and enforced, including appropriate provisions for review by the Administrator and the Secretary of such statements; and

Report to Congress.

(B) report to the Congress on June 1, 1978, and on June 1 of each year thereafter with respect to such statements and the actions taken in regard thereto during the preceding calendar year.

(3) The Administrator may by rule identify specific positions with the Environmental Protection Agency, and the Secretary may by rule identify specific positions with the Department of Health, Education, and Welfare, which are of a nonregulatory or nonpolicymaking nature, and the Administrator and the Secretary may by rule provide that officers or employees occupying such positions shall be exempt from the requirements of paragraph (1).

(4) This subsection does not supersede any requirement of chapter 11 of title 18, United States Code.

Penalty.

(5) Any officer or employee who is subject to, and knowingly violates, this subsection or any rule issued thereunder, shall be fined not more than $2,500 or imprisoned not more than one year, or both.

(f)   Statement of Basis and Purpose.—Any final order issued under this Act shall be accompanied by a statement of its basis and purpose. The contents and adequacy of any such statement shall not be subject to judicial review in any respect.

Appointment.

(g)   Assistant Administrator.—(1) The President, by and with the advice and consent of the Senate, shall appoint an Assistant Administrator for Toxic Substances of the Environmental Protection Agency. Such Assistant Administrator shall be a qualified individual who is, by reason of background and experience, especially qualified to direct a program concerning the effects of chemicals on human health and the environment. Such Assistant Administrator shall be responsible for (A) the collection of data, (B) the preparation of studies, (C) the making of recommendations to the Administrator for regulatory and other actions to carry out the purposes and to facilitate the administration of this Act, and (D) such other functions as the Administrator may assign or delegate.

5 USC app. II.

(2) The Assistant Administrator to be appointed under paragraph (1) shall (A) be in addition to the Assistant Administrators of the Environmental Protection Agency authorized by section 1 (d) of Reorganization Plan No. 3 of 1970, and (B) be compensated at the rate of pay authorized for such Assistant Administrators.

Sec. 27. Development and Evaluation of Test Methods

Consultation.
15 USC 2626.

(a)   In General.—The Secretary of Health, Education, and Welfare, in consultation with the Administrator and acting through the Assistant Secretary for Health, may conduct, and make grants to public and nonprofit private entities and enter into contracts with public and private entities for, projects for the development and evaluation of inexpensive and efficient methods (1) for determining and evaluating the health and environmental effects of chemical substances and mixtures, and their toxicity, persistence, and other characteristics which affect health and the environment, and (2) which may be used for the development of test data to meet the requirements of rules promulgated under section 4. The Administrator shall consider such methods in prescribing under section 4 standards for the development of test data.

(b) Approval by Secretary.—No grant may be made or contract entered into under subsection (a) unless an application therefor has been submitted to and approved by the Secretary. Such an application shall be submitted in such form and manner and contain such information as the Secretary may require. The Secretary may apply such conditions to grants and contracts under subsection (a) as the Secretary determines are necessary to carry out the purposes of such subsection. Contracts may be entered into under such subsection without regard to sections 3648 and 3709 of the Revised Statutes (31 U.S.C. 529; 41 U.S.C. 5). *Grants or contracts, application.*

(c) Annual Reports.—(1) The Secretary shall prepare and submit to the President and the Congress on or before January 1 of each year a report of the number of grants made and contracts entered into under this section and the results of such grants and contracts. *Report to President and Congress.*

(2) The Secretary shall periodically publish in the Federal Register reports describing the progress and results of any contract entered into or grant made under this section. *Publication in Federal Register.*

Sec. 28. State Programs

(a) In General.—For the purpose of complementing (but not reducing) the authority of, or actions taken by, the Administrator under this Act, the Administrator may make grants to States for the establishment and operation of programs to prevent or eliminate unreasonable risks within the States to health or the environment which are associated with a chemical substance or mixture and with respect to which the Administrator is unable or is not likely to take action under this Act for their prevention or elimination. The amount of a grant under this subsection shall be determined by the Administrator, except that no grant for any State program may exceed 75 per centum of the establishment and operation costs (as determined by the Administrator) of such program during the period for which the grant is made. *15 USC 2627.*

(b) Approval by Administrator.—(1) No grant may be made under subsection (a) unless an application therefor is submitted to and approved by the Administrator. Such an application shall be submitted in such form and manner as the Administrator may require and shall— *Grants, application.*

(A) set forth the need of the applicant for a grant under subsection (a),

(B) identify the agency or agencies of the State which shall establish or operate, or both, the program for which the application is submitted,

(C) describe the actions proposed to be taken under such program,

(D) contain or be supported by assurances satisfactory to the Administrator that such program shall, to the extent feasible, be integrated with other programs of the applicant for environmental and public health protection,

(E) provide for the making of such reports and evaluations as the Administrator may require, and

(F) contain such other information as the Administrator may prescribe.

Application approval.

(2) The Administrator may approve an application submitted in accordance with paragraph (1) only if the applicant has established to the satisfaction of the Administrator a priority need, as determined under rules of the Administrator, for the grant for which the application has been submitted. Such rules shall take into consideration the seriousness of the health effects in a State which are associated with chemical substances or mixtures, including cancer, birth defects, and gene mutations, the extent of the exposure in a State of human beings and the environment to chemical substances and mixtures, and the extent to which chemical substances and mixtures are manufactured, processed, used, and disposed of in a State.

Report to Congress.

(c) **Annual Reports.**—Not later than six months after the end of each of the fiscal years 1979, 1980, and 1981, the Administrator shall submit to the Congress a report respecting the programs assisted by grants under subsection (a) in the preceding fiscal year and the extent to which the Administrator has disseminated information respecting such programs.

(d) **Authorization.**—For the purpose of making grants under subsection (a) there are authorized to be appropriated $1,500,000 for the fiscal year ending September 30, 1977, $1,500,000 for the fiscal year ending September 30, 1978, and $1,500,000 for the fiscal year ending September 30, 1979. Sums appropriated under this subsection shall remain available until expended.

Sec. 29. Authorization for Appropriations

15 USC 2628.

There are authorized to be appropriated to the Administrator for purposes of carrying out this Act (other than sections 27 and 28 and subsections (a) and (c) through (g) of section 10 thereof) $10,100,000 for the fiscal year ending September 30, 1977, $12,625,000 for the fiscal year ending September 30, 1978, $16,200,000 for the fiscal year ending September 30, 1979. No part of the funds appropriated under this section may be used to construct any research laboratories.

Sec. 30. Annual Report

Report to President and Congress. 5 USC 2629.

The Administrator shall prepare and submit to the President and the Congress on or before January 1, 1978, and on or before January 1 of each succeeding year a comprehensive report on the administration of this Act during the preceding fiscal year. Such report shall include—

(1) a list of the testing required under section 4 during the year for which the report is made and an estimate of the costs incurred during such year by the persons required to perform such tests;

(2) the number of notices received during such year under section 5, the number of such notices received during such year under such section for chemical substances subject to a section 4 rule, and a summary of any action taken during such year under section 5 (g) ;

(3) a list of rules issued during such year under section 6;

(4) a list, with a brief statement of the issues, of completed or pending judicial actions under this Act and administrative actions under section 16 during such year;

(5) a summary of major problems encountered in the administration of this Act; and

(6) such recommendations for additional legislation as the Administrator deems necessary to carry out the purposes of this Act.

Recommendations.

Sec. 31. Effective Date

Except as provided in section 4 (f), this Act shall take effect on January 1, 1977.
  Approved October 11, 1976.

15 USC 2601 note.

LEGISLATIVE HISTORY:

HOUSE REPORTS: No. 94-1341 accompanying H.R. 14032 (Comm. on Interstate and Foreign Commerce) and No. 94-1679 (Comm. of Conference).
SENATE REPORTS: No. 94-698 (Comm. on Commerce) and No. 94-1302 (Comm. of Conference).
CONGRESSIONAL RECORD, Vol. 122 (1976):
Mar. 26, considered and passed Senate.
Aug. 23, considered and passed House, amended, in lieu of H.R. 14032.
Sept. 28, Senate and House agreed to conference report.
WEEKLY COMPILATION OF PRESIDENTIAL DOCUMENTS, Vol. 12, No. 42:
Oct. 12, Presidential statement.

Index

Italicized page numbers refer to material contained in statutory text found in appendixes.

Abatement period, in OSHAct, 92-93
Access
  citizen or corporate, to adminis-
    trative process, 61-68
  by Congress, to data submitted
    under TSCA, *575-576*
  to employee records, 347-350
  to employer's premises or records
    by OSHA, Fourth Amend-
    ment as limitation of, 69-73,
    96
  medical. *See* Medical Access Rule
  restrictions on disclosure of
    toxics information following
    granting of, 358-362
ACGIH. *See* American Conference
    of Governmental Industrial
    Hygienists
Action(s)
  agency. *See* Agency action(s)
  citizens' civil, under TSCA,
    194-195, *580-582*
  unilateral, versus duty to bargain,
    259
  See also Private right of action
Actual damages, under tort law, 467
Acute toxic exposure, compensation
    for, 464
Adams, Larry T., 251
Ad Hoc Committee on the Eval-
    uation of Low Levels of
    Environmental Chemical
    Carcinogens, 114
Adjudication, distinction between
    rulemaking and enforce-
    ment and, 58-59
Adler, P., 15
Administration
  of OSHAct, 89-90, *516-518*
  of TSCA, *586-588*
Administrative common law, and
    judicial limitations on
    agency authority, 69
Administrative law, 39-86
Administrative Office of the United
    States Courts, 55

Administrative Procedure Act
    (APA), 44, 60, 61, 62, 63, 64,
    68, 73-74, 78, 79, *80-86,* 90,
    97, 195
  partial text of, 80-86
  rulemaking under, 59
Administrative process, citizen or
    corporate access to, 61-68
Administrative rulemaking, 58-61
Administrator, EPA
  approval by, *589-590*
  Assistant, *588*
  laws administered by, *570*
  laws not administered by,
    569-570
  notice to, of substantial risks,
    568-569
Administrator's Toxic Substances
    Advisory Committee
    (ATSAC), 195
Advisory committees
  access to, 62-63
  in EPA, 60-61, 195
  in OSHA, 60-61, 90, 95, *511,
    516-518*
  in regulatory decisionmaking,
    60-61
Advisory opinions, by Federal
    courts, 51
AFL-CIO, 256, 279, 337, 371. *See
    also* American Federation of
    Labor; Congress of Industrial
    Organizations
Age discrimination, protection
    against, 450-451
Age discrimination in Employment
    Act, 24, 450
Agency(ies)
  access to employee records by,
    348-349
  bypassing of, through private right
    of action, 68
  Federal. *See* Federal agency *entries;
    see also* specific agencies by
    name in general. *See*
    Administrative law

rules and procedures of, as
    limitations on agency
    authority, 69
Agency action(s)
  final, 63
  new chemical, under TSCA, 198
  with respect to categories of
    chemicals, under TSCA, *587*
  reviewable, 85
  review of, role of courts in, 69-70
  statement of reasons for not taking,
    under Section 5 of TSCA, *557*
Agency authority. *See* Authority,
    agency
Agency decisions not to act, judicial
    review of, 79
Agency proceedings and records,
    access to, 62
Agency rulemaking. *See* Rulemaking
AIDS, 483
Allar, 483
Allegheny County, Pennsylvania
    Steelworkers, 28
Amalgamated Clothing and Textile
    Workers Union, 26, 196
Amendment, statutory, 45
American Association for Labor
    Legislation, 454
American Cancer Society, 337
American Civil Liberties Union, 371
American Conference of Govern-
    mental Industrial Hygienists
    (ACGIH), 99, 120, 320
American Cyanamid Company,
    421-425
Americans with Disabilities Act
    (ADA), 452
American Federation of Labor, 7, 8.
    See also AFL-CIO
American industrial relations, trans-
    formation of, 251-252
American Lung Association, 337
American National Standards
    Institute, 90-91, 99
American Petroleum Institute (API),
    122

Ammonia fumes, refusal to work because of, 376–382
Analyzing regulatory framework, questions in, 40–41
Annual report(s)
  under OSHAct, *534–535*
  under TSCA, *589, 590–591*
Anti-discrimination laws, monitoring data and exclusion under, 448–452
Appeals, courts of. *See* Court(s) of Appeals
Appropriations
  under OSHAct, *538*
  under TSCA, *590*
Approval of grants
  by Administrator, under TSCA, *589–590*
  by Secretary, under TSCA, *589*
Arbitrary and capricious standard, of judicial review, 73–79, *86*
Arbitration, of labor disputes, 285–286
Arbitration clauses, and right to strike, 283–287
Article III, U.S. Constitution, 50
Asbestos
  compensation for exposure to, 466, 467–470, 477–482
  OMB and, 46–47
  Paterson, New Jersey plant for, 28
  permanent OSHA standard for, 102–113
Ashford, N. A., 2, 26, 27, 32, 34, 60, 78, 89, 137, 180, 215, 218, 226, 228, 236, 239, 240, 242, 243, 247, 275, 288, 301, 306, 310, 322, 325, 346, 351, 372, 446, 447, 464, 475, 499, 501, 503
Aspirin, and Reye's syndrome, OMB and, 47–48
Assessment of Toxic Agents, 306, 307
Assistance, from Small Business Administration, under OSHAct, 94
Assistance office, under TSCA, *587*
Associated Industries of Massachusetts (AIM), 300, 360
Assumption of risk, 225, 484
AT&T, 3, 24, 26
  hotel billing and information systems (HOBIS) of, 281
Atomic Energy Act of 1954, 88, 406, *512, 543*
Attorney General, U.S., *525, 547, 548, 551, 558, 579, 581*
Auchter, Thorne, 97
Audits, under OSHAct, *534*
Authority, agency
  judicial limitations on, 69–73
  to regulate, under TSCA, 190–193
  to require generation of information, under OSHAct, 310–313

to require testing, under TSCA, 193, 197–199, *545–551*
  for research and development, under TSCA, *570*
Authorization for appropriations, under TSCA, *590*
Automation, 10–11
  worker involvement in decisions regarding, 275–282
Axelson, O., 466
Ayres, C., 26, 27, 236, 242, 247, 275

Backpay award, in unfair labor practice cases, 402
Bacow, L., 263, 266, 279
Bagnara, 279
Bargaining
  collective. *See* Collective bargaining *entries*
  effects, decision versus, 266–275
Bargaining representative, recognition as under NLRA, 253
Barth, P. S., 456, 507
BASF Corporation, 200
Bayer, Ronald, 446
Bazelon, D., 78–79
Bendix, 252
Benefits
  income, for scheduled injuries, 458
  public policy, problems in estimation of, 238–240
  See also Cost-benefit analysis
"Benzene decision," under OSHAct, 122–137
  standard setting after, 137–157
Berlin, 306
Berman, D., 9
Bernstein, 48
Bertin, J., 446
Bethlehem Steel, 8
Bingham, Eula, 97
Biological monitoring, 307–308
Biomarkers, potential utility in toxic substance exposure compensation proceedings, 504–506
Birnbaum, Sheila, 474
Black Lung Benefits Reform Act of 1977, 461
Blacks, employment of, in low-paid occupations, 18–19
Blackstone, W., 365, 366
Blank, Rome, Comisky & McCauley, 369
BNA. *See* Bureau of National Affairs
Boden, L., 455
Boeing Company, 281
Bona fide occupational qualification (BFOQ), in discrimination cases, 425, 426, 431, 434, 435, 437, 440, 441, 442, 445, 446, 452
Boston College Law School, 274
"Bottom-line" myopia, 245
Boulding, K. E., 244

Bowles, S. D., 2, 24, 26, 28
Braverman, H., 2
Breach of implied covenant of good faith and fair dealing, 409
Breach of implied promise of job security, 409
Brody, D., 2, 6, 7
Brooks, H., 2
Brown, M. S., 2
Brown lung disease, and OSHA cotton dust standard, 137–144
Bruchey, S., 2
Burden of proof
  in discrimination cases, 395, 432, 451, 452
  in tort and worker compensation cases, 457–459, 466–467, 503–507
  under the APA, *81–82*
Burdick, Walton E., 369
Bureau of Labor Standards, U.S., 87, 88
Bureau of Labor Statistics (BLS), 22, 252, 262, 263–264, 265
  Annual Survey of, 31
Bureau of National Affairs (BNA), 2, 32, 262–263, 264, 279, 288, 466
Bureau of Occupational Safety and Health, 90
Bush administration, 195
Business necessity, in discrimination cases, 426
Business Week, 367, 369
Byington, John, 216
Bypassing of agency, through private right of action, 68

Cadmium, OMB and, 47
Cahan, V., 367
Caldart, C. C., 78, 215, 244, 300, 301, 306, 310, 322, 325, 346, 351, 446, 447, 475, 499, 501, 503, 506
California Chamber of Commerce, 371
California Department of Industrial Relations; 265
California Occupational Safety and Health Act, 383
California Supreme Court, 483
Cancer standard, generic, under OSHAct, 100–102
Carter administration, 49, 166, 195, 196
Case(s)
  early, under OSHAct, 102–121
  rule of the, 58
  types of, heard in federal court, 51, 52
Case system, and how to read judicial decisions, 55–58
Castleman, B. I., 100
Catler, S., 372

Causation, proof of, in occupational disease cases growing out of exposure to toxic substances, 503–507
Center for Science in the Public Interest, 48
Centers for Disease Control, 47, 90
Certiorari, 53
CFCs, 196, 199
Challenger shuttle, 407
Chamber of Commerce
    California, 371
    Ohio, 474
    U.S., 458, 459, 464
Chamot, 279
Chandler, A. D., Jr., 2, 3, 4
Change(s)
    postwar, in the American workplace, 10–32
    technological, 2, 10–12
        collective bargaining over issues of, 275–287
Chemical Industry Institute of Technology, 215
Chemical Manufacturers Association, 300
Chemical substances
    categories of, action with respect to, under TSCA, *587*
    hazardous
        microelectronic-based technologies and, 32
        and mixtures, regulation of, under TSCA, 191–194, 196–197, 559–564
    and mixtures, testing of, under TSCA, 193, 197–199, 545–551
    new, under TSCA, 190
        actions on, 198
    for which OSHA has promulgated health standards, 100
Chicago Bears, 502
Children of workers, suits by, 475
Chown, P., 264
Chrysler Corporation, 280
Citations, under OSHAct, *520*
Citizen access to administrative process, 61–68
Citizen enforcement through private right of action, 68
Citizens' civil actions, under TSCA, 194–195, *580–582*
Citizenship of parties, and effect on federal court jurisdiction, 51, 52
Citizens' petitions, under TSCA, 195, *582–583*
Citizen suits. *See* Citizens' civil actions, under TSCA, *and* Private right of action
Civil litigation, representation of OSHA in, *525*
Civil penalties, 93, 200, *525–526, 576–577*

Civil Rights Act of 1964, 42
    Title VII of, 24, 259, 289, 395, 396, 421, 450, 452
        and age discrimination, 450–451
        and exclusion of pregnant or fertile female employees, 425–447
Classification study, under TSCA, *586*
Clay, Don, 216, 217
Clean Air Act, 29, 41, 42, 194, 247, 407
Clean Air Act Amendments of 1977, 24
Clean Water Act, 29, 41, 55, 194, 407
Cleveland Citizen, 8
Coal Mine Safety and Health Act of 1969. *See* Mine Safety and Health Act
Code of Federal Regulations, 58
Cohen, S., 17
Coke oven emissions, 121–122
Cole, R., 12
Colgate-Palmolive Co., 330
Collective bargaining, 250, 256–287, *293–295*
    duty to engage in, 258–260, *294–295*
        timing of, "decision" versus "effects" bargaining, 266–275
    general overview of, 256–258
    over issues of technological change, 275–287
    over safety and health, 260–266
Collective bargaining agreement, duty of care independent of, 495
Colton, J., 2
Columbia University, 55, 370
Commerce clause, 42
Commitment gap, between workers and their jobs, 23
Committee on the Evolution of Work, AFL–CIO, 371
Committee on Science, Engineering and Public Policy, 33
Committees, health and safety, labor–management, 287–289. *See also* House of Representatives, U.S.; Senate, U.S.
Commoner, Barry, 11
Common law
    administrative, and judicial limitations on agency authority, 69
    monitoring data and worker exclusion under, 447–448
    protection against worker discharge, 408–430
    See also Tort law *and* Suits in tort
Communication Workers of America (CWA), 26, 280–281
Community right-to-know laws, 335, 338

Company policies   *See* Speaking out about company policies, 364
Comparative negligence, 483
Compensation
    for exposure to toxic substances, 463–466
    for occupational injury and disease, 453–507
        level and adequacy of, 461–466
    workers'. *See* Workers' compensation
Competition, imperfect, 230–231
Compliance Operations Manual, 180
Comprehensive Assessment Information Rule (CAIR), 325–326
Comprehensive Environmental Response, Compensation and Liability Act, 194, 407. *See also* Superfund
Compromise settlements, in workers' compensation cases, 459–460
Comptroller General of the United States, *534*
Concerted activity, under Section 7, NLRA, 382–385
Conclusiveness, in APA formal hearing, *82–84*
Concurrent jurisdiction, between federal and state courts, 51, 52
Concurring opinions, 53
Conference Report, 45, 180
Confidentiality
    employee, 345–350
    of trade secrets
        in general, 351–358
        under NLRA, 332–334
        under OSHAct, 317–318, *525*
        under OSHA Hazard Communication Standard, 317–318, 321
        under TSCA, 358, *574–575*
Congress, U.S., 12, 39, 41, 42, 43, 44, 45, 46, 49, 51, 54, 58, 60, 67, 73–74, 75, 78, 79, 80, 84, 88, 90, 96, 98, 189, 217, 247, 258, 291, 292, 311, 336, 338, 339, 385, 408, 448, 452, 454, 460, 461–463, *509–510, 532, 533, 534, 535, 536, 541, 542, 543, 570, 586, 589, 590*
    access by, under TSCA, 575–576
    See also House of Representatives, U.S.; Legislative branch; Senate, U.S.
Congress of Industrial Organizations, 8. *See also* AFL–CIO
Congressional Record
    on OSHAct, *539*
    on TSCA, *591*
Consensus standards, under OSHAct, 90–91, 99–100
    authority to require medical surveillance on, 312
    judicial interpretation of, 166–171

Consolidation, under TSCA, *565-566, 582*
Constitution, U.S., 39, 42, 49, 79, 291, 346, 348, 484
 and judicial limitations on agency authority, 69
Constitutional basis for labor, health, and safety regulation, 41-58
Constitutional courts, 50-55
Constitutional support, of employment at will doctrine, 364
Consumer Product Safety Act, 483
Consumer Product Safety Commission (CPSC), 194, 219
Contingent work force, growth of, 19-20
Contract, "yellow-dog," 250, 364
Contractor(s)
 health and safety, suits against, 495-498
 independent, company physician as, 502
Contributory negligence, 225, 483
Controversies, types of, heard in federal court, 51, 52
Cooperation of Federal agencies, under TSCA, *586*
Coordination, under TSCA, *570*
Corporate access to administrative process, 61-68
Cost-benefit analysis, 40, 236-247
 feasibility and, 137-145
 politicization of, 46-48, 245-246
Cost-effectiveness analysis, contrasted with health-effectiveness analysis, 238
Costs
 award of, under TSCA, *580*
 public policy, problems in estimation of, 241-243
Cotton dust, OSHA standard governing, 137-144
Council on Environmental Quality, 550, 586
Council on Wage and Price Stability, 48
Court(s)
 access to, 63-64
 appellate. *See* Court(s) of Appeals
 Constitutional, 50-55
 and dispute before it, 56
 district. *See* District Court(s)
 Federal, jurisdiction of, 51, 52
 role of, in reviewing agency action, 69-79
 trial, 50
 U.S. Supreme. *See* Supreme Court, U.S.
Court(s) of Appeals, 50, 53-54, 55, 91, 92, *528, 584*
 California, 409
 D.C. Circuit, 74-75, 78, 79, 97, 102, 122, 137, 145, 171, 179, 181, 195, 200, 331, 349, 384-385, 391, 421, 448, 451, 521-522, 576, 578-579

Eighth Circuit, 180
Eleventh Circuit, 452
 for Federal Circuit, 55
Fifth Circuit, 27, 122, 137, 215, 259, 282
First Circuit, 345, 395
Fourth Circuit, 96, 97, 166-167, 451
New York, 366
Ninth Circuit, 97
Ohio, 475
Second Circuit, 114, 122, 181, 345
Seventh Circuit, 181, 484
Sixth Circuit, 345, 349
 of State of Washington, 484
Third Circuit, 97, 113, 313, 323, 338-339, 345, 356
Court of Claims, 55
Court officers, 55
Court of International Trade, 55
Creation of new agency, 41
Criminal laws, violation of, 80, 415
Criminal penalty(ies), under TSCA, *577*
 for wrongful disclosure, *575*
Crisis in the Workplace, 34, 89, 288
Cullen, M., 32
Customs territory of the United States, entry into, under TSCA, *573*
Cyert, R. M., 2, 13, 16, 32, 33
Cytogenetic monitoring, 307

Daedalus, 256
Dallas, 53
Damages, under tort law
 actual, 467
 emotional, 506-507
 in proportion to attributable risk, 506
Dana, 252
Dangers
 to health and safety. *See* Health and safety *entries*
 imminent, procedures to counteract, under OSHAct, 92, *524-525*
 See also Hazards
Daniels, Dominick, 180
Data
 collection, dissemination and utilization of, under TSCA, *570-571*
 disclosure of, under TSCA, *574-576*
 monitoring, exclusion of employees on basis of, 447-452
 test
  development of, petition for standards for, under TSCA, *551*
  submission of, under TSCA, *552-553*
Data systems, under TSCA, *570-571*
Davis, D. L., 466
Debevoise, Judge Dickinson R., 357

Decision(s)
 contents of, APA formal hearing, *82-84*
 initial, APA formal hearing, *82-84*
 record as basis of formal hearing, APA and, *81-82*
 of Supreme Court, 53
Decision bargaining, versus effects bargaining, 266-275
Decision making, management
 arbitrary, regarding job security, 364
 employee participation in, 25-27, 33-34, 279-281
Decision points, for workers' compensation claims, 457
Decision rule, cost-benefit criterion as, 246
Delay, in unfair labor practice cases, 403-405
 1960-1980, 404
Delegation doctrine, 42-43
Deoxyribonucleic acid (DNA), 307, 505
Department of Commerce, *550*
Department of Defense, 364, 407
Department of Health and Human Services, 90
Department of Health, Education and Welfare, 23, 93, 94, 311, *570, 571, 586, 587-588.* See *also* Secretary of Health, Education and Welfare
Department of Justice, 171, 337
Department of Labor, 2, 17, 18, 26, 88, 89, 195, 229, 287, 395, 398, 461, 464, *526, 534, 538, 550*
Department of Occupational Safety, Health and Social Security, AFL-CIO, 337
Department of Transportation, 74, 75, 384
Descotes, J., 505
Designation of confidential data, under TSCA, *575*
Deutsch, Steven, 31-32, 281
Development
 of information, regulation pending, under TSCA, *554-556*
 research and, under TSCA, *570-571*
 of test data, petition for standards for, under TSCA, *551*
 of test methods, under TSCA, *588-589*
Dibromo chloropropane (DBCP), 330
Dicta, 53, 56
Direction, to administrative agencies
 from executive branch, 46-49
 from judicial branch, 49-58
 from legislative branch, 41-46
Disability, occupational, compensation for, 453-507

Discharge, wrongful, 412. *See also* Protections from retaliatory employment practices.
Disclosure(s)
of data
potential barriers to, 345-362
restrictions on, once access has been granted, 358-362
under TSCA, *574-576*
financial, under TSCA, *587-588*
Discretionary function, and governmental torts, 484-485
Discrimination
age, 450-451
handicap, 448-450
impermissible, under NLRA, 289-290
as prohibited by Section 210 of Energy Reorganization Act, and other environmental statutes, 406
proof of, under OSHAct, 393-396
in retaliation for employee actions, 363-420
on the basis of employee physical condition, 420-452
under the Civil Rights Act, 425-447, 450-452
Diseases, occupational. *See* Occupational disease(s)/illness(es)
Disease-exposure risk matrix, 504
Disparate impact, in discrimination cases, 426, 430-431, 440, 450-451
defined, 426, 450
Disparate treatment, in discrimination cases, 450
Dispute before court, 56
Dissemination of data, under TSCA, *570-571*
Dissenting employee, rights under NLRA, 289-291
Dissenting opinions, 53
District Court(s), U.S., 54, 55, 74, 92, 194-195, 200, 207, 357, 398, *522, 551, 555, 556, 577, 581*
Division of Labor Standards, 9
DNA, 307, 505
Dobbs, D., 483
Dodge, 9
Doll, R., 464, 466
Dominick, Peter H., 311
Donohue, Thomas R., 256
Dow Chemical, 120
Dual capacity doctrine, in tort law, 475-476, 502
Dubofsky, M., 5
Duchin, F., 2
Dunlop, John T., 21, 256, 368
du Pont, 2
Dust, cotton, OSHA standard governing,137-144
Duty(ies)
APA and, formal hearings, *81-82*
of care, independent of collective bargaining agreement, 495

of employers and employees, under OSHAct, 89, 512
to engage in collective bargaining, 258-260
timing of, 266-275
of fair representation, 289-290
to generate and/or retain information, 301
to inform, 301
See also General duty clause
Dyer, L., 265
Dymmel, 279

Early, S., 287, 288
Early cases, under OSHAct, 102-121
Early period of twentieth century: 1900-1945, 3-9
background of, 3-4
health and safety in, 8-9
labor-management relations in, 6-8
structure of industry and occupations in, 5-6
technology and organization of work in, 4-5
Easterbrook, Judge Frank, 446
Eastman, Crystal, 454
Eastman Kodak, 3
Eaton, 252
Eau Claire, Wisconsin, plant, Uniroyal, 28
Economic assistance to small businesses, under OSHAct, *537*
Economic feasibility, 40, 107-108, 137-145, 155-156, 221-222
Economic issues, in occupational health and safety, 221-247
Economic rationale, for right-to-know, 304
"Economic" strike, under NLRA, 282-283
Economy, postindustrial, 16
Education programs, under OSHAct, 94, *531*
Effects bargaining
decision versus, 266-275
in light of *First National Maintenance*, 274-275
Egg Products Inspection Act, *543*
Eleventh Amendment, 51, 484
Elkins, Charles, 196
Emergency locator beacons, under OSHAct, *538*
Emergency Planning and Community Right to Know Act, 335-336, 338
Emergency temporary standards (ETS), under OSHAct, 90, 120, 157-163
Emotional damages, under tort law, 506-507
Employee(s)
black, in low-paid occupations, 18-19
coercion of, to join union, 290

discrimination against. *See* Discrimination
dissenting, under NLRA, 289-291
duties of, under OSHAct, 89, 96, *512*
exclusion of, on basis of human monitoring, 447-452
female
exclusion of pregnant or fertile, 421-447
in low-paid occupations, 18-19
prospective, exclusion of, 446
relations between management and. *See* Labor-management relations
under Section 11(c) of OSHAct, 392-393
See also Worker(s)
Employee confidentiality, 345-350
Employee education, under OSHAct, 94, *531*
Employee freedoms, potential, 363-364
Employee information, in OSHA Hazard Communication Standard, 321
Employee monitoring. *See* Monitoring, of worker
Employee participation in management decision making, 25-27, 33-34, 279-281
Employee protection. *See* Protection(s)
Employee records, access to, 347-350
Employee right(s)
to confidentiality, 347-350
under NLRA, in general, 253-255, *293*, 400-405
under OSHAct, in general, 94-95, 392-400
to refuse hazardous work, 286, 364, 372-392
revolution in, 367-371
to strike, under NLRA
arbitration clauses and, 283-287
in general, 282-283
over unsafe working conditions, 372-375
Employee Stock Ownership Plans, 21, 25
Employee training, in OSHA Hazard Communication Standard, 321
Employers
access to employee records by, 350
attitudes of, toward workers' compensation, 454
duties of, under OSHAct, 89, *512*
suits against, 467-476
unfair labor practices by. *See* Unfair labor practice *entries*
Employment
of females and blacks, in low-paid occupations, 18-19
hazardous
right to refuse, 286, 364, 372-392

Employment *(continued)*
  hazardous *(continued)*
    wealth effect on worker deci-
      sions to accept, 232
    See also Work
Employment at will doctrine/rule,
  25
  decline of, 364-372
  development of, 365-367
Employment effects, under TSCA,
  585-586
Employment growth by sector:
  1979-1987, 16
Employment practices, retaliatory,
  363-452
  See also Protections from retaliatory
    employment practices
Employment Retirement Income
  Security Act of 1974, 24
Enabling legislation, 41
Energy Reorganization Act, Section
  210 of, text of, 406-407
Enforcement
  citizen, through private right of
    action, 68
  distinction between rulemaking
    and adjudication and, 58-59
  under OSHAct, 91-92
    procedure for, *520-521*
  under TSCA
    Section 23, *585*
    specific, *577-578*
Enforcing standard, promulgating
  versus, as pertains to discretionary government
  functions, 486
Entry into customs territory of the
  United States, under TSCA,
  573
Environmental and safety statutes,
  federal, prohibiting discrimination, 406-407
Environmental Defense Fund, 196
Environmental monitoring, 308-309
Environmental Protection Agency
  (EPA), 41, 42-43, 45, 46-47,
  189-219, 335, 346, 348, 351,
  352-356, *543, 550, 551, 555,
  556, 559, 566, 569, 570, 587,
  588*
  advisory committee of, 61
  informal rulemaking by, 60
  and information transfer, 325-326
Environmental Research, Development, and Demonstration
  Authorization Act, 61
Epstein, S., 466
Equal Access to Justice Act of 1984,
  67-68
Equal Employment Opportunity
  Commission, 370, 426, 446
Equilibrium level(s) of occupational
  risk, 223-224
  as determined by new and exist-
    ing technologies, comparison
    of, 242

Equity, problems of, in policy
  decisions, 243-244
Estimating
  of public policy benefits, prob-
    lems in, 238-240
  of public policy costs, problems
    in, 241-243
Estreicher, Samuel, 370-371
Ethical considerations
  in cost-benefit analysis, 243-244
  and refusal to work on projects,
    364, 415-420
Ethylene oxide (EtO), 164-166
  OMB and, 47
Evaluation of test methods, under
  TSCA, *588-589*
Evidence
  in APA formal hearings, *81-82*
  substantial
    in general, 73-74, *86*
    under OSHAct, 102-121
    under TSCA, 200-202
Examinations, medical, 313
Exchange of research and development results, under *TSCA*,
  571
Exclusion
  under TSCA, Section 23 (a), *585*
  from workplace, because of pre-
    sumed susceptibility to
    workplace hazards, 420-452
  anti-discrimination law protection
    against
    under Age Discrimination in
      Employment Act, 450-452
    under Americans with Dis-
      abilities Act, 452
    under handicap discrimination
      laws, 448-450
    under OSHAct, 421-425, 448,
      451-452
    under Title VII of Civil Rights
      Act, 425-447, 450-452
  common law protection against,
    447-448, 451
Exclusive jurisdiction, of federal
  courts, 51, 52
Executive branch, direction to
  administrative agency from,
  46-49
Executive Order(s), 245
  11821, 48
  12044, 49
  12291, 48, 49
Exemption(s)
  under OSHAct, *525*
  under TSCA, 193, *547-548,
    557-559, 578*
Exhaustion of administrative
  remedies, as condition for
  judicial review of administrative agency, 63
Existing Chemicals Program, under
  TSCA, 196
Expansion, of individual rights,
  24-25

Expected social benefit or loss, 236
Exports, under TSCA, *572-573*
Exposure
  monitoring worker for, 306-309
  to noise, protection of workers
    from, 166-171
  to toxic substances. *See* Toxic
    substances, exposure to
  See also Short-term exposure limit(s)
Exposure records, access to, 302,
  322-325, 333-335, 336-338,
  347-350
Extension of notice period, under
  TSCA, *553*
Externalities, and job market,
  228-230, 234

Facts at hand, application of law to,
  59
Factual review, scope of, 73-79
Fair Labor Standards Act of 1938,
  97, 250
Fair representation, duty of, 289-290
Farber, Henry S., 369
Farmington, West Virginia, mine
  explosion, 28
Fear of job loss, 32
Feasibility. *See* Economic feasibility
  and Technological feasibility
Federal Advisory Committee Act
  (FACA), 44, 62
Federal agency(ies)
  cooperation of, under TSCA,
    586
  creation of new, 41
  granting of new powers and
    responsibilities to, 41
  review by, informal rulemaking,
    82-84
Federal agency safety programs and
  responsibilities, under
  OSHAct, *528-529*
Federal Aviation Act of 1958, 538
Federal Coal Mine Health and
  Safety Act. *See* Mine Safety
  and Health Act
Federal courts, jurisdiction of, 51,
  52
Federal environmental and safety
  statutes, prohibiting discrimi-
  nation, 406-407
Federal Food, Drug and Cosmetic
  Act, 192, *526, 543*
Federal Highway Administration,
  384
Federal information transfer laws,
  310-338
Federal Insecticide, Fungicide, and
  Rodenticide Act (FIFRA),
  192, 194, 352-356, 357-358,
  543
Federal judges, 50-51
Federal judicial system, 50
Federal labor statutes, right to refuse
  hazardous work under, 364,
  372-392

Federal laws, relationship of TSCA to other, 194, *569-570*
Federal magistrates, 54-55
Federal Meat Inspection Act, *543*
Federal Mediation and Conciliation Service, 297, 298
Federal Panel on Formaldehyde, 215
Federal preemption, and role of state and local laws, 40, 79-80, 313
of state and local right to know laws, 313, 338-345
under TSCA, *578*
Federal questions, 51
Federal Railroad Administration, 347
Federal Register, 58, 59, 61, 81, 90, 91, 192, 194, 216, 217, *513, 515, 516, 548, 549, 550, 551, 553-554, 557, 559, 563, 569, 582, 583, 589*
Federal regulation, of occupational safety and health, history of, 87-88
Federal Rules of Civil Procedure, *523, 524*
Federal Tort Claims Act, 484-486
Federal Trade Commission, *547, 548, 558*
Fees, under TSCA, *580, 586-587*
Feinman, J. M., 365
Fellow-servant rule, 225
Female employees
exclusion of pregnant or fertile, 421-447
in low-paid occupations, 18-19
Fetus, protection of, 421-447
Field Operations Manual, OSHA, 97
Fifth Amendment, 364
Final agency action, 63
Final Rule on Occupational Exposures to Hazardous Chemicals in Laboratories, 321
Final Rule to Expand Scope of Hazard Communication Standard, 320
Financial disclosures, under TSCA, *587-588*
Firestone Rubber, 256
Firm, economics of, 40
First Amendment, 292
"First principle" of union representation, 289
Fischer, G. W., 239
Fischhoff, B., 227, 239
Flamm, Gary, 48
Florini, Karen, 196
Food and Drug Administration (FDA), 47-48
Ford, Henry, 4
Ford administration, 48
Ford Motor Co., 4-5, 8, 26
Formaldehyde, regulatory history of, 215-219

Formaldehyde Institute, 216
Formal rulemaking, under the APA, 59, *81-84*
judicial review of, 73, 84-86
Form and venue of proceeding, in judicial review, 85
Forming union, right of, 253
Forrestiere, L., 466
Fortune 500 firms, 26
Fourth Amendment, 91, 96, 346, 347
Freedom of Information Act (FOIA), 44, 62
Freedoms, workplace, four potential, 363-364
Freeman, Richard B., 2, 251, 369
Freon, 182-186
Frumin, Eric, 196

Gary Works, U.S. Steel, 6
Gaydos, Representative, 336
General Accounting Office (GAO), 195, 586
General duty clause, OSHAct, 89
monitoring data and exclusion under, 448, 452
obligation, evolution of, 179-186, 421-425
General Electric, 22, 252
General Motors, 2, 23, 26, 280
Generation of information, 310-313
duty of, 301
Generic cancer standard, OSHA, 100-102
Genetic monitoring, 307
Genetic screening, 309
for promulgated health standards, 312-313
Georgetown University, 425
Getman, J., 401
Gillespy, Robert, 474
Girard-diCarlo, David F., 369
Gissel bargaining order, 254-255, 403
Glaberson, W. G., 367
Gordon, D. M., 2, 4, 5, 6
Gorsuch, Anne, 216
Government, suits against, 484-486
Government Accountability Project (GAP), 408
Government intervention
market imperfections as basis for, 226-236
types of, 233-236
Government Printing Office, U.S., 291, 461-463
Government in the Sunshine Act, 44, 62
Graber, Glenn, 303
Granting new powers and responsibilities to agency, 41
Grants to states, under OSHAct, *532-533*
Great Britain, and employment at will, 365
Great Crash of 1929, 7
Great Depression, 3, 6, 7, 9

Green, J. R., 2, 5, 6, 11, 23
Green, M., 484
Greenfield, P., 265
Gross, J., 265
Grossman, R. L., 9
Growth and Opportunity Committee, 280
Grozuczak, J., 31
Guest, R. H., 13
Guile, B., 2

Hamburger, F., 463
Hand, Judge Learned, 482
Handicap discrimination, 448-450
Harm
increased risk of, 506
intention of, 474, 503
possibility of, 216-217
Harris Poll, 369
Hartmann, H., 2, 19
Harvard University, 21, 266, 367, 368, 369
Business School, 13
School of Public Health, 280
Hastings Center, 446
Hattis, D., 306, 310, 322, 325, 346, 447, 503, 506
Hawkins, Representative, 336
Hazard Communication Standard, OSHA, 80, 300, 313-321, 322, 323, 335-336, 337, 338-339, 356, 358
definitions in, 320
employee information and training in, 321
purpose of, 319-320
requirements in, 320-321
"tailored" provision in, 321
trade secrets, protection of, in, 317-318, 321
Hazardous chemical substances
microelectronics-based technologies and, 32
and mixtures, regulation of, under TSCA, 191-194, 196-197 *559-564*
Hazardous work
right to refuse, 286, 364, 372-392
wealth effect on worker decisions to accept, 232
Hazards
chemical, from microelectrical-based technologies, 32
imminent, under TSCA, 191-192, 197, 207-208, 564-566
reproductive, excluding women for, 421-447
safety, information in job market and, 227
susceptibility to, 309, 313, 420-451
video display terminal, 32, 277, 278
See also Dangers
Health and safety
bargaining over, 260-266
in early period: 1900-1945, 8-9

Health and Safety *(continued)*
  market for, 222–226
  occupational, economic issues in,
    221–247
  in postwar period: 1945–present,
    27–32
  in shaping attitudes toward work, 2
  strikes over, 372–375
Health and safety committees, labor-
    management, 287–289
Health and safety contractors, suits
    against, 495–498
Health and safety crisis of 1960s,
    27–29
Health and safety protection, under
    LMR Act, 372–375
Health and safety studies, under
    TSCA, *568*
  disclosure of, *574–575*
Health-effectiveness analysis, cost-
    effectiveness analysis
    contrasted with, 238
Health risk. *See* Risk(s) *entries*
Health standards. *See* Standard(s)
Healy, James J., 257
Hearings
  in administrative rulemaking
    formal, 59, *81–82*
    informal, 59
    under OSHAct, 60, 91
    under TSCA, 60, 191
  oversight, 45
Heaton, G., 483
Heckscher, C. C., 2, 22, 24, 26
Henman, 306
Hernandez, John, 216
Hewlett-Packard Co., 369
High-risk manufacturing industries,
    injury and illness incidence
    rates in, 31
High risk notification, 336–338
High Risk Occupational Disease
    Notification and Prevention
    Act, 336
Hirschhorn, L., 2
History
  of federal regulation of
    occupational health and
    safety, 87–88
  of standard setting under OSHAct,
    99–102
Hoerr, J., 367
Hoffa, Jimmy, 367
Homogenization of labor, 4
House of Representatives, U.S., 44,
    88, 336, 337, 509, 529, 541
  Armed Services Committee of,
    583
  Committee of Conference of,
    591
  Committee on Education and
    Labor of, 539
    Subcommittee on Safety and
      Health of, 337
  Committee on Interstate and For-
    eign Commerce of, 591

Subcommittee on Employment
    and Housing of Committee
    on Government Operations
    of, 281
House Report(s), 44–45
  on OSHAct, *539*
  on TSCA, *591*
House–Senate Conference Com-
    mittee, on OSHAct, 180
Hunt, H. A., 456
Hybrid rulemaking, 60

Idealized job market, 223–224
Illinois Supreme Court, 80
Illness, occupational. *See* Occupa-
    tional disease(s)/illness(es)
Illness tax, 234
Immerwahr, J., 23
Imminent dangers, procedures to
    counteract, under OSHAct,
    92, *524–525*
Imminent hazards, under TSCA,
    191–192, 197, 207–208,
    564–566
Immunity, sovereign, 484
Impact, disparate, in discrimination
    cases, 426, 430–431, 440,
    450–451
  defined, 426, 450
Impermissible discrimination,
    under NLRA, 289–290
Implementation, of TSCA, reasons
    for slow, 199
Implied covenant of good faith and
    fair dealing, breach of, 409
Implied promise of job security,
    breach of, 409
Incidence rates, injury and illness, 31
Income benefits, for scheduled
    injuries, 458
Indemnification study, under TSCA,
    586
Independent contractor, company
    physician as, 502
Independent union, right to form,
    253
Individual, sensitivity of, to
    chemicals, 309
Individual employee concern,
    versus concerted activity,
    under NLRA, 384
Individual rights, expansion of, 24–25
Industrial relations, American, trans-
    formation of, 251–252
Industrial relations system, 21
Industry(ies)
  American, early period of. *See*
    Early period
  economics of firm versus, 40
  high-risk manufacturing, injury
    and illness incidence rates
    in, 31
  major, union membership by, 22
  and occupations, structure of, 2,
    5–6, 16–20
  part-time work by, 20

Inequities, market-transmitted,
    231–233
Informal controls, by Congress over
    agencies, 45–46
Informal rulemaking,
  by OSHA and EPA, 60
  in general, 59, *80–81*
  judicial review of, 73–79, *84–86*
Information
  employee, in Hazard Communi-
    cation Standard, 321
  exposure, access to, 302, 322–325,
    333–335, 336–338, 347–350
  generation of, 310–313
    duty of, 301
  imperfect, in job market, 226–228,
    234
  medical, access to, 302, 322–325,
    333–335, 336–338, 347–350
  regulation pending development
    of, under TSCA, *554–556*
  reporting and retention of, under
    TSCA, 325–326, *566–569*
  withholding of, by employee,
    346
  See also Duty to inform
Information transfer. *See* Toxics infor-
    mation transfer
Initial decisions, in APA formal
    rulemaking, *82–84*
Initiation of rulemaking, 61
Injunction, 53
Injunction petitions, for unfair labor
    practices, 1960–1980, 405
Injuries, occupational. *See* Occupa-
    tional injury(ies)
Injury tax, 234
Injustices, market-transmitted,
    231–233
Inspections
  under OSHAct, 69–73, 91–92, 96,
    518–520
  under TSCA, *571–572*
Institute of Medicine, 33
Intentional harm, under tort law,
    474, 503
Interagency Regulatory Liaison
    Group (IRLG), 215
Interagency Testing Committee (ITC),
    193, 197–198
Interim relief, in unfair labor
    practice cases, 404–405
Interim standards, under OSHAct.
    See Consensus standards
Internal Revenue Code of 1954, 55,
    543
International Agency for Research
    on Cancer, 320
International Association of Machin-
    ists and Aerospace Workers,
    26, 265, 281
International Brotherhood of Elec-
    trical Workers, 26
International Business Machines
    Corp., 369
International Harvester, 3

International Union of Bricklayers, 368
International Union, United Automobile, Aerospace and Agricultural Implement Workers of America, 280
Interpretations, judicial, of standard setting under OSHAct, 102–179
Intervention, government
market imperfections as basis for, 226–236
types of, 233–236
Intrafirm intertemporal externalities, 228, 230
Inventory, under TSCA, *567–568*
Investigations
under OSHAct, *518–520*
under TSCA, *585–586*

Job loss and fear of job loss, 32
Job market
idealized, 223–224
performance of, in U.S., 224–225
Job market imperfections, as basis for government intervention, 226–236
Job Opportunity Bank-Security (JOB Security) Program, 280
Job security, arbitrary management decisions regarding, 364, 408–409
Jobs, growth in, by type, 20
Johns-Manville, 465
Johnson, 464
Johnson administration, 28, 88
Johnston, W. B., 2
Joyce, John T., 368
Jubak, J., 301
Judges, federal, 50–51
Judicial branch, direction to administrative agencies from, 49–58
Judicial interpretation(s)
of standard setting under OSHAct, 102–179
of TSCA, 200–215
Judicial limitations on agency authority, 69–73
Judicial review of agency decisions
of agency decisions not to act, 79
under APA, 59, 73–79, *84–85*
in general, 63–68, 69–79
monetary limitations on availability of, 67–68
under OSHAct, 60, *521–522*
scope of factual review in, 73–79
threshold limitations on availability of, and standing to seek, 63
under TSCA, 60, 195, *578–580, 584*
See also Judicial interpretations
Judicial system, Federal, 50–55
Judicial writs, common, 53
Judiciary Act of 1789, 53

Jurisdiction, of Federal courts, 51–52
state, under OSHAct, *527–528*
Jurisdictional standards, of NLRB, 292

Kahneman, D., 239
Kaldor-Hicks criterion, 243
Katz, H., 251
Katz, J. I., 372
Kazis, R., 1, 9
Keeton, R., 483
Keeton, W. P., 483, 484
Knapp, Andria S., 370
Kochan, T. A., 2, 22, 251, 265
Kohler, Thomas, 274

Labor, homogenization of, 4. *See also* Department of Labor
Labor Law Journal, 369
Labor-management health and safety committees, 287–289
Labor-management relations, 2
in early period: 1900–1945, 6–8
under National Labor Relations Act, 249–298
in postwar period: 1945–present, 21–27
transformation of, 251–252
Labor Management Relations Act (LMRAct). *See* National Labor Relations Act (NLRA)
Labor practice, unfair, *See* Unfair labor practice *entries*
Labor Reform Act of 1978, 401
Labor regulation, Constitutional basis for, 41–58
Labor revolt, 23
Labor statutes, federal, right to refuse hazardous work under, 364, 372–392
Labor unions. *See* Union(s) *entries*
LaDou, J., 12
Landrum-Griffin Act of 1959, 21, 250
Larson, L. W., 475
Lauwarys, Robert R., 307
Law(s)
administrative. *See* Administrative law *entries*
anti-discrimination, monitoring data and exclusion under, 448–452
common. *See* Common law *entries*
local
federal preemption and role of, 79–80
information transfer, 338–345
right-to-know. *See* Right-to-know laws
state
effect of TSCA on, 578
federal preemption and role of, 79–80
information transfer, 338–345
tort, use of, to augment workers' compensation, 466–503

toxics information transfer
federal, 310–338
state and local, 338–345
See also Legislative *entries*
Lead, 145–156, 171–179, 421–425, 426–446
Legislation
enabling, 41
See also Law
Legislative branch
direction to administrative agency from, 41–46
See also Congress, U.S.
Legislative history, 44–45
of OSHAct, *539*
of TSCA, *591*
Leontief, W., 2
Leventhal, H., 79
Levy, B. S., 30, 455
Lewin, Tamar, 475
Liability, for injury, under tort law, 466–467
Lichtenstein, S., 227
Limitation, to civil action, under TSCA, *581*
Lipsky, D., 265
Litigation, civil, representation of OSHA in, 525
Livernash, E. Robert, 257
Llewellyn, K. N., 55
Local emergency planning commission (LEPC), 335
Local information transfer laws, 338–345
Local laws, federal preemption and role of, 79–80
Lockout, 286
Loperamide, 416–420
Low, J. O., 5
Low-paid occupations, employment of females and blacks in, 18–19

Machina, M. J., 239
MacLaury, J., 454
Magistrates, Federal, 54–55
Maier, P., 483
Major industry, union membership by, 22
Majority opinions, 53
Management decision making
arbitrary, regarding job security, 364
employee participation in, 279–281
Management, scientific, 5. *See also* Labor-management *entries*
Mandamus, writ of, 53, 68
Manufacturers, suits against, 476–484
Manufacturing and processing notices, under TSCA, *551–559*
Manufacturing industries, high-risk, injury and illness incidence rates in, 31
Manville Corporation, 300, 466

Manville Property Damage Settlement Trust, 466
Market
  for health and safety, 222–226
  job. See Job market entries
Market-transmitted injustices or inequities, 231–233
Markowitz, G., 9
Marquis, J. K., 463
Martin, D., 499
Maskin, A., 505
Massachusetts Institute of Technology (MIT), 266, 369
Massachusetts Special Legislative Commission on Liability for Releases of Oil and Hazardous Materials, 506
Massachusetts Supreme Judicial Court, 250, 465–466
Massachusetts Toxics Use Reduction Law, 282
Mass production, 4
Material safety data sheets (MSDS), 320–321
Materials Research Corp., 370
Matrix
  disease-exposure risk, 504
  of policy consequences, 237
Mattson,   , 446
McCulloch, Kenneth J., 370, 371
McGarity, T. ,78
McKensie, R., 251
Meany, George, 256
Medical Access Rule, 322–325, 337
Medical information, access to, 302, 322–325, 333–335, 336–338, 347–350
Medical professionals, suits against, 501–503
Medical record. See Record, employee medical
Medical removal protection (MRP), 171–179, 336, 347, 391, 447, 448
Medical surveillance, 306–307, 310, 312
Medoff, J. L., 2
Mentzer, M., 330
Metzenbaum, Senator H., 336
Meyers Industries, 383–384
Meyers standard, 384
Michigan State University, School of Labor & Industrial Relations, 370
Michigan Supreme Court, 80
Michigan Whistleblowers' Protection Act, 408
Microelectronic-based products, as postwar change, 12
Microelectronics-based technologies, areas of concern emerging from, 32
Miller, C. S., 32
Mine Safety and Health Act, 9, 27, 29, 88, 375

Minnesota Mining and Manufacturing Co. (3M), 22, 330
Mintz, B. W., 97
Mishan, E. J., 232
Mishel, Lawrence, 16, 20
Misiti, 279
Mix of industries and occupations, 2, 5–6, 16–20
Mixtures
  chemical substances and, testing of, under TSCA, 193, 197–199, 545–551
  hazardous, regulation of, under TSCA, 191–194, 196–197, 559–564
Modifications of rule, during judicial review, under TSCA, 579
Monetary limitations on availability of judicial review. 67–68
Monetary terms, estimates of social benefit or loss expressed in, 236
Monitoring
  under OSHA, 310–313, 346–347
  right to refuse, 346–347
  under TSCA, 325–326, 346–347, 577
  of worker, 305, 313, 325–326, 346–347
    biological, 307–308
    genetic, 307
  workplace, 305–309
  environmental, 308–309
  See also Confidentiality and Surveillance
Monitoring data, exclusion of employees on basis of, 447–452
Monsanto, 22
Montgomery, J., 484
Moon, Richard G., 370
Moral rationale, for right-to-know, 302–304
Morris, C., 259
Morton Thiokol, 407
Moskowitz, D. B., 367
Mowery, D. C., 2, 13, 16, 32, 33
Musculoskeletal problems, 32
Mutagens, regulating of, under TSCA, 192

Nabisco, 3
Nader, Ralph, 164, 336
National Academy of Engineering, 33
National Academy of Sciences, 13, 33
National Advisory Committee on Occupational Safety and Health (NACOSH), 60–61, 90, 95, 511, 516–518
National Aeronautics and Space Administration (NASA), 407
National Association of Manufacturers, 454

National Association of Working Women, 369
National Cancer Institute, 98, 550
National Civic Federation, 9, 454
National Commission on State Workmen's Compensation Laws, 511, 535–537
  recommendations of, 460–461
National Congress of Employees, 371
National defense waiver, under TSCA, 583
National Enforcement Investigation Center, Office of Compliance Monitoring, EPA, 200
National Environmental Policy Act, 29, 44
National Fire Protection Association, 90
National Football League, 502
National Foundation on Arts and Humanities Act, 512
National Highway Traffic Safety Administration, 75–77
National Industrial Recovery Act of 1933, 7
National Institute for Occupational Safety and Health (NIOSH), 30, 32, 90, 91, 94, 99–100, 102, 114, 310–313, 336–337, 346, 348, 349, 511, 513, 531–532, 546, 550
National Institute of Environmental Health Services, 550
National Joint Committee on New Technology, 280
National Labor Relations Act (NLRA), 7, 42, 79, 179, 249–298, 304, 305, 310, 334, 365, 368, 486, 524
  employee rights under, 372–375
  partial text of, 293–298
  and right to refuse hazardous work, 372–385
  Section 8 of, 400–406
  self-organization rights under, 401–405
  and toxics information transfer, 327–335
National Labor Relations Board (NLRB), 7, 27, 253–275, 276, 279, 285, 294, 296, 297, 327, 330–331, 349, 373–375, 383–385, 401–405
  role of, in administering NLRA, 291–293
National Occupational Exposure Survey, 309
National Research Council of Italy, 279
National Safety Council, 31, 88, 225
National Science Foundation, 550
National Toxicology Program, 320
Natural Resources Defense Council (NRDC), 196, 215, 218

Navy, U.S., 486
Negligence, comparative and contributory, 483
Negotiated testing program, under TSCA, 211, 213-214
Nelkin, D., 2
Network of rights and obligations, in right to know laws, 301
New Business Venture Development Group, of U.A.W. and G.M., 280
New chemical actions, TSCA and, 198
New chemical substances, TSCA and, 190
New Deal, 3, 7, 9, 21
New Directions program, under OSHAct, 98
New Jersey State Chamber of Commerce, 339-344, 357-358
New Jersey Supreme Court, 368
New Venture Fund, of U.A.W. and G.M., 280
New York Times, 474-475
New York University, 370
Nichols, A. L., 229
9 to 5, National Association of Working Women, 369
Ninth Amendment, 346
Nixon administration, 27, 28, 29, 88, 311
Noble, C., 23, 28, 29
Noise, protection of workers from exposure to, 166-171
Noncytogenetic monitoring, 307
Norris-LaGuardia Act, 250
No-strike clause, 283-285
Nothstein, G., 446
Notice(s) under TSCA,
    to Administrator, of substantial risks, *568-569*
    manufacturing and processing, *551-559*
    of proposed rulemaking (NPR), 191
    regarding exports, *573*
    of test data, *548*
Notification
    high risk, 336-338
    premanufacturing (PMN), under TSCA, 192
        exemption from, 193
    under Section 210 of Energy Reorganization Act, 406-407
Nuclear power plants, 406-407
Nuclear Regulatory Commission (NRC), 74-75
Nulty, 281
Nussbaum, Karen, 17, 18, 20, 23

Obiter dicta, 53, 56
Occidental Chemical, 330
Occupational disability, compensation for, 453-507

Occupational disease(s)/illness(es)
    compensation for, 453-507
        level of, 461-466
    general duty obligation and, 180-181
    imperfect information in job market and, 227-228
    incidence rates for, 31
    leading, 30
    magnitude and severity of, 461-462
    monitoring worker for, 305-313, 325-326, 346-347
    recordkeeping on
        under OSHAct, 93, 312, 322-323, *518-520*
        under TSCA, 193-194, 325-326, *568*
    reporting on
        under OSHAct, 312
        under TSCA, 193-194, 325-326, *566-569*
Occupational disease cases, growing out of exposure to toxic substances, proof of causation in, 503-507
Occupational disease claims, time limits on filing of, 459
Occupational Health Advisory Board, 280
Occupational health and safety. *See* Health and safety *entries*
Occupational Health Guidelines for Chemical Hazards, 306-307, 312
Occupational illnesses. *See* Occupational disease(s)/illness(es)
Occupational injury(ies)
    compensation for, 453-507
        level of, 461-466
    general duty obligation and, 180-181
    incidence rates for, 31
    income benefits for scheduled, 458
    leading, 30
    recordkeeping on, under OSHAct, 93, 312, *518-520*
    reporting on, under OSHAct, 312
Occupational risk. *See* Risk(s)
Occupational safety and health. *See* Health and safety *entries*
Occupational Safety and Health Act (OSHAct) of 1970, 9, 24, 27, 29, 34, 43, 59, 60, 80, 87-187, 266, 310, 322, 323, 382, 406, 451, 452, *509-539*, 550
    additional positions under, *537-538*
    advisory committees: administration of, 89-90, *516-518*
    annual report under, *534-535*
    applicability of, *511-512*
    appropriations under, *538*

audits under, *534*
citations under, *520*
conditions following passage of, 29-32
confidentiality of trade secrets under, 317-318, *525*
Congressional findings and purpose of, *509-510*
definitions in, *510-511*
description and summary of, 89-99
duties under, 89, *512*
economic assistance to small businesses under, *537*
effective date of, *538*
emergency locator beacons under, *538*
employee rights under, 94-95, 372-375
enforcement under, 91-92
    procedure for, *520-521*
and exclusion of pregnant or fertile female employees, 421-425
general duty clause of, 89, 179-186, 421-425
    monitoring data and exclusion under, 448, 452
grants to States under, *532-533*
inspections, investigations, and recordkeeping under, *518-520*
judicial review under, procedure and standard for, 60, *521-522*
occupational safety and health standards under, 90-91, *512-516*
    judicial review and interpretation of, 102-179
penalties under, 93, *525-526*
procedures to counteract imminent dangers under, 92, *524-525*
refusal of hazardous work under, 286, 385-392
representation of OSHA in civil litigation under, *525*
research and related activities under, *529-531*
Section 11(c) of, 385-400, *522*
separability under, *538*
standard setting under
    history of, 99-102
    judicial interpretation of, 102-179
state jurisdiction and state plans under, *527-528*
statistics under, 94, *533-534*
text of, *509-539*
and toxic information transfer, 310-325
training and employee education under, *531*
TSCA and, *570*
variations, tolerances, and exemptions under, *525*

Occupational Safety and Health
Administration (OSHA), 43,
47, 58, 88–100, 101–131,
136–187, 264, 306–307,
310–313, 338–339, 344–345,
349, 350, 356, 385–391,
392–398, 421, 451, 486
advisory committee of, 60–61, 90,
95, *511, 516–518*
chemical substances for which
health standards have been
promulgated by, 100
informal rulemaking by, 60
Medical Access Rule of, 322–325,
337
See also Hazard Communication
Standard
Occupational Safety and Health Reporter,
300
Occupational Safety and Health
Review Commission
(OSHRC), 59, 90, 92, 97, 180,
181, 421–425, *510, 523–524,
538*
Occupational safety and health
standards, under OSHAct,
90–91, *512–516*
Occupations
industries and, structure of, 2,
5–6, 16–20
low-paid, employment of females
and blacks in, 18–19
Odell, R., 11
Office of Compliance Monitoring,
EPA, 200
Office of General Counsel, EPA,
217
Office of Industrial Hygiene and
Sanitation, U.S. Public Health
Service, 87
Office of Management and Budget
(OMB), 46–48, 49, 145, 164,
245–246
Office of Technology Assessment
(OTA), 2, 12, 32, 196, 225,
454
Office of Toxicological Sciences,
FDA, 48
Office of Toxic Substances, EPA,
195, 196, 199, 215–216
Official citation, of federal statutes, 42
Ohio Chamber of Commerce, 474
Oil, Chemical & Atomic Workers
(OCAW), 330, 331–334,
421–425
Olson, C., 265
Omnibus Judgeship Act of 1978, 50
Ontario worker compensation
system, 464
On the record decision, in formal
rulemaking under the APA,
59
Opinion Research Corporation
(ORC), 23–24

Opinions, of Supreme Court, nature
and publication of, 53
Ordstown, Ohio, GM plant, 23
O'Reilly, J. T., 301
Organization of work, 2
in early period: 1900–1945, 4–5
in postwar period: 1945–present,
12–16
Organizing process, union, 254–255,
401–405
Originating statute, 41, 69
OSHA Log, 312
Osterman, P., 2, 13
Oversight hearings, 45
Owen, D., 483, 484

Panel on Technology and Employ-
ment, of Committee on
Science, Engineering and
Public Policy, National
Academy of Sciences and
Engineering and Institute of
Medicine, 33
Paperwork Reduction Act (PRA),
49
Pareto, Vilfredo, 232
Pareto optimal (pareto efficient),
232, 243
Participation, employee, in management decision-making,
25–27, 33–34, 279–281
Parties, submissions by, in formal
hearing under APA, 82–84
Part-time work, by industry, 20
Patent and Trademark Office, 55
Paterson, New Jersey, asbestos plant,
28
Penalty(ies)
civil
under OSHAct, 93, *525–526*
under TSCA, *576–577*
criminal
under OSHAct, 93, *526*
under state law, 80
under TSCA, *575, 577*
Percentage composition, of chemical mixture, trade secret protection for, 351
Performance, of job market, in U.S.,
224–225
Perkins, Frances, 88
Permanent standards, under
OSHAct, 90–91
chemical substances for which
there are, 100
genetic screening for, 312–313
judicial interpretation of, 102–157,
164–166, 171–179
medical surveillance required by,
310–311
Permissible exposure limits (PELs),
99–100, 186, 187, 307, 320
for noise, 166
for vinyl chloride, 120

Personnel, supervisory, suits against,
499–501
Personnel policies, violation of,
408–409
Petition(s)
citizens', under TSCA, 194–195,
582–583
injunction, for unfair labor
practices, 1960–1980, 405
for standards for development of
test data, under TSCA, *551*
Peto, R., 464, 466
Petrochemical-based materials, as
postwar change, 11
Philadelphia Project on Occupational Safety and Health
(PHILAPOSH), 29
Phosphorus, 485–486
Physical integrity, as grounds for
refusal of worker monitoring,
346–347
Physician, role of, in worker's
compensation, 456–457, 458
Piore, M. J., 2, 13
"Pittsburgh Survey," 454
PMN. *See* Premanufacturing
notification
Polaroid Corporation, 351
Policy(ies)
company, speaking out against,
364
Congressional, general, under-
lying promulgation of statute
in OSHAct, 88, *509–510*
in TSCA, 190, *542–543*
OSHA, resolution of scientific
issues as matter of, in generic
cancer standard, 100–101
personnel, violation of, 408–409
public. *See* Public policy *entries*
Policy consequences, matrix of,
238
Policy decisions, and judicial review
of agency action, 78
Political rationale, for right-to-know,
304–305
Politicization of cost-benefit analysis,
46–48, 245–246
Pollack, S., 31
Pollock, M. A., 367
Polychlorinate biphenyls (PCBs), 29
regulations on, 196, 197, 199,
200–207
spills of, 207–208
under TSCA, *563–564*
Poor Laws, 365
Population-based data, on correlation between disease and
exposure, as proof of
causation, 504
Portsmouth Naval Shipyard, 486
Posner, Judge Richard, 452
Possibility of harm, under TSCA
Section 4 (f), 216–217

Postal Reorganization Act of 1970, 293
Postindustrial economy, 16
Postol, 502
Postwar changes: 1945–present, 10–32
Potlach Corp., 97
Poultry Products Inspection Act, *543*
Powers
    of administrative law judge, under APA, *81–82*
    granting of new, to agency, 41
Preemption, federal, and role of state and local laws, 40, 79–80
    of state and local right to know laws, 313, 338–345
    under TSCA, *578*
Pregnancy Discrimination Act of 1978, 425
    and fetus protection policies, 425–447
Pregnant or fertile female employees, exclusion of, 421–447
Premanufacturing notification (PMN), under TSCA, 192, 197, 198, 200
    exemption from, 193
Presentations, additional, during judicial review under TSCA, *579*
President, U.S., 46, 50–51, 90, 291, *523, 529, 532, 533, 535, 536, 583, 588, 589, 590*
Presidential Executive Orders. *See* Executive Order(s)
Presiding employees, in formal hearing under APA, *81–82*
Prill, Kenneth, 383–384, 385
"Primer on TSCA 4(f)," 217
Priority list, under TSCA, *549–551*
Priority Review Level 1 (PRL–1), under TSCA, 216
Privacy Act, 44, 62
Private right of action, citizen enforcement through, 68
    lack of, under section 11(c) of OSHAct, 398
Procedural mandate, governing agency action, 43–44
Procedural requirements, under Section 11(c), OSHAct, 396–400, *522*
Procedures
    agency, as limitations on agency authority, 69
    for petitions, under TSCA, *582–583*
Proceeding, form and venue of, in judicial review under APA, *85*
Process, chemical manufacturing, trade secret protection for, 351
Processing notices, under TSCA, *551–559*

Products liability, 476–484
Programs
    education, under OSHAct, 94
    federal agency safety, under OSHAct, *528–529*
    state, under TSCA, *589–590*
    training. *See* Training
Progressive era, 3, 8, 454
Prohibited acts, under TSCA, *576*
Prohibition of union shop, 291
Promulgated health standards, genetic screening for, 312–313
Promulgation
    of standard, versus enforcing of, as pertains to discretionary government functions, 486
    of subsection (a) rules, of Section 6, TSCA, *561–563*
Proof
    burden of. *See* Burden of proof
    of causation, in occupational disease cases growing out of exposure to toxic substances, 503–507
    of discrimination, 393–396
Prospective workers/employees, exclusion of
    under fetus protection policies, 446
    under Section 11(c) of OSHAct, 392–393
Protection(s)
    from exposure to noise, 166–171
    of fetus, 421–447
    medical removal, 171–179, 336, 347, 391, 447, 448
    from retaliatory employment practices, 363–452
        under common law, 408–420
        under federal environmental and safety statutes, 406–407
        under general whistleblower statutes, 407–408
        under NLRA, 400–406
        under Section 11(c) of the OSHAct, 392–400, *522*
        under Section 210 of Energy Reorganization Act, 406–407
        under TSCA, *583–585*
        See also Right to refuse hazardous work *and* Exclusion from workplace because of presumed susceptibility to workplace hazards
    against unreasonable risks, under TSCA, 190, 191, 200–207, 219, *556–557*
Proximate cause, 467
    in toxic substance exposure cases, 503–507
Public Citizen Health Research Group, 164–166, 336
Public health agencies, access to employee records by, 348

Public Health Service, U.S., 87, 114
Public policy benefits, problems in estimating of, 238–240
Public policy costs, problems in estimating of, 241–243
Public Works and Economic Development Act of 1965, 537

Quality control, under TSCA, *560–561*
Quality of Work Life (QWL), 280
Quantification bias, 244–245
Questions, federal, and federal court jurisdiction, 51

Railway Labor Act, 249–250, 365
Raising wages, without bargaining, 260
Reagan administration, 30, 46–47, 48, 49, 61, 97, 98, 166, 195, 246, 323
Record(s)
    of agency hearing
        formal hearing under APA, 59, *82–84:* as basis of agency decison, 59, *81–82*
        in standard setting under TSCA, 191, *561–562*
    agency, public access to, 62
    employee
        exposure, access to, 302, 322–325, 335–336, 336–338, 347–350
        medical, access to, 302, 322–325, 335–336, 347–350
    rulemaking, under TSCA, *579*
    See also Record-keeping
Record-keeping
    under OSHAct, 93, 312, *518–520*
    under TSCA, 193–194, 325–326, *566–569*
Reeve, Tapping, 366
Refusal
    to carry out action contravening articulated public policy, 415–420
    of hazardous work, right of, 286, 364, 372–392
    to work on unethical or socially irresponsible projects, 364, 415–420
Registry of Toxic Effects of Chemical Substances, 323
Regulation
    cost-benefit analysis and, generally, 236–247
    economic analysis of, generally, 221–248
    federal, history of, 87–88
    of formaldehyde, reluctance to engage in, under TSCA, 215–219
    health and safety, Constitutional basis for, 41–58

Regulation *(continued)*
    of labor–management relations.
        See Labor–Management
        Relations Act
    market imperfections as basis for,
        226–236
    questions to consider in eval-
        uating, 40
    types of, 233–236
    See also Administrative law *and
        entries for specific statutes,
        agencies, and regulatory subjects*
Regulatory decisionmaking, advisory
    committees in OSHA and
    EPA in, 60–61
Regulatory framework, analyzing of,
    questions in, 40–41
Regulatory history of formaldehyde,
    215–219
Rehabilitation Act of 1973, 449, 450
Rehnquist, Justice William, 43
Reinstatement, of employees in un-
    fair labor practices cases,
    402–403
Release of confidential data, under
    TSCA, *575*
Relief
    authorized against imminent
        hazards, under TSCA, 565
    pending review, under APA, *85*
Reluctance to regulate, formalde-
    hyde, under TSCA, 215–219
Remedial philosophy of NLRA, with
    regard to unfair labor
    practices, 402
Remedy, for retaliatory discrimi-
    nation, under TSCA, *584*
Report(s)
    annual. *See* Annual report(s)
    of Congressional committees, 44–45
Reporting
    of occupational injuries and
        illnesses, under OSHAct,
        312
    and retention of information,
        under TSCA, 193–194,
        325–326, *566–569*
Representation
    of EPA, under TSCA, *566*
    fair, duty of, under NLRA, 289–290
    of OSHA, in civil litigation, *525*
Required action(s), under TSCA
    under Section 4 (f), 217–218,
        551
    under Section 7, *564–565*
Research
    under OSHAct, 94, *529–531*
    under TSCA, *570–571*
Resource Conservation and Recov-
    ery Act (RCRA), 45, 194, 196,
    407
Responsibility(ies), granting of new,
    to agency, 41
Rest, K., 137, 218

Restatement (Second) of Torts, 476,
    483, 484
Restrictions on disclosure of toxics
    information once access has
    been granted, 358–362
Retaliatory employment practices,
    363–452. *See also* Protections
    from retaliatory employment
    practices
Retrieval study, under TSCA, 586
Reuther, Walter, 367
Review
    by agency, of decision of hearings
        officer, under APA, *82–84*
    of agency action, by courts. *See*
        Judicial review
Review Commission. *See* Occupa-
    tional Safety and Health
    Review Commission
Revised Statutes, *532, 570, 571,
    589*
Reye's syndrome, aspirin and, OMB
    and, 47–48
Riding circuit, 54
Right(s)
    of access to information, 62–63,
        299–362
    to bargain collectively, under
        NLRA. *See* Collective
        bargaining
    employee. *See* Employee right(s)
    individual, expansion of, 24–25
    of judicial review, under APA,
        63–68, 69–79, *84–85*
    and obligations, network of, in
        right to know laws, 301
    to refuse hazardous work
        under NLRA, 372–385
        under OSHAct, 385–392
    to self-organization, under NLRA,
        253–255, *293*, 401–405
    to strike
        over "abnormally dangerous
            conditions," 373–374
        in general, 282–283
        over unsafe working conditions,
            372–375
        arbitration clauses and, 282–287
    See also Private right of action
Right-to-know laws, 299–362
    conceptual basis of, 302–305
    defined, 299
    federal, 310–338
    framework for understanding,
        301–302
    state, and local, 299–300, 313,
        338–345
Ripeness, as condition for judicial
    review of administrative
    agency, 63–64
Risk(s)
    damages in proportion to attrib-
        utable, as basis for compar-
        ison, 506

equilibrium level(s) of occupa-
    tional, 223–224
    as determined by new and
        existing technologies,
        comparison of, 242
    of harm, increased, as basis for
        comparison, 506
    health, monitoring for. *See*
        Monitoring
    high. *See* High risk *entries*
    substantial, notice to Administra-
        tor of, under TSCA, 194, 326,
        568–569
    unreasonable, under TSCA, 190,
        191, 200–207, 219, *556–557*
    See also Significant risk requirement
Risk formulation, as element of
    regulatory statute, 40
Risk matrix, disease-exposure, 504
Roberts, 279
Robinson, J. C., 22, 27, 29, 262, 287
Rom, W., 30
Roosevelt, Franklin D., 88
Rosner, D., 9
Rothman, R., 505
Rothstein, M., 498, 503
Rudkin, Reba, 467–470, 474, 475
Ruggerio, C., 219
Rule(s)
    agency, as limitation on agency
        authority, 69
    of the case, 58
    decision, cost-benefit criterion as,
        246
    employment at will, 25, 364–372
    Medical Access, 322–325, 337
    significant new use, under TSCA,
        193, 195, 197
    subsection (a), of Section 6, TSCA,
        promulgation of, *561–563*
    testing requirement, under TSCA,
        209–215, *545–547*
Rulemaking
    administrative, 58–61
    under Administrative Procedure
        Act, 59, *80–81*
    distinction between adjudication
        and enforcement and,
        58–59
    hybrid, 60
    informal, by OSHA and EPA, 60
    initiation of, 61
    under Section 6 of OSHAct, 91,
        512–516
    under Section 6 of TSCA, 191,
        559–564
Ruttenberg, R., 25,·287
Ryan, C. W., 78, 215

Sabel, C. F., 2, 13
"Safer" technology, collective bar-
    gaining and, 278
Safety, health and. *See* Health and
    safety *entries*

Safety hazards, information in job market and, 227
Safety programs, Federal agency, under OSHAct, 528–529
Safety statutes, federal environmental and, prohibiting discrimination, 406–407
St. Antoine, Theodore J., 368
Salary workers, union membership as percentage of, 251
Santa Barbara oil rig explosion, 28
Satisfaction, worker, historical analysis of, 23–24
Saturn Project, 280
Scale, of technological change, as determinant of working conditions, 2
Schneiderman, 466
Schultz, William B., 46, 49
Schwartz, B., 122
Science Advisory Board, EPA, 61
Scientific management, of industrial production, 5
Scope
  of factual review, 73–79
  of inspection, under TSCA, *572*
  of judicial review, under APA, *85–86*
  of regulation, under TSCA, *559–560*
  of technological change, as determinant of working conditions, 2
Screening
  sensitivity, 309
  techniques of, for effects of chemicals, under TSCA, *571*
Secretary of Commerce, *550, 586*
Secretary of Health, Education, and Welfare, *513, 515, 516, 517, 518–519, 520, 526, 529–531, 532, 533, 534, 535, 570, 571, 586, 587–588*
  See also Secretary of Health and Human Services
Secretary of Health and Human Services (HHS), 90, 93, 94
Secretary of Interior, *511*
Secretary of Labor, 90–91, 406–407, *510, 538, 550*
  approval by, under TSCA, *589*
  Assistant, under OSHAct, *537*
Section numbers, in federal statutes, 42
Security fee, paid to unions, 290
Seizure, of chemical, under TSCA, *577–578*
Self-organization, rights of, under NLRA, 253–255, *293*, 401–405
Senate, U.S., 44, 46, 88, 291, *509, 541, 529, 588*
  Armed Services Committee of, *583*

Committee of Conference of, *591*
Committee on Commerce of, *591*
Committee on Labor and Public Welfare of, 180, 181, *539*
Committee on OSHAct of, 463
Senate Report(s), 44–45
  on OSHAct, 463, *539*
  on TSCA, *591*
Sensitivity screening, 309
Separability, under OSHAct, *538*
Separation of powers, constitutional doctrine of, 57
Service Contract Act of 1965, *512*
Service Employees International Union, 369
Service sector, growth of, 1947–1987, 17
Settlements, compromise, in worker's compensation cases, 459–460
Shell, 215
Sherman Antitrust Act of 1890, 250
Short-term exposure limit(s) (STELs), 99, 164–166, 307
  for vinyl chloride, 120
Signature diseases, 227
Significant new use rule(s) (SNURs), under TSCA, 193, 195, 197
Significant risk requirement, under OSHAct, 102, 122–137, 147–148, 162, 219
Simon, Jaqueline, 16, 20
Singleman, J., 2
Slichter, Sumner H., 257, 367, 368
Slovic, P., 227
Slow implementation, of TSCA, reasons for, 199
Small Business Act, 94, 537
Small Business Administration, 567, 587
  assistance from, under OSHAct, 94
Small businesses, economic assistance to, under OSHAct, 537
Smith, Charles Manly, 366
Smith, R. S., 234
SNUR. *See* Significant new use rule(s)
Socially irresponsible projects, refusal to work on, 364, 415–420
Social Security, 229, 453, 463
Solicitor General, 96
Solicitor of Labor, 525
Solid Waste Disposal Act, 45, 407. *See also* Resource Conservation and Recovery Act
Sovereign immunity, 484
Spadafor, C., 306, 310, 322, 325, 346, 447, 503
Speaking out about company policies, 364. *See also* Whistleblowers
  protections against reprisals for
    under common law, 412–415
    under federal environmental safety and statutes, 406–407

under general whistleblower statutes, 407–408
  under OSHAct, 392–400
  under NLRA, 400–406
Special Task Force, HEW, 23
Specific enforcement, under TSCA, 577–578
Speed, of technological change, 2
Spenner, K., 15
Stafford, Senator, 336
Standard(s)
  of judicial review
    under APA, 59, 73–79, *85–86*
    under OSHAct, 60, *521–522*
    under TSCA, 60, 195, *579–580*
    See also Arbitrary and capricious standard, Judicial interpretations, Judicial review, *and* Substantial evidence test
  jurisdictional, of NLRB, 292
  Meyers, 384
  regulatory
    chemical regulation and testing, under TSCA: authority for, 190–193; judicial interpretation of, 200–215; promulgation of, 195–200, 215–219
    occupational safety and health, under OSHAct: consensus, 90–91, 99–100 (authority to require medical surveillance on, 312; judicial interpretation of, 166–171); emergency temporary, 90–91, 120, 157–163; generic cancer, 100–102; hazard communication. *See* Hazard Communication Standard; interim. *See* consensus, *this heading;* permanent, 90–91 (chemical substances for which there are, 100; genetic screening for, 312–313; judicial interpretation of, 102–157, 164–166, 171–179; medical surveillance required by, 310–311)
  statutory, questions to consider when analyzing, 40
Standard Oil, 3
Standard setting
  under OSHAct
    history of, 99–102
    judicial interpretation of, 102–179
  under TSCA
    history of, 195–200, 215–219
    judicial interpretation of, 200–215
  See also Rulemaking *and* Standards
Standing doctrine. *See* Standing to seek judicial review

Standing to seek judicial review, 63, 64–67
Standing to sue. *See* Standing to seek judicial review
State Bar of California, 371
State emergency response commission (SERC), 335
State information transfer laws
federal preemption of, 313, 338–345
history of, 299–300
State jurisdiction and state plans, under OSHAct, *527–528*
State law(s), federal preemption and role of, 40, 79–80, 313, 338–345
under TSCA, *578*
Statement
of Congressional policies underlying statute
under OSHAct, 88, *509–510*
under TSCA, 190, *542–543*
of reasons for not taking action. under Section 5 of TSCA, *557*
State programs, under TSCA, *589–590*
State and local right-to-know laws. *See* State information transfer laws
State workers' compensation. *See* Workers' compensation
States, grants to, under OSHAct, *532–533*
Statistical correlation between disease and exposure, as proof of causation, 504
Statistics, required by OSHAct, 94, *533–534*
Statute(s)
environmental and safety, prohibiting discrimination, 406–407
federal, right to refuse hazardous work under, 364, 373–392
originating, 41, 69
"whistleblower," 407–408
See also specific statutes by name
Statutes of Labourers, 365
Statutory amendment, 45
Statutory mandate
defined, 41
interpreting of, 44–45
See also specific statutes and subject matter headings
Steiger, William, 180
STEL. *See* Short-term exposure limit(s)
Stieber, Jack, 25, 370
Stillman, N., 446
Stone, Robert F., 221, 226, 236, 239, 242
Storage study, under TSCA, *586*
Stress, from microelectronics-based technology, 32

Strike(s)
purpose of, 282, 382
right to, under NLRA
arbitration clauses and, 282–287
in general, 282–283
over unsafe working conditions, 372–375
types of, 282–283
Structure of industry and occupations, 2
as determinant of working conditions, 2
in early period: 1900–1945, 5–6
in postwar period: 1945–present, 16–20
Studies, under TSCA
health and safety, 326, *568*
disclosure of, 358, *574–575*
other, *586*
Subchapters, in federal statutes, 42
Suboptimization, 244
Subpoena, under TSCA, *572*
Substantial evidence test, as standard of judicial review, 59, 60, 73–74, *86*, 195
See also Judicial interpretations
Substantial risks, notice to Administrator of, under TSCA, 194, 326, *568–569*
Suits
citizen. *See* Citizens' civil actions, under TSCA, *and* Private right of action
in tort, for occupational injury or disease, 225, 406–503
against employers, 467–476
against health and safety contractors, 495–498
against the government, 484–486
against manufacturers, 476–484
against medical professionals, 501–503
against supervisory personnel, 499–501
against unions, 486–495
level of compensation under, 464–465
proof of causation under, 503–507
in tort, for wrongful discharge, 368–371, 408–420
Summers, C. W., 371–372
Summons, 53
Sunshine Mine Disaster, 487–494, 495
Superfund, 194. *See also* Comprehensive Environmental Response, Compensation, and Liability Act
Superfund Amendment and Reauthorization Act (SARA), 199, 300, 326, 335–336, 338

Supervisory personnel, suits against, 499–501
Supplier-induced externalities, 229–230
Supreme Court, U.S., 42, 43, 44, 49, 54, 74, 75, 78, 80, 91, 96, 101, 120, 123, 136, 137, 145, 250, 258, 266, 276, 281, 282, 286, 289, 290, 331, 346, 347, 351–352, 357, 364, 373, 374, 375, 384, 395, 406, 407, 425, 426, 448, 450, 451, *522, 525, 528, 580*
decisions and opinions of, 53
See also individual state supreme courts
Surface Transportation Assistance Act (STAA), 385
Surgeon General, U.S., 114
Surveillance, medical, 306–307, 310, 312
Susceptibility to occupational disease or injury
biological markers of, 309, 313
presumed, exclusion from workplace because of, 420–452
Swados, Harvey, 23
Sweeney, John J., 17, 18, 20, 23
Syracuse University, 464

Taft-Hartley Act of 1947, 21, 250, 373–374
"Tailored" provisions, in Hazard Communications Standard, 321
Target site, chemical, 307
Tariff Schedules of the United States, *544, 573*
Tasini, J., 367
Tax, injury or illness, 234
Tax Court, U.S., 55
Taylor, Frederick W., 5
Taylorism, 13
Technological change, collective bargaining over issues of, 275–287
Technological feasibility, 117–122, 149–157
Technology, workplace
as determinant of working conditions, 2
in early period: 1900–1945, 4–5
in postwar period: 1945–present, 10–12
"safer," collective bargaining and, 278
worker participation in decisions affecting, 26–27, 33–34, 279–281
Technology and Employment, 33
Technology and the American Economic Transition (OTA study), 12

Technology-forcing
    concepts behind, 241-242
    by government regulation. *See*
        Technological feasibility
    by tort law, 481, 482
Tennessee Public Service Com-
    mission, 384
Tennessee Valley Authority (TVA),
    257
Tenth Amendment, 346
Teratogens, regulating of, under
    TSCA, 192
Test data
    development of, petition for stan-
        dards for, under TSCA, *551*
    submission of, under TSCA,
        552-553
Testing, of chemical substances and
    mixtures, under TSCA,
    197-199, 209-215, *545-551*
    authority to require, 193
Testing program, negotiated, under
    TSCA, 211
Test methods, development and
    evaluation of, under TSCA,
    588-589
Threshold limitations on availability
    of judicial review, 63
Threshold limit values (TLVs), 99,
    120, 320
Time limits, on filing occupational
    disease claims, 459
Timing, of duty to bargain, under
    NLRA, 266-275
Title III, of SARA, 199, 335-336,
    338
Title VII, of Civil Rights Act. *See*
    Civil Rights Act
Titles, in federal statutes, 42
TLVs. *See* Threshold limit values
Todhunter, John, 216, 217
Tolerances, under OSHAct, *525*
Tomlinson, Warren L., 369
Tort law, use of, to augment workers'
    compensation, 466-503. *See
    also* Suits in tort
Townley & Updike, 370, 371
Toxic Release Inventory, 199
Toxics information transfer
    unions and, 330-331
    in workplace, 299-362
Toxics information transfer laws,
    299-362
    See also Right-to-know laws
Toxic substances, exposure to
    compensation for, 463-465
    proof of causation in occupational
        disease cases growing out of,
        503-507
Toxic Substances Control Act (TSCA)
    of 1976, 27, 39, 41, 42-43, 60,
    61, 63, 68, 74, 79, 302, 305,
    310, 323, 326, 351, 358, 407,
    483

administration of, *586-588*
authorization for appropriations
    under, *590*
citizens' civil actions under,
    194-195, *580-582*
citizens' petitions under, *582-583*
definitions under, 190, *543-545,
    559, 566, 569*
development and evaluation of
    test methods under, *588-589*
disclosure of data under, 358,
    574-576
effective date of, *591*
effective date of rules under, *563*
employee protection under, 190,
    191, 200-207, 219, *583-585*
employment effects under,
    585-586
entry into customs territory of the
    United States under, *573*
exemptions under, *547-548,
    578*
exports under, *572-573*
findings, policy and intent of,
    190, *542-543*
formaldehyde, regulation under,
    215-219
implementation of, 195-200
information transfer and, 193-194,
    325-326, *566-571, 574-576*
inspections and subpoenas under,
    571-572
judicial interpretations of, 200-215
judicial review under, 60, 195,
    578-580, 584
national defense waiver under,
    583
notice of receipt of test data
    under, *548*
penalties under, 200, *576-577*
petition for standards for devel-
    opment of test data under,
    551
preemption under, 578
premanufacturing notice (PMN)
    under. *See* Premanufacturing
    notice
priority list under, *549-551*
prohibited acts under, *576*
purposes of, 190, *542-543*
regulation under, 190-194,
    195-199
relationship of, to other Acts/
    Federal laws, 194, *569-570*
and reluctance to regulate for-
    maldehyde, 215-219
reporting and retention of infor-
    mation under, 193-194,
    325-326, *566-569*
required actions under, *551,
    564-565*
    nature of, 217-218
research, development, collection,
    dissemination, and utilization

of data under, 193-194,
    325-326, *570-571*
short title and table of contents
    of, *541-542*
specific enforcement and seizure
    under, *577-578*
studies under. *See* Studies, under
    TSCA
testing of chemical substances
    and mixtures under,
    197-199, 209-215, *545-551*
    authority to require, 193
text of, *541-591*
Toyota, 26
"Trade-off" analysis, 247
Trade secrets
    confidentiality of
        in general, 351-358
        under NLRA, 332-334
        under OSHAct, 317-318,
            525
        under OSHA Hazard Com-
            munication Standard,
            317-318, 321
        under TSCA, 358, 574-575
    definition of, 317-318, 353
Training
    employee, under OSHA Hazard
        Communication Standard,
        321
    under OSHAct, 94, *531*
    under TSCA, *571*
Transfer, information. *See* Toxics
    information transfer
Trauberman, J., 506
Treatment, disparate, in discrimi-
    nation cases, 450
Trial courts, 50
Tribe, L. H., 244
Tversky, A., 239

Unfair labor practice(s), under
    NLRA
    cases, delay in processing,
        1960-1980, 404
    by employers, 1950-1980, 401
    explained, 253-254, *294-298*
    injunction petitions for,
        1960-1980, 405
    protections against, effectiveness
        of, 254-255, 400-406
    refusal to bargain as, 258-260.
        See also Collective bargaining
    strikes and, 282-283
"Unfair labor practice" strikes, under
    NLRA, 282-283
Unilateral action, by management,
    259
Union(s)
    access to employee records by,
        333-334, 349-350
    involvement of, in occupational
        safety and health, 1957-1987,
        262-265

Union(s) *(continued)*
  reactions of, to workers' compensation, 454
  right of, to information about occupational health hazards, under NLRA, 327–335
  role of
    after OSHA, 29–30
    as driving force in industrial relations, 21–23, 250–253, 367–371
  suits against, 486–495
Union membership, as percentage of wage and salary workers, 251
Union representation, "first principle" of, 289
Union security agreement, 290
Union shop, 290
  prohibition of, 291
Uniroyal, 28
United Auto Workers (UAW), 9, 26, 279–280, 281, 337, 350
United Mine Workers (UMW), 7, 252, 276
United States, entry into customs territory of, under TSCA, *573*
United States Code, 42
United States Judicial Conference, 54
United States Reports, 53
U.S. Steel, 3, 6, 9, 12, 263
United States Tax Court, 55
United Steelworkers of America, 26, 95, 263
University of Michigan, 368
University of Pittsburgh, 370
University of Tennessee, 303
University of Wisconsin, 265
Unreasonable risk(s), under TSCA, 190, 191, 219, 200–207, *556–557*
Uptake, versus exposure, 308
Urban Environmental Conference, 31
Urethane, OMB and, 48
Utilization of data, under TSCA, *570–571*

Variations, under OSHAct, *525*
Venue
  in judicial review, under APA, 85
  under OSHAct, 521, 524
  under TSCA, *565–566, 577, 579, 581, 584*
Video display terminal (VDT) hazards, 32, 277, 278
Vietnam War, 28
Vinyl chloride, 114–120, 121
  PELs for, 120
Violation. *See* Penalty(ies)
Viscusi, W. K., 229, 230

Vladeck, David C., 46, 49
Vocational Rehabilitation Act, 24

Wagener, 466
Wages, raising of, without bargaining, 260
Wage workers, union membership as percentage of, 251
Wagner Act, 3, 7, 8, 21, 26, 401
Wald, R., 78–79
Walkaround right, under OSHAct, 92
Walker, C. R., 13
Walsh-Healy Act of 1936, 88, 250, 512
Walton, Richard, 13, 24
Warner, W. L., 5
Warrant, 53
Warren, Jacqueline, 196
Washington Post, 46
Wealth effect, on worker decisions to accept hazardous employment, 232
Weaver, P. H., 233
Weekly Compilation of Presidential Documents, on TSCA, *591*
Wegman, D. H., 30, 455
Weil, D., 2, 29, 30, 266, 289
Weiler, Paul C., 254, 368, 401, 406, 464
Welfare, as a source of compensation for workplace disease or injury, 463
Westin, Alan F., 370
West Virginia miners, and black lung disease, 28–29
Whirlpool Corporation, 385, 386–391
Whistleblowers, 364, 368, 407–408, 412–415
  defined, 407
  protections for. *See* Speaking out about company policies, protections against reprisals for
White Lung Association, 466
Williams, Wendy W., 425–426, 446
Wilson, W. J., 19
Wintersberger, 279
Witt, M., 287, 288
Wolfe, Sidney, 336
Wood, Horace Gray, 366–367
Work
  hazardous
    right to refuse, 286, 364, 372–392
    wealth effect on worker decision to accept, 232
  organization of. *See* Organization of work
  part-time, by industry, 20
  on unethical or socially irresponsible project, refusal of, 364
See also Employment

Worker(s)
  monitoring of. *See* Monitoring, of worker
  protection of. *See* Protection(s)
  rights of. *See* Employee right(s)
  wage and salary, union membership as percentage of, 251
  See also Employee(s)
Worker decisions to accept hazardous employment, wealth effect on, 232
Worker participation, in management decision-making, 25–27, 33–34, 279–281
Worker satisfaction, historical analysis of, 23–24
Workers' compensation, 264, 455–461, 462
  access to data generated by, 264
  compromise settlements in, 459–460
  description of, 455–456
  employers' attitudes toward, 454
  establishment of work-relatedness for, 457–459, 503–507
  "exclusivity" clauses in, 467, 225
  and externalities, 228–230
  in general, 225, 453–461
  history of, 454
  level of compensation under, 461–462, 463–466
  role of physician in, 456–457, 458
  union reaction to, 454
  use of tort law to augment, 466–503
Workers' compensation claims, decision points for, 457
Work force, contingent, growth of, 19–20
Work-force strategies, 14–15
Work in America, 23
Workmens' compensation. *See* Workers' compensation
Workmen's Compensation Commission. *See* National Commission on State Workmen's Compensation Laws
Workplace
  illness in. *See* Occupational disease(s)/illness(es)
  injury in. *See* Occupational injury(ies)
  monitoring of, 305–309
    environmental, 308–309
    toxics information transfer in, 299–362
Workplace freedoms, four potential, 363–364
Workplace hazards, susceptibility to biological markers of, 309, 313
  presumed, exclusion from workplace because of, 420–452

Workplace health and safety. *See*
    Health and safety *entries*
Workplace Health Fund, 281
Work-relatedness, establishment of,
    for compensation, 457–459,
    503–507
World War I, 3, 6–7
World War II, 3, 6, 8, 9, 10, 11, 18,
    28, 250, 251
Wright, J. S., 79

Writ of ejectment, 53
Writ of execution, 53
Writ of mandamus, 53, 68
Writs, judicial, common, 53
Wrongful discharge, 412. *See also*
    Protections from retaliatory
    employment practices

Xerox, 26

Yankelovich, D., 23
"Yellow-dog" contract, 250, 364
Yodaiken, R. E., 306

Zeckhauser, R., 229
Ziem, G. E., 100
Zuboff, S., 2, 13
Zysman, J., 17

Index of Cases

Cases excerpted in this text appear in italics; case excerpts appear at italicized page numbers.

ABC Trans-National Transport, Inc. v. NLRB, 272

Abilities and Goodwill, Inc. [NLRB], 378

Abood v. Detroit Board of Education, 290

Action for Children's Television v. FCC, 74

Adair v. United States, 364

AFL–CIO v. Brennan, 121, 150, 151

AFL–CIO v. Marshall, 146, 149, 150, 152, 173, 203

Albemarle Paper Co. v. Moody, 397, 450

Alleluia Cushion Co., 383, 384

Allen-Bradley Local v. Wisconsin Employment Relations Bd., 494

Allen v. Seidman, 438

Allis-Chalmers Corp. v. Lueck, 488, 489, 490, 492, 493

Almeida-Sanchez v. United States, 71

American Iron and Steel Institute v. OSHA, 121, 150, 151, 153

American Petroleum Institute v. OSHA, 123, 148, 150, 155, 160, 161. *See also* Industrial Union Department, AFL–CIO v. American Petroleum Institute.

American Smelting and Refining Co. v. Occupational Safety and Health Review Commission, 180

American Textile Manufacturers Institute v. Donovan, 43, 77, *137–144,* 168, 317

Anaconda Aluminum [NLRB], 374

Andrus v. Allard, 354

Arnold Hansen, d.b.a. Hansen Brothers Logging, 181

Asbestos Information Association/North America v. OSHA, 114, 157–163

Associated Industries of Massachusetts v. Snow, 345

Associated Industries v. United States Dept. of Labor, 102

Association of Data Processing Service Organizations, Inc. v. Camp, 66

Atlantic & Gulf Stevedores, Inc. v. OSHRC, 152, 154, 155, 174

Atlas Roofing Co. v. Occupational Safety and Health Review Commission, 168, 389

Automotive Parts & Accessories Association v. Boyd, 106

Avco Corp. v. Machinists, 488

Baker v. Carr, 64

Baltimore Gas & Electric Co. v. Natural Resources Defense Council, Inc., 170

Barker v. Lull Engineering Co., 476–477, 483

Barley v. Harrison Manufacturing, 471, 473

Beck v. C.W.A., 290

Begay v. United States, 485, 486

Berman v. Parker, 355

Beshada v. Johns-Manville Products Corporation, 477–482, 483

Bethlehem Steel v. OSHRC, 97

Black v. Hirsh, 355

Blankenship v. Cincinnati Milacron Chemicals, 470–474, 475

Blessing v. United States, 497

Block v. Community Nutrition Institute, 66

Board of Regents v. Roth, 353

Boliden Metech Inc. v. United States, 200

Borden Chemical [NLRB], 330. *See also* Oil, Chemical & Atomic Workers Local Union No. 6-418, AFL–CIO v. National Labor Relations Board.

Borel v. Fibreboard Paper Products, 478

Botany Mills v. United States, 411

Bowman Transp. Inc. v. Arkansas-Best Freight System, 75

Boyle v. Breme, 502

Boys Markets, Inc. v. Retail Clerks Union, Local 770, 283–285, 286, 374, 375

Brawner v. Sanders, 494

Breininger v. Sheet Metal Workers, 490

Brennan v. Occupational Safety and Health Review Commission and Reilly Tar and Chemical Corp., 181

Brennan v. Occupational Safety and Health Review Commission and Vy Lactos Laboratories, Inc., 180

Brisk Waterproofing Co., 185

Brock v. Emerson Electric Co., 96

Brock v. Roadway Express, Inc., 385

Brockway Motor Trucks v. NLRB, 268, 269

Brotherhood of R. R. Trainmen v. Chicago River & Ind. R. R., 284

Brown v. Harris, 214

Bryant v. Fox, 502

Burlington Truck Lines v. United States, 75

Burns Ford, Inc. [NLRB], 272

Bussey v. Travelers Insurance Co., 498

Buttner v. American Bell Tel. Co., 469

Byrd v. Fieldcrest Mills, Inc., 177

Caldwell v. United States, 486

California v. Sierra Club, 68

Calumet Industries v. Brock, 65–67

Camara v. Municipal Court, 70, 71, 72, 73

Camp v. Pitts, 75

Canipe v. National Loss Control Service Corp., 495–498

Carroll v. United Steelworkers of America, 493

Cepeda v. Cumberland Engineering Company, Inc., 478, 479
Chadsey v. United States of America, 399-400
Chamberlin v. Isen, 485
Charles Dowd Box Co. v. Courtney, 488
Chemical & Alkali Workers v. Pittsburgh Plate Glass Co., 269, 271
Chevron Chemical Co. v. Costle, 317
Chevron, U.S.A., Inc. v. NRDC, 44, 184
Chicago Teachers Union Local No. 1 v. Hudson, 290
Chrysler Corp. v. Brown, 354
Chrysler Corp. v. Dept. of Transportation, 118
Citizens to Preserve Overton Park v. Volpe, 74-75, 78, 133, 168
Clark v. Uebersee Finanz-Korporation, 143
Clemons v. Mississippi, 492
Collopy v. Newark Eye and Ear Infirmary, 417
Colonnade Catering Corp. v. United States, 71, 73
Commonwealth Edison Co. v. Montana, 342
Con. Agra, Inc. McMillan Co. Division, 185, 186
Condon v. Local 2944, United Steelworkers of America, 494
Consolidated Edison Company v. NLRB, 203
Consolidated Edison Company v. Public Service Commission of New York, 361
Coopage v. Kansas, 364
Copper & Brass Fabricators Council, Inc. v. Department of the Treasury, 66
Corning Glass Workers v. Brennan, 423
Corn Products Refining Co. v. Eddy, 354
Cranberry Lake Quarry Co. v. Johnson, 344
Cunningham v. United States, 485-486

Dalehite v. United States, 485, 486
Daly v. General Motors, 483-484
Davis v. Stover, 502
Day-Brite Lighting, Inc. v. Missouri, 354
De Arroyo v. Sindicato De Trabajadores Packinghouse, 290
Delamotte v. Midland Ross, 471, 473
Del-Costello v. Teamsters, 490
Detroit Edison Co. v. NLRB, 331, 332, 333
DiGiovanni v. Pessel, 418
DiLuzio v. United Electrical, Radio, and Machine Workers of America, 494

Dockery v. Lambert Table Co., 411
Donovan, Secretary of Labor, v. Anheuser-Busch, Inc., 381, 382
Donovan v. R. D. Anderson Constr. Co., 393
Donovan v. Square D Company, 396-398
Dothard v. Rawlinson, 434, 435-436, 437, 441
Douglas v. E. J. Gallo Winery, 475-476
Drivers' Union v. Lake Valley Co., 284
Dry Colors Mfrs. Assn., Inc. v. Dept. of Labor, 130, 158, 160
Dube v. Pittsburgh Corning, 486
Ducote v. Albert, 502
Dunbar v. United Steelworkers of America, 487
Duprey v. Shane, 502
Du-Tri Displays [NLRB], 378

E. E. Black, Ltd. v. Marshall, 449-450
Electrical Products Div. of Midland-Ross Corp. [NLRB], 272
Electrical Workers v. Hechler, 488, 489, 490, 492, 493
Ellis v. Brotherhood of Railways, Airline and Steamship Clerks, 290
English v. College of Medicine and Dentistry of N.J., 411
Environmental Defense Fund v. Environmental Protection Agency, 191, 200-207
Environmental Encapsulating Corp. v. New York City, 345
Escola v. Coca Cola Bottling Co., 476

Farmer v. Carpenters, 494
Farmworker Justice Fund, Inc. v. Brock, 165
Farwell v. Boston and Worcester Rail Road Co., 465-466
Fawcett Printing Corp. [NLRB], 332, 334
Feldman v. Lederle Laboratories, 483
Fibreboard Paper Products Corp. v. NLRB, 261, 269, 270, 281
Fidelity Federal Savings and Loan Association v. de la Cuesta, 340
Fire Equipment Manufacturers' Association, Inc. v. Marshall, 66
Fireman's Fund American Insurance Co. v. Coleman, 497
First National Bank of Arizona v. Cities Service Co., 432
First National Bank of Omaha v. National Labor Relations Board, 379
First National Maintenance Corporation v. National Labor Relations Board, 266-275, 276, 277, 279, 282, 286, 406

Florida Peach Growers Assn., Inc. v. Dept. of Labor, 130, 160
Florida Steel Corporation [NLRB], 379
Ford Motor Co. v. Hoffman, 491, 494
Ford Motor Co. v. NLRB, 269
Ford Motor Credit Co. v. Milhollin, 389
Forging Industry Association v. Secretary of Labor, 167-171
FPC v. Florida Power & Light Co., 135
Frampton v. Central Indiana Gas Company, 411
Frank Diehl Farms v. Secretary of Labor, 168
Freund v. Cellofilm Properties, Inc., 478-480, 482
Furnco Construction Corp. v. Waters, 429, 433, 435
F. W. Woolworth Co. [NLRB], 379

Gateway Coal v. United Mine Workers, 374-375, 381, 390
Geary v. United State Steel Co., 418, 420
Geduldig v. Aiello, 440
General Electric Co. v. Gilbert, 431, 440
General Motors Corp., GMC Truck & Coach Div. [NLRB], 273
General Motors Corp. v. NIOSH, 349
General Motors Leasing Corporation v. United States, 70
George Hyman Constr. Co. v. OSHRC, 97
Glanzer v. Shepard, 493
Greenman v. Yuba Power Products, Inc., 477
Griggs v. Duke Power Co., 426, 451
Gulf Light v. IBEW, 27
Gulf Mississippi Marine Corp. v. George Engine Co., 495
Guy v. Arthur H. Thomas Co., 502

Harrington v. National Outdoor Advertising Co., 353
Hartmann Luggage Co. [NLRB], 272
Hawaii Housing Authority v. Midkiff, 355, 356
Hayes v. Shelby Memorial Hospital, 426, 430-431, 438, 441, 442, 445, 446-447, 452
Herber v. Johns-Manville Corp., 505
Herbert v. Lando, 361
Hercules, Inc. v. EPA, 147, 206
Hitchman Coal and Coke Co. v. Mitchell, 250
Hodel v. Virginia Surface Mining and Recl. Assn., 353
Holloman v. Watt, 485
Hulahan v. Sheehan, 494
Humphrey v. Moore, 490

Industrial Union Department, AFL-CIO v. American Petroleum Institute, 43, 77, 120, *123–136*, 138, 141, 142, 143, 147–151, 152, 155, 156, 158, 160, 161, 162, 168, 169, 444

Industrial Union Department, AFL-CIO v. Hodgson, *102–113*, 117, 131, 143, 151, 155, 173, 178, 222

Ingalls Shipbuilding Corp., 333

Inglis v. Operating Engineers Local Union No. 12, 494

International Association of Machinists v. Street, 290

International Brotherhood of Teamsters. *See* Teamsters

International Offset Corp., 272

International Union of Electrical, Radio & Machine Workers v. NLRB, 331, 332, 333, 334

International Union, United Automobile, Aerospace and Agricultural Implement Workers of America v. General Dynamics Land Systems Div., *182–186*

Isis Plumbing & Heating Co. [NLRB], 379

Jackson v. Muhlenberg Hospital, 419

Jersey Shore Medical Center—Pitkin Hospital v. Baum, 417

Jeter v. St. Regis Paper Co., 177

J. I. Case Co. v. NLRB, 332

John B. Kelly, Inc. [OSHRC], 181

John Hancock Mutual Life Insurance Co. v. NLRB, 397

Johns-Manville Products Corporation v. Contra Costa Superior Court, *467–470*, 474, 475

Johnson v. Murph Metals, Inc., 483

Joiner v. Benton Community Bank, 413

Jones v. Truck Drivers Local Union No. 299, 494

Kaczmarek v. N.J. Turnpike Authority, 410

Kaiser Aetna v. United States, 353

Kaplan v. Tenth Judicial District, 500

Katz v. United States, 71

Kelsay v. Motorola, Inc., 411, 412, 413–414

Kewanee Oil Co. v. Bicron Corp., 318, 353

Knitz v. Minister Machine Co., 483

Kottis v. United States Steel Corp., 476

Lally v. Copygraphics, *410–411*, 412, 420

Lawlor v. Shannon, *359–362*

Law v. Dartt, 469

Lead Industries Ass'n v. Donovan, 165

Lever Brothers Co. v. International Chemical Workers Union, 272

Lewis v. American Hoist & Derrick Co., 477

Lewis v. Benedict Coal Corp., 491

Lilly v. Grand Trunk R. Co., 389

Lincoln Mills, 284

Lingle v. Norge Division of Magic Chef, Inc., 492, 493

Local No. 93, International Association of Fire Fighters v. City of Cleveland, 431

Local 24, Int'l Brhd of Teamsters v. Oliver, 177

Longever v. Revere Copper & Brass, Inc., 476

Los Angeles Dept. of Water & Power v. Manhart, 436

Loucks v. Star City Glass Co., 411

Luckett v. Orange Julep Co., 353

Maher v. Roe, 439

Manhattan General Equipment Co. v. The Commissioner of Internal Revenue, 214

Manufacturers' Association of Tri-County v. Knepper, 345

Marbury v. Madison, 49

Marcucci v. Johns-Manville Sales Corp., 480

Marshall v. American Petroleum Institute, 101

Marshall v. Barlow's, Inc., *69–73*, 96

Marshall v. Certified Welding Corp., 387

Marshall v. Commonwealth Aquarium, *393–395*

Marshall v. Daniel Construction Co., 387, 395

Marshall v. Gibson's Products, Inc. of Plano, 174

Marshall v. Intermountain Electric Co., 398

Marshall v. International Longshoremen's and Warehousemen's Union, 494

Marshall v. Lumco, 399–400

Marshall v. Springville Poultry Farm, 393

Marshall v. Whirlpool Corp., 387, 392, 395. *See also* Whirlpool Corp. v. Marshall

Martin v. Althoff, 411

Martin v. Tapley, 411

Mastro Plastics Corp. v. NLRB, 284

Matsushita Electric Industrial Co. v. Zenith Radio Corp., 432

Maynard v. Revere Copper Products, 494

McAlister v. Methodist Hospital, 502

McCown v. International Harvester Co., 483

McDonnell Douglas Corp. v. Green, 429

McLaughlin v. A. B. Chance Co., 96

McLaughlin v. Kings Island, Division of Taft Broadcasting Co., 96

Mercer v. Uniroyal Inc., 475

Meyers Industries Inc. v. Prill, 383, 384, 385. *See also* Prill v. NLRB

Micallef v. Miehle Co., 484

Midgley v. S. S. Kresge Co., 477

Mid-Louisiana Gas Co. v. Federal Energy Regulatory Commission, 214

Mitchell v. Robert De Mario Jewelry, Inc., 397

Mobil Oil Corp. v. FPC, 106

Monsanto Co. v. Acting Administrator, United States Environmental Protection Agency, 352

Moran v. Raymond Corp., 484

Motor Vehicle Manufacturers' Association v. State Farm Mutual Automobile Insurance Company, *75–77*

Mt. Healthy City Board of Education v. Doyle, 394

Mourning v. Family Publications Service, Inc., 178, 389

Muller v. Oregon, 440–441

Nabors v. NLRB, 397

National Ass'n of Recycling Industries, Inc. v. Secretary of Labor, 170

National Labor Relations Board. *See* NLRB

National Petroleum Refiners Ass'n v. FTC, 174

National Realty & Constr. Co. v. OSHRC, 184

Natural Gas Pipeline Co. v. United States, 485

Natural Resources Defense Council. *See* NRDC

New Jersey State Chamber of Commerce v. Hughey, *339–344*, *357–358*

Nicoletta v. No. Jersey District Water Supply Comm'n, 411

NLRB v. Adams Dairy, Inc., 271

NLRB v. American Nat. Ins. Co., 261, 269

NLRB v. Babcock & Wilcox Co., 71

NLRB v. City Disposal Systems, 384–385

NLRB v. Gissel Packing Co., 254–255

NLRB v. Gulf Power Co., *27, 177, 259, 260–261, 277, 282, 329, 331*

NLRB v. International Harvester Co., 272

NLRB v. Jones & Laughlin Steel Corp., 268, 365

NLRB v. Katz, 268

NLRB v. Knight Morley Corp., 373–374

NLRB v. Lyon & Ryan Ford, Inc., 382

NLRB v. Mars Sales & Equipment Co., 382
NLRB v. Mayes, 261
NLRB v. Royal Plating & Polishing Co., 271, 272
NLRB v. Scrivener, 397
NLRB v. Tamara Foods, Inc., 376–382
NLRB v. Thompson Transport Co., 272
NLRB v. Trident Seafoods Corp., 382
NLRB v. Reed & Prince Mfg. Co., 272
NLRB v. Washington Aluminum Company, 261, 372–373, 377, 379, 380, 383, 390
NLRB v. Winn-Dixie Stores, Inc., 268
NLRB v. Wooster Division of Borg-Warner Corp., 258, 269
Northwestern National Gas Co. v. McNulty, 472
NRDC v. Costle, 79, 193, *209–211*, 213, 219
NRDC v. EPA, 193, *211–214*
NRDC v. SEC, 79

Occidental Life Insurance Co. v. EEOC, 397, 398
Ohio Manufacturers' Association v. City of Akron, 345
Oil, Chemical and Atomic Workers International Union v. American Cyanamid Company, *421–425*, 448, 451–452
Oil, Chemical and Atomic Workers Local Union No. 6-418, AFL–CIO v. National Labor Relations Board, *331–334*
Olympic Medical Corporation [NLRB], 379
Oxford Consumer Discount Co. v. Stefanelli, 344
Ozark Trailers, Inc. [NLRB], 267, 269

Palmateer v. International Harvester Co., *412–415*
Palmer v. Massey-Ferguson, Inc., 484
Pariseau v. Wedge Products, Inc., 471, 473
Payton v. Abbot Laboratories, 507
Penn Central Transportation Co. v. New York City, 353
Pennsylvania Coal Co. v. Mahon, 353, 354
Penn Terra Ltd. v. Department of Environmental Resources, 342
People v. Chicago Magnet Wire Corporation, 80
People v. Hegedus, 80
Percival v. General Motors Corp., 414
Perry v. Sindermann, 411

Peterson v. Air Line Pilots Assn., International, 494
Pierce v. Ortho Pharmaceutical Corp., *410, 411, 415–420*
Pike v. Frank G. Hough Co., 476
Pime v. Loyola University, 438
Planned Parenthood League of Massachusetts v. Bellotti, 362
Plastilite Corporation [NLRB], 377
Prill v. NLRB, 384
Pruneyard Shopping Center v. Robins, 353
Public Citizen Health Research Group v. Auchter, 160, 164, 165
Public Citizen Health Research Group v. Brock, 79, *164–166*
Public Citizen Health Research Group v. Food and Drug Administration, 318
Public Citizen Health Research Group v. Tyson, 164, 165, 166
Public Citizen v. Young, 444
Pugh v. See's Candies, Inc., 409

Rawson v. United Steelworkers of America, 488, 492. *See also* United Steelworkers of America v. Rawson
REA Express, Inc. v. Brennan and Occupational Safety and Health Review Commission, 181
Reddi-Whip, Inc. v. Lemay Value Co., 353
Redwing Carriers [NLRB], 374
Republic Aviation Corp. v. NLRB, 71
Republic Creosoting Co., Division of Reilly Tar and Chemical Corp. *See* Brennan v. OSHRC and Reilly Tar and Chemical Corp.
Republic Steel Corp. v. Maddox, 491
Retail Clerks v. Schermerhorn, 494
Richardson v. United Steelworkers of America, 494
Richards v. United States, 284
Richmond Block, Inc. [OSHRC], 181
Rindge Co. v. Los Angeles, 355
Robinson v. Lorillard Corp., 433, 451
Rogers Materials Co. v. Ind. Acc. Com., 469
Rosenfeld v. Southern Pacific Co., 441
Ross v. Schubert, 502
Rostker v. Goldberg, 446
R. T. Vanderbilt Co. v. OSHRC, 66
Ruckelshaus v. Monsanto Co., *317*, *352–356*, *357–358*
Ruzick v. General Motors, 290

Safeway Stores, 334
Seattle Times v. Rhinehart, 360–361
SEC v. Chenery Corp., 75

Secretary of Transportation v. Railway Labor Executives Association, 347
See v. City of Seattle, 70, 71, 73
Segal v. Arrow Industries Corp., 411
Shell Chemical v. EPA, 215
Shell Oil Co. [NLRB], 272
S. H. Kress & Co. v. Godman, 493
Sierra Club v. Morton, 64
Silkwood v. Kerr-McGee Corporation, 80, 340
Simpson v. Young, 444
Sinclair Refining Co. v. Atkinson, 283, 284, 285
Skidmore v. Swift & Co., 388, 431
The Society of the Plastics Industry, Inc. v. Occupational Safety and Health Administration, *114–120*, 121, 131, 150, 151, 152, 153, 160
SOCMA v. Brennan, 113, 119, 120, 122
South Central Timber Development, Inc. [NLRB], 377
Southern Ry. v. OSHRC, 423
Stacy v. Aetna Casualty & Surety Co., 497
Stanlick v. Kaplan Industries, 499
State, ex rel. Crawford, v. Indus. Comm., 472
State, ex rel. Miller, v. Mead Corp., 473
Steele v. Louisville & Nashville R. Co., 493
Steelworkers. *See* United Steelworkers of America
Steelworkers Trilogy, 284
Steiner Corp. v. American District Telegraph, 493
Stephens v. Justiss-Mears Oil Co., 411
Sterling v. Velsicol Chemical Corp., 507
Streeter v. Sullivan, *499–501*, 503
Suter v. San Angelo Foundry & Machine Company, 478, 479, 481
Sweeney v. Board of Trustees of Keene State college, 394
Synthetic Organic Chemical Mfrs. Assn. v. Brennan. *See* SOCMA v. Brennan

Tameny v. Atlantic Richfield Co., 413
Tatrai v. Presbyterian University Hospital, 502
Tax Analysts and Advocates v. Blumenthal, 67
Taylor Diving & Salvage v. Department of Labor, 158, 160
Taylor v. Brighton Corp., 398
Teamsters v. Lucas Flour Co., 284, 488, 491
Teamsters v. Oliver, 269
Teamsters v. United States, 429

Texas & New Orleans Railroad v. Brotherhood of Railway Clerks, 365

Textile Workers v. Darlington Mfg. Co., 255, 269, 271

Textile Workers v. Lincoln Mills of Alabama, 488

Thorpe v. Housing Authority of the City of Durham, 178

T. J. Hooper, 482

Todd Shipyards Corp. v. Secretary, 97

Torres v. Wisconsin Dept. of Health & Social Services, 434–435, 436, 437, 441

Transamerica Mortgage Advisors v. Lewis, 411

Triff v. National Bronze & Aluminum Foundry, 473

United Auto Workers v. Johnson Controls, Inc., 426–446, 452

United Mine Workers Health & Retirement Funds v. Robinson, 490

United Parcel Service, Inc. v. Mitchell, 490

United States v. Bacto-Unidisk, 389

United States v. Biswell, 71, 73

United States v. Chadwick, 70

United States v. Commonwealth Edison Co., 192, 207–208

United States v. General Motors Corp., 353

United States v. Hayes, 395

United States v. Hutcheson, 284

United States v. O'Brien, 361

United States v. Orleans, 485

United States v. S. A. Empresa de Viacao Aerea Rio Grandeuse ("Varig Airlines"), 485, 486

United Steelworkers of America, AFL–CIO-CLC v. Auchter, 313–321, 338, 341, 342, 356–357

United Steelworkers of America v. Dunbar, 487. See also Dunbar v. United Steelworkers of America

United Steelworkers of America, AFL–CIO-CLC v. Marshall and Bingham, 145–156, 171–179

United Steelworkers of America v. Marshall, 165, 170, 171, 203, 316, 331, 422, 444

United Steelworkers of America v. OSHA, 49

United Steelworkers of America v. Pendergras, 166

United Steelworkers of America v. Rawson, 290, 487–494, 495

United Steelworkers of America v. Warrior and Gulf Navigation Co., 286

Usery v. Whirlpool Corp., 387, 394–395. See also Whirlpool Corp. v. Marshall

Vaca v. Sipes, 289, 490, 494

Vasquez v. Glassboro Services, Inc., 418

Vermont Yankee Power Corp. v. Natural Resources Defense Council, 74, 75, 76, 134, 146

Vic Tanney International [NLRB], 378

Vy Lactos Laboratories, Inc. See Brennan v. OSHRC and Vy Lactos Laboratories, Inc.

Wards Cove Packing Co. v. Atonio, 395, 431–432, 433, 435, 438, 442, 451, 452

Watson v. Fort Worth Bank & Trust, 433

Webb's Fabulous Pharmacies, Inc. v. Beckwith, 353, 354

Weeks v. Southern Bell Telephone & Telegraph Co., 435, 441

Weinberger v. Hynson, Westcott & Dunning, Inc., 143

Welch v. Schneider National Bulk Carriers, 194

Westinghouse Electric Corp., 331, 332

Westinghouse Electric Corporation v. NLRB, 261

Westinghouse Electric Corp. v. United States Nuclear Regulatory Commission, 354, 356

West v. Jessop, 500

Whalen v. Roe, 348, 349, 350

Whirlpool Corp. v. Marshall, 130, 175, 380–381, 386–391, 392, 448

Whiting v. United States, 395

Winona Industries, Inc. and International Chemical Workers Union, Local 893, AFL–CIO, 327–329, 330, 331

Wright v. FMC Corporation, 469

Wright v. Olin Corp., 426, 429, 430, 431, 433, 438, 440, 441, 442, 445

Zurich Insurance Co. v. Scofi, 500